Behavioural models in psychopharmacology are used for different purposes. The main concern of industrial psychopharmacologists is to develop new and improved drugs for the treatment of mental disorders; basic scientists use animal models to investigate the underlying nature of such conditions. The important distinction between these different perspectives is made explicit for the first time in this book. By considering anxiety, depression, mania and schizophrenia, feeding disorders, dementia, and drug dependence, this book provides a comprehensive and critical review of the adequacy of the behavioural procedures used by psychopharmacologists to model psychiatric disorders. In addition to the theoretical and industrial perspective, each section of the book also assesses the contribution of behavioural models from the point of view of the clinician.

Graduate students and research workers in psychopharmacology, from both academic and industrial spheres, as well as clinicians, will find this book of considerable interest.

**Behavioural models in psychopharmacology:
theoretical, industrial and clinical perspectives**

Behavioural models in psychopharmacology: theoretical, industrial and clinical perspectives

Edited by
PAUL WILLNER
Professor of Psychology
City of London Polytechnic, London, England

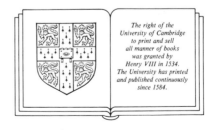

The right of the
University of Cambridge
to print and sell
all manner of books
was granted by
Henry VIII in 1534.
The University has printed
and published continuously
since 1584.

CAMBRIDGE UNIVERSITY PRESS

Cambridge
New York   Port Chester
Melbourne   Sydney

Published by the Press Syndicate of the University of Cambridge
The Pitt Building, Trumpington Street, Cambridge CB2 1RP
40 West 20th Street, New York, NY 10011, USA
10 Stamford Road, Oakleigh, Melbourne 3166, Australia

First published 1991

Printed in Great Britain at the University Press, Cambridge

British Library cataloguing in publication data

Behavioural models in psychopharmacology.
1. Drugs. Action. Behavioural aspects
I. Willner, Paul
615.78

Library of Congress cataloguing in publication data

Behavioural models in psychopharmacology : theoretical,
industrial, and clinical perspectives / edited by Paul Willner.
   p. cm.
ISBN 0-521-39192-X
1. Psychopharmacology – Research – Methodology. 2. Mental
illness – Animal models. 3. Mental illness – Chemotherapy –
Research – Methodology. 4. Human behaviour – Animal
models. I. Willner, Paul.
[DNLM: 1. Behaviour – drug effects. 2. Disease Models, Animal.
3. Drug Screening. 4. Models, Psychological. 5. Psychotropic
Drugs – pharmacology. QV 77 B4185]
RM315.B367   1991
616.89′18–dc20   90-1779   CIP

ISBN 0 521 39192 X hardback

CE

For Matthew and Jessica

Contents

Contributors

Sven Ahlenius, *Professor, Department of Neuropharmacology, Astra Alab AB, Sodertalje, Sweden.*

Harvey J. Altman, *Director, Department of Behavioral Animal Research, Lafayette Clinic, Detroit, Michigan, U.S.A.*

John S. Andrews, *Research Scientist, Department of Neuropsychopharmacology, Schering AG, Berlin, F.R.G.*

Trevor Archer, *Formerly: Head, Department of Neuropharmacology, Astra Alab AB, Sodertalje, Sweden. Now: Professor, Department of Psychology, University of Gothenburg, Göteborg, Sweden.*

Timothy J. Barth, *Research Fellow, Department of Psychology, University of Texas, Austin, Texas, U.S.A.*

John Cookson, *Consultant Psychiatrist and Honorary Senior Lecturer, Department of Psychiatry, The London Hospital, London, U.K.*

Donald V. Coscina, *Head, Section of Biopsychology and Professor, Departments of Psychiatry and Psychology, University of Toronto, Toronto, Canada.*

Wojciech Danysz, *Docent, Department of Pharmacology and Physiology of the Nervous System, Psychoneurological Institute, Warszawa, Poland.*

J.F.W. Deakin, *Senior Lecturer, Department of Psychiatry, University Hospital of South Manchester, Manchester, U.K.*

Steven B. Dunnett, *Lecturer, Department of Experimental Psychology, University of Cambridge, Cambridge, UK..*

Christopher J. Fowler, *Adjunct Professor, Department of Geriatric Medicine, Karolinska Institutet, Huddinge, Sweden.*

Paul E. Garfinkel, *Psychiatrist in Chief, Toronto General Hospital, and Professor, Department of Psychiatry, University of Toronto, Toronto, Canada.*

Samuel Gershon, *Vice-President for Research, Medical and Health Care Division, Western Psychiatric Institute and Clinic, University of Pittsburgh, Pittsburgh, Pennsylvania, U.S.A.*

Andrew J. Goudie, *Reader, Department of Psychology, University of Liverpool, Liverpool, U.K.*

Simon Green, *Senior Lecturer, Department of Psychology, Birkbeck College, London, U.K.*

Richard Hartnoll, *Senior Research Fellow and Director, Drugs Indicators Project, Birkbeck College, University of London, London, U.K.*

Helen Hodges, *Research Fellow, Department of Psychology, Institute of Psychiatry, University of London, London, U.K.*

Malcolm Lader, *Professor, Department of Psychiatry, Institute of Psychiatry, University of London, London, U.K.*

Melvin Lyon, *Associate Professor, Department of Psychiatry and Behavioral Science, University of Arkansas for Medical Sciences, 4301 West Markham, Slot 554 Little Rock, Arkansas 72205, U.S.A.*

Anthony M.J. Montgomery, *Senior Lecturer, Department of Psychology, City of London Polytechnic, London, U.K.*

Howard J. Normile, *Assistant Professor, Department of Behavioral Animal Research, Lafayette Clinic, Detroit, Michigan, U.S.A.*

John D. Salamone, *Formerly: Project Leader, Merck, Sharp and Dohme Research Laboratories, Harlow, U.K.*

Now: Assistant Professor, Department of Psychology, University of Connecticut, Storrs, Connecticut, U.S.A.

David J. Sanger, *Head, CNS Pharmacology, Synthelabo Recherche (L.E.R.S), 31 Ave P.V. Couturier, 92220 Bagneux, France.*

Jerry Sepinwall, *Director, Department of Neurobiology and Obesity Research, Hoffman-La Roche Inc., Nutley, New Jersey, U.S.A.*

D. N. Stephens, *Head, Department of Neuropsycho-pharmacology, Schering AG, Berlin, F.R.G.*

Ann C. Sullivan, *Group Director, Pharmacology and Chemotherapy, Hoffman-La Roche Inc., Nutley, New Jersey, U.S.A.*

Paul Willner, *Professor, Department of Psychology, City of London Polytechnic, London, U.K.*

Preface

The problem of using animal behaviour to model human mental disorders is, explicitly or implicitly, the central preoccupation of psychopharmacology. The idea that psychiatric disorders might be modelled in animals provides the basis for a substantial proportion of current research; and research that does not employ behavioural models directly is usually justified by reference to its eventual benefits, in terms of an understanding of the human brain and the development of more effective and safer therapies. This book sets out to evaluate the extent to which behavioural models of psychiatric disorders succeed in meeting their objectives.

There are a number of compelling reasons for assessing the status of behavioural models at this time. First, there has been a considerable recent growth in the number of such models, in all areas of psychopharmacology, and it serves a useful purpose to bring these developments together within a single volume, as a source for future reference. It is a regrettable, if wholly understandable, feature of new paradigms that their adherents tend to be more enthusiastic than critical; with this in mind, a second aim has been to examine objectively the strengths, weaknesses and potential of established and newer models, in order to provide a basis and a context for future research. A third reason for undertaking this exercise is that behavioural models are used for a variety of different purposes, and the importance of this simple observation is not widely recognized. In particular, academic and industrial psychopharmacologists have rather different research objectives, which place different requirements on the models they use. This has led to some confusion that we have attempted to clarify. Finally, at a time when animal experimentation is coming under increasing criticism, there are strong ethical reasons to evaluate the extent to which work with animal models is truly applicable to the human condition. In addition to rigorous self-assessment, the opinions of clinicians provide an important further source of feedback, which experimental psychopharmacologists should value.

All of the areas covered in this volume have been reviewed previously; however, our present treatment differs from earlier reviews in three important respects. First, there is no comparable attempt to provide comprehensive coverage; the six substantive sections of this book include all significant areas of behavioural modelling (with the single exception of pain). This has allowed us to meet a second important objective, the application of a common set of critical standards. All of the contributors have worked within a common analytical framework, which is outlined in Chapter 1. As a result, it is possible for the reader to make realistic comparisons between models, both within a single area of the literature, and between different areas. Finally, we have, perhaps for the first time, recognized explicitly that psychopharmacology is a coalition of different professional groups, who use the same paradigms to different ends. Therefore, in each part of the book, the theoretical, industrial, and clinical perspectives on behavioural models have been examined separately, so enabling a clearer and less ambiguous presentation of the issues.

The original suggestion for this book came from Erich Lehr, of Boehringer Ingelheim, F.R.G., and I am grateful to him for it. I am indebted to Heather Reid for her support and encouragement (and for many other things besides). I wish also to record a debt of gratitude to the contributing authors, without whom this book would not exist: having accepted a difficult and unusual brief, they tolerated my frequent and sometimes demanding editorial interventions with good grace and without complaint. If this book succeeds in its purpose, the credit is theirs.

Paul Willner
London, September 1989

PART I

Introduction

I

Behavioural models in psychopharmacology

PAUL WILLNER

Most psychopharmacologists who work with animals do so in the belief that their research will ultimately be relevant to people. The behavioural paradigms and procedures that they use therefore serve, in some sense, as models of human behaviour. However, the precise nature of this relationship is often left implicit. If we attempt to define what, in this context, is meant by the term 'model', it rapidly becomes apparent that its meaning is as varied as the interests of psychopharmacologists, who come from a diversity of academic backgrounds and work in a variety of institutional settings. Inevitably, attempts to define the term 'model' succeed in describing only some of its uses. Take a recent example:

Animal models represent experimental preparations developed in one species for the purpose of studying phenomena occurring in another species. In the case of animal models of human psychopathology one seeks to develop syndromes in animals which resemble those in humans in certain ways in order to study selected aspects of human psychopathology. (McKinney, 1984)

This seems a perfectly reasonable definition. But is it? Consider, for example, one of the older animal models of depression, the muricide test (Horovitz, 1965). In this model, potential antidepressant drugs are tested for their ability to prevent rats from killing mice. The model has been used quite widely, despite being inefficient as a way of detecting antidepressant activity and ethically objectionable (Willner, 1984a). But does it fall within the scope of our definition of 'animal models'? If animal models depend upon a resemblance between the model and the condition being modelled, it clearly does not: depressed people do

not typically kill mice, and it is difficult to see any other respect in which muricide and depression are similar.

One way to preserve the integrity of the definition is to argue that the muricide test is not really an animal model of depression, but rather, a screening test for antidepressant drugs. We would then wish to distinguish between 'animal models' and 'screening tests'. However, this has not been the position taken in reviews of the literature on 'Animal Models of Depression', for they invariably include the muricide test (see, e.g. Katz, 1981; McKinney, 1984; Willner, 1984a; Jesberger & Richardson, 1985). Should we then accept the implicit view of these reviewers that antidepressant screening tests are a type of animal model of depression? If we do, we are tempted to extrapolate models like muricide beyond their proper domain, and all manner of falacious claims are likely to arise as a result. The dopamine hypothesis of schizophrenia provides a very obvious example of the conceptual confusion that arises when a theory of the mechanism of drug action (antischizophrenic drugs are antagonists at dopamine receptors . . .) becomes a theory of the nature of the disorder being treated (. . . therefore schizophrenics suffer from excessive transmission at dopamine synapses).

It is clearly necessary to consider carefully, in every case, the domain in which the use of a particular model is justified. However, this prospect is not so daunting as it may appear, since some general principles may be elaborated that apply to broad classes of behavioural model. This chapter will describe three such classes, and consider what the models in each class are designed to achieve and how they may be assessed. The term 'behavioural model', rather than 'animal model', is used deliberately, to allow an initial consideration of as wide a range of models as possible.

1.1 A TAXONOMY OF BEHAVIOURAL MODELS

A straightforward taxonomy of behavioural models emerges from simply specifying the field in which the models are to be deployed. It is clear from the most cursory glance at any of the relevant journals or textbooks, that the name 'psychopharmacology' is actually a misnomer. The subject matter of psychopharmacology is not simply the effects of drugs on mental processes, but also the underlying physiological mechanisms. The name 'neuropsychopharmacology', which is clumsier but more accurate, reminds us that we are dealing with the meeting point of three disciplines: neuroscience, psychology, and pharmacology. Correspondingly, there are three classes of behavioural model: behaviour may be used to model brain function, psychological processes, or drug actions. These three classes of model will be referred to as behavioural bioassays, simulations, and screening tests. Screening tests are 'drug centred': they are designed solely to expedite the discovery of new drugs, and therefore form a branch of pharmacology. Behavioural bioassays fall within the realm of neuroscience, in that they are 'brain centred': their purpose is to use behaviour to measure activity in a specified brain system. Simulations attempt to use animals to further our understanding of human mental processes. Thus, they are 'mind centred', and fall within psychology.

1.1.1 Screening tests

A screening test models a drug action: the search for novel psychotropic agents is based upon the actions of known drugs, which serve as reference points against which to compare the performance of new candidates. Two types of screening procedures may be distinguished, which correspond to different strategies for drug development. Traditionally, the object of screening tests has been to identify agents likely to have a specific type of clinical action: neuroleptic, antidepressant, anxio-lytic, and so on. Tests of this kind may be capable of identifying clinically effective drugs that vary widely in their chemical structure. However, they carry no guarantee, and may sometimes fail to identify structurally novel compounds. This problem is perhaps best recognized in relation to the socalled atypical antidepressants, several of which were discovered accidentally during testing in humans after failing all the standard traditional screening tests (see Danysz *et al.*, this vol., Ch. 6).

A second strategy is to identify specific biochemical actions as targets for drug development. This strategy becomes appropriate once we understand the mode of action of existing treatments. Anxiolytic drug development, for example, now proceeds primarily via the initial identification of compounds that interact with the GABA/benzodiazepine receptor complex (Squires & Braestrup, 1977). Screening for specific biochemical actions is carried out primarily by biochemical tests, but behavioural techniques are also used for this purpose. The advantage of screening for a specific neurochemical action is that the usefulness of the resulting compounds may transcend traditional diagnostic boundaries. Serotonin (5-HT) uptake inhibitors are probably the best example: developed primarily as antidepressants, these drugs may also be of value as anorectics (see Sepinwall & Sullivan, this vol., Ch. 9), as cognitive enhancers in dementia (Bergman *et al.*, 1983), as anxiolytics (Westenberg & den Boer, 1988), and as analgesics (Ogren & Holm, 1980).

The biochemical strategy has the major disadvantage of precluding the discovery of chemically novel modes of treatment. One of the most popular ways of screening for neuroleptic activity, for example, is to test the ability of drugs to antagonize apomorphine-induced stereotyped behaviour. The potency of drugs in this test correlates almost perfectly with their clinical potency in schizophrenia, and also with their ability to bind to dopamine receptors (Creese & Snyder, 1978). But why is it that virtually all drugs effective in the treatment of schizophrenia are dopamine receptor antagonists? It may be that the dopamine synapse is indeed a crucial site for intervention in schizophrenia. However, it seems equally likely that the apparent specificity of neuroleptics for the dopamine receptor simply reflects the degree to which neuroleptic screening tests succeed in identifying dopamine receptor antagonism (see Ahlenius, this vol., Ch. 12).

If the objective is to discover drugs having a specific clinical indication, then screening tests may attempt to simulate the condition for which drugs are being sought. Many of the screening tests for anxiolytic drugs, for example, embody procedures involving some degree of conflict and a presumption that the animal is anxious (see Stephens & Andrews, this vol., Ch. 3). However, simulation is in no way a requirement in a screening test, and many screening tests have an exclusively empirical foundation; their rationale is simply that agents known to have the desired property are found to alter behaviour in the test. The literature abounds with empirically based screening tests: the muricide test for antidepressant activity has already been mentioned; suppression of apomorphine-induced emesis, one of the classic tests for neuroleptic activity (Niemegeers & Janssen, 1979), is another. The increasingly popular 'drug discrimination' procedures fall within this category. In drug discrimination experiments, animals are rewarded for making one response if they have been injected with a reference drug, and for a different response if they have not; the similarity between a novel agent and the reference drug is then assessed by measuring the degree to which the animals respond to the novel agent as though it were the reference drug (Colpaert & Slangen, 1982). Thus these procedures require the animal to discriminate its internal state. However, at the behavioural level, the consequence of that judgement is always the same: the rat presses one of two levers.

Irrespective of the manner in which they are constructed, screening tests are subject primarily to one very simple requirement: the test should predict accurately the desired activity; it should accept drugs that are effective and reject those that are ineffective. However, if a screen is less than perfect (as most of them are) then the two types of error are not of equal importance. If a screening test accepts an ineffective compound (false positive) the error will eventually come to light in further testing, and no permanent damage will have been done. However, if the test wrongly rejects an effective compound (false negative) a potentially beneficial drug will be lost irretrievably.

When assessing the ability of a screening test to discriminate active from inactive compounds, it should be borne in mind that failure to discriminate may not necessarily be the fault of the screening test: it is always possible that the clinical effects of a drug have been incorrectly classified. Many antidepressant screens, for example, admit anticholinergic drugs as 'false positives'. But are they really false? Anticholinergics have never been subjected to properly controlled clinical trials, but there are numerous reports from the 1940s and 1950s that they have antidepressant activity (Janowsky & Risch, 1984). Conversely, a 'false negative' could reflect the fact that a drug considered to be clinically effective, actually, is not. A false impression of clinical effectiveness can easily arise in early, uncontrolled clinical trials. What is less frequently recognized is that this impression can continue through into controlled studies, if they compare the new drug against a reference drug that is known to be active. In such a trial, it is possible that the new agent might perform worse than the reference drug, but for a variety of reasons, the difference might not be statistically significant. The new drug is then said to be 'as good as drug X'. However, if a placebo condition had been included, we might have observed that in addition to being 'no worse than drug X', the new drug was also 'no better than no drug at all'. This seems to be what happened in the case of the antidepressant iprindole, which has caused havoc by challenging antidepressant researchers to explain how it works, but actually appears not to be clinically effective in properly designed, placebo-controlled tests (Zis & Goodwin, 1979).

In view of the large number of candidate drugs entered into screening programmes, a screening test should ideally respond to a single administration of an active compound (though it will also be necessary in later testing to confirm that the drug retains its activity following chronic administration). There appears to be a degree of conflict between the logistical requirement that a test respond to acute treatment and the clinical observation that many psychotherapeutic agents act slowly over a period of days, weeks, or even months. However, while this is an important consideration in assessing the validity of a simulation (see Section 1.2.2), it plays no part in the assessment of a screening procedure, which is concerned solely with the ability of the test to make accurate predictions.

In addition to fulfilling the scientific imperative of efficient prediction, and the logistic imperative of speed, a good screening test will also meet a number of other, essentially economic, criteria. It should be cheap, simple, and reliable; it should allow adequate experimental design as well as efficient data processing and statistical analysis; it should not be labour intensive; and it should not rely unduly on the goodwill or sustained concentration of the operator. For several of these reasons, automated procedures have much to recommend them.

1.1.2 Behavioural bioassays

Behavioural bioassays model a physiological action: the whole animal is used as a measuring device to assess the functional state of an underlying physiological system, in exactly the same way that such measurements might be made in an isolated tissue or in a test tube. Examples are the increases in locomotor activity and stereotyped behaviour induced by psychomotor stimulants, which may be used to measure the responsiveness of dopamine receptors in nucleus accumbens and striatum, respectively (Kelly *et al.*, 1975). One use of behavioural assays is as screening tests to identify a specified biochemical action. More commonly, behavioural assays are used to study the mechanisms responsible for changes in brain function, typically those resulting from chronic drug administration, brain lesions, or other similar experimental manipulations.

If the object is to measure a physiological action, why use behavioural methods? There at least are four reasons why indirect behavioural measurement might at times be preferable to a direct biochemical assay. First, behavioural methods are non-destructive. They do not require removal of the brain, which has a number of advantages both for the experimenter and for the experimental subjects. Second, the use of a behavioural measure guarantees that the agents being tested are actually reaching the brain: the poor bioavailability of many novel drugs is only apparent in the whole animal. Third, behavioural measures are functional measures. Many biochemical indices are notoriously unreliable as guides to functional changes. This applies particularly to the parameters derived from receptor binding studies: to

take just one example, the well established supersensitivity of dopamine receptors in the nucleus accumbens that follows chronic antidepressant administration has proved extremely difficult to confirm using receptor binding methods (Martin-Iverson *et al.*, 1983). Fourth, behaviour integrates the activity of the whole brain. Behavioural measures therefore take into account any corrective changes occurring 'downstream' from the point of measurement. As a result, a behavioural assay may measure a change in brain function less accurately than a biochemical assay, but nevertheless, may give a clearer indication of its functional significance.

Like screening tests, behavioural assays are subject primarily to a single, simple requirement: in this case, that they should measure what they claim to measure. In practice, this may actually be rather difficult to demonstrate, and demonstration is essential, since 'obvious' assumptions have sometimes been shown to be false. It tends to be assumed, for example, that behavioural effects of the 5-HT precursor 5-HTP provide a measure of activity at central 5-HT synapses. It is true that most effects of 5-HTP are indeed mediated by 5-HT, but it can often be demonstrated that the 5-HT systems responsible are actually outside the brain (Carter *et al.*, 1978).

Additionally, if behaviour is being used as a measuring instrument, it is important that some attention be paid to basic principles of measurement (Martin & Bateson, 1986). One issue of obvious importance is the need to ascertain that the behaviour is sensitive to changes in the underlying physiological variable, and is not subject to 'floor' or 'ceiling' effects. A second pertinent issue is the level of measurement achieved: quantitative comparisons require that measurements be at least at the level of ordinal scaling (i.e. rank ordering). For many purposes, interval scaling (i.e. arithmetically linear relationships) would be preferable, and this desideratum is frequently translated into an unjustified assumption that interval scaling has been achieved. Interval scaling is implicit in the use of parametric statistical tests, such as *t*-tests or analysis of variance; if parametric analyses are applied to non-linear data the 'obvious' interpretation of the results may well be incorrect. A monotonic dose–response function, though not essential, is a third desirable feature in a behavioural assay. The head-twitch response to 5-HT agonists provides a useful measure of (spinal) 5-HT receptor function. However, the response wanes at high doses, with the emergence of other forms of abnormal behaviour (Drust *et al.*, 1979). There is no problem in interpreting an increase in head-twitching as an increase in responsiveness, but a decrease in head-twitching could result either from a decrease in responsiveness or from a large increase, and in published studies the direction of change is not always obvious.

1.1.3 Simulations

Having discussed models of drug action (screening tests) and models of brain function (behavioural assays), it is clear that the term 'animal model of anxiety/mania/obesity/. . .' properly applies to the third class of model, the simulation of human behaviour. In principle, any facet of behaviour is open to simulation. However, simulations of 'normal' behaviour usually take the form of a demonstration that effects first characterized in animals may also be observed in human subjects: examples include classical conditioning (Prokasy, 1965), the control of operant behaviour by schedules of reinforcement (Bradshaw *et al.*, 1976; Davison & Morley, 1987), or the application of animal learning theory to educational practice (Gagne, 1970). It would be more accurate to view this literature as consisting of studies of human models of animal behaviour, rather than the reverse.

While simulations of normal human behaviour are far from rare, simulations of abnormal behaviours predominate, for a variety of social, financial, ethical and legal reasons. An animal model of an abnormal behaviour attempts to simulate a symptom of the disorder, a group of symptoms, or exceptionally, a complete syndrome. Methods of constructing the simulation vary greatly; they include brain damage, selective breeding, selection of extreme individuals, and the application of a variety of factors assumed to be implicated in the etiology of mental disorders, such as stressors, social isolation, or ageing. The object of these manipulations is to produce a behavioural state that can be used as a tool to study aspects of the disorder being modelled. Broadly, four facets of the disorder may be addressed in a model: its

etiology, its treatment, its physiological basis, and the physiological mechanisms underlying its successful treatment. A given model may be appropriate to all or to only some of these problems.

In using an animal model to study a human disorder, the overriding consideration is the validity of the simulation. While it is relatively straightforward to assess the validity of a screening test or a behavioural assay, assessing the validity of a simulation is far more complex. However, the value of data derived from simulations depends crucially on their validity as models of human behaviour. For this reason, the problem of validating simulations will be discussed in some detail.

1.2 ASSESSING THE VALIDITY OF SIMULATIONS OF HUMAN BEHAVIOUR

Models are tools. As such, they have no intrinsic value; the value of a tool derives entirely from the work one can do with it. Conclusions arising from the use of a simulation of abnormal behaviour are essentially hypotheses, that must eventually be tested against the clinical state. An assessment of the validity of a simulation gives no more than an indication of the degree of confidence that we can place in the hypotheses arising from its use. However, the fact that a simulation appears to be valid carries no guarantee that such predictions will be fulfilled. We may feel that a model has a high degree of validity and yet see a prediction falling flat when tested in the real world; on the other hand, a model of apparently low validity may occasionally hit the jackpot. The validity of a simulation is a matter of judgement, rather than measurement.

Against this background, there are a number of yardsticks on which a judgement may be based. They address three broad aspects of a model, from which a picture may be built up of its overall validity. By analogy with the procedures used for validating psychological tests, I have called these three facets predictive validity, face validity and construct validity. Predictive validity means that performance in the test predicts performance in the condition being modelled; face validity means that there are phenomenological similarities between the two; and construct validity means that the model has a sound theoretical rationale (Willner, 1984*a*, 1986).

Earlier attempts to develop criteria for validating animal models of human behaviour have tended to concentrate largely on the assessment of face validity (McKinney & Bunney, 1969; Abramson & Seligman, 1977). The identification of two further categories reflects two ways in which the literature has developed in recent years. First, there has been a considerable expansion in the literature dealing with the pharmacological exploitation of animal models, much of which contributes to the assessment of predictive validity. Second, there has been significant growth in our understanding of the psychological mechanisms underlying psychopathological states, and examination of construct validity provides a convenient way of bringing animal models into contact with this very relevant literature. The exercise of distinguishing different types of validity has practical value, in that it allows ready identification of areas in which information about a particular model is weak or missing, and ensures that comparisons between different models are made on the basis of comparable data.

1.2.1 Predictive validity

In principle, questions of predictive validity can be addressed to a number of features of simulations, including their etiology and physiological basis. In practice, the predictive validity of the behavioural models used in psychopharmacology is determined primarily by their response to therapeutic drugs. Assessing the predictive validity of a simulation is therefore similar to assessing a screening test: the first question is whether the model discriminates efficiently between those agents that are clinically effective and those that are not. However, there are a number of important differences; in some areas we would ask more of a simulation than of a screening test, while in others we would ask less.

There are three major areas in which a simulation might be expected to go further than a screening test. One is that whereas a screening test is designed to identify drugs that improve the clinical condition, a simulation should respond also to treatments that make the clinical condition worse. It should be possible to show, for example, that the behavioural abnormalities apparent in a simulation of schizophrenia are increased by acute administration of amphetamine or l-dopa, drugs that exacerbate schizophrenic symptoms (Angrist *et al.*, 1973); this demonstration would have no relevance in a screening test for anti-schizophrenic drugs. A second difference is that a simulation should be responsive to all of the classes of drugs that are useful in treating the disorder, whereas a screening test might embody a strategic decision to look for drugs with particular chemical properties; this would then be recognized as an explicit limitation of the test. At the present time, for example, it makes good strategic sense to search for anxiolytic agents among the benzodiazepines and related compounds, and to design screening tests to minimize their undesirable side-effects; however, a simulation of anxiety might be expected to respond not only to benzodiazepines, but also to beta-blockers and various drugs that interact with the 5-HT system (see Lader, this vol., Ch. 4).

A third effect that should be demonstrable in a simulation, but is superfluous in a screening test, is that the relative potencies of different agents in the model correlate positively with their potencies in clinical use. This is potentially a very powerful test: witness, for example, the almost perfect correlations between the clinical potencies of neuroleptic drugs in schizophrenia, their ability to inhibit apomorphine-induced stereotyped behaviour, and their affinity for dopamine D2 receptors (Creese & Snyder, 1978). However, this test should be applied cautiously, having regard to species differences in drug absorption and metabolism. Furthermore, it can only be applied if drugs are chosen that vary widely in their clinical potency. This condition is satisfied for the neuroleptics, which vary over four orders of magnitude in their clinical potency. It is much less satisfactory for antidepressants, for example, most of which are used clinically within a very narrow dosage range; nevertheless, even in this case, there are sufficient outliers to allow a correlation of potencies to provide useful information (Willner, 1984*b*).

The major point of divergence between a screening test and a simulation is that while the value of a screening test is totally undermined by a failure to predict efficiently, the presence of false positive or false negative responses would not automatically invalidate a simulation. In discussing screening tests, we have already observed that a lack of agreement between a model and the clinical condition might arise from the incorrect classification of drugs as clinically active or inactive. False positives or negatives can also arise from species differences in drug kinetics. In particular, species differences in the rate or the route of drug metabolism can have profound effects on the concentration of the drug reaching its site of action; discrepancies arising in this way are especially likely in the case of a drug which has active metabolites. A third source of variability is the regime of drug administration. Clinical trials almost always assess the efficacy of drugs after a period of chronic treatment, and experiments carried out using acute drug administration may not accurately reproduce the appropriate conditions. The development of tolerance on chronic treatment can obliterate a 'false positive' response; conversely, the development of sensitization may overcome a 'false negative'.

In addition to these factors that can apply equally to false positives and false negatives, there are two further considerations. In the case of a false positive, it is important to consider whether a drug that appears to be clinically ineffective might reveal a positive effect if higher doses had been used. In practice, clinical trials at adequate dosage might prove impossible to carry out, owing to the emergence of unacceptable side-effects, which in a simulation would be far less of a deterrent. It is possible, for example, that the very poor performance of cholinergic agonists in dementia, in contrast to their positive effects in many animal models (see Dunnett & Barth, this vol., Ch. 14), might reflect a difficulty in achieving an appropriate dose level in the clinical studies (Summers *et al.*, 1986). False negatives, on the other hand, raise the interesting possibility of clinical subgroups, distinguishable by their differential drug response. Unlike other antidepressants, for example, monoamine oxidase inhibitors fail to normalize behaviour in olfactory bulbectomized rats; some workers have suggested that this may be related to the fact that these drugs are useful clinically only in an atypical group of patients (Jesberger & Richardson, 1985).

It should by now be clear that the process of validating an animal model cannot proceed mechanistically to a yes/no answer. A model may have predictive validity despite failing to discriminate efficiently between agents that are clinically active and those that are not. It is a matter of judgement to decide on the significance of discrepancies between drug effects in the model and in the clinical state. Inefficient prediction may not be incompatible with predictive validity, provided that we can understand the origin of the discrepancy.

1.2.2 Face validity

Face validity refers to a phenomenological similarity between the model and the disorder it simulates: on one hand, the model should resemble the disorder, while on the other, there should be no major dissimilarities. In a seminal paper that for the first time endowed the topic of modelling psychopathology in animals with a degree of scientific respectability, McKinney & Bunney (1969) suggested that to be valid, a model should resemble the condition it models in four respects: etiology, symptomatology, treatment, and physiological basis. Though admirable in principle, this requirement is in fact too stringent. In practice, we cannot require a simulation to correspond in every particular with the disorder it simulates, for the very good reason that there are major gaps in our understanding of the disorders themselves. It is not reasonable, for example, to require similarity of etiology in a simulation of schizophrenia, when the etiology of schizophrenia is virtually a closed book. Indeed, the main objective in setting up a simulation is precisely to fill out the missing pages.

We can, however, reasonably ask that a simulation be compared with the disorder it models in those areas where we do understand something of the disorder. A corollary to this position is that in areas where more is known about the disorder, we should examine the model more carefully for the degree of phenomenological similarity. A second corollary is that as the understanding of a disorder develops, criteria for evaluating the face validity of an animal model will automatically change. Where facts have been established about the etiology of a disorder or its physiological basis, they should obviously be included in an assessment of face validity. In general, however, the etiology and physiological basis of psychiatric disorders are poorly understood; assessment of face validity will usually be based primarily on symptomatology, about which most is known, and to some extent, on treatment.

A number of relevant questions may be asked

about treatment, additional to those issues raised under the heading of 'predictive validity'. Do abnormalities in the model respond to effective behavioural modes of treatment, where they exist? Are drug effects in the model achieved at reasonable doses, that produce tissue concentrations of the drug comparable to those found clinically? However, the most important of these questions concerns the treatment regime. Most prescriptions of psychotropic drugs envisage that the drug will be taken regularly over an extended period of weeks or months, and consequently, the drugs must continue to exert their effects after chronic treatment. It is therefore essential to establish that drug effects in the model are also demonstrable after a period of chronic administration. It is also necessary to establish that therapeutic effects are actually present while the drug is being administered: it is often expedient, following chronic drug treatment, to precede testing by a period of drug 'wash-out', and in these circumstances confusion between drug effects and withdrawal symptoms can easily arise (Noreika *et al.*, 1981; see also Section 1.3 below).

A related issue is less straightforward. In clinical use, not only must drugs continue to act after chronic treatment, but also, their onset of action is frequently delayed. This fact has been used to argue against the validity of models in which therapeutic effects can be demonstrated using acute drug treatment. The problem with accepting this argument is that delayed onsets of clinical action are poorly understood, and may in part reflect a delay in delivering an adequate concentration of drug to the brain (see, e.g., Willner, 1989). As with so many of the guidelines for validating animal models, an acute drug effect, in a model of a disorder that appears to require chronic treatment, may or may not indicate a lack of validity.

The central activity in an assessment of face validity is likely to be a comparison of symptoms displayed in the disorder with behavioural and other abnormalities apparent in the model. Since most disorders are syndromes that include a number of different symptoms, it should usually be possible to make multiple comparisons between the disorder and the model; the symptom check-list approach adopted in the DSM-III system of diagnosis provides a useful starting point for this exercise (American Psychiatric Association, 1987). In

practice, models often tend to focus on a single behaviour. In this case, the significant question is that of its centrality in the disorder; again, the DSM-III distinction between essential symptoms and optional extras may be helpful. However, the question of centrality cannot be answered fully at the level of phenomenology. Even though a reduction in food intake is the essential feature of anorexia nervosa, it remains a moot point whether anorexia is properly described as a disorder of feeding or a disorder of personality (see Montgomery, this vol., Ch. 8); if the latter, then simulations based on a reduction in food intake may be missing the target. A related question is the specificity of symptoms for a particular disorder (i.e. their value in differential diagnosis); clearly, a model will be less valid if the symptoms it simulates are common to a variety of disorders (Abramson & Seligman, 1977).

Where a number of points of similarity may be adduced between a model and the disorder it simulates, it becomes necessary to ask whether the identified cluster of symptoms forms a coherent grouping that might realistically be seen in a single patient, or whether they are drawn from a variety of diagnostic subgroupings. It is important in this context not to rely too heavily upon the diagnostic categories refined in DSM-III and related systems, since these categories were derived consensually rather than empirically, and themselves stand in need of validation (see, e.g., Carroll, 1983). Simulations offer a way of relating clusters of symptoms to specific etiological factors, specific treatment modalities, and specific physiological abnormalities. It remains an optimistic possibility that through this process, valid simulations of psychiatric disorders might themselves contribute to the clearer definition of diagnostic boundaries.

1.2.3 Construct validity

There is one important natural limitation to the attempt to establish face validity by mapping a point-to-point correspondence between a disorder and an animal model: there is no good reason to suppose that a given condition will manifest itself in identical ways in different species (Hinde, 1976) and, in fact, there are many obvious instances of divergence. For example, rearing on the hind legs is a prominent component of stimulant-induced stereotyped behaviour in rats but not in primates, while the reverse is true of scratching (Randrup & Munkvad, 1970). The physical topography of these behaviours is quite different; nevertheless, we are able to say that they are homologous across species; the judgement of homology arises from an understanding that the two behaviours have the same physiological substrate. Patterns of maternal behaviour, which vary widely across species, provide another example of homology. In the case of a human mother cradling her baby and a female rat retrieving her pups, the judgement of homology arises primarily from an understanding of the psychosocial environment in which the behaviours take place. In both instances, the decision that different behaviours in different species reflect different manifestations of a similar underlying process is based upon a theoretical rationale, which derives from consideration of factors other than the behaviours themselves.

The theoretical rationale behind a model requires evaluation, and a satisfactory outcome endows the model with construct validity. In advancing this concept it is assumed implicitly that it is possible to construct theories of psychopathology that have some application to species other than people. As this has in the past been a source of some confusion (exemplified by the extreme difficulty of defining what might be meant by a schizophrenic rat) it may be helpful to consider briefly the general shape of psychopharmacological theories.

I have argued elsewhere (Willner, 1984c, 1985) that it is inadequate for a psychopharmacological theory simply to describe a link between a set of biochemical events (e.g. antagonism of benzodiazepine receptors) and a change in human experience (relief of anxiety). An adequate theory must also interpose an account at two intermediate levels. At the first, it must explain the consequences of the identified biochemical changes for functional activity within the brain (which pathways are affected, and how), while at the second, it must explain how these changes in functional activity affect the brain's ability to process information (which requires some understanding of the role of the affected systems in the cognitive activity of the normal brain). It then becomes necessary to consider, in addition, the manner in which those underlying cognitive changes are incorporated into

subjective experience, and so to arrive at a proper understanding of the experiential effects of a drug.

Having distinguished between these various levels of analysis we are now in a position to consider which aspects of a theory of psychopathology may be addressed in animals, and which may not. For most practical purposes, we do not possess experimental tools that will allow us to address queries to animals about their subjective state; the questions we can ask them are strictly limited by the necessity that their replies must be behavioural rather than verbal. Even in the case of the drug discrimination procedure, the data from human experiments caution against equating internal states with subjective states (Chait *et al.*, 1986; see also Goudie, this vol., Ch. 17). We do not, therefore, attempt to simulate the experiential aspects of drug action or of psychopathology. What we are attempting in a simulation is to model the constraints on information processing and behaviour that underly the experiential changes.

From this perspective, the assessment of construct validity is a three-stage process (Willner, 1986). The first step is to identify the behavioural variable that is being modelled; the second is to assess the degree of homology between the identified variable and behaviour in the simulation; the third is to assess the significance of the identified variable in the clinical picture.

Each of these stages is fraught with difficulty. It might, for example, seem a straightforward matter to decide what is being modelled, and this may indeed be so in some cases, where the behavioural objectives of the model are specified in advance. Many models of dementia, for example, embody an explicit choice to simulate specific disorders of learning and memory (see Dunnett & Barth, this vol., Ch. 14). However, there are many other starting points for a simulation, such as the application of a supposed etiological factor, or the use of drugs to induce a supposedly relevant physiological state; the behavioural state simulated may then be shrouded in obscurity. It remains a total mystery what aspect of depression was being addressed by experiments in which monkeys were isolated in the dark, in a vertical cylinder nicknamed 'the well of despair' (Harlow & Suomi, 1971).

Having established what is being modelled, the next step is to assess whether the two behaviours are homologous. However, this can only be done if both behaviours are well understood, and frequently they are not. The learned helplessness model of depression provides an excellent example of this problem. In this model, prior exposure to uncontrollable aversive events causes a subsequent learning deficit; this effect may be demonstrated very readily in both rats and people (Miller *et al.*, 1977). However, despite a very substantial research effort, the nature of the underlying behavioural mechanisms remains unclear. The explanation originally proposed was that subjects exposed to uncontrollable events learn that events are uncontrollable: a homologous process in rats and people (Seligman, 1975). However, in rats there is evidence that motor disabilities and stress-induced analgesia contribute to the behavioural deficits (Jackson *et al.*, 1978, 1979), while in people helplessness effects are sensitive to very minor changes in the experimental procedure (Buchwald *et al.*, 1978). Returning to the original examples of maternal behaviour and drug-induced stereotypies, we are able to see that for two behaviours to be homologous, they should share a similar physiological basis and occur in a similar behavioural context. In learned helplessness, the growth of information about the two behaviours has, if anything, reduced the likelihood of homology.

The final step in the assessment of construct validity is an evaluation of the significance in the overall clinical picture of the behaviour modelled by the simulation. Many animal models of drug dependence, for example, focus on the rewarding properties of drugs. However, while rewarding effects are of undoubted importance in the early stages of drug use, some theoretical formulations emphasize the fear of withdrawal as a major factor in dependence (see Goudie and Hartnoll, this volume, Chapters 17 and 19). If this is correct, then models that simulate rewarding effects may be of only limited relevance. A similar problem concerning the significance of low food intake in anorexia nervosa has already been remarked: the anorectic's low food intake should not be mistaken for a reduced appetite for food (see Montgomery, this vol., Ch. 8). Because debates of this kind take place entirely within the human literature, these problems do not figure prominently in discussions of animal models. However, it is obviously of some considerable importance to know whether a simulation is asking the correct question.

Of the three roads to validity, construct validity is the most fundamental: while a model may, for a variety of reasons, fail to meet criteria for predictive or face validity and still survive, it would be difficult to retain confidence in a model whose theoretical rationale had been exploded. Construct validity is the most difficult aspect of a model to establish, but also, the most challenging.

1.3 USING MODELS

Of course, while it is important to distinguish clearly between screening tests, behavioural assays and simulations, in practice the three types of model interact, in a variety of ways. One obvious point at which models of different types meet is through the cross-talk in their scientific development. Screening tests, for example, are frequently developed as offshoots of attempts to simulate a disorder, or as a result of insight into the physiological mechanisms revealed by a simulation. This process can also work in the opposite direction: the early neuroleptic screening tests threw up a range of active agents that were later discovered to be dopamine receptor antagonists; as a result, later screening tests were based on bioassays for dopamine antagonism, and chronic overstimulation of dopamine receptors forms the basis of some current attempts to simulate schizophrenia (see Lyon, this vol., Ch. 11).

A second significant point of intersection is that the same procedure may be used for two quite different purposes. In such cases, assessing the model's suitability for its two uses might involve radically different sets of criteria. The use of the same procedure either as a screening test or as a bioassay has already been discussed. If the procedure is being used as an assay then it must have good metric properties; however, if it is being used as a screening test all we ask is that it should discriminate accurately between agents that are active and those that are inactive. Similarly, many procedures may serve either as screening tests or as simulations. The suppression of lever pressing by electric shock, for example (Geller & Seifter, 1960), may be used either as a simulation of anxiety, or as a screening test for anxiolytic drugs (see Green & Hodges, and Stephens & Andrews, this vol., Ch. 2, 3). Again, the important point is that the criteria employed to decide whether or not this should be considered a 'good' model depend upon the purpose for which it is being used. If the Geller-Seifter conflict procedure is being used as a screening test, then its validity as a simulation is irrelevant: the only requirement is that it should efficiently predict anxiolytic activity. However, if the test is being used to investigate the brain mechanisms of anxiety, then the overriding concern is with its validity as a simulation; its predictive power is relatively unimportant since, as discussed above, there are a number of factors that might turn a valid simulation into an inefficient predictor of clinical activity.

A third point of contact is that two different types of model may be used within a single experiment. This applies particularly to the simultaneous use of a simulation and a behavioural assay. A case study will serve to illustrate both the development of a behavioural assay and its use in conjunction with a simulation procedure. The problem addressed in these experiments was whether antidepressant drugs, chronically administered, reduce the sensitivity of presynaptic dopamine autoreceptors, and whether this effect contributes to their clinical action; the literature was equivocal with respect to the first question and silent on the second. The model used to assay the responsiveness of dopamine autoreceptors was the suppression of food intake by a low dose of the dopamine agonist apomorphine. To validate this model we demonstrated that the response was blocked by centrally acting dopamine receptor antagonists, but not by a dopamine antagonist that does not enter the brain, or by a variety of other types of drug; that the response could be elicited by administering apomorphine directly to the ventral tegmental area of the midbrain, where the cells of origin of the mesolimbic dopamine system are located; and that administration of a dopamine antagonist to the ventral tegmental area blocked the effect of systemically administered apomorphine (Willner *et al.*, 1985; Muscat *et al.*, 1986; Towell *et al.*, 1986*a*). Then we examined the effect of apomorphine on food intake during chronic treatment with antidepressant drugs, and found that there was indeed a reduced response, indicative of dopamine autoreceptor subsensitivity. However, the effect could only be demonstrated following withdrawal from the antidepressants (Towell *et al.*, 1986*b*). Could it be relevant to the clinical action? To

answer this, we turned to a simulation of depression. In this model, which simulates the cardinal symptom of endogenous depression, the inability to experience pleasure, rats are exposed chronically to a variety of mild stressors; over a period of weeks, they gradually lose their preference for a highly rewarding sweet solution (Willner *et al.*, 1987; see also Willner, this vol., Ch. 5). On administering apomorphine, to measure dopamine autoreceptor sensitivity, we found that the attenuated response typical of withdrawal from antidepressants was also present in the stressed animals. Antidepressant administration restored normal preference behaviour in the stressed animals, but caused no further desensitization of dopamine autoreceptors. We therefore concluded that this effect is not, in fact, one of the mechanisms by which antidepressants exert their clinical action (Muscat *et al.*, 1988).

I have presented these experiments in some detail as relatively few studies have exploited the possibilities of using behavioural assays within the context of a simulation procedure. However, studies of this kind must inevitably increase in the future. Numerous effects of psychotherapeutic drugs on brain function have been described, but almost invariably in 'normal' animals. Before accepting that such effects are responsible for clinical improvement, it will be necessary in every case to demonstrate that the effect is also present in simulations of the disorder (or in clinical trials). As the above example shows, there is no guarantee of success in the attempt to demonstrate the clinical relevance of drug effects described in 'normal' animals.

1.3.1 Models and their users: three caricatures

Before moving from general principles to a more detailed examination of the behavioural models used in specific areas of psychopharmacology, we need to ask two further questons. Who uses behavioural models, and why? Unlike many disciplines, psychopharmacology contains not one, but three scientific communities: academics, industrial pharmacologists, and clinicians. These three groups relate to the discipline in different ways; they have different objectives, which give rise to different scientific priorities, and lead them to different uses in the available behavioural tools.

The major concern of an academic psycho-pharmacologist is (or should be) the construction, development and evaluation of theories. I have already discussed what is required of a theory in psychopharmacology. To reiterate briefly, an adequate theory of psychotropic drug action must provide a coherent account of how the drug interacts with the brain, the specific anatomical sites of action, the resulting physiological changes, the consequent changes in cognitive and behavioural abilities, and the relationship between cognitive change and subjective experience. The job of the academic psychopharmacologist is to provide explanations in some or all of these domains: to elucidate the biochemical, physiological and psychological mechanisms that underlie psycho-pathological behaviour and its treatment by psy-chotherapeutic drugs. The major behavioural tool deployed to these ends is the simulation, and assessment of the validity of simulations therefore falls within this area of work. Behavioural assays are also used extensively, in the investigation of physiological mechanisms. However, the academic psychopharmacologist has very little scientific interest in screening tests, whose only contribution to the development of explanatory theory is that new drugs often prove to be of value as experimental tools.

The industrial psychopharmacologist, on the other hand, is concerned primarily to discover new drugs. The screening test is therefore the major behavioural tool for this group of workers. They also make heavy use of behavioural assays, both as screening tests and as tools to investigate the physiological actions of clinically active drugs. However, industrial psychopharmacologists are interested in simulations (and their validity) only to the extent that a valid simulation might serve as the starting point from which to develop a superior screening test.

The major concern of clinical psychopharmacologists is to improve the care of their patients. Their involvement with animal models is usually less direct than that of the other two groups; they do not themselves use the models, but rather, apply and test clinically the fruits of others' work. Clinicians are obviously interested in simulations, since the main object of a simulation is to throw light on the nature of the disorder. However, because clinical psychopharmacologists are pri-marily clinicians and only secondarily pharmaco-logists, unlike the other two groups they have no commitment to drugs, if other forms of treatment or prevention should prove more effective. Their interest is primarily in the behavioural aspects of a model rather than its physiological mechanisms: many of the behavioural techniques in psycho-therapy derived initially from simulations (Eysenck & Martin, 1987). Even though their contact with simulations is somewhat distant, the interests of clinicians extend more widely than that of the other groups, to include areas in which there are behavioural simulations of disorders that are not treated pharmacologically (Keehn, 1986). Clinical psychopharmacologists are also interested in screening tests, and look to them to provide compounds that will be more efficaceous than existing drugs, and/or have fewer side-effects. However, as clinicians, they have little interest in using behavioural models as assay systems. Information of this kind becomes important only when it leads to new treatments or to diagnostic improvements.

Obviously, these sketches of the academic, the industrial, and the clinical psychopharmacologist are grossly oversimplified caricatures, and many psychopharmacologists do not fit neatly into these stereotypes. Many industrial pharmacologists are academics at heart, whose wish to contribute to the development of scientific theory is at least the equal of their desire to develop new products; some pharmaceutical companies are sufficiently enlightened to recognize this as a legitimate aspir-ation, and make appropriate provision for 'pure' research. Equally, many academics are attracted by the possibility of making an immediate practical impact in the real world, by developing screening tests for the pharmaceutical industry. Scientific collaboration between the academic and industrial sectors is highly developed, to the benefit of both: industry takes theoretical advances and uses them to develop both new products and new routes to product development; conversely, novel agents produced by the industry are among the most powerful analytical tools available to conduct 'pure' research. Clinical psychopharmacology plays a central role in guiding preclinical research, through the essential function of testing the hypo-theses arising from the use of animal models; at the same time, the agenda for clinical psychopharmacology is largely determined by preclinical research, which identifies the physiological questions to be asked in the clinic and provides the novel agents that form the subject matter of clinical trials. Furthermore, there is considerable movement of personnel, at all levels, between the academic world and industry; the adoption of a new persona appropriate to the new institutional context can be a long drawn-out process. In addition, many clinicians take time out from their clinical responsibilities to do basic research, and in so doing slip into the other two roles.

1.3.2 Theoretical, industrial and clinical perspectives

Despite the blurring of distinctions between the three sections of the psychopharmacological community, it is important to recognize that their interests are not identical, and that these differences are reflected in the direction of research and in the choice of experimental methodologies. In this book, we have tried for the first time to identify the particular perspectives of academic, industrial and clinical psychopharmacologists working in different areas of psychopharmacology. In the six areas covered, we have attempted to examine the ways in which the three perspectives interact, and to clarify the contribution of behavioural models to progress in each of the three domains.

Of the three caricatures outlined above, that of the academic psychopharmacologist is probably the most accurate: the chapters that follow demonstrate a remarkable creativity in generating new ideas for modelling psychiatric disorders in animals. However, in most of the areas surveyed the emphasis has been largely on the development of models, rather than their use to probe the mechanisms underlying the disorders modelled. In large part, this reflects the sad fact that while great vigour has been applied to the development of a wide variety of models, many have little to recommend them. Nevertheless, it is clear that the framework outlined in this chapter (Section 1.2) provides a straightforward means of assessing the strengths and weaknesses of a model; it is also clear that in some instances, converging guidance from a number of models can be informative despite the weakness of each model individually. Thus, while the actual contribution of simulations to our

understanding of psychiatric disorder has been limited, their potential is considerable.

The least accurate of the three sketches is that of the industrial psychopharmacologist. It is apparent in reading accounts of the screening process that industrial psychopharmacologists do in fact share the academic preoccupation with the validity of models as simulations of psychiatric disorders. It is instructive to examine why this should be, given that the theoretical perspective is not a necessary feature of a screening programme. One reason is more political than scientific: the development of a new drug represents a major investment of resources, which can more easily be justified to the company decision makers by evidence derived from a test that has face validity, and so makes intuitive sense (Broekkamp, personal communication). However, there are also important considerations within the scientific context of drug development, which broadly reflect the belief that a better understanding of the disorder may lead to new avenues for drug development. For some disorders, such as dementia or obesity, there are no well established treatments, and therefore there is no basis for an empirical predictive test. In these cases the only alternative to seeking valid simulations is to take promising compounds directly into the clinic; this is a high-risk strategy on which few companies would wish to rely exclusively. At the other extreme, the established antidepressants have a great variety of biochemical actions, which have formed the basis for the development of a bewildering variety of novel agents that may have antidepressant properties (see Danysz *et al.*, this vol., Ch. 6). The number of potential antidepressants undergoing clinical trials has therefore seen rapid expansion. An unforseen consequence is that as we do not at present know which of the new agents work and which do not, it has become impossible to assess with any degree of accuracy how well the available screening tests succeed in predicting clinical activity. For other disorders, most notably schizophrenia, a particular treatment modality is so successful that drug development is based primarily around particular chemical structures with specified biochemical actions. However, this strategy by its nature precludes the discovery of compounds acting through different mechanisms. The example of anxiolytic drugs illustrates perfectly the dangers of this preemptive reliance on medicinal chemistry: no sooner had screening programmes settled down to searching for compounds active at the benzodiazepine receptor, than several new families of potential non-benzodiazepine anxiolytics were discovered, that interact in various ways with 5-HT receptors (see Part 2). The fact that these new agents differ significantly from benzodiazepines, and from one another, in traditional anxiolytic-sensitive behavioural tests, highlights an urgent need to clarify which aspects of anxiety the different tests model.

It is clear from this discussion that the quest for valid simulations is actually central to industrial psychopharmacology, since such models may be used to identify desirable biochemical properties. However, as discussed earlier (Section 1.1), screening tests and simulations have very different requirements. This essential tension at the heart of industrial psychopharmacology is most commonly resolved by an uneasy compromise: a layered screening programme in which rough and ready primary screens are followed by tests of greater complexity that approximate more closely to valid simulations.

The clinical perspective on behavioural models, in reality, is twofold. On the one hand, clinicians view animal models of psychiatric disorders with a healthy scepticism, and provide a valuable critical service in pointing out their many shortcomings, when measured against the complexities of human behaviour. On the other hand, the literature generated from animal models encourages clinicians to confront their own limitations, and these are many. Psychiatric diagnosis is to a large extent descriptive and consensual, and there are few indications that this situation is likely to improve in the near future. Ideas of the etiology of psychiatric disorders are still in their infancy (note, for example, the pervasive role of 'stress'). Even treatment, the jewel in the crown of biological psychiatry, remains largely empirical. The absence of a sound theoretical basis is illustrated by the way in which treatments find application outside their original indications: for example, antidepressant drugs as anxiolytics or anorectics, and neuroleptic drugs as anxiolytics or antidepressants. (Consider also that mild dementias may be cured by treating an underlying depression, but depression is exacerbated by cholinergic agonists, the current front-runners in the race to find an anti-dementia drug.)

As one of the 'clinical' contributors to this volume notes, because behavioural models are explicitly related to a broader body of theory, they fulfill a valuable function in forcing clinicians to examine critically their own assumptions. It is all too easy, from a clinical standpoint, to denigrate animal models of psychiatric disorders. However, it could equally be seen as an area of strength that the models lend themselves so readily to critical analysis. Ultimately, the major contribution of animal models may well be that they encourage clinicians to apply a similar rigour to the analysis of disordered human behaviour.

REFERENCES

Abramson, L.Y. & Seligman, M.E.P. (1970). Modelling psychopathology in the laboratory. History and rationale. In: J.D. Maser & M.E.P. Seligman (eds.), *Psychopathology: Animal Models*, pp. 1–26. Freeman: San Francisco.

American Psychiatric Association (1987). *Diagnostic Criteria from DSM-III-R*. APA, Washington D.C.

Angrist, B., Sathananthan, G.S. & Gershon, S. (1973). Behavioural effects of L-dopa in schizophrenic patients. *Psychopharmacologia* **31**, 1–12.

Bergman, I., Brane, G., Gottfries, C.G., Jostell, K.-G., Karlsson, I. & Svennerholm, L. (1983). Alaproclate: A pharmacokinetic and biochemical study in patients with dementia of the Alzheimer type. *Psychopharmacology* **80**, 279–83.

Bradshaw, C.M., Szabadi, E. & Bevan, P. (1976). Behaviour of humans in variable interval schedules of reinforcement. *Journal of the Experimental Analysis of Behavior* **27**, 275–9.

Buchwald, A.M., Coyne, J.C. & Cole, C.S. (1978). A critical evaluation of the learned helplessness model of depression. *Journal of Abnormal Psychology* **87**, 180–93.

Carroll, B.J. (1983). Neurobiologic dimensions of depression and mania. In: J. Angst (ed.), *The Origins of Depression: Current Concepts and Approaches*, pp. 163–86. Springer: Berlin.

Carter, R.B., Dykstra, L.A., Leander, J.D. & Appel, J.B. (1978). Role of peripheral mechanisms in the behavioral effects of 5-hydroxytryptophan. *Pharmacology, Biochemistry and Behavior* **8**, 249–53.

Chait, L.D., Uhlenluth, E.H. & Johanson, C.E. (1986). The discriminative stimulus and subjective effects of phenylpropanolamine, mazindol and d-amphetamine in humans. *Pharmacology, Biochemistry and Behavior* **24**, 1665–72.

Colpaert, F. & Slangen, J.L. (1982). *Drug*

Discrimination: Applications in CNS Pharmacology. Elsevier: Amsterdam.

Creese, I. & Snyder, S.H. (1978). Behavioral and biochemical properties of the dopamine receptor. In: M.A. Lipton, A. DiMascio & K.F. Killam (eds.), *Psychopharmacology: A Generation of Progress*, pp. 377–88. Raven Press: New York.

Davison, M. & Morley, D. (1987). *The Matching Law: A Research Review* Lawrence Erlbaum: Hillsdale, N.J.

Drust, E.G., Sloviter, R.S. & Connor, J.D. (1979). Effect of morphine on 'wet dog' shakes caused by cerebroventricular injections of serotonin. *Pharmacology* **18**, 299–305.

Eysenck, H.J. & Martin, I. (1987), *Theoretical Foundations of Behaviour Therapy.* Plenum Press: New York.

Gagne, R.M. (1970). *Conditions of Learning.* Holt, Reinhart & Winston: New York.

Geller, I. & Seifter, J. (1960). The effects of meprobamate, barbiturates, D-amphetamine and promazine on experimentally induced conflict in the rat. *Psychopharmacologia* **1**, 482–92.

Harlow, H.F. & Suomi, S.J. (1971). Production of depressive behaviours in young monkeys. *Journal of Autism and Child Schizophrenia* **1**, 246–55.

Hinde, R.A. (1976). The uses of similarities and differences in comparative psychopathology. In: G. Serban & A. Kling (eds.), *Animal Models in Human Psychobiology*, pp. 187–202. Plenum Press: New York.

Horovitz, Z.P. (1965). Selective block of mouse killing by antidepressants. *Life Science* **4**, 1909–12.

Jackson, R.L., Maier, S.F. & Rapoport, P.M. (1978). Exposure to inescapable shock produces both activity and associative deficits in rats. *Learning and Motivation* **9**, 69–98.

Jackson, R.L., Maier, S.F. & Coon, D.J. (1979). Long-term analgesic effect of inescapable shock and learned helplessness. *Science* **206**, 91–3.

Janowsky, D.S. & Risch, S.C. (1984), Cholinomimetic and anticholinergic drugs used to investigate an acetylcholine hypothesis of affective disorders and stress. *Drug Development Research* **4**, 125–42.

Jesberger, J.A. & Richardson, J.S. (1985). Animal models of depression: Parallels and correlates to severe depression in humans. *Biological Psychiatry* **20**, 764–86.

Katz, R.J. (1981). Animal models and human depressive disorders. *Neuroscience and Biobehavioral Reviews* **5**, 231–46.

Keehn, J.D. (1986). *Animal Models for Psychiatry.* Routledge: London.

Kelly, P.H., Seviour, P.W. & Iversen, S.D. (1975). Amphetamine and apomorphine responses in the rat following 6-OHDA lesions of the nucleus accumbens septi and corpus striatum. *Brain Research* **94**, 507–22.

Martin, P. & Bateson, P. (1986). *Measuring Behaviour: An Introductory Guide.* Cambridge University Press: Cambridge.

Martin-Iverson, M., Leclere, J.F. & Fibiger, H.C. (1983). Cholinergic–dopaminergic interactions and the mechanisms of action of antidepressants. *European Journal of Pharmacology* **94**, 193–201.

McKinney, W.T. (1984). Animal models of depression: An overview. *Psychiatric Developments* **2**, 77–96.

McKinney, W.T. & Bunney, W.E. (1969). Animal model of depression: Review of evidence and implications for research. *Archives of General Psychiatry* **21**, 240–8.

Miller, W.R., Rosellini, R.A. & Seligman, M.E.P. (1977). Learned helplessness and depression. In: J.D. Maser & M.E.P. Seligman (eds.), *Psychopathology: Animal Models*, pp. 104–30. Freeman: San Francisco.

Muscat, R., Willner, P. & Towell, A. (1986). Apomorphine anorexia: A further pharmacological characterization. *European Journal of Pharmacology* **123**, 123–31.

Muscat, R., Towell, A. & Willner, P. (1988). Changes in dopamine autoreceptor sensitivity in an animal model of depression. *Psychopharmacology* **94**, 545–50.

Niemegeers, C.J.E. & Janssen, P.A.J. (1979). A systematic study of the pharmacological activities of dopamine antagonists. *Life Science* **24**, 2201–16.

Noreika, L., Pastor, G. & Liebman, J. (1981). Delayed emergence of antidepressant efficacy following withdrawal in olfactory bulbectomized rats. *Pharmacology, Biochemistry and Behavior* **15**, 393–8.

Ogren, S.-O. & Holm, A.-C. (1980). Test-specific effects of the 5-HT uptake inhibitors alaproclate and zimelidine on pain sensitivity and morphine analgesia. *Journal of Neural Transmission* **47**, 253–71.

Prokasy, W.F. (ed.) (1965). *Classical Conditioning: A Symposium.* Appleton, Century, Crofts: New York.

Randrup, A. & Munkvad, I. (1970). Biochemical and psychological investigations of stereotyped behaviour. In: E. Costa & S. Garattini (eds.), *Amphetamines and Related Compounds*, pp. 695–713. Raven Press: New York.

Seligman, M.E.P. (1975). *Helplessness: On Depression, Development and Death.* Freeman: San Francisco.

Squires, R.F. & Braestrup, C. (1977). Benzodiazepine receptors in the brain. *Nature* **266**, 732–4.

Summers, W.K., Majovski, L.V., Marsh, G.M., Tachiki, K. & Kling, V. (1986). Oral tetrahydroaminoacridine in long-term treatment of senile dementia. *New England Journal of Medicine* **315**, 1241–5.

Towell, A., Muscat, R. & Willner, P. (1986*a*). Apomorphine anorexia: The role of dopamine cell body autoreceptors. *Psychopharmacology* **89**, 65–8.

Towell, A., Willner, P. & Muscat, R. (1986*b*). Behavioural evidence for autoreceptor subsensitivity in the mesolimbic dopamine system during withdrawal from antidepressant drugs. *Psychopharmacology* **90**, 64–71.

Westenberg, H.G.M. & den Boer, J.A. (1988). Serotonin uptake inhibitors and agonists in the treatment of panic disorder. *Psychopharmacology* **96** (Suppl.), 56.

Willner, P. (1984*a*). The validity of animal models of depression. *Psychopharmacology* **83**, 1–16.

Willner, P. (1984*b*). The ability of antidepressant drugs to desensitize beta-adrenergic receptors is not correlated with their clinical potency. *Journal of Affective Disorders* **7**, 53–8.

Willner, P. (1984*c*). Drugs, biochemistry and subjective experience: Towards a theory of psychopharmacology. *Perspectives in Biology and Medicine* **28**, 49–64.

Willner, P. (1985). *Depression: A Psychobiological Synthesis*. Wiley: New York.

Willner, P. (1986). Animal models of human mental disorders: Learned helplessness as a paradigm case. *Progress in Neuropsychopharmacology and Biological Psychiatry* **10**, 677–90.

Willner, P. (1989). Sensitization to the actions of antidepressant drugs. In: M.W. Emmett-Oglesby & A.J. Goudie (eds.), *Psychoactive Drugs: Tolerance and Sensitization*, pp. 407–59. Humana: Clifton, N.J.

Willner, P., Towell, A. & Muscat, R. (1985). Apomorphine anorexia: A behavioural and neuropharmacological analysis. *Psychopharmacology* **87**, 351–6.

Willner, P., Towell, A., Sampson, D., Sophokleous, S. & Muscat, R. (1987). Reduction of sucrose preference by chronic mild unpredictable stress, and its restoration by a tricyclic antidepressant. *Psychopharmacology* **93**, 358–64.

Zis, A.P. & Goodwin, F.K. (1979). Novel antidepressants and the biogenic amine hypothesis of depression. The case for iprindole and mianserin. *Archives of General Psychiatry* **36**, 1097–107.

PART II

Models of anxiety

2

Animal models of anxiety

SIMON GREEN and HELEN HODGES

2.1 INTRODUCTION

Animal models have been used extensively in psychopharmacology but with little discussion of their validity (see Willner, this vol., Chapter 1). Consideration of the validity of animal models of anxiety has usually been restricted, implicitly or explicitly, to the success or failure of a given model in predicting the clinical anxiolytic potency of pharmacological agents. Although screening may be the most valuable short-term function of models, their contribution to brain research as bioassays and, particularly, as simulations cannot be ignored. In fact the long-term significance of animal models may well lie in their contribution to the unravelling of human psychopathology. At present, psychiatric diagnosis, classification, and treatment are imprecise. The fact that substantial diagnostic categories, such as 'neurotic disorder', have shifted around between DSMII and DSM-III (i.e. between 1968 and 1980; American Psychiatric Association, 1980) is solid evidence that classificatory systems are still evolving. For no major psychopathology does a true nosology of the disorder exist to relate systematically aetiology, symptoms, treatment, and outcome. Given this fuzziness at the human level, we can be optimistic and conclude that animal models of anxiety may well be able to help clarify some of these human issues.

The underlying motivation of the academic psychopharmacologist is to construct or modify theories of anxiety and anxiolytic drug action; we examine some of these theories later. Therefore, in general terms the academic psychopharmacologist does not need to discover a new and successful anxiolytic tomorrow (although we would not ignore one if it turned up) and can be concerned more with the use of animal models as simulations rather than as bioassays or screens. As has been pointed out in Chapter 1, these categories are not mutually exclusive, but we hope to show that simulations of complex human psychopathologies must themselves be complex, which often makes them too expensive and time consuming to be useful as screening tests. This position also assumes that simulations with adequate construct validity add significantly to the sum of knowledge about a given psychopathology; i.e. that findings will feed forward to the analysis of human anxiety and, possibly, backward to the design of screening tests. If this assumption is false, then all animal models can be reduced to the simplest and cheapest screening test.

In this chapter we review a number of animal models of anxiety. As a later chapter (Stephens & Andrews, Ch. 3), deals with screening tests in detail, we shall take the opportunity to concentrate upon the role of models as simulations of anxiety. To anticipate our conclusions, and perhaps reflecting a novelty in our approach, it now seems clear that no one model is completely adequate as either a screen or a simulation. So we shall not be producing a 'best-buy' list (cf. Lal & Emmett-Oglesby, 1983; Pellow *et al.*, 1985; Treit, 1985), but rather presenting arguments in favour of a broad-based approach to anxiolytic psychopharmacology.

2.1.1 Anxiety – definitions and diagnosis

The subjective experience of anxiety is ubiquitous; it is a rare individual who survives a day without occasional bouts of anxiety, precipitated perhaps by physical danger, social interaction, or problems with the family or work. Clinically anxious people, then, are to a large extent self-defined as those for whom anxiety has become extreme and for whom coping responses have become ineffective, leading to them seeking professional help (Lader, 1983; this vol., Ch. 4).

The psychiatric diagnosis of anxiety disorders is not straightfoward, although there is broad agreement about major categories (American Psychiatric Association, 1980). Anxiety presents as autonomic hyperactivity, with raised blood pressure and heart-rate, increased sweating, and dryness of mouth and throat, along with cognitive and affective states of increased vigilance, distractibility, and apprehension. Motor fatigue and insomnia are common. Anxiety disorders of childhood include separation anxiety, avoidance disorder (of strangers), and overanxious disorder, a more general apprehension of the world. In adults, anxiety disorders (which have an overall point prevalence of 2–4% of the population) are divided into 'anxiety states' and 'phobic disorders'.

Anxiety states are further subdivided into: 'panic disorder', in which the autonomic and cognitive/affective signs are exaggerated into attacks of palpitations, choking, dizziness, and fainting, with intense fear of death, insanity, or uncontrollability; 'generalised anxiety disorder', a free-floating anxiety syndrome, with no obvious precipitant and not consistently attached to any particular object or situation; 'obsessive-compulsive disorder', which involves recurrent obsessions (thoughts, ideas, images, impulses) and compulsive behaviours such as hand-washing that seem to evolve in an attempt to reduce the anxiety level; acute or chronic 'post-traumatic stress disorder'; and a residual category of 'atypical anxiety disorder'.

Phobic disorders are also subdivided, into agoraphobia, social phobia, and a residual category of simple phobia. One complication of this particular subclassification, involving conditions where the anxiety state has become attached to a particular object or situation, is a division of agoraphobia depending upon the presence or absence of associated panic attacks. This obviously muddies the distinction between phobic disorders and the panic disorders categorized under anxiety states, and it has been suggested (Marks, 1987) that the diagnosis of 'panic' as an entity separable from severe anxiety is unjustified (see also Lader, this vol., Ch. 4).

Anxiety has been studied extensively as a dimension of normal personality. Attempts to demonstrate significant and consistent psychophysiological differences between subjects high and low in trait anxiety have in general been unsuccessful (reviewed in Eysenck & Eysenck, 1985). However, consistent cognitive differences have been identified (Eysenck, 1987), which will presumably be reflected in some aspects of brain function. At present, however, even the most detailed model of the brain substrates of anxiety (Gray, 1982) cannot incorporate such findings.

Whether animal models have a contribution to make in this particular area is uncertain; one would hope that individual differences in susceptibility to anxiogenic stimuli, and perhaps variable responses to different types of threat, could be modelled in animals (e.g. Blanchard *et al.*, 1974). Genetic and pharmacological studies of Maudsley reactive and non-reactive rat strains may be important in this regard (e.g. Robertson *et al.*, 1978).

2.1.2 Drug treatment of anxiety

Drug treatment of anxiety is dominated by the benzodiazepines (BZs) of which chlordiazepoxide (librium) and diazepam (valium) are the classic examples. Other drugs prescribed include meprobamate, barbiturates, triazolopyridazines; and various non-prescribed drugs are also used, of which alcohol is the most popular. However, there is as yet no convincing evidence that any anxiolytic as effective clinically as the BZs acts through brain mechanisms independent of those influenced by the BZs, although the recent interest in serotonin 5-HT$_{1A}$ agonists or mixed agonist/antagonists such as buspirone may alter this view (Chopin & Briley, 1987; Stephens & Andrews, this vol., Ch. 3); even then it is possible that common substrates will emerge.

The application of animal models to human anxiety has not been consistent across the range of diagnostic categories. This is because the models are drug-driven, and the different categories are not equally drug-responsive. Phobic disorders do not on the whole respond significantly to drug therapy alone, although BZs may be a useful adjunct to the more usual treatment involving one of the behaviour or cognitive therapies (Rickels, 1985; Davison & Neale, 1986). Similarly, in obsessive-compulsive disorders drugs may be used as an adjunct to forms of behaviour or psycho-therapy, but are rarely effective when given alone (Rickels *et al.*, 1978). Of the major anxiety dis-orders, panic attacks do not respond consistently to standard anxiolytic drugs, although there are reports that high doses of diazepam may be effective (Noyes *et al.*, 1984), while more effective still are the tricyclic antidepressant imipramine and the mixed anxiolytic/antidepressant BZ alpra-zolam (reviewed in Rickels, 1985; also Lader, this vol., Ch. 4). Panic attacks are consistently induced in panic disorder patients by caffeine. There has been some interest over recent years in the inter-actions between methylxanthines (such as caf-feine), brain adenosine, and established neuro-transmitters such as dopamine (e.g. Jarvis & Williams, 1987). However, glucose, lactate, and the alpha$_2$-adrenergic antagonist yohimbine can also produce 'panic' in anxious patients. As one would predict, the effect of yohimbine is blocked by the alpha$_2$ agonist clonidine, but not by alprazolam, and although the clinical anxiolytic potency of clonidine is not well established, the yohimbine/clonidine interaction suggests some noradrenergic involvement in anxiety states (see also Redmond, 1979; Gray, 1982; and Section 2.1.3). In general, though, the pharmacology of panic disorder is poorly understood.

Attempts have been made to produce animal models of some of these drug-resistant anxiety states. Particular use has been made of the monkey's natural fear of snakes in the learning theory-based analysis of phobias (e.g. Mineka, 1986), but the lack of effective drug therapy at the human or animal level leaves them outside the scope of psychopharmacology.

The predominant use of anxiolytic drugs is asso-ciated with the diagnostic category of generalized anxiety disorder. Prognosis is best where the neurotic anxiety has no psychotic or depressive overtones, where somatic involvement is high, pre-morbid functioning and social adjustment are good, and illness is acute rather than chronic. However, recent work suggests that for chronic generalized anxiety states where drug therapy is

indicated, antidepressants may be more effective than BZs (reviewed in Tyrer & Murphy, 1987). If this becomes established, it will increase the pressure to broaden the psychopharmacology of anxiety beyond the narrow study of BZs.

Patient response in general is influenced by a host of specific and non-specific factors relating to drug and patient characteristics, and in any group of treated subjects up to 30% may remain sympto-matic and essentially resistant to drug therapy (see e.g. Rickels *et al.*, 1978). This situation has a superficial resemblance to the responsive/resistant subgroups of schizophrenics which, while ignored for many years, eventually gave rise to the Type I/Type II distinction and an interest in organic pathology (Crow, 1984; see also Cookson, this vol., Ch. 13). There is no suggestion that the same may happen in regard to anxiety; nevertheless, it is apparent that any drug-driven research, such as animal modelling, applies only to the anxiety con-dition that responds to drug therapy.

2.1.3 Benzodiazepines and behaviour

Animal models of anxiety are effectively models of BZ psychopharmacology. This position was reinforced by the discovery and analysis of a speci-fic BZ receptor in the brain associated with the GABA–BZ supramolecular complex (Squires & Braestrup, 1977; Schofield *et al.*, 1987). The cur-rently accepted hypothesis is that the immediate action of BZs and other established clinical anxio-lytics is to facilitate GABA neurotransmission (Haefely, 1985), with a consequent enhancement of GABA-mediated inhibition of other neurotrans-mitter systems. Behavioural and other effects of BZs would then emerge as a consequence of this primary action.

'Anxiety' is usually equated with 'fear' in both humans and animals, an equation that seems intui-tively convincing. Fear as a motivating force is ubiquitous throughout the phylogenetic scale, as danger or the anticipation of danger has powerful implications for survival. The presence of BZ receptors across the vertebrate animal kingdom (Nielsen *et al.*, 1978) supports the view that brain mechanisms dedicated to fear or anxiety evolved relatively early in evolution, and, incidentally, powerfully justifies the use of animal models in analyzing BZ effects. However, the ubiquity of 'fear' and its central role in anxiety states means that any procedure that induces fear is *ipso facto* a possible animal model of anxiety. Any paradigm involving shock, food or water deprivation, novel environments, changes in environmental conditions or experimental parameters, conditions of very high or very low certainty (Hennessey *et al.*, 1977), i.e. virtually any behavioural test devised by experimental psychopharmacologists, can induce mild or severe fear. So many models are able to claim face validity.

BZs have significant effects across a whole range of behavioural tests (for review, see Gray, 1982), with only such tasks as locomotion in a familiar environment and simple discriminations being relatively immune. The position is further complicated by the range of behavioural and physiological actions exerted by BZs. In line with what is known of their human psychopharmacology, BZs, in rats, have anxiolytic, anticonvulsant, and sedative/muscle-relaxant properties, plus direct effects on food and water consumption (Cooper, 1983). Finer-grained analysis of cognitive effects in humans and animals also reveals drug-induced impairments of stimulus discrimination and memory (Frith *et al.*, 1984; Ghoneim *et al.*, 1984; Hodges *et al.*, 1986).

However, anxiety involves affective, cognitive, and motivational changes, and drugs may in theory modulate anxiety by direct or indirect effects on any of these elements. Thus, there is no *a priori* basis for excluding any of the range of behavioural actions of BZs as being irrelevant to their anxiolytic potency. However, experimental work showing the development of tolerance to anticonvulsant and sedative/muscle-relaxant properties whilst the anxiolytic profile is retained (e.g. File, 1981) means that those properties can be eliminated from consideration, though the puzzle remains as to how behaviours mediated through the same receptor can show different profiles of tolerance development (see also Stephens & Andrews, this vol., Ch. 3). Direct BZ-induced increases in food and water consumption also seem unlikely to provide a basis for anxiety reduction, *per se*, although they may influence patient reaction to prolonged drug therapy, and may also reflect changes in motivation which do affect anxiety. Cognitive effects of BZs are harder to eliminate. As the cognitive mediation of anxiety is still subject to debate and controversy, it is impossible to conclude that a drug-induced impairment in, say, stimulus discrimination, is irrelevant to a patient prone to interpret ambiguous stimuli as threatening (Eysenck, 1987).

So for psychopharmacologists the development of animal-behavioural models which are consistently sensitive to effects of anxiolytic or anxiogenic drugs is only the beginning. The main purpose is to use these models in two ways: to investigate the neurochemical systems involved, and to develop theories about how drugs affect behaviour. The assumption, implicit or explicit, is that the effects of the drugs will tell us something about the substrates and behavioural processes involved in human anxiety.

Despite these problems, there have been attempts to generate theoretical models of anxiolytic drug action. Different theories use rather similar behavioural paradigms, or reinterpret established paradigms, so that differences in the relative validity of animal models do not necessarily help us choose between theories. Theories do, however, differ in their range of application; some encompass complex systems of behaviour and associated neural circuits, while others focus on what might be happening in particular behavioural tests. The most comprehensive neuropsychological account of anxiety (Gray, 1982) argues that a septo-hippocampal 'behavioural inhibition system' (BIS) monitors environmental events and feedback from the animal's own responses, and is particularly sensitive to signals associated with novelty, non-reward or punishment. When these are detected, ongoing behaviour is suppressed, while attention and arousal are increased. Thus anxiety *is* 'activity in the BIS'. Gray hypothesizes that anxiolytics act indirectly to impair the BIS through GABAergic modulation of the ascending noradrenergic and serotonergic pathways to the hippocampus which are responsible for the registration and evaluation of aversive stimuli. He uses three main types of evidence to support this theory: the effects of anxiolytic drugs across the range of tests involving novelty, non-reward and punishment that comprise the bulk of this chapter; the similar effects of septal and/or hippocampal lesions when tested in these models; and the effects of anxiolytic drugs on hippocampal theta rhythms (see Section 2.2.5.1). Thus the theory draws together an impressive variety of models and evidence, and it is consistent, in general terms, with pharmacological and receptor-binding evidence about anxiolytic drug action. It is also capable of generating specific lines of research; indeed most current theories of anxiolytic drug action are variations or extensions of Gray's model. The model argues, essentially, that anxiolytics impair attention to certain types of stimuli. Subsequent theorists have been concerned to show that the cognitive effects are more general than Gray claims, and/or that BZs also affect emotional processes and that this may have as much to do with their behavioural and clinical actions as their effects on information processing.

For example, Ljungberg *et al.* (1987) suggest that BZs impair decision making in rats in a very general way, by altering their evaluation of the aversive significance of stimuli. Using a procedure developed from optimum foraging theory they showed that BZ-treated rats persisted much longer in making non-rewarded responses before trying alternative courses of action. Indeed there is much evidence to suggest that BZs and other anxiolytics impair information processing in a wide variety of situations, since they affect learning of new material and performance of complex discrimination tasks in both animals and humans (Frith *et al.*, 1984; Ghoneim *et al.*, 1984; Hodges *et al.*, 1986). Thus their effects may not be restricted to the three classes of stimuli that Gray proposes.

Thiebot *et al.* (1985; Thiebot, 1986) regard BZ-induced increases in responding in situations of non-reward or punishment as evidence that anxiolytics affect motivational processes – they undermine an animal's ability to wait for rewards. Unlike control animals, BZ-treated rats chose small immediate rewards in preference to larger delayed rewards, and could only refrain from lever pressing for short (<1 min) non-rewarded or punished periods. This approach would be consistent with findings that BZs powerfully increase the incentive or hedonic properties of food, particularly preferred food (Cooper, 1980). Thus BZ effects on appetite, which at first sight seem remote from their effects on anxiety, may be relevant to changes in motivational states which alter the normal balance between approach and avoidance, so that the increased hedonic value of rewards come to outweigh the cost of obtaining them. An alternative motivational theory has been put forward by Graeff and co-workers (Graeff & Rawlins, 1980; Schutz *et al.*, 1985) who suggest that BZs impair the functioning of a periaqueductal gray (PAG) 'brain aversive system': BZs, GABA and 5-HT, administered centrally, increased the latency to escape following aversive stimulation within the PAG (see Section 2.2.5). The septo–hippocampally-mediated cognitive effects central to Gray's theory would be of only marginal importance in this interpretation of BZ action.

Recent evidence suggests that cognitive and emotional effects of BZs may be dissociated (Hodges *et al.*, 1987). BZs given peripherally increased both non-rewarded and punished responding in a multiple component conflict procedure (see Section 2.2.2.4). In contrast BZs infused into the amygdala (which is part of Graeff's aversive circuit) selectively increased punished responding. Though response suppression in both components must include cognitive elements such as attention to cues, these and related findings (Hodges & Green, 1987) do suggest that two effects of BZs, resistance to punishment and disruption of discrimination, involve different systems, possibly linked to reduction of fear and impaired information processing respectively. 'Fear' may in theory be reduced by changes in cognitive appraisal of threat or in the emotional reaction to it. Therefore experiments which dissociate various behavioural or neurochemical effects of anxiolytics are potentially useful, in providing more selective targets for either behavioural analysis or therapy (Hodges & Green, 1987).

Given the multifaceted nature of anxiety, there are many different ways in which it may be influenced by drugs. Indeed, as File (1987) points out, the search for novel anxiolytics that will avoid the side effects and tolerance/dependence problems of BZs may well unearth clinically effective compounds which appear not to be anxiolytic in the current animal models based on BZ psychopharmacology (see also Stephens & Andrews, this vol., Ch. 3). This has obvious implications for interpreting possible false negatives in these tests, and for current theories, and further justifies the use of more complex behavioural models to analyze the psychopharmacology of the brain's anxiety system.

2.2 ANIMAL MODELS OF ANXIETY

Many paradigms are used in anxiolytic psychopharmacology which, as discussed earlier, can claim at least a face validity, as they are fear-inducing, and/or a predictive validity, in that they are relatively selective for clinically useful anxiolytics. Even without claiming to be inclusive, it is difficult not to describe and discuss, albeit briefly, at least twenty such models chosen on the basis of popularity and/or theoretical utility. They can be divided broadly into models using unconditioned (spontaneous) behaviour, models involving simple or complex classical and/or operant conditioning procedures, models concentrating upon drug-induced discriminative states, and brain stimulation paradigms. After a brief consideration of each in turn, the closing section will attempt an overall evaluation in terms of the validation criteria proposed by Willner (this volume, Chapter 1; see also Willner, 1984, 1986).

2.2.1 Unconditioned responses

A number of tests involve variations upon the theme of exploratory locomotion in novel environments. Although the pattern of activity exhibited by a rat in an unfamiliar maze or arena can look straightforward, the motivational basis is complicated and often unpredictable (Leyland *et al.*, 1976; Rosenfeld *et al.*, 1978); for much of this century psychologists have debated exactly why a rat 'explores', (for reviews, see Whimbey & Denenberg, 1967; Russell, 1973). At its simplest, exploratory activity appears to be a function of general activity level, impulsivity (the need to become familiar with strange territory), and anxiety (the desire to escape from the unknown and unpredictable); if offered the opportunity to leave a novel maze, rats usually take it (Welker, 1959; Montgomery, 1958; Blanchard *et al.*, 1974; see also Zuckerman, 1982).

Given this ambiguous response to novelty, the effects of various manipulations (e.g. of situational factors such as size, illumination, elevation, or of drug state) are not always predictable. 'Fear', as influenced by the novelty of the environment and other situational variables, is often assumed to be inversely related to locomotor exploration, but under certain circumstances can be directly related

or apparently independent of exploratory activity (Halliday, 1966; Whimbey & Denenberg, 1967). So the effects of treatments supposedly reducing levels of fear become in turn unpredictable, especially as in this area of psychopharmacology data are typically reported with only a minimal theoretical interpretation.

2.2.1.1 Open-field test

In the simplest animal model of anxiety, rats or mice are placed in a novel and relatively large arena, and drug effects assessed on measures of exploration (locomotion, sniffing, rearing) and 'fearfulness' (defaecation, freezing, grooming, urination). Anxiolytics tend to increase locomotion (Simon & Soubrie, 1979), decrease rearing (Hughes, 1972; Thiebot *et al.*, 1973), and reduce fearfulness indicators (Fukuda & Iwahara, 1974). There is evidence that effects on locomotion and rearing are specific to novel environments, and time-dependent, in that initially locomotion is reduced and rearing increased (Simon & Soubrie, 1979). However, contrary to expectations drug effects have been found to be independent of the animal's fear level as manipulated by shock-presentation in the apparatus before drug testing (Kumar, 1971).

As a relatively non-specific exploratory response, locomotion is increased by stimulants such as amphetamine, and anticholinergics such as scopolamine (Anisman *et al.*, 1976). Rearing may be reduced by drug-induced motor incoordination, as well as by possible effects on fearfulness, suggesting that open-field exploration is not a pure test of anxiolytic potency *per se*. In any case, the differential effects on two measures of exploratory activity give the procedure an unwelcome inconsistency. The lack of any means of analyzing the behaviour to any great degree makes the interpretation of unexpected drug effects impossible (Walsh & Cummins, 1976). However, the basic simplicity of testing neophobia via exploratory activity has encouraged the development of procedures more sophisticated than the open-field.

2.2.1.2 Hole-board test

In the hole-board apparatus rats or mice are allowed to explore a square board with up to sixteen holes. They naturally 'head dip' into the holes (a response which can be encouraged by positioning novel objects beneath them), whilst locomotion around the board can give a separate measure of general activity (File & Wardill, 1975). BZs given acutely consistently reduce head dipping, rearing, and locomotor activity (File & Lister, 1982). However, the comparison with the open-field, where BZs increase the supposedly exploratory locomotor response, again emphasizes that in order to go beyond the simple screen and predictive validity, there must be some means of unpacking the behavioural effects and locating them in a theoretical framework that focuses upon 'fear' as a central construct. It is notable that drug effects on measures of anxiety in the elevated plus-maze (see Section 2.2.1.5) do not correlate with their effects on hole-board activity (Pellow *et al.*, 1985).

2.2.1.3 Staircase test

In this test, which we consider briefly as an example of the pragmatic approach to animal models, mice are placed in a chamber containing a five-step staircase. They climb up and down and rear, i.e. they explore. It has been proposed (Simiand *et al.*, 1984) that anxiolytics selectively reduce rearing while leaving the number of steps climbed unaffected. The precise relationship between stairclimbing and conventional exploration is unclear, while rearing, as in the open field, would seem more sensitive to motor incoordination effects than to anything else. Such an animal model would appear to have little theoretical (face and construct) validity, and would survive only as an efficient screen. Unfortunately, and, to the sceptic, predictably, the test can produce significant false negatives and false positives; it is also insensitive to anxiogenic treatments (Pollard & Howard, 1986).

2.2.1.4 Light/dark crossing

Crawley (1981; Crawley *et al.*, 1984) has suggested that an emergence test involving crossings between a dark (safe) and a brightly-lit (aversive) compartment provides a more reliable model of neophobia and exploration. She has shown that anxiolytics consistently and selectively increase light/dark

crossings, which are in theory normally inhibited by fear of the lit area; the effect was specific to the exploratory response rather than involving a general motor disinhibition. The potencies of BZs in this test are roughly proportional to their clinical potency, and the effect is selectively blocked by BZ antagonists and inverse agonists.

This would indicate a model with more behavioural and pharmacological validity than the open-field and hole-board. However, as Treit (1985) points out, we still do not know whether, for instance, impulsive or aversive aspects of exploratory behaviour are being affected. This does not imply that the model is not useful, only that its usefulness is limited; until the precise psychological nature of exploratory behaviour is known, any model based on it has difficulty in relating drug effects to particular dimensions of anxiety.

2.2.1.5 Elevated plus-maze

A problem often encountered in the use of animal models is that basic parametric data from individual laboratories are not always subjected to replication by other groups. This is by way of making a general observation that, with the proliferating range of available models, certain laboratories become practised with certain methods and naturally feel that these are the most suitable for a given purpose. The outside observer has then to gauge the objective value of any paradigm, and of the rationale behind it. This becomes much easier when the laboratory itself has attempted to validate a model using objective measures. The elevated plus-maze (and also the social interaction model: see Section 2.2.1.7) are particularly associated with File and colleagues, who are rare in that they have tried to validate their models according to certain basic criteria.

The elevated plus-maze itself is raised 50 cm off the ground, and is arranged as a cross with opposing pairs of arms either open, or enclosed by sides 40 cm high. The underlying principle (Montgomery, 1958; Halliday, 1966) is that the open arms are more anxiety-provoking and that the ratio of either time spent on open/closed arms or entries into open/closed arms reflects the relative 'safety' of closed arms compared to the relative 'fearfulness' of open arms. Non-drugged control animals spend significantly more time in closed arms. Anxiolytics increased the proportion of entries into, and time spent on, open arms, in a five-minute trial. Amphetamine, caffeine, neuroleptics, and antidepressants, did not increase entries into, or time spent on, open arms, while anxiogenic compounds such as yohimbine and pentylenetetrazol had mixed effects but did tend to reduce exploration of the open arms. Blood corticosterone levels, which were used to assay the stress response, were significantly greater in animals confined to open arms, as were levels of defaecation and freezing behaviour (Pellow *et al.*, 1985).

These validation procedures, and a similar analysis of their social interaction test (Section 2.2.1.7; File & Hyde, 1978, 1979), have given these two models some prominence in the literature (e.g. Chopin & Briley, 1987), and it is certainly true that few other models have been so systematically analyzed. Our final section discusses in more detail the significance of the validation process, but at this stage we can at least conclude from File's work that the elevated plus-maze appears to measure fear, that it is at least as selective for anxiolytics as other models, and that it is economical in terms of time and effort. In addition it is one of the few bidirectional tests, sensitive to anxiogenic as well as to anxiolytic effects. In passing we should however mention that we ourselves have found some problems with the plus-maze, predominantly a large variance, that sometimes obscures possibly significant results. In any group of control or treated animals a combination of active and freezing subjects produces a very wide range of scores, although anxiolytic effects are consistently in the predicted direction.

Using as it does spontaneous exploration, the plus-maze does not lend itself to a fine-grained analysis of the behaviour itself or of drug effects thereon. If two animal models of anxiety respond differentially to putative anxiolytics, it is virtually impossible to debate the issue in relation to the complex nature of anxiety if one of the tests involves exploratory behaviour (see closing comments in File, 1987).

2.2.1.6 Exploration and cognition

We have considered a number of animal models of anxiety involving exploratory behaviours. Leaving

aside a general discussion (see Section 2.3) it is relevant at this point, and in line with one of the purposes of this chapter, to mention a recent shift of emphasis from motivational variables to the cognitive aspects of maze exploration (O'Keefe & Nadel, 1978; Olton, 1979). Rats learning a radial maze in which subsets of arms are either rewarded with food pellets or left unrewarded, utilize a long-term, between-trial reference memory for the location of rewarded arms and a short-term, within-trial working memory for arms already explored (Olton, 1979). Specific lesion or drug-induced deficits can then be assigned to impairments of particular memory components; Olton & Papas (1979) for instance found that fornix-fimbria lesions did not affect the within-trial avoidance of non-rewarded arms, but disinhibited re-entries into rewarded arms already explored i.e. they caused a specific working memory deficit.

This approach allows a much tighter analysis of drug-induced variations in spontaneous behaviour. It does not appear to be a model of anxiety, as BZ effects suggest general impairments of the early stages of information processing and a consequent disruption of both working and reference memory (Willner & Birbeck, 1986; Hodges & Green, 1986). These impairments may be relevant to the clinical anxiolytic and/or side-effects of BZs, but either way, if they are part of the psychopharmacological profile of BZs, they need to be investigated. Only a substantial theoretical context and suitable experimental paradigms will allow this to happen.

2.2.1.7 Social interaction

File & Hyde (1978, 1979) have shown that the amount of social interaction (sniffing, nipping, grooming, following, kicking, boxing, wrestling, jumping on, crawling over or under the partner) between two male rats placed in a test chamber for ten minutes is inhibited by increasing 'anxiety', by using an unfamiliar chamber, and increasing light levels. This inhibition is selectively antagonized by chronic treatment with anxiolytics. Exploration of the chamber similarly declines with unfamiliarity and more intense light, but is less affected by drug treatment, suggesting that social interaction *per se* is the more sensitive 'anxiety' measure.

Under appropriate conditions, then, the social

interaction model matches the elevated plus-maze in terms of basic behavioural and pharmacological validity. File points out that interaction is sensitive to individual variations in weight and to time-of-day effects, and that drug-induced increases do not wholly reflect reduced anxiety, though these points may well apply to the majority of animal models in psychopharmacology; it takes systematic analysis to reveal them, which is usually lacking. Even in rats, social behaviour is a complex process (Krsiak, 1986) with multiple motivations. This need not be a drawback, since human anxiety as reflected, for example, in social phobias and avoidance disorders of childhood, is equally complex, if not more so. However, it becomes a problem when we need to unpack the anxiety component in social interaction to give a clearer account of anxiolytic drug action; for instance, we do not know whether the drugs affect the emission or reception of social cues, particularly as either two drugged or two non-drugged rats are always tested together; this represents a serious procedural flaw.

The social interaction model has become popular as a preliminary screen for known and putative anxiolytics (e.g. Chopin & Briley, 1987; Traber & Glaser, 1987), although there are some logistical drawbacks to its use in screening (see Stephens & Andrews, this vol., Ch. 3), and as with the plus-maze, other laboratories have found it difficult to use because of the large variability of responses. Nevertheless the valuable work of File's group in attempting to validate these models gives them a substance not possessed by most other animal models of anxiety.

2.2.1.8 Aggression

Aggressive behaviour has been a popular topic for psychopharmacological investigation in its own right. An awareness of the various distinctive categories of aggression – in terms of eliciting stimuli, species-specific purpose, and topology (Moyer, 1968) – and their heterogeneous neurochemical substrates (Avis, 1974) would immediately suggest that drug effects would either be inconsistent or, if consistent, highly specific to a particular subset of aggressive responses. In line with this prediction, and despite early hopes that anxiolytics, especially BZs, would generally reduce aggressiveness, results have indeed been inconsistent. BZs may increase

(Di Mascio, 1973; Beck & Cooper, 1986; Miczek *et al.*, 1986) or decrease (Valzelli, 1973; Malick, 1978) aggressive behaviour, depending upon treatment parameters and the particular response under study. There has been some recent interest in the analgesia induced by the stress of intraspecific social conflict, and the possibility that one subtype of this analgesia may be mediated through the BZ receptor and therefore modifiable by BZ receptor-active agents (Rodgers, 1986). It remains to be seen whether this line of research has any contribution to make to the study of anxiety and anxiolytics.

2.2.1.9 Hyponeophagia

Eating, in rats and other species, is inhibited by novel aspects of the environment and of the food. The amount of food consumed in these situations is increased by anxiolytics, and this hyponeophagic effect has been presented as an animal model of anxiolytic drug action (Poschel, 1971; Cole, 1983; Shephard & Broadhurst, 1982).

A particular and serious problem for this model is that BZs directly stimulate feeding (e.g. Cooper, 1980), which is bound to confound a test based entirely on amounts consumed. Although it has been shown that a low dose stimulation of feeding can become a genuine high dose anxiolytic action (Hodges *et al.*, 1981), this test seems unsatisfactory for the study of anxiety. Besides the feeding problem, the potential sources of anxiety are themselves too heterogeneous: novel food, novel food pots, novel test chamber, etc. In addition, effects on the consumption of appetizing food, which are bidirectional for BZs and inverse agonists (Cooper, 1986), suggest that incentive motivation rather than feeding *per se* may be an important feature of BZ action, making it necessary to tease apart novelty and preference. Results vary as these aspects are varied, but in no consistent pattern. It is of course perfectly valid to study anxiolytic effects on appetite and relate findings to the clinical situation; and the use of BZ receptor agonists, inverse agonists, and antagonists can reveal much about appetite control (Cooper, 1982; Cooper & Estall, 1985). But as an animal model of anxiety hyponeophagia seems to have little to recommend it, in the absence of systematic efforts to isolate the factors involved.

2.2.2 Conditioned avoidance responses

The advantage of conditioning, and in particular, operant, procedures is the level of experimenter control over the animal's behaviour and the possibility of isolating confounding variables by suitable experimental design. This is of course in addition to any predictive, face, or construct validity any given model may possess.

Most experimental procedures used in this area involve aversive stimuli such as mild footshock. Conditioning responses to these stimuli can involve both classical and instrumental processes, and often both, and an introductory statement must point out that learning-theory interpretations of aversive conditioning are extremely complicated (Olton, 1973; Anisman *et al.*, 1979; Gray, 1982). Behaviour in avoidance paradigms is also subject to a host of subject and situational variables (Bignami, 1976, 1986).

So these models have positive and negative features. The first group to be considered use the aversive effects of footshock to condition the inhibition of normally ongoing behaviour; it is hypothesized that the inhibition of behaviour in anticipation of punishment is mediated by the hypothetical construct 'fear' or 'anxiety', and that such inhibition should be reduced by anxiolytic treatments.

2.2.2.1 Four-plate test

Boissier *et al.* (1968) introduced this modified locomotion test as a screen for anxiolytics. Animals are shocked for running between four metal plates, and, predictably, this inhibits the behaviour. Anxiolytics reinstate crossing responses, but the specificity of the test has been questioned (Aron *et al.*, 1971) and it has not proved particularly popular. Reactions to aversive stimuli are psychologically and neurochemically complicated. The four-plate test involves active avoidance, passive avoidance, and escape responses, and one would not therefore expect consistent drug effects in what is a relatively uncontrolled environment. Alternative paradigms exist with a much greater degree of experimenter control and which tell us much more about anxiolytic psychopharmacology in relation to shock-induced behavioural inhibition.

2.2.2.2 Conditioned suppression

Also known as the conditioned emotional response (CER), the conditioned suppression procedure involves the suppression of ongoing, usually appetitive, operant responding in the presence of a light or tone (conditioned stimulus, CS) previously paired with an electric shock (unconditioned stimulus, UCS). The CS-UCS pairings can take place in the operant chamber during the performance of the operant (on-the-baseline), or in a separate apparatus (off-the-baseline). On-the-baseline procedures have the disadvantage that the animal may learn a spurious association between responding and the footshock; the paradigm would therefore involve instrumental conditioning (punishment) as well as the classical conditioning of 'fear'.

Both Gray (1982) and Treit (1985), in reviewing the data, emphasize the inconsistency of anxiolytic effects on the CER. Gray is slightly more optimistic, pointing out that on-the-baseline conditioned suppression does seem vulnerable to anxiolytic-induced disruption, but in no reliable pattern. This inconsistency and the ability of non-anxiolytics to disrupt the CER make it easy to agree with Treit that this particular paradigm is not suitable as a model for anxiolytic psychopharmacology.

2.2.2.3 Passive and active avoidance

Although not in the mainstream of animal models of anxiety, avoidance learning paradigms are susceptible to anxiolytic effects in a theoretically predictable manner. In passive avoidance the animal is punished with footshock for making a response, usually locomotion or stepping down from a raised platform. When it is later replaced in the apparatus, the response is inhibited. Passive avoidance depends upon classically conditioned secondary aversive stimuli (i.e. the learned association between situational cues and the footshock), and the acquired instrumental contingency between the response and the footshock. Anxiolytics disinhibit punished responding in passive avoidance tasks, and in a thorough analysis Gray (1982) concludes that their action is on the control of behavioural inhibition by secondary aversive signals, rather than upon response/footshock contingencies.

This analysis is particularly cogent when

extended to anxiolytic drug effects on tasks in which an animal has to respond actively to avoid footshock. In a one-way active-avoidance task, one area is always shocked and one area always safe. The animal acquires the association between situational cues and shock, learns therefore to anticipate shock, and responds accordingly. In the two-way task, shocked and safe areas are constantly reversed. Typically, in order to avoid shock, the animal has to shuttle repeatedly from one end of a box to the other. The experienced subject can therefore only avoid shock by approaching cues that have previously been associated with shock. Anxiolytics typically are ineffective in one-way tasks, but facilitate performance in two-way tasks, in theory by reducing the impact of signals previously associated with shock in line with their effects on passive avoidance (Gray, 1982).

Passive and active avoidance tasks are complex and influenced by a range of procedural and pharmacological manipulations (e.g. Bignami, 1976). This makes them unsuitable as screens for anxiolytic activity; however, the level of experimental control possible, the extensive theoretical context, and their role as major examples of the inhibitory effects of aversive stimulation (mediated through conditioned 'fear' or 'anxiety') makes them valuable in the analysis of anxiolytic psychopharmacology.

2.2.2.4 Geller-Seifter conflict

The Geller-Seifter paradigm has been a popular model for the study of anxiolytic drug effects. The paradigm typically consists of several operant components cued by different light signals. A period of positive reinforcement, usually on a variable interval schedule, is followed by a short time-out phase, when responses are neither reinforced nor punished, and then by a conflict component in which all responses are simultaneously rewarded with food and punished with a mild footshock. The suppression of responding in the conflict component is attenuated by anxiolytics, relatively specifically and with a potency proportional to their clinical potency (Sepinwall & Cook, 1978).

This has made Geller-Seifter conflict a valuable model for the study of anxiety, although it has

recently been subject to criticism (Lal & Emmett-Oglesby, 1983; Treit, 1985; File, 1987; Stephens & Andrews, this vol., Ch. 3). The procedure typically involves food reward, and therefore could be confounded by, for example, BZ effects on appetite. However, the separate components allow for the isolation of drug-induced appetitive responding, and typically increases in conflict responding are much greater than any increases in the rewarded component, either as analyzed by analysis of covariance (Hodges *et al.*, 1986), or with response levels in the two components equalized before drug administration (Jeffrey & Barrett, 1979). In our experience, and contrary to the comments of Treit (1985) and Lal & Emmett-Oglesby (1983), animals are not hard to train (although, as with all operant paradigms, this is a time-consuming procedure and therefore unsuitable as a simple screen), and response rates in all components remain relatively stable over long periods. It is in fact this stability of baseline responding between treatments that allows repeated drug-testing if required, for instance, to demonstrate reliable and repeatable responses to anxiolytics over time in individual subjects.

BZs also tend to increase responding in the time-out period (Sepinwall & Cook, 1978; Hodges & Green, 1984). Although this could represent a BZ-induced attenuation of the aversive consequences of non-reward, procedural variations, such as randomizing the order in which components are presented, show that BZs impair the ability of animals to discriminate between the separate components (Hodges *et al.*, 1986). This implies that the Geller-Seifter paradigm involves much more than anxiety. However, it should be remembered that while the BZ-induced increase in conflict responding may consist of a genuine antianxiety action plus a discrimination impairment, BZ effects in human anxiety states may also involve more than one behavioural and pharmacological substrate.

As with any procedure involving behavioural suppression in control conditions, the Geller-Seifter paradigm is not convenient for detecting anxiogenic effects, i.e. treatments which decrease conflict responding. Although baseline conflict rates can be controlled at a high enough level to detect decreases (Petersen & Jensen, 1984; Hodges *et al.*, 1987), they usually become much more variable over time and therefore more difficult to work with. This will become a major problem as the study of the mechanisms of anxiety increasingly comes to involve anxiogenic manipulations.

2.2.2.5 *Vogel water-lick conflict*

Originally introduced by Vogel *et al.* (1971), the water-lick paradigm has become a popular conflict model as it eliminates the prolonged training necessary for more elaborate schedules such as Geller-Seifter conflict. Rats or mice first learn to lick from a water spout in an operant chamber. Then, usually after a period of unpunished licking, responses are punished with mild footshock, such that licking in controls is suppressed. Anxiolytics release this suppressed behaviour while non-specific effects on, for example, drinking *per se*, can be assessed via non-punished drinking or *ad-lib* drinking in home cages.

The lack of a systematic analysis of drug effects on non-conflict behaviour is a partial drawback of this procedure, especially as Leander (1983) has shown that BZ effects on punished drinking correlate more with effects on deprivation-induced drinking than with anti-conflict potency in Geller-Seifter-type paradigms, suggesting that direct stimulation of drinking mechanisms may be an important source of contamination. In contrast, the finding that displacement of ^{3}H-diazepam binding does correlate with potency in the water-lick conflict test, but not with potency in a test of neophobia-induced suppression of drinking (Malick & Enna, 1979), points to a BZ-selectivity for water-lick conflict independent of an action on consummatory responses.

More recent developments of the water-lick paradigm by Petersen and colleagues (Petersen & Scheel-Kruger, 1982; Scheel-Kruger & Petersen, 1982; Petersen & Jensen, 1984) have been aimed at improving replicability. By pre-selecting rats that water-lick and are also sensitive to shock-induced suppression they produced more consistent data and were also able to demonstrate anxiogenic actions of BZ-inverse agonists and pentylenetetrazol. Despite the possible confound of direct effects on drinking, the relative procedural simplicity of water-lick conflict makes it a popular animal model of anxiety and anxiolytic psychopharmacology.

2.2.3 **Other conditioning models**

2.2.3.1 *Potentiated startle*

The potentiated startle paradigm has been particularly associated with the work of Davis and colleagues. The startle response to a loud tone is potentiated by simultaneous presentation of a light that has been previously paired with footshock (i.e. an off-the-baseline procedure). Anxiolytics produce a dose-dependent reduction in potentiated startle, an effect not found with, for example, antidepressants (Davis, 1979*a*; Cassella & Davis, 1985). The paradigm is also sensitive to putative anxiolytics such as clonidine (Davis *et al.*, 1979), buspirone (Davis *et al.*, 1988), and 5-HT$_3$ receptor antagonists (Glenn & Green, 1989). Morphine is also effective against this 'conditioned fear' enhancement of startle, an effect blocked by naloxone (Davis, 1979*b*). This latter is an interesting observation, as opiates do have euphorigenic (anti-anxiety?) effects in humans, but are not active in, for example, Geller-Seifter conflict (Geller *et al.*, 1963; Sepinwall & Cook, 1978), and are not used routinely as clinical anxiolytics (see also Stephens & Andrews, this vol., Ch. 3).

Perhaps the major advantage of potentiated startle is that the brain pathways underlying the basic startle response have been mapped out in detail (Davis *et al.*, 1982), while more recent work has implicated the amygdala in mediating the potentiation of startle by conditioned fear (Hitchcock & Davis, 1986). This allows a tighter analysis of drug and lesion effects, with implications for models of the brain mechanisms of conditioned fear. As the basic brainstem response is reflexive, there would appear to be fewer problems of confounding by drug effects on other aspects of behaviour, although in the long run this could be a disadvantage when modelling human anxiety.

It remains to be seen whether forebrain modulatory mechanisms, such as the amygdala, are important in anxiety states (Gray, 1982), but if they are, then potentiated startle could be a useful model, especially as the anxiogenic stimulus evokes an increase in behavioural output, implying that the model should also be sensitive to anxiogenic treatments (Davis *et al.*, 1979). The paradigm does need to be used extensively by other laboratories before its value can be fully assessed.

2.2.3.2 *Partial reinforcement extinction effect (PREE)*

Animals trained on operant or runway appetitive tasks using partial reinforcement schedules show an increased resistance to extinction when compared with controls trained with continuous reinforcement during acquisition. Anxiolytics such as chlordiazepoxide and sodium amylobarbitone, given during training, block the PREE (Gray, 1982; see also Section 2.2.5.1).

The PREE has more theoretical than practical significance. It is embedded in a theoretical context involving signals of frustrative non-reward, and its sensitivity to anxiolytics provides valuable data for theories of drug action and the brain mechanisms of anxiety (Gray, 1982). However, it is itself a sensitive measure, with parameters such as inter-trial interval and number of acquisition trials critical both to its development and to drug actions upon it (Gray, 1982; Rawlins, 1982). This sensitivity is theoretically consistent, but makes it unsuitable as an anxiolytic screen or even as a simulation, in comparison with models which match it in terms of behavioural complexity and drug specificity but are methodologically more robust. In addition, and like other models closely associated with individual groups, a more widespread usage would help to evaluate its general contribution.

2.2.3.3 *Conditioned defensive burying*

Rodents appear to possess a species-typical defensive reaction to noxious stimuli (e.g. to aversive smell or taste) or to threatening situations of burying the stimulus object or burying the entrance to their burrows (e.g. Jackson & Allgeyer, 1985; Fanselow *et al.*, 1987). This has given rise to various paradigms, some of which seem to be sensitive to anxiolytic-induced disruption (reviewed in Treit, 1985). Treit and colleagues (1981; Treit, 1985; Pinel & Treit, 1983) have introduced more control into the basic situation by shocking the rat when it touches a prod mounted in the test chamber; the rat subsequently buries the prod with material made available in the chamber. The shock-test interval can be as long as 20 days,

and the prod-shock association can be made in another chamber or in the home cage, rather than in the test chamber (i.e. it can be on- or off-the-baseline).

Conditioned defensive burying is dose-relatedly reduced by anxiolytics such as chlordiazepoxide, diazepam, and sodium pentobarbital, with a potency roughly equal to their clinical potency. Some non-anxiolytics, such as d-amphetamine, morphine, and picrotoxin, were ineffective, but chlorpromazine did suppress defensive burying. However, increasing shock levels eliminated the effect of diazepam but left chlorpromazine-induced suppression intact, which may represent a real qualitative distinction between neuroleptics and anxiolytics; this remains to be fully established. There is no possibility of conditioned defensive burying being confounded by appetitive behaviours, and it has some attraction as a more ecologically-valid, 'prepared' and species-specific behaviour (Seligman, 1970). However, it is unsuitable for examining possible anxiogenic treatments, and does use electric shock to establish conditioned fear, which detracts from its ecological validity. Obviously those variations not using shock do not have this problem, but also lack the high degree of experimental control provided by this simple conditioning paradigm, which can be used, for example, to vary the aversiveness of the stimulus. A further problem is that the basic dependent variable (depth of sawdust etc. piled up) is likely to prove unsuitable for fine-grained dosage or drug comparisons.

Unlike those models using spontaneous behaviours reviewed earlier, there is some possibility of isolating other effects of drugs in this model. Motor impairment and analgesic side-effects appear not to contribute to BZ-induced suppression of conditioned defensive burying (reviewed in Treit, 1985), nor does any impairment of the original conditioning. Beyond this, the behaviour is difficult to analyze. As we discuss later, a 'pure' model of anxiety awaits the psychological unpacking of human anxiety. But if anxiety states are associated with perceptual and other cognitive changes, then paradigms which reflect these and allow them to be analyzed will be more useful as simulations than those that do not. However, as a model which, although at a preliminary stage, appears to select out anxiolytic actions on a phylogenetically established substrate of 'fear' or 'anxiety', conditioned defensive burying seems to have some promise.

2.2.3.4 Contrast effects

Contrast effects, which have played an important part in the development of learning theory, make use of an animal's own expectations about the size or quality of a reward. Elliot (1928) and Crespi (1942) first demonstrated that the running speed of rats decreased when they were shifted from a larger or preferred reward to a smaller or less appetizing one, whereas rats that had been switched to a larger or more favoured reward ran faster. Such findings forced the recognition that learning involves expectancies and preferences over and above the stamping in of stimulus–response associations by mere repetition.

Contrast effects have been studied in three main paradigms; simultaneous, anticipatory and successive (Flaherty & Rowan, 1986). In simultaneous contrast two rewards, such as sucrose solutions, are given either together, or in close proximity. Rats given a 4% solution accompanied or immediately followed by a 32% solution rapidly develop a slower lick rate for the former and a faster rate for the latter, relative to rats given only one concentration. This procedure therefore demonstrates both positive and negative contrast effects. Anticipatory contrast occurs when one reward is systematically followed by a second, so that it comes to predict the second reinforcement. If the first reward is meagre (e.g. 4% sucrose) and the second is appetizing (e.g. 32% sucrose), the first will be consumed slowly. This behaviour does not occur if the two concentrations are the same. Unlike simultaneous contrast, anticipatory contrast takes several days to acquire, and affects only the lick rate for the first solution, if non-preferred. Thus only negative contrast effects are clearly demonstrable using this procedure. The third paradigm, successive contrast, is based on the original Crespi procedure. After a stable lick or run rate has been developed for a particular reward, rats are shifted to another while the reward for controls remains unchanged. In general it has proved easier to demonstrate 'depression' (negative contrast) effects when the rat is shifted to a less attractive reward than to show 'elation' (positive contrast) when the reward

is enhanced, because of ceiling effects on lick rate or run speed. However with manipulations such as delay of reinforcement, there have been some reports of positive contrast effects (for a review see Flaherty, 1982).

Contrast effects have face validity as a model for investigating anxiolytic drug action because they involve evaluative and emotional responses to the quality of rewards. Also, they potentially provide bidirectional tests for use with both anxiolytic and anxiogenic compounds. The major experimental thrust has been in the work of Flaherty's group, which has gone far to validate the use of successive negative contrast as a test of anxiety. These workers find that rats shifted from 32% to 4% sucrose sharply reduce their lick rate, with a gradual recovery to pre-shift levels over several days. Control rats given only the 4% solution maintain a stable rate. Anxiolytic drugs (e.g. BZs, alcohol) reliably increase the lick rate from the second post-shift day, though they do not necessarily prevent the initial depression (Flaherty et al., 1980, 1982; Becker & Flaherty, 1982, 1983). Flaherty et al. (1986) argue that this initial depression involves sensory discrimination, but that after experience with a less preferred solution, lick rate is reduced by devaluation of the reward, which involves stress and conflict. Indeed corticosterone elevation correlated with contrast on the second post-shift day, but there was no elevation on the first post-shift day (Flaherty et al., 1985). To support this hypothesis Flaherty et al. (1986) have shown that with a longer lick period, BZs disinhibit licking in the second, but not the first, five minutes even on the first post-shift day. The abolition of Day 2 negative contrast appears specific to anxiolytics, since other classes of drug (antipsychotics, antidepressants, cholinergic and noradrenergic treatments) do not affect successive negative contrast (Flaherty et al., 1977, 1987; Flaherty & Meinrath, 1979).

Interestingly, anxiolytic drugs have not been found to disrupt either simultaneous or anticipatory contrast (Flaherty et al., 1977, 1979). These paradigms may involve different processes from those engaged when a less appetising reward is substituted for a preferred one. Neither simultaneous nor anticipatory contrast involve a reduction in reinforcement; in simultaneous contrast the rat chooses to lick more avidly at the preferred solution, whereas in successive contrast it reduces lick rate for the first solution when it has learned that something more tasty is coming next. Thus the sensory processes involved in simultaneous contrast and associative processes involved in anticipatory contrast do not seem to be a target for anxiolytic drug action, which strengthens the case for the specificity of the conflict phase of successive negative contrast to anxiety.

Day 2 successive negative contrast effects appear to be robust and selective for anti-anxiety drugs, which makes them a potentially useful tool with which to investigate both the neurochemical and anatomical substrates of anxiolytic action. Some 5-HT antagonists (cyproheptadine and cinanserin) have been found to abolish or diminish Day 2 contrast, consistent with reports of anti-conflict effects of these drugs (Becker, 1985). However neither buspirone nor methysergide were effective, though both these compounds have anxiolytic effects in other procedures (Hodges *et al.*, 1987; Davis *et al.*, 1988). Negative contrast was eliminated by medial amygdalectomy and reduced by lateral amygdala lesions to a degree comparable with effects of anxiolytic drugs (Becker *et al.*, 1984). However, neither septal nor hippocampal lesions were found to affect contrast (Flaherty *et al.*, 1979). These findings are in agreement with the evidence from other paradigms for the involvement of the central (Hitchcock & Davis, 1986) and lateral/basolateral amygdaloid nuclei (Petersen & Scheel-Kruger, 1982; Thomas & Inversen, 1985; Hodges *et al.*, 1987) in the effects of anti-anxiety drugs.

The use of successive negative contrast has some practical and theoretical drawbacks. Since control post-shift baseline rates increase steadily back to preshift levels over five days or so, there is a relatively narrow window (days 2–3) during which to observe the maximal effects of anxiolytic drugs. The method also involves comparisons between groups of drugged and control animals tested under shifted and non-shifted conditions, so large numbers are involved and repeated testing at different drug doses is not possible. Flaherty's hypothesis that Day 2 effects are specifically related to conflict and anxiety is plausible, and has received some confirmation. However it remains to be seen whether positive successive contrast effects can be reliably obtained and then used as a

putative model for testing the effects of anxiogenic drugs. So far the model has proved most successful with drugs active at the GABA–BZ receptor complex. Though sensitive to the effects of some 5-HT antagonists, it appears unable to detect effects of buspirone and clonidine which have been found to be clinically effective as anxiolytics in some circumstances. Thus the extent to which the procedure can pick up non-BZ anxiolytics is still debatable.

2.2.3.5 *Learned helplessness*

Animals exposed to unavoidable and uncontrollable stressors, such as electric footshock, may show deficits in subsequent learning tasks, as well as reduced locomotion, decreased appetite, and weight loss. Reasoning that the animal learns on exposure to uncontrollable stress that responding is futile, Seligman and co-workers (e.g. Seligman, 1975; Abramson *et al.*, 1978) have suggested that the cognitive and motivational deficits produced in the helplessness paradigm are parallel to the human behaviour defined as clinical depression. These parallels are discussed in detail by Willner (this vol., Ch. 5; see also Willner, 1986).

However, it seems intuitively plausible that helplessness might model anxiety rather than depression. The animal subjected to inescapable uncontrollable shock becomes fearful and anxious, which may be independent of the development of helplessness but is more likely to be aetiological (Willner, this vol., Ch. 5). Antidepressants are often effective in human clinical anxiety (see Section 1.2; Lader, this vol., Ch. 4), and it has been argued (Gray, 1982) that anxiety and neurotic depression, while not identical, are overlapping syndromes.

Because learned helplessness has long been considered an analogue of human depression, with a particular sensitivity to antidepressant drugs (see Willner, this vol., Ch. 5; Sherman *et al.*, 1982), studies on anxiolytics have been limited. Sherman *et al.* (1982) showed that chronic treatment with benzodiazepines (diazepam, lorazepam, chlordiazepoxide) between training and the test session did not attenuate helplessness in a paradigm sensitive to similar treatment with antidepressants. However, lorazepam or chlordiazepoxide given before training did prevent the development of helplessness (Sherman *et al.*, 1979; Drugan *et al.*, 1984). A likely mechanism for these effects is the prevention by anxiolytics of stress-induced NA depletion (Lidbrink *et al.*, 1973), which appears an important physiological substrate of helplessness (see e.g. Gray, 1982). Drugan *et al.* (1985) used the anxiogenic beta-carboline FG-7142 as the training stressor, and found that it produced a helplessness-like learning deficit 24 h later indistinguishable from that induced by training with inescapable tailshock. The FG-7142 effect was blocked by the BZ-receptor antagonist Ro 15-1788 (flumazenil).

Thus, while learned helplessness, once established, shows a reasonable specificity for antidepressant treatments, it is also clear that anxiolytics can prevent the inception of helplessness. Perhaps the paradigm can best be considered a good example of the difficulty of separating out behaviours relevant to depression and anxiety. There is no easy *a priori* distinction between fear, stress, and anxiety. As the human clinical picture accepts overlapping syndromes and antidepressant therapy for anxiety states, then perhaps our animal modelling should also accept that a perfect distinction between models of anxiety and models of depression may be neither possible nor desirable (see also Willner, this vol., Ch. 5).

2.2.4 Drug-induced discriminative states

Anxiolytic and anxiogenic drugs induce internal states sufficiently intense for human subjects to be able to describe them unequivocally. Perhaps most impressive are 'fear' states induced by agents such as pentylenetetrazol (PTZ; reviewed in Lal & Emmett-Oglesby, 1983), yohimbine (Holmberg & Gerson, 1961), and beta-carbolines (Braestrup *et al.*, 1982). Anxiogenic reactions may also be observed in primates (e.g., see Insel *et al.*, 1984). Lal (1979; Lal & Emmett-Oglesby, 1983) has suggested that such drug-induced discriminative states can serve to cue choice behaviour in rats, and that they may therefore serve as models of anxiety states.

Using PTZ as a standard anxiogenic, Lal has shown that rats trained to choose a rewarded lever under PTZ choose the same lever when given beta-carbolines or yohimbine, but the other, unrewarded lever when given convulsants or anticonvulsants. Anxiolytics block the preference for the PTZ-lever; the BZ antagonist Ro15-1788 in turn antagonizes the BZ-induced blockade, but leaves the action of non-BZ anxiolytics such as meprobromate intact (Lal & Emmett-Oglesby, 1983). The latter finding is interesting as it suggests that PTZ-induced anxiety is not solely dependent upon the BZ receptor; in fact one could argue that PTZ, as a potent convulsant, is too messy a drug anyway to use as an anxiogenic agent, and that its main value may be as a screen for anticonvulsant activity (see Stephens & Andrews, this vol., Ch. 3). There is some evidence (Bennett, 1986) that interoceptive stimuli associated with activity at the BZ receptor involve muscle-relaxant properties of BZ agonists, which would further confuse the issue. The blockade of PTZ-induced choice behaviour could also represent non-specific drug effects such as perceptual, memory, or motor deficits.

More useful are procedures investigating BZ-induced discriminative states (e.g. Stolerman *et al.*, 1986). These states do seem to represent BZ-receptor activity, in that they react predictably to agonists and antagonists (Colpaert, 1977; Herling & Shannon, 1982; Stolerman *et al.*, 1986). The central problem remains; which of BZs' numerous behavioural effects is in fact being assayed?

These discrimination paradigms are interesting in that they represent the closest approximation in anxiolytic psychopharmacology to a bioassay model. Particularly when BZs are used as the discriminative stimulus, the internal state presumably represents the outcome of a selective activation of the BZ receptor. Of course, the downflow from the BZ–GABA receptor complex potentially encompasses a whole range of neurotransmitter systems, in addition to BZ actions at receptors not linked to the GABA complex. However, if the investigation is directed at the broader reaches of BZ behavioural actions then, in its limited way, the drug discrimination paradigm has a role to play in identifying the relative contributions to the discriminated state of anxiolytic, anticonvulsant, muscle-relaxant, and other properties.

2.2.5 Brain stimulation

Stimulation of many different sites can produce behavioural indications of fear in animals. These include the amygdala (Applegate *et al.*, 1983), the locus coeruleus (Redmond *et al.*, 1976; Redmond, 1979), the median raphe nucleus (Graeff & Silveira

Filho, 1978), and the dorsal periaqueductal grey of the midbrain (DPAG) (De Molina & Hunsperger, 1959). Interactions of brain stimulation-induced 'fear' and drugs or lesions have not so far produced a particularly coherent body of data, probably because stimulation at different sites engages different aspects of the brain's aversive systems. The thrust behind much of the work is the delineation of these systems. For instance, Graeff & Rawlins (1980) found that whereas septal lesions disinhibited responding suppressed by footshock but not responding suppressed by DPAG stimulation, chlordiazepoxide disinhibited both. They conclude that chlordiazepoxide acts on at least two separable mechanisms – a behavioural inhibition system and an aversive or punishment system, which implies that BZ effects will be more potent when punishment (e.g. footshock) is actually present (e.g. in conflict procedures) than when only signals of punishment are present (e.g. in conditioned suppression paradigms) (Brandao *et al.*, 1982). 5-HT also has an anti-aversive action on DPAG-stimulation-induced escape (Schutz *et al.*, 1985). On the face of it, this would contradict the conventional view that reduction of 5-HT transmission has anxiolytic and anti-aversive effects (Gardner, 1986; Johnstone & File, 1986; Hodges *et al.*, 1987; see also Deakin, this vol., Ch. 7). It seems clear that analysis of the brain's aversive and behavioural inhibitory systems has some way to go.

Brain stimulation paradigms have not so far seen extensive use in psychopharmacological studies. Their considerable value would appear to lie much more in the unravelling of those pathways mediating behavioural and affective reactions to aversive stimuli, which would be a crucial contribution to understanding the varieties of fear-induced responses and their relation to the concept of anxiety.

2.2.5.1 *Medial septal driving of the hippocampal theta rhythm*

The theta rhythm recorded from electrodes in the hippocampus can be driven by electrodes placed in the medial septum, where the pacemaker cells are thought to be located; the threshold current required to drive theta shows a characteristic curve which in the male rat falls to a minimum at 7.7 Hz.

All anxiolytics tested have selectively increased the threshold current required to drive theta at around 7.7 Hz (Gray 1982). Gray has therefore argued that this 7.7-Hz theta is specifically related to the processing of novelty, non-reward, and punishment, the effect being to switch on the behavioural inhibition system (see Section 2.1.3).

This account predicts that driving theta at 7.7 Hz should affect responses to anxiogenic stimuli. This hypothesis has been tested in various PREE paradigms (Section 2.2.3.2) and the data are complex. However, animals trained in partial reinforcement schedules following several sessions of theta driving at 7.7 Hz were found to extinguish even more slowly than partially reinforced controls (Williams *et al.*, 1989). This effect is opposite to that found with anxiolytics administered during acquisition (see Section 2.2.3.2). These bidirectional effects of theta driving and anxiolytics have been demonstrated most clearly using a runway, but similar changes in extinction after PR training in the Skinner box have also been reported (Gray, 1982; Holt & Gray, 1983).

Lesions of the ascending NA pathway and pharmacological depletion of NA mimic the action of anxiolytics in elevating the threshold driving current at 7.7 Hz minimum, as does the GABA agonist muscimol (Gray *et al.*, 1975; McNaughton *et al.*, 1977; Mellanby *et al.*, 1981). Conversely GABA antagonists counteract this effect of anxiolytics (Quintero *et al.*, 1985). Effects of novel serotonergic anxiolytics and BZ-receptor antagonists and partial agonists have not been examined in the theta driving model, but the pharmacological assessment of the model in terms of agents active at the GABA–BZ receptor complex is broadly consistent, so far, with evidence from other procedures. The model is one of the few available which is responsive to noradrenergic manipulations, and thus may prove sensitive to a wider variety of agents than procedures which are sensitive to BZs. However, effects of theta driving appear to be very sensitive to the precise frequency of stimulation. Thus Williams (personal communication) has found that while driving theta at 7.7 Hz increased resistance to extinction, stimulation at 7.5 Hz actually facilitated extinction after PR training. On the other hand, increased resistance to extinction and tolerance to both punishment and conditioned suppression have been reported at theta driving frequencies of both 7.7 Hz and 8.3 Hz (Holt & Gray, 1985).

Thus theta driving has generated research on the relationships between hippocampal electrical activity and behaviour, and on the neurochemical mechanisms of anxiolytic drug action. However, there is evidence to link hippocampal theta both to different types of motor activity (Vanderwolf *et al.*, 1975) and to memory processes (Deuprée *et al.*, 1982), and we should remember that the anxiolytics tested in this model also affect activity levels, attention and memory. Nevertheless findings that behavioural effects of theta driving are revealed in those paradigms shown to be sensitive to anxiolytics, and that the proactive behavioural effects of theta driving itself are substantial and complex, indicate that this may be a powerful tool for investigating brain systems related to aspects of anxiolytic drug action.

This said, the model is not well suited to serve either as a bioassay or a simulation. Theta driving is complicated to set up, the 'anxiety' is internally generated by electrical stimulation, and inferred from complex, learned, behavioural responses. The phenomena appear to be so sensitive to precise experimental features such as stimulation frequency that they demand replication in other laboratories. In addition there has been no study of any natural variation in hippocampal electrical activity, though the wide variation in behavioural responses to anxiolytics within a single strain of animals is well known (File, 1983). Furthermore, there may well be individual differences in the occurrence of theta, to add to the wide range of interspecies differences (Winson, 1972).

2.3 ANIMAL MODELS AND CONCEPTS OF VALIDITY

As Willner (this volume, Chapter 1) points out, there can be tremendous overlap between the use of animal models as screens, bioassays, and simulations. This is particularly the case with anxiety, where an approach making conditioned or unconditioned fear responses central to the clinical syndrome means that any screening procedure eliciting such responses is also a simulation; fear-induced behaviour in animals would represent an hypothesized aetiology and symptomatology of human anxiety, and can be used to study

treatments and physiological bases. Only in drug-induced discriminative states and brain-stimulation models is the state pharmacologically or electrophysiologically elicited rather than depending upon situational cues analogous to those thought to be important in the human condition.

Drug-discrimination paradigms, subject to the problems outlined in Section 2.2.4, have some utility as screens, particularly for agents active at the GABA–BZ receptor complex. In addition, and as previously discussed, they can serve as a bioassay for activation of central BZ receptors, although it is not clear which of the behavioural and physiological effects of BZs determines choice discrimination behaviour. At best this is a useful paradigm for identifying agents active at BZ receptors.

The remainder of the models can all lay claims to being simulations of at least some aspects of human anxiety, and are therefore subject to evaluation in terms of predictive, face, and construct validity.

2.3.1 Concepts of validity

The dominant thrust in the use of animal models has been directed at screening current and putative clinical anxiolytics. Although lip-service may be paid to notions of validity beyond the predictive, assessment of models is largely in terms of success or failure in identifying novel anxiolytic drugs (Chopin & Briley, 1987). The predictive validity of animal models of anxiety has now entered a critical phase.

2.3.1.1 Predictive validity

BZs are currently the most popular of treatments for generalized anxiety disorder, presumably because they have been assumed to be the most effective over the last twenty years or so (Tyrer & Murphy, 1987). It is therefore reasonable for animal models to be validated on them, especially as a specific BZ receptor exists in the brain which presumably evolved in relation to the animal's need to respond to fearful or anxiety-provoking stimuli. No single model unambiguously selects out BZs without a false positive or false negative, but several (e.g. elevated plus-maze, social interaction, Geller-Seifter conflict, and possibly potentiated startle) have proved highly selective for a range of compounds, and have also been shown to respond to non-BZ anxiolytics, such as meprobamate, acting via the GABA–BZ receptor complex (see also Stephens & Andrews, this vol., Ch. 3).

The current dilemma is that the search for novel drugs which reduce anxiety without the side-effects and dependency problems of BZs has produced a range of compounds acting on 5-HT neurotransmission, either as 5-HT_{1A} mixed agonist/antagonists, as 5-HT_2 antagonists, or as 5-HT_3 antagonists, whose potency in our current range of animal models is at best patchy and at worst absent (File, 1987; Chopin & Briley, 1987; Traber & Glaser, 1987; Jones *et al.*, 1988). Modulations of brain 5-HT via traditional methods using PCPA-induced depletion, neurotoxic lesions of the raphe nuclei, or relatively non-specific drugs such as the antagonist methysergide, have repeatedly demonstrated some activity in classical models, but to a lesser and more variable degree than with BZs and probably via separable mechanisms (Hodges & Green, 1984; Thiebot *et al.*, 1984; Green & Hodges, 1986). Older theories of the role of 5-HT in anxiety (Stein *et al.*, 1973) have evolved into the hypothesis that 5-HT may have a role in impulsivity rather than in anxiety *per se* (Soubrie, 1986; Thiebot, 1986), although several recent reviews emphasize how methodological problems prevent clear conclusions emerging (Gardner, 1986; Johnstone & File, 1986). Given that a major effect of BZs is the reduction of brain 5-HT turnover (Thiebot, 1986), an involvement of 5-HT in the anxiolytic actions of BZs would not be unexpected, although the work of Graeff *et al.* (see Section 2.2.5) suggests that the profile of 5-HT involvement in response to aversive stimuli varies with the type of stimulus and the point in the brain's aversive network under investigation.

If clinically potent anxiolytics emerge – and only buspirone has undergone any systematic testing so far – which are as effective as BZs and which are relatively inactive in current animal models of anxiety, then severe problems arise (see also Lader, this vol., Ch. 4). Such agents would presumably be acting on separate neurochemical substrates; in fact, if they prove to be specific to 5-HT, it would reinforce the idea that the behavioural actions of

BZs are not mediated via 5-HT pathways (although Davis *et al.* (1988) have demonstrated an anxiolytic effect of buspirone on potentiated startle which did not appear to be mediated by 5-HT). The idea that at least two relatively independent mechanisms can modulate anxiety would have profound implications for models of the disorder, emphasizing its heterogeneous nature and the fundamental need to analyze its psychological dimensions at the human level. It would also have implications for the validity of animal models, although it is worth reiterating that BZs can be effective anxiolytics and are efficiently screened by our current models; it is still not certain that agents specific to 5-HT, or other agents, will emerge to challenge their supremacy, and if they do not, the problem disappears. A stronger view (Stephens & Andrews, this vol., Ch. 3) is that either non-BZs will prove effective anxiolytics, in which case their non-selection by our current models shows how those tests have retarded research, or they will not, in which case we have to explain the occasional positive results from our current models. Setting this particular debate aside, we turn to other aspects of validity.

2.3.1.2 Face validity

The McKinney & Bunney (1969) criteria for face validity – that an animal model should faithfully represent the aetiology, symptoms, treatment, outcome, and physiological bases of a disorder – are clearly too stringent, as Willner (this volume, Chapter 1) points out. Insufficient is known of human anxiety for the attempt to be made, and so the emphasis has been on symptoms and treatment, and to an extent on aetiology, with the hope that data from animal models will feed forward to the human condition.

Behavioural dimensions of human anxiety vary from the bizarre and obvious (e.g. phobias) to the ordinary and often unnoticed (e.g. agitation). Crucial to diagnosis are patient reports of subjective experience, with peripheral physiological arousal a reliable correlate. Dominating subjective reports are feelings of anxiety and uncertainty about current or anticipated events, i.e. a sensitivity to potential or actual aversive aspects of the world around.

As animal models cannot handle the subjective

aspects of the human experience they have concentrated on reactions to anticipated or actual aversive stimuli, mediated by the theoretical construct 'fear'. In this respect, all the models discussed in Sections 2.2.1, 2.2.2, and 2.2.3 can claim a minimal face validity in terms of the precipitating conditions; all involve aversive stimuli of some sort, ranging from straightforward environmental novelty through interactions with conspecifics in unfamiliar surroundings, to loud noises and footshock. Behavioural responses are varied and usually unaccompanied by any attempt to relate them unambiguously to 'fear'. Spontaneous behaviours in particular are subject to unjustified assumptions about their motivational bases, although File's work on the elevated plus-maze, and to a lesser extent Crawley's analysis of the light/dark crossing paradigm, are notable exceptions. By demonstrating systematic variations in the dependent behavioural variable with situational manipulations designed to increase aversiveness, File helps justify her claim that the behaviour reflects fear or anxiety, without exclusive recourse to the effects of anxiolytic drugs. That these effects are predictable and that the models are relatively efficient screens (at least for BZ receptor-active agents) gives them an additional pharmacological validity, while the correlation of hypothesized fear levels with blood corticosteroid levels provides physiological support.

Conditioned behaviours have not been systematically analyzed to the same degree, but simple observation does reduce this potential 'validity gap'. Behavioural inhibition in conflict procedures is a straightforward function of shock intensity, i.e. the behavioural measure assumed to reflect 'fear' varies predictably with situational manipulations designed to increase aversiveness. The effects of anxiolytic drugs are predictable and the Geller-Seifter and water-lick conflict in particular serve as relatively accurate screening tests (see Stephens & Andrews, this vol., Ch. 3). We know that the adrenomedullary and adrenocortical stress systems are sensitive to the intensity of aversive stimuli (Mason, 1968) and, although it needs to be demonstrated experimentally, we would be surprised if blood corticosteroid levels did not discriminate conflict from non-conflict periods. Finally, piloerection, defaecation, and lever approach-avoidance are apparent during the conflict component of the Geller-Seifter procedure (E. Lehr, personal communication).

The use of potentiated startle and conditioned defensive burying has not been sufficiently extensive for similar arguments to be advanced at this stage, and for various reasons discussed in the relevant Sections, other models involving conditioned behaviours have basic drawbacks besides any relating to face validity. One notable exception is the use of successive negative contrast, which Flaherty has tried to validate using pharmacological, physiological, and lesion studies (see Section 2.2.3.4).

Phenomenological similarity is central to face validity, and the assumption for animal models of anxiety is that the expression of fear in animals represents the basic symptom of human anxiety. We shall discuss in the next section whether this is an acceptable equation, but a point that arises here concerns the precipitating conditions. In models based either on spontaneous behaviours or on conditioned responses the expressions of fear – behavioural inhibition, defecation, freezing, etc. – seem identical. However, fear or anxiety induced by environmental novelty and assessed during spontaneous exploration or social interaction can be seen as involving the continued presence of aversive stimuli, whilst the state induced by signals conditioned to aversive stimuli involves the anticipation of their occurrence rather than their actual presence. There are also differences in intensity, with novelty *per se* being less aversive than footshock. The peripheral issue of footshock being an alien and artificial experience for the rat seems to us misplaced. Experimental psychopharmacology by its nature deals in artificial situations; rats do not as a matter of course explore elevated plus-mazes or suddenly engage with strange rats in unfamiliar territory. Their reactions to footshock are easily conditioned, suggesting that the brain mechanisms for evaluating and adapting to it are well developed, probably because footshock is categorized along with a range of other powerful aversive stimuli that the rat's nervous system is designed to code and process.

Whether the differences in fear-eliciting procedures between spontaneous and conditioned paradigms has wider implications is uncertain. Behaviour in the continual presence of mild aversive stimuli as opposed to behaviour (in the broadest sense) in anticipation of relatively intense aversive stimuli may well involve overlapping but distinctive patterns of neurochemical activity, as suggested by the correlation between ^{3}H-diazepam binding and shock-induced but not neophobia-induced suppression of drinking (Malick & Enna, 1979). It is also possible to speculate on relationships between different eliciting conditions and different categories of human anxiety. Phobias, for instance, involve innate or acquired avoidance responses, usually to discrete identifiable stimuli. A plausible case can be made (Gray, 1982) for considering them as instances of passive avoidance and analogous to passive avoidance learning in animals. On a more general level, phobias involve the anticipated occurrence of noxious stimuli, a feature of the conditioned paradigms such as passive avoidance learning, conflict procedures, and potentiated startle. Generalized anxiety states, on the other hand, are not associated with one or two specific events and reflect instead a chronic state of foreboding about any number of potentially aversive situations. A limited case can be made for equating this with the prolonged exposure of rats to unfamiliar and therefore unpredictable environments, as in models using spontaneous behaviours. A critical problem, of course, is that whereas human anxiety states are differentially responsive to drug treatment (see Section 2.1.2; Lader, this vol., Ch. 4), the pharmacological validity of the most acceptable animal models from conditioned and spontaneous categories is comparable; conversely, the sensitivity of these models to recently-developed putative anxiolytics is equally patchy and unimpressive (Chopin & Briley, 1987). So, if the anxiety states induced in animal models by different situational manipulations are qualitatively distinct, this has not yet become apparent in any consistent fashion.

An account of the aetiology of human anxiety must encompass many factors. There are elaborate conditioning models (Eysenck & Eysenck, 1985) involving the generalization of classically-conditioned fear to inappropriate neutral objects or situations, and leading directly to behavioural therapeutic intervention, especially for phobic disorders. Individual differences in state or trait anxiety (see Lader, this vol., Ch. 4) are important endogenous variables. The predominance of female patients being treated for anxiety disorders may reflect diagnostic biases, for instance in self-report, but equally, may represent real biological variation; by contrast, in the rat it is the male that shows the greatest sensitivity to experimental manipulations of fear, and to early environmental stress (Gray, 1971).

This variety of factors has not been addressed systemically in animal models. Conditioning paradigms would in theory relate more to the drug-resistant phobic states than to the drug-sensitive generalized anxiety disorder, to which it is difficult to apply a straightforward conditioning model. Individual differences have occasionally been studied, using, for instance, the Maudsley reactive and non-reactive strains (Robertson *et al.*, 1978), but sex differences in incidence have been ignored. Individual variability in response to anxiolytics has also been studied occasionally (File, 1983; Hodges & Green, 1984), and this would have a broader relevance to the substantial number of patients resistant to drug therapy; however, it has no particular implications for any one animal model. A final underlying point is that many anxiety states and treatments are chronic, whilst the majority of studies using animal models are acute.

So, in terms of face validity assessed via aetiology, symptoms, and treatment it is hard to discriminate between five or six models from spontaneous and conditioned behaviour categories. Aetiology is reduced to aversive stimuli, symptoms become apparently simple behavioural responses and autonomic arousal, and while none is a perfect screen for currently-used anxiolytics, all are relatively efficient. So the choice for a screen should emphasize logistical investment, i.e. time, money, and effort (see Stephens & Andrews, this vol., Ch. 3). We consider the choice for a simulation in the final section.

2.3.1.3 Construct validity

In the introductory chapter Willner (this vol., Ch. 1) outlined three stages in establishing the construct validity of animal simulations; identifying the behavioural variable being modelled; assessing homology (in terms of common physiological substrates) between the variable and behaviour in the simulation; and assessing the significance of the variable in the clinical picture.

For animal models of anxiety the picture is

reasonably clear. The behavioural variable modelled is the observable expression of an implied central state of fear or anxiety; only in relation to BZs do we have a clear idea of the physiological substrates in animals and humans; we do not know the precise role of fear in the genesis and maintenance of clinical anxiety.

As with most clinical syndromes there is a shortage of human data. Although we know of many sites in the animal brain from which 'fearful' behaviour can be elicited (see Section 2.2.5; also Valenstein, 1973; Thompson, 1978), which appear to concentrate in brainstem, diencephalic, and limbic areas, there is little in the way of matching human research. There is an assumption that brain circuitry derived from animals can be extrapolated to the human brain, and that has been very much the basis of the perenially popular limbic model of human emotion (MacLean, 1949), although direct supporting evidence has accumulated only slowly (e.g. Kelly, 1973).

The pharmacological approach to homology has been more successful. BZs reduce human anxiety (fear?), and antagonize behavioural indices of fear in animal models. They act on a specific receptor in the brain, found in humans and animals with comparable distributions (Nielsen *et al.*, 1978), and which presumably represent a system whose behavioural functions must then be homologous in animals and humans. Until the endogenous agonist/antagonist of the BZ receptor is isolated, identified, and studied, the precise role of the system in behaviour will remain uncertain; but it does seem reasonable to assume that it will be related to fear and anxiety. However, the system does not act independently, but via a heterogeneous range of pathways downstream from the GABA–BZ receptor complex, and itself must depend upon ascending sensory and descending 'cognitive' inputs.

Clinical anxiety has affective, behavioural, cognitive, motivational, and physiological aspects, and is in no sense homogeneous (Rachman, 1978; Hugdahl, 1981). The failure of BZs to ameliorate major categories of anxiety disorders suggests that the endogenous BZ system does not mediate all aspects of anxiety, and that the 'fear' it appears to mediate is not central to all aspects of anxiety. If human anxiety is multifaceted, and can be modulated by interference with any one of those facets,

then there are potentially many ways of reducing it. Thus even the most rigid and detailed theoretical approach – via learning theory – has had to incorporate notions of cognitive mediation, informational transmission, and 'preparedness' (Seligman, 1970) in trying to formulate a conditioning account of anxiety and anxiety-reduction (Rachman, 1977, 1981).

Earlier we discussed the development of potential non-BZ anxiolytics and the apparent but important failure of current animal models to detect them reliably. If their clinical activity is eventually confirmed, it will reinforce the view that not only can anxiety be reduced via non-BZ mechanisms, but that any notion that elicited fear in animals is homologous with and central to human anxiety states is misguided. This would be a major shock to animal models of anxiety, but stranger things have happened. It would, in parallel, put great pressure on to the analysis of human anxiety and the development of new animal models; it is hard to predict what these would look like.

As clinical anxiety is a heterogeneous syndrome, animal research must not define itself too narrowly. The identification of new anxiolytics is important, as is the search for animal models with construct validity. At present this latter search, for the reasons just outlined, seems fruitless. What is important, perhaps more so for the academic psychopharmacologist, is the broad psychopharmacological analysis of anxiolytic and anxiogenic drugs. If human anxiety can, in theory, be influenced by manipulations of cognitive, affective, and motivational systems, then actual and potential anxiolytics should be analyzed for such effects. This can only be done by using a wide range of animal models, selected for their relative efficiency in isolating particular psychological variables. Only by collating data from a number of models and a number of drugs can an overall picture of the neurochemistry of anxiety emerge (Gray, 1982). This, we suggest, is as important an aim as the development of an efficient screening test; in fact, it is fundamental to the concept of an animal model of anxiety claiming construct validity.

We would like to acknowledge the helpful comments of Mike Snape, Brok Glenn, and Paul Willner, on earlier drafts of this chapter.

REFERENCES

Abramson, L., Seligman, M.E.P. & Teasdale, J. (1978). Learned helplessness in humans: critique and reformulation. *Journal of Abnormal Psychology* **84**, 49–94.

American Psychiatric Association. (1980). *Diagnostic and Statistical Manual of Mental Disorders*, 3rd edn. Washington, D.C.: A.P.A.

Anisman, H., Kokkinidis, L., Glazier, S. & Remington, G. (1976). Differentiation of response biases elicited by scopolamine and d-amphetamine: effects on habituation. *Behavioral Biology* **18**, 401–17.

Anisman, H., Remington, G. & Sklar, L.S. (1979). Effect of inescapable shock on subsequent escape performance: catecholaminergic and cholinergic mediation of response initiation and maintenance. *Psychopharmacology* **61**, 107–24.

Applegate, C.D., Kapp, B.S., Underwood, M.D. & McNall, C.L. (1983). Autonomic and somatomotor effects of amygdala central n. stimulation in awake rabbits. *Physiology and Behavior* **31**, 353–60.

Aron, C., Simon, P., Larousse, C. & Boissier, J.R. (1971). Evaluation of a rapid technique for detecting minor tranquilizers. *Neuropharmacologia* **10**, 459–69.

Avis, H.H. (1974). Neuropharmacology of aggression: a critical review. *Psychological Bulletin* **81**, 47–63.

Beck, C.H.M. & Cooper, S.J. (1986). β-carboline FG 7142 – reduced aggression in male rats: reversed by the benzodiazepine receptor antagonist Ro 15-1788. *Pharmacology, Biochemistry and Behavior* **24**, 1645–9.

Becker, H.C. (1985). Comparison of the effects of the benzodiazepine midazolam and three serotonin antagonists on a consummatory conflict paradigm. *Pharmacology, Biochemistry and Behavior* **24**, 1057–64.

Becker, H.C. & Flaherty, C.F. (1982). The influence of ethanol on contrast in consummatory behaviour. *Psychopharmacology* **77**, 253–8.

Becker, H.C. & Flaherty, C.F. (1983). Chlordiazepoxide and ethanol additively reduce gustatory negative contrast. *Psychopharmacology* **80**, 35–7.

Becker, H.C., Jarvis, M.F., Wagner, G.C. & Flaherty, C.F. (1984). Medial and lateral amygdalectomy differentially influences consummatory negative contrast. *Physiology and Behavior* **33**, 707–12.

Bennett, D.A. (1986). Comparison of discriminative stimuli produced by full and partial benzodiazepine agonists: pharmacological specificity. *Psychopharmacology* **89**, S41.

Bignami, G. (1976). Nonassociative explanations of behavioral changes induced by central anticholinergic drugs. *Acta Neurobiologiae Experimentalis* **36**, 5–90.

Bignami, G. (1986). Limitations of current models for benzodiazepine and antimuscarinic effects on punishment suppression and extinction. *Psychopharmacology* **89**, S9.

Blanchard, R.J., Kelley, M.J. & Blanchard, D.C. (1974). Defensive reactions and exploratory behavior in rats. *Journal of Comparative and Physiological Psychology* **87**, 1129–33.

Boissier, J.R., Simon, P. & Aron, C. (1968). A new method for rapid screening of minor tranquilizers in mice. *European Journal of Pharmacology* **4**, 145–51.

Braestrup, C., Schmiechen, R., Neff, G., Nielsen, M. & Petersen, E.N. (1982). Interaction of convulsive ligands with benzodiazepine receptors. *Science* **216**, 1241–3.

Brandao, M.L., De Aguiar, J.C. & Graeff, F.G. (1982). GABA mediation of the anti-aversive action of minor tranquillizers. *Pharmacology, Biochemistry and Behavior* **16**, 397–402.

Cassella, J.V. & Davis, M. (1985). Fear-enhanced acoustic startle is not attenuated by acute or chronic imipramine treatment in rats. *Psychopharmacology* **87**, 278–82.

Chopin, P. & Briley, M. (1987). Animal models of anxiety: the effect of compounds that modify 5-HT neurotransmission. *Trends in Pharmacological Sciences* **8**, 383–8.

Cole, S.O. (1983). Combined effects of chlordiazepoxide treatment and food deprivation on concurrent measures of feeding and activity. *Pharmacology, Biochemistry and Behavior* **18**, 369–72.

Colpaert, F.C. (1977). Discriminative stimulus properties of benzodiazepines and barbiturates. In *Discriminative Stimulus Properties of Drugs*, H. Lal (ed.), pp. 93–106. New York: Plenum.

Cooper, S.J. (1980). Benzodiazepines as appetite-enhancing compounds. *Appetite* **1**, 7–19.

Cooper, S.J. (1982). Specific benzodiazepine antagonist Ro 15-1788 and thirst-induced drinking in the rat. *Neuropharmacology* **21**, 483–6.

Cooper, S.J. (1983). Benzodiazepine–opiate antagonist interactions in relation to feeding and drinking behavior. *Life Sciences* **32**, 1043–51.

Cooper, S.J. (1986). Bidirectional effects on behaviour of β-carbolines acting at central benzodiazepine receptors (Meeting review). *Trends in Pharmacological Sciences* **7**, 210–12.

Cooper, S.J. & Estall, L.B. (1985). Behavioural pharmacology of food, water and salt intake in relation to drug action at benzodiazepine receptors. *Neuroscience and Biobehavioral Reviews* **9**, 5–19.

Crawley, J.N. (1981). Neuropharmacologic specificity

of a simple animal model for the behavioral actions of benzodiazepines. *Pharmacology, Biochemistry and Behavior* **15**, 695–9.

Crawley, J.N., Skolnick, P. & Paul, S.M. (1984). Absence of intrinsic antagonistic actions of benzodiazepine antagonists on an exploratory model of anxiety in the mouse. *Neuropharmacology* **23**, 531–7.

Crespi, L.P. (1942). Quantitative variation in incentive and performance in the white rat. *American Journal of Psychology* **55**, 467–517.

Crow, T.J. (1984). A re-evaluation of the viral hypothesis. *British Journal of Psychiatry* **145**, 243–53.

Davis, M. (1979*a*). Diazepam and flurazepam: effects on conditioned fear as measured with the potentiated startle paradigm. *Psychopharmacology* **62**, 1–7.

Davis, M. (1979*b*). Morphine and naloxone: effects on conditioned fear as measured with the potentiated startle paradigm. *European Journal of Pharmacology* **54**, 341–7.

Davis, M., Cassella, J.V. & Kehne, J.H. (1988). Serotonin does not mediate anxiolytic effects of buspirone in the fear-potentiated startle paradigm: comparison with 8-OH-DPAT and ipsapirone. *Psychopharmacology* **94**, 14–20.

Davis, M., Gendelman, D.S., Tischler, M.D. & Gendelman, P.M. (1982). A primary acoustic startle circuit: lesion and stimulation studies. *Journal of Neuroscience* **2**, 791–805.

Davis, M., Redmond, D.E., Jr. & Baraban, J.M. (1979). Noradrenergic agonists and antagonists: effects on conditioned fear as measured by the potentiated startle paradigm. *Psychopharmacology* **65**, 111–18.

Davison, G.C. & Neale, J.M. (1986). *Abnormal Psychology*, 4th edn. New York: Wiley.

De Molina, A. & Hunsperger, R.W. (1959). Central representation of affective reactions in forebrain and brainstem: electrical stimulation of amygdala, stria terminalis, and adjacent structures. *Journal of Physiology* (London) **145**, 251–65.

Deupree, D., Coppock, W. & Willer, H. (1982). Pre-training septal driving of hippocampal rhythmic slow activity facilitates acquisition of visual discrimination. *Journal of Comparative and Physiological Psychology* **96**, 557–62.

Di Mascio, A. (1973). The effects of benzodiazepines on aggression: reduced or increased? *Psychopharmacology* **30**, 95–102.

Drugan, R.C., Maier, S.F., Skolnick, P., Paul, S.M. & Crawley, J.N. (1985). An anxiogenic benzodiazepine receptor ligand induces learned helplessness. *European Journal of Pharmacology* **113**, 453–7.

Drugan, R.C., Ryan, S.M., Minor, T.R. & Maier, S.F.

(1984). Librium prevents the analgesia and shuttlebox escape deficit typically observed following inescapable shock. *Pharmacology, Biochemistry and Behavior* **21**, 749–54.

Elliot, M.H. (1928). The effect of change of reward on the maze performance of rats. *University of California Publications in Psychology* **4**, 19–30.

Eysenck, H.J. & Eysenck, M.W. (1985). *Personality and Individual Differences*. London: Plenum.

Eysenck, M.W. (1987). Trait theories of anxiety. In *Personality Dimensions and Arousal*, ed. J. Strelau & H.J. Eysenck, pp. 79–97. New York: Plenum.

Fanselow, M.S., Sigmundi, R.A. & Williams, J.L. (1987). Response selection and the hierarchial organization of species-specific defence reactions: the relationship between freezing, flight, and defensive burying. *Psychological Review* **37**, 381–6.

File, S.E. (1981). Rapid development of tolerance to the sedative effects of lorazepam and triazolam in rats. *Psychopharmacology* **73**, 240–5.

File, S.E. (1983). Variability in behavioral responses to benzodiazepines in the rat. *Pharmacology, Biochemistry and Behavior* **18**, 303–6.

File, S.E. (1987). The search for novel anxiolytics. *Trends in Neurosciences* **10**, 461–3.

File, S.E. & Hyde, J.R.G. (1978). Can social interaction be used to measure anxiety? *British Journal of Pharmacology* **62**, 19–24.

File, S.E. & Hyde, J.R.G. (1979). A test of anxiety that distinguishes between the actions of benzodiazepines and those of other minor tranquilisers and of stimulants. *Pharmacology, Biochemistry and Behavior* **11**, 65–9.

File, S.E. & Lister, R.G. (1982). β-CCE and chlordiazepoxide reduce exploratory head-dipping and rearing: no mutual antagonism. *Neuropharmacology* **21**, 1215–18.

File, S.E. & Wardill, A.G. (1975). Validity of head-dipping as a measure of exploration in a modified holeboard. *Psychopharmocologia* **44**, 53–9.

Flaherty, C.F. (1982). Incentive contrast: a review of behavioural changes following shifts in rewards. *Animal Learning and Behaviour* **10**, 409–40.

Flaherty, C.F. & Meinrath, A.B. (1979). The influence of scopolamine on sucrose intake under absolute and relative test conditions. *Physiological Psychology* **7**, 412–18.

Flaherty, C.F. & Rowan, G.A. (1986). Successive simultaneous and anticipatory contrast in the consumption of saccharin solutions. *Journal of Experimental Psychology* **12**, 381–93.

Flaherty, C.F., Lombardi, B.R., Kapnot, J. & D'Amato, M.R. (1977). Incentive contrast undiminished by extended testing with imipramine or chlordiazepoxide. *Pharmacology, Biochemistry and Behavior* **7**, 315–22.

Flaherty, C.F., Powell, G. & Hamilton, L.W. (1979). Septal lesions, sex and incentive shift effects in the open field. *Physiology and Behavior* **22**, 903–9.

Flaherty, C.F., Lombardi, B.R., Wrightson, J. & Deptula, D. (1980). Conditions under which chlordiazepoxide influences gustatory negative contrast. *Psychopharmacology* **67**, 269–77.

Flaherty, C.F., Becker, H.C. & Driscoll, C. (1982). Conditions under which amobarbital sodium influences contrast on consummatory behavior. *Physiological Psychology* **10**, 122–8.

Flaherty, C.F., Becker, H.C. & Pohorecky, L. (1985). Correlation of corticosterone elevation and negative contrast varies as a function of post-shift day. *Animal Learning and Behavior* **13**, 309–14.

Flaherty, C.F., Grigson, P.S. & Rowan, G.A. (1986). Chlordiazepoxide and the determinants of negative contrast. *Animal Learning and Behavior* **14**, 315–21.

Flaherty, C.F., Grigson, P.S., Demetrikopoulos, P.S. & Demetrikopoulos, M.K. (1987). Effect of clonidine on consummatory negative contrast and on novelty-induced stress. *Pharmacology, Biochemistry and Behavior* **27**, 659–64.

Frith, C.D., Richardson, J.T.E., Samuel, M., Crow, T.J. & McKenna, P.J. (1984). The effects of intravenous diazepam and hyoscine upon human memory. *Quarterly Journal of Experimental Psychology* **36A**, 133–44.

Fukuda, S. & Iwahara, S. (1974). Dose effects of chlordiazepoxide upon habituation of open-field behavior in white rats. *Psychologia* **17**, 82–90.

Gardner, C.R. (1986). Recent developments in 5HT-related pharmacology of animal models of anxiety. *Pharmacology, Biochemistry and Behavior* **24**, 1479–85.

Geller, I., Bachman, E. & Seifter, J. (1963). Effects of reserpine and morphine on behavior suppressed by punishment. *Life Sciences* **4**, 226–31.

Ghoneim, M.M., Hinrichs, J.V. & Mehwaldt, S.P. (1984). Dose–response analysis of the behavioural effects of diazepam. (I) Learning and memory. *Psychopharmacology* **82**, 291–5.

Glenn, B. & Green, S. (1989). Anxiolytic profile of GR 38032F in the potentiated startle paradigm. *Behavioural Pharmacology* **1**, 91–4.

Graeff, F.G. & Rawlins, J.N.P. (1980). Dorsal periaqueductal gray punishment, septal lesions and the mode of action of minor tranquilizers. *Pharmacology, Biochemistry and Behavior* **12**, 41–5.

Graeff, F.G. & Silveira Filho, N.G. (1978). Behavioral inhibition induced by electrical stimulation of the median raphe nucleus in the rat. *Physiology and Behavior* **21**, 477–84.

Gray, J.A. (1971). *The Psychology of Fear and Stress*. London: Weidenfeld & Nicolson.

Gray, J.A. (1982). *The Neuropsychology of Anxiety*. Oxford: Oxford University Press.

Gray, J.A., James, D.T.D. & Kelly, P.H. (1975). Effect of minor tranquillisers on hippocampal theta rhythm mimicked by depletion of forebrain noradrenaline. *Nature* **258**, 424–5.

Green, S.E. & Hodges, H. (1986). Differential effects of dorsal raphe lesions and intraraphe GABA and benzodiazepines on conflict behavior in rats. *Behavioral and Neural Biology* **46**, 13–29.

Haefely, W. (1985). The biological basis of benzodiazepine actions. In *The Benzodiazepines: Current Standards for Medical Practice*, D.E. Smith & D.R. Wesson (eds.), pp. 7–41. Lancaster: MTP Press.

Halliday, M.S. (1966). Exploration and fear in the rat. *Symposia of the Zoological Society of London* **18**, 45–59.

Hennessey, J.W., King, M.G., McClure, T.A. & Levine, S. (1977). Uncertainty, as defined by the contingency between environmental events, and the adrenocortical response of the rat to electric shock. *Journal of Comparative and Physiological Psychology* **91**, 1447–60.

Herling, S. & Shannon, H.E. (1982). Ro 15-1788 antagonizes the discriminative stimulus effects of diazepam in rats but not similar effects of pentobarbital. *Life Sciences* **31**, 2105–12.

Hitchcock, J. & Davis, M. (1986). Lesions of the amygdala, but not of the cerebellum or red nucleus, block conditioned fear as measured with the potentiated startle paradigm. *Behavioral Neuroscience* **100**, 11–22.

Hodges, H. & Green, S. (1984). Evidence for the involvement of brain GABA and serotonin systems in the anticonflict effects of chlordiazepoxide in rats. *Behavioral and Neural Biology* **40**, 127–54.

Hodges, H. & Green, S. (1986). Effects of chlordiazepoxide on cued radial maze performance in rats. *Psychopharmacology* **88**, 460–6.

Hodges, H. & Green, S. (1987). Are the effects of benzodiazepines on discrimination and punishment dissociable? *Physiology and Behavior* **41**, 257–64.

Hodges, H.M., Green, S.E., Crewes, H. & Mathers, I. (1981). Effects of chronic chlordiazepoxide treatment on novel and familiar food preference in rats. *Psychopharmacology* **75**, 311–14.

Hodges, H., Baum, S., Taylor, P. & Green, S. (1986). Behavioural and pharmacological dissociation of chlordiazepoxide effects on discrimination and punished responding. *Psychopharmacology* **59**, 155–61.

Hodges, H., Green, S. & Glenn, B. (1987). Evidence that the amygdala is involved in benzodiazepine and serotonergic effects on punished responding but not on discrimination. *Psychopharmacology* **92**, 491–504.

Holmberg, G. & Gerson, S. (1961). Autonomic and psychic effects of yohimbine hydrochloride. *Psychopharmacology* **2**, 93–106.

Holt, L. & Gray, J.A. (1983). Septal driving of the hippocampal theta rhythm produces a long-term proactive and nonassociative increase in resistance to extinction. *Quarterly Journal of Experimental Psychology* **35B**, 92–118.

Holt, L. & Gray, J.A. (1985). Proactive behavioral effects of theta-driving septal stimulation on conditioned suppression and punishment in the rat. *Behavioral Neuroscience* **99**, 60–74.

Hugdahl, K. (1981). The three-systems-model of fear and emotion – a critical examination. *Behaviour Research and Therapy* **19**, 75–85.

Hughes, R.N. (1972). Chlordiazepoxide-modified exploration in rats. *Psychopharmacology* **24**, 462–9.

Insel, T.R., Ninan, P.T., Aloi, J., Jimerson, D.C., Skolnick, P. & Paul, S.M. (1984). A benzodiazepine receptor-mediated model of anxiety. *Archives of General Psychiatry* **41**, 741–50.

Jackson, R.L. & Allgeyer, R.L. (1985). On the nature of processes governing defensive burying of aversive flavors and odors in rats. *Learning and Motivation* **16**, 315–33.

Jarvis, M.F. & Williams, M. (1987). Adenosine and dopamine function in the CNS. *Trends in Pharmacological Sciences* **8**, 330–2.

Jeffrey, D.R. & Barrett, J.E. (1979). Effects of chlordiazepoxide on comparable rates of punished and unpunished responding. *Psychopharmacology* **64**, 9–11.

Johnstone, A.L. & File, S.E. (1986). 5-HT and anxiety: promises and pitfalls. *Pharmacology, Biochemistry and Behavior* **24**, 1467–70.

Jones, B.J., Costall, B., Domeny, A.M., Kelly, M.E., Naylor, R.J., Oakley, N.R. & Tyers, M.B. (1988). The potential anxiolytic activity of GR 38032F, a 5-HT receptor antagonist. *British Journal of Pharmacology* **93**, 985–93.

Kelly, D. (1973). Therapeutic outcome in limbic leucotomy in psychiatric patients. *Psychiatrica Neurologica Neurochirurgica* (Amsterdam) **76**, 353–63.

Krsiak, M. (1986). Social behaviour in animals. *Psychopharmacology* **89**, S2.

Kumar, R. (1971). Extinction of fear. II. Effects of chlordiazepoxide and chlorpromazine on fear and exploratory behaviour in rats. *Psychopharmacology* **19**, 297–312.

Lader, M. (1983). Anxiety and depression. In *Physiological Correlates of Human Behaviour. Vol. III: Individual Differences and Psychopathology*, A. Gale & J.A. Edwards (eds.), pp. 155–67. London: Academic Press.

Lal, H. (1979). Interoceptive stimuli as tools of drug development. *Drug Development and Industrial Pharmacology* **5**, 133–49.

Lal, H. & Emmett-Oglesby, M.W. (1983). Behavioural analogues of anxiety. *Neuropharmacology* **22**, 1423–41.

Leander, J.D. (1983). Effects of punishment-attenuating drugs on deprivation-induced drinking: implications for conflict procedures. *Drug Development Research* **3**, 185–92.

Leyland, M., Robbins, T. & Iversen, S.D. (1976). Locomotor activity and exploration: the use of traditional manipulators to dissociate these two behaviors in the rat. *Animal Learning and Behavior* **4**, 261–5.

Lidbrink, P., Corrodi, H., Fuxe, K. & Olson, L. (1973). The effects of benzodiazepines, meprobamate, and barbiturates on central monoamine neurons. *The Benzodiazepines*, S. Garattini, E. Mussini & L.O. Randall (eds.), pp. 203–24. New York: Raven Press.

Ljungberg, T., Lidfors, L., Enquist, M. & Ungerstedt, U. (1987). Impairment of decision making in rats by diazepam: implications for the 'anticonflict' effects of benzodiazepines. *Psychopharmacology* **92**, 416–23.

McKinney, W.T. & Bunney, W.E. (1969). Animal model of depression: review of evidence and implications for research. *Archives of General Psychiatry* **21**, 240–8.

MacLean, P.D. (1949). Psychosomatic disease and the 'visceral brain': recent developments bearing on the Papez theory of emotion. *Psychosomatic Medicine* **11**, 338–53.

McNaughton, N., James, D.T.D., Stewart, J., Gray, J.A., Valero, I. & Drewnowski, A. (1977). Septal driving of hippocampal theta rhythm as a function of frequency in the male rat: effects of drugs. *Neuroscience* **2**, 1019–27.

Malick, J.B. (1978). Selective antagonism of isolation-induced aggression in mice by diazepam following chronic administration. *Pharmacology, Biochemistry and Behavior* **8**, 497–9.

Malick, J.B. & Enna, S.S. (1979). Comparative effects of benzodiazepines and non-benzodiazepine anxiolytics on biochemical and behavioural tests predictive of anxiolytic activity. *Communications in Psychopharmacology* **3**, 245–52.

Marks, I. (1987). Agoraphobia, panic disorder and related conditions in the DSM-IIIR and ICD-10. *Journal of Psychopharmacology* **1**, 6–12.

Mason, J.W. (1968). A review of psychoendocrine research on the pituitary-adrenal cortical system. *Psychosomatic Medicine* **30**, 576–607.

Mellanby, J., Gray, J.A., Quintero, S., Holt, L. & McNaughton, N. (1981). Septal driving of hippocampal theta rhythm: a role for GABA in the effects of minor tranquillizers. *Neuroscience* **6**, 1413–21.

Miczek, K.A., DeBold, J.F. & Winslow, J.T. (1986). Ethological analysis of aggression: mechanisms for effects of alcohol and benzodiazepines. *Psychopharmacology* **89**, S3.

Mineka, S. (1986). A primate model of phobic fears. *Bulletin of the British Psychological Society* **39**, A81.

Montgomery, K.C. (1958). The relation between fear induced by novel stimulation and exploratory behaviour. *Journal of Comparative and Physiological Psychology* **48**, 254–60.

Moyer, K.E. (1968). Kinds of aggression and their physiological basis. *Communications in Behavioral Biology* **2**, 65–87.

Nielsen, M., Braestrup, C. & Squires, R.F. (1978). Evidence for a late evolutionary appearance of brain-specific benzodiazepine receptors: an investigation of 18 vertebrate and 5 invertebrate species. *Brain Research* **141**, 342–6.

Noyes, R., Anderson, D.G., Clancy, J., Crowe, R.R., Slyman, D.J., Ghoneim, M.M. & Hinrichs, J.V. (1984). Diazepam and propranolol in panic disorder and agoraphobia. *Archives of General Psychiatry* **41**, 287–92.

O'Keefe, J. & Nadel, L. (1978). *The Hippocampus as a Cognitive Map*. Oxford: Oxford University Press.

Olton, D.S. (1973). Shock-motivated avoidance and the analysis of behavior. *Psychological Bulletin* **79**, 243–51.

Olton, D.S. (1979). Mazes, maps and memory. *American Psychologist* **34**, 583–96.

Olton, D.S. & Papas, B.C. (1979). Spatial memory and hippocampal function. *Neuropsychologia* **17**, 669–82.

Pellow, S., Chopin, P., File, S.E. & Briley, M. (1985). Validation of open:closed arm entries in an elevated plus-maze as a measure of anxiety in the rat. *Journal of Neuroscience Methods* **14**, 149–67.

Petersen, E.N. & Jensen, L.H. (1984). Proconflict effects of benzodiazepine receptor inverse agonists and other inhibitors of GABA function. *European Journal of Pharmacology* **103**, 91–7.

Petersen, E.N. & Scheel-Kruger, J. (1982). The GABAergic anticonflict effects of intra-amygdaloid benzodiazepines demonstrated by a new waterlick conflict paradigm. In *Behavioral Models and the Analysis of Drug Action*, M.Y. Spiegelstein & A. Levy (eds.), pp. 467–73. Amsterdam: Elsevier.

Pinel, J.P.J. & Treit, D. (1983). The conditioned defensive burying paradigm and behavioral neuroscience. In *Behavioral Approaches to Brain Research*, T. Robinson (ed.), pp. 212–34. Oxford: Oxford University Press.

Pollard, G.T. & Howard, J.L. (1986). The staircase test: some evidence of nonspecificity for anxiolytics. *Psychopharmacology* **89**, 14–19.

Poschel, B.P.H. (1971). A simple and specific screen for benzodiazepine-like drugs. *Psychopharmacology* **19**, 193–8.

Quintero, S., Mellanby, J., Thompson, M.K., Nordeen, H., Nult, D., McNaughton, N. & Gray, J.A. (1985). Septal driving of hippocampal theta rhythm: role of γ-aminobutyrate–benzodiazepine receptor complex in mediating effects of anxiolytics. *Neuroscience* **16**, 875–84.

Rachman, S. (1977). The conditioning theory of fear-acquisition: a critical examination. *Behaviour Research and Therapy* **15**, 375–87.

Rachman, S. (1978). *Fear and Courage*. San Francisco: Freeman.

Rachman, S. (1981). The primacy of affect: some theoretical implications. *Behaviour Research and Therapy* **19**, 279–90.

Rawlins, J.N.P. (1982). The relationship between memory and anxiety. *Behavioral and Brain Sciences* **3**, 498–9.

Redmond, D.E., Jr. (1979). New and old evidence for the involvement of a brain norepinephrine system in anxiety. In *Phenomenology and Treatment of Anxiety*, W.G. Fann, I. Karacan, A.D. Pokorny & R.L. Williams (eds.), pp. 153–203. New York: Spectrum.

Redmond, D.E., Huang, Y.H., Snyder, D.R. & Maas, J.W. (1976). The behavioral effects of stimulation of the locus coeruleus in the stumptail monkey (*Macaca arctoides*). *Brain Research* **116**, 502–10.

Rickels, K. (1985). Benzodiazepines in emotional disorders. In *The Benzodiazepines: Current Standards for Medical Practice*, D.E. Smith & D.R. Wesson (eds.), pp. 97–109. Lancaster: MTP Press.

Rickels, K., Downing, R.W. & Winokur, A. (1978). Antianxiety drugs: clinical use in psychiatry. In *Handbook of Psychopharmacology, Vol. 13: Biology of Mood and Antianxiety Drugs*, L.L. Iversen, S.D. Iversen & S.H. Snyder (eds.), pp. 395–430. New York: Plenum.

Robertson, H.A., Martin, I.L. & Candy, J.M. (1978). Differences in benzodiazepine receptor binding in Maudsley reactive and Maudsley non-reactive rats. *European Journal of Pharmacology* **50**, 455–7.

Rodgers, R.J. (1986). A pharmacoethological approach to intrinsic analgesia mechanisms. *Psychopharmacology* **89**, S30.

Rosenfeld, J., Lasko, L.A. & Simmel, E.C. (1978). Multivariate analysis of exploratory behavior in gerbils. *Bulletin of the Psychonomic Society* **12**, 239–41.

Russell, P.A. (1973). Relationships between

exploratory behaviour and fear: a review. *British Journal of Psychology* **64**, 417–33.

Scheel-Kruger, J. & Petersen, E.N. (1982). Anticonflict effect of the benzodiazepines mediated by a GABAergic mechanism in the amygdala. *European Journal of Pharmacology* **83**, 115–16.

Schofield, P.R., Darlison, M.G., Fujita, N., Burt, D.R., Stephenson, F.A., Rodriguez, H., Rhee, L.M., Ramachandran, J., Reale, V., Glencorse, T.A., Seeburg, P.H. & Barnard, E.A. (1987). Sequence and functional expression of the GABAa receptor shows a ligand-gated super-family. *Nature* **328**, 221–7.

Schutz, M.T.B., De Aguiar, J.C. & Graeff, F.G. (1985). Anti-aversive role of serotonin in the dorsal periaqueductal gray matter. *Psychopharmacology* **85**, 340–5.

Seligman, M.E.P. (1970). On the generality of the laws of learning. *Psychological Review* **77**, 406–18.

Seligman, M.E.P. (1975). *Helplessness: On Depression, Development and Death*. San Francisco: Freeman.

Sepinwall, J. & Cook, L. (1978). Behavioural pharmacology of anti-anxiety drugs. In *Handbook of Psychopharmacology, Vol. 13: Biology of Mood and Antianxiety Drugs*, L.L. Iversen, S.D. Iversen & S.H. Snyder (eds.), pp. 345–93. New York: Plenum.

Shephard, R.A. & Broadhurst, P.L. (1982). Effects of diazepam and picrotoxin on hyponeophagia in rats. *Neuropharmacology* **21**, 771–3.

Sherman, A.D., Allers, G.L., Petty, F. & Henn, F.A. (1979). A neuropharmacologically-relevant animal model of depression. *Neuropharmacology* **18**, 891–93.

Sherman, A.D., Sacquitne, J.L. & Petty, F. (1982). Specificity of the learned helplessness model of depression. *Pharmacology, Biochemistry and Behavior* **16**, 449–54.

Simiand, J., Keane, P.E. & Morre, M. (1984). The staircase test in mice: a simple and efficient procedure for primary screening of anxiolytic agents. *Psychopharmacology* **84**, 48–53.

Simon, P. & Soubrie, P. (1979). Behavioral studies to differentiate anxiolytic and sedative activity of the tranquilizing drugs. In *Modern Problems of Pharmacopsychiatry, Vol. 14: Differential Psychopharmacology of Anxiolytics and Sedatives*, J.R. Boissier (ed.), pp. 99–143. Basel: S. Karger.

Soubrie, P. (1986). Reconciling the role of central serotonin neurons in human and animal behavior. *Behavioral and Brain Sciences* **9**, 319–64.

Squires, R.F. & Braestrup, C. (1977). Benzodiazepine receptors in rat brain. *Nature* **266**, 732–4.

Stein, L., Wise, C.D. & Berger, B.D. (1973). Anti-anxiety action of benzodiazepines: decrease in activity of serotonin neurons in the punishment system. In *The Benzodiazepines*, S. Garattini,

E. Mussini & L.O. Randall (eds.), pp. 299–326. New York: Raven Press.

Stolerman, I.P., Garcha, H.S. & Rose, I.C. (1986). Midazolam cue in rats: effects of Ro 15-1788 and picrotoxin. *Psychopharmacology* **89**, 183–8.

Thiebot, M.H. (1986). Are serotonergic neurons involved in the control of anxiety and in the anxiolytic activity of benzodiazepines? *Pharmacology, Biochemistry and Behavior* **24**, 1471–7.

Thiebot, M.H., Soubrie, P., Hamon, M.M. & Simon, P. (1984). Evidence against the involvement of serotonergic neurons in the antipunishment activity of diazepam in the rat. *Psychopharmacology* **82**, 355–9.

Thiebot, M.H., Soubrie, P. & Simon, P. (1985). Is delay of reward mediated by shock-avoidance behavior a critical target for antipunishment effects of diazepam in rats? *Psychopharmacology* **87**, 473–9.

Thiebot, M.H., Soubrie, P., Simon, P. & Boissier, J.R. (1973). Dissociation de deux composants du comportement chez le rat sous l'effet de psychotropes: application a l'étude des anxiolytiques. *Psychopharmacology* **31**, 77–90.

Thomas, S.R. & Iversen, S.D. (1985). Correlation of 3H-diazepam binding with anxiolytic locus in the amygdaloid complex of the rat. *Brain Research* **342**, 85–90.

Thompson, R. (1978). Localization of a 'passive avoidance memory system' in the white rat. *Physiological Psychology* **6**, 263–74.

Traber, J. & Glaser, T. (1987). 5-HT1a receptor-related anxiolytics. *Trends in Pharmacological Sciences* **8**, 432–7.

Treit, D. (1985). Animal models for the study of anti-anxiety agents: a review. *Neuroscience and Biobehavioral Reviews* **9**, 203–22.

Treit, D., Pinel, J.P.J. & Fibiger, H.C. (1981). Conditioned defensive burying: a new paradigm for the study of anxiolytic agents. *Pharmacology, Biochemistry and Behavior* **15**, 619–26.

Tyrer, P. & Murphy, S. (1987). The place of benzodiazepines in psychiatric practice. *British Journal of Psychiatry* **151**, 719–23.

Valenstein, E.S. (ed.) (1973). *Brain Stimulation and Motivation*. Glenview, Illinois: Scott, Foresman.

Valzelli, L. (1973). Activity of benzodiazepines on aggressive behavior in rats and mice. In *The Benzodiazepines*, S. Garattini, E. Mussini & L.O. Randall (eds.), pp. 405–18. New York: Raven Press.

Vanderwolf, C.H., Kramis, R., Gillespie, L.A. & Bland, B.H. (1975). Hippocampal rhythmical slow activity and neocortical low voltage fast activity: relations to behavior. In *The Hippocampus, Vol. 2. Neurophysiology and Behavior*, R.L. Isaacson & K.H. Pribram (eds.), pp. 101–127. New York: Plenum.

Vogel, J.R., Beer, B. & Clody, D.E. (1971). A simple and reliable conflict procedure for testing anti-anxiety agents. *Psychopharmacology* **21**, 1–7.

Walsh, R.N. & Cummins, R.A. (1976). The open-field test: a critical review. *Psychological Bulletin* **83**, 482–7.

Welker, W.I. (1959). Escape, exploratory and food-seeking responses of rats in a novel situation. *Journal of Comparative and Physiological Psychology* **52**, 106–11.

Whimbey, A.E. & Denenberg, V.H. (1967). Two independent behavioural dimensions in open field performance. *Journal of Comparative and Physiological Psychology* **63**, 500–4.

Williams, J., Gray, J.A., Snape, M. & Holt, L. (1989). Long-term effects of septohippocampal stimulation on behavioural responses to anxiogenic stimuli. In *The Psychopharmacology of Anxiety*, P. Tyrer (ed.), pp. 80–108. Oxford: Oxford University Press.

Willner, P. (1984). The validity of animal models of depression. *Psychopharmacology* **83**, 1–16.

Willner, P. (1986). Validation criteria for animal models of human mental disorders: learned helplessness as a paradigm case. *Progress in Neuro-Psychopharmacology and Biological Psychiatry* **10**, 677–90.

Willner, P. & Birbeck, K.A. (1986). Effects of chlordiazepoxide and sodium valproate in two tests of spatial behaviour. *Pharmacology, Biochemistry and Behavior* **25**, 747–51.

Winson, J. (1972). Interspecies differences in the occurrence of theta. *Behavioral Biology* **7**, 479–87.

Zuckerman, M. (1982). Leaping up the phylogenetic scale in explaining anxiety: perils and possibilities. *Behavioral and Brain Sciences* **3**, 505–6.

3

Screening for anxiolytic drugs

D.N. STEPHENS and J.S. ANDREWS

Anxiolytic drugs account for 40% of the world sales of psychoactive drugs in cash terms with an annual turnover of £2,500 M. They are thus one of the most important areas of activity for pharmaceutical companies. The bulk of these prescriptions is for benzodiazepines (BZs), and of these, although it was introduced to the clinic nearly 20 years ago diazepam still represents more than 15%. The BZs have been extraordinarily successful drugs in commercial terms, and have had an enormous influence on medical practice from the general practitioner level upwards.

3.1 BENZODIAZEPINES

It is not only in the clinic that BZs have exerted their influence. A great deal of our understanding of the neuropsychology of anxiety is based on an understanding and description of the mode of action of these drugs (e.g. Gray, 1982), and our entire methodology in terms of developing animal models of anxiety is heavily reliant on the BZs for its validity (see Green & Hodges, Chapter 2 of this volume). The success of the BZs owes itself to the fact that they are both extremely effective anxiolytics, and very safe drugs. A new drug must show considerable benefits over them to be either ethically justifiable or commercially successful. Thus in designing a screening programme for a new anxiolytic drug it is necessary to start out from the strengths and weaknesses of the BZs. BZ anxiolytics do indeed have properties which can be improved upon. Firstly, they have a wide spectrum of effects apart from anxiolysis; they are anticonvulsant, sedative, hypnotic, induce muscle relax-

ation and induce amnesia; and while all these properties find their clinical applications, for the use of BZs as anxiolytics, all must be regarded as unwanted or unnecessary side-effects. To this list may also be added the ability of BZs to potentiate the effects of alcohol and barbiturates.

Additionally, BZs have won a growing reputation for bringing problems during withdrawal at the end of treatment. Withdrawal symptoms include intense anxiety, sleeplessness and even, in the worst cases, convulsions (Levy, 1984; Breir et al., 1984). Such problems are not universally observed and are often associated with short half-lives, long term treatments and high doses. However they bring about a legitimate concern amongst clinicians and patients alike.

A perhaps related problem is the development of tolerance to the therapeutic actions of the BZs so that following repeated administration, either the therapeutic effect wears off, or higher doses are required to maintain the same effect. Such problems of tolerance are of special importance in the use of BZs for the treatment of epilepsy where the therapeutic effect declines after some weeks or months of treatment, but the clinician feels the necessity to maintain treatment in the fear that withdrawal may give rise to an increase in the frequency of convulsions. Whether tolerance develops to the anxiolytic effect of BZs is hotly disputed. Many clinicians believe that tolerance does not occur but others are equally convinced that it does. Certainly in the laboratory it has been possible to demonstrate tolerance to the effects of several BZs in animal models of anxiety (e.g. Vellucci & File, 1979;

Stephens & Schneider, 1985; Emmet-Oglesby *et al.*, 1987).

A further problem with BZs, though much sensationalised, is their abuse potential, that is, their use for non-therapeutic purposes. Since measurement of abuse potential, dependence and withdrawal, is the subject of other chapters (see Goudie and Sanger, Chapters 17 and 18, this volume), it will not be dealt with further here, even though it is of great importance for our theme.

From the above it will be clear that it is as important for a new anxiolytic drug to exclude side effects as to establish anxiolytic potency. A screening programme for novel anxiolytics must therefore contain not only tests capable of identifying potential anxiolytic properties, but also at some stage, tests for possible side effects and problems of tolerance, dependence, withdrawal and abuse.

On the other hand, the BZs are extraordinarily safe drugs in terms of their lack of effects on organ systems other than the brain. Thus a new anxiolytic must not only pass the tests of toxicology which are standard for all novel compounds planned for testing in humans, it must rival the BZs in its relative lack of effects on, for instance the cardiovascular or reproductive system. Since this question of 'safety pharmacology' and toxicology is common to all the classes of drugs discussed in this volume (see e.g. Ahlenius, Chapter 12), it, too, will not be dealt with further here.

3.2 NON-BEHAVIOURAL SCREENING PROCEDURES

In the past it was common in pharmaceutical research to investigate novel compounds in a variety of tests for a variety of indications. By employing rather simple models, an experienced pharmacologist might identify a novel compound as behaviourally active, in (e.g.) the Irwin screen (Irwin, 1968), or as affecting the cardiovascular system or as possessing another action. This approach, which requires a high capacity, and is rather hit and miss in the first stages, has become almost extinct in psychopharmacology.

Today we understand more about the actions of drugs, from the receptor level to their clinical effects and it is therefore possible, and more efficient, to synthesise novel compounds with a direct aim in mind. Thus one approach is to search for compounds with an action at the BZ receptor. A novel approach, based to an extent on the apparent therapeutic effect of buspirone (Rakel, 1987) but also on theoretical considerations (Gray, 1982) is to identify substances acting on the serotonin (5-HT) system; we will return to this approach later. The nature of the screening programme will thus depend on the approach decided upon at the outset.

It should also be appreciated that the initial aim of a screening programme is not immediately to identify a drug with the required properties for the clinic out of the hundreds or thousands produced by chemists. Research in the pharmaceutical industry is a team effort with constant feedback (and forward) at each stage of development. At the beginning of a project it is more useful to produce a small amount of information on a large number of compounds in order to allow hypotheses of structure–activity relationships to be formed. This is achieved by making systematic changes in the molecular structure of a 'lead' molecule and then recording alterations in the pharmacological properties. The lead molecule itself need only have some of the properties desired, e.g. it may bind to the receptor but only weakly and have no anticonflict activity. However, systematic changes may lead to compounds with diverging properties and pharmacological interest. This development of an active compound from an inactive one will be discussed later. At this early stage of drug-finding a single measure of the biological activity of a large number of related compounds may be of great importance. Later, when several substances have been identified with a good activity in this single measure or in a few different tests, it becomes necessary to identify advantages of these compounds over one another, or over existing standard compounds. At this stage, a breadth of tests applied for a small number of compounds will become necessary. In practice, these two procedures usually progress in parallel and a screening programme, consisting of primary, secondary and tertiary level screens of increasing breadth and complexity, is designed to cover both the need to understand structure–activity relationships, and to identify possible drugs for development. However, at different stages of the project the emphasis of the screening programme will be shifted between these approaches.

3.2.1 Biochemical screening

BZs achieve their pharmacological effects by enhancing the efficacy of γ-amino-butyric acid (GABA) in the central nervous system (CNS) (Costa *et al.*, 1975; Haefely *et al.*, 1975). In the last few years enormous strides have been made in understanding this interaction and it is now clear that specific binding sites for BZs exist on a membrane protein complex which also possesses receptor sites for the inhibitory neurotransmitter, GABA. The BZ binding site interacts allosterically with the GABA site to modulate the ability of GABA to influence chloride ion flux through a related ion channel in the nerve membrane (e.g. Study & Barker, 1981; Gee *et al.*, 1986). By enhancing the inhibitory effects of GABA on neuronal firing, BZs are able to influence the activity of many types of central neurones accounting for their diverse effects in anxiety, epilepsy, sedation, muscle relaxation, memory function, sleep and so on. (See Schneider *et al.*, 1989, for a recent review).

This knowledge of BZ receptor function allows a great deal about the likely properties of BZ receptor ligands to be inferred from their biochemical pharmacological properties, without having to perform an initial screen in behavioural tests. Thus novel substances not displacing labelled BZs from brain membranes in simple receptor assays might be rejected simply by using an *in vitro* test system (but see the discussion below on tracazolate). Not all substances which bind to the BZ receptor possess anxiolytic properties, and there exist BZ antagonist ligands for the receptor and even ligands which induce anxiety (e.g. Dorow *et al.*, 1983; Corda *et al.*, 1983; Stephens *et al.*, 1987). Nevertheless, by studying the allosteric interaction between the GABA and BZ receptor it is possible to predict certain pharmacological effects of novel compounds from biochemical observations.

The primary action of anxiolytic BZs is to enhance the affinity of GABA; this allosteric enhancement of GABA binding can be demonstrated by incubating brain membranes with tritiated GABA in the presence of BZ receptor ligands (Tallman *et al.*, 1978). In point of fact, this method is technically not as accessible as its logical mirror-image. Since the BZ and GABA binding sites are allosterically linked, the binding of BZs is also altered in the presence of GABA. The direction and extent to which GABA modifies the affinity of BZs reflects their efficacy at the receptor. Thus, in the presence of GABA or GABA agonists such as muscimol, the binding of BZs such as diazepam and alprazolam is markedly enhanced whereas the affinity of convulsant and anxiogenic ligands at the BZ receptor is actually reduced (e.g. Braestrup & Nielsen, 1981). The affinities of BZ receptor ligands which exhibit only weak intrinsic activity but which antagonise the effects of both BZs and the convulsant ligands BZ i.e. antagonists are hardly altered in the presence of GABA (Stephens *et al.*, 1987).

Thus on the basis of *in vitro* biochemical experiments it is already possible to suggest where in the anxiolytic–anxiogenic and anticonvulsant–convulsant spectrum new compounds may be located. It may also be possible to separate sedative anxiolytics from non-sedative anxiolytics by the magnitude of the change in affinity induced by GABA (see Stephens *et al.*, 1987, for a discussion of the correlation between biochemical and behavioural predictors of anxiolytic activity).

The GABA/BZ receptor complex contains binding sites not only for these two types of ligand but also for barbiturates, for picrotoxin (Leeb-Lundberg *et al.*, 1980; Olsen, 1981; Huidobro-Toro *et al.*, 1987) and for cage-convulsants of the t-butylbicyclophosphorothionate (TBPS) type (Squires *et al.*, 1983; Havoundijian & Skolnick, 1986). All of these sites exhibit allosteric interactions with the BZ site which predict to some degree the pharmacological effect of the BZ ligand in question. This receptor complex is summarised in Fig. 3.1. To an extent, these different allosteric effects correlate with each other as must be expected from current models of BZ receptor function, and since, for instance, the effect of BZ ligands on binding of ^{35}S-TBPS is technically easier to perform than GABA stimulation of BZ binding, this measure offers a useful primary screen for predicting BZ-like effects (see Stephens *et al.*, 1987; Young & Glennon, 1987).

Other possibilities for studying the effect of BZ receptor ligands *in vitro* include electrophysiological measures on ion flux through chloride channels, in patch clamp studies (see Study & Barker, 1981; Mathers, 1987). However, these methods are too complex and labour intensive for drug screen-

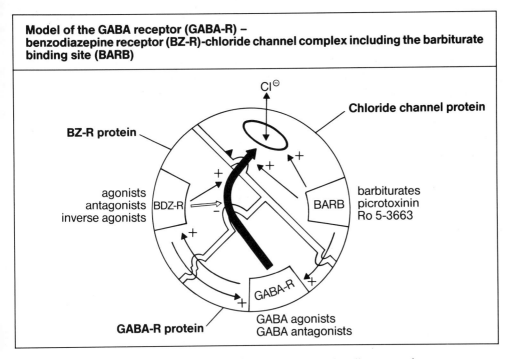

Model of the GABA receptor (GABA-R) – benzodiazepine receptor (BZ-R)-chloride channel complex including the barbiturate binding site (BARB)

Cl⁻

Chloride channel protein

BZ-R protein

agonists
antagonists
inverse agonists

BDZ-R

barbiturates
picrotoxinin
Ro 5-3663

BARB

GABA-R

GABA-R protein

GABA agonists
GABA antagonists

Fig. 3.1. Schematicised diagram of the GABA/BZ receptor complex illustrating the most important interactions between the various binding sites. All three types of binding site shown regulate the chloride channel. The BZ and barbiturate binding sites can enhance (or inhibit) the binding of GABA. This type of interaction is the basis for the GABA ratio test and TBPS ratio test described in the text as one way of using a biochemical screen to predict where in the anxiolytic/sedative–anxiogenic/convulsant spectrum a BZ ligand may lie. (Adapted from Polc et al., 1982.)

ing. A recent advance allows the measurement of chloride flux through synaptosomal membranes by using radioactive chloride ions (Majewska & Schwartz, 1987). This method offers a direct measure of the ability of BZ receptor ligands to influence the function of the BZ/GABA receptor/chloride channel complex, and it may thus be a useful screening method for identifying BZ/GABA receptor modulators. *In vivo* electrophysiological measures of the ability of novel compounds to modulate the effects of GABA on neuronal firing rates (e.g. McNaughton *et al.*, 1986; Smith & Gallager, 1987; Mereu *et al.*, 1987) belong only to intense characterisation of drug action.

In vitro methods offer a number of advantages over *in vivo* procedures. They allow a high through-put of compounds, and exclude factors such as poor oral bioavailability, or brain penetration, which might otherwise cause an active sub-

stance to be excluded. A consideration not to be made light of is the reduction in the number of experiments on animals used in drug research.

A further advantage of an *in vitro* approach is that it offers a degree of sensitivity which is missing from the *in vivo* models. Thus structures which exhibit only weak binding might provide a 'lead' substance which would not be picked up by anxiety models. Such leads can then be modified and improved in interplay between research chemists and pharmacologists to provide a substance which is also active in the behavioural tests.

Hesp & Resch (1987) describe such an interactive process in a chemical and series based on the non-BZ anxiolytic tracazolate. Unlike tracazolate itself, a decomposition product, desbutyltraczaolate displaced labelled BZs from brain membranes, indicating an activity at the BZ receptor, though with a low affinity. This compound exhibited no behavioural influence. On the basis of the BZ receptor affinity, however, a series of modifications were made to produce the orally active compound ICI 174 329. This compound not only had a higher receptor affinity than desbutyltracazolate, but was equipotent with diazepam in several anxiety models. A screening programme based on behavioural methods would have had no access to the finally interesting substance.

A second example from our own laboratories also illustrates this point. Following the discovery of binding sites for BZs in the brain (Squires & Braestrup, 1977; Mohler & Okada, 1977), Braestrup and his coworkers began a search for possible endogenous ligands for this receptor. As a result of this effort a novel structure, β-carboline carboxylic acid ethyl ester (β-CCE), was isolated from human urine and found to bind with an affinity similar to that of BZs to ^{3}H-diazepam labelled binding sites (Braestrup *et al.*, 1980). Although β-CCE was in fact an artefact of the isolation procedure, it provided a lead for a series of derivatives. Interestingly, the lead structure possessed none of the properties of BZs in animal models, and indeed in many models exhibited an activity opposite to those of the BZs. That is, it is proconvulsant (Braestrup *et al.*, 1982) and anxiogenic (Corda *et al.*, 1983). Such a compound could not have been identified with existing animal models of anxiety as a likely lead for a series of anxiolytic substances. On the basis of receptor binding studies, however, it was possible to identify other β-carbolines with activity at the BZ receptor, including substances with anxiolytic and other BZ-like properties, selective compounds with anxiolytic but no sedative/ataxic properties (Petersen *et al.*, 1984), antagonists of BZ action (Jensen *et al.*, 1984), and compounds with properties opposite to those of the BZs.

3.2.2 Pharmacological screening for anxiolytics

Although such *in vitro* methods provide a great deal of easily won information, they can only partially replace direct measures of the effects of a new drug in the whole animal.

The basic requirements of a primary *in vivo* screen is a high throughput, with a high predictive ability and using little substance. A certain incidence of false positives (substances showing activity in the test though not actually possessing

the required property) can be tolerated since they will be excluded later by more sophisticated tests. False negatives are less acceptable: once an active compound has been excluded from testing, it is lost to science (and economics!).

The choice of a primary screen is always difficult and in the field of anxiolytic research many different models are employed. Not all of them have anything very obvious to connect them with anxiety, and some, such as the ability of a substance to antagonise convulsions induced by pentylenetetrazole (PTZ), are related to other qualities of BZs. This test has been a useful one because BZs which possess anxiolytic properties are also anticonvulsant. The disadvantage of the method is that the model has no predictive ability for anxiolytics which are not at the same time anticonvulsant. Although certain anticonvulsants such as valproic acid, which are not used for their anxiolytic effects, are also active in this test, they would be excluded at a later stage of screening. The task here is to provide feedback to the chemist that a particular line of thinking is worth following or should be abandoned.

3.3 TESTS OF ANXIOLYTIC ACTIVITY

Only at this already advanced stage in the screening procedure does it become necessary to introduce tests which are thought to test specifically the anxiolytic properties of the novel substance. The number of actual tests available is large, (see e.g. Treit, 1985*b*; Green & Hodges, Chapter 2 of this volume, for comprehensive reviews). Green & Hodges discuss fully almost all of the models available and their theoretical base. We shall discuss only a few of the major models used today in the pharmaceutical industry as screens for anxiolytic drugs rather than as models of anxious behaviour. Therefore we shall discuss only some of the most popular methods, and their advantages and disadvantages in an industrial setting. Additionally we shall provide one possible rationale for a screening programme.

3.3.1 Behavioural inhibition

Tests based on behavioural inhibition provide the main armoury of the animal pharmacologist seeking novel compounds. The principle is straightforward. Animals are punished, usually with electric shock, for carrying out some spontaneous or rewarded behaviour. As a result of the punishment the frequency of occurrence of the behaviour is reduced, but administration of known anxiolytics reinstates the punished behaviour.

3.3.1.1 The four-plate test

The four-plate test was introduced by Boissier and coworkers (Boissier *et al.*, 1968) as one of a series of ingenious, simple tests for the rapid screening of minor tranquilisers. In this procedure drug- and test-naive mice are introduced to a novel test chamber whose floor is divided into four metal plates. Crossing from one plate to the next is punished by electric shock. A short observation time of one minute is adequate to reveal effects of anxiolytic compounds.

The advantages of this method are several: no lengthy training is required; drugs need only be administered once so that independent groups of animals can be used for each drug dosage; the low body weight of mice demands only small quantities of drug so that the chemists do not need to synthesise large quantities of expensive or difficult compounds; the method is quick and complete dose–response curves can be established in a few hours, giving rapid feedback as to whether a new structure is a step in the right or wrong direction. Furthermore, the method can be used as part of a test battery so that a single mouse can give rise to information not only on the possible anxiolytic effects of a compound but also; say, its anticonvulsant, sedative or muscle-relaxing activities, resulting in a saving in both animals and material.

This efficiency, however, is at the expense of specificity and clarity. Table 3.1 shows some of the substances which are known to be active in increasing punished activity in the four-plate test. As is necessary for such a test, clinically active anxiolytics including BZs, barbiturates and also ethanol are all effective. Also active are certain β-carbolines which although not yet tested in the clinic, show activity in other animal models of anxiety. Certain antihistamines, and the tricyclic antidepressant amitriptyline, also increase punished activity although they are not known to possess anxiolytic activity in the clinic. Additionally, a

number of drugs, including d-amphetamine, cocaine and caffeine, increase both punished and unpunished activity (Aron *et al.*, 1971). Since in our experience (e.g. Stephens & Kehr, 1985) BZs also increase both punished and unpunished crossings, these observations indicate the possibility of obtaining false positives in this test.

3.3.1.2 Vogel test

The Vogel test (Vogel *et al.*, 1971) is based on the punishment of drinking. Drug and test-naive thirsty rats are allowed to drink from a drinking tube; electric shock delivered through the tube reduces the number of licks compared to that observed in controls. The rate of punished responding is enhanced by clinically active anxiolytics such as chlordiazepoxide, diazepam, meprobamate and pentobarbital, but not by stimulants such as amphetamine, nor by scopolamine. Thus the test is pharmacologically specific, easy to perform, does not require lengthy training and is carried out with drug-naive animals. Modifications of the lick-suppression test (Kilts *et al.*, 1981; Petersen & Buus-Lassen, 1981), however, revert to the use of pre-trained rats, which can be used repeatedly for drug-testing. This procedure was introduced to provide more stable baseline behaviour than is sometimes seen in the Vogel test, but by reintroducing a period of training, and repeated drug treatments, sacrifices the Vogel test's primary advantages over operant procedures.

3.3.1.3 Geller–Seifter type tests

In the original Geller procedure (Geller & Seifter, 1960), rats were trained in operant chambers to operate a lever to obtain food. Presentation of a tone signalled that the reinforcement contingencies had been changed and that further responding would result in both an increased availability of food, and in foot-shock. Response rates during the 'conflict' component of this and similar procedures (Cook & Davidson, 1973) were enhanced by existing clinically active anxiolytics such as chlordiazepoxide and barbiturates, but not by drugs such as amphetamine. Responding in the simple food-rewarded component is not increased by the anxiolytics.

The main practical problem with the Geller

Table 3.1. *Summary table indicating the most commonly reported effects (+ = active; − = inactive) of various classes of psychoactive drugs in different tests believed to measure anxiolytic efficacy.*

	Four-plate	Geller	Vogel	CER	Plus maze	Social interaction	BZ cue	PTZ cue	Potentiated startle
Benzodiazepine	+	+	+	(±)	+	+	+	+	+
But e.g.									
lorazepam	−								
triazolam						−			
BZ receptor ligands									
e.g.									
CGS 9896	+	+	+		+	+	+	+	
CL 218 872	+	+	+		+	+	+	+	
ZK 93 423	+	+	+				+	+	
ZK 91 296	+	(+)	−		−	+	+	(+)	
Barbiturates									
e.g.									
Pentobarbital	+	+	+			+	+	+	+
Neuroleptics									
e.g.									
Haloperidol	−	−	−		−		−	−	−
Stimulants									
e.g.									
Amphetamine	+	−	−		−		−	−	
Antidepressants									
e.g.									
Imipramine	(−)	−			−				
5-HT anxiolytics*									
e.g.									
Buspirone	−	(−)	(+)	+	−	−	−	−	+
Opiates									
e.g.									
Morphines	−	(−)	−	(+)	−		−	−	+

In general, all non-convulsive ligands for the GABA/BZ receptor complex (see Fig. 3.1) are active in all tests (noteworthy exceptions are indicated). Some tests such as the four-plate test give false positive results with stimulants. Weak or contradictory reports of an effect are indicated by parentheses.
*for comprehensive analysis, see Chopin & Briley (1987).

procedure is that it requires a great deal of effort. Firstly, it is necessary to train the animals (usually rats) to perform the two components of the operant task; this in itself requires several weeks' training. Secondly, the individual sensitivity of rats to the foot-shock punishment varies considerably so that some animals virtually ignore a shock of, say 0.5 mA, whereas others react so strongly to the same shock that they may stop operating the lever permanently. Furthermore, if the degree of suppression of the operant behaviour is very great, then even potent anxiolytics may be ineffective. On the other hand, if the suppression is only small, it may be difficult to find statistical evidence for increases in responding in the presence of the drug. So it is often necessary to adjust the shock level for each individual animal, and at intervals to re-establish the optimum level. Even when the animals are well trained and the individual shock levels titrated, however, the problems are not at an end. It is a common (though not universal) experience that newly trained, previously drug-naive rats do not show increases in punished responding when first treated with a BZ; rather, repeated treatments are often necessary to reveal an anti-conflict effect (Margules & Stein, 1968; Sepinwall & Cook, 1978). However, once the anti-conflict effect of one drug (usually diazepam) has been established, then the animals are available for testing novel compounds. This means, of course, that novel compounds are routinely tested in drug-experienced animals, something which pharmacologists normally try to avoid.

This problem is usually compounded because the trained animals are used to test many different doses of many compounds, and one can never be sure what influence experience with one compound might have on the effect of another. The alternative, to train a new batch of animals for each dose of each drug, is impractical because of the long training time required.

In addition to the practical problems which arise from use of the Geller procedure and its close relatives, there are a number of theoretical ones (see also Green & Hodges, Chapter 2 of this volume). Perhaps the most problematic is that during the long training, animals acquire a good understanding of the relationship between the signal and punishment and may show no physical signs of fear or anxiety (Andrews, unpublished observations). If the animals are not anxious or fearful, it is not clear that anxiolytics act by reducing fear or anxiety, or whether they otherwise disrupt the control of behaviour by the operant schedule. It is well known that BZs reduce vigilance (e.g. Ott, 1984; Stephens & Sarter, 1988) and it has been proposed that they impair ability to process relevant external stimuli, and dissociate external stimuli from their emotional associations (Gray, 1982). While this might on the one hand be a description of *how* tranquillisers achieve their anxiolytic effects, it might also be a reason to think that disruption of stimulus control rather than anxiolysis forms the basis for their actions in operant conflict tests.

The evidence on these points is not large and only a few workers have considered this possibility (e.g. Hodges & Green, 1987; Ljungberg *et al.*, 1987). If it is true, then the Geller type of test would remain valuable for drug screening, but lose much of its apparent face validity as a model of anxiolytic activity.

3.3.2 Drug discrimination

Drug discrimination is a procedure which to an extent cuts through the Gordian knot of establishing a theoretical basis for animal models of anxiolytic activity. In drug discrimination experiments the response required to obtain a reinforcement is signalled by whether the animal has been treated with a given drug or not. Thus animals learn to operate one lever in the presence of the training drug and a second lever in the presence of placebo (Colpaert *et al.*, 1976).

3.3.2.1 Benzodiazepine cue

If a novel compound resembles the training drug, then at appropriate doses the animal selects the lever associated with the training drug; if the test substance does not resemble the training drug the animal selects the placebo lever. This methodology offers a very reliable test for identifying substances with properties similar to those of existing drugs. Thus drug discrimination procedures using BZs as the standard training drug have found widespread favour in screening for substances with similar properties (Bennett, 1985; Bennet *et al.*, 1985; Ator & Griffiths, 1986; Sanger, 1987).

The problem is that BZs have many properties, some desirable (e.g. anxiolytic) and some not (e.g. muscle relaxant, amnestic). We do not know how the animal recognises the BZ cue, i.e. if the BZ discriminative stimulus is related to its anxiolytic, sedative or other properties and theoretically, we could as easily end up selecting a sedative, amnesic, but non-anxiolytic compound, as an anxiolytic lacking side-effects. In point of fact, drug-discrimination procedures based on BZs do not seem to reflect their sedative or muscle-relaxant properties (Colpaert *et al.*, 1976). For instance, zolpidem is a sedative BZ-receptor ligand without clear anxiolytic effects, but a cue based on this compound does show a different pharmacological profile from

BZ cues (Sanger & Zivkovic, 1986; Sanger *et al.*, 1987); additionally a number of compounds, including the β-carbolines ZK 91 296 and ZK 95 962, and the pyrazoloquinoline CCS 9896 are recognised as BZ-like in drug-discrimination procedures even though they do not possess sedative effects in other animal models (Stephens *et al.*, 1987). Since in other animal tests these compounds act as anxiolytics, it seems at least a reasonable working hypothesis that what they have in common with BZs is their anxiolytic activity. It is worth remembering, however, that novel types of anxiolytic such as buspirone are not identified as being BZ-like in such drug discrimination procedures.

3.3.2.2 Pentylenetetrazol (PTZ) cue

A second drug discrimination model (Shearman & Lal, 1980) employs the ability of animals to discriminate PTZ. PTZ at high doses gives rise to convulsions; at lower doses, however, it has been reported that PTZ induces feelings of anxiety in human subjects (Hildebrandt, 1937; Rodin & Calhoun, 1970). Thus, drugs, which are able to antagonise the effects of non-convulsant doses of PTZ in animals may possess anxiolytic properties. Pharmacological evidence suggests that known anxiolytics block the ability of rats to discriminate PTZ, and that anticonvulsants without known anxiolytic activities do not, suggesting that the PTZ discriminative stimulus is not based on its convulsant properties, (Shearman & Lal, 1980); other convulsants such as strychnine do not generalise to the PTZ cue while anxiogenic substances such as cocaine and the anxiogenic β-carbolines methyl-β-carboline-3-carboxylate (BCCM) (Lal & Emmett-Oglesby, 1983) and FG 7142 (Stephens *et al.*, 1984*b*) do. The evidence appears consistent with the PTZ discriminative stimulus depending on its anxiogenic properties. However, a more conservative interpretation might be that the ability to antagonise or to mimic the PTZ discriminative stimulus reflects a drug's ability to influence the GABA/BZ receptor complex. This interpretation removes the test's apparent face validity, but leaves it as one useful model for screening drugs acting at this complex. Whether the test offers any more in this respect than, say, the PTZ convulsion model is still open to question.

There are several disadvantages of the PTZ cue model that have become apparent in our laboratory. Firstly the length of time it takes to train the animals (80–100 sessions) means that it is a costly screen to run. Secondly, animals become unstable in responding from time to time without any apparent reason. And finally, we have noticed that all the drugs we have tested that generalise to PTZ, also induce seizures or kindled seizures in rats. That is, repeated administration of subconvulsive doses of such drugs leads eventually to seizures. Furthermore we have observed that all drugs that antagonise the PTZ cue antagonise PTZ-kindled seizures, regardless of their ability or inability to antagonise seizures brought about by an acute dose of PTZ (Andrews *et al.*, 1989; see Table 3.2). This suggests that the PTZ-cue may be a pre-kindled seizure cue.

Recently, the β-carboline FG 7142, which is a partial inverse agonist at the BZ receptor has been reported to induce attacks of intense anxiety in human volunteers (Dorow *et al.*, 1983). An FG 7142 cue has been established but not yet fully characterised pharmacologically (Leidenheimer *et al.*, 1987; Leidenheimer & Schechter, 1988). However, it has been possible to demonstrate that animals exposed to acute stress choose the lever associated in training with FG 7142 even when their test-day treatment has been saline (Leidenheimer & Schechter, 1988).

This would seem to be *a priori* evidence that the FG 7142 cue is associated with its stress-inducing or anxiogenic properties and suggests that a drug discrimination test based on FG 7142 might, used with caution, provide a useful anxiolytic screening method. The problem here is that many substances binding to the BZ receptor, including the antagonists such as flumazenil and ZK 93 426 would be expected to antagonise the FG 7142 cue while not possessing anxiolytic activity. Furthermore, FG 7142 also causes kindled seizures in rats and generalises to PTZ (Stephens *et al.*, 1984*b*). It seems likely that, as with PTZ, any anxiogenic properties would be inextricably linked with seizure activity and therefore would provide no advantage over simpler seizure models of BZ activity.

3.3.3 Social interaction

A model of anxiety based on increases in social interaction between male rats in a novel, well-lit

Table 3.2. *The relationship between activity in the PTZ cue of full and partial BZ receptor agonists and clinically effective anti-epileptics, and activity against clonic convulsions induced by an acute injection of PTZ and against PTZ kindled seizures (adapted from Andrews et al., 1989).*

Substance	PTZ cue	Antagonism Acute PTZ clonic	Kindled PTZ seizure
BZ Agonists			
Diazepam	0.37	1.86	1.84
Chlordiazepoxide	1.06	3.1	1.87
Lorazepam	0.05	0.36	0.18
Lormetazepam	0.04	0.22	0.24
Triazolam	0.02	0.25	0.17
Midazolam	1.25	2.49	0.82
Clonazepam	0.32	0.6	0.24
Non-BZ Agonists			
ZK 91 296	0.83	10.8	1.86
ZK 93 423	0.08	0.66	0.32
ZK 95 962	0.18	12.87	0.59
Zolpidem	0.96	6.81	1.83
CGS 9896	1.47	> 80	1.53
CL 218 872	2.16	4.32	4.86
Antiepileptics			
Valproate	126.84	293.6	99.99
Phenytoin	> 60	> 100	> 100
Carbamazepine	> 40	> 80	> 80
Ethosuximide	107.69	> 212	176
New/Potential anxiolytics			
Buspirone	> 2.5	> 30	> 30
MK-801	> 0.32	> 0.5	> 0.5
2-APH	> 80	—	> 80

All values stated are ED_{50} (mg/kg) in the rat.
Note that a compound's ability to antagonise the PTZ discriminative stimulus is predicted by its ability to prevent PTZ kindled seizures, but not necessarily by its action against seizures induced following a single high dose of PTZ. Of the clinically effective anti-epileptics, only those with the ability to antagonise PTZ convulsive effects were active in the cue. For discussion, see text under Section 3.2.2; for full details of the experimental procedures see Andrews et al., 1989.

environment has been established and popularised by File and her colleagues (File & Hyde, 1979; File, 1980). This important member of the test battery of the psychopharmacologist interested in anxiety is dealt with extensively by Green & Hodges (Chapter 2, this volume) and these considerations will not be repeated here. For an industrial pharmacologist, however, the social interaction test described by File has two major disadvantages. Firstly, the test requires observation by well trained personnel, and is therefore labour intensive and more or less impossible to automate satisfactorily. Thus the test is unsuitable for routine drug screening procedures. Secondly, the version of the test developed by File uses chronic drug administration. This not only increases the work load, but also introduces the problem of tolerance. A modification of the social interaction test proposed by Guy & Gardner (1987) overcomes this latter problem, but not the former. An additional problem is that certain anxiolytic BZs such as triazolam are not active in the test (File, 1980).

3.3.4 Exploratory models of anxiety

A number of tests make use of the tendency of rodents to explore a novel environment. In such tests, two factors seem to be important, a tendency to explore and at the same time a tendency to avoid novel situations; these two tendencies are somehow in balance with one another. Anxiolytics, by dampening the second component, might be expected to enhance exploratory activity and this idea has been exploited in a number of test models including the staircase test (e.g. Simiand et al., 1984) and the emergence test of Crawley et al. (1984).

In the former, naive mice or rats are tested in an enclosed staircase. A reduction in the number of steps climbed gives a measure of sedative properties of the drug, while anxiolytics are said to reduce the amount of rearing. Although there is no obvious connection between either behaviour and anxiolysis, and it seems at least as likely that the reduction in rearing is also a measure of sedation, this test is sometimes used in industry as a primary screen, where it has the advantage of simplicity, a high throughput, and can be combined with other primary screens. It also appears to be rather specific for clinically useful anxiolytics, though once again, buspirone appears inactive as does alprazolam, while morphine give a false positive in this test (Pollard & Howard, 1986).

The test described by Crawley *et al.* (1984) uses the aversion of mice to areas of high illumination. Anxiolytic drugs increase the frequency of entries into the light compartment of a two-compartment box. This test appears to have some face validity and the fact that it gives rise to false positives such as with amphetamine need not be too important in a primary screen. However, it must be noted that a recent study has suggested that the effects of diazepam in this test are better described by changes in general locomotor stimulation rather than by an anxiolytic action (Carey & Fry, 1988).

A refinement of a similar principle is found in the plus-maze (Handley & Mithani, 1984; Pellow *et al.*, 1985; Pellow, 1986). The apparatus consists of a plus-shaped maze with two open and two enclosed arms. Naive rats (Pellow *et al.*, 1985) and mice (Stephens *et al.*, 1986) spend more time in the enclosed than in the open arms, but following treatment with anxiolytic drugs, the animal divides its activity more equally between the two types of arm. Pellow and her colleagues (1986) have characterised this test behaviourally, physiologically and pharmacologically (see Green & Hodges, Chapter 2 of this volume). It is fast, simple, does not involve the use of expensive equipment, and can be easily automated. Furthermore, no training is required, nor is the use of noxious stimuli or food or water deprivation.

Since mice can be employed and effects seen on acute administration the test requires only minimal amounts of test compound. The pharamcological specificity also appears to be good (Table 3.1) though the test failed to reveal an anxiolytic activity of busipirone or the putative anxiolytic ZK 91 296, a β-carboline partial agonist at BZ receptor, or of TVX-Q7821 (ipsapirone) a 5-HT$_{1A}$ agonist (Critchley & Handley, 1987). In our own laboratory, using mice, the plus maze has enjoyed some success. However, recently we have tried to establish the plus maze with rats as subjects. This has proved extremely difficult and because of large variances in individual rats' performance significant drug effects have not been obtained. Other laboratories have also experienced this difficulty (D. Sanger, pers. comm.) and it is not clear whether the problem lies with species (rat vs mouse), strain (hooded vs albino), or as yet unidentified but necessary test conditions. With these reservations the plus-maze may be a useful addition to the battery of tests available for anxiolytic drug screening.

3.3.5 Other tests

The tests mentioned up to now by no means exhaust the possibilities, and the interested reader is referred to Chapter 2 by Green & Hodges (this volume) for information on a variety of other models, including the conditioned defensive burying test (Treit *et al.*, 1981) which exploits the observation that rodents bury objects which they associate with noxious stimuli, and the potentiated startle response (Davis, 1986). A new screening model, the ability of anxiolytics to reduce the frequency of ultrasonic vocalisation by rat pups on separation from their mother (Gardner, 1985), is too recent to evaluate.

3.4 SCREENING FOR SIDE EFFECTS

Drug therapies can be improved upon in two ways: the effectiveness of the therapy itself can be improved, or, by removing a drug's unwanted properties, side effects can be reduced. In the case of BZ anxiolytics, clinicians and patients alike appear to be satisfied with their effectiveness. Therefore, the major area for improvement is in side effects.

It is well known that BZs are sedative, muscle relaxant and amnestic. All three of these properties are more or less problematic for patients taking BZs for the treatment of anxiety, though it is often said that at the beginning of their therapy agitated patients welcome a degree of sedation. Nevertheless, sedation and muscle relaxation prevent the patient under therapy from, for instance, driving or operating machinery. The amnestic effect of BZs is a particular problem when the drugs are employed as hypnotics, since the patients may forget they have already taken the evening's sleeping pill and take a second dose; forgetting appointments, etc., also occurs and impairs the patient's normal life. To a lesser extent (since the doses are usually lower), similar effects also occur with anxiolytic therapy (see Lister, 1985, for a review). All three side effects are grossly potentiated by alcohol and barbiturates.

3.4.1 Sedation and ataxia

Muscle relaxation, sedation and ataxia are sometimes difficult to distinguish from one another in simple animal screens, which usually employ some measure of muscle coordination. For instance, mice or rats may be required to maintain their position on a rotating cylinder by walking backwards, or to escape from a narrow chimney by climbing up it backwards, or to climb on to a wire which they are allowed to grip with their forepaws. BZs impair performance in all such tests at doses not far above those in which they exert effects in anxiolytic tests. Interestingly, a number of novel BZ receptor ligands which are active in the common models of anxiolytic activity are not active in such tests. Examples include a pyrazoloquinoline, CGS 9896, the β-carbolines ZK 91 296 and ZK 95 962, and the BZ Ro 17-1812.

These results indicate the possibility of dissociating anxiolytic from side effects. Such a dissociation has also been confirmed in human pharmacology (Dorow *et al.*, 1987*b*) and one such non-sedative anxiolytic, ZK 95 962, has even been demonstrated to antagonise the sedative/hypnotic effects of a benzodiazepine hypnotic (Dorow *et al.*, 1987*b*).

Sedative effects of BZ can also be readily demonstrated in simple measures of locomotor activity, though it is not clear that this effect models sedation, in the sense of reduced vigilance, in patients. A more relevant test using signal detection analysis of a continuous attention task as a measure of vigilance in animals was described by Warburton & Brown (1972). Although such tasks readily detect vigilance-decreasing effects of BZs (Francis & Cooper, 1979; Jensen *et al.*, 1987), ZK 91 296 was found to be inactive (Stephens & Sarter, 1988), consistent with observations in people (Dorow *et al.*, 1987*b*).

Despite the theoretical attractiveness of signal-detection tasks they demand too much time and effort to find a role in a screening programme. Nevertheless, there may be a case for using complex methods. The BZ Ro 17-1812 shows a good anxiolytic profile without ataxic or locomotor-sedative properties in animal models (Haefely, 1984). Subsequent clinical experience, however, suggests that the substance is extremely sedative in patients (Merz, 1984). Although this discrepancy between the animal and human findings may simply reflect a species difference, it may alternatively indicate weaknesses in currently used animal screens for sedative action. The cost of running an animal test, however expensive, is invariably several orders of magnitude cheaper than the cost of developing a substance through to the clinic. Therefore, it is sensible to employ the best available animal model for clinical questions, almost regardless of cost. Whether the signal-detection test may have been more worthwhile is not clear, but the experience with Ro 17-1812 points to the need for more sophisticated tests even of side effects.

3.4.2 Amnestic properties of benzodiazepines

The amnestic properties of BZs are receiving growing attention (Hindmarch & Ott, 1988). These effects may be directly linked to their sedative properties. Whatever the cause, the effect has been noted in several human studies (e.g. Roth *et al.*, 1980; Dorow *et al.*, 1987*a*). Unfortunately the most common task used to evaluate amnestic activity in animals, passive avoidance, is also the most flawed and inappropriate. In these procedures the animal is punished for making a response, e.g. walking from a lighted runway into a dark compartment, so that when later reintroduced to the same apparatus the response is inhibited. Certainly anxiolytic effects might be expected to interfere with the acquisition process. Perhaps the single largest problem is the failure to dissociate performance from the various states of arousal. If BZ-induced amnesia is only a consequence of sedation and not a direct effect on memory *per se* this may not seem to be important. However, the test itself then becomes irrelevant as any test of BZ sedation would also be a test of BZ amnestic properties; this seems a little naive. Indeed, the most commonly used single-trial passive avoidance task has been so severely criticised as mnemonic tests that its survival even in preliminary screens is surprising. A full discussion of the problems associated with this test is unnecessary here (see Sahgal, 1984). Carew (1970) and Sahgal & Mason (1985) have provided alternatives that are as simple to run but require sophisticated statistics to analyse properly.

It is accepted clinical experience that certain BZs are particularly problematic with respect to their amnestic properties. Prominent among these are the hypnotics triazolam and midazolam. On the other hand, among hypnotics, lormetazepam has gained a reputation for its relative lack of amnestic properties (Subhan, 1984). The basis of this difference is not clear but may reflect the extreme rapidity of onset of action of the former compounds. Whether animal models employing passive avoidance procedures are capable of predicting such differences is very questionable, especially in view of the species differences in the kinetics of the different compounds.

3.4.3 Questions of kinetics

A further problem for industrial pharmacologists which hardly comes to the notice of their academic colleagues is that of pharmacokinetics. To be a possible candidate for development, a new drug must not only possess the necessary therapeutic activity and lack of side effects, it must also exhibit an acceptable resorption and half-life. Many interesting compounds are degraded rapidly by the liver so that they do not reach the circulating blood (and hence the brain) in high enough concentrations. This phenomenon is known as the first-pass effect. Thus the route of administration can be an important factor in determining the pharmacological effect of a drug. Drugs administered orally or intraperitoneally all have to pass through the liver to reach their target organ. On the other hand, those administered intravenously or subcutaneously avoid the liver during their first passage around the animal and reach the brain before they enter the hepatic circulation.

Many pharmacologists argue that since anxiolytics are likely, eventually, to be administered orally to patients, then oral administration should also be the preferred route in screening tests. However, the metabolism of drugs varies considerably from species to species and there is no guarantee that a drug which is active when given orally to the rat, will also be orally active in people. More importantly, since rats are notoriously fast metabolisers of drugs, a compound which is not active in the rat following oral administration may nevertheless show a good bioavailability in humans. Thus in deciding on route of administration, the question must be asked whether the screen is a test of pharmacodynamics (i.e. anxiolytic activity) or of pharmacokinetic parameters. If the idea is to find a pharmacologically active lead substance, even though the substance class is known to be subject to a high rate of peripheral metabolism, whether in gut, liver or blood, then even intracerebroventricular administration may be the preferred route for the screen.

A second important kinetic factor is the drug's biological half-life. Indeed, whether a novel BZ is to be developed as an anxiolytic or an hypnotic will depend to a large extent on how quickly the substance is metabolised. This is typically measured by the half-life: the time for half of the drug relative to peak values to disappear from the blood. Modern hypnotics exhibit half-lives of six to eight hours or even less, whereas the anxiolytic diazepam has a half-life of 20–70 hours, and its active metabolite, desmethyldiazepam has a half-life of 36–96 hours (Greenblatt & Shader, 1987). The advantage of a long half-life, of course, is that the drug need not be administered so frequently, but half-lives also influence factors such as intensity of withdrawal and hangover effects. It is also hypothesized that within a pharmacological class, drugs with a rapid onset of action and short half-life are more likely to be abused and lead to more severe withdrawal symptoms (Jaffe, 1985). Thus drugs which are rapidly absorbed may give rise to more problems of this sort. A compound's pharmacokinetic properties are therefore of interest at two levels. Firstly, interesting compounds should not be accidentally excluded because they have the wrong pharmacokinetic properties in the rat or mouse; and secondly, kinetics may have an important bearing on the pharmacodynamic properties in the clinic.

3.4.4 Tolerance development

Although tolerance clearly develops to the sedative, muscle relaxant and anticonvulsant properties of BZs, it is hotly disputed whether their anxiolytic potency declines with repeated administration. Many of the most prominent clinical researchers in the pharmacology of anxiolytics hold that potency is maintained over weeks or months (e.g. Rickels *et al.*, 1983) while on the other hand, the prevailing opinion among psychiatrists, particularly in the U.K., is that it is not useful to maintain BZ anxiolytic therapy for more than three months (Committee for the Review of Medicines, 1980; Lader, 1980). Wilson & Gallager (1987) have recently presented evidence that the rate of development of tolerance to BZs differs among different brain structures, and this may help to explain why tolerance develops more rapidly to some effects than to others.

For many years it has also been held that tolerance does not develop to the action of BZs in animal models of anxiolytic activity (Margules & Stein, 1967). In the last several years, however, tolerance to various BZs has been demonstrated in such diverse models as the social interaction test (Vellucci & File, 1979), an operant conflict test (Gonzalez *et al.*, 1984), the four-plate test (Stephens & Schneider, 1985), a test of conditioned defensive burying (Treit, 1985*a*), and the plus-maze (Fritz *et al.*, 1986; Stephens *et al.*, 1988), i.e. in virtually all classes of test used to evaluate the anxiolytic properties of drugs.

The success in demonstrating tolerance in these tests is something of an embarrassment. If the clinical viewpoint that the anxiolytic effects of BZs do not undergo tolerance is correct, then all of the above tests fail in an important respect as animal models. In that case, whatever the mechanism by which BZs achieve their effects in such tests it cannot be the same as that by which they achieve their anxiolytic effects in the clinic, and therefore the tests have no more validity as models than, say, the PTZ convulsant test. On the other hand, if the animal models reflect the anxiolytic action of BZs, then the clinical trial designs were apparently inappropriate for measuring tolerance.

3.5 A RATIONAL SCHEME FOR SCREENING BZ-TYPE ANXIOLYTICS ACTING AT THE BZ/GABA RECEPTOR COMPLEX

The foregoing sections illustrate some of the variety of tests available for identifying compounds acting (for the most part) at the BZ-GABA receptor.

Thus a typical screening procedure for anxiolytics acting at the BZ receptor will involve biochemical measures of receptor binding activity, as well as a series of behavioural measures for anxiolytic activity and side effects. A summary of this approach can be seen in Fig. 3.2. Many compounds are tested in the primary screens in order to

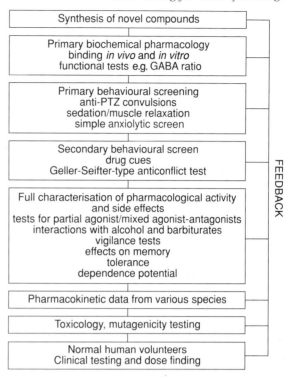

Fig. 3.2. Schematicised summary of procedure for identifying a BZ anxiolytic compound. Many compounds are tested in the primary screens in order to develop feedback to the chemists on structure–function relationships. Although this schema is essentially for identifying BZ ligands, the flow can be applied to a search for any compound with anxiolytic properties. A substance should not necessarily be rejected because it does not bind to the targeted receptor or is active in only some of the tests. See the text for a full discussion of the suitability of the available models for non-BZ anxiolytics.

provide rapid feedback for the chemists, and a full characterisation is limited to only a few substances.

Within such a test scheme it is always necessary to consider the possible redundancy of tests. If, for instance, activity in the four-plate test always predicts activity in the Geller conflict test, then one of them can be spared. On the other hand, the most interesting compound may well be that which, unlike all existing compounds, is inactive in one particular sort of test. Such a compound can be anticipated as having a novel profile in the clinic.

That such selectivity does exist may be illustrated by comparing the potencies of some stan-

Table 3.3. *Potencies of some BZ receptor ligands in four animal models sensitive to anxiolytic drugs, and* in vivo *potency in displacing ^{3}H-lormetazepam from rat forebrain membranes.*

	Displacement of ^{3}H-BZ *in vivo* ED$_{50}$ (mg/kg)	PTZ cue: Antagonism ED$_{50}$ (mg/kg)	CDP cue: Generalisation ED$_{50}$ (mg/kg)	Lick Suppression Test: antipunishment MED (mg/kg)	Protection against stress-induced increases in plasma CS levels (MED; mg/kg)
ZK 93 426	0.2	> 40	> 40	> 40	> 3.0
Ro 15-1788		> 40	> 40	> 40	—
ZK 91 296	2.5	(2.5)	10	> 60	1.0
ZK 95 962	0.8	0.04	2.5	> 50	> 3.0
CGS 9896		1.1	0.8	30	—
CL 218 872		2.5	3.8	30	—
Diazepam	6.0	0.5	1.5	3	1.0
Clonazepam	0.4	0.2	0.3	3	0.1
Lorazepam		0.04	0.4	0.3	—
Lormetazepam	0.8	0.04	0.4	0.3	0.3
ZK 93 423	1.5	0.1	0.3	< 0.3	
Triazolam	0.9	0.01	0.8	—	—

All drugs were given i.p. 30 min before the test. Data from Stephens *et al.*, 1987; see discussion under Section 3.5.

dard and novel benzodiazepine receptor ligands in animal models of anxiolytic activity. We carried out this exercise for four tests of anxiolytic activity which we have in the past routinely used (Stephens *et al.*, 1987). Table 3.3 illustrates there was a good qualitative correspondence between activities in the chlordiazepoxide drug discrimination test, in antagonising the PTZ discriminative stimulus, in enhancing punished licking in thirsty rats, and in protecting against the rise in plasma corticosterone level induced by stress. However, the potencies of the substances in the four tests did not correlate highly. Whereas activity in the CDP-cue correlated reasonably with potency in the lick-suppression test and in the anti-stress test, activity in antagonising the PTZ-cue correlated poorly with all the other measures (Table 3.3). Limited data from another test, the social interaction test of File, suggested that potency in this test is not predicted by activity in the other models. This form of analysis suggests two conclusions – firstly that the different models of anxiety may relate to different brain processes and perhaps, therefore, to different aspects of anxiety, and secondly that BZ receptor ligands may be differentially active in such processes. However, it is equally possible that some of

the tests are not legitimate tests of anxiety but model some other aspect of BZ activity.

It is also interesting to look at the predictive value of the *in vitro* measures described above for potency in the *in vivo* models. As will be already clear, potency in displacing labelled BZs from their binding sites is of only limited value in predicting pharmacological activity. Although the activity of BZs in antagonising PTZ convulsions correlates well with their potency in displacing ^{3}H-BZs *in vivo* (Duka *et al.*, 1979) as does their potency in the four-plate test (Stephens *et al.*, 1984a), the existence of antagonist and inverse agonist ligands which bind but do not possess the required pharmacological activity make such correlations dangerous.

This problem can be corrected by employing measures of efficacy at the receptor in addition to information on binding, and both the GABA ratio and TBPS ratio allow a classification of substances into anxiolytics, antagonists and anxiogenic substances (see e.g. Stephens & Kehr, 1985). Furthermore, we were able to show with a limited number of substances that those with high TBPS or GABA ratios required a lower receptor occupancy to achieve their action in the CDP-cue and in a water-lick conflict test than substances with lower GABA or TBPS ratios (Stephens *et al.*, 1987). Indeed, for the series of seven BZ receptor ligands employed, potency in the water-lick conflict test was as well predicted by the TBPS ratio as by potency in generalising to the CDP-cue (Stephens *et al.*, 1987). However, neither the TBPS nor the GABA ratio predicted the degree of receptor occupancy necessary to achieve activity in antagonising the PTZ discriminative stimulus. Furthermore, within this very limited sample of substances active in anxiolytic tests, there was only a poor quantitative relationship between the two biochemical measures. It should be noted, however, that the two biochemical assays were made using membranes from different brain areas and it is not clear whether a better correlation would have been found within a single structure.

Thus *in vitro* test models, used with caution, may provide very useful information predicting activity in *in vivo* tests, though their predictive ability is clearly not sufficient to replace animal tests.

3.6 NON-BENZODIAZEPINE ANXIOLYTICS

The test schema outlined above is specifically designed to identify compounds whose anxiolytic mechanism resembles that of the BZs, but which are superior to existing BZs in terms of side effects. It is an example of a strategy to exploit and improve an already existing therapy. However, major advances in therapeutics often depend on leaps which subsequently require a rethinking of screening procedures. Strategies for the development of new anxiolytic therapies are currently undergoing such a sea-change since the introduction into the clinic of buspirone as a novel anxiolytic (Rakel, 1987).

Buspirone is structurally unrelated to the BZs and does not bind to the BZ receptor. Although its mode of action is not yet clear, it has been suggested to act at DA, NA and, most recently, at 5-HT$_{1A}$ receptors (see Eison, 1984; Taylor *et al.*, 1984). It is this latter effect which most likely contributes to buspirone's activity. The fact that buspirone has been reported in clinical trials to be a useful therapy in the case of mild to moderate anxiety has caused animal pharmacologists to think deeply about their animal models.

Buspirone (and its derivatives gepirone and ipsapirone) not only do not bind to BZ receptors, they are also inactive in protecting against PTZ-induced or other convulsions. Thus the biochemical and primary pharmacological screens outlined above are of no use in identifying this sort of compound. This is perhaps not surprising since these screens simply exploited the fact that all anxiolytic BZs bind to BZ receptors and have anticonvulsant properties. They are, as already discussed, screens for BZ-like compounds, not screens for anxiolytics.

More surprisingly, however, buspirone is rather inactive in traditional animal models of anxiety. Thus in tests employing the release of punished behaviour as a measure of anxiolytic activity, buspirone is variously reported as inactive (Goldbert *et al.*, 1983) or only weakly active (McClosky *et al.*, 1987) or comparable to BZs (Taylor *et al.*, 1985). A clue to this inconsistency may be the very poor oral bioavailability of buspirone in mammals (less than 4% in man), and McClosky *et al.* (1987) report marginally better effects following subcutaneous administration. In the pigeon with presumably a different metabolism, buspirone apparently exhibits a marked anti-punishment effect comparable to that of standard BZs (Barret *et al.*, 1986).

It is not only in punishment procedures that the anxiolytic action of buspirone is difficult to identify, since File's group (Pellow *et al.*, 1987) reported a lack of effect in the plus-maze and Pollard & Howard (1986) a lack of effect in the staircase test. On the other hand, buspirone is reported to be active in protecting against stress-induced increases in plasma corticosterone levels (Urban *et al.*, 1986), in decreasing fear-potentiated startle (Davis, 1986), in increasing exploration (Pich & Samanin, 1986) and in increasing social interaction between pairs of rats (Guy & Gardner, 1985), all tests in which BZs are also active.

This pattern of results can have many kinds of explanation. Perhaps the insensitivity of our acute tests reflects (or predicts) the fact that buspirone appears to be active only after chronic treatment (Pecknold *et al.*, 1985; Tyrer *et al.*, 1985), or perhaps it reflects the possibility that our models are measuring different kinds of anxiety, or different aspects of anxiety for which the BZ receptor ligands and the buspirone type of compound are differentially selective. Most worrying is the possibility that our models are not models of anxiety at all, but simply models of BZ action. If this is the case, it is no wonder that buspirone is not active in all of them. A further possibility, which it is still too early to rule out, is that compounds of the buspirone type will not become a useful and generally accepted therapy for anxiety. Indeed at the present time there are several clinical studies in which buspirone fails to show any anxiolytic activity, usually in patients who have previously been treated with BZs (e.g. Olajide & Lader, 1987). If this is the case, then the problem will be to explain why so many animal tests have sometimes given rise to false positives in identifying buspirone and its analogues as possessing anxiolytic properties.

Recently, further confusion has been added in the continuing debate as to the exact mode of action of buspirone. It has been reported that although buspirone does not itself bind to BZ receptors it does enhance BZ receptor binding (Goeders *et al.*, 1988). Moreover, Davis *et al.* (1988) have suggested that the effects of buspirone on the fear-potentiated startle response are not due to an action at either 5-HT_{1A} or BZ receptors.

The initial success with buspirone in the clinic may have come about as a result of serendipity but it does not come entirely out of the blue. Since its discovery there has been a return of interest to the role of the 5-HT system in anxiety. A function of 5-HT in the action of BZs was proposed by Stein and his colleagues (Wise *et al.*, 1972) who demonstrated that 5-HT transmission was reduced in the presence of BZs, and Gray has proposed a central role for 5-HT in behavioural inhibition systems in the brain (Gray, 1982 see also Green & Hodges, Chapter 2, this volume). In keeping with these views, lesions of central 5-HT systems (Tye *et al.*, 1979) or a blockade of 5-HT synthesis with *p*-chlorophenylalanine have anxiolytic, or at least anticonflict effects in animal tests (Robichaud & Sledge, 1969; Hartman & Geller, 1971). Furthermore, a recent review of the effects of putative 5-HT anxiolytics in a wide range of tests, supported the suggestion that reducing 5-HT transmission results mostly in anxiolytic effects and increasing 5-HT transmission in anxiogenic effects (Chopin & Briley, 1987). Exceptions to this rule exist, for example the 5-HT_{1A} receptor agonist 8-hydroxy-N,N-dipropylaminotetralin

(8-OHDPAT) has also been claimed to have an anxiolytic-like effect in the Vogel lick-suppression test (Engel *et al.*, 1984) and the plus-maze (Critchley & Handley, 1987), but these findings are sometimes difficult to replicate (e.g. Pellow *et al.*, 1987; Deacon & Gardner, 1986). A problem here may be that 5-HT_{1A} receptors appear to exist both pre- and post-synaptically, so that whether an agonist ligand at this receptor subtype acts to facilitate or to dampen 5-HT neurotransmission will depend on the balance between pre- and postsynaptic activity. Such a hypothesis could account for the marked inconsistency of findings with all the 5-HT_{1A} receptor ligands in animal models of anxiety. Nevertheless, these results would seem to indicate that an assault on the 5-HT system might give rise to a novel group of anxiolytics, and indeed, a number of pharmaceutical companies are investigating modulation of 5-HT transmission as a means of identifying anxiolytic activity.

While evidence that 5-HT_{1A} agonists possess anxiolytic properties either in animal tests or in the clinic is still controversial (see Johnston & File, 1986; Chopin & Briley, 1987), especially as occasional anxiogenic-type effects have been reported (Pellow *et al.*, 1987), there appears to be a growing body of evidence that 5-HT antagonists other than buspirone may indeed have some anxiolytic properties. Ritanserin, for example, an antagonist at 5-HT_2 receptors, has been reported to reduce the symptoms of general anxiety disorder (Ceulemans *et al.*, 1984) and to be active in animal models of conflict and of emergence (Colpaert *et al.*, 1985; Amrick & Bennett, 1986). In addition, the relatively selective 5-HT_{1A} partial agonist ipsapirone is reported to be active not only in several models of anxiety and depression, but also in the treatment of clinical anxiety according to recently completed clinical trials featuring a placebo-controlled double-blind design (quoted in Glaser, 1988). A ligand for yet a further 5-HT receptor subtype has also been reported to have anxiolytic properties. GR 38032F, a 5-HT_3 antagonist is reported to be active in the social interaction test but not in a lick-suppression test (Jones *et al.*, 1988). It is too early to assess these new compounds realistically, but it does appear that several pharmaceutical companies take the role of 5-HT in anxiety sufficiently seriously to invest in drug-finding programmes. Such a switch away from BZ-receptor ligands will provide an important feedback for our animal tests – are they able to identify these new compounds? And if they are, will this activity be reflected in the clinic?

The existence of this new type of ligand has also promoted the development of several novel 'models' of anxiolytic activity, including the use of fear responses in the common marmoset (Jones *et al.*, 1988) which both benzodiazepines and 5-HT$_3$ antagonists are capable of suppressing. The development of such a new model is to be welcomed since it offers a test involving a natural fear response, with a higher species and with the ability to identify compounds from outside the group acting at the BZ receptor. However, a note of caution must be sounded. The main justification for the novel test is that it identifies 5-HT$_3$ antagonists: since it is not yet known whether such substances will be active in the clinic in the treatment of generalised anxiety, the validity of the test remains unproven. Furthermore, as with the range of tests identifying the benzodiazepine anxiolytics, a novel test based on the behavioural properties of a particular class of pharmacological substances may lead to the exclusion of substances with another mode of action. Lastly, it is not clear what effect non-anxiolytic substances such as neuroleptics might have in this test. Only time will provide an answer to these problems.

There is another class of drugs which are effective anxiolytics but which also fail to be identified consistently as such by most behavioural models of anxiety, the opiates (Millan & Duka, 1981). Interestingly many positive results with morphine in animal models have been dismissed as false positives. Nevertheless, several authors have convincingly discussed the anxiolytic properties of opiates (see Shephard, 1986; Redmond, 1985) in humans. Although the anxiolytic action is pronounced it is short-lived as tolerance rapidly develops to this effect. It is also interesting to note that intense anxiety, symptoms of panic, feelings of depression and dysphoria are some of the classical symptoms of opiate withdrawal (Jaffe, 1985). Given the severity of opiate side effects, and their dependence liability, it seems unlikely that they could become drugs of choice for the treatment of anxiety outside the surgical setting. However, just as we are now developing BZ ligands without the major side effects of the BZs for the treatment of

anxiety, it cannot be ruled out that specifically anxiolytic opiates might be developed.

Finally, one other class of drugs is currently receiving attention as a possible source for new anxiolytics: steroid hormones. Reports of an anxiolytic action of both sex and adrenal hormones in animals (Rodriguez-Sierra *et al.*, 1984; Crawley *et al.*, 1986) and in humans (Nott *et al.*, 1976) are not new. Recently however, it has been shown that these anxiolytic effects appear to be via an action at the GABA receptor complex (Majewska *et al.*, 1986; Smith *et al.*, 1987). It remains to be seen whether steroids can be realistically developed as anxiolytics, whether the hormonal side effects can be controlled, and whether they really offer something different from other compounds acting at the GABA complex such as BZs or barbiturates.

3.7 RETROSPECTIVE

It is apparent that we are at a crossroads in the search for new anxiolytics. The current tests of anxiolytic activity have all been developed subsequent to the introduction of BZs. Furthermore, they rely for their validity almost exclusively on the fact that BZs work in them. Moreover, the development of more sophisticated animal models of anxiolytic action in the 25 years since chloridazepoxide was introduced has not led to the identification of a single anxiolytic outside the class of substances acting at the BZ receptor. Surely if these were true models of anxiolytic activity at least one new class of anxiolytics would have been discovered. Unfortunately, it is also all too easy to argue that the development of tests for identifying BZ-like anxiolytics has actually hindered the finding of anxiolytics with other pharmacological actions. The lack of a firm theoretical basis for the behavioural tests currently available becomes clear when a new substance with apparent anxiolytic properties appears, as with the discovery of buspirone and its derivatives. In the case of this new class of compounds, the fact that they are not active in a particular animal model of anxiety has not been thought a sufficient reason to reject the compound. Rather, the results from that test have been ignored. Clearly the theoretical bases of such models cannot have been particularly persuasive.

No model of anxiety currently used in screening

procedures derives from an attempt to model the symptoms of anxiety itself. This is not entirely surprising given that current diagnostic criteria separate anxiety into several different classes of symptoms and causes, a differentiation largely ignored in the laboratory. Attempting to classify existing animal models of anxiety according to whether they model phobic disorders, anxiety states such as panic, or generalised anxiety disorder, described by the DSM III diagnostic criteria is impossible (see also Green & Hodges, Chapter 2 of this volume).

The problem here is that the symptoms which are so important for the diagnosis of patients are not measured (and to an extent not measurable) in the laboratory animal where most available methods are essentially reducible to readily observed forms of avoidance behaviour. In conventional conflict tasks, for instance, the animal learns to avoid bar pressing in the presence of a stimulus predicting punishment. However, by the time anxiolytic drugs come to be tested the animals are performing so well they are in complete control of the situation and may experience no anxiety.

3.7.1 Is there an alternative?

Curiously, the behavioural literature describing attempts to model anxiety based disorders has had little impact on psychopharmacology, even though experimental neuroses in animals were described by Pavlov (1927), and from the classical conditioning school; Wolpe (1958, 1971) specifically argued for similarities between cats' phobic responses to the boxes in which they had been shocked, and anxiety-based disorders in humans. This similarity incorporates both a common etiology (classical conditioning of fear or anxiety) and a common therapy (extinction of anxiety through counter-conditioning). Wolpe also argued that a simple classical conditioning model like that for phobias was also applicable to explain generalised anxiety states. In the case of generalised anxiety, however, the anxiety is conditioned to a series of environmental stimuli, giving rise to a multiplicity of situations in which anxiety is experienced. Although neither Wolpe's hypothesis of phobic nor of generalised anxiety is widely accepted, learning theory models of anxiety do have the advantage of attempting to integrate animal

models and human disorders with a common theoretical framework.

The problem for the animal psychopharmacologist in attempting to model generalised anxiety is the long-lasting nature of the disorder. In contrast, current animal models at the very best mimic short-term, situational anxiety. One phenomenon which appears to model a more persistent anxiety in animals is the experimental neuroses induced during classical conditioning experiments with monkeys (see Mineka, 1985, for discussion). Among the symptoms reported in experimental neurosis are extreme agitation, restlessness, rapid heart rate and respiration, muscle tension, piloerection, enhanced vigilance and distractability. Such symptoms were also to be seen outside the experimental situation and persisted for months after the end of the experiment. These effects are extremely interesting from the point of view of developing a model of the clinical condition, but present several difficulties in application to a drug-finding programme. The effects are clearly seen, but require intense care and observation as well as a species, monkeys, not commonly found in screening labs. An additional factor in the genesis of persistent anxiety may be unpredictability and uncontrollability of aversive events. Although Seligman's work is best known for its application to theories of depression (see Willner, Chapter 5 of this volume), he has also argued for a role of unpredictability in the genesis of anxiety (Seligman, 1974). Again, the problem for the industrial pharmacologist is to translate this idea into a suitable method for use with, preferably, rodents.

We should note here that there are two animal models which identify as effective both buspirone and morphine as well as the BZs. It is especially interesting to our theme that both of these tests involve classical conditioning: the conditioned emotional response (e.g. Hill *et al.*, 1966) and potentiated startle (Davis, 1986; see also Green & Hodges, Chapter 2 of this volume for discussion). Although these two methods may be useful as screens they are unlikely to answer all of our problems. The potentiated startle is a conditioned fear response. However, the measured response is a reflex action which quickly undergoes habituation. Thus, this is also a test of situational as opposed to generalised chronic anxiety.

At the present time there is no fully adequate anxiety model. Thus, the industrial pharmacologist has no choice but to employ the existing tools in as rational and as critical manner as possible. The approach is to rely not on one test, but on a profile in several, and, given the initial success of buspirone, to watch carefully for anomalies. It may well be that some of the present tests are perfectly adequate to identify non-BZ anxiolytics, but that we have been conditioned from experience with the BZ anxiolytics to expect an immediate result from a single dosing. In the clinical studies showing buspirone to be effective, it appears that the anxiolytic effect is not immediate but develops after 10 days or so of treatment.

3.8 CONCLUSIONS

This chapter has attempted to sketch out the problem of screening for novel anxiolytic drugs as seen from the standpoint of an industrial pharmacologist. We have avoided phobias and panic attacks to concentrate on generalised anxiety. In the case of phobias, behaviour therapies may be useful (e.g. Marks *et al.*, 1971), and for panic attacks antidepressants (Klein, 1964), though several BZs are currently undergoing clinical trials for this indication (see Lader, Chapter 4 of this volume). However, each of these topics would require a chapter in itself. As a review of screening approaches, the chapter has frequently veered away from the question of animal models of anxiety and their validity to concentrate on matters of little interest to the theoretician. We make no apology for giving as much weight to screens based on biochemical methodology, to screening for side effects, and to a brief discourse on kinetic problems as to a discussion of animal models of anxiety themselves. Neither have we spent much time in analysing the theoretical basis of the animal models (see Hodges & Green, 1987, and this vol., Ch. 2), though we hope that our scepticism as to their theoretical validity is clear. Rather we have attempted to address the various models from a practical viewpoint. The only justification that any particular screening model or procedure needs is its ability to deliver successful drugs. From this standpoint, current screens for anxiolytic activity must be judged an enormous

success since they have provided clinical psycho-pharmacologists with some of their most popular drugs.

And yet, one is left wondering whether the very success of our screening procedures has led to a restriction in the development of genuinely novel anxiolytic agents. Both chlordiazepoxide and buspirone were discovered serendipitously: anxiolytic activity was noted whilst they were being tested for something completely different. Neither of these compounds was identified by the available screens of the time. We are therefore forced to ask, do the present tests really improve our chances of identifying novel anxiolytics, or do they just prevent accidents from happening?

We gratefully acknowledge the cooperation of the chemists, biochemists, pharmacologists and pharmacokineticists of Schering AG and A/S Ferrosan without whom this work would have been considerably more naive. In particular Herbert Schneider, Ralph Schmiechen, Theodora Duka and Rainer Dorow are thanked for their free exchange of ideas. The assistance of Bettina Fogel and Evelin Weber in preparation of the manuscript is also gratefully acknowledged.

REFERENCES

Amrick, C. & Bennett, D. (1986). A comparison of the anticonflict activity of serotonin agonists and antagonists in rats. *Society for Neuroscience Abstracts* **12**, 907.

Andrews, J.S., Turski, L.A. & Stephens, D.N. (1989). Does the pentylenetetrazol discriminative stimulus reflect kindling rather than anxiogenic properties? *Drug Development Research* **16**, 247–56.

Aron, C., Simon, P., Larousse, C. & Boissier, J.R. (1971). Evaluation of a rapid technique for detecting minor tranquilizers. *Neuropharmacologia* **10**, 459–69.

Ator, N.A. & Griffiths, R.R. (1986). Discriminative stimulus effects of atypical anxiolytics in baboons and rats. *Journal of Pharmacology and Experimental Therapeutics* **237**, 393–403.

Barrett, J.E., Witkin, J.M., Mansbach, R.S., Skolnick, P. & Weissman, B.A. (1986). Behavioral studies with anxiolytic drugs. III Antipunishment actions of buspirone in the pigeon do not involve benzodiazepine receptor mechanisms. *Journal of Pharmacology and Experimental Therapeutics* **238**, 1009–13.

Bennett, D.A. (1985). The non-sedating anxiolytic CGS 9896 produces discriminative stimuli that may be related to an anxioselective effect. *Life Sciences* **37**, 703–9.

Bennett, D.A., Amrick, C.L., Wilson, D.E., Bernard, P.S., Yokoyama, N. & Liebman, J.M. (1985). Behavioral pharmacological profile of CGS 9895: a novel anxiomodulator with selective benzodiazepine agonist and antagonist properties. *Drug Development Research* **6**, 313–25.

Boissier, J.R., Simon, P. & Aron, C.A. (1968). New method for rapid screening of minor tranquilizers in mice. *European Journal of Pharmacology* **4**, 145–51.

Braestrup, C. & Nielsen, M. (1981). GABA reduces binding of ^{3}H-methyl β-carboline-3-carboxylate to brain benzodiazepine receptors. *Nature* **294**, 472–4.

Braestrup, C., Nielsen, M. & Olsen, C.E. (1980). Urinary and brain β-carboline-3-carboxylates as potent inhibitors of brain benzodiazepine receptors. *Proceedings of the National Academy of Science* **77**, 2288–92.

Braestrup, C., Schmiechen, R., Neef, G., Nielsen, M. & Petersen, E.N. (1982). Interaction of convulsive ligands with benzodiazepine receptors. *Science* **26**, 1241–3.

Breir, A., Charney, D.S. & Nelson, J.C. (1984). Seizures induced by abrupt discontinuation of alprazolam. *American Journal of Psychiatry* **141**, 1606–7.

Carew, T.J. (1970). Do passive avoidance tasks permit assessment of retrograde amnesia in rats? *Journal of Comparative and Physiological Psychology* **72**, 269–71.

Carey, M.P. & Fry, J.P. (1988). An evaluation of light aversion as a measure of anxiolytic benzodiazepine activity in the mouse. *Abstract of the 11th Annual Meeting of the European Neuroscience Association*, Zürich 4–8 September 1988 (Supplement to the *European Journal of Neuroscience*) Abstr. 90.1, p. 343.

Ceulemans, D., Hoppenbrouwers, M.L., Gelders, Y. & Reyntjens, A. (1984). Serotonin blockade. In: Benzodiazepines: what kind of anxiolysis? *Proceeding of 14th CINP, Florence*, p. 727.

Chopin, P. & Briley, M. (1987). Animal models of anxiety: the effect of compounds that modify 5-HT neurotransmission. *Trends in Pharmacological Science* **38**, 383–8.

Colpaert, F.C., Desmedt, L.K.C. & Janssen, P.A.J. (1976). Discriminative stimulus properties of benzodiazepines, barbiturates and pharmacologically related drugs: relation to some intrinsic and anticonvulsant effects. *European Journal of Pharmacology* **37**, 113–23.

Colpaert, F.C., Meert, T.F., Niemegeers, C.J.E. & Janssen, P.A.J. (1985). Behavioural and 5-HT

antagonistic properties of ritanserin: a pure and selective antagonist of LSD discrimination in the rat. *Psychopharmacology* **86**, 45–54.

Committee for Review of Medicine (1980). Benzodiazepines. *Drug Therapeutics Bulletin* **18**, 97–8.

Cook, L. & Davidson, A.B. (1973). Effects of behaviorally active drugs in a conflict-punishment procedure in rats. In: S. Garattini, E. Mussini & L.O. Randall (eds.), *The Benzodiazepines*, pp. 327–45. Raven Press: N.Y.

Corda, M.G., Blaker, W.D., Mendelson, W.B., Guidotti, A. & Costa, E. (1983). Beta-carbolines enhance shock-induced suppression of drinking rats. *Proceedings of the National Academy of Science* **80**, 2072–6.

Costa, E., Guidotti, A., Mao, C.C. & Suria, A. (1975). New concepts on the mechanism of action of benzodiazepines. *Life Sciences* **17**, 167–86.

Crawley, J.N., Skolnick, P. & Paul, S.M. (1984). Absence of intrinsic antagonist actions of benzodiazepine antagonists on an exploratory model of anxiety in the mouse. *Neuropharmacology* **23**, 531–7.

Crawley, J.N., Glowa, J.R., Majewska, M.D. & Paul, S.M. (1986). Anxiolytic activity of an endogenous adrenal steroid. *Brain Research* **398**, 382–5.

Critchley, M.A.E. & Handley, S.L. (1987). Effects in the X-maze anxiety model of agents acting at 5-HT$_1$, and 5-HT$_2$ receptors. *Psychopharmacology* **93**, 502–6.

Davis, M. (1986). Pharmacological and anatomical analysis of fear conditioning using the fear-potentiated startle paradigm. *Behavioral Neuroscience* **100**, 814–24.

Davis, M., Cassella, J.V. & Kehne, J.H. (1988). Serotonin does not mediate anxiolytic effects of buspirone in the fear-potentiated startle paradigm: comparison with 8-OH-DPAT and ipsapirone. *Psychopharmacology* **94**, 14–20.

Deacon, R. & Gardner, C.R. (1986). Benzodiazepines and 5-HT ligands in a rat conflict test. *British Journal of Pharmacology* **88**, 330 P.

Dorow, R., Horowski, R., Paschelke, G., Amin, M. & Braestrup, C. (1983). Severe anxiety induced by FG 7142, a β-carboline ligand for benzodiazepine receptors. *Lancet* **9**, 98–9.

Dorow, R., Berenberg, D., Duka, T. & Sauerbrey, N. (1987a). Amnestic effects of lormetazepam and their reversal by the benzodiazepine antagonist RO 15-1788. *Psychopharmacology* **93**, 507–14.

Dorow, R., Duka, T., Höller, L. & Sauerbrey, N. (1987b). Clinical perspectives on β-carbolines from first studies in humans. *Brain Research Bulletin* **19**, 319–26.

Duka, T., Höllt, V. & Herz, A. (1979). In vivo receptor occupation by benzodiazepines and correlation with the pharmacological effect. *Brain Research* **179**, 147–56.

Eison, M.S. (1984). Use of animal models: toward anxioselective drugs. *Psychopharmacology* **17** Suppl. 1, 37–44.

Emmett-Oglesby, M.W., Mathis, D.A. & Lal, H. (1987). Diazepam tolerance and withdrawal assessed in an animal model of subjective drug effects. *Drug Development Research* **11**, 145–56.

Engel, J.A., Hjorth, S., Svensson, K., Carlson, A. & Lindquist, S. (1984). Anticonflict effect of the putative serotonin receptor agonist 8-hydroxy-2 (M-n-propylamino) tetralin (8-OHDPAT). *European Journal of Pharmacology* **105**, 365–8.

File, S.E. (1980). The use of social interaction as a method for detecting anxiolytic activity of chloridazepoxide-like drugs. *Journal of Neuroscience Methods* **2**, 219–38.

File, S.E. & Hyde, J.R.G. (1979). A test of anxiety that distinguishes between the actions of benzodiazepines and those of other minor tranquilisers. *Pharmacology, Biochemistry and Behavior* **11**, 65–9.

Francis, R.L. & Cooper, S.J. (1979). Chlordiazepoxide-induced disruption of discrimination behaviour: a signal detection analysis. *Psychopharmacology* **63**, 307–10.

Fritz, S., Schneider, H.H., Stephens, D.N. & Weidmann, R. (1986). Tolerance to the anxiolytic action of diazepam in rats. *British Journal of Pharmacology* **88**, 336.

Gardner, C.R. (1985). Distress vocalisation in rat pups: a simple screening method for anxiolytic drugs. *Journal of Pharmacological Methods* **14**, 181–7.

Gee, K.W., Lawrence, L.J. & Yamamura, H.I. (1986). Modulation of the chloride ionophore by benzodiazepine receptor ligands: influence of γ-aminobutyric acid and ligand efficacy. *Molecular Pharmacology* **30**, 218–25.

Geller, I. & Seifter, J. (1960). The effects of meprobamate, barbiturates d-amphetamine and promazine on experimentally induced conflict in the rat. *Psychopharmacologia* **9**, 482–92.

Glaser, T. (1988). Ipsapirone, a potent and selective 5-HT$_{1A}$-receptor ligand with anxiolytic and antidepressant properties. *Drugs of the Future* **13**, 429–39.

Goeders, N.E., Ritz, M.C. & Kuhar, M.J. (1988). Buspirone enhances benzodiazepine receptor binding in vivo. *Neuropharmacology* **27**, 275–80.

Goldberg, M.E., Salama, A.I., Patel, J.B. & Malik, J.B. (1983). Novel non-benzodiazepine anxiolytics. *Neuropharmacology* **22**, 1499–504.

Gonzalez, J.P., McCullock, A.J., Nicholls, P.J., Sewell, R.D.E. & Tekle, A. (1984). Subacute benzodiazepine treatment: observations on

behavioural tolerance and withdrawal. *Alcohol and Alcoholism* **19**, 325–32.

Gray, J.A. (1982). *The Neuropsychology of Anxiety*. Clarendon Press/Oxford University Press: N.Y.

Greenblatt, D.J. & Shader, R.I. (1987). Pharmacokinetics of anti-anxiety agents. In: H.Y. Meltzer (ed.) *Psychopharmacology: The Third Generation of Generation of Progress*, pp. 1377–86. Raven Press, N.Y., pp. 1377–86.

Guy, A.P. & Gardner, C.R. (1985). Pharmacological characterisation of a modified social interaction model of anxiety in the rat. *Neuropsychobiology* **13**, 194–200.

Haefeley, W. (1984). Pharmacological profile of two benzodiazepine partial agonists: Ro 16-6028 and Ro 17-1812. *Clinical Neuropharmacology* **7**, Suppl. 1, 670–4.

Haefeley, W., Kulcsar, A., Mohler, H., Pieri, L., Polc, P. & Schaffner, R. (1975). Possible involvement of GABA in the central actions of benzodiazepines. In: E. Costa and P. Greengard (eds.), *Mechanism of Action of Benzodiazepines*, pp. 131–51. Raven Press: N.Y.

Handley, S.L. & Mithani, S. (1984). Effects of alpha-adrenoceptor agonists and antagonists in a maze-exploration model of 'fear-motivated' behavior. *Naunyn Schmiedeberg's Archives of Pharmacology* **327**, 1–5.

Hartman, R.J. & Geller, I. (1971). p-Chlorophenylalanine effects on a conditioned emotional response in rats. *Life Science* **10**, 927–33.

Havoundjian, H. & Skolnick, P. (1986). A quantitative relationship between Cl^{-} enhanced [^{3}H]flunitrazepam and [^{35}S]t-butylbicyclophosphorothionate binding to the benzodiazepine/GABA receptor chloride ionophore complex. *Molecular Brain Research* **1**, 281–7.

Hesp, B. & Resch, J.F. (1987). The medicinal chemist's approach to CNS drug discovery. In: H.Y. Meltzer (ed.) *Psychopharmacology: The Third Generation of Progress*, pp. 1637–47. Raven Press: N.Y.

Hildebrandt, F. (1937). Pentamethylenetetrazol (cardioxol). *Handbook of Experimental Pharmacology*, Julius Springer, Berlin, **5**, 151–83.

Hill, H.E., Belleville, R.E., Pescor, F.T. & Wikler, A. (1966). Comparative effects of methadone, meperidine and morphine on conditioned suppression. *Archives Internationales de Pharmacodynamie et de Thérapie* **163**, 341–52.

Hindmarch, I. & Ott, H. (eds.) (1988). *Benzodiazepine Receptor Ligands: Memory and Information Processing*. Springer, Berlin.

Hodges, H. & Green, S. (1987). Are the effects of benzodiazepines on discrimination and punishment dissociable? *Physiology and Behavior* **41**, 257–64.

Huidobro-Toro, P., Bleck, V., Allan, A.M. & Adron Harris, R. (1987). Neurochemical actions of anesthetic drugs on the aminobutyric acid receptor-chloride channel complex. *Journal of Pharmacology and Experimental Therapeutics* **242**, 963–9.

Irwin, S. (1968). Comprehensive observational assessment 1a. A systematic, quantitative procedure for assessing the behavioral and physiologic state of the mouse. *Psychopharmacologia* **13**, 222–57.

Jaffe, J.H. Drug addiction and drug abuse. (1985). In: A.G. Gilman, L.S. Goodman, T.W. Rall and F. Murad (eds.), *The Pharmacological Basis of Therapeutics* (7th edn), pp. 532–81. Macmillan: N.Y.

Jensen, L.H., Stephens, D.N., Sarter, M. & Petersen, E.N. (1987). Bidirectional effects of β-carbolines and benzodiazepines on cognitive processes. *Brain Research Bulletin* **19**, 359–64.

Jensen, L.H., Petersen, E.N., Braestrup, C., Honore, T., Kehr, W., Stephens, D.N., Schneider, H.H., Seidelmann, D. & Schmiechen, R. (1984). Evaluation of ZK 93 426 as a benzodiazepine receptor antagonist. *Psychopharmacology* **83**, 249–56.

Johnston, A.L. & File, S.E. (1986). 5-HT and anxiety: promises and pitfalls. *Pharmacology, Biochemistry and Behavior* **24**, 1467–70.

Jones, B.J., Costall, B., Domeney, A.M., Kelly, M.E., Naylor, R.J., Oakley, N.R. & Tyers, M.B. (1988). The potential anxiolytic activity of GR 38032F, a 5-HT$_3$ receptor antagonist. *British Journal of Pharmacology* **93**, 985–93.

Kilts, C.D., Commissaris, R.L. & Rech, R.H. (1981). Comparison of anticonflict drug effect in three experimental models of anxiety. *Psychopharmacologia* **74**, 290–6.

Klein, D.F. (1964). Delineation of two drug-responsive anxiety syndromes. *Psychopharmacologia* **5**, 397–408.

Lader, M.A. (1980). The present status of benzodiazepines in psychiatric medicine. *Arzneimittelforschung* **30**, 851–916.

Lal, H. & Emmett-Oglesby, M.W. (1983). Behavioral analogues of anxiety. *Neuropharmacology* **22**, 1423–41.

Leeb-Lundberg, F., Snowman, A. & Olsen, R. (1980). Barbiturate receptors are coupled to benzodiazepine receptor. *Proceedings of the National Academy of Science* **77**, 7468–72.

Leidenheimer, N.J. & Schechter, M.D. (1988). Discriminative stimulus control by the anxiogenic β-carboline FG 7142: generalization to a physiological stressor. *Pharmacology, Biochemistry and Behavior* **30**, 351–5.

Leidenheimer, N.J., Yamamoto, B.K. & Schechter, M.D. (1987). Behavioral and neurochemical evidence for serotonergic modulation by the anxiogenic β-carboline FG 7142. *Society for Neuroscience Abstracts* **13**, 454.

Levy, A.B. (1984). Delirium and seizures due to abrupt alprazolam withdrawal: case report. *Journal of Clinical Psychiatry* **45**, 38–9.

Lister, R.G. (1985). The amnesic action of benzodiazepines in man. *Neuroscience and Biobehavioural Reviews* **9**, 87–94.

Ljungberg, T., Lidfors, L., Enquist, M. & Ungerstedt, U. (1987). Impairment of decision making in rats by diazepam: implications for the 'anticonflict' effects of benzodiazepines. *Psychopharmacology* **92**, 416–23.

Majewska, M.D. & Schwartz, R.D. (1987). Prenenolone-sulphate: an endogenous antagonist of the γ-aminobutyric acid receptor complex in brain? *Brain Research* **404**, 355–60.

Majewska, M.D., Harrison, N.L., Schwartz, R.D., Barker, J.L. & Paul, S.M. (1986). Steroid hormone metabolites are barbiturate-like modulators of the GABA receptor. *Science* **232**, 1004–7.

Margules, D.L. & Stein, L. (1967). Neuroleptics and tranquilizers: evidence from animal studies of mode and site of action. In: H. Brill, J.O. Cole, P. Deniker, H. Hippius & P.B. Bradley (eds.), *Neuropsychopharmacology*, pp. 108–20. Excerpta Medica Foundation: Amsterdam.

Margules, D.L. & Stein, L. (1968). Increase of 'antianxiety' activity and tolerance of behavioral depression during chronic administration of oxazepam. *Psychopharmacologia* **13**, 74–80.

Marks, I.M., Boubgouris, J.C. & Marset, P. (1971). Flooding versus desensitization in the treatment of phobic patients: cross-over study. *British Journal of Psychiatry* **119**, 353–75.

Mathers, D.A. (1987). The GABA$_A$ receptor: new insights from single-channel recording. *Synapse* **1**, 96–101.

McCloskey, T.C., Paul, B.K. & Commissaris, R.L. (1987). Buspirone effects in an animal conflict procedure: comparison with diazepam and phenobarbital. *Pharmacology, Biochemistry and Behavior* **27**, 171–5.

McNaughton, N., Richardson, J. & Gore, C. (1986). Reticular elicitation of hippocampal slow waves: common effects of some anxiolytic drugs. *Neuroscience* **19**, 899–903.

Mereu, G., Corda, M.G., Carcangiu, P., Giogi, O. & Biggio, G. (1987). The β-carboline ZK 93 423 inhibits reticulata neurons: an effect reversed by benzodiazepine antagonists. *Life Sciences* **40**, 1423–30.

Merz, W.A. (1984). Partial benzodiazepine agonists: initial results in man. *Clinical Neuropharmacology* **7**, Suppl. 1, 672–3.

Millan, M. & Duka, T. (1981). Anxiolytic properties of opiates and endogenous opioid peptides and their relationship to the actions of benzodiazepines. *Modern Problems of Pharmacopsychiatry* **17**, 123–41.

Mineka, S. (1985). Animal models of anxiety-based disorders: their usefulness and limitations. In: A. Hussain Tuna and J.D. Maser, (eds.), *Anxiety and Anxiety Disorders*, pp. 199–244. Lawrence Erlbaum Associates: N.J.

Mohler, H. & Okada, T. (1977). Benzodiazepine receptor: demonstration in the central nervous system. *Science* **198**, 849–51.

Nott, P.N., Franklin, M., Armitage, C. & Gelder, M.G. (1976). Hormonal changes and mood in puerperium. *British Journal of Psychiatry* **128**, 379–83.

Olajide, D. & Lader, M. (1987). A comparison of buspirone, diazepam, and placebo in patients with chronic anxiety states. *Journal of Clincal Psychopharmacology* **7**, 148–52.

Olsen, R.W. (1981). GABA-benzodiazepine-barbiturate receptor interactions. *Journal of Neurochemistry* **37**, 1–13.

Ott, H. (1984). Are electroencephalographic and psychomotor measures sensitive in detecting residual sequelae of benzodiazepine hypnotics? In: I. Hindmarch, H. Ott & T. Roth (eds.), *Sleep, Benzodiazepines and Performance*, pp. 133–51. Springer: Berlin.

Pavlov, I.P. (1927). *Conditioned Reflexes* (Translated G. Anrep). Oxford University Press: N.Y.

Pecknold, J.C., Familamin, P., Chang, H., Wilson, R., Alarcia, J. & McClure, D.J. (1985). Buspirone: anxiolytic? *Progress in Neuro-Psychopharmacology and Biology Psychiatry* **9**, 639–42.

Pellow, S. (1986). Anxiolytic and anxiogenic drug effects in a novel test of anxiety: are exploratory models of anxiety in rodents valid? *Methods and Findings in Experimental Clinical Pharmacology* **8**, 557–65.

Pellow, S., Chopin, P., File, S.E. & Briley, M. (1985). Validation of open: closed arm entries in an elevated plus-maze as a measure of anxiety in the rat. *Journal of Neuroscience Methods* **14**, 149–67.

Pellow, S., Johnston, A.L. & File, S.E. (1987). Selective agonists and antagonists for 5-hydroxytryptamine receptor subtypes, and interactions with yohimbine and FG F7142 using the elevated plus-maze test in the rat. *Journal of Pharmacy and Pharmacology* **39**, 917–28.

Petersen, E.N. & Buus Lassen, J. (1981). A water lick conflict paradigm using drug experienced rats. *Psychopharmacology* **75**, 236–9.

Petersen, E.N., Jensen, L.H., Honore, T., Braestrup, C., Kehr, W., Stephens, D.N., Wachtel, H., Seidelmann, D. & Schmiechen, R. (1984). ZK 91 296 a partial agonist at benzodiazepine receptors. *Psychopharmacology* **83**, 240–8.

Pich, E.M. & Samanin, R. (1986). Disinhibitory effects of busipirone and low doses of sulpiride and haloperidol in two experimental anxiety models in rats: possible role of dopamine. *Psychopharmacology* **89**, 125–30.

Polc, P., Bonetti, E.P., Schaffner, R. & Haefeley, W. (1982). A three-state model of the benzodiazepine receptor explains the interactions between the benzodiazepine antagonist RO 15-1788, benzodiazepine tranquilizers, β-carbolines, and phenobarbitone. *Naunyn-Schmiedeberg's Archives of Pharmacology* **321**, 260–4.

Pollard, G.T. & Howard, J.C. (1986). The staircase test: some evidence for non-specificity for anxiolytics. *Psychopharmacology* **89**, 14–19.

Rakel, R. (1987). Assessing the efficacy of antianxiety agents. In: *Anxiety: Quest for Improved Therapy. American Journal of Medicine* **82**, Suppl. 5A, 1–6.

Redmond, D.E. (1985). Neurochemical basis for anxiety and anxiety disorders: evidence from drugs which decrease human fear or anxiety. In: A. Hussain Tuna and J.D. Maser (eds.), *Anxiety and Anxiety Disorders*, pp. 533–55. Lawrence Erlbaum Associates: N.J.

Rickels, K., Case, W.G., Downing, R.W. & Winokur, A. (1983). Long term diazepam therapy and the clinical outcome. *Journal of the American Medical Association* **250**, 767–71.

Robichaud, R.C. & Sledge, K.L. (1969). The effects of p-chlorophenylalanine on experimentally induced conflict in the rat. *Life Science* **8**, 965–9.

Rodin, E.A. & Calhoun, H.D. (1970). Metrazol tolerance in a 'normal' volunteer population. *Journal of Nervous and Mental Disorders* **150**, 438–50.

Rodrigues-Sierra, J.F., Howard, J.L., Pollard, G.T. & Hendrickes, S.E. (1984). Effect of ovarian hormones on conflict behavior. *Psychoneuroendocrinology* **9**, 293–300.

Roth, T., Hartse, K.M., Saab, P.G., Piccione, P.M. & Kramer, M. (1980). The effects of flurazepam, lorazepam and triazolam on sleep and memory. *Psychopharmacology* **70**, 231–7.

Sahgal, A. (1984). A critique of the vasopressin-memory hypothesis. *Psychopharmacology* **83**, 215–28.

Sahgal, A. & Mason, J. (1985). Drug effects on memory: assessment of a combined active and passive avoidance task. *Behavioural Brain Research* **17**, 251–5.

Sanger, D.J. (1987). Further investigation of the stimulus properties of chlordiapoxide and zolipidem: agonism and antagonism by two novel benzodiazepines. *Psychopharmacology* **93**, 365–8.

Sanger, D.J. & Zivkovic, B. (1986). The discriminative stimulus properties of zolipidem, a novel imidazopyridine hypnotic. *Psychopharmacology* **89**, 317–22.

Sanger, D.J., Perrault, G., Morel, E., Joly, D. & Zivkovic, B. (1987). The behavioral profile of Zolipidem a novel hypnotic drug of imidazopyridine structure. *Physiology and Behavior* **41**, 235–40.

Schneider, H.H., Turski, L.A. & Stephens, D.N. (1989). Modulators of the GABA receptor complex. In: Bowery, N. (ed.) *GABA: From Basic Research to Clinical Practice.* Pythagoras Press: Rome. (In press.)

Sepinwall, J. & Cook, L. (1978). Behavioral pharmacology of antianxiety drugs. In: L.L. Iversen, S.D. Iversen and S.H. Snyder (eds.), *Handbook of Psychopharmacology*, vol. 13: *Biology of Mood and Antianxiety Drugs*, pp. 345–93. Plenum Press: N.Y.

Seligman, M.E.P. (1974). Depression and learned helplessness'. In: R.J. Friedman and M.M. Katz (eds.), *The Psychology of Depression: Contemporary Theory and Research*, pp. 83–111. Winston-Wiley: Washington.

Shearman, G.T. & Lal, H. (1980). Generalization and antagonism studies with convulsant, GABAergic and anticonvulsant drugs in rats trained to discriminate pentylenetetrazol from saline. *Neuropharmacology* **19**, 473–9.

Shephard, R.A. (1986). Neurotransmitters, anxiety and benzodiazepines: a behavioral review. *Neuroscience and Biobehavioral Reviews* **10**, 449–61.

Simiand, J., Keane, P.E. & Morre, M. (1984). The staircase test in mice: a simple and efficient procedure for primary screening of anxiolytic agents. *Psychopharmacology* **84**, 48–53.

Smith, D. & Gallager, D. (1987). GABA, benzodiazepine and serotonergic receptor development in the dorsal raphe nucleus: electrophysiological studies. *Developmental Brain Research* **35**, 191–8.

Smith, S.S., Waterhouse, B.D., Chapin, J.K. & Woodward, D.J. (1987). Progesterone alters GABA and glutamate responsiveness: a possible mechanism for its anxiolytic action. *Brain Research* **400**, 353–9.

Squires, R.F. & Braestrup, C. (1977). Benzodiazepine receptors in rat brain. *Nature* **266**, 732–4.

Squires, R., Casida, J., Richardson, M. & Saederup, E. (1983). [^{35}S]t-Butylbicyclophosphorothionate binds with high affinity to brain specific sites coupled to gamma-aminobutyric acid-A and ion recognition sites. *Molecular Pharmacology* **23**, 326–36.

Stephens, D.N. & Kehr, W. (1985). β-carbolines can enhance or antagonize the effects of punishment in mice. *Psychopharmacology* **85**, 143–7.

Stephens, D.N. & Sarter, M. (1988). Bidirectional nature of benzodiazepine receptor ligands extends to effects on vigilance. In: I. Hindmarch, H. Ott & T. Roth (eds.), *Benzodiazepine Receptor Ligands: Memory and Information Processing*, pp. 205–17. Springer: N.Y.

Stephens, D.N. & Schneider, H.H. (1985). Tolerance to the benzodiazepine diazepam in an animal model of anxiolytic activity. *Psychopharmacology* **87**, 322–7.

Stephens, D.N., Kehr, W., & Schneider, H.H. & Braestrup, C. (1984a). Bidirectional effects on anxiety of β-carbolines acting as benzodiazepine receptor ligands. *Neuropharmacology* **23**, 879–80.

Stephens, D.N., Shearman, G.T. & Kehr, W. (1984b). Discriminative stimulus properties of β-carbolines characterized as agonists at benzodiazepine receptors. *Psychopharmacology* **83**, 233–9.

Stephens, D.N., Meldrum, B.S., Weidmann, R., Schneider, C. & Grützner, M. (1986). Does the excitatory amino acid receptor antagonist 2-APH exhibit anxiolytic activity? *Psychopharmacology* **90**, 166–9.

Stephens, D.N., Schneider, H.H., Kehr, W., Jensen, L.H., Petersen, E. & Honore, T. (1987). Modulation of anxiety by β-carbolines and other benzodiazepine receptor ligands: relationship of pharmacological to biochemical measures of efficacy. *Brain Research Bulletin* **19**, 309–18.

Stephens, D.N., Schneider, H.H., Weidmann, R. & Zimmermann, L. (1988). Decreased sensitivity to benzodiazepine receptor agonists and increased sensitivity to inverse agonists following chronic treatment: evidence for separate mechanisms. In: Biggio, G. & E. Costa (eds.), *Chloride Channels and Their Modulation by Neurotransmitters and Drugs. Advances in Biochemical Psychopharmacology* **43**, pp. 337–54. Raven Press, N.Y.

Study, R.E. & Barker, J.L. (1981). Diazepam and (−)pentobarbital: fluctuation analysis reveals different mechanisms for the potentiation of gamma-aminobutyric acid responses in cultured central neurons. *Proceedings of the National Academy of Science* **78**, 7180–4.

Subhan, Z. (1984). The effects of benzodiazepines on short-term memory and information and processing. In: I. Hindmarch, H. Ott & T. Roth (eds.), *Sleep, Benzodiazepines and Performance*, pp. 173–81. Springer: Berlin.

Tallman, J.F., Thomas, J.W. & Gallager, D.W. (1978). Modulation of benzodiazepine binding site sensitivity. *Nature* **274**, 383–5.

Taylor, D.P., Allen, L.E., Becker, J.A., Crane, M.,

Hyslop, D.K. & Riblet, L.A. (1984). Changing concepts of the biochemical action of the anxioselective drug buspirone. *Drug Development Research* **4**, 95–108.

Taylor, D.P., Eison, M.S., Riblet, L.A. & van der Maelen, C.P. (1985). Pharmacological and clinical effects of buspirone. *Pharmacology, Biochemistry and Behavior* **23**, 687–94.

Treit, D. (1985*a*). Evidence that tolerance develops to the anxiolytic effect of diazepam in rats. *Pharmacology, Biochemistry and Behavior* **22**, 383–7.

Treit, D. (1985*b*). Animal models for the study of anti-anxiety agents: a review. *Neuroscience and Biobehavioral Reviews* **9**, 203–22.

Treit, D., Pinel, J.P.J. & Fibiger, H.C. (1981). Conditioned defensive burying: a new paradigm for the study of anxiolytic agents. *Pharmacology, Biochemistry and Behavior* **17**, 359–61.

Tye, N.C., Iversen, S.D. & Green, A.R. (1979). The effects of benzodiazepines and serotonergic manipulations on punished responding. *Neuropharmacology* **18**, 689–95.

Tyrer, P., Murphy, S. & Owen, R.T. (1985). The risk of pharmacological dependence with buspirone. *British Journal of Clinical Practice* **39** (Suppl. 38), 91–3.

Urban, J.H., VandeKar, L.D., Lorens, S.A. & Bethea, C.L. (1986). Effect of the anxiolytic drug Buspirone on prolactin and corticosterone secretion in stressed and unstressed rats. *Pharmacology, Biochemistry and Behavior* **25**, 457–62.

Velluci, S.V. & File, S.E. (1979). Chlordiazepoxide loses its anxiolytic action with long-term treatment. *Psychopharmacology* **62**, 61–5.

Vogel, J.R., Beer, B. & Clody, D.I. (1971). A simple and reliable conflict procedure for testing antianxiety agents. *Psychopharmacology* **21**, 1–7.

Warburton, D.M. & Brown, K. (1972). The facilitation of discrimination performance by physostigmine sulphate. *Psychopharmacology* **27**, 275–84.

Wilson, M.A. & Gallager, D.W. (1987). Effects of chronic diazepam exposure on GABA sensitivity and on benzodiazepine potentiation of GABA-mediated responses of substantia nigra pars reticulata neurons of rats. *European Journal of Pharmacology* **136**, 333–43.

Wise, C.D., Berger, B.D. & Stein, L. (1972). Benzodiazepines: anxiety reducing activity by reduction of serotonin turnover in the brain. *Science* **177**, 180–3.

Wolpe, J. (1958). *Psychotherapy by Reciprocal Inhibition.* Stanford University Press: Stanford, CA.

Wolpe, J. (1971). The behavioristic conception of neurosis: a reply to two critics. *Psychological Review* **78**, 341–3.

Young, R. & Glennon, R.A. (1987). Stimulus properties of benzodiazepines: correlations with binding affinities, therapeutic potency, and structure activity relationships (SAR). *Psychopharmacology* **93**, 529–33.

4

Animal models of anxiety: a clinical perspective

MALCOLM LADER

4.1 INTRODUCTION

Clinicians, by and large are much more concerned to have effective treatments for their patients than to be given elegant explicatory hypotheses for their patients' ills. Consequently, they prefer animal models of, say, anxiety to be more useful in developing new drugs rather than to provide fundamental insights into the mechanisms of that condition. Naturally, in a perfect world, the rational approach is preferable to the empirical one. Unfortunately, most of current psychiatry is so inchoate that the rational approach is still a goal for the dim, distant future.

Nevertheless, research must have some direction. To that end various models of anxiety have been proposed, usually on the basis of covert anthropomorphism, but occasionally on an ethological basis which takes into account simple facts such as the rat being a nocturnal creature. Attempts are then made to validate those models using accepted human treatments as the main criterion.

Although the validity or even the usefulness of models of complex psychiatric conditions such as schizophrenia could be challenged, anxiety presents features which should be reproducible in animals. Thus, it seems reasonable to expect animals to experience a state akin to anxiety, or at least show behaviour homologous to the human state. Situations likely to elicit anxiety in people such as uncertainty or anticipation of unpleasant situations or events would be expected to have similar effects in animals and to produce behavioural responses and physiological changes

in parallel. The only exception might be the more severe, apparently spontaneous anxiety states in which complex cognitive factors, usually covert ('unconscious'), might operate.

The main tenor of this review will be to examine the assumptions underlying this approach and to scrutinise two aspects in particular, namely which human conditions the animal models model and secondly, whether the benzodiazepines (BZs) are selective anxiolytics.

4.2 SYNDROMES OF ANXIETY

Anxiety is both a normal emotion and a pathological condition. Furthermore, both types of anxiety may result in the individual experiencing the emotion seeking medical or other professional advice. In the U.K., the help sought in over 80 per cent of instances is that of the general practitioner, not of a psychiatrist or psychologist. But, let us first look at the semantics of the word 'anxiety'.

The first meaning is that of being solicitous or eager to do something, and need concern us no further. The second is 'uneasy with fear and desire regarding something doubtful' (Chambers Dictionary). The essence of this anxiety is uneasiness and doubt. Anxiety is an ineffable feeling of apprehension and expectation, that is directed towards the future. In its normal form this is related to a fairly specific possibility, such as loss of employment or a physical illness and merges into the kindred emotion of fear which relates to a very specific object or situation. This type of normal anxiety is a natural and understandable reaction to a threat or potential threat. Nevertheless, it may cause sufficient symptomatic concern for the sufferer to present to his or her general practitioner who often prescribes anxiolytic medication. Such drug usage has been criticised as being inappropriate or even hazardous. Simple counselling to help a patient cope with the threat is regarded as appropriate and, indeed, has been shown to be as effective as anxiolytics in the majority of patients (Catalan *et al.*, 1984a, b).

Abnormal or pathological anxiety is diagnosed when the intensity of the emotion felt and complained of (i.e. 'state anxiety') seems quite disproportionate to its putative source (Lewis, 1967). This entails a value judgment on the part of the patient and therapist and, again, no clear demarcation exists between normal and pathological anxiety. Most often, the exaggeration of the anxiety is a habitual mode of response of the individual (i.e. 'trait anxiety'), who is likely to be dubbed an 'anxious personality'. Sometimes, a previously calm person develops an anxiety state and this is the only one of these conditions which constitutes a true disorder in the medical sense. It is then postulated that the mechanism of the overreaction lies in some internal symbolic meaning of external circumstances that would not normally constitute a threat (Freud, 1962; Fenichel, 1971).

To recapitulate, most patients complaining of anxiety are reacting to external events of threat, the degree of reaction being determined by that threat and its perception by the individual, the latter being in part a personality attribute. In some cases, however, the threat is minimal or unidentifiable and previous coping behaviour is quite normal so that a disorder must be postulated.

This all seems a far cry from the syndromes defined in nosological schemata such as those of the American Psychiatric Association (1987) (DSM-IIIR) or the World Health Organisation (ICD-10). The former lays down rigid criteria such as unrealistic anxiety and worry about two or more life circumstances for six months or more, before the diagnosis of Generalized Anxiety Disorder (GAD) can be made (Table 4.1). These requirements exclude from the category all but severe, chronic anxiety states. The diagnosis of Generalized Anxiety Disorder under DSM-IIIR will cover only a small proportion of patients who at present are routinely receiving anxiolytics. One can predict that the residual category, Anxiety Dis-

Table 4.1. *Diagnostic criteria for generalized anxiety disorder (from Diagnostic Criteria from DSM-III-R, American Psychiatric Association, 1987)*

(A) Unrealistic or excessive anxiety and worry (apprehensive expectation) about two or more life circumstances, e.g., worry about possible misfortune to one's child (who is in danger) and worry about finances (for no good reason), for a period of six months or longer, during which the person has been bothered more days than not by these concerns. In children and adolescents, this may take the form of anxiety and worry about academic, athletic, and social performance.

(B) If another Axis I disorder is present, the focus of the anxiety and worry in (A) is unrelated to it, e.g., the anxiety or worry is not about having a panic attack (as in Panic Disorder), being embarrassed in public (as in Social Phobia), being contaminated (as in Obsessive Compulsive Disorder), or gaining weight (as in Anorexia Nervosa).

(C) The disturbance does not occur only during the course of a mood disorder or a psychotic disorder.

(D) At least six of the following 18 symptoms are often present when anxious (do not include symptoms present only during panic attacks):

Motor tension
(1) trembling, twitching, or feeling shaky
(2) muscle tension, aches, or soreness
(3) restlessness
(4) easy fatigability.

Autonomic hyperactivity
(5) shortness of breath or smothering sensations
(6) palpitations or accelerated heart rate (tachycardia)
(7) sweating, or cold clammy hands
(8) dry mouth
(9) dizziness or lightheadedness
(10) nausea, diarrhea, or other abdominal distress
(11) flushes (hot flashes) or chills
(12) frequent urination
(13) trouble swallowing or 'lump in throat'

Vigilance and scanning
(14) feeling keyed up or on edge
(15) exaggerated startle response
(16) difficulty concentrating or 'mind going blank' because of anxiety
(17) trouble falling or staying asleep
(18) irritability.

(E) It cannot be established that an organic factor initiated and maintained the disturbance, e.g., hyperthyroidism, caffeine intoxication.

Table 4.2. *Classification of anxiety disorders (or anxiety and phobic neuroses) (from Diagnostic Criteria from DSM-III-R, American Psychiatric Association, 1987)*

Panic disorder
  with agoraphobia
    Specify current severity of agoraphobic avoidance
    Specify current severity of panic attacks
  without agoraphobia
    Specify current severity of panic attacks

Agoraphobia without history of panic disorder
    Specify with or without limited symptom attacks

Social phobia
    Specify if generalized type

Simple phobia

Obsessive compulsive disorder (or neurosis)

Post-traumatic stress disorder
    Specify if delayed onset

Generalized anxiety disorder

Anxiety disorder not otherwise specified

order Not Otherwise Specified, will become increasingly popular!

Anxiety is a symptom in a wide variety of related disorders; the DSM-IIIR system of classification includes nine conditions under the rubric of anxiety disorders (Table 4.2). Currently, the relationship of panic to anxiety is being hotly debated (Carr & Sheehan, 1984; Roth, 1984; Consensus Statement, 1987). The null hypothesis regards it as a symptom of severe anxiety which does not justify consideration as a separate disorder. However, many authorities regard evidence from genetic and familial studies, epidemiological surveys, and response patterns to anxiolytic and antidepressant drugs to be sufficiently compelling to warrant separation of panic disorders from generalised anxiety disorder. The relationship of agoraphobia to panic and anxiety is also controversial and no clear consensus has yet emerged. Thus, the DSM-IIIR scheme includes panic disorder with agoraphobia, and agoraphobia without panics. To complicate matters further agoraphobia can include limited symptom attacks, a sort of incipient panic.

The World Health Organization's recent development, the ICD-10, is rather more pragmatic and less doctrinaire. This scheme divides anxiety disorders into two primary classes of phobic disorders and other anxiety disorders. Phobic anxiety disorders are further divided into simple phobia, social phobia (Liebowitz *et al.*, 1985) and agoraphobia (Hallam, 1978; Foa *et al.*, 1984); other anxiety disorders are split up into generalised anxiety disorder, panic disorder and anxiety-depressive disorder. This scheme acknowledges the close relation between anxiety and depression, and the presence of mixed disorders, which are yet another complication.

A drawback of both DSM-IIIR and ICD-10 is the failure to take a natural historical rather than a cross-sectional approach. Although G.A.D. in DSM-IIIR requires a six-month history, no recognition is made of the fact that syndromes change, merge, alter and remit over time. Indeed, especially in primary care, a Venn diagram (Fig. 4.1) is held by many to better accord with clinical observations (Angst *et al.*, 1984; Angst & Dobler-Mikola, 1985) and genetic data (Jardine *et al.*, 1984). Not only are mixed syndromes common but a patient may shift among the symptomatic clusters over the course of months or years (Goldberg & Simpson, 1986).

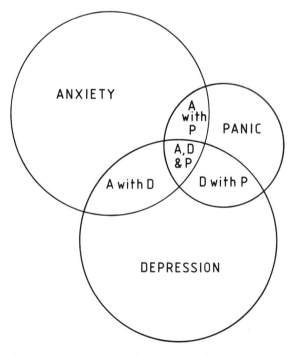

Fig. 4.1. *Venn diagram of relationships between syndromes of anxiety (A), depression (D), and panic (P).*

4.3 ANIMAL MODELS OF ANXIETY

As described in earlier chapters, animal tests of anxiety fall into two main groups (Green & Hodges; Stephens & Andrews, Chapters 2 and 3, respectively of this volume; see also Johnston & File, 1988). The first are tests based on conflict or conditioned fear such as the Geller-Seifter, the Vogel punished drinking and the punished loco-motion tests. All involve a reduction in the amount of behaviour emitted during punishment. It is assumed that the anticipation of punishment causes reduced responding and that anxiolytic drugs should therefore return the rate of respond-ing to normal. Control measures, e.g. of unpu-nished responding, are needed to assess the effects of any non-specific drug effects such as sedation or stimulation or of changes in food or water intake. Of these tests, most reliance is placed on the Geller-Seifter procedure although it is time-consuming and difficult to establish reproducible baselines (Iversen, 1980).

The second group of tests are those of uncertainty generated by placing the animal in an unfamiliar environment or unnatural conditions as bright light. The social interaction test, for example, measures the disruption of normal social intercourse between two male rats caused by such a procedure (File, 1980). Anxiolytics prevent the decline in social interaction in the unfamiliar brightly-lit arena. Other behavioural measures such as overall level of motor activity can act as controls for general sedation and yet others such as defaecation as further measures of anxiety. The elevated plus-maze generates anxiety by placing animals on an elevated open arm, the height con-stituting the relevant stimulus. Anxiolytics increase the percentage of time the animals spend in the open arms (Pellow *et al.*, 1985).

Another approach is to use reactive and non-reactive strains of rats (Robertson *et al.*, 1978). Whether reactivity equates to anxiety and non-reactivity to calmness is uncertain. Little sys-tematic work has been carried out evaluating anxiolytic effects on these strains.

What about validity? It is very difficult and probably misleading to put much store by face validity as a criterion for relating an animal model to a human condition. Often the procedures involved in training the animal bear little resem-blance to human experience as recounted by patients with anxiety disorders. Indeed, many of the operant conditioning models could be con-strued to be akin to Post-traumatic Stress Disorder in humans. (Probably the best animal example of post-traumatic stress disorder, however, was Pavlov's dogs who nearly drowned when his labor-atory was flooded.)

Other models seem more closely related to phobic situations or just to heightened alertness or arousal, in a non-specific way. The social inter-action test is probably not a model of social phobias, rather of heightened alertness in an unfa-miliar situation interfering with normal social behaviour. The elevated plus-maze is similarly sampling the reaction to an unfamiliar and hazard-ous situation.

Reactive strains of rats may bear some resem-blance to humans who have high levels of 'trait' anxiety, i.e., over-respond to situations. But the model may be picking up primarily physiological responsivity.

Another criterion which is widely advocated is whether the model responds in the predicted way to drugs known to lower anxiety levels in humans. But here the question of selectivity of action has too often been overlooked. This question is dis-cussed in the next Section.

4.4 SPECIFICITY OF ANXIOLYTICS

4.4.1 Benzodiazepines

In the 1960s and early 1970s, the benzodiazepines (BZs) steadily replaced the barbiturates as the treatment of choice for anxiety disorders because they were perceived as much safer drugs. Indeed, in many countries, tranquilliser usage expanded appreciably as doctors prescribed them for more and more trivial conditions and patients demanded them to deaden normal responses to everyday occurrences. As this era coincided with the develop-ment of behavioural pharmacology into a discipline in its own right, it was inevitable that BZs became the yardstick against which animal models of anxiety were established. Tests were refined until they became very sensitive to the effects of BZs.

In all this, there was an assumption that the BZs were specific anxiolytics and in this respect differed from their predecessors which were general sedatives (Haefely, 1978). This notion was fostered by the pharmaceutical industry who always have a vested (and legitimate) interest in promoting their latest products. Symptomatic of this was the way the term 'sedative' which originally meant 'anxiety-allaying' took on connotations of unwanted drowsiness and torpor, a condition previously known as 'over-sedation'. Unfortunately, neither clinicians nor behavioural psychopharmacologists questioned sufficiently vigorously this replacement of scientific data by advertising copy. BZs were synonymous with tranquillisers – anxiety reduction without sedation – and it was (and is) widely believed that the key to understanding anxiety lay with unravelling the mode of action of these drugs (Insel *et al.*, 1984).

The latter supposition is, of course, a hoary old chestnut of misplaced logic. Because a treatment may assuage a symptom, or even a syndrome, it does not follow that the mechanism of the condition itself are being altered. For example, diuretics are an effective treatment for cardiac failure, but working out their actions on the kidney throws no light on the mechanisms of the failing heart. Psychopharmacology has numerous examples of similar misconceptions, the best one being the dopamine hypothesis of schizophrenia. Because neuroleptics lessen some symptoms of schizophrenia, vast research programmes were mounted in an attempt to demonstrate abnormalities of dopamine mechanisms in schizophrenic patients (see Cookson, Chapter 13 of this volume). That neuroleptics lessen symptoms of mania, agitated depression, organic brain syndromes and so on was quietly forgotten. The most, surely, that could be expected was that elucidation of the actions of neuroleptic drugs might tell us something of the way dopaminergic transmission was involved in some non-specific psychiatric symptoms and signs, particularly those involving motor overactivity. A second and even worse consequence of the reliance on neuroleptics was the establishment of animal models for putative new neuroleptics which merely detected extrapyramidal effects. Only recently, 35 years after the introduction of chlorpromazine, have compounds been developed which may be antipsychotic without inducing extrapyramidal effects (see Ahlenius, this vol., Ch. 12).

It can be argued that a parallel situation developed with respect to models of anxiety and the anxiolytics. BZs undoubtedly allay anxiety in many patients and can be shown to lessen normal anxieties in us all. But several studies, backed by a wealth of clnical experience, attest to their equal efficacy in lessening aggression and other emotions. In fact, the first BZ, chlordiazepoxide, was discovered because of its anti-aggression properties in animals (Cohen, 1970). But even wider properties are obvious. Hundreds of studies have shown that the BZs impair a wide range of psychological functions. In higher doses, drowsiness ensues, the basis of the use of benzodiazepines as hypnotics. In addition the BZs have well-known anticonvulsant and muscle-relaxant properties. The only conclusion can be that benzodiazepines are fairly nonspecific general cerebral depressants, the only major difference from their predecessors, the barbiturates, chloral, paraldehyde, ethanol and so on, being safety in overdose.

How, then, could the BZs have become established as the anxiolytic gold standard? Although qualitative differences from general cerebral depressants are unconvincing, BZs probably have less steep dose-effect curves so that doses can be found which, clinically at least, lessen anxiety without undue general cerebral depression. However, that dosage 'window' is quite narrow. But the onus of responsibility lies with the behavioural psychopharmacologists to convince the clinician that animal models sensitive to BZs are picking up anxiolytic effects and not something else, for example, sedative effects, amnesiogenic properties, or effects on appetitive behaviour (Iversen, 1980; see also Green & Hodges, this vol., Ch. 2).

The elucidation over the past decade of the biochemical properties of the BZs has reinforced scepticism concerning selectivity of action. BZs act at specific receptors in the central nervous system and potentiate the effects of GABA at $GABA_A$ sites, modulating the opening of chloride ionophores (Paul *et al.*, 1981; Haefely, 1985). GABA is such an ubiquitous inhibitor neurotransmitter that it is difficult to see how much selectivity of action could be present unless clearly distinguishable subpopulations of BZ receptors exist (Braestrup & Nielsen, 1980). Although claims have been made for such divisions, e.g. BZ_1 and BZ_2 receptors (Lippa *et al.*, 1982; Langer & Arbilla, 1988), the situation remains unclear and there is certainly no evidence that a specific group of BZ receptors subserve solely an anxiolytic function (Petersen *et al.*, 1986). However, much recent research has involved the investigation of compounds having partial agonist or mixed agonist/antagonist actions at BZ receptors, and such compounds may have promise as anxiolytics with few side effects (Gardner, 1988; see also Stephens & Andrews, Chapter 3 of this volume.)

4.4.2 Other anxiolytics

A wide range of drugs has been used to combat anxiety. For millenia, alcohol, opiates and cannabis were popular and it was only in the last century that they were superseded by simple organic chemicals such as chloral and paraldehyde. More recently, a series of compounds, such as the barbiturates, the propanediols such as meprobamate, and chlormethiazole, have been introduced, to be largely replaced by the BZs. All seem to have actions on the GABA receptor–chloride ionophore complex. All show some degree of cross-tolerance with the others. Despite unwanted effects such as toxicity in overdose and the risk of dependence, all seemed to be effective anxiolytics and one or two, e.g. meprobamate, enjoyed enormous, albeit ephemeral, popularity. Animal models of anxiety are usually sensitive to these compounds.

However, other drugs used in the treatment of anxiety include the antidepressants, both the tricyclic group (Kahn *et al.*, 1987) and the monoamine oxidase inhibitors (MAOIs), antipsychotic drugs in low dosage and the beta-adrenoceptor antagonists (Shader & Greenblatt, 1983). The antidepressants seem particularly effective in panic disorders: both tricyclics and MAOIs are fairly efficacious, certainly more so than the BZs (Robinson *et al.*, 1978; Zitrin *et al.*, 1980; Kahn *et al.*, 1986). Onset of action is usually fairly slow, taking three to four weeks with the TCAs and up to six weeks with the MAOIs (Tyrer *et al.*, 1973). The improvement in anxious mood appears to be a combination of heightened self-esteem and more self-confidence, greater optimism, and then reduced anxiety and fear. The clinical change is often sudden. As well as blocking panics, some studies have shown that antidepressants have a useful therapeutic effect in generalised anxiety disorders (Shader & Greenblatt, 1983).

This often-overlooked clinical fact introduces yet more complications for animal models of anxiety, their interpretation, and their practical utility as screening techniques for new anxiolytic drugs. Care is usually taken to separate animal models of anxiety from those purporting to model depression. However, clinically, not only are anxiety and depressive syndromes closely related and even overlapping but the anxiolytic and antidepressant treatments are not clearly demarcated (Schatzberg & Cole, 1978; Gelenberg, 1979; Liebowitz, 1984; Angst & Dobler-Mikola, 1984; Angst *et al.*, 1985; Goldberg & Simpson, 1986). Highly specific models may therefore be throwing the baby out with the bathwater. Mixed models may be more appropriate.

The antipsychotic drugs such as trifluoperazine have also enjoyed wide popularity as anxiolytics. Their usage has waned as the danger of inducing tardive dyskinesia on long-term use could never be completely discounted. Nevertheless, many patients have obtained relief from anxiety symptoms while taking low doses of one of these drugs so that they have undoubted anxiolytic properties. The beta-blockers are more limited in their actions. By and large, they only help patients who have marked somatic symptoms, such as, palpitations, gastrointestinal upset, and tremor, and whose somatic symptoms are pivotal to the syndrome of anxiety disorder (Tyrer, 1980; Lader, 1988). Patients with other somatic symptoms or psychological symptoms are less likely to be helped. Clonidine, an alpha-2 agonist, seems ineffective (Hoehn-Saric *et al.*, 1981).

Thus, the insistence by some model-makers that their behavioural test should discriminate between BZs (and similar anxiolytics) and all other psychotropic drugs may have been misplaced. As shown with newer anxiolytics in the next section, failure to detect anxiolytic properties of non-benzodiazepines may vitiate the validity of the models available.

It must also be remembered that a range of non-pharmacological treatments for anxiety are being developed and tested. These include both behavioural and cognitive methods (Beck & Emery, 1985; Clark, 1986). Most acute responses to stress and adjustment disorders can be managed successfully by simple counselling. Exposure treatment, derived originally from animal studies (Wolpe, 1958), is appropriate for phobic anxiety disorders, used alone for simple phobic disorders and combined with appropriate cognitive techniques for agoraphobia and social phobia. Cognitive therapies might be a challenge to animal psychopharmacologists to model in the laboratory but behavioural techniques are eminently feasible. The validity of animal models of anxiety would be greatly strengthened if they could be adapted to show the efficacy of non-pharmacological treatments. This would parallel the use of electroconvulsive shocks to validate animal models of depression.

4.4.3 Newer anxiolytics

The gravest reservations concerning current animal models of anxiety are raised by the newer anxiolytics which are believed not to involve the BZ/GABA/chloride ionophore complex. A remarkable range of different compounds which are supposed to act on one or other of the 5-HT$_1$, 5-HT$_2$ and 5-HT$_3$ receptor systems have been synthesised recently and evaluated for therapeutic properties. The first of these drugs is buspirone, which, among other pharmacological effects, is a 5-HT$_{1A}$ agonist or partial agonist (Eison & Temple, 1986). Buspirone has been marketed as an anxiolytic in West Germany from 1985, in the U.S.A. from late 1986 and in the U.K. from early 1988. It is as effective as BZs, except in patients recently taking BZs in whom for some unknown reason it appears to be ineffective (Schweizer *et al.*, 1986; Olajide & Lader, 1987); it has few if any sedative properties, and it promises to be devoid of abuse and dependence potential (Goa & Ward, 1986). However, the therapeutic effects of buspirone come on more slowly than those of the BZs (Tyrer *et al.*, 1985). In animal models of anxiety it is unimpressive (Riblet *et al.*, 1982, 1984). Positive results have on occasion been reported in most models, but the results are highly controversial, difficult to replicate, and seemingly dependent on a multitude of methodological variables. Buspirone is usually reported to be active in the Vogel lick-suppression test. However, positive and negative reports have been equally frequent in the Geller-Seifter test; and in the social interaction test and the elevated plus-maze, buspirone is usually inactive (Johnston & File, 1988; Stephens & Andrews, this vol., Ch. 3).

The delay in attainment of the full therapeutic effects of buspirone is reminiscent of that associated with both main types of antidepressants, the tricyclics and the MAOIs, although rather less in extent. As with the antidepressants, it may be that the effects of single doses of buspirone are not informative and that repeated doses are necessary in animals to mimic the therapeutic situation in human patients.

Another group of drugs acting on 5-HT mechanisms are the 5-HT$_2$ antagonists such as ritanserin and ICI 169,369. Both have shown anxiolytic properties in the clinic although the evidence is only preliminary in both cases (Ceulemans *et al.*, 1985). In animal tests, anxiolytic effects are quite weak and in fact ritanserin seems anxiogenic in the elevated plus-maze (Pellow *et al.*, 1987).

Very recently, 5-HT$_3$ receptors have been identified in the brain (Kilpatrick *et al.*, 1988) so there is now more rationale to the claimed anxiolytic (and antipsychotic) properties of 5-HT$_3$ antagonists such as GR 38032F (Jones *et al.*, 1988), zacopride (Costall *et al.*, 1988), BRL 43694 (Fake *et al.*, 1987) and ICS 20593 (Donatsch *et al.*, 1984). The reports to date of the activity of these various compounds in animal test models have been somewhat inconsistent. In general, they do appear to be effective, in the social interaction test and the elevated plus-maze, but not in the Vogel lick suppression test (Tyers *et al.*, 1987). Thus, the profile of action of these compounds in animal models of anxiety is almost exactly opposite to that of buspirone. However, reports of clinical efficacy had not yet accrued at the time of writing so these compounds have yet to be established as anxiolytics. Nevertheless, the development of these compounds has given a great fillip to the study of 5-HT and anxiety (Conn & Saunders-Bush, 1987).

An interesting group of compounds are the direct GABA agonists such as progabide and fengabine (Bartholini *et al.*, 1986). They have anticonvulsant properties, progabide being marketed for this purpose in France. Somewhat unexpectedly, they have no anxiolytic actions in the clinic but appear to possess definite antidepressant properties. In animal models of anxiety they are mostly but not entirely inactive; in animal models of depression they are mostly but not entirely active (Bartholini *et al.*, 1986). These data might be construed as supporting the validity of animal models of both depression and anxiety, at least with respect to drugs involving GABA mechanisms.

4.5 Models of other drug properties

Although interest and activity have focussed on animal models of anxiety and of anxiolytic properties of drugs, other aspects can be profitably pursued in animals and appropriate models set up. As 'sedation' is regarded as an unwanted effect, screening programmes usually include a test or tests which are believed to reflect sedative rather than anxiolytic properties. These tests are also regarded as predictive of hypnotic qualities in a New Chemical Entity. Such tests include general locomotor activity, food-motivated lever-pressing, and the rota-rod; some researchers regard the last as indicative of general central depression and the others of sedation, but to the clinician the distinction is unconvincing.

The need for sedation-free tranquillisers can be challenged. Many patients are so distressed by the hypervigilance and over-reactivity associated with a state of anxiety that a modicum of sedation – of feeling a little distant, languid or 'spaced-out' – is quite welcome. This is not the case, of course, with all patients, particularly those who need to drive or operate dangerous machinery.

Another aspect of anxiolytic drug properties which has been modelled in animals is abuse/dependence liability. Much confusion reigns here, partly due to clinicians not sufficiently clarifying their terms. Abuse signifies the occasional or regular use of high doses of a drug in a non-medical context. It has been documented with the BZs but tends to be sporadic and typically confined to polydrug users who may resort to benzodiazepines when preferred illicit drugs such as diamorphine and amphetamines become scarce and expensive. Animal models involve self-administration of the drug, usually intravenously, in response to some repeated operant behaviour such as lever-pressing (see Goudie and Sanger, Chapters 17 and 18 of this volume). BZs are not very active in this test whereas newer anxiolytics such as buspirone seem totally inactive.

Place preference and similar tests in which the animal signifies behaviourally whether it can distinguish between the cues provided by administration of one drug or another are more difficult to interpret for their predictive value in people (Goudie and Sanger, this vol., Ch. 17, 18). Patients are sometimes astonishingly precise in their ability to distinguish not only between drugs of different types but even between drugs of similar types such as the BZs.

Another set of models can be used to evaluate physical dependence. This is manifested in people by a characteristic withdrawal syndrome when a drug is discontinued. In animals, changes in spontaneous behaviour can be observed when a drug of dependence is stopped, or disruptions can be induced in learned behaviour such as lever-pressing (Goudie, this vol., Ch. 17). After the repeated administration of a BZ, the BZ antagonist flumazenil (Ro 15-1788) can be used to precipitate an immediate withdrawal reaction. Cross-tolerance can be investigated by seeing whether the test compound can suppress the withdrawal reaction from a standard anxiolytic such as a barbiturate or benzodiazepine, and vice versa. The predictive validity of many of these tests is claimed to be good and they undoubtedly have some face validity.

4.5.1 Human models

It can be seen that, despite some appealing face validity, most animal models fall short of being entirely convincing and are analogue rather than homologue models. Somewhat unexpectedly, many human models of anxiety are also uncompelling and examining the reasons for this might shed light on the shortcomings of the animal models. Human models can be divided into the non-pharmacological and the pharmacological and have been developed to reproduce elements of panic and phobic states as well as of generalised anxiety disorders (Lader, 1983).

Stressful tasks include naming a rapid series of differently-coloured strips, the Stroop test, difficult precision tasks with or without white noise and so on. Electric shocks or loud noises, predictable or unpredictable, have also been used. The problem is that anxiety is not reliably engendered: annoyance, hostility, frustration, or resignation and amusement are often reported by normal subjects (Wing, 1964).

Real-life situations have also been exploited such as sports parachuting, racing driving and military training. Again, anxiety is not reproducible, and is often mixed with euphoria (Berkun *et al.*, 1962). Many normal people have idiosyncratic fears, e.g. of lifts or snakes, which do not inconvenience them but can be used to induce fear responses (Tyrer & Lader, 1974). The latter seem to be good models for phobic responses in patients, although direct extrapolation is probably unwise. Confirmation of findings in these normal subjects should be sought in phobic patients seeking treatment.

A range of physiological stimuli is available. The cold-pressor test, in which the subject immerses a limb in ice-cold water, is an old favourite. Hyperventilation is another procedure (Thyer *et al.*, 1984), which can produce panic attacks in susceptible normal subjects, and much more frequently in agoraphobic patients. Inhalation of carbon dioxide can also induce anxiety and panic attacks. The mechanisms of such effects are unclear: it is curious that hyperventilation, which produces respiratory alkalosis, and breathing carbon dioxide, which produces acidosis, may both induce anxiety.

Several pharmacological techniques are also available (Griez, 1984; Shear, 1986; Lader & Bruce, 1986). The earliest was the infusion of catecholamines such as adrenaline or noradrenaline of mixtures of the two (Basowitz *et al.*, 1956; Frankenhaeuser & Järpe, 1963). Unfortunately, these substances do not readily pass the blood–brain barrier, so their effects are attributable mainly to perception of peripheral changes such as palpitations and tremor and the secondary induction of anxiety. The quality of the feeling induced is unconvincing to the subject who reports its 'as-if' nature. In patients, however, the anxiety may closely resemble spontaneous attacks. It has been suggested that anxious patients have learned to associate peripheral physiological changes with central feelings of anxiety whereas normal subjects have not. Catecholamine infusions therefore constitute anxiety-induction agents of limited validity.

Yohimbine is a more satisfactory sympathomimetic agent but it has complex actions affecting both alpha$_1$ and alpha$_2$ adrenoceptors. It can induce a range of anxiety symptoms and is quite a convincing model (Charney *et al.*, 1983, 1984). Reactivity is greater in anxious patients than in normal subjects. Conversely, lactate infusions which recently enjoyed a vogue are often not followed by convincing and typical anxiety symptoms. Also the mode of action of lactate infusions remains a mystery (Van der Molen *et al.*, 1986).

The development of adrenaline, yohimbine and lactate models of anxiety stemmed from a peripheral physiological approach. The BZs have provided another starting point. Inverse agonists at the BZ receptor, such as certain of the beta-carbolines, induce syndromes resembling anxiety in primates, with profound physiological changes (Ninan *et al.*, 1982). Fragmentary accounts in human volunteers suggest major anxiogenic properties (Dorow *et al.*, 1983). However, the reservations expressed earlier about the non-specificity of benzodiazepine anxiolysis could be applied also to beta-carboline anxiogenesis.

The most widely-used anxiogenic substance is caffeine (Bruce & Lader, 1986). Case reports indicate that caffeine in high dose may produce symptoms of anxiety (Greden, 1974; Lee *et al.*, 1985), and panic attacks may be precipitated in the susceptible. Spontaneous anxiety levels can be lessened by advising the patient to reduce or stop caffeine consumption. Unfortunately, the basic pharmacology of caffeine is complex (Boulenger *et al.*, 1982) and the model of anxiety it produces cannot be interpreted into pharmacological terms. Nevertheless, it is the most convincing model phenomenologically.

An interesting offshoot of these induction models has been the literature relating cognitive and arousal elements (Tyrer, 1976), the upshot of which was the theory that emotions arise from an interaction between a state of physiological arousal and cognitive information derived from the situation (Schachter, 1966). Three assumptions underlie the theory:

(1) Individuals label an inexplicable state of physiological arousal according to situational cognitive clues.

(2) If an immediate and totally congruent explanation is forthcoming, the person will not feel an emotion.

(3) Even if cognitive clues are forthcoming, an emotion will result only insofar as the underlying physiological arousal is raised.

To these can be added a personal speculation that might help to throw light on both human and animal models of anxiety:

(4) If an inexplicable state of physiological arousal is unaccompanied by cognitive clues, i.e., remains unexplained, the individual will experience the emotion of anxiety.

Thus, anxiety can be induced by raising arousal with or without cognitive clues. The former represents a model of specific anxiety, the latter of non-specific anxiety. Similarly, anxiolytics might be postulated to work specifically on the cognitive mechanisms subserving anxiety or non-specifically on the arousal mechanisms. Animal tests developed to detect BZ actions, arguably non-specific, could be expected to be relatively insensitive to more specific anxiolytics, as they would detect drugs lessening arousal rather than direct anti-anxiety effects.

4.6 CONCLUSIONS

The success of prediction from animals to humans depends on the degree of success with which a screening test detects drug effects, and the pragmatic usefulness of a screening test as judged by the confirmatory tests on humans, both normal and disordered. A battery of tests should possess a useful measure of predictive validity, i.e., both false positives and false negatives should be minimal. The reliability of any test should be high and for standard drugs must yield reproducible results.

Several recognised methods exist for selecting behavioural and other animal models. One is the analysis of criterion behaviour. First, a human aberrant behaviour pattern, such as phobic avoidance, is analysed for its basic characteristics. Next, apparently similar elements of animal behaviour are selected. Then, chemical compounds are sought which modify the animal behaviour. Finally, these compounds, if active, are tested on the original abnormal human behaviour. Unfortunately, because of our ignorance of the mechanisms of abnormal human behaviour, the predictive validity of such tests is usually low.

The second approach is through the analysis of criterion drugs. Drugs known to be effective in the clinic are investigated in animal tests and a battery of the most sensitive and discriminatory tests assembled. New drugs are tested on the same battery and their profiles of action compared with the standards. This method is the most widely used but suffers the drawback that new types of compounds may be missed. Whereas typical drugs used to be the most studied, with intense competition for novel compounds, atypical drugs are now also being carefully scrutinised.

A third approach is to try to codify behaviour and then to seek compounds with different profiles of action. This academic approach has not been favoured.

Although the academician might advocate the first of these three approaches, the second is still far and away the most widely used. It is therefore hardly surprising that new psychotropic drugs in general and new anxiolytic drugs in particular are still being discovered serendipitously. What has changed over the past decade or two is the attitude to animal models. They are now regarded as interesting behavioural systems in their own right and as limited but nevertheless indispensable screening tests.

The role of the clinician remains that of the eminence grise. We are the final arbiters of whether a drug will be a commercial success. This decision is hopefully based on adequate clinical evidence concerning the risk/benefit ratio of the drug. But we can also play a role in defining abnormal human states more clearly without conferring on psychiatry the pseudo-precision exemplified by some nosological schemata. Our main function will be to assess quickly, accurately and dispassionately putative new anxiolytics and to provide the feedback concerning efficacy that the behavioural psychopharmacologist needs. If we also provide new insights into anxiety, that facilitate new, reliable, sensitive and valid animal models of anxiety, this will be perceived as a bonus. What we must stop doing is ascribing specificity to anxiolytic drugs where none exists, and distinct syndromes to anxiety disorders where the complexities, psychiatric and pharmacological, are too great.

REFERENCES

American Psychiatric Association. (1987). *Diagnostic Criteria from DSM-III-R*. Washington, DC: APA.
Angst, J. & Dobler-Mikola, A. (1985). The Zurich study – a prospective epidemiological study of depressive, neurotic and psychosomatic syndromes. IV: Recurrent and nonrecurrent brief

depression. *European Archives of Psychiatry and Neurological Sciences* **234**, 408–16.

Angst, J., Dobler-Mikola, A. & Binder, J. (1984). The Zurich study – a prospective study of depressive, neurotic and psychosomatic syndromes. I: Problem, methodology. *European Archives of Psychiatry and Neurological Sciences* **234**, 13–20.

Bartholini, G., Lloyd, K.G. & Morselli, P.L. (eds.) (1986). *GABA and Mood Disorders: Experimental and Clinical Research*. New York: Raven Press.

Basowitz, H., Korchin, S.J., Oken, D., Goldstein, M.S. & Gussack, H. (1956). Anxiety and performance changes with a minimal dose of epinephrine. *Archives of Neurology and Psychiatry* **76**, 98–105.

Beck, A.T. & Emery, G. (1985). *Anxiety Disorders and Phobias: A Cognitive Perspective*. New York: Basic Books.

Berkun, M.M., Bialek, H.M., Kern, R.P. & Yagi, K. (1962). Experimental studies of psychological stress in man. *Psychological Monographs* **76**, No. 15.

Boulenger, J.P., Patel, J. & Marangos, P.J. (1982). Effects of caffeine and theophylline on adenosine and benzodiazepine receptors in human brain. *Neuroscience Letters* **30**, 161–6.

Braestrup, C. & Nielsen, M. (1980). Multiple benzodiazepine receptors. *Trends in Neuroscience* **3**, 301–3.

Bruce, M. & Lader, M. (1986). Caffeine: clinical and experimental effects in humans. *Human Psychopharmacology* **1**, 63–82.

Carr, D.B. & Sheehan, D.V. (1984). Panic anxiety: a new biological model. *Journal of Clinical Psychiatry* **45**, 323–30.

Catalan, J., Gath, D., Edmonds, G. & Ennis, J. (1984*a*). The effects of non-prescribing of anxiolytics in general practice. I. Controlled evaluation of psychiatric and social outcome. *British Journal of Psychiatry* **144**, 593–602.

Catalan, J., Gath, D., Bond, A. & Martin, P. (1984*b*). The effects of non-prescribing of anxiolytics in general practice. II. Factors associated with outcome. *British Journal of Psychiatry* **144**, 603–10.

Ceulemans, D.L.S., Hoppenbrouwers, M.-L.J.A., Gelders, Y.G. & Reyntjens, A.J.M. (1985). The influence of ritanserin, a serotonin antagonist, in anxiety disorders: a double-blind placebo-controlled study versus lorazepam. *Pharmacopsychiatry* **18**, 303–5.

Charney, D.S., Heninger, G.R. & Redmond, D.E. (1983). Yohimbine induced anxiety and increased noradrenergic function in humans: effects of diazepam and clonidine. *Life Science* **33**, 19–26.

Charney, D.S., Heninger, G.R. & Breier, A. (1984). Noradrenergic function in panic anxiety, effects of yohimbine in healthy subjects and patients with agoraphobia and panic disorder. *Archives of General Psychiatry* **41**, 751–63.

Clark, D.M. (1986). A cognitive approach to panic. *Behaviour Research and Therapy* **24**, 461–70.

Cohen, I.M. (1970). The benzodiazepines. In: *Discoveries in Biological Psychiatry*, ed. F.J. Ayd & B. Blackwell, pp. 130–41. Philadelphia: Lippincott.

Conn, P.J. & Sanders-Bush, E. (1987). Central serotonin receptors: effector systems, physiological roles and regulation. *Psychopharmacology* **92**, 267–77.

Consensus Statement. (1987). Panic disorder. *British Journal of Psychiatry* **150**, 557–8.

Costall, B., Domeney, A.M., Gerrard, P.A., Kelley, M.E. & Naylor, R.A. (1988). Zacopride: Anxiolytic profile in rodent and primate models of anxiety. *Journal of Pharmacy and Pharmacology* **40**, 302–5.

Donatsch, P., Engel, G., Richardson, B.P. & Stadler, P. (1984). ICS 205-930: a highly selective and potent antagonist at peripheral neuronal 5-hydroxytryptamine (5-HT) receptors. *British Journal of Pharmacology* **81**, 34P.

Dorow, R., Horowski, R., Paschelke, G., Amin, A. & Braestrup, C. (1983). Severe anxiety induced by FG 7142, a beta-carboline ligand for benzodiazepine receptors. *Lancet* **ii**, 98–9.

Eison, A.S. & Temple, D.L. (1986). Buspirone: Review of its pharmacology and current perspectives on its mechanism of action. *American Journal of Medicine* **80** (3B, Suppl.), 1–9.

Fake, C.S., King, F.D. & Sanger, G.J. (1987). BRL 43694: a potential novel 5-HT$_3$ antagonist. *British Journal of Pharmacology* **91**, 335P.

Fenichel, O. (1971). *The Psychoanalytic Theory of Neurosis*. London: Routledge & Kegan Paul.

File, S.E. (1980). The use of social interaction as a method for detecting anxiolytic activity of chlordiazepoxide-like drugs. *Journal of Neuroscience Methods* **2**, 219–38.

Foa, E.B., Steketee, G. & Young, M.C. (1984). Agoraphobia: Phenomenological aspects, associated characteristics and theoretical considerations. *Clinical Psychology Review* **4**, 431–57.

Frankenhaeuser, M. & Järpe, G. (1963). Psychophysiological changes during infusions of adrenaline in various doses. *Psychopharmacology* **4**, 424–32.

Freud, S. (1962). On the grounds for detaching a particular syndrome from neurasthenia under the description 'anxiety neurosis'. In: *the Standard Edition of the Complete Psychological Works of Sigmund Freud*, vol. 3, ed. J. Strachey. London: Hogarth Press.

Gardner, C.R. (1988). Pharmacological profiles in vivo

of benzodiazepine receptor ligands. *Drug Development Research* **12**, 1–28.

Gelenberg, A.J. (1979). Prescribing antidepressants: the rational use of psychotropic drugs. *Drug Therapy* **9**, 95–8, 103–6, 109–12.

Goa, K.L. & Ward, A. (1986). Buspirone: A preliminary review of its pharmacological properties and therapeutic efficacy as an anxiolytic. *Drugs* **2**, 114–29.

Goldberg, D. & Simpson, N. (1986). The diagnosis of anxiety in primary care settings. In: *Drug Treatment of Neurotic Disorders*, eds. M.H. Lader & H.C. Davies, pp. 76–82. London: Churchill Livingstone.

Greden, J.F. (1974). Anxiety or caffeinism: a diagnostic dilemma. *American Journal of Psychiatry* **131**, 1089–92.

Griez, E. (1984). Experimental models of anxiety. Problems and perspectives. *Acta Psychiatrica Belgica* **84**, 511–32.

Haefely, W. (1978). Behavioral and neuropharmacological aspects of drugs used in anxiety and related states. In: *Psychopharmacology: A Generation of Progress*, eds. M.A. Lipton, A. DiMascio & K.F. Killam, pp. 1359–74. New York: Raven Press.

Haefely, W. (1985). Biochemistry of anxiety. *Annals of the Academy of Medicine (Singapore)* **14**, 81–3.

Hallam, R.S. (1978). Agoraphobia: A critical review of the concept. *British Journal of Psychiatry* **133**, 314–19.

Hoehn-Saric, R., Merchant, A.F. & Keyser, M.L. (1981). Effects of clonidine on anxiety disorders. *Archives of General Psychiatry* **38**, 1278–82.

Insel, T.R., Ninan, P.T., Aloi, J., Jimerson, D.C., Skolnick, P. & Paul, S.M. (1984). A benzodiazepine receptor-mediated model of anxiety. Studies in non-human primates and clinical implications. *Archives of General Psychiatry* **41**, 741–50.

Iversen, S.D. (1980). Animal models of anxiety and benzodiazepine actions. *Arzneimittel-Forschung/Drug Research* **30**, 862–8.

Jardine, R., Martin, N.G. & Henderson, A.S. (1984). Genetic covariation between neuroticism and the symptoms of anxiety and depression. *Genetic Epidemiology* **1**, 89–107.

Johnston, A.L. & File, S.E. (1988). Effects of ligands for specific 5-HT receptor sub-types in two animal tests of anxiety. In: *Buspirone – A New Introduction to the Treatment of Anxiety*, ed. M. Lader, *Royal Society of Medicine Symposium Series* **133**, 31–41.

Jones, B.J., Costall, B., Domeney, A.M., Kelley, M.E., Naylor, R.J., Oakley, N.R. & Tyers, M.B. (1987). The anxiolytic activity of GR38032F, a 5-HT$_3$ receptor antagonist, in the rat and cynomolgus monkey. *British Journal of Pharmacology* **93**, 985–93.

Kahn, R.J., McNair, D.M., Lipman, R.S., Covi, L., Rickels, K., Downing, R., Fisher, S. & Frankenthaler, L.M. (1986). Imipramine and chlordiazepoxide in depressive and anxiety disorders. 2. Efficacy in anxious patients. *Archives of General Psychiatry* **43**, 79–85.

Kahn, R.J., McNair, D.M. & Frankenthaler, L.M. (1987). Tricyclic treatment of generalized anxiety disorder. *Journal of Affective Disorders* **13**, 145–51.

Kilpatrick, G.J., Jones, B.J. & Tyers, M.B. (1988). Direct labelling of 5-HT$_3$ receptors in rat brain using [^{3}H]-GR65630: characterisation and distribution. *British Journal of Pharmacology* **96**, 871 P.

Lader, M. (1983). Anxiety and depression. In: *Physiological Correlates of Human Behaviour*, ed. A. Gale & J.A. Edwards, pp. 155–67. London: Academic Press.

Lader, M. (1988). Beta-adrenoceptor antagonists in neuropsychiatry: an update. *Journal of Clinical Psychiatry* **6**, 213–23.

Lader, M. & Bruce, M. (1986). States of anxiety and their induction by drugs. *British Journal of Clinical Pharmacology* **22**, 251–61.

Langer, S.Z. & Arbilla, S. (1988). Limitations of the benzodiazepine receptor nomenclature: A proposal for a pharmacological classification as omega receptors. *Fundamental and Clinical Pharmacology* **2**, 159–70.

Lee, M.A., Cameron, O.G. & Greden, J.F. (1985). Anxiety and caffeine consumption in people with anxiety disorders. *Psychiatry Research* **15**, 211–17.

Lewis, A. (1967). Problems presented by the ambiguous word 'anxiety' as used in psychopathology. *Israel Annals of Psychiatry and Related Disciplines* **5**, 105–21.

Liebowitz, M.R. (1984). The efficacy of antidepressants in anxiety disorders. In: *Psychiatry Update*, ed. L. Grinspoon, pp. 503–19, Washington, DC: American Psychiatric Press.

Liebowitz, M.R., Gorman, J.M. & Fryer, A.J. (1985). Social phobia: Review of a neglected anxiety disorder. *Archives of General Psychiatry* **42**, 729–36.

Lippa, A.S., Meyerson, L.R. & Beer, B. (1982). Molecular substrates of anxiety: clues from the heterogeneity of benzodiazepine receptors. *Life Science* **31**, 1409–17.

Ninan, P.T., Insel, T.R., Cohen, R.M., Cook, J.M., Skolnick, P. & Paul, S.M. (1982). Benzodiazepine receptor-mediated experimental 'anxiety' in primates. *Science* **218**, 1332–4.

Olajide, D. & Lader, M.H. (1987). A comparison of buspirone, diazepam and placebo in patients with chronic anxiety states. *Journal of Clinical Psychopharmacology* **7**, 148–52.

Paul, S.M., Marangos, P.J. & Skolnick, P. (1981). The benzodiazepine-GABA-chloride ionophore receptor complex: Common site of minor tranquilizer action. *Biological Psychiatry* **16**, 213–29.

Pellow, S., Chopin, P., S.E. & Briley, M. (1985). Validation of open–closed arm entries in an elevated plus-maze as a measure of anxiety in the rat. *Journal of Neuroscience Methods* **14**, 149–67.

Pellow, S., Johnstone, A. & File, S.E. (1987). Selective agonists and antagonists for 5-hydroxytryptamine receptor subtypes, and interactions with yohimbine and FG 7142 using the elevated plus-maze test in the rat. *Journal of Pharmacy and Pharmacology* **39**, 917–28.

Petersen, E.N., Jensen, L.H., Drejer, J. & Honore, T. (1986). New perspectives in benzodiazepine receptor pharmacology. *Pharmacopsychiatry* **19**, 4–6.

Riblet, L.A., Taylor, D.P., Eison, M.S. & Stanton, H.C. (1982). Pharmacology and neurochemistry of buspirone. *Journal of Clinical Psychiatry* **43** (Section 2), 11–16.

Riblet, L.A., Eison, M.S., Taylor, D.P., Temple, D.L. & VanderMaelen, C.P. (1984). Neuropharmacology of buspirone. *Psychopathology* **17** (Suppl. 3), 69–78.

Robertson, H.A., Martin, O.L. & Candy, J.M. (1978). Differences in benzodiazepine receptor binding in Maudsley reactive and Maudsley non-reactive rats. *European Journal of Pharmacology* **50**, 455–7.

Robinson, D.S., Nies, A., Ravaris, C.L., Ives, J.O. & Bartlett, D. (1978). Clinical pharmacology of phenelzine. *Archives of General Psychiatry* **35**, 629–35.

Roth, M. (1984). Agoraphobia, panic disorder and generalized anxiety disorder: some implications of recent advances. *Psychiatric Developments* **2**, 31–52.

Schachter, S. (1966). The interaction of cognitive and physiological determinants of emotional state. In: *Anxiety and Behavior*, ed. C.D. Spielberger, pp. 193–224. New York: Academic Press.

Schatzberg, A.F. & Cole, J.O. (1978). Benzodiazepines in depressive disorders. *Archives of General Psychiatry* **24**, 509–14.

Schweizer, E., Rickels, K. & Lucki, I. (1986).

Resistance to the anti-anxiety effect of buspirone in patients with a history of benzodiazepine use. *New England Journal of Medicine* **314**, 719–20.

Shader, R.I. & Greenblatt, D.J. (1983). Some current treatment options for symptoms of anxiety. *Journal of Clinical Psychiatry* **44** (11, Section 2), 21–9.

Shear, M.K. (1986). Pathophysiology of panic: A review of pharmacologic provocative tests and naturalistic monitoring data. *Journal of Clinical Psychiatry* **47:6** (Suppl.), 18–26.

Thyer, B.A., Papsdorf, J.D. & Wright, P. (1984). Physiological and psychological effects of acute intentional hyperventilation. *Behaviour Research and Therapy* **22**, 587–90.

Tyers, M.B., Costall, B., Domeney, A., Jones, B.J., Kelley, M.E., Naylor, R.J. & Oakley, N.R. (1987). The anxiolytic activities of $5HT_3$ antagonists in laboratory animals. *Neuroscience Letters* **29**, 568.

Tyrer, P. (1976). *The Role of Bodily Feelings in Anxiety*. London: Oxford University Press.

Tyrer, P.J. (1980). Use of beta-blocking drugs in psychiatry and neurology. *Drugs* **20**, 300–8.

Tyrer, P.J. & Lader, M.H. (1974). Physiological and psychological effects of ± propranolol and diazepam in induced anxiety. *British Journal of Clinical Pharmacology* **1**, 379–85.

Tyrer, P., Candy, J. & Kelly, D. (1973). A study of the clinical effects of phenelzine and placebo in the treatment of phobic anxiety. *Psychopharmacologia* **32**, 237–54.

Tyrer, P., Murphy, S. & Owen, R.T. (1985). The risk of pharmacological dependence with buspirone. *British Journal of Clinical Practice* **39**, (Suppl. 38), 91–3.

Van der Molen, G.M., Van den Hout, M.A., Vroemen, J., Lousberg, H. & Griez, E. (1986). Cognitive determinants of lactate-induced anxiety. *Behaviour Research and Therapy* **24**, 677–80.

Wing, L. (1964). Physiological effects of performing a difficult task in patients with anxiety states. *Journal of Psychosomatic Research* **7**, 283–94.

Wolpe, J. (1958). *Psychotherapy by reciprocal inhibition*. Palo Alto: Stanford University Press.

Zitrin, C.M., Klein, D.R. & Woerner, M.G. (1980). Treatment of agoraphobia with group exposure in vivo and imipramine. *Archives of General Psychiatry* **37**, 63–72.

PART III

Models of depression

5

Animal models of depression

PAUL WILLNER

5.1 INTRODUCTION

As recently as the mid-1970s, animal models of depression were of interest almost exclusively within the pharmaceutical industry for their (limited) value as antidepressant screening tests. However, many clinicians would now be prepared to accept, at least in principle, that animal models might serve as simulations within which to investigate aspects of depression (see Deakin, this vol., Ch. 7). Recent years have seen a major expansion in the number of behavioural paradigms that might serve this purpose, and several review articles have been published that have brought many of them to the attention of a wider audience (Katz, 1981a; McKinney, 1984; Willner, 1984; Jesberger & Richardson, 1985).

The object of this chapter is to survey the models that are currently popular, or have been so in the past, as well as some newer models that have not yet been widely utilized. A distinction is drawn between those models that are in reality behavioural assays for the effects of antidepressants on specific neurotransmitter systems, and those (the majority) that simulate some aspect of depression.

5.1.1 Symptomatology of depression

Depression is a multifaceted disorder: a variety of symptoms may be present, but none of them, including depressed mood, is essential. In the DSM-III classification system (American Psychiatric Association, 1980), the diagnosis of major depression requires the presence for at least two weeks of either a dysphoric mood (usually, though not necessarily, depression), or loss of interest or pleasure in activities that would usually be enjoyed. In addition, four of the following eight symptoms must be present (or three with both of the cardinal symptoms): psychomotor retardation or agitation; feelings of worthlessness, self-reproach or excessive or inappropriate guilt; thoughts of suicide; decreased ability to concentrate; loss of energy; decreased sex drive; sleep disturbance; appetite disturbance.

As subjective feelings are not amenable to animal modelling, what symptoms of depression might animal models attempt to address? Of the two cardinal symptoms, one, dysphoric mood, is excluded; so, too, are feelings of guilt, thoughts of suicide, and loss of energy. The other symptoms translate readily into measurable behaviours: loss of motivation and insensitivity to rewards emerge as the major features, along with psychomotor change, decreased persistence, decreased sexual activity, and disturbances of sleep or food intake. It is important to note that psychomotor, sleep and appetite changes may be in either direction, which allows considerable flexibility to the proponents of an animal model: for example, support may be claimed from both decreases and increases in locomotor activity. This lack of precision, together with the fact that the clinical phenomena of psychomotor retardation and agitation are considerably more complex than gross changes in locomotor activity, and indeed, may co-exist (Nelson & Charney, 1981), suggests that simulations in which a change in locomotor activity is the major, or only, behavioural feature should not be taken too seriously.

5.1.1.1 *Subtypes of depression*

For many years, there has been a broad consensus that there are two types of depression, though only recently has a measure of agreement been reached over how they differ. The important distinction seems to be between a reactive depression, which responds with mood improvements to attention and reassurance, and an endogenous depression, or melancholia, which does not (Klein, 1974; Kendell, 1976; Nelson & Charney, 1981; Willner, 1985). The defining feature of an endogenous depression is the inability to experience pleasure, which has been conceptualized as a decrease in the sensitivity of brain reward systems (Klein, 1974). It is important to be aware that the distinction between endogenous and reactive depressions is based on their clinical features; endogenous and reactive depressions do not differ in the degree to which external precipitants can be identified (see e.g. Lewinsohn *et al.*, 1977; Paykel, 1979). This point is widely misunderstood by clinicians, and also misinforms many discussions of animal models of depression.

The inability to experience pleasure is strongly correlated with severity of depression (Nelson & Charney, 1981), though severe, non-endogenous depressions do also exist (Feinberg & Carroll, 1982). Almost invariably, severe depressions also involve some degree of psychomotor change. Indeed, in many studies, psychomotor retardation has emerged as the symptom most characteristic of severe endogenous depression (Nelson & Charney, 1981; Jablensky *et al.*, 1981). Psychomotor agitation tends to be more closely associated with psychotic features such as delusions of guilt (Frances *et al.*, 1981; Nelson & Charney, 1981). Nevertheless, there is considerable overlap between these two groups of symptoms, and agitated endogenous depressions are not uncommon.

However, two biological markers are reliably associated with endogenous, but not non-endogenous depressions: a decrease in the latency to enter the first period of rapid eye movement (REM) sleep (Kupfer & Thase, 1983) and an increased secretion of cortisol, usually detected by the dexamethasone suppression test (DST) (Carroll, 1982). REM latency has not yet been measured in an animal model of depression, but abnormally high blood cortisol levels are frequently reported

(see below). As this abnormality is a specific feature of endogenous depression, it should co-exist, in a valid simulation, alongside a decreased sensitivity to reward.

One aspect of depression that up to now has been ignored entirely in animal models is the fact that depression is a periodic disorder, which typically recurs throughout life at gradually decreasing intervals (Angst *et al.*, 1973). Furthermore, in a significant minority (10–15%) of patients, depressive episodes are interspersed with episodes of mania and hypomania. The unipolar/bipolar distinction is clinically useful: bipolar disorders have a clear genetic component (Allen, 1976; Robertson, 1987), and bipolar depressions tend to show a more homogenous pattern of endogenous and retarded symptoms (Beigel & Murphy, 1971). However, there have been no attempts to model bipolar manic-depressive disorder in animals. The only lead in this direction is chronic amphetamine intoxication, which has been proposed as an animal model of mania (see Lyon, this vol., Ch. 11), and withdrawal from which has been advanced as an animal model of depression (see Section 5.3.2.8).

5.1.2 The response to antidepressants

Responsiveness to antidepressant drugs is usually taken to be a basic requirement for an animal model of depression (though this has not always been the case – c.f. Klinger, 1975). This requirement is frequently considered not merely to be necessary, but also to be sufficient: the demonstration that a behaviour responds to antidepressants but not to neuroleptics and/or anxiolytics generally leads without more ado to the proposal that the behaviour constitutes a potential model. This tendency should be resisted. A substantial minority of depressed patients fail to respond to antidepressants, and in addition to their antidepressant effects, tricyclics appear to be highly effective anxiolytics (see Deakin and Lader, Chapters 7 and 4 of this volume). Thus, depression cannot be defined by the response to antidepressants. Neither, therefore, can animal models of depression. Indeed, after chronic treatment, tricyclics appear to be effective in at least one animal model of anxiety (Bodnoff *et al.*, 1988).

The issue of whether a model responds acutely

to antidepressants or responds only to chronic treatment has also been a source of some confusion. It is widely believed that because the pharmacotherapy of depression requires chronic drug treatment, for a period of weeks, the validity of an animal model is called into question by an acute antidepressant response. In fact, the clinical requirement for chronic treatment may be more apparent than real: there is some evidence of very early antidepressant responses in clinical studies designed explicitly to detect them (e.g. Frazer *et al.*, 1985; see also Willner, 1989*a*). The real test for a simulation of depression is that tolerance must not develop to the antidepressant response: irrespective of how it responds to acute antidepressant treatment, the model must respond to chronic treatment. This test has not been applied universally to animal models of depression, but in general, those models to which the test has been applied have passed it. Indeed, it is usually found that if a test does respond acutely to antidepressants, the response is potentiated by chronic treatment (Willner, 1989*a*). However, tolerance to antidepressant effect has been reported in some behavioural paradigms (Niesink & van Ree, 1982; Cuomo *et al.*, 1983).

In practice, judgements about the desirable time course of drug action in an animal model of depression are largely pragmatic. If the intention is to use a simulation to investigate the physiological mechanisms of antidepressant action, it clearly makes sense to choose a model in which the antidepressant effects increase with chronic treatment. However, if a model is to be used as an antidepressant screening test, then an acute drug response is much to be preferred for logistical reasons (and should not be a source of embarrassment!).

It is also important not to be too hasty in rejecting a model simply because it responds to non-antidepressants, or fails to respond to known antidepressants. The most frequently encountered false positives in animal models of depression are anticholinergics, stimulants, and opioids (see Willner, 1984). But are these positive responses really false? It is all too easy to forget that prior to the development of antidepressants, drugs from all three of these classes were regularly prescribed for the relief of depression (see Willner, 1985). Some recent studies have confirmed that certain opiates (e.g. buprenorphine) have antidepressant activity (Emrich *et al.*, 1983), but anticholinergics and stimulants have never been properly assessed, as their use was discontinued prior to the introduction of blind clinical trials. It would be quite wrong to infer from the fact that these drugs were prescribed as antidepressants, that therefore they were effective. Equally, however, when animal models respond to stimulants or anticholinergics there is sufficient doubt about the status of these compounds to return an open verdict.

Three classes of drug giving rise to false negative response also require special attention. First are the monoamine oxidase inhibitors (MAOIs). These drugs appear to be effective clinically in a small group of 'atypical depressions' characterized by moderately depressed mood, high levels of anxiety, and usually, overeating and hypersomnia (Tyrer, 1979; Paykel *et al.*, 1983). As the MAOIs are relatively ineffective in severe depressions (Tyrer, 1979), their lack of effect in an animal model may actually strengthen its validity. However, the converse reasoning does not apply to tricyclics. It now appears that, contrary to earlier impressions, tricyclics are in fact effective in 'atypical depressions' (Paykel *et al.*, 1983). If MAOIs are more effective than tricyclics in these conditions, their superiority probably reflects a greater anti-anxiety effect (Ravaris *et al.*, 1980).

Secondly, it must be remembered that the clinical efficacy of many of the newer 'atypical antidepressants' is not well established. Iprindole is a particular case in point. Great effort has been exerted in attempts to reach some understanding of the mode of action of this drug, when in fact, it may well be ineffective (Zis & Goodwin, 1979). Finally, serotonin (5-HT) uptake inhibitors present a different problem. These drugs are clearly efficacious as antidepressants (Asberg *et al.*, 1986), though perhaps inferior to tricyclics (see Deakin, this vol., Ch. 7). 5-HT uptake inhibitors are particularly effective in certain animal models, yet are inactive in others (see below). Rather than rejecting these latter models, it may be more fruitful to attempt to relate differences in drug response to patterns of depressive symptoms. This amounts, in effect, to using differences between groups of animal models to assist in the development of a theory of depression.

5.1.3 Theories of depression

Theoretical approaches to depression tend to be either strongly biological or strongly psychological in nature. The former tend to emphasize neurochemical abnormalities, with scant attention to how such abnormalities arise or what their functional significance might be (e.g. Schildkraut, 1965: Lapin & Oxenkrug, 1969), while the latter tend to emphasize personality differences and maladaptive modes of thinking (e.g. Beck, 1967; Seligman, 1975). Although it has been pointed out frequently that these two levels of analysis are by no means incompatible (Gray, 1982; Maier, 1984), there have been few attempts at synthesis (but see Willner, 1985).

5.1.3.1 *Modelling the etiology of depression*

It is now clear that a variety of different factors are implicated in the etiology of depression. 'Psychological' factors include undesirable life events, chronic mild stress, adverse childhood experiences, and personality traits such as introversion and impulsiveness; 'biological' factors include genetic influences, and a variety of physical illnesses and medications (see Akiskal & Tashjian, 1983; Akiskal, 1985, 1986; Willner, 1985, 1987 for reviews). In certain cases, the immediate precipitant of a depression may be clearly identified; seasonal affective disorder (Rosenthal *et al.*, 1985) and post-partum depression are good examples. More usually, however, the etiology of depression is better understood as the result of an accumulation of a number of different risk factors (see, e.g. Brown & Harris, 1978; Aneshensel & Stone, 1982; Akiskal, 1985). This point has been largely overlooked in the construction of animal models of depression, which in general have assumed a single causal factor. (Indeed, as will be apparent below (Sections 5.3.1 and 5.3.2), assumptions about causality provide a convenient basis for classification of the models.) The attempt to simulate depression using a single physiological or (more usually) behavioural manipulation may be counterproductive, since few of the identified etiological factors appear sufficiently potent to precipitate depression in an otherwise risk-free individual. Nevertheless, some of the attempts to simulate depression have enjoyed a measure of success, so offering the opportunity to examine the effect of specific etiolo-

gical factors on, for example, symptom patterns, treatment efficacy and physiological substrates.

The major group of animal models of depression are based on responses to stressors of various kinds (see Section 5.3.2). The theoretical rationale underlying these models is usually derived from the well-established finding that the risk of depression is increased substantially by a stressful life event (Brown & Harris, 1978, 1988; Lloyd 1980*b*). However, it is important to recognize that the adverse consequences of life events are present for a prolonged period of six to twelve months. A number of psychological processes have been described that mediate these long-lasting effects. In particular, life events may undermine the victim's sense of identity and can exacerbate ongoing life difficulties (Brown & Harris, 1988; Brown, 1989). Thus, life events should not be viewed as acute stressors. Indeed, in the case of bereavement, a DSM-III diagnosis of depression is explicitly excluded during the period of acute loss. From this perspective, it may be more appropriate to use chronic stress regimes, rather than acute stressors, to model the etiological role of life events.

Additionally, a number of factors have been identified that confer a long-lasting vulnerability to depression. The major vulnerability factor is an inadequate level of social support (Brown & Harris, 1978, 1988; Aneshensel & Stone, 1982; Brown, 1989), which to a large extent arises from inadequate socialization, owing to parental loss during childhood or to an unloving and discordant family environment (Rutter & Madge, 1976; Brown & Harris, 1978; Quinton *et al.*, 1984). These experiences can result in a lack of social skill and a low level of self-esteem, both of which increase the likelihood that stress will precipitate a depressive episode, and also impede recovery (Beck, 1967; Lewinsohn, 1974; Becker, 1979). From these starting points, a number of animal models of depression have been developed that are based on the adverse effects of social isolation (see Section 5.3.1). These models have largely ignored the complexity of the psychological consequences of childhood social deprivation. Neither have they explored the effects of early social deprivation on later social relationships, which appear to play a central role in the expression of vulnerability to depression. However, the development of more realistic models may not be impossible. Berger and

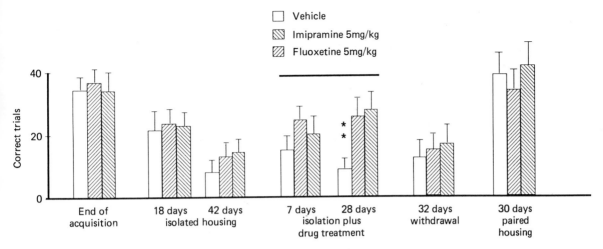

Fig. 5.1. *Male hooded rats were housed in pairs during training on the social cooperation task of Berger & Schuster (1982). At the end of acquisition, they were transferred to single housing and tested in their original pairs at 7–10 day intervals for 42 days, followed by 30 days of antidepressant treatment, 30 days of withdrawal, and a final 30 days during which they were returned to paired housing. The disruptive effect of social isolation on cooperative behaviour was reversed by drug treatment or by returning the animals to social housing (from Willner* et al., *1989).*

colleagues have described a model of social cooperation, in which pairs of rats are rewarded for running in a shuttle box, but only if they run together. This behaviour is readily learned by group-housed animals, but is disrupted by housing the animals singly (Berger & Schuster, 1982, 1987). We have recently shown that cooperative behaviour may be restored in singly housed animals by returning them to social housing, or alternatively, by treatment for four weeks with imipramine or the 5-HT uptake inhibitor fluoxetine (Willner *et al.*, 1989: see Fig. 5.1). The full potential of this model has yet to be explored.

In this review, I have attempted to indicate aspects of a model that detract from its validity, but not in any systematic way to identify the 'best' model. Given the heterogeneity of depression, which is manifest in its etiology and symptomatology, it comes as no surprise to find that this disorder may be simulated in a variety of different ways. Indeed, the diversity of animal models of depression may prove to be a particularly valuable source of theoretical insights. It follows that while there might be good reasons to reject certain models, in general, models based on widely different etiological assumptions should be seen as complementary, rather than as competitors for a 'best simulation' prize.

5.1.3.2 Monoamines and depression

Psychobiological research on depression has traditionally concentrated on the neurotransmitters noradrenaline (NA) and 5-HT (see Willner, 1985). Despite the prominence of NA in theoretical accounts of depression our understanding of the relevant role of NA in brain function remains woefully inadequate. It has not helped matters that there has been a prolonged controversy over whether NA function in depression is decreased, as originally proposed (Schildkraut, 1965), or increased, as suggested by the evidence that chronic treatment with antidepressants 'downregulates' beta-adrenergic receptors (Sulser, 1978). However, the electrophysiological and neuroendocrine markers of depression described earlier (Section 5.1.1.1) are compatible with a decrease in NA function (see Willner, 1987), and there is little evidence that beta-receptor 'downregulation' is of functional significance in the action of antidepressants (see Willner, 1989a).

This said, the relevance to depression of a decrease in NA function remains more obscure than one would wish. The major forebrain NA projection, the dorsal bundle, appears to be involved primarily in attention and the efficient processing of sensory information (Oades, 1985).

An early suggestion that NA may function as a 'reward transmitter' (Stein, 1968) has been largely discounted (see, e.g., Gray, 1982), though recent evidence does suggest that the dorsal NA bundle may play a more subtle role in the perception of reward (Morley *et al.*, 1987). The major link between NA and depression is found in the stress literature: there is considerable evidence that stress can cause a substantial depletion of brain NA, and that this effect mediates in part the adverse behavioural consequences of stress (reviewed by Anisman & Zacharko, 1982; Willner, 1987); attentional deficits seem at least in part responsible (Minor *et al.*, 1988). Under some circumstances brain NA levels may be decreased by relatively mild stressors (Anisman & Zacharko, 1982; Minor *et al.*, 1988). However, NA depletion usually occurs only after the administration of unrealistically high levels of stress (see, e.g., Weiss *et al.*, 1981), which suggests that this effect may actually be less relevant to depression than to post-traumatic stress disorder.

The role of 5-HT in depression is also problematic, although the difficulty here is more one of developing an appropriate theory from among several available alternatives. Treatments that reduce central 5-HT function have frequently been found to cause anxiolytic-like effects (reviewed by Soubrie, 1986), leading to the hypothesis that 5-HT mediates responses to aversive stimuli (Wise *et al.*, 1970). The anxiolytic activity of 5-HT_2 receptor antagonists has been confirmed clinically (see Lader, this vol., Ch. 4). However, recent evidence suggests that in addition to their antidepressant activity, 5-HT uptake inhibitors, which increase 5-HT transmission (Willner, 1985), may also be effective anxiolytics, at least in panic disorder (Westenberg & Den Boer, 1988) and obsessive-compulsive disorder (Goodwin *et al.*, 1988). Deakin (1989; this vol., Ch. 7) has suggested that an interaction between different subtypes of 5-HT receptor might provide a resolution of this paradox, such that stimulation of 5-HT_2 receptors is aversive, but stimulation of 5-HT_1 receptors is anti-aversive. There is some empirical support for this view (Giral *et al.*, 1988; Soubrie, 1989).

However, this line of argument is clearly of greater immediate relevance to anxiety rather than to depression. In fact, the clinical evidence suggests strongly that similar disorders of 5-HT function cut a broad swathe across diagnostic boundaries, which is probably better summarized as 'pathologically impulsive behaviour' than as any particular disorder (Soubrie, 1986; Willner, 1989*b*). Impulsive behaviour may be viewed as an inability to defer gratification, which is usually associated with an immature level of socialization. The fact that social isolation in animals reduces forebrain 5-HT function (Garattini *et al.*, 1967; Segal *et al.*, 1973), raises the intriguing possibility that the 5-HT dysfunctions observed in depressed patients might be related to their social inadequacies (Willner, 1989*b*). It is striking that performance in the rat social cooperation paradigm is severely impaired by lesions of the forebrain 5-HT projections (Berger & Schuster, 1987) or by the 5-HT antagonist metergoline (Willner *et al.*, 1989*a*) and may be restored by fluoxetine, a 5-HT uptake inhibitor (Willner *et al.*, 1989*a*; see Fig. 5.1) or by fluprazine, a 5-HT_1 agonist (Berger & Schuster, 1987).

Although largely absent from earlier formulations, the potential importance of a third neurotransmitter, dopamine (DA) has more recently been highlighted (Randrup *et al.*, 1975; Willner, 1983, 1985). It is now well established that brain DA systems, particularly the mesolimbic projections from the ventral tegmental area of the midbrain to the 'limbic forebrain' and prefrontal cortex, are crucially involved in motivated behaviour. Although it is not yet possible to specify precisely the functions of the mesolimbic DA system, it is clear that this system plays a central role in reward processes and/or in incentive motivation (Wise, 1982; Beninger, 1983; Salamone, 1987). These properties make the mesolimbic DA system a prime suspect in a disorder characterized by anhedonia and apathy.

There appear to be two major reasons why DA has previously been overlooked in theoretical accounts of depression. Firstly, tricyclic antidepressants appeared to have little effect on the functioning of DA neurons. However, we now know that after chronic administration, tricyclics do significantly influence DA function in the mesolimbic system, by increasing the responsiveness of postsynaptic receptors (see Willner, 1989*a*; Danysz *et al.*, this vol., Ch. 6). Secondly, when the precursor of DA, L-DOPA, was administered to depressed patients, there were frequent switches to mania or hypomania, but little evidence of mood elevation (Murphy *et al.*, 1971). However, the poor antidepressant effect of L-DOPA may have been over-interpreted, since, unlike tricyclics, L-DOPA does not increase DA function preferentially in the mesolimbic system. Thus, it is possible that a potential antidepressant effect of L-DOPA may be swamped by motor-stimulant effects mediated through the nigro-striatal DA system. A very selective DA uptake inhibitor, GBR-12909 (van der Zee *et al.*, 1980), which is active in the learned helplessness model of depression (Nielsen *et al.*, 1987) is currently undergoing clinical trials as an antidepressant, the results of which may resolve this issue.

Other neurotransmitter systems have received little theoretical attention in relation to depression. However, anticholinergics, opiate agonists and GABAergic drugs appear to be active in some animal models of depression, and in each case, there is clinical support for the suggestion that these compounds may possess some antidepressant activity (see Willner, 1985, and this vol., Ch. 18).

5.2 NON-SIMULATIONS

A number of the models reviewed in this chapter are, effectively, non-starters, as far as simulating depression is concerned (though they might well be of value as antidepressant screening tests). These models are behaviours that have been discovered empirically to respond to antidepressants, but which, considered as simulations of depression, either are totally lacking in face and construct validity, or have claims to validity that do not stand up to critical examination. They will be considered here briefly; those models that in practice are useful as screening tools are examined in greater detail in Chapter 6 (Danysz *et al.*).

5.2.1 Pharmacological interactions

One group of non-simulations involves the interaction of antidepressants with drugs of other pharmacological classes. Numerous tests of this kind have been proposed as antidepressant screening procedures (see Danysz *et al.*, this vol., Ch. 6); indeed, for many years, the only animal models of depression were of this type (see, e.g., Hill &

Tedeschi, 1971). In general, these procedures serve as bioassays for specific neurochemical actions of antidepressants, and it has not been seriously proposed that they simulate depression. However, this claim has been made in three cases, which therefore merit discussion.

5.2.1.1 Reserpine reversal

The earliest animal model of depression was the reversal of the behavioural and physiological effects of the monoamine-depleting drugs reserpine and tetrabenazine. The syndrome induced by reserpine-like drugs is characterized by ptosis (eye closure), hypothermia and catalepsy. The reversal of ptosis and hypothermia by pretreatment with tricyclic antidepressants and MAOIs is very well established, and was the first clear demonstration of a difference in pharmacological activity between tricyclic antidepressants and neuroleptics (which potentiate reserpine) (Costa *et al.*, 1960; Howard *et al.*, 1981). However, the test fails to detect some newer antidepressants, which differ structurally from the tricyclics and MAOIs, such as mianserin, trazodone or the selective 5-HT uptake inhibitors. Conversely, a wide range of non-antidepressants are detected by the test (see, e.g. Colpaert *et al.*, 1975; Danysz *et al.*, this vol., Ch. 6).

If the predictive validity of this test is poor, its face validity is worse. The claim to face validity rests upon two foundations; that reserpine induces depression in people, and that reserpine-induced catatonia is normalized by antidepressant drugs. The first of these claims is questionable: despite the many published studies of supposed reserpine-induced depressions that appeared in the 1950s and 1960s, the incidence of true depressions in these studies may actually have been as low as the 5–10% of patients who had a prior history of depression (Goodwin *et al.*, 1972). The second claim, that reserpine-induced catatonia is normalized by antidepressants, is simply false. While reversal of the physiological effects of reserpine-like drugs is always reported, antidepressants frequently fail to reverse the catatonia (e.g. Colpaert *et al.*, 1975). And when antidepressant-treated animals do awaken from their drug-induced stupor, it is to take up a highly stereotyped and abnormal behaviour, consisting of continuous sniffing and incessant, inexorable forward locomotion, which continues unabated for a period of hours (Willner & Clark, 1978).

It has been clear for some time that reserpine reversal is actually a bioassay for adrenergic stimulant effects of antidepressants. In fact, the test may include three independent bioassays, since reserpine-induced ptosis, hypothermia and catalepsy are reversed, respectively, by alpha-adrenergic or serotonergic agonists, beta-adrenergic agonists, and DA agonists (Bourin *et al.*, 1983).

5.2.1.2 *Amphetamine potentiation*

The potentiation of a variety of behavioural and physiological effects of amphetamine is another classic antidepressant screening test (Halliwell *et al.*, 1964). Lever pressing for brain stimulation reward is among the behaviours so potentiated, and it is this effect which appears to lend the test some degree of face validity (Stein, 1962). However, this appearance is spurious, since the effects depend primarily on the fact that antidepressants impair the metabolism of amphetamine in the liver, which increases the dose of amphetamine reaching the brain (Sulser *et al.*, 1966). Not surprisingly, this action is shared by representatives of numerous other classes of drug, while many newer antidepressant agents, structurally dissimilar to the tricyclics, do not potentiate amphetamine. It has been claimed that while drug effects on the duration of methamphetamine-induced locomotor stimulation are extremely variable, only antidepressants amplified the intensity of the response (Dall'Olio *et al.*, 1986). However, it has not been demonstrated that this more specific interaction is of central origin, or that similar results may be obtained using brain stimulation reward.

5.2.1.3 *Reversal of 5-HTP-induced behavioural depression*

A more recent drug interaction model involves the reversal by antidepressant drugs of the suppression of operant responding by 5-HTP, the precursor of 5-HT. 'Behavioural depression' in rats working for milk reinforcement was attenuated by acute pretreatment with imipramine, amitriptyline, iprindole, mianserin or trazodone (Nagayama *et al.*, 1980, 1981; Aprison *et al.*, 1982). However, chlori-

mipramine, DMI and zimelidine were ineffective (Hintgen *et al.*, 1985; Nagayama *et al.*, 1986), and behavioural depression was potentiated by the 5-HT uptake inhibitor fluoxetine (Aprison *et al.*, 1982), which like the other ineffective drugs, is an effective antidepressant (Asberg *et al.*, 1986); conversely, the most potent blocker of behavioural depression was the 5-HT-receptor antagonist methysergide, which is not known to have antidepressant properties (Nagayama *et al.*, 1980, 1981). This model has been advanced as a simulation of depression, on the basis that animals achieve fewer rewards when pretreated with 5-HTP. However, the suppression of operant performance by 5-HTP is secondary to a general behavioural sedation, and in fact, like many other effects of 5-HTP, is mediated peripherally (Carter *et al.*, 1978; Leander, 1986). It seems more appropriate to see this model as simply a biossay for the antagonistic effects of antidepressants at (peripheral) 5-HT receptors.

5.2.2 Miscellaneous models

The second set of non-simulations consists of a diverse group of unrelated, somewhat bizarre, and largely unconvincing, behavioural procedures.

5.2.2.1 *Predatory behaviour*

One of the earliest empirical models of depression was based on the discovery that the tricyclic antidepressant imipramine blocked the mouse-killing behaviour of rats (muricide), which occurs spontaneously in a small proportion of rats, but with a far higher incidence following various forms of brain damage (Horovitz, 1965; Albert & Walsh, 1982). In a related model, tricyclics have also been found to block the attack on an anaesthetized rat elicited by hypothalamic stimulation in cats (Dubinsky & Goldberg, 1971). The effect of imipramine on muricide is shared by other tricyclics, MAOIs, electroconvulsive shock (ECS), and some atypical antidepressants, including mianserin, which is missed by many of the pharmacological screening procedures (Delina-Stula & Vassout, 1979; Ueki, 1982). However, one study found that only a minority of antidepressants blocked muricide at doses significantly lower than those which caused motor debilitation (Sofia, 1969), and a number of other classes of drug are also effective, including psychomotor stimulants, and some anti-cholinergics and antihistamines (Horovitz, 1965; Horovitz *et al.*, 1966; Barnett & Taber, 1971). The test therefore has only moderate predictive value.

In the original study, tricyclics were found to suppress muricide when applied to the medial amygdala (Horovitz, 1965). Chronic treatment has been found to potentiate the anti-muricidal effects of tricyclics, applied systemically or to the amygdala (Shibata *et al.*, 1984). While some studies suggest a noradrenergic mechanism (Shibata *et al.*, 1984), most of the evidence indicates that tricyclics suppress muricide by a stimulant effect at 5-HT receptors (Albert & Walsh, 1982; Eisenstein *et al.*, 1982).

5.2.2.2 *Amygdaloid seizure activity*

Following the initial observations in the muricide test, a focus on the amygdala as a site of anti-depressant action has been a recurring theme in a number of other models. In a more direct examination of antidepressant effects on the amygdala, it has been claimed that tricyclic antidepressants suppress electrical and behavioural seizure activity elicited by electrical stimulation of the amygdala, at doses lower than those required to suppress cortical seizures. Anticonvulsant and anxiolytic drugs also suppressed seizure activity but did not discriminate between amygdala and cortex (Babington & Wedeking, 1973). However, iprindole, mianserin and MAOIs were ineffective in this test (Babington, 1981), and the selective action of tricyclics in the amygdala has also been questioned (Knobloch *et al.*, 1982).

5.2.2.3 *Waiting behaviour*

Thiebot and colleagues have shown that if rats are offered a choice, in a T-maze, between a small immediate reward and a larger delayed reward, they are more likely to wait for the larger reward if pretreated acutely with a variety of antidepressants, including tricyclics, 5-HT uptake inhibitors and an MAOI (Bizot *et al.*, 1988*a*). Anxiolytics had the opposite effect (Thiebot *et al.*, 1985). As in other models, inhibition of 5-HT synthesis had an anxiolytic-like effect (decreased waiting capacity), while certain 5-HT$_1$ receptor agonists were antidepressant-like (Bizot *et al.*, 1988*b*).

A related paradigm measures waiting capacity using the 'differential reinforcement of low rate' (DRL) operant schedule. In this paradigm, rats must space their responding, because only responses that follow a predetermined delay will be rewarded. If the delay is sufficiently long (72 s), antidepressants, administered acutely, are found to increase reinforcement rates: this effect is shared by tricyclics, MAOIs, ECS, and many atypical antidepressants including iprindole, mianserin, trazodone, fluoxetine and zimelidine, though the stimulant antidepressants nomifensine and buproprion were ineffective (O'Donnell & Seiden, 1982, 1983; Seiden *et al.*, 1985). While it is claimed by Seiden and colleagues that this procedure is insensitive to various non-antidepressants, there is some evidence, from studies with the neuroleptic haloperidol, that a similar effect might occur with any drug that causes a modest decrease in response rate (Pollard & Howard, 1986). However, Seiden's group have been unable to repeat these observations (Seiden, personal communcation).

The effects of antidepressants in this model may be blocked by neurotoxic lesion of forebrain 5-HT pathways, or by nonspecific 5-HT receptor antagonists. However, specific 5-HT$_2$ antagonists did not block the effects of antidepressants, and in fact, were themselves antidepressant-like. These results suggest that the antidepressant effects were mediated through 5-HT$_1$ receptors, and indeed, 5-HT$_1$ agonists were antidepressant-like in this test (Seiden, 1988). Thus, the data from both of these 'waiting' paradigms are consistent in suggesting that antidepressant drugs may have 5-HT$_1$-mediated 'anti-impulsive' effects.

5.2.2.4 *Circadian rhythm models*

A number of studies have reported effects of antidepressants on circadian rhythmicity. The first of these was the observation that following reversal of the light-dark cycle, rats readjusted to a normal circadian cycle of locomotor activity more rapidly if they received chronic antidepressant treatment (imipramine, maprotiline or pargyline) before and after the phase-shift; chlordiazepoxide, chlorpromazine, reserpine and amphetamine were all ineffective (Baltzer & Weiskrantz, 1973). Imipramine and clorgyline have also been found to cause a lengthening of the circadian period of hamsters shifted from a normal light-dark cycle to constant darkness (Goodwin *et al.*, 1982). A third model is based on the suppression by a variety of anti-depressants of rapid eye movement (REM) sleep in cats (Scherschlicht *et al.*, 1982); amphetamine and morphine had a similar effect, but also suppressed non-REM sleep (though the only other non-antidepressant for which results were reported, phenobarbital, had an effect similar to the anti-depressants).

Little work has been carried out using these models, and at first sight, their relevance appears obscure. However, disturbance of circadian rhythms does seem to be a characteristic feature of depression; a decrease in the latency of the first period of REM sleep is well established (Kupfer, 1976; Akiskal, 1980), and a phase-advance of many other circadian rhythms has also been observed (Goodwin *et al.*, 1982). Some workers have proposed that changes in circadian rhythms may be of etiological significance in depression (Wehr & Wirz-Justice, 1982), or even that depression should be considered to be primarily a disorder of circadian rhythmicity (Healy & Williams, 1988). This point of view has not gained wide acceptance – but should it do so, then circadian rhythm models would acquire a more prominent status.

5.3 SIMULATIONS OF DEPRESSION

We now turn to those models that to some extent do succeed in simulating depression. Some of these models are also suitable for consideration as anti-depressant screening tests, while in others, the complexity of the procedures rules them out for this purpose. Three groups of simulations may conveniently be distinguished. The first group consists of models in which abnormal behaviours are engendered by social separation. The second group consists of stress-induced impairments of locomotor activity or of behaviour in aversive situations; in certain of these models responding for rewards is also reduced. In the final group, abnormal behaviour is engendered by brain damage.

5.3.1 Separation models

For many authors (e.g. Everitt & Keverne, 1979; Howard *et al.*, 1981), the only worthwhile animal models of depression are those involving separa-tion phenomena in non-human primates; the evolutionary proximity of primate species seems to afford intuitive insights into their behaviour that are lacking in less closely related animals. In fact, separation phenomena of 'protest' followed by 'despair' are present to some extent in many species, including cats, dogs, rodents and precocial birds (reviewed by McKinney & Bunney, 1969; Katz, 1981*a*). In addition to the classic primate studies, separation phenomena have been used as a basis for the development of animal models of depression in three other species: rats, hamsters and chicks.

5.3.1.1 *Primate models*

In infant monkeys, the response to maternal separ-ation consists of an initial stage of 'protest', char-acterized by agitation, sleeplessness, distress calls and screaming, followed after one or two days by 'despair', characterized by a decrease in activity, appetite, play and social interaction, and the assumption of a hunched posture and 'sad' facial expression (Kaufman & Rosenblum, 1967; McKinney & Bunney, 1969; Suomi, 1976; Hinde *et al.*, 1978). Similar phenomena are also observed when group-reared animals are isolated from their peers (Suomi *et al.*, 1970; Bowden & McKinney, 1972; Kraemer & McKinney, 1979). The nature of the separation response is sensitive to the environ-ment in which the experiments are carried out, however (e.g. Suomi, 1976; Reite *et al.*, 1981), and the incidence of 'depressive' behaviours may in some experiments be as low as 15% (Lewis *et al.*, 1976).

Very few published studies have attempted to use antidepressant treatments to modify primate separation behaviour. Chronic DMI has been found to increase social contact and decrease dis-tress vocalization and self-oriented behaviours in maternally-separated infant macaques (Hrdina *et al.*, 1979). Similarly, chronic imipramine was found to decrease self-clasping in peer-separated infant rhesus monkeys; however, acute treatment had the opposite effect, and other separation-induced behavioural changes were unaffected (Suomi *et al.*, 1978). A partial response to ECS in isolated rhesus monkeys has also been reported (Lewis & McKinney, 1976). Trifluoperazine, amphetamine and diazepam were not found to affect responses to social isolation in chimpanzees (Menzel *et al.*, 1963; Turner *et al.*, 1969), but some therapeutic effects of chlorpromazine were seen in rhesus monkeys (McKinney *et al.*, 1973). All in all, the results of these pharmacological studies have not been impressive: though it should be added that while antidepressant drugs are frequently used in childhood depression, their efficacy has not been convincingly demonstrated (Kashani *et al.*, 1981; Pearce, 1981). While further pharmacological data are essential if these models are to be validated, the expense of using primates, and the consequent small group sizes in most studies, make these paradigms unsuitable for most practical purposes, such as the investigation of the mechanisms of antidepressant action.

The primate separation response shows a marked similarity to the state of 'anaclitic depression', first described by Spitz (1946) and Robertson & Bowlby (1952): institutionalized children showed the same sequence of protest (agitation, crying, insomnia, and oral stereotypies) followed in approximately 15–20% of cases by despair (retardation, self-clasping, withdrawal from social contact, and an increase in the likelihood of succumbing to disease). Unfortunately, these similarities are no guarantee of face validity, since the relationship between infantile anaclitic depression and adult depression remains uncertain (Schulterbrandt & Raskin, 1977). Furthermore, while it is now reasonably well established that under certain circumstances, the loss of a loved one in childhood predisposes to adult depression (Lloyd, 1980*a*), it is not clear that this fact, in itself, helps to validate the primate separation models, since these models examine behaviour in the immediate aftermath of loss. However, it is of interest in this context that in animals with a history of separation, the catecholamine synthesis inhibitor alpha-methylparatyrosine (AMPT) produced 'depressive' responses at doses substantially lower than those needed in control subjects (Kraemer & McKinney, 1979).

The assumption that separation from a loved one is a significant precipitant of adult depression is also open to question. A clear relationship exists in the case of bereavement, but the incidence of clinical depression following the mourning period may be as low as 5% (Parkes, 1972), and bereavement also precipitates a wide range of other psychiatric disorders (Brown *et al.*, 1973), as well as a

variety of non-psychiatric medical conditions (Schmale, 1973), particularly coronary thrombosis (Parkes, 1972). In the case of marital breakdown, it has not been established whether separation precipitates depression, or conversely, whether a prior depression in one partner was the cause of separation (Briscoe & Smith, 1975). Thus, the theoretical formulation of depression as a phase of the protest-despair cycle, which is assumed by the model, is tenuous. (This criticism also applies, of course, to non-primate separation models.) An alternative formulation is that social separation models the absence of social support (see Willner, 1989*b*), which has been identified as another significant predisposing factor in adult depression (e.g. Williams *et al.*, 1981; Aneshensel & Stone, 1982). What is clear is that the construct validity of separation models is less well-established than has often been assumed.

5.3.1.2 *Distress calling in isolated chicks*

A very simple 'protest–despair' model has recently been described in week-old chicks. On isolation, chicks produce distress calls; typically, these subside in frequency after the first hour. Distress calling at a high frequency was prolonged by a wide range of antidepressants, including tricyclics, MAOIs, 5-HT uptake inhibitors, and most other atypical antidepressants. The only false negative was the antidepressant beta-receptor agonist salbutamol (Lecrubier *et al.*, 1980), while the only false positive was chlorpromazine. Distress calling was not reactivated by anxiolytics, anticholinergics or stimulants (Lehr, 1986, 1989, and personal communication).

5.3.1.3 *Separation in pair-bonded hamsters*

In pair-bonded Siberian dwarf hamsters, separation has been reported to produce a characteristic syndrome of increased body weight, decreased exploratory behaviour, and decreased social interaction with an unfamiliar opposite-sex animal; the effects were greater in males than in females. However, it has not yet been established whether this 'separation syndrome' results from disruption of the pair bond, or simply from social isolation. Imipramine reversed some, but not all, of the behavioural changes; the syndrome was unaffected by chronic treatment with clorgyline, a specific inhibitor of MAO-A, but totally reversed by tranylcypromine, a non-specific MAOI (Crawley, 1984, 1985).

5.3.1.4 Social isolation in rats

Social isolation also has marked effects in rats. Isolation at two to three weeks of age causes hyperactivity, relative to group-reared controls (Sahakian *et al.*, 1975), whereas isolation of two-month-old animals results in aggression towards the experimenter and muricide (Valzelli & Bernasconi, 1971). It has been found that the activity difference after isolation for a year between (early) isolated and group-reared animals was abolished by acute treatment with a range of tricyclic antidepressants, MAOIs, atypical antidepressants (mianserin, iprindole, nomifensine, viloxazine, trazodone), and the beta-receptor agonist salbutamol. The 5-HT-receptor blocker and antihistamine cyproheptadine, for which antidepressant activity has not been reported, and seems unlikely, was also effective. However, neuroleptics and anxiolytics did not abolish the activity difference between isolated and group-reared animals except at neurotoxic doses (Garzon *et al.*, 1979; Garzon & Del Rio, 1981). Anticholinergic drugs have not been tested in this model, while conflicting data have been obtained with amphetamine and the beta-blocker propranolol (Sahakian *et al.*, 1975; Weinstock *et al.*, 1978; Garzon *et al.*, 1979; Garzon & Del Rio, 1981). Otherwise, the model appears to respond specifically to antidepressants. However, the need to isolate animals for a year prior to testing, combined with the fact that the effects of isolation were abolished by handling (Garzon & Del Rio, 1981), severely restricts the usefulness of this model as a screening test.

Apart from hyperactivity, which is seen in a significant proportion of depressions (Kupfer & Detre, 1978), there is little information on which to judge the validity of this model as a simulation of depression. One potential problem is that in operant tasks, isolated animals may show greater persistence (Morgan *et al.*, 1975), which depressed people certainly do not (Weingartner & Silberman, 1982). However, as noted earlier (Section 5.1.3.1), chronic (4–6 week) social isolation of mature rats also causes a disruption of cooperative social

behaviour (Berger & Schuster, 1982), reminiscent of the poor social performance of depressed people (Lewinsohn, 1974), which may be reversed by chronic antidepressant treatment (Willner *et al.*, 1989; see Fig. 5.1). In a preliminary study, the effect of imipramine in this model was abolished by the 5-HT antagonist metergoline (Willner *et al.*, 1989). At a higher dose, metergoline also impaired cooperative performance in group-housed animals (Willner *et al.*, 1989). However, the specific 5-HT$_2$ antagonists ritanserin and ketanserin did not (Willner, Muscat & Turner, unpublished observations). These observations suggest a specific role for 5-HT$_1$ receptors in cooperative behaviour; the therapeutic effect in isolated animals of the 5-HT$_1$ agonist fluprazine (Berger & Schuster, 1987) provides support for this view.

5.3.2 Stress models

The largest group of animal models of depression involve abnormal behaviours engendered by stress. A number of theoretical approaches have been developed which attempt to explain the supposed etiological effect of stress in depression (see, e.g. Depue, 1979; Anisman & Zacharko, 1982). It should be remembered, however, that the definition of 'stress' in relation to human studies remains a topic of endless debate (e.g., Anisman & Zacharko, 1982), and a number of fundamental questions are unanswered, including the accuracy with which depressed individuals recall stressful life events (Lishman, 1972), the possibility that stress may simply provoke hospitalization in already depressed people (Hudgens *et al.*, 1967), and the provocative thought that the life style of depressives may be responsible for many of the stresses they experience (Beck & Harrison, 1982).

With these qualifications, there is general agreement that the likelihood of entering an episode of depression is increased five- or six-fold in the six months following the occurrence of stressful 'life events' (Paykel, 1979; Lloyd, 1980*b*; Fava *et al.*, 1981). Evidence is also accumulating to support the commonsense notion that chronic low grade stress may be a powerful predisposing factor (Kanner *et al.*, 1981; Billings *et al.*, 1983). Unemployment, for example, carries a high risk of depression (Jahoda, 1979; Winefield *et al.*, 1987);

and in another study, almost 40% of subjects reporting a high level of marital, financial or work-related stress were clinically depressed (Anashensel & Stone, 1982). The following review covers four models based on the behavioural sequelae of acute stress, and four models in which behavioural abnormalities are induced by chronic stress.

One of the most significant effects of stress, observed in a number of the models described below, is a reduction in the performance of rewarded behaviour. The hypothesis that depression results from a reduction in the activity of reward systems is central to a number of theories of depression (Stein, 1962; Costello, 1972; Ferster, 1973; Lewinsohn, 1974), and a failure to respond to normally pleasurable events is the defining feature of endogenous depression (melancholia) (Klein, 1974; American Psychiatric Association, 1980; Nelson & Charney, 1981). The demonstration of a reduced sensitivity to reward therefore invests a model with some degree of construct validity.

5.3.2.1 *Learned helplessness*

The learned helplessness model is the most familiar simulation of depression, and also the most controversial. The phenomenon of learned helplessness, originally described by Seligman and co-workers in dogs, and subsequently extended to a large number of other species, including people, is that exposure to uncontrollable stress produces performance deficits in subsequent learning tasks, which are not seen in subjects exposed to the identical stressor but able to control it (reviewed by Seligman, 1975; Maier & Seligman, 1976; Garber *et al.*, 1979; Maier, 1984; Willner, 1986). In addition to their passivity in the face of stress, 'helpless' animals (typically, rats or mice) show a variety of other behavioural changes, including decreased locomotor activity, poor performance in appetitively motivated tasks (including intracranial self-stimulation (ICSS)), decreased aggression, and loss of appetite and weight (reviewed by Weiss *et al.*, 1982; Maier, 1984; Willner, 1986). Indeed, the range of symptomatic parallels between learned helplessness and depression is so striking that it has been suggested that 'helpless' animals would meet DSM-III criteria for major depression (Weiss *et al.*, 1982).

If we take this claim seriously, it is probably incorrect, for most of the effects of uncontrollable shock are transient, with recovery within two to three days (Weiss *et al.*, 1982; Maier, 1984), whereas a DSM-III diagnosis of depression requires at least two weeks of abnormal symptomatology (American Psychiatric Association, 1980). Recovery from learned helplessness, which appears to result from activation of the processes responsible for adaptation to stress (Anisman & Zacharko, 1982; Willner, 1987), severely limits the usefulness of this model. Nevertheless, under certain circumstances, uncontrollable shock does have long-lasting effects (Glazer & Weiss, 1976). Very little of the learned helplessness research has used chronic paradigms. However, Desan *et al.* (1988) recently reported that inescapable shock causes a long-term (seven-week) depression of home-cage locomotor activity, which could be reversed by chronic treatment with DMI.

For a model that has received so much attention, learned helplessness has until recently attracted surprisingly few studies of antidepressant action. The learning deficit is reversed by subchronic treatment (three to seven days) with a variety of antidepressants, including tricyclics, MAOIs, atypical antidepressants and ECS (Dorworth & Overmeier, 1977; Leshner *et al.*, 1979; Petty & Sherman, 1980; Sherman *et al.*, 1982), while chronic treatment with neuroleptics, stimulants, sedatives and anxiolytics was ineffective (Sherman *et al.*, 1982). One important study reported that the suppression of intracranial self-stimulation by inescapable shock (Zacharko *et al.*, 1983) was also prevented by chronic treatment with the tricyclic DMI (Zacharko *et al.*, 1984). Acute antidepressant treatment sometimes fails to reverse helplessness (Petty & Sherman, 1980; Zacharko *et al.*, 1984), but in other studies, acute treatment was effective (Kametani *et al.*, 1983), particularly at high doses (Martin *et al.*, 1986a). The requirement for prolonged treatment appears, in this model, to represent nothing more than the delay in achieving adequate drug tissue concentrations (Petty & Sherman, 1980). Versions of the learned helplessness model using subchronic (two to three day) treatment regimes are becoming increasingly popular as antidepressant screening tests.

The symptomatic richness of the learned helplessness model has led to some discussion of just what is being modelled. Learned helplessness was originally advanced as a model of reactive depression (Seligman, 1975). However, the clinical features of helpless animals, including psychomotor retardation, loss of appetite, lack of aggression, sleep disturbance and corticosteroid elevation, are far more similar to the symptoms of endogenous depression (Nelson & Charney, 1981), perhaps the depressive phase of a bipolar illness (Depue & Monroe, 1978). The main problem with this formulation is that the defining feature of endogenous depressions is their failure to respond to psychosocial intervention (Depue & Monroe, 1978; Nelson & Charney, 1981). Helplessness does respond to psychosocial intervention, in that performance deficits may be overcome by demonstrating to the animal that its responding does produce shock termination (Seligman *et al.*, 1975; Williams & Maier, 1977). The solution to this problem is not yet clear, though endogenous depressions do apparently respond to cognitive therapy (Beck, 1983). Furthermore, anxiolytic treatment prior to the helplessness induction procedure prevented the development of behavioural deficits, while learned helplessness was induced by an anxiogenic drug (Drugan *et al.*, 1984, 1985), suggesting that anxiety may play a crucial etiological role (Jackson & Minor, 1988). Again, this is difficult to reconcile with a bipolar depression – though there is some evidence that bipolar depressives may suffer a particularly high level of chronic stress (Depue & Monroe, 1979).

The theoretical rationale of learned helpessness as a model of depression has usually been assumed to lie within the 'learned helplessness hypothesis of depression' (Seligman, 1975), and consists, in effect, of three assertions: that animals exposed to uncontrollable aversive events do become helpless; that a similar state is induced in people by uncontrollability; and that helplessness in people is the central symptom of depression. Each of these assumptions has been the source of intense controversy, and overall, the relevant literature tends not to validate the model. As this evidence has recently been discussed in detail elsewhere (Willner, 1986), it will be summarized here briefly.

First, do animals exposed to uncontrollable shock become 'helpless': do they perform badly in subsequent tasks because they have learned that their responses fail to control their environment (Seligman, 1975; Maier & Seligman, 1976)? A number of alternatives, and simpler, accounts of the performance deficits shown by inescapably shocked animals have also been advanced. Broadly, they argue that (for a variety of reasons) the performance deficits of inescapably shocked animals are caused by a reduction of motor activity (Glazer & Weiss, 1976; Anisman *et al.*, 1979). A number of studies have demonstrated convincingly that animals subjected to inescapable shock do suffer learning difficulties (Maier & Testa, 1975; Jackson *et al.*, 1978), which appear to result from an attentional impairment (Minor *et al.*, 1984, 1988). It is equally clear, however, that uncontrollable shock does also cause motor deficits (Maier & Jackson, 1979), and these remain a plausible explanation of the performance impairments seen in the vast majority of 'learned helplessness' experiments. Furthermore, the effects of uncontrollable shock may be overcome simply by providing signals at the onset or offset of shock; this suggests that the inability to predict shock (and shock-free) periods may be rather more important, as a cause of behavioural impairments, than the inability to exercise control (Jackson & Minor, 1988).

Secondly, do people exposed to uncontrollable stress become 'helpless'? Exposure to uncontrollable stress – typically, loud noise – or to insoluble problems – typically, anagrams – has been found to induce subsequent performance deficits in human subjects (Garber *et al.*, 1979; Garber & Seligman, 1980), but again, the interpretation is unclear. Following numerous discrepant findings, the learned helplessness theory as applied to people was reformulated. In the new version, it is asserted that subjects exposed to uncontrollable aversive events come to see themselves as helpless only if they attribute their failure to causes that are internal to themselves, global in scope, and stable in time (e.g. 'I am stupid') (Abramson *et al.*, 1978). The reformulated theory is able to explain most of the findings of human helplessness induction experiments (though not all – see Buchwald *et al.*, 1978; Wortman & Dintzer, 1978). However, the cost is a considerable increase in the difficulty of testing the theory, and an insuperable distancing from the animal literature.

Thirdly, does 'helplessness' cause depression? Since the reformulation of the learned helplessness hypothesis, this question has usually been investigated by asking whether people who tend to make 'depressive attributions' are at greater risk of depression. The evidence is clear; they are not (Lewinsohn *et al.*, 1981; Manly *et al.*, 1982; Winefield *et al.*, 1987).

We may conclude that the learned helplessness hypothesis of depression does not confirm the construct validity of the learned helplessness model. However, the alternative account of 'helplessness' as stress-induced motor inactivation (Weiss *et al.*, 1982) remains viable. The proper evaluation of this hypothesis must await a more formal theoretical conceptualization of the relationship between stress and depression.

In fact, two aspects of helplessness research may contribute to the development of such a theory. One is the overwhelming importance of uncontrollability (failure to cope) and/or unpredictability in determining the deleterious behavioural consequences of stress (Maier, 1984); the protective effect of control appears to result from lower levels of fear, suggesting an important role for anxiety in the genesis of the behavioural abnormalities in this animal model of depression (Jackson & Minor, 1988). The other is the considerable progress towards understanding the neurochemical bases of stress-induced inactivation (see Maier, 1984; Willner, 1987). Briefly, the debilitating effect of uncontrollable stress on later performance may be reversed by agonists at DA, alpha-1 and beta-adrenergic receptors or by anticholinergic drugs; conversely, helplessness may be simulated pharmacologically by drugs that reduce DA and/or NA function, or by drugs that increase cholinergic function (Anisman & Zacharko, 1982; Willner, 1987). Alpha-1 and beta-adrenergic antagonists (as well as the opiate antagonist naloxone) have also been found to block the therapeutic effect of tricyclics (Martin *et al.*, 1986*b*). In contrast, studies of the role of 5-HT in learned helplessness are inconsistent (Anisman *et al.*, 1979; Petty & Sherman, 1980; Brown *et al.*, 1982). 5-HT$_1$ agonists are effective in this model, as are 5-HT uptake inhibitors (within a narrow dose range (Giral *et al.*, 1988; Martin *et al.*, 1989); however, 5-HT appears not to be involved in mediating the actions of tricyclic antidepressants in this model (Soubrie *et al.*, 1986).

5.3.2.2 *Exhaustion stress*

Female rats, reared in revolving cages, show a cyclical activity pattern tied to the estrous cycle. Forced running in the wheel, to the point of exhaustion, killed about half the animals. Of the survivors, half resumed running within several days; the others, however, showed a very low spontaneous locomotor activity, with no cyclicity, for several weeks, accompanied by constant diestrous. Normal activity was restored by daily imipramine treatment (Hatotani *et al.*, 1982). Two features of the model are potentially of interest: the effect was all-or-none, only appearing in some of the subjects, and long-lasting. However, the face validity of the model is seriously compromised by the severity of the procedure. Also, the model is clearly unacceptable on ethical grounds.

5.3.2.3 *'Behavioural despair'*

In the very well known 'behavioural despair' test, mice or rats are forced to swim in a confined space; after an initially frenzied attempt to escape, they assume an immobile posture, and on subsequent immersion, the onset of immobility is much more rapid. This state has been named 'behavioural despair', on the assumption that the animals have 'given up hope of escaping' (Porsolt *et al.*, 1977, 1979; Porsolt, 1981), though some authors prefer the more theoretically neutral term 'the forced swim test'. The onset of immobility in the second swimming test is delayed by pretreatment with a wide variety of antidepressants, usually in a subacute treatment schedule consisting of three injections in 24 h. Effective agents include tricyclics, MAOIs, most atypical antidepressants (e.g. maprotiline, iprindole, buproprion, mianserin, nomifensine, viloxazine), ECS, and deprivation of REM sleep (see Willner, 1984; Borsini & Meli, 1988 for references). There is, in fact, a significant correlation between the potency of antidepressants in the 'behavioural despair' test and their clinical potency, which has not been demonstrated with any other animal model of depression (Willner, 1984).

However, certain antidepressants are ineffective in this test: for example, chlorimipramine, trazodone and the beta-receptor agonist salbutamol did not reduce immobility in the rat (Porsolt *et al.*,

1979; Porsolt, 1981). This probably indicates a lack of sensitivity to antidepressants that act predominantly as indirect 5-HT agonists: 5-HT uptake inhibitors are also usually ineffective (Porsolt *et al.*, 1979; Gorka *et al.*, 1979; Satoh *et al.*, 1984), while chlorimipramine, a tricyclic relatively selective for 5-HT, did reduce immobility in the mouse, in which species (as in people) it is metabolized to a potent NA uptake inhibitor (Nagy, 1977).

A further problem arises from the large number of non-antidepressants that also reduce immobility. While the test successfully discriminates antidepressants from neuroleptics and anxiolytics (Porsolt *et al.*, 1977), false positives have been reported for stimulants, convulsants, anticholinergics, antihistamines, pentobarbital, opiates and various other brain peptides, and a number of other drugs (see Willner, 1984, and Borsini & Meli, 1988 for references). Some of these effects are non-specific, however: stimulants and anticholinergics appear to reduce immobility by an indiscriminate stimulation of motor activity rather than by delaying the onset of immobility, and could be distinguished from antidepressants simply by prolonging the period of the test. Additionally, the responses to an antihistamine (Kitada *et al.*, 1981) and an anticholinergic (Kawashima *et al.*, 1986) were found to disappear on chronic administration. The generality of these effects has not yet been established.

The fact that the 'behavioural despair' test responds to acute drug treatment has been a frequent, but unjustified, source of criticism. While small antidepressant effects may be present after a single high dose (Kitada *et al.*, 1981; Zebrowska-Lupina, 1980), large effects are observed after chronic treatment at lower doses (Kitada *et al.*, 1981; Nomura *et al.*, 1984; Araki *et al.*, 1985; Kostowski, 1985). Potentiation by chronic administration of an antidepressant effect detectable early in treatment appears also to describe the clinical situation (Frazer *et al.*, 1985; Willner, 1989*a*). Unlike the situation in the learned helplessness model, the potentiation of the antidepressant effect in the 'behavioural despair' test by repeated treatment cannot be explained simply by the elevation of brain drug concentrations (Kitada *et al.*, 1981; Poncelet *et al.*, 1986; Mancinelli *et al.*, 1987).

While the predictive validity of the 'behavioural despair' test may be better than has sometimes been assumed, its face and construct validity are minimal. The only symptomatic resemblance to depression, other than the temporal characteristics of the antidepressant response, lies in the observation that 'behavioural despair' reflects an inability or reluctance to maintain effort, rather than a generalized hypoactivity; it is in tests that require the sustained expenditure of effort that depressed subjects show their most pronounced psychomotor impairments (Weingartner & Silberman, 1982). The theoretical rationale of this test derives entirely from its supposed relationship to learned helplessness: consequently, the problems discussed in relation to the construct validity of learned helplessness apply equally to the 'behavioural despair' model. In addition, the relationship between the two models is unclear. The role of inescapability in the genesis of immobility has been touched on in two studies (Altenor *et al.*, 1977; O'Neill & Valentino, 1982), but inconclusively (see Willner, 1984). Prior inescapable, but not escapable, shock has been found to increase immobility in the 'behavioural despair' test (Nomura *et al.*, 1982; Weiss *et al.*, 1981), but in view of the consistent finding of decreased motor activity following inescapable shock (see Anisman & Zacharko, 1982), it would be surprising were this not the case. The reciprocal finding has not been demonstrated: forced swimming in the 'behavioural despair' procedure has not been found to impair subsequent escape performance in a water maze (Porsolt, 1981), or in a shock-avoidance task (pressing a lever three times) in which performance deficits are typically seen following inescapable shock (O'Neill & Valentino, 1982). 'Behavioural despair' might be a milder version of learned helplessness, but this has not yet been demonstrated.

Nevertheless, the two tests do seem to share similar physiological substrates. Like learned helplessness, 'behavioural despair' is also reversed by stimulating DA or alpha-1 adrenergic receptors (Porsolt *et al.*, 1979; Plaznik *et al.*, 1985*a*, *b*) or by anticholinergics (Browne, 1979), and potentiated by treatments that decrease the activation of DA or NA systems (Porsolt *et al.*, 1979; Kitada *et al.*, 1983; Parale & Kulkarni, 1986) or increase cholinergic transmission (Overstreet *et al.*, 1986). Similarly, the therapeutic actions of antidepressants in this model are blocked by DA receptor antagonists (Borsini *et al.*, 1985; Pulvirenti & Samanin, 1986;

Cervo & Samanin, 1987) and by treatments that reduce NA function, including neurotoxic destruction of the ascending NA pathways (Araki *et al.*, 1984; Plaznik *et al.*, 1985*a*) and alpha-1 receptor antagonists (Borsini *et al.*, 1981; Kitada *et al.*, 1983). And as in the learned helplessness test, 5-HT does not appear to be involved either in 'behavioural despair' (Porsolt *et al.*, 1979; Plaznik *et al.*, 1985*a*, *b*), or in its reversal by antidepressants (Borsini *et al.*, 1981; Kitada *et al.*, 1983; Araki *et al.*, 1985).

5.3.2.4 *Variants on 'behavioural despair'*

A number of modifications of the 'behavioural despair' procedure have been described, including the provision of an escape route (a rope to climb) following the inescapable swim (Cools, 1980; Thornton *et al.*, 1985), and automation of the recording procedure by means of a water wheel, onto which the animals unsuccessfully attempt to climb (Shimizu *et al.*, 1984). However, the most significant of these developments is the 'tail suspension test' (Steru *et al.*, 1985; Porsolt *et al.*, 1986). In this model, mice suspended by the tail show a temporal pattern of struggling followed by immobility, similar to that seen in the forced swimming test. Antidepressants, at strikingly low doses, have been shown to increase the duration of mobility and also, in an automated version of the test, to increase the power of the movements (though the latter effect is rather less convincing). Effective agents include tricyclics, MAOIs, and atypical antidepressants; the latter include mianserin and, significantly, 5-HT uptake inhibitors, which are usually ineffective in the forced swim test. Immobility was also reduced by the stimulants amphetamine and caffeine, but was potentiated by neuroleptics or anxiolytics (Steru *et al.*, 1985; Porsolt *et al.*, 1986).

5.3.2.5 *Failure to adapt to stress*

The major problem in using chronic stress to generate a simulation of depression, is that rats will usually adapt, if stressors are presented repeatedly. However, the degree of adaptation may be subject to sex differences. Male rats responded to a single restraint stress by a decrease in locomotor activity and food intake, but adapted over a five-day test period; females, however, while having a smaller response to the first stress, failed to adapt (Kennett *et al.*, 1986). Potentially, this model addresses the important issue of the two- to three-fold greater incidence of depression in women (Weissman & Klerman, 1977); it is the only animal model of depression to do so.

Few pharmacological studies have been conducted using this model. In male rats, chronic treatment with DMI reversed some of the open field deficits, but did not reverse stress-induced anorexia; acute treatment with DMI or with anxiolytic drugs was without effect (Kennett *et al.*, 1987). Adaptation to restraint stress could, however, be promoted by inhibition of corticosterone synthesis (Kennett *et al.*, 1985). The suggestion that elevated corticosterone levels might be of etiological significance in depression finds clinical support in the observation that reducing corticosteroid levels is an effective treatment of the depressions associated with Cushing's disease (Kelly *et al.*, 1983). As in the learned helplessness model, 5-HT$_1$ agonists also promoted adaptation to stress (Kennett *et al.*, 1987; see also Deakin, this vol., Ch. 7).

5.3.2.6 *Chronic unpredictable stress*

Another model prevents adaptation to repeated stress by presenting a variety of stressors in an unpredictable sequence. Following three weeks of exposure to electric shocks, immersion in cold water, immobilization, reversal of the light/dark cycle and other stressors, rats were exposed to loud noises and bright lights, followed immediately by an open field test; chronically stressed animals failed to show the usual increase in locomotor activity. However, the activating effect of an acute stress was maintained in animals receiving daily antidepressant treatment during the chronic stress period. This effect was observed with tricyclics, an MAOI (pargyline), atypical antidepressants (iprindole, mianserin, buproprion) and ECS. A neuroleptic, an anxiolytic, an antihistamine, an anticholinergic and a stimulant were ineffective; so, also, however, was the MAOI tranylcypromine. In addition to causing changes in open field activity, chronic stress also increased plasma corticosteroid levels. This effect showed the same spectrum of pharmacological sensitivity, with the exception that the anticholinergic scopolamine was also effective (Katz, 1981*b*; Katz & Hersh, 1981; Katz *et al.*, 1981*a*, *b*; Roth & Katz, 1981; Katz & Baldrighi, 1982; Katz & Sibel 1982*a*, *b*). Similar effects have also been reported in mice; corticosteroid levels and the response to an acute stress were normalized by tricyclic antidepressants, but not by fluoxetine (Soblosky & Thurmond, 1986). A further effect observed in rats after chronic stress was a failure to increase fluid consumption when saccharin was added to the drinking water; this deficit was partially restored by imipramine (Katz, 1982).

This model would appear to have a substantial degree of face validity, since the effects observed – increased corticosteroid levels, a lack of reactivity to an acute stress, and a failure to respond to a (presumably) pleasurable stimulus are all central symptoms of endogenous depression (Carroll, 1978; American Psychiatric Association, 1980). Additionally, the stress regime employed in these experiments, though unnecessarily severe, appears a somewhat more realistic analogue of the stress of living than a single session of either electric shock or water immersion. However, the model involves prophylactic treatment; it has not been demonstrated that antidepressants can reverse an established behavioural deficit in this model.

There are two sources from which the model might derive construct validity. One is the learned helplessness literature; to the extent that the model is related to learned helplessness, it is subject to the problems previously discussed in relation to that model. However, an alternative theoretical rationale derives from the literature relating depression to psychological stress: in particular, the evidence that while the contribution of major 'life events' to the etiology of depression may be relatively small (Lloyd, 1980*b*), the contribution of a high level of chronic stress may be substantial (Aneshensel & Stone, 1982).

5.3.2.7 *Chronic unpredictable mild stress*

A variant of the chronic stress model employs very mild stressors, such as periods of food and water deprivation, small temperature reductions, changes of cage mates, and other similarly innocuous manipulations. Over a period of weeks of exposure to the mild stress regime, rats gradually reduced their consumption of a highly preferred

sucrose solution, and in untreated animals, this deficit persisted for several weeks following the cessation of stress. Antidepressant treatment had no effect in nonstressed animals, but following the reduction of sucrose intake by stress, normal behaviour was restored by chronic treatment (two to five weeks) with DMI, imipramine or amitriptyline. The results could not be explained by stress-induced motor deficits or by a 'sweet tooth' effect of the antidepressants; rather, the behavioural changes appear to reflect changes in sensitivity to the reward (Willner *et al.*, 1987, 1988; Muscat *et al.*, 1988; Sampson *et al.*, 1990). This model is unique among the stress models in demonstrating reversal of an established behavioural deficit during the continued presence of the stressor. If, as seems likely, chronic stress does play a role in the etiology of depression (see, e.g. Aneshensel & Stone, 1982), its continued presence during antidepressant therapy would usually be the norm.

The mild stressors used in this procedure were intended to represent a closer analogue of life stress than the more severe stressors used in the parent model. As a decrease in sensitivity to reward is assumed to simulate an endogenous depression, this model predicts that endogenous depressions should involve a higher level of mild stress. A recent test of this hypothesis found no difference in the frequency of mild stressors between endogenous and reactive patients, and normal volunteers. However, the endogenous depressives (but not the reactive group), experienced the stresses they encountered as being considerably more severe (Willner *et al.*, 1990).

The effects of antidepressants in the chronic mild stress model appear to be mediated by an increase in the sensitivity of postsynaptic DA receptors. The effect of DMI was blocked by a low dose of the DA receptor antagonist pimozide, which had no effect either in non-stressed animals, or in stressed, non-treated animals (Willner *et al.*, 1988). The effects of imipramine and amitriptyline were similarly blocked by SCH-23390 and sulpiride, which are specific antagonists at the D1 and D2 subtypes of DA receptor (Sampson *et al.*, 1990: see Fig. 5.2).

5.3.2.8 *Amphetamine withdrawal*

A number of studies have reported that responding for brain stimulation reward was reduced during

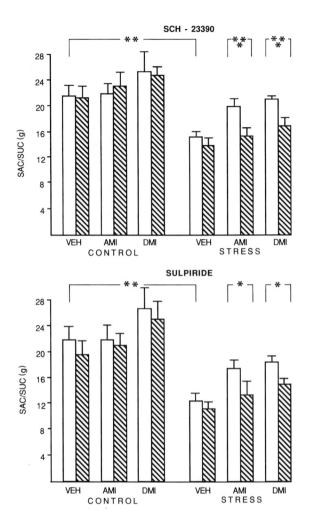

Fig. 5.2. *Male hooded rats were subjected to chronic mild unpredictable stress throughout this experiment, and consumption of a 1% sucrose solution was tested weekly. Treatment with vehicle (VEH), amitriptyline (AMI), or DMI commenced after two weeks of stress, and lasted nine weeks (three weeks each at 7.5 mg/kg, 5 mg/kg and 2.5 mg/kg). All animals received SCH-23390 (12.5 μg/kg) on day 6 after drug withdrawal and sulpiride (40 mg/kg) on day 13 of withdrawal. SCH-23390 and sulpiride reduced sucrose intake in the antidepressant-treated stressed animals, but did not affect intake in any of the other groups (from Sampson* et al., *1990).*

chronic amphetamine treatment and for a period of weeks following withdrawal; the threshold for brain stimulation reward was elevated during this period, confirming that the rate reduction reflects a subsensitivity to reward rather than a depression of motor activity (Leith & Barrett, 1976, 1980; Simpson & Annau, 1977; Barrett & White, 1980; Kokkinidis & Zacharko, 1980). In a single pharmacological study, this effect was alleviated by two days of imipramine or amitriptyline treatment, and with continued treatment, normal responding was restored (Kokkinidis *et al.*, 1980).

Amphetamine withdrawal is included here as another type of chronic stress model, since a number of studies have assumed stimulant drug treatment to be a form of stress (Post, 1975; Antelman *et al.*, 1980). In this case, however, the model gains additional face validity from the obvious parallel with the depressions that frequently follow the cessation of chronic amphetamine use (Watson *et al.*, 1973; Schick *et al.*, 1973). The construct validity of the model depends largely on the assumption that ICSS does activate natural reward pathways (Hoebel, 1976). This assumption is supported by experiments showing that, while the precise effects vary from electrode to electrode (see Gallistel & Beagley, 1971), responding for brain stimulation reward may be increased by food deprivation or by a variety of hormonal manipulations (Olds, 1958), and decreased by pre-loading with food or water (Hoebel & Teitelbaum, 1962). Stimulation of electrodes implanted in medial forebrain sites has been found to evoke pleasurable sensations in human subjects (Heath *et al.*, 1968; Valenstein, 1973), though a variety of other reasons for stimulating, such as curiosity, are also reported (Valenstein, 1973; Sem-Jacobsen, 1976).

5.3.3 Brain damage models

One influential review has argued that because brain function is demonstrably altered in severe depressions, a valid animal model of depression should be based on the infliction of brain damage (Jesberger & Richardson, 1985). It is difficult to see the force of this argument, particularly given that neurobiological abnormalities are readily demonstrable in many of the environmentally determined models discussed in earlier sections. Models in which behaviour is abnormal for a prolonged period do, however, offer a considerable advantage for studies in which a model is to be used to investigate mechanisms of antidepressant action. This condition is met in only a minority of the models considered in earlier sections, but might be expected in a model engendered by brain damage.

5.3.3.1 Olfactory bulbectomy

The major brain damage model involves bilateral lesions of the olfactory bulbs. Rats subjected to this operation display a variety of behavioural changes, including irritability, hyperactivity, and an elevation of circulating levels of plasma corticosteroids; as a result of their hyperactivity, the animals also are deficient in passive avoidance learning. All of these changes are reversed by antidepressant drugs (Cairncross *et al.*, 1977, 1978, 1979). The specificity of the effects is variable, however: while all of the effects of bulbectomy were reversed by amitriptyline, mianserin and viloxazine, irritability and the hormonal changes were also reversed by the neuroleptic chlorpromazine and the anxiolytic chlordiazepoxide. Of the three changes, therefore, the passive avoidance deficit appears to be the only one that responds specifically to antidepressants. The effectiveness of antidepressants in the passive avoidance paradigm has been widely confirmed, and extended to other tricyclic and many atypical antidepressants (Broekkamp *et al.*, 1980; Noreika *et al.*, 1981; Leonard, 1982; Lloyd *et al.*, 1982; Joly & Sanger, 1986). The model has reasonably good specificity for antidepressants. The 5-HT agonist quipazine did reverse the passive avoidance deficit (Lloyd *et al.*, 1982), but amphetamine and the anticholinergic atropine were ineffective (Cairncross *et al.*, 1978, 1979; Noreika *et al.*, 1981; Lloyd *et al.*, 1982). However, the only MAOI to have been tested, tranylcypromine, does not reverse the passive avoidance deficit (Cairncross *et al.*, 1978, 1797; Noreika *et al.*, 1981; Jesberger & Richardson, 1986), though in one study, tranylcypromine did reverse irritability and corticosteroid elevations (Jesberger & Richardson, 1986). The therapeutic effect of antidepressants persists for at least five days following the cessation of treatment (Jesberger & Richardson, 1986). In one study, however, antidepressant effects were only apparent following withdrawal (Noreika *et al.*, 1981).

In most cases, effects of antidepressant in this model are only seen after subchronic treatment (five to ten days), but 5-HT uptake inhibitors were effective after a single injection (Noreika *et al.*,

1981; Lloyd *et al.*, 1982; Joly & Sanger, 1986), and tricyclics have been shown to act acutely if applied to the medial amygdala (Garrigou *et al.*, 1981). In addition to the effects already described, olfactory-bulbectomy also induces muricide in non-killer strains of rat (Ueki, 1982), and both spontaneous and bulbectomy-induced muricide were blocked by amygdala lesions (Horovitz, 1967; Ueki, 1982), which further implicates the amygdala in the effects of olfactory bulbectomy. Bulbectomy has also been found to reduce the threshold for kindling of amygdaloid seizure activity (Watanabe *et al.*, 1979).

In fact, while bulbectomy causes numerous neurochemical changes (see Jesberger & Richardson, 1986), the behavioural and hormonal effects appear to depend on a deficiency of 5-HT in the amygdala. The effects may be mimicked by 5-HT lesions, or reversed by the application to the medial amygdala of 5-HT agonists or of 5-HT (but not NA) (Cairncross *et al.*, 1979; Broekkamp *et al.*, 1980; Garrigou *et al.*, 1981; Lloyd *et al.*, 1982). Similarly, the 5-HT antagonist metergoline has been found to block the effect of antidepressants in this model, whether administered to the medial amygdala or systemically (Broekkamp *et al.*, 1980; Garrigou *et al.*, 1981; Joly & Sanger, 1986).

This model differs from the stress models in representing an agitated rather than a retarded depression; behavioural output in the model is increased rather than decreased. Although the learning of a passive avoidance task is the usual paradigm for studying the bulbectomized rat, it is likely that the learning deficit simply reflects hyperactivity, since antidepressants reduce locomotor activity in bulbectomized animals at doses that are ineffective in sham-operated controls (van Riezen *et al.*, 1977; Leonard, 1982). While a significant proportion of depressed patients are hyperactive (Kupfer & Detre, 1978; Nelson & Charney, 1981), psychomotor changes of both kinds, agitation and retardation, are strongly diagnostic of endogenous depression (Nelson & Charney, 1981). The inactivity of MAOIs in this model, and the elevation of plasma corticosteroids, also point in the same direction: MAOIs are effective in 'atypical' depression, but there is little or no evidence that they are effective in endogenous depression (Tyrer, 1979); similarly, abnormalities of the pituitary-adrenal system are seen in the majority of endogenous depressions, but not in non-endogenous depressions (Carroll, 1978). It would appear, therefore, that the bulbectomized rat models a specific subgroup of depressions – endogenous depressions with psychomotor agitation. It is difficult to find any theoretical rationale for this degree of specificity, except to note that clinically, low 5-HT turnover is usually associated with psychomotor agitation (Asberg *et al.*, 1980; Banki *et al.*, 1981).

5.3.3.2 *Internal capsule lesion*

A final model utilizes a lesion of the internal capsule, 'in the region of the telencephalic–diencephalic border' to reduce rates of responding for brain stimulation reward. This deficit was alleviated by subchronic treatment (five to nine days) with tricyclics, MAOIs and various atypical antidepressants. Morphine, which may have antidepressant properties (Emrich *et al.*, 1983) was also effective; diazepam, yohimbine, propranolol and 'other non-antidepressants' were not. The results of these studies have only been reported in abstract form (Cornfeldt *et al.*, 1982; Szewczak *et al.*, 1982). While the behavioural changes may reflect altered sensitivity to reward rather than non-specific sedative effects, as in the amphetamine withdrawal model, this has not been confirmed.

5.4 COMMENTARY

The approach taken in this and related papers (Willner, 1984, 1986) has been to deploy an extensive set of criteria with which to judge areas of strength and weakness, and thereby to assess the suitability of a model for a particular purpose. The reader should now have the information with which to carry out such an assessment of the models reviewed in this chapter. My own impression after conducting this survey, is that while no one model is particularly convincing, the overall picture gives grounds for cautious optimism. It remains to draw out some common threads and to point to some future directions.

5.4.1 Limitations

First, some commonly encountered limitations of animal models of depression.

(i) False positive and false negative drug responses seem to be inevitable whenever a test is extensively investigated, and although, as discussed earlier, it is important not to read too much into false responses, their absence would be preferable. This goal presents a number of interesting scientific challenges. First, to clinicians: are false responses really false? The persistent response of animal models of depression to anticholinergics, for example (see also Willner, 1985, Chapter 14), identifies a need for proper clinical trials of these compounds. It is unlikely that an anticholinergic could ever be developed as an antidepressant for clinical use, but the theoretical importance of this question (see also Janowsky & Risch, 1984) suggests that these studies should be carried out. Secondly, to users of a model: can false responses be explained away? Adherents of the 'behavioural despair' model have been particularly successful in devising procedures to reduce (or otherwise explain) the incidence of false responses; their example should be adopted more widely. Thirdly, to industrial pharmacologists: if false responses can be minimized, then screening procedures could be improved: for example by using a two-stage version of the 'behavioural despair' test in which false positives identified using acute drug treatment are later screened out by the use of chronic treatment (Kitada *et al.*, 1981).

(ii) It is striking how little attention has been paid to the breadth of symptoms simulated in an animal model of depression. The face validity of the majority of models is far from impressive, and while there are some exceptions, notably learned helplessness and olfactory bulbectomy, most modellers are content to demonstrate one or two behavioural changes and their reversal by antidepressants. From the example of learned helplessness, it seems likely that many further points of correspondence between animal models and depression might be revealed if only the appropriate questions were asked. Such studies could, in principle, assist in improving the classification of depressive subgroups. It has been taken for granted that models require extensive pharmacological characterization. The need for extensive behavioural characterization should be equally recognized.

(iii) One of the principal aims of simulating depression is to provide experimental paradigms within which to study the physiological bases of depressive disorders and the mechanisms of action of antidepressant drugs. It is widely recognized that antidepressants show few signs of their clinical activity in 'normal' people. However, there is an understandable reluctance to follow through the logic and accept that the results of studies of antidepressants in 'normal' animals cannot be trusted until they are also shown to hold true in simulations of depression. However, here we encounter another problem, for very few animal models of depression have a time course appropriate for studying chronic antidepressant administration. Those models that do involve prolonged behavioural changes, such as the chronic mild stress and olfactory bulbectomy models, can be expected to play a particularly important role in the future.

5.4.2 Contributions

For many years, the thought that animal models of depression might make some scientific contribution in addition to their use in screening has been little more than wish and a hope. However, it is now possible to discern some of the directions from which such contributions might arise.

(i) Antidepressant research has identified a wide range of physiological changes consequent on chronic treatment (albeit in 'normal' animals). But which of the effects are relevant to their clinical actions? A powerful method for discriminating between irrelevant and relevant effects (between actions and mechanisms of action) is to determine how the effects may be prevented. As antidepressants have few behavioural effects of any interest in 'normal' animals (Tucker & File, 1986), these questions can only be asked in simulations of depression. Studies of antidepressant antagonism in the 'behavioural

despair' (e.g. Plaznik *et al.*, 1985*a*; Pulverenti & Samanin, 1986), learned helplessness (Martin *et al.*, 1986*b*), and chronic mild stress (Willner *et al.*, 1988; Sampson *et al.*, 1989) models have shown that antidepressants restore suppressed behaviour in these models by increasing DA and/or alpha-adrenergic transmission. Conversely in the olfactory bulbectomy, muricide, and waiting models, antidepressants reduce impulsive behaviour by increasing 5-HT transmission (Broekkamp *et al.*, 1980; Eisenstein *et al.*, 1982; Seiden, 1988). In addition to the specific information they provide, these studies suggest that the pharmacological blockade of antidepressant action is a powerful analytical method that could be used clinically (c.f. Shopsin *et al.*, 1975, 1976).

(ii) Despite the fact that their validity as simulations of depression is still largely unproven, animal models are beginning to provide some insights into the brain mechanisms underlying depression. As indicated by the antidepressant data, the models that have undergone extensive physiological investigations appear to fall into two groups. In one (e.g. learned helplessness, behavioural despair) a decrease in behavioural output is caused by a decrease in transmission through NA and DA synapses, while in the second (e.g. olfactory bulbectomy) an increase in behavioural output is caused by a decrease in transmission through 5-HT synapses. Clinically, low DA turnover is strongly associated with psychomotor retardation, while low 5-HT turnover is strongly associated with psychomotor agitation (see Willner, 1985, Chapters 9, 16, for review). The association in animal models of particular symptoms with specific neurochemical mechanisms supports the view that depression may consist of two relatively independent syndromes, overlapping in many of their symptoms but of differing etiologies (Klein, 1974; Willner, 1985).

(iii) The elucidation of physiological mechanisms in animal models can also help to resolve disputes about the significance of neurochemical changes in depression. There is current debate, for example, as to whether 5-HT transmission in depression is decreased, as suggested by low 5-HT turnover (van Praag, 1980), or increased by virtue of postsynaptic receptor supersensitivity (Fuxe *et al.*, 1983). The role of 5-HT depletion in bulbectomized animals, together with the symptomatic correspondence between these animals and patients with evidence of low 5-HT turnover, supports the hypothesis of a 5-HT deficiency in depression, rather than an excess.

(iv) One of the most striking observations arising from the present review is the consistency with which antidepressant-reversible effects of stress may be identified in different paradigms. One of these effects, seen in the learned helplessness, chronic stress, chronic mild stress, and amphetamine-withdrawal models, is the antidepressant-reversible suppression of rewarded behaviour. This finding has considerable implications for a specific role of stress in the etiology of endogenous depressions (see also Willner, 1987), which deserves further concerted investigation both in animal models and in the clinic.

(v) The consistency between the various stress models also highlights the need to clarify the role of anxiety in the genesis of depression. Equally, the fact that other procedures (particularly social isolation) can also generate abnormal behaviours reminiscent of depression, and with apparently different physiological mechanisms, argues against unitary explanations of depression and in favour of multifactorial models (see Willner, 1987). More generally, as the etiological factors in depression are progressively identified, animal models offer an opportunity of establishing causal relationships between risk factors and patterns of symptoms, and interactions between different risk factors (see, e.g. Kennett *et al.*, 1986).

5.4.3 Postscript

At the end of the day, animal models of depression are only of value to the extent that they expand our understanding of the nature of depression or help to expedite its treatment. Their role in the search for new antidepressants has been substantial (though the extent to which the newer agents are an improvement on existing drugs is questionable).

Currently, clinical trials of 5-HT$_1$ agonists are strongly indicated by data from a number of diverse models, and a re-examination of the possibility of developing dopaminergic antidepressants would also seem warranted. However, the contribution of animal models of depression to a better understanding of depression has up to now been slight. While the emphasis of this chapter has been on assessing the validity of simulations of depression, it should never be forgotten that validity is only a means to an end: the models must ultimately be shown to be useful, and in the final analysis, they must be judged by their ability to supply clinically productive hypotheses. The study of animal models of depression has advanced to a point at which a more intensive dialogue between simulations and the clinic should now prove mutually beneficial.

REFERENCES

Abramson, L.Y., Seligman, M.E.P. & Teasdale, J.D. (1978). Learned helplessness in humans: Critique and reformulation. *Journal of Abnormal Psychology* **87**, 49–74.

Akiskal, H.S. (1980). External validating criteria for psychiatric diagnosis: Their application in affective disorders. *Journal of Clinical Psychiatry* **41**, 12, Sec. 2, 6–15.

Akiskal, H.S. (1985). Interaction of biologic and psychologic factors in the origin of depressive disorders. *Acta Psychiatrica Scandinavica* **71**, 131–9.

Akiskal, H.S. (1986). A developmental perspective on recurrent mood disorders: A review of studies in man. *Psychopharmacological Bulletin* **22**, 579–86.

Akiskal, H.S. & Tashjian, R. (1983). Affective disorders: II. Recent advances in laboratory and pathogenetic approaches. *Hospital Communications in Psychiatry* **34**, 822–30.

Albert, D.J. & Walsh, M.L. (1982). The inhibitory modulation of agonistic behaviour in the rat brain: A review. *Neuroscience and Biobehavioral Reviews* **6**, 125–43.

Allen, M.G. (1976). Twin studies of affective illness. *Archives of General Psychiatry* **33**, 1476–8.

Altenor, A., Kay, E. & Richter, M. (1977). The generality of learned helplessness in the rat. *Learning and Motivation* **8**, 54–61.

American Psychiatric Association (1980). *DSM III – Diagnostic and Statistical Manual of Psychiatric Disorders*, A.P.A., Washington.

Aneshensel, C.S. & Stone, J.D. (1982). Stress and depression: A test of the buffering model of social support. *Archives of General Psychiatry* **39**, 1392–6.

Angst, J., Baastrup, P., Grof, P., Hippius, H., Poldinger, W. & Weis, P. (1973). The course of monopolar depression and bipolar psychoses. *Psychiatrica Neurologica Neurochirurgica* **76**, 489–500.

Anisman, H., Irwin, J. & Sklar, L.S. (1979). Deficits of escape performance following catecholamine depletion: implications for behavioural deficits induced by uncontrollable stress. *Psychopharmacology* **64**, 163–70.

Anisman, H.A. & Zacharko, R.M. (1982). Depression: The predisposing influence of stress. *Behavioral and Brain Sciences* **5**, 89–137.

Antelman, S.M., Eichler, A.J., Black, C.A. & Kocan, D. (1980). Interchangeability of stress and amphetamine sensitization. *Science* **207**, 329–31.

Aprison, M.H., Hintgen, J.N. & Nagayama, H. (1982). Testing a new theory of depression with an animal model: Neurochemical-behavioural evidence for postsynaptic serotonergic receptor involvement. In: S.Z. Langer, R. Takahashi, T. Segawa & M. Briley (eds.), *New Vistas in Depression*, pp. 171–8. Pergamon Press, New York.

Araki, H., Kazuaki, K., Uchiyama, Y. & Aihara, H. (1985). Involvement of amygdaloid catecholaminergic mechanism in suppressive effects of desipramine and imipramine on duration of immobility in rats forced to swim. *European Journal of Pharmacology* **113**, 313–18.

Asberg, M., Bertilsson, R., Rydin, E., Schalling, D., Thoren, P. & Traskman-Bendz, L. (1980). Monoamine metabolites in cerebrospinal fluid in relation to depressive illness, suicidal behaviour and personality. In B. Angrist, G.D. Burrows, M. Lader, O. Lingjaerde, G. Sedvall & D. Wheatley (eds.), *Recent Advances in Neuropsychopharmacology*, pp. 257–71. Pergamon: New York.

Asberg, M., Eriksson, B., Martensson, B., Traskman-Bendz, L. & Wager, A. (1986). Therapeutic effects of serotonin uptake inhibitors in depression. *Journal of Clinical Psychiatry* **47**:4 (Suppl.) 23–5.

Babington, R.G. (1981). Neurophysiologic techniques and antidepressive activity. In: S.J. Enna, J.B. Malick & E. Richelson (eds.), *Antidepressants: Neurochemical, Behavioral and Clinical Perspectives*, pp. 157–73. Raven Press: New York.

Babington, R.G. & Wedeking, P.W. (1973). The pharmacology of seizures induced by sensitization with low intensity brain stimulation. *Pharmacology, Biochemistry and Behavior* **1**, 461–7.

Baltzer, V. & Weiskrantz, L. (1973). Antidepressant agents and reversal of diurnal activity cycles in the rat. *Biological Psychiatry* **10**, 199–209.

Banki, C.M., Molnar, G. & Vojnik, M. (1981). Cerebrospinal fluid amine metabolites, tryptophan and clinical parameters in depression. II. Psychopathological symptoms. *Journal of Affective Disorders* **3**, 91–9.

Barnett, A. & Taber, R.I. (1971). Antidepressant agents. In R.A. Turner & P. Hebborn (eds.), *Screening Methods in Pharmacology*, pp. 209–26, Academic Press: New York.

Barrett, R.J. & White, D.K. (1980). Reward system depression following chronic amphetamine: antagonism by haloperidol. *Pharmacology, Biochemistry and Behavior* **13**, 555–9.

Beck, A.R. (1967). *Depression: Clinical, Experimental and Theoretical Aspects*. Harper and Row: New York.

Beck, A.T. (1983). Cognitive therapy of depression: New perspectives. In P.J. Clayton & J.E. Barrett (eds.), *Treatment of Depression: Old Controversies and New Approaches*, pp. 265–90. Raven: New York.

Beck, A.T. & Harrison, R.P. (1982). Stress, neurochemical substrates and depression: Concommitants are not necessarily causes. *Behavioral and Brain Sciences* **5**, 101–2.

Becker, J. (1979). Vulnerable self-esteem as a predisposing factor in depressive disorders. In R.A. Depue (ed.), *The Psychobiology of the Depressive Disorders: Implications for the Effects of Stress*, pp. 317–34. Academic Press: New York.

Beigel, A. & Murphy, D.L. (1971). Unipolar and bipolar affective illness: Differences in clinical characteristics accompanying depression. *Archives of General Psychiatry* **24**, 215–20.

Beninger, R.S. (1983). The role of dopamine in locomotor activity and learning. *Brain Research Reviews* **6**, 173–96.

Berger, B.D. & Schuster, R. (1982). An animal model of social interaction: Implications for the analysis of drug action. In M.H. Spiegelstein & A. Levy (eds.), *Behavioral Models and the Analysis of Drug Action*, pp. 415–28. Elsevier: Amsterdam.

Berger, B.D. & Schuster, R. (1987). Pharmacological aspects of social cooperation in animals and humans. In B. Olivier, S. Mos & P. Brain (eds.), *Ethopharmacology of Agonistic Behaviour*, pp. 14–32, Martenus Nijkoff: Dordrecht.

Billings, A.G., Cronkite, R.C. & Moos, R.H. (1983). Social-environmental factors in unipolar depression: Comparisons of depressed patients and non-depressed controls. *Journal of Abnormal Psychology* **92**, 119–33.

Bizot, J.C., Thiebot, M.H., Le Bihan, C., Soubrie, P. & Simon, P. (1988a). Effects of imipramine-like drugs and serotonin uptake blockers on delay of reward in rats: possible implication in the behavioral mechanism of action of antidepressants. *Journal of Pharmacology and Experimental Therapeutics* **246**, 1144–51.

Bizot, J.C., Thiebot, M.H. & Puech, A.J. (1988*b*). Effects of 5-HT-related drugs on waiting capacities in rats. *Psychopharmacology* **96**, S5.

Bodnoff, S.R., Suranyi-Codotte, B., Aitken, D.H., Quirion, R. & Meaney, M.Y. (1988). The effects of chronic antidepressant treatment in an animal model of anxiety. *Psychopharmacology* **95**, 298–302.

Borsini, F. & Meli, A. (1988). Is the forced swimming test a suitable model for revealing antidepressant activity. *Psychopharmacology* **94**, 147–60.

Borsini, F., Bendotti, G., Velkov, V., Rech, R. & Samanin, R. (1981). Immobility test: effects of 5-hydroxytryptaminergic drugs and role of catecholamines in the activity of some antidepressants. *Journal of Pharmacy and Pharmacology* **33**, 33–7.

Borsini, F., Pulvirenti, L. & Samanin, R. (1985). Evidence of dopamine involvement in the effect of repeated treatment with various antidepressants in the behavioural 'despair' test in rats. *European Journal of Pharmacology* **110**, 253–6.

Bourin, M., Poncelet, M., Chermat, R. & Simon, P. (1983). The value of the reserpine test in psychopharmacology. *Arneimittel-Forschung: Drug Research* **33**, 1173–6.

Bowden, D.M. & McKinney, W.T. (1972). Behavioral effects of peer separation, isolation and reunion on adolescent male rhesus monkeys. *Developmental Psychobiology* **5**, 353–62.

Briscoe, C.W. & Smith, J.B. (1975). Depression in bereavement and divorce. *Archives of General Psychiatry* **32**, 439–43.

Broekkamp, C.L., Garrigou, D. & Lloyd, K.G. (1980). Serotonin-mimetic and antidepressant drugs on passive avoidance learning by olfactory bulbectomized rats. *Pharmacology, Biochemistry and Behavior* **13**, 643–6.

Brown, G.W. (1989). A psychosocial view of depression. In: D.H. Bennett & H. Freeman (eds.), *Community Psychiatry*. Churchill-Livingstone: London. (In press.)

Brown, G.W. & Harris, T. (1978). *Social Origins of Depression*. Tavistock: London.

Brown, G.W. & Harris, T. (eds.) (1988). *Life Events and Illness*. Guilford Press: New York.

Brown, G.W., Sklair, F., Harris, T.O. & Birley, J.L.T. (1973). Life events and psychiatric disorders. I. Some methodological issues. *Psychological Medicine* **3**, 74–87.

Brown, L., Rosellini, R.A., Samuels, O.B. & Riley, E.P. (1982). Evidence for a serotonergic mechanism of the learned helplessness phenomenon. *Pharmacology, Biochemistry and Behavior* **17**, 877–83.

Browne, R.G. (1979). Effects of antidepressants and anticholinergics in a mouse 'behavioural despair' test. *European Journal of Pharmacology* **58**, 331–4.

Buchwald, A.M., Coyne, J.C. & Cole, C.S. (1978). A critical evaluation of the learned helplessness model of depression. *Journal of Abnormal Psychology* **87**, 180–93.

Cairncross, K.D., Wren, A.F., Cox, B. & Schnieden, H. (1977). Effects of olfactory bulbectomy and domicile on stress induced corticosterone release in the rat. *Physiology and Behavior* **19**, 485–7.

Cairncross, K.D., Cox, B., Forster, C. & Wren, A.F. (1978). A new model for the detection of antidepressant drugs: Olfactory bulbectomy in the rat compared with existing models. *Journal of Pharmacological Methods* **1**, 131–43.

Cairncross, K.D., Cox, B., Forster, C. & Wren, A.F. (1979). Olfactory projection systems, drugs and behaviour: A review. *Psychoneuroendocrinology* **4**, 253–72.

Carroll, B.J. (1978). Neuroendocrine function in psychiatric disorders. In M.A. Lipton, A. DiMaschio & K.F. Killam (eds.), *Psychopharmacology: A Generation of Progress*, pp. 487–97. Raven Press: New York.

Carroll, B.J. (1982). The dexamethasone suppression test for melancholia. *British Journal of Psychiatry* **140**, 292–304.

Carter, R.B., Dykstra, L.A., Leander, J.D. & Appel, J.B. (1978). Role of peripheral mechanisms in the behavioral effects of 5-hydroxytryptophan. *Pharmacology, Biochemistry and Behavior* **9**, 249–53.

Cervo, L. & Samanin, R. (1987). Evidence that dopamine mechanisms in the nucleus accumbens are selectively involved in the effect of desipramine in the forced swim test. *Neuropharmacology* **26**, 1469–72.

Colpaert, F.C., Lenaerts, F.M., Niemegeers, C.J.E. & Janssen, P.A.J. (1975). A critical study on RO-4-1284 antagonism in mice. *Archives Internationales de Pharmacodynamie et Thérapie* **215**, 40–90.

Cools, A.R. (1980). Role of the neostriatal dopaminergic activity in sequencing and selecting behavioural strategies: Facilitation of processes involved in selecting the best strategy in a stressful situation. *Behavioral Brain Research* **1**, 361–78.

Cornfeldt, M., Fisher, B. & Fielding, S. (1982). Rat internal capsule lesion: a new test for detecting antidepressants. *Federation Proceedings* **41**, 1066.

Costa, E., Garattini, S. & Valzelli, L. (1960). Interactions between reserpine, chlorpromazine and imipramine. *Experientia* **16**, 461–3.

Costello, C.G. (1972). Depression: loss of reinforcers or loss of reinforcer effectiveness? *Behavior Therapy* **3**, 240–7.

Crawley, J.N. (1984). Preliminary report of a new rodent separation model of depression. *Progress in Neuro-Psychopharmacology and Biological Psychiatry* **8**, 447–57.

Crawley, J.N. (1985). A monoamine oxidase inhibitor reverses the 'separation syndrome' in a new hamster separation model of depression. *European Journal of Pharmacology* **112**, 129–33.

Cuomo, V., Cagiano, R., Brunello, N., Fumagalli, R. & Racagni, G. (1983). Behavioural changes after acute and chronic administration of typical and atypical antidepressants in rat: Interactions with reserpine. *Neuroscience Letters* **40**, 315–19.

Dall'Olio, R., Vaccheri, A., Gandolfi, O. & Montanaro, N. (1986). Behavioral differentiation between pharmacokinetic and pharmacodynamic components of the interaction of antidepressants or neuroleptics with methamphetamine. *Psychopharmacology* **90**, 18–23.

Deakin, J.F.W. (1989). Role of 5HT receptor subtypes in depression and anxiety. In P. Bevan, A. Cools & T. Archer (eds.), *Behavioural Pharmacology of 5HT*, pp. 179–204. Erlbaum: New York.

Delina-Stula, A. & Vassout, A. (1979). Differential effects of psychoactive drugs on aggressive responses in rats and mice. In M. Sandler (ed.), *Psychopharmacology of Aggression*, pp. 41–60. Raven Press: New York.

R.A. Depue (ed.) (1979). *The Psychobiology of the Depressive Disorders: Implications for the Effects of Stress*. Academic Press: New York.

Depue, R.A. & Monroe, S.M. (1978). Learned helplessness in the perspective of the depressive disorders: Conceptual and definitional issues. *Journal of Abnormal Psychology* **87**, 3–20.

Depue, R.A. & Monroe, S.M. (1979). The unipolar–bipolar distinction in the depressive disorders: implications for stress-onset interactions. In R.A. Depue (ed.), *The Psychobiology of the Depressive Disorders: Implications for the Effects of Stress*, pp. 25–53. Academic Press: New York.

Desan, P.H., Silbert, L.H. & Maier, S.F. (1988). Long-term effects of inescapable stress on daily running activity and antagonism by desipramine. *Pharmacology, Biochemistry and Behavior* **30**, 21–9.

Dorworth, T.R. & Overmeier, J.B., (1977). On 'learned helplessness': The therapeutic effects of electroconvulsive shock. *Physiology and Behavior* **4**, 355–8.

Drugan, R.C., Ryan, S.M., Minor, T.R. & Maier, S.F. (1984), Librium prevents the analgesia and shuttlebox escape deficits typically observed following inescapable shock. *Pharmacology, Biochemistry and Behavior* **21**, 749–54.

Drugan, R.C., Maier, S.F., Skolnick, P., Paul, S.M. & Crawley, J.N. (1985). An anxiogenic benzodiazepine receptor ligand induces learned helplessness. *European Journal of Pharmacology* **113**, 453–7.

Dubinsky, B. & Goldberg, M.E. (1971). The effect of imipramine and selected drugs on attack elicited by hypothalamic stimulation in the cat. *Neuropharmacology* **10**, 537–45.

Eisenstein, N., Iorio, L.C. & Clody, D.E. (1982). Role of serotonin in the blockade of muricidal behaviour by tricyclic antidepressants. *Pharmacology, Biochemistry and Behavior* **17**, 847–9.

Emrich, H.M., Vogt, P. & Herz, A. (1983). Possible antidepressant effects of opioids: Action of buprenorphine. *Annals of the New York Academy of Science* **398**, 108–12.

Everitt, B.J. & Keverne, E.B. (1979). Models of depression based on behavioural observations of experimental animals. In E.S. Paykel & A. Coppen (eds.), *Psychopharmacology of Affective Disorders*, pp. 41–59. Oxford University Press: Oxford.

Fava, G.A., Munari, F., Pavan, L. & Kellner, R. (1981). Life events and depression: A replication. *Journal of Affective Disorders* **3**, 159–65.

Feinberg, M. & Carroll, B.J. (1982). Separation of subtypes of depression using discriminant analysis. I. Separation of unipolar endogenous depression from non-endogenous depression. *British Journal of Psychiatry* **140**, 384–91.

Ferster, C.B. (1973). A functional analysis of depression. *American Psychologist* **28**, 857–70.

Frances, A., Brown, R., Koskis, J. & Mann, J. (1981). Psychotic depression: A separate entity? *American Journal of Psychiatry* **183**, 829–33.

Frazer, A., Lucki, I. & Sills, M. (1985). Alterations in monoamine-containing neuronal function due to administration of antidepressants repeatedly to rats. *Acta Pharmacologica et Toxicologica* **56**, Suppl. 1, 21–34.

Fuxe, K., Ogren, S.O., Agnati, L.F., Benfenati, F., Fredholm, B., Andersson, K., Zini, I. & Eneroth, P. (1983). Chronic antidepressant treatment and central 5-HT synapses. *Neuropharmacology* **22**, 389–400.

Garattini, S., Giacalone, R. & Valzelli, L. (1967). Isolation, aggressiveness and brain 5-hydroxytryptamine turnover. *Journal of Pharmacy and Pharmacology* **19**, 338–9.

Gallistel, C.R. & Beagley, G. (1971). Specificity of brain stimulation reward in the rat. *Journal of Comparative and Physiological Psychology* **76**, 199–205.

Garber, J. & Seligman, M.E.P. (eds.) (1980). *Human Helplessness: Theory and Applications*. Academic Press: New York.

Garber, J., Miller, W.R. & Seaman, S.F. (1979). Learned helplessness, stress and the depressive disorders. In R.A. Depue (ed.), *The Psychobiology of the Depressive Disorders: Implications for the Effects of Stress*, pp. 335–63. Academic Press: New York.

Garrigou, D., Broekkamp, C.L.E. & Lloyd, K.G. (1981). Involvement of the amygdala in the effects of antidepressants on the passive avoidance deficit in bulbectomized rats. *Psychopharmacology* **74**, 66–70.

Garzon, J. & Del Rio, J. (1981). Hyperactivity induced in rats by long-term isolation: further studies on a new animal model for the detection of antidepressants. *European Journal of Pharmacology* **74**, 287–94.

Garzon, J., Fuentes, J.A. & Del Rio, J. (1979). Antidepressants selectively antagonize the hyperactivity induced in rats by long-term isolation. *European Journal of Pharmacology* **59**, 293–6.

Giral, P., Martin, P., Soubrie, P. & Simon, P. (1988). Reversal of helpless behaviour in rats by putative 5HT$_{1A}$ agonists. *Biological Psychiatry* **23**, 237–42.

Glazer, H.I. & Weiss, J.M. (1976). Long-term interference effect: an alternative to 'learned helplessness'. *Journal of Experimental Psychology: Animal Behavior Processes* **2**, 201–13.

Goodwin, F.K., Ebert, M.H. & Bunney, W.E. (1972). Mental effects of reserpine in man: A review. In R.I. Shader (ed.), *Psychiatric Complications of Medical Drugs*, pp. 73–101. Raven Press: New York.

Goodwin, F.K., Wirz-Justice, A. & Wehr, T.A. (1982). Evidence that pathophysiology of depression and the mechanism of action of antidepressant drugs both involve alterations in circadian rhythms. In E. Costa & G. Racagni (eds.) *Typical and Atypical Antidepressants: Clinical Practice*, pp. 1–11. Raven Press: New York.

Goodwin, F.K., Price, L.H. & Charney, D.S. (1988). Fluvoxamine in OCD. *Psychopharmacology* **96** (Suppl.) 234.

Gorka, Z., Wojtasik, E., Kwiatek, H. & Maj, J. (1979). Action of serotoninmimetics in the behavioural despair test in rats. *Communications in Psychopharmacology* **3**, 133–6.

Gray, J.A. (1982). *The Neuropsychology of Anxiety: An Enquiry into the Functions of the Septo-Hippocampal System*. Oxford University Press: Oxford, New York.

Halliwell, G., Quinton, R.M. & Williams, F.E. (1964). A comparison of imipramine, chlorpromazine and related drugs in various tests involving autonomic functions and antagonism of reserpine. *British Journal of Pharmacology* **23**, 330–50.

Hatotani, N., Nomura, J. & Kitayama, I. (1982). Changes of brain monoamines in the animal model for depression. In S.Z. Langer, R. Takahashi, T.

Segawa & M. Briley (eds.), *New Vistas in Depression*, pp. 65–72. Pergamon Press: New York.

Healy, D. & Williams, J.M.G. (1988). Dysrhythmia, dysphoria and depression: The interaction of learned helplessness and circadian dysrhythmia in the pathogenesis of depression. *Psychological Bulletin* **103**, 163–78.

Heath, R.G., John, S.B. & Fontana, C.J. (1968). The pleasure response: Studies by stereotaxic techniques in patients. In N.S. Kline & E. Laska (eds.), *Computer and Electronic Devices in Psychiatry*, pp. 178–89. Grune and Stratton: New York.

Hill, R.T. & Tedeschi, D.H. (1971). Animal testing and screening procedures in the evaluation of psychotropic drugs. In R.H. Rech & K.E. Moore (eds.), *An Introduction to Psychopharmacology*, pp. 237–88. Raven Press: New York.

Hinde, R.A., Leighton-Shapiro, M.E. & McGinnis, L. (1978). Effects of various types of separation experience on rhesus monkeys 5 months later. *Journal of Child Psychology and Psychiatry* **19**, 199–211.

Hintgen, J.N., Fuller, R.W., Mason, N.R. & Aprison, M.H. (1985). Blockade of a 5-hydroxytryptophan-induced animal model of depression with a potent and selective 5-HT$_2$ receptor antagonist (LY53857). *Biological Psychiatry* **20**, 592–7.

Hoebel, B.G. (1976). Brain-stimulation reward and aversion in relation to behaviour. In A. Wauquier & E. Rolls (eds.), *Brain-Stimulation Reward*, pp. 331–72. North Holland/American Elsevier: New York.

Hoebel, B.G. & Teitelbaum, P. (1962). Hypothalamic control of feeding and self-stimulation. *Science* **135**, 375–7.

Horovitz, Z.P. (1965). Selective block of rat mouse killing by antidepressants. *Life Science* **4**, 1909–12.

Horovitz, Z.P. (1967). The amygdala and depression. In S. Garattini & M.N.G. Dukes (eds.), *Antidepressant Drugs*, pp. 121–9. Excerpta Medica: Amsterdam.

Horovitz, Z.P., Piala, J.J., High, J.P., Burke, J.C. & Leaf, R.C. (1966). Effects of drugs on the mouse killing (muricide) test and its relationship to amygdaloid function. *International Journal of Neuropharmacology* **5**, 405–11.

Howard, J.L., Soroko, F.E. & Cooper, B.R. (1981). Empirical behavioral models of depression, with emphasis on tetrabenazine antagonism. In S.J. Enna, J.B. Malick & E. Richelson (eds.), *Antidepressants: Neurochemical, Behavioral and Clinical Perspectives*, pp. 107–20. Raven Press: New York.

Hrdina, P.D., Von Kulmiz, P. & Stretch, R. (1979).

Pharmacological modification of experimental depression in infant macaques. *Psychopharmacology* **64**, 89–93.

Hudgens, H., Morrison, J. & Barcha, R. (1967). Life events and onset of primary affective disorders. *Archives of General Psychiatry* **16**, 134–45.

Jablensky, A., Sartorius, N., Gulbinat, W. & Ernberg, G. (1981). Characteristics of depressive patients contacting psychiatrists in four cultures. *Acta Psychiatrica Scandinavica* **63**, 367–83.

Jackson, R.L. & Minor, T.R. (1988). Effects of signaling inescapable shock on subsequent escape learning: Implications for theories of coping and 'learned helplessness'. *Journal of Experimental Psychology: Animal Behavior Processes* **14**, 390–400.

Jackson, R.L., Maier, S.F. & Rapoport, P.M. (1978). Exposure to inescapable shock produces both activity and associative deficits in rats. *Learning and Motivation* **9**, 69–98.

Jahoda, M. (1979). The impact of unemployment in the 1930s and the 1970s. *Bulletin of the British Psychological Society* **32**, 309–14.

Janowsky, D.S. & Risch, S.C. (1984). Cholinomimetic and anticholinergic drugs used to investigate an acetylcholine hypothesis of affective disorders and stress. *Drug Development Research* **4**, 125–42.

Jesberger, J.A. & Richardson, J.S. (1985). Animal models of depression: Parallels and correlates to severe depression in humans. *Biological Psychiatry* **20**, 764–84.

Jesberger, J.A. & Richardson, J.S. (1986). Effects of antidepressant drugs on the behavior of olfactory bulbectomized and sham-operated rats. *Behavioral Neuroscience* **100**, 256–74.

Joly, D. & Sanger, D.J. (1986). The effects of fluoxetine and zimelidine on the behavior of olfactory bulbectomized rats. *Pharmacology, Biochemistry and Behavior* **24**, 199–204.

Kametani, H., Nomura, S. & Shimizu, J. (1983). The reversal effect of antidepressants on the escape deficit induced by inescapable shock in rats. *Psychopharmacology* **80**, 206–8.

Kanner, A.D., Coyne, J.C., Schaefer, C. & Lazarus, R.S. (1981). Comparison of two modes of stress measurement: Daily hassles and uplifts versus major life events. *Journal of Behavioral Medicine* **4**, 1–39.

Kashani, J.H., Husain, A., Shekim, W.O., Hodges, K.K., Cytryn, L. & McKnew, D.H. (1981). Current perspectives on childhood depression: An overview. *American Journal of Psychiatry* **138**, 143–53.

Katz, R.J. (1981*a*). Animal models and human depressive disorders. *Neuroscience and Biobehavioral Reviews* **5**, 231–46.

Katz, R.J. (1981*b*). Animal model of depression:

Effects of electroconvulsive shock therapy. *Neuroscience and Biobehavioral Reviews* **5**, 273–7.

Katz, R.J. (1982). Animal model of depression: Pharmacological sensitivity of a hedonic deficit. *Pharmacology, Biochemistry and Behavior* **16**, 965–8.

Katz, R.J. & Baldrighi, G. (1982). A further parametric study of imipramine in an animal model of depression. *Pharmacology, Biochemistry and Behavior* **16**, 969–72.

Katz, R.J. & Hersh, S. (1981). Amitriptyline and scopolamine in an animal model of depression. *Neuroscience and Biobehavioral Reviews* **5**, 265–71.

Katz, R.J. & Sibel, M. (1982*a*). Animal model of depression: tests of three structurally and pharmacologically novel antidepressant compounds. *Pharmacology, Biochemistry and Behavior* **16**, 973–7.

Katz, R.J. & Sibel, M. (1982*b*). Further analysis of the specificity of a novel animal model of depression: Effects of an antihistaminic, antipsychotic and anxiolytic compound. *Pharmacology, Biochemistry and Behavior* **16**, 979–82.

Katz, R.J., Roth, K.A. & Carroll, B.J. (1981*a*). Acute and chronic stress effects on open field activity in the rat: implications for a model of depression. *Neuroscience and Biobehavioral Reviews* **5**, 247–51.

Katz, R.J., Roth, K.A. & Schmaltz, K. (1981*b*). Amphetamine and tranylcypromine in an animal model of depression: pharmacological specificity of the reversal effect. *Neuroscience and Biobehavioral Reviews* **5**, 259–64.

Kaufman, I.C. & Rosenblum, L.A. (1967). The reaction to separation in infant monkeys: Anaclitic depression and conservation-withdrawal. *Psychosomatic Medicine* **29**, 648–75.

Kawashima, K., Araki, H. & Aihara, H. (1986). Effect of chronic administration of antidepressants on duration of immobility in rats forced to swim. *Japanese Journal of Pharmacology* **40**, 199–204.

Kelly, W.F., Checkley, S.A., Bender, D.A. & Mashiter, K. (1983). Cushing's syndrome and depression – A prospective study. *British Journal of Psychiatry* **142**, 16–19.

Kendell, R.E. (1976). The classification of depression: A review of contemporary confusion. *British Journal of Psychiatry* **129**, 15–28.

Kennett, G.A., Dickinson, S.L. & Curzon, G. (1985). Central serotonergic responses and behavioural adaptation to repeated immobilisation: the effect of the corticosterone synthesis inhibitor metapyrone. *European Journal of Pharmacology* **119**, 143–52.

Kennett, G.A., Chaouloff, F., Marcou, M. & Curzon, G. (1986). Female rats are more vulnerable than males in an animal model of depression: the possible role of serotonin. *Brain Research* **382**, 416–21.

Kennett, G.A., Dourish, C.T. & Curzon, G. (1987). Antidepressant and antianorexic properties of 5HT$_{1A}$ agonists and other drugs in animal models. In C.T. Dourish, S. Ahlenius & P.H. Hutson (eds.), *Brain Serotonergic Mechanisms: The Pharmacological, Biochemical and Potential Therapeutic Actions of 8-OH-DPAT and Other 5HT$_{1A}$ Agonists*, pp. 243–60. Ellis Horwood: Chichester.

Kitada, Y., Miyauchi, T., Satoh, A. & Satoh, S. (1981). Effects of antidepressants in the rat forced swimming test. *European Journal of Pharmacology* **72**, 145–52.

Kitada, Y., Miyauchi, T., Kanazawa, Y., Nakamichi, H. & Satoh, S. (1983). Involvement of alpha- and beta-adrenergic mechanisms in the immobility-reducing action of desipramine in the forced swimming test. *Neuropharmacology* **22**, 1055–60.

Klein, D.F. (1974). Endogenomorphic depression: A conceptual and terminological revision. *Archives of General Psychiatry* **31**, 447–54.

Klinger, E. (1975). Consequences of commitment to and disengagement from incentives. *Psychological Review* **82**, 1–24.

Knobloch, L.C., Goldstein, J.R. & Malick, J.B. (1982). Effects of acute and subacute antidepressant treatment on kindled seizures in rats. *Pharmacology, Biochemistry and Behavior* **17**, 461–5.

Kokkinidis, L. & Zacharko, R.M. (1980). Response sensitization and depression following long-term amphetamine treatment in a self-stimulation paradigm. *Psychopharmacology* **68**, 73–6.

Kokkinidis, L., Zacharko, R.M. & Predy, P.A. (1980). Post-amphetamine depression of self-stimulation responding from the substantia nigra: reversal by tricyclic antidepressants. *Pharmacology, Biochemistry and Behavior* **13**, 379–83.

Kostowski, W. (1985). Possible relationship of the locus coeruleus–hippocampal noradrenergic neurons to depression and mode of action of antidepressant drugs. *Polish Journal of Pharmacology and Pharmacy* **37**, 727–43.

Kraemer, G.W. & McKinney, W.T. (1979). Interactions of pharmacological agents which alter biogenic amine metabolism and depression. *Journal of Affective Disorders* **1**, 33–54.

Kupfer, D.J. (1976). REM latency: A psychobiological marker for primary depressive disease. *Biological Psychiatry* **11**, 159–74.

Kupfer, D.J. & Detre, T.P. (1978). Tricyclic and monoamine-oxidase inhibitor antidepressants: Clinical use. In L.L. Iversen, S.D. Iversen & S.H. Snyder (eds.), *Handbook of Psychopharmacology*, vol. *14*, pp. 199–232. Plenum Press: New York.

Kupfer, D.J. & Thase, M.E. (1983). The use of the sleep laboratory in the diagnosis of affective disorders. *Psychiatric Clinics of North America* **6**, 3–25.

Lapin, I.P. & Oxenkrug, G.F. (1969). Intensification of the central serotonergic processes as a possible determinant of the thymoleptic effect. *Lancet* **1**, 132–6.

Leander, D.J. (1986). Peripheral action of serotonin as a model of depression. *Biological Psychiatry* **21**, 842–4.

Lecrubier, Y., Puech, A.J., Jouvent, R., Simon, P. & Widlocher, D. (1980). A beta adrenergic stimulant (salbutamol) versus imipramine in depression: A controlled study. *British Journal of Psychiatry* **136**, 354–8.

Lehr, E. (1986). Distress call activation in isolated chicks: A new behavioural model for antidepressants. *Psychopharmacology* **89**, S21.

Lehr, E. (1989). Distress-call reactivation in isolated chicks: A behavioural indicator with high selectivity for antidepressants. *Psychopharmacology* **97**, 145–6.

Leith, N.J. & Barrett, R.J. (1976). Amphetamine and the reward system: evidence for tolerance and post-drug depression. *Psychopharmacology* **46**, 19–25.

Leith, N.J. & Barrett, R.J. (1980). Effects of chronic amphetamine or reserpine on self-stimulation; animal model of depression? *Psychopharmacology* **72**, 9–15.

Leonard, B.E. (1982). On the mode of action of mianserin. In E. Costa & G. Racagni (eds.), *Typical and Atypical Antidepressants: Molecular Mechanisms*, pp. 301–129. Raven Press: New York.

Leshner, A.I., Remler, H., Biegon, A. & Samuel, D. (1979). Effects of desmethylimipramine (DMI) on learned helplessness. *Psychopharmacology* **66**, 207–13.

Lewinsohn, P.M. (1974). A behavioural approach to depression. In R.J. Friedman & M.M. Katz (eds.), *The Psychology of Depression: Contemporary Theory and Research*, pp. 157–85. Winston/Wiley: New York.

Lewinsohn, P.M., Zeiss, A.M., Zeiss, R.A. & Haller, R. (1977). Endogenicity and reactivity as orthogonal dimensions in depression. *Journal of Nervous and Mental Disorders* **164**, 327–32.

Lewinsohn, P.M., Steinmetz, J.L., Larson, D.W. & Franklin, J. (1981). Depression-related cognitions: Antecedent or consequence? *Journal of Abnormal Psychology* **90**, 213–19.

Lewis, J.K. & McKinney, W.T. (1976). Effects of electroconvulsive shock on the behavior of normal and abnormal rhesus monkeys. *Behavioral Psychiatry* **37**, 687–93.

Lewis, J.K., McKinney, W.T., Young, L.D. & Kraemer, G.W. (1976). Mother–infant separation in rhesus monkeys as a model of human depression: A reconsideration. *Archives of General Psychiatry* **33**, 699–705.

Lishman, W.A. (1972). Selective factors in memory. *Psychological Medicine* **2**, 248–53.

Lloyd, C. (1980*a*). Life events and depressive disorder reviewed. I. Events as predisposing factors. *Archives of General Psychiatry* **37**, 529–35.

Lloyd, C. (1980*b*). Life events and depressive disorders reviewed. II. Events as precipitating factors. *Archives of General Psychiatry* **37**, 541–8.

Lloyd, K.G., Garrigou, D. & Broekkamp, C.L.E. (1982). The action of monoaminergic, cholinergic and gabaergic compounds in the olfactory bulbectomized rat model of depression. In S.Z. Langer, R. Takahashi, T. Segawa & M. Briley (eds.), *New Vistas in Depression*, pp. 179–86. Pergamon Press: New York.

Maier, S.F. (1984). Learned helplessness and animal models of depression. *Progress in Neuropsychopharmacology and Biological Psychiatry* **8**, 435–46.

Maier, S.F. & Jackson, R.L. (1979). Learned helplessness: All of us were right (and wrong): Inescapable shock has multiple effects. In G. Bower (ed.), *The Psychology of Learning and Motivation*, vol. 13, pp. 155–218. Academic Press: New York.

Maier, S.F. & Seligman, M.E.P. (1976). Learned helplessness: Theory and evidence. *Journal of Experimental Psychology*: General **1**, 3–46.

Maier, S.F. & Testa, T.J. (1975). Failure to learn to escape by rats previously exposed to inescapable shock is partly produced by associative interference. *Journal of Comparative and Physiological Psychology* **88**, 554–64.

Mancinelli, A., D'Aranno, V., Borsini, F. & Meli, A. (1987). Lack of relationship between effect of desipramine on forced swimming test and brain levels of desipramine or its demethylated metabolite in rats. *Psychopharmacology* **92**, 441–3.

Manly, P.C., McMahon, R.J., Bradley, C.F. & Davidson, P.O. (1982). Depressive attributional style and depression following childbirth. *Journal of Abnormal Psychology* **91**, 245–54.

Martin, P., Soubrie, P. & Simon, P. (1986*a*). Shuttle-box deficits induced by inescapable shocks in rats: reversal by the beta-adrenoreceptor stimulants clenbuterol and salbutamol. *Pharmacology, Biochemistry and Behavior* **24**, 177–81.

Martin, P., Soubrie, P. & Simon, P. (1986*b*). Noradrenergic and opioid mediation of tricyclic-induced reversal of escape deficits caused by inescapable shock pretreatment in rats. *Psychopharmacology* **90**, 90–4.

Martin, P., Laporte, A.M., Soubrie, P., El Mestikawy, S. & Hamon, S. (1989). Reversal of helpless behaviour in rats by serotonin uptake inhibitors. In P. Bevan, A. Cools & T. Archer (eds.), *Behavioural Pharmacology of 5-HT*, pp. 231–4. Lawrence Erlbaum: New York.

McKinney, W.T. (1984). Animal models of depression: An overview. *Psychiatric Developments* **2**, 77–96.

McKinney, W.T. & Bunney, W.E. (1969). Animal model of depression: Review of evidence and implications for research. *Archives of General Psychiatry* **21**, 240–8.

McKinney, W.T., Young, L.D. & Suomi, S.J. (1973). Chlorpromazine treatment of disturbed monkeys. *Archives of General Psychiatry* **29**, 490–4.

Menzel, E.W., Davenport, R.K. & Rogers, C.M. (1963). Effects of environmental restriction upon the chimpanzee's responsiveness to objects. *Journal of Comparative and Physiological Psychology* **56**, 78–85.

Minor, T.R., Jackson, R.L. & Maier, S.F. (1984). Effects of task-irrelevant cues and reinforcement delay on choice-escape learning following inescapable shock: Evidence for a deficit in selective attention. *Journal of Experimental Psychology: Animal Behavior Processes* **10**, 543–56.

Minor, T.R., Pelleymounter, M.A. & Maier, S.F. (1988). Uncontrollable shock, forebrain norepinephrine, and stimulus selection during choice escape learning. *Psychobiology* **16**, 135–45.

Morgan, M.J., Einon, D.F. & Nicholas, D. (1975). The effects of isolation rearing on behavioural inhibition in the rat. *Quarterly Journal of Experimental Psychology* **27**, 615–34.

Morley, M.J., Bradshaw, C. & Szabadi, E. (1987). DSP4 alters the effect of d-amphetamine on variable-interval performance: analysis in terms of Herrnstein's equation. *Psychopharmacology* **92**, 247–53.

Murphy, D.L., Brodie, H.K.H., Goodwin, F.K. & Bunney, W.E. (1971). Regular induction of hypomania by L-dopa in 'bipolar' manic-depressive patients. *Nature* **229**, 135–6.

Muscat, R., Towell, A. & Willner, P. (1988). Changes in dopamine autoreceptor sensitivity in an animal model of depression. *Psychopharmacology* **94**, 545–50.

Nagayama, H., Hintgen, J.N. & Aprison, M.H. (1980). Pre- and postsynaptic serotonergic manipulations in an animal model of depression. *Pharmacology, Biochemistry and Behavior* **13**, 575–9.

Nagayama, H., Hintgen, J.N. & Aprison, M.H. (1981). Postsynaptic action by four antidepressive drugs in an animal model of depression. *Pharmacology, Biochemistry and Behavior* **15**, 125–30.

Nagayama, H., Akiyoshi, J. & Tobo, M. (1986). Action of chronically administered antidepressants on the serotonergic postsynapse in a model of depression. *Pharmacology, Biochemistry and Behavior* **25**, 805–11.

Nagy, A. (1977). Blood and brain concentrations of imipramine, clomipramine and their monomethylated metabolites after oral and intramuscular administration in rats. *Journal of Pharmacy and Pharmacology* **29**, 104–7.

Nelson, J.C. & Charney, D.S. (1981). The symptoms of major depression. *American Journal of Psychiatry* **138**, 1–13.

Niesink, R.J.M. & van Ree, J.M. (1982). Antidepressant drugs normalize the increased social behaviour of pairs of male rats induced by short term isolation. *Neuropharmacology* **21**, 1343–8.

Nomura, A., Shimizu, J., Kamateni, H., Kinjo, M., Watanabe, M. & Nakazawa, T. (1982). Swimming mice: In search of an animal model for human depression. In S.Z. Langer, R. Takahashi, T. Segawa & M. Briley (eds.), *New Vistas in Depression*, pp. 203–10. Pergamon: New York.

Nomura, S., Shimizu, J., Ueki, N., Sakaida, S. & Nakazawa, T. (1984). The activation of mice's behavior and reduction of brain beta-adrenergic receptor binding following repeated administration of antidepressant drugs. *Japanese Journal of Psychopharmacology* **4**, 237–41.

Noreika, L., Pastor, G. & Liebman, J. (1981). Delayed emergence of antidepressant efficacy following withdrawal in olfactory bulbectomized rats. *Pharmacology, Biochemistry and Behavior* **15**, 393–8.

Oades, R.D. (1985). The role of noradrenaline in tuning and dopamine in switching between signals in the CNS. *Neuroscience and Biobehavioral Reviews* **9**, 261–82.

O'Donnell, J.M. & Seiden, L.S. (1982). Effects of monoamine oxidase inhibitors on performance during differential reinforcement of low response rate. *Psychopharmacology* **78**, 214–18.

O'Donnell, J.M. & Seiden, L.S. (1983). Differential-reinforcement-of-low-rate 72-second scehedule: Selective effects of antidepressant drugs. *Journal of Pharmacology and Experimental Therapeutics* **224**, 80–8.

Olds, J. (1958). Effects of hunger and male sex hormone on self-stimulation of the brain. *Journal of Comparative and Physiological Psychology* **51**, 320–4.

O'Neill, K.A. & Valentino, D. (1982). Escapability and generalization: Effect on behavioural despair. *European Journal of Pharmacology* **78**, 379–80.

Overstreet, D.H., Janowsky, D.S., Gillin, J.C., Shiromani, P.J. & Sutin, E.L. (1986). Stress-induced immobility in rats with cholinergic supersensitivity. *Biological Psychiatry* **21**, 657–64.

Parale, M.P. & Kulkarni, S.K. (1986). Clonidine-induced behavioural despair in mice: Reversal by antidepressants. *Psychopharmacology* **89**, 171–4.

Parkes, C.M. (1972). *Bereavement: Studies of Grief in Adult Life*. International University Press: New York.

Paykel, E.S. (1979). Recent life events in the development of the depressive disorders. In R.A. Depue (ed.), *The Psychobiology of the Depressive Disorders: Implications for the Effects of Stress*, pp. 245–62. Academic Press: New York.

Paykel, E.S., Rowan, R.R., Rao, B.M. & Bhat, A. (1983). Atypical depression: Nosology and response to antidepressants. In P.J. Clayton & J.E. Barrett (eds.), *Treatment of Depression: Old Controversies and New Approaches*, pp. 247–52. Raven Press: New York.

Pearce, J.B. (1981). Drug treatment of depression in children. *Acta Paedopsychiatrica* **46**, 317–28.

Petty, F. & Sherman, A.D. (1980). Reversal of learned helplessness by imipramine. *Communications in Psychopharmacology* **3**, 371–3.

Plaznik, A., Danysz, W. & Kostowski, W. (1985*a*). Mesolimbic noradrenaline but not dopamine is responsible for organization of rat behavior in the forced swim test and an anti-immobilizing effect of desipramine. *Polish Journal of Pharmacology and Pharmacy* **37**, 347–57.

Plaznik, A., Danysz, W. & Kostowski, W. (1985*b*). A stimulatory effect of intraaccumbens injections of noradrenaline on the behavior of rats in the forced swim test. *Psychopharmacology* **87**, 119–23.

Pollard, G.T. & Howard, J.L. (1986). Similar effects of antidepressant and non-antidepressant drugs on behavior under an interresponse-time > 72-s schedule. *Psychopharmacology* **89**, 253–8.

Poncelet, M., Gaudel, G., Danti, S., Soubrie, P. & Simon, P. (1986). Acute versus repeated administration of desipramine in rats and mice: Relationships between brain concentrations and reduction of immobility in the swimming test. *Psychopharmacology* **90**, 139–41.

Porsolt, R.D. (1981). Behavioural despair. In S.J. Enna, J.B. Malick & E. Richelson (eds.), *Antidepressants: Neurochemical, Behavioral and Clinical Perspectives*, pp. 121–39. Raven Press: New York.

Porsolt, R.D., LePichon, M. & Jalfre, M. (1977). Depression: A new animal model sensitive to antidepressant treatment. *Nature* **266**, 730–2.

Porsolt, R.D., Bertin, A., Blavet, M., Deniel, M. & Jalfre, M. (1979). Immobility induced by forced swimming in rats: Effects of agents which modify central catecholamine and serotonin activity. *European Journal of Pharmacology* **57**, 201–10.

Porsolt, R.D., Chermat, R., Simon, P. & Steru, L. (1986). The tail suspension test: Computerized device for evaluating psychotropic activity profiles. *Psychopharmacology* **89**, S28.

Post, R.M. (1975). Cocaine psychoses: A continuum model. *American Journal of Psychiatry* **132**, 225–31.

Pulvirenti, L. & Samanin, R. (1986). Antagonism by dopamine, but not noradrenaline receptor blockers of the anti-immobility activity of desipramine after different treatment schedules in the rat. *Pharmacological Research Communications* **18**, 73–80.

Quinton, D., Rutter, M. & Liddle, C. (1984). Institutional rearing, parenting difficulties and marital support. *Psychological Medicine* **14**, 107–24.

Randrup, A., Munkvad, J., Fog, R., Gerlach, J., Molander, L., Kjellberg, B. & Scheel-Kruger, J. (1975). Mania, depression and brain dopamine. In W.B. Essman & L. Valzelli (eds.), *Current Developments in Psychopharmacology*, vol. 2, pp. 206–48. Spectrum: New York.

Ravaris, C.L., Robinson, D.S., Ives, J., Nies, A. & Bartlett, D. (1980). Phenelzine and amitriptyline in the treatment of depression. *Archives of General Psychiatry* **37**, 1075–80.

Reite, M., Short, R., Seiler, C. & Pauley, J.D. (1981). Attachment, loss and depression. *Journal of Child Psychology and Psychiatry* **22**, 141–69.

Robertson, J. & Bowlby, J. (1952). Responses of young children to separation from their mothers. *Cour du Centre Internationale de l'Enfance* **2**, 131–42.

Robertson, M. (1987). Molecular genetics of the mind. *Nature* **325**, 755.

Rosenthal, N.E., Sack, D.A., Carpenter, C.J., Parry, B.L., Mendelson, W.B. & Wehr, T.A. (1985). Antidepressant effects of light in seasonal affective disorder. *American Journal of Psychiatry* **142**, 163–70.

Roth, K.A. & Katz, R.J. (1981). Further studies on a novel animal model of depression: therapeutic effects of a tricyclic antidepressant. *Neuroscience and Biobehavioral Reviews* **5**, 253–8.

Rutter, M. & Madge, N. (1976). *Cycles of Disadvantage*. Heinemann: London.

Sahakian, B.J., Robbins, T.W., Morgan, M.J. & Iversen, S.D. (1975). The effects of psychomotor stimulants on stereotypy and locomotor activity in socially deprived and control rats. *Brain Research* **84**, 195–205.

Salamone, J.D. (1987). The actions of neuroleptic drugs on appetitive instrumental behaviours. In L.L. Iversen, S.D. Iversen & S.H. Snyder (eds.), *Handbook of Psychopharmacology*, vol. 19, pp. 575–608. Plenum Press: New York.

Sampson, D., Muscat, R. & Willner, P. (1990). Reversal of antidepressant action by dopamine antagonists in an animal model of depression. *Psychopharmacology*, in press.

Satoh, H., Mori, J., Shimomura, K., Ono, T. & Kikuchi, H. (1984). Effect of zimelidine, a new antidepressant, on the forced swimming test in rats. *Japanese Journal of Pharmacology* **35**, 471–3.

Schershlicht, R., Polc, P., Schneeberger, J., Steiner, M. & Haefely, W. (1982). Selective suppression of rapid eye movement sleep (REMS) in cats by typical and atypical antidepressants. In E. Costa & G. Racagni (eds.), *Typical and Atypical Antidepressants: Molecular Mechanisms*, pp. 359–84. Raven Press: New York.

Schick, J.F.E., Smith, D.E. & Wesson, D.R. (1973). An analysis of amphetamine toxicity and patterns of use. In D.E. Smith & D.R. Wesson (eds.), *Uppers and Downers*, pp. 23–61. Prentice Hall: Englewood Cliffs, N.J.

Schildkraut, J.J. (1965). The catecholamine hypothesis of depression: A review of supporting evidence. *American Journal of Psychiatry* **122**, 509–22.

Schmale, A.H. (1973). Adaptive role of depression in health and disease. In J.P. Scott & E. Senay (eds.), *Separation and Depression*, pp. 187–214. American Association for the Advancement of Science, Washington.

Schulterbrandt, J.G. & Raskin, A. (1977). *Depression in Childhood: Diagnosis, Treatment and Conceptual Models*. Raven Press: New York.

Segal, D.S., Knapp, S., Kuczenski, R. & Mandell, A.J. (1973). The effects of environmental isolation on behavior and regional rat brain tyrosine hydroxylase and tryptophan hydroxylase activities. *Behavioral Biology* **8**, 47–53.

Seiden, L.S. (1988). DRL-72 as a possible drug screen for antidepressants. Paper presented to the European Behavioural Pharmacology Society, Athens, September 1988.

Seiden, L.S., Dahms, J.L. & Shaughnessy, R.A. (1985). Behavioral screen for antidepressants: The effects of drugs and electroconvulsive shock on performance under a differential-reinforcement-of-low-rate schedule. *Psychopharmacology* **86**, 55–60.

Seligman, M.E.P. (1975). *Helplessness: On Depression, Development and Death*. Freeman: San Francisco.

Seligman, M.E.P., Rossellini, R.A. & Kozak, M. (1975). Learned helplessness in the rat: Reversibility, time course and immunization. *Journal of Comparative and Physiological Psychology* **88**, 542–7.

Sem-Jacobsen, C.W. (1976). Electrical stimulation and self-stimulation in man with chronic implanted electrodes. Interpretation of results and pitfalls. In A. Wauquier & E. Rolls (eds.), *Brain-Stimulation*

Reward, pp. 508–20. North Holland/American Elsevier: New York.

Sherman, A.D., Sacquitne, J.L. & Petty, F. (1982). Specificity of the learned helplessness model of depression. *Pharmacology, Biochemistry and Behavior* **16**, 449–54.

Shibata, S., Nakanishi, H., Watanabe, S. & Ueki, S. (1984). Effects of chronic administration of antidepressants on mouse-killing behavior (muricide) in olfactory bulbectomized rats. *Pharmacology, Biochemistry and Behavior* **21**, 225–30.

Shimizu, J., Nomura, S., Kuno, H. & Nakazawa, T. (1984). Effects of water temperature and weight of wheel load on water wheel turning behavior of imipramine administered mice. *Psychopharmacology* **84**, 20–1.

Shopsin, B., Friedman, E., Goldstein, M. & Gershon, S. (1975). The use of synthesis inhibitors in defining a role for biogenic amines during imipramine treatment in depressed patients. *Psychopharmacological Communications* **1**, 239–49.

Shopsin, B., Friedman, E. & Gershon, S. (1976). Parachlorophenylalanine reversal of tranylcypromine effects in depressed patients. *Archives of General Psychiatry* **33**, 811–19.

Simpson, D.M. & Annau, Z. (1977). Behavioral withdrawal following several psychoactive drugs. *Pharmacology, Biochemistry and Behavior* **7**, 59–64.

Soblosky, J.S. & Thurmond, J.B. (1986). Biochemical and behavioral correlates of chronic stress: Effects of tricyclic antidepressants. *Pharmacology, Biochemistry and Behavior* **24**, 1361–8.

Sofia, R.D. (1969). Structural relationship and potency of agents which selectively block mouse killing (muricide) behaviour in rats. *Life Science* **8**, 1201–10.

Soubrie, P. (1986). Reconciling the role of central serotonin neurons in human and animal behaviour. *Behavioral and Brain Science* **9**, 319–64.

Soubrie, P. (1989). Role of serotonin-containing neurones and 5HT receptors in animal models of anxiety and depression. In P. Bevan, A. Cools & T. Archer (eds.), *Behavioural Pharmacology of 5HT*, pp. 337–52. Erlbaum: New York.

Soubrie, P., Martin, P., El Mestikawy, S., Thiebot, M.H., Simon, P. & Hamon, M. (1986). The lesion of serotonergic neurons does not prevent antidepressant-induced reversal of escape failures produced by inescapable shocks in rats. *Pharmacology, Biochemistry and Behavior* **25**, 1–6.

Spitz, R. (1946). Anaclitic depression. *Psychoanalytical Studies of the Child* **2**, 113–17.

Stein, L. (1962). New methods for evaluating stimulants and antidepressants. In J.H. Nodine & J.H. Moyer (eds.), *The First Hahnemenn*

Symposium on Psychosomatic Medicine, pp. 297–301. Lea and Fibiger: Philadelphia.

Stein, L. (1968). Chemistry of reward and punishment. In D.H. Efron (ed.), *Psychopharmacology: A Review of Progress. 1957–1967*, pp. 105–23. U.S. Govt. Printing Office: Washington D.C.

Steru, L., Chermat, R., Thierry, B. & Simon, P. (1985). The tail suspension test: A new method for screening antidepressants in mice. *Psychopharmacology* **85**, 367–70.

Sulser, F. (1978). New perspectives on the mode of action of antidepressant drugs. *Trends in Pharmacological Science* **1**, 92–4.

Sulser, F., Owens, M.L. & Dingell, J.V. (1966). On the mechanism of amphetamine potentiation by desipramine. *Life Science* **5**, 2005–10.

Suomi, S.J. (1976). Factors affecting responses to social separation in rhesus monkeys. In G. Serban & A. Kling (eds.), *Animal Models in Human Psychobiology*, pp. 9–26. Plenum Press: New York.

Suomi, S.J., Harlow, H.F. & Domek, C.J. (1970). Effect of repetitive infant-infant separation of young monkeys. *Journal of Abnormal Psychology* **76**, 161–72.

Suomi, S.J., Seaman, S.F., Lewis, J.K., DeLizio, R.B. & McKinney, W.T. (1978). Effects of imipramine treatment on separation-induced social disorders in rhesus monkeys. *Archives of General Psychiatry* **35**, 321–5.

Szewczak, M.R., Fielding, S. & Cornfeldt, M. (1982). Rat internal capsule lesion: further characterization of antidepressant screening potential. *Society for Neuroscience Abstracts* **8**, 465.

Thiebot, M.H., Le Bihan, C., Soubrie, P. & Simon, P. (1985). Benzodiazepines reduce the tolerance to reward delay in rats. *Psychopharmacology* **86**, 147–52.

Thornton, E.W., Evans, J.A.C. & Harris, C. (1985). Attenuated response to nomifensine in rats during a swim test following lesion of the habenula complex. *Psychopharmacology* **87**, 81–5.

Tucker, J.C. & File, S.E. (1986). A review of the effects of tricyclic and atypical antidepressants on spontaneous motor activity in rodents. *Neuroscience and Biobehavioral Reviews* **10**, 115–21.

Turner, C., Davenport, R. & Rogers, C. (1969). The effect of early deprivation on the social behaviour of adolescent chimpanzees. *American Journal of Psychiatry* **125**, 1531–6.

Tyrer, P. (1979). Clinical use of monoamine oxidase inhibitors. In E.S. Paykel & A. Coppen (eds.), *Psychopharmacology of Affective Disorders* pp. 159–78. Oxford University Press: Oxford.

Ueki, S. (1982). Mouse-killing behaviour (muricide) in the rat and the effect of antidepressants. In S.Z.

Langer, R. Takahashi, T. Segawa & M. Briley (eds.), *New Vistas in Depression*, pp. 187–94. Pergamon Press: New York.

Valenstein, E.S. (1973). *Brain Control: A Critical Examination of Brain Stimulation and Psychosurgery*. Wiley: New York.

Valzelli, L. & Bernasconi, S. (1971). Psychoactive drug effects on behavioural changes induced by prolonged socio-environmental deprivation in rats. *Psychological Medicine* **6**, 271–6.

van der Zee, P., Koger, H.S., Gootjes, J. & Hespe, W. (1980). Aryl 1,4-dialle(en)ylpiperazines as selective and very potent inhibitors of dopamine uptake. *European Journal of Medicinal Chemistry – Chimica Therapeutica* **15**, 363–70.

van Praag, H.M. (1980). Central monoamine metabolism in depression. I. Serotonin and related compounds. *Comprehensive Psychiatry* **21**, 30–43.

van Riezen, H., Schnieden, H. & Wren, A.F. (1977). Olfactory bulb ablation in the rat: Behavioural changes and their reversal by antidepressant drugs. *British Journal of Pharmacology* **60**, 521–8.

Watanabe, S., Inoue, M. & Ueki, S. (1979). Effects of psychotropic drugs injected into the limbic structures on mouse-killing behaviour in the rat with olfactory bulb ablation. *Japanese Journal of Pharmacology* **29**, 493–6.

Watson, R., Hartman, E. & Schildkraut, J.J. (1973). Amphetamine withdrawal: Affective state, sleep patterns and MHPG excretion. *American Journal of Psychiatry* **129**, 263–9.

Wehr, T.A. & Wirz-Justice, A. (1982). Circadian rhythm mechanisms in affective illness and in antidepressant drug action. *Pharmacopsychiatry* **15**, 31–9.

Weingartner, H. & Silberman, E. (1982). Models of cognitive impairment: Cognitive changes in depression. *Psychopharmacological Bulletin* **18**, 27–42.

Weinstock, M., Speiser, Z. & Ashkenazi, R. (1978). Changes in brain catecholamine turnover and receptor sensitivity induced by social deprivation in rats. *Psychopharmacology* **56**, 205–9.

Weiss, J.M., Goodman, P.A., Losito, B.G., Corrigan, S., Charry, J.M. & Bailey, W.H. (1981). Behavioral depression produced by an uncontrollable stressor: relationship to norepinephrine, dopamine, and serotonin levels in various regions of rat brain. *Brain Research Reviews* **3**, 167–205.

Weiss, J.M., Bailey, W.H., Goodman, P.A., Hoffman, L.J., Ambrose, M.J., Salman, S. & Charry, J.M. (1982). A model for neurochemical study of depression. In M.Y. Spiegelstein & A. Levy (eds.), *Behavioral Models and the Analysis of Drug Action*, pp. 195–223. Elsevier: Amsterdam.

Weissman, M.S. & Klerman, G.L. (1977). Sex differences and the epidemiology of depression. *Archives of General Psychiatry* **34**, 98–111.

Westenberg, H.G.M. & Den Boer, J.A. (1988). Serotonin uptake inhibitors and agonists in the treatment of panic disorder. *Psychopharmacology* **96** (Suppl.), 56.

Williams, A.W., Ware, J.E. & Donald, C.A. (1981). A model of mental health, life events and social support applicable to general populations. *Journal of Health Social Behaviour* **22**, 324–36.

Williams, J.L. & Maier, S.F. (1977). Transsituational immunization and therapy of learned helplessness in the rat. *Journal of Experimental Psychology: Animal Behavior Processes* **3**, 240–52.

Willner, P. (1983). Dopamine and depression: A review of recent evidence. *Brain Research Reviews* **6**, 211–46.

Willner, P. (1984). The validity of animal models of depression. *Psychopharmacology* **83**, 1–16.

Willner, P. (1985). *Depression: A Psychobiological Synthesis*. Wiley: New York.

Willner, P. (1986). Validating criteria for animal models of human mental disorders: Learned helplessness as a paradigm case. *Progress in Neuropsychopharmacology and Biological Psychiatry* **10**, 677–90.

Willner, P. (1987). The anatomy of melancholy: The catecholamine hypothesis of depression revisited. *Reviews in Neuroscience* **1**, 77–99.

Willner, P. (1989a). Sensitization to the actions of antidepressant drugs. In M.W. Emmett-Oglesby & A.J. Goudie (eds.), *Psychoactive Drugs: Tolerance and Sensitization*, pp. 157–78. Humana Press: New Jersey.

Willner, P. (1989b). Towards a theory of serotonergic dysfunction in depression. In P. Bevan, A. Cools & T. Archer (eds.), *Behavioural Pharmacology of 5-HT*, pp. 157–78. Lawrence Erlbaum: New York.

Willner, P. & Clark, D. (1978). A reappraisal of the interaction between tricyclic antidepressants and reserpine-like drugs. *Psychopharmacology* **58**, 55–62.

Willner, P., Towell, A., Sampson, D., Sophokleous, S. & Muscat, R. (1987). Reduction of sucrose preference by chronic mild unpredictable stress, and its restoration by a tricyclic antidepressant. *Psychopharmacology* **93**, 358–64.

Willner, P., Muscat, R., Sampson, D. & Towell, A. (1988). Dopamine and antidepressant drugs: the antagonist challenge strategy. In H. Belmaker, M. Sandler & A. Dahlstrom (eds.), *Progress in Catecholamine Research, Part C*, pp. 275–9. Alan R. Liss: New York.

Willner, P., Sampson, D., Phillips, G., Fichera, R., Foxlow, P. & Muscat, R. (1989). Effects of isolated housing and chronic antidepressant treatment on cooperative social behaviour in rats. *Behavioural Pharmacology*, **1**, 85–90

Willner, P., Wilkes, M. & Orwin, A. (1990*b*). Attributional style and perceived stress in endogenous and reactive depression. *Journal of Affective Disorders*, **18**, 281–7.

Winefield, A.H., Tiggeman, M. & Smith, S. (1987). Unemployment, attributional style and psychological well-being. *Personality and Individual Differences* **8**, 659–65.

Wise, C.D., Berger, B.D. & Stein, L. (1970). Serotonin: A possible mediator of behavioral suppression induced by anxiety. *Diseases of the Nervous System*, Suppl. 31, 34–7.

Wise, R.A. (1982). Neuroleptics and operant behaviour: The anhedonia hypothesis. *Behavioral and Brain Sciences* **5**, 39–87.

Wortman, C.B. & Dintzer, L. (1978). Is an attributional analysis of the learned helplessness phenomenon viable? A critique of the Abramson-Seligman-Teasdale reformulation. *Journal of Abnormal Psychology* **87**, 75–90.

Zacharko, R.M., Bowers, W.J., Kokkinidis, L. & Anisman, H. (1983). Region-specific reductions of intracranial self-stimulation after uncontrollable stress: Possible effects on reward processes. *Behavioral Brain Research* **9**, 129–41.

Zacharko, R.M., Bowers, W.J. & Anisman, H. (1984). Responding for brain stimulation: Stress and desmethylimipramine. *Progress in Neuro-Psychopharmacology and Biological Psychiatry* **8**, 601–6.

Zebrowska-Lupina, I. (1980). Presynaptic alpha-adrenoceptors and the action of tricyclic antidepressant drugs in behavioural despair in rats. *Psychopharmacology* **71**, 169–72.

Zis, A.P. & Goodwin, F.K. (1979). Novel antidepressants and the biogenic amine hypothesis of depression. The case for iprindole and mianserin. *Archives of General Psychiatry* **36**, 1097–107.

6

Screening for new antidepressant compounds

WOJCIECH DANYSZ, TREVOR ARCHER and CHRISTOPHER J. FOWLER

6.1 INTRODUCTION

Since the introduction of the monoamine oxidase inhibitors (MAOIs) and the tricyclic antidepressant compounds into clinical practice for the treatment of depression, a large number of agents with different pharmacological profiles have been shown to possess antidepressant properties. However, available tests and methods used for both the prediction and selection of new antidepressant agents (ADs) are generally unsatisfactory. Indeed the suitability of many of these methods may be questionable (see Willner, 1984*a*). This dilemma represents an important and pressing problem that psychopharmacologists must address, since there is a clear clinical requirement for more effective and, at least as important, less toxic, ADs. In the present chapter, some of the different biochemical and behavioural methods used both for the study of AD action and for the selection of new compounds as potential ADs are reviewed. These two areas of AD research are closely interconnected, and it is not merely difficult but also inadvisable to dissociate the search for the mechanism of action of ADs from the practical aspects of drug screening.

One of the difficulties inherent in animal models for AD action is that they represent, at best, a model for drugs active in the disorder, rather than the disorder itself. This is often forgotten and even more often overlooked, leading to the appearance in the literature of a large body of articles with titles in the style 'XXXX, a new potential antidepressant', where the key word 'potential' means active in selected animal models. Such a practice should be discouraged, since it blurs the distinction between an agent with pharmacological properties similar to those of established ADs and an agent with AD actions. Such AD actions can only be established in clinical trials, designed to provide answers to the following questions:

(a) Is the test compound effective in depression?
(b) What is the efficacy of the test compound with respect to reference compounds (particularly amitriptyline)?
(c) What is the spectrum of activity of the test compound?
(d) Does the test compound have fewer side-effects than the reference compound?
(e) What is the speed of action of the test compound?

The aim of the industrial psychopharmacologist is to provide animal drug screens which maximise the chances of a good performance of the test compound in these clinical trials. Up to now, this has not generally been the case: many compounds have been discovered that are clinically effective, but none of them has demonstrated convincing superiority as ADs over traditional therapies (see Zis & Goodwin, 1979; Cording-Tommel & Von Zerssen, 1982; Van Dorth, 1983; Rickels *et al.*, 1985). The main achievements over the last 20 years have been the discovery of compounds (in particular, the serotintin (5-HT) uptake inhibitors) that are almost as effective as the original ADs, but which are less toxic and may prove to have a broader spectrum of clinical indications. The introduction of such compounds has in general been the result of one of four approaches:

New Drug 'XAD'

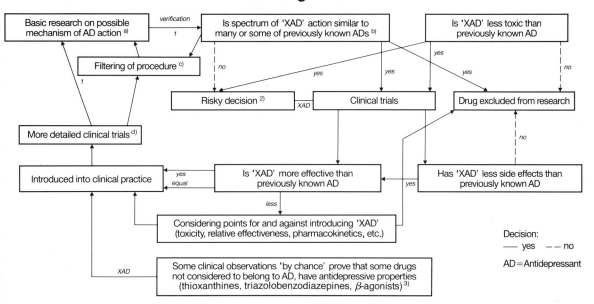

Fig. 6.1. Procedure in the search for new antidepressant compounds, XADs.

1. *Clinical trials select effective ADs which in consequence create our opinions and hypotheses on the mechanism of AD action. This changes the criteria that new AD should fulfil, which is in fact self-limiting. However, 'different new' drugs may be introduced where the abbreviation XAD is given.*

2. *It is extremely risky to make clinical trials when we are not completely sure that 'XAD' is effective since always the possibility of suicide attempts should be considered.*

3. *It should be considered that 'antidepressive' properties of some of these compounds may result from relieving some of the symptoms (e.g. anxiety) and thus elevating mood indirectly (see Cording-Tömmel & von Zerssen, 1982).*

a) Drug screens, bioassays and animal models. b) Status of 'predictive validity'. c) Considerations of 'construct validity'. d) Clinical trials should not be continued (as is often the case) in relatively young patients but should also include phase I, II and III studies in elderly depressed patients.

(1) The chance (or serendipitous) finding in a clinical trial of the mood elevating action of a compound. Such serendipity led to the discovery of the antidepressant properties of iproniazid in trials for tuberculostatic activity (Bosworth, 1959), and of imipramine during an on-going study of antihistamines as potential hypnotic agents (see Kuhn, 1970). More recently, mianserin was synthesised with the aim of developing an anti-asthmatic anti-migraine agent, due to its anti-histaminergic and antiserotoninergic activity. Clinical trials demonstrated efficacy for these indications, but the compound was found in addition to have certain central 'side-effects', such as sedation and a mood-enhancing action (Itil *et al.*, 1972; Vossenaar, 1980).

Such observations led in each case to clinical trials that demonstrated the antidepressant properties of the compounds.

(2) The clinical trial of structurally related compounds on the basis of their similarity to known ADs in animal models (the 'me-too' approach).

(3) The development of new compounds not structurally related to the existing ADs, but active in the screening tests.

(4) The screening of new compounds based on a novel research hypothesis of AD action.

These approaches are shown schematically in Fig. 6.1.

We now know that many compounds with widely differing pharmacological profiles have AD action (see Table 6.1). This finding has had both a negative and a positive effect on attitudes towards screening tests for new ADs. The negative effect has been disenchantment with existing tests, in which many of the compounds may not be active. Such tests are generally those which select for particular pharmacological actions of compounds, such as noradrenaline (NA) or 5-HT uptake inhibition. The positive effect has been a search for screening procedures that are 'catch-all', such that ADs will give a positive response in the test, regardless of the primary pharmacological mechanism of action (for a review of the behavioural screens with respect to their predictive and 'catch-all' ability, see Willner, 1984*a*).

6.2 TREATMENT PARADIGMS IN PRECLINICAL AD RESEARCH

When interpreting results with established ADs and new compounds in animal experiments a large number of variables (i.e. dose, duration, etc.) must be taken into account. These are summarised briefly below.

(a) *Dosage*
 The general tendency of the psychopharmacologist is to give the AD at a single dose, often 5 or 10 mg/kg. Such an approach is understandable, but differs somewhat from the considerable variation in dosage of ADs given to patients. Ideally, several doses of the AD should be given in the animal experiment to determine whether observed results are dose-related, and whether the effects are observed at dosages relevant to the situation in people. Relevant to this is the question of whether single or multiple daily doses should be given, and whether the compound should be given orally, subcutaneously or intraperitoneally (for further discussion, see Danysz *et al.*, 1987*a*).

(b) *Selection of reference ADs*
 Since the ultimate aim of the psychopharmacologist is either to use a 'catch-all' screen or alternatively to find a common mechanism of action of AD, it is advisable to select a number of ADs with widely differing pharmacological profiles for reference. This may not, however, be as easy as it sounds,

given that conflicting clinical evidence causes difficulty in establishing beyond doubt that a given compound is an effective AD (Morris & Beck, 1974). Nevertheless, there are a wide range of compounds and/or treatments from which to choose, ranging from the standard AD such as amitriptyline and electroconvulsive treatment, to a variety of compounds for which antidepressant properties have been claimed. These include neuroleptics (flupenthixol, chlorprothixene), anxiolytics (alprazolam, adinazolam), beta-adrenoceptor agonists (salbutamol, clenbuterol, fenoterol), compounds interacting with GABA-ergic systems (carbamazepine), dopaminergic agonists (pizotifen), alpha$_2$-adrenoceptor antagonists (idazoxan), angiotensin convertase inhibitors (captopril), and phosphodiesterase inhibitors (rolipram) (Standal, 1977; Zebrowska-Lupina, 1980; Zebrowska-Lupina & Kozyrska, 1980; Lahti *et al.*, 1982; Robertson & Trimble, 1982; Emrich *et al.*, 1983; Przegalinski & Bigajska, 1983; Post *et al.*, 1982, 1983, 1984; Simon *et al.*, 1984; Maj *et al.*, 1985; Von Voigtlander & Straw, 1985; Deicken, 1986).

(c) *Duration of treatment*
 In the majority of AD screening tests, single doses are employed, which is convenient because of the simplicity of this procedure. Although this type of procedure may have less significance than repeated administration studies for the later therapeutic action, it may still offer some not inconsiderable predictive value in the search for new ADs. For repeated administration studies, two to three weeks of treatment is generally given, since this is the sort of period usually required for significant therapeutic responses in clinical trials (see Oswald *et al.*, 1972). In biochemical studies, a 24–48 hour wash-out time following the last administration is usually used so that direct effects of the compounds on the receptor systems under study do not obfuscate interpretation of the results. A potential problem concerning withdrawal effects exists, as seen for example, in the behavioural studies of 5-HT receptor function after chronic AD treatment (see Willner, 1985)

and studies on the desensitization of DA autoreceptors (Towell *et al.*, 1986). The interval after the last dose of the compound is also relevant in studies where receptor function is estimated *in vivo* by administration of receptor agonists (see Tables 6.3–6.6), since misleading results can be obtained if pharmacokinetic interactions are present. Thus, for example, it has been shown that desipramine both increases plasma levels of morphine and elevates morphine analgesia (Goldstein *et al.*, 1982), an effect found as late as 24 hours after the final dose. However, when the pharmacokinetic interaction was avoided by direct central administration of morphine, the opposite effect, i.e. an attenuation of morphine analgesia, was found (Kellstein *et al.*, 1984). Similarly, some of the tricyclic ADs have been found to decrease the metabolism of amphetamine in the liver, and consequently may increase some responses related to this drug (Sulser *et al.*, 1966).

6.3 BIOCHEMICAL SCREENING PROCEDURES

6.3.1 Acute biochemical effects

The finding that the monoamine oxidase inhibitors (MAOIs) were active antidepressants led to the synthesis of a large number of such compounds. Initial compounds, such as phenelzine, were rather unselective, in that they inhibited both -A and -B forms of the enzyme. In addition these drugs suffer the problem of the 'cheese effect', an interaction with tyramine in foodstuffs that can cause hypertensive crises, occasionally with fatal outcome. The finding that MAO existed in two forms, MAO-A and MAO-B (Johnston, 1968) led to the development of a large series of inhibitors selective for one of the forms. Several MAO-A inhibitors appear to be effective ADs, of which the reversible inhibitors have relatively mild 'cheese effects' (for review, see Fowler & Ross, 1984). A similar pattern has been followed for the monoamine reuptake inhibitors, where the relatively unselective compounds produced initially have been followed up by compounds selective for NA, 5-HT or dopamine (DA) reuptake (see Table 6.1). Both these areas have used predominantly biochemical screening

methods, such as: *in vitro* inhibition of MAO/monoamine reuptake, or *ex vivo* monitoring of monoamine and monoamine metabolite levels. This approach was backed up by 'traditional' behavioural models of monoamine function such as potentiation of 5-hydroxytryptophan-induced behaviours or prevention of reserpine-induced ptosis (for an example of a screening project for the reversible MAO-A selective inhibitor amiflamine, see Florvall *et al.*, 1978; Ask *et al.*, 1982). While acute biochemical tests of the type listed above are well suited for screening purposes, they have two important disadvantages.

First, the approach is self-limiting. If the aim is to produce compounds that are more efficacious than the compounds presently available, the odds of success in this respect will decrease with time. On the other hand, this approach has had some success in producing compounds that are as efficacious as compounds presently available but with lower toxicity. Thus, for example, inclusion of a screen assessing antihistaminic and anticholinergic properties of monoamine reuptake inhibitors can easily be used to 'screen out' compounds where potential toxicity problems could rise (Hall & Ögren, 1981, 1984). Similarly, a tyramine-potentiating test is often included at a fairly early stage in the development of new MAO inhibitors (Ask *et al.*, 1982).

Second, the screen may not be relevant. In the clinic, it is generally thought that two to three weeks of treatment are required before therapeutic effects of antidepressants are seen. This may be an oversimplification – some studies have reported mood improvements after shorter treatment periods with compounds such as amoxapine, amitryptyline, clenbuterol and salbutamol (Haskell *et al.*, 1975; Prusoff *et al.*, 1981; Simon *et al.*, 1984; see also Willner, this vol., Ch. 5). Nevertheless, this general observation would suggest that it is not the acute pharmacological effects of the AD but rather the changes in neurotransmission following repeated administration that are of primary relevance to the therapeutic actions.

6.3.2 Biochemical effects following repeated treatment

The finding that repeated AD treatment results in a reduction of the number of beta-adrenoceptors in

Table 6.1. *A selection of some recently developed antidepressant agents*

Name of compound	Primary acute pharmacological effect	Example reference
Adinazolam	GABA potentiation	Dunner *et al.*, 1987
Alaproclate	5-HT uptake inhibition	Frost *et al.*, 1984; Ögren *et al.*, 1985
Amoxapine	NA uptake inhibition	Ban *et al.*, 1982
AHR-9377	NA uptake inhibition	Kinnier *et al.*, 1984
BAY R 1531	5-HT$_{1A}$ agonist	Schuurman *et al.*, 1987
Bupropion	DA uptake inhibition	Soroko *et al.*, 1977; Ferris *et al.*, 1981
Buspirone	5-HT$_{1A}$ agonist, DA antagonist	Glaser & Traber, 1983 Hjorth & Carlsson, 1982
Citalopram	5-HT uptake inhibition	Hyttel, 1983
Diclofensine	NA, DA, 5-HT uptake inhibition	Keller *et al.*, 1982
Dothiepin	(weaker anticholinergic activity)	Stratas, 1984
Febarbamate	NA uptake inhibition	Brenner *et al.*, 1984
Femoxetine	5-HT uptake inhibition	Dahl *et al.*, 1982
Fluvoxamine	5-HT uptake inhibition	Claasen *et al.*, 1977; Claasen, 1983
Gepirone	5-HT$_{1A}$ agonist	Eison & Vocca, 1985
Indalpine	5-HT uptake inhibition	Sechter *et al.*, 1984
Ipsapirone	5-HT$_{1A}$ agonist	Cunningham *et al.*, 1985
Mianserine	5-HT antagonist	Brogden *et al.*, 1978
Minaprine	Increase in NA, DA transmission, MAO-A inhibition	Biziere *et al.*, 1985
Meclobemide	MAO-A inhibition	Da Prada *et al.*, 1984; Stefanis *et al.*, 1984
Nomifensine	DA uptake inhibition	Brogden *et al.*, 1979
Oxaprotiline	NA uptake inhibition	Delini-Stula *et al.*, 1982
Paroxetine	5-HT uptake inhibition	Oswald & Adam, 1986
Pirlindol	NA potentiation	Martorana & Nitz, 1979
Progabide	GABA-mimetic	Mondrup *et al.*, 1983
Ritanserin	5-HT$_2$ antagonist	Colpaert *et al.*, 1985; Leysen *et al.*, 1985
Rolipram	cAMP phosphodiesterase inhibitor	Zeller *et al.*, 1984
RO-11-2465	5-HT uptake inhibition	Avento *et al.*, 1984
RU-24969	5-HT$_1$ agonist	Euvrard & Boissier, 1980
Salbutamol	Beta$_2$-agonist	Jouvent *et al.*, 1977; Simon *et al.*, 1978
Sertraline	5-HT uptake inhibition	Koe *et al.*, 1983
SL 79 229	GABA agonist	Bartholini *et al.*, 1984
Sulpiride	DA antagonist	Montanaro *et al.*, 1982
Tiflucarbine	5-HT uptake inhibition	Glaser *et al.*, 1987
Tolaxatone	MAO-A inhibition	Strolin Benedetti *et al.*, 1983
Trazodone	5-HT uptake inhibition	Silvestrini, 1982
Viqualine	5-HT uptake inhibition	Fourtillan *et al.*, 1985
YM-08054-1	5-HT, NA uptake inhibition	Tachikawa *et al.*, 1979
Zimeldine	5-HT uptake inhibition	Coppen *et al.*, 1983

All the compounds in this list (which is by no means complete) have been reported to possess antidepressant properties in at least one clinical study.

the rat brain with a concomitant reduction in response to beta-adrenoceptor stimulation (Vetulani & Sulser, 1975), had important consequences. A large number of studies have been undertaken to determine whether or not antidepressants with widely different acute pharmacological profiles (including electroconvulsive shock) have common effects on central neurotransmission following repated treatment. Initial studies concentrated on beta-adrenoceptors and 5-HT$_2$ receptors, but in recent years a vast literature has accumulated documenting the effects of antidepressants on other receptor systems in the brain. As an example, some of the effects of repeated treatment with the antidepressant MAO-A inhibitor clorgyline are given in Table 6.2, indicating that repeated clorgyline treatment results in profound changes in the brain.

Since this area of AD research has been the subject of an excellent recent review (Green, 1987), we shall restrict ourselves to a brief discussion of the following points:

(a) As pointed out by Green (1987), the search for a common mechanism of action of antidepressants in an animal model may be misplaced, since patients classified as 'non-responders' with respect to a given AD may show significant clinical improvement when another compound is administered.

(b) The case that the change in receptor numbers is directly related to AD action is not proven. This is perhaps best encapsulated in the title of a paper by Ferris & Beaman (1983): 'Bupropion: a new antidepressant drug, the mechanism of action of which is not associated with down-regulation of post-synaptic beta-adrenergic, serotonergic (5-HT$_2$), alpha$_2$-adrenergic, imipramine and dopaminergic receptors in brain'. The functional relevance of various changes in receptor number reported from binding experiments is sometimes hard to elucidate (see Danysz *et al.*, 1987*a*). The 'down-regulation' of beta-adrenoceptors is perhaps the most promising with respect to screening, since it is the most common effect observed with many different ADs, although the lack of 'down-regulation' with some ADs including the tricyclic AD trimipramine and 5-HT uptake inhibitors,

raises the possibility of false negatives (Hauser *et al.*, 1985). As against this is the finding that the ability of ADs to down-regulate beta-adrenoceptors is negatively rather than positively correlated to their clinical efficacies (Willner, 1984*b*). Thus, a screen based on what is perhaps a secondary adaptive response may not give rise to the 'ideal' antidepressant.

(c) Screening capacity may be a problem: the repeated treatment 'down-regulation' approach is time consuming and not particularly well suited to screening of large numbers of compounds.

From the above discussion, it would appear (at least to the present authors) that a biochemical approach to AD screening tests may well have outlived its usefulness as a source of novel antidepressants. On the other hand, it is vital that compounds emerging from behavioural screens (see below) are investigated biochemically both for their acute pharmacological profiles (since this could lead to new biochemical screens to improve the profiles of ADs) and for the changes in neurotransmission after repeated treatment. Biochemical tests for potential toxic properties of ADs, such as tyramine potentiation in the case of MAO inhibitors and anticholinergic properties of ADs in general, also remain a vital ingredient of preclinical screening programmes.

6.4 BIOBEHAVIOURAL ASSAYS AND SCREENING TESTS

Biobehavioural assays and screening tests can be divided into two main groups: drug-induced, and those based on other treatments, such as interaction with the environment, surgical lesion or electrical stimulation of the brain. These two main groups are discussed in turn below.

6.4.1 Effects of ADs on drug-induced behaviours

This paradigm is used to investigate the effect of both acute and repeated AD treatment upon behaviours induced by directly and indirectly acting receptor agonists. Directly acting agonists include both neurotransmitters and synthetic agonists, such as 5-methoxy-N,N-dimethyltrypta-

Table 6.2. *Effect of the irreversible monoamine oxidase-A inhibitor clorgyline on central and peripheral neurotransmission – some recent findings (data for rat unless otherwise stated)*

Neurotransmitter system/ dose	Finding	Reference
Noradrenergic neurons		
alpha$_1$-adrenoceptors		
1 mg/kg/day, 21 days	Decreased cortical ^{3}H-WB4101 binding, partially reversed by i.c.v. 6-OHDA; decreased pressor response to yohimbine in pithed rat	1, 2
alpha$_2$-adrenoceptors		
1 mg/kg/day, 10–21 days	Decreased cortical ^{3}H-clonidine binding, reversed by i.c.v. 6-OHDA; decreased sensitivity of NA release from rat cortical microsomes to inhibitory effects of clonidine	1, 3, 4
1 mg/kg/day, 28 days	No change in inhibitory effects of clonidine on K$^+$-stimuilated NA release from cortical synaptosomes; increased Ca^{2+}-sensitivity of NA release process	5
1 mg/kg/day, 28 days	Decreased inhibitory effect of clonidine on motor activity	6
2 mg/kg/day, 21 days	Decreased sensitivity of vas deferens to effects of clonidine	7
beta-adrenoceptors		
1 mg/kg/day, 14–21 days	Decreased B$_{max}$ of cortical ^{3}H-dihydroalprenolol binding (blocked by 6-OHDA), decreased binding in brain stem; decreased cortical cAMP response to NA	1, 3, 4, 8, 9
Electrophysiology		
4 mg/kg/day, 21–28 days	Decreased locus coeruleus cell firing	10
1–4 mg/kg/day, 21–28 days	No change in responsiveness of hippocampal cell firing to iontophoresed NA	11
Dopaminergic neurons		
1 mg/kg/day, 21–28 days	Increased apomorphine-induced stereotypies; no change in striatal ^{3}H-spiperone binding	12
Serotonergic neurons		
1 mg/kg/day, 4 days	Decreased ^{3}H-5-HT binding in cortex; not found in 5,7-DHT or PCPA-treated animals	13, 14
1 mg/kg/day, 21 days	Decreased effects of m-chlorophenylpiprazine on food intake, limb abduction and head shakes	15
0.1 mg/kg/day, 7 days	No change in ejaculation and behavioural responses produced by 5-MeODMT; decreased resonses when clorgyline and p-chloramphetamine treatment combined	10
Electrophysiology		
4 mg/kg/day, 21–28 days	No change in hippocampal pyramidal cell	10
1 mg/kg/day, 21 days	Decreased hippocampal pyramidal cell responsiveness to iontophoresed 5-HT	11
Circadian rhythms		
2 mg/kg/day, 14 days	Slows free-running circadian rest–activity cycle in hamsters	17
4 mg/kg/day, 14 days	Changes in apparent rhythm of number of ^{3}H-WB4101 binding sites in forebrain membranes	18
4 mg/kg/day, 14 days	Phase change of alpha-melanotropin rhythm in forebrain	19
2 mg/kg/day, 10–11 days	Dampens rhythm of disk-shedding from visual cell outer segments of retina	20
Other changes		
1 mg/kg/day, 21 days	Changes in brain protein concentrations	21

mine (5-MeODMT) or clonidine. Indirectly acting agonists include monoamine precursors, releasers and uptake inhibitors. Although there are a number of potential pitfalls associated with these methods (for discussion, see Danysz *et al.*, 1987*a*), the information obtained from such studies has a number of advantages over purely biochemical studies (e.g. receptor binding): these bio-behavioural assays given an indication of the functional activity of given neurotransmitter systems and can more easily be related to some of the behavioural parameters that show dysfunction in affective disorders (motivation, cognition, emotions, pain arousal, reinforcement, sleep and EEG changes, etc.).

Some of the tests used to study ADs based on pharmacological interactions are shown in Tables 6.3–6.6 for the monoamines, acetylcholine and histamine. The Tables summarise the responses found after treatment with ADs. In the following section, some of these tests are discussed. However, it should be pointed out that these tests are essentially *in vivo* methods selecting either for a given pharmacological profile, or else establishing whether or not a change in the function of a given receptor system is found after repeated treatment. Thus, they are subject to the same criticisms as the

Table 6-2 (*cont.*)
References:
1, Cohen *et al.*, 1985;
2, Finberg & Kopin, 1986;
3, Cohen *et al.*, 1982*a*;
4, Cohen *et al.*, 1982*b*;
5, Campbell & McKernan, 1986;
6, Aulakh *et al.*, 1983*a*;
7, Finberg & Tal, 1985;
8, Mishra *et al.*, 1983;
9, Youdim & Finberg, 1983;
10, Campbell *et al.*, 1985*a*;
11, Blier *et al.*, 1986;
12, Campbell *et al.*, 1985*b*;
13, Savage *et al.*, 1980*a*;
14, Savage *et al.*, 1980*b*;
15, Cohen *et al.*, 1983;
16, Rényi, 1986;
17, Wehr & Wirz-Justice, 1982;
18, Kafka *et al.*, 1983;
19, O'Donohue *et al.*, 1982;
20, Remé *et al.*, 1984;
21, Sills *et al.*, 1986.

purely biochemical tests discussed above, namely the questionable relevance of the receptor regulation studies, and the self-limiting nature of the acute screens.

6.4.1.1 Tests related to the noradrenergic system (Table 3).

The earliest tests introduced were based on antagonism of the actions of reserpine, Ro-4-1284 or tetrabenazine, such as monoamine depletion, catalepsy, ptosis, sedation and hypothermia, the last test being the most common. The hypothermia test, in spite of the many false positives (e.g. stimulants, beta-adrenergic receptor blockers, LSD) (Sigg *et al.*, 1965; Sigg & Hill, 1967; Colpaert *et al.*, 1975) and false negatives (iprindole, mianserin, trazadone, selective 5-HT uptake inhibitors) (Van Riezen, 1972; Silvestrini, 1982), is still often used to screen for the NA-potentiating action of new compounds. The actions of tetrabenazine and Ro-4-1284 are of shorter duration than those of reserpine.

AD and test compounds are usually given before the reserpine/tetrabenazine, so that 'protection' is indexed rather than reversal. Thus, for example, amitriptyline protects against reserpine-induced hypothermia, but, when given 18 hours after the reserpine, potentiates the hypothermia response (Danysz, unpublished observations). Note also that amitriptyline is weak in protecting against tetrabenazine actions in rats, but not mice, indicating the crucial importance of the species used (Howard *et al.*, 1981). The tetrabenazine test seems to produce the fewest false negatives, although it does not select mianserin or 5-HT uptake inhibitors (see Malick, 1981). A related test is the antagonism of NA (noradrenaline) depletion after 6-hydroxydopamine lesion, proposed by Von Voigtlander & Losey (1976, 1978) as a procedure for screening uptake inhibitory properties.

Enhancement by ADs and test compounds of L-DOPA-induced central actions (Everett, 1967) has also been widely used as a noradrenergic screening procedure, although obviously this test also involves dopaminergic mechanisms. In this procedure, the ADs and test compounds are given before the L-DOPA, together with an MAOI. The responses include parameters such as piloerection, salivation, excitation, irritability, jumping, squeak-

Table 6.3. *Biobehavioural assays and screening tests for antidepressant drugs based on interactions with target drugs: noradrenergic system*

Drug	Effect	Spec.	AC	RP	Remarks (mechanism etc.)	Reference
Reserpine	Hypothermia	m, r	−	?		Garrantini & Jori, 1967; Askew, 1963
	Ptosis	m, r	−	?		Garrantini et al., 1960
	Sedation	m, r	−	−		Sulser et al., 1962
	NA, 5-HT depletion	r	−	?		Von Voigtlander & Losey, 1976; Chen & Bohner, 1961
Tetrabenazine	Ptosis	m	−	?		Vernier et al., 1962
	Sedation	m	−	?		Howard et al., 1981
	Catalepsy	m	−	?		Giurgea & Dauby, 1966
Ro-4-1284	Sedation	m	−	?		Garratini et al., 1960
L-DOPA	Excitation	m	+	?		Everett, 1967; Sigg & Hill, 1967
	Aggression	m	+	?		
	Piloerection	m	+	?		
Yohimbine	Toxicity	m	+	?	MAOI ineffective	Quinton, 1963; Halliwell et al., 1964; Malick, 1981
	Rise in blood pressure	d	+	?		Lang & Gershon, 1963; Sanghvi et al., 1969; Sanghvi & Gershon, 1969
TRH	Hypothermia	m	+	?		Desiles & Rips, 1981
6-OHDA	NA, A depletion	r	−	?		Von Voigtlander & Losey, 1976
NA	Rise in blood pressure	d	+	?	NA given i.v. TAD	Halliwell et al., 1964
	Constriction of nictitating membrane	c	+	?		Haefely et al., 1964
	Hyperthermia	r	+	?	NA given centrally	Von Voigtlander et al., 1983
Clonidine	Sedation	r	0 −	−	α-2 down regulation	Spyraki & Fibiger, 1980
		m	?	0		Von Voigtlander et al., 1978
		r	0	−	Intrahippocampal injection	Plaznik et al., 1984
	EEG-synchronization	r	0	−		Passarelli & Scotti de Carolis, 1983; Kostowski & Dyr, 1984
	Hypothermia	m	0 +	−	α-2 postsynaptic	Von Voigtlander et al., 1978; Danysz et al., 1987b
		r	0 +	0 −		
	Decrease in exploration	g	0	−	Relatively selective Selecting NA-uptake inhibitors	Kostowski & Malatynska, 1983
	Decrease in domination	r	0 +	−		Malatynska & Kostowski, 1984
	Decrease in startle	r	0	0 −		Davis & Menkes, 1982
Clonidine	Decrease in ICSS	r	0	−		Aulakh et al., 1983b
	Hyperphagia	r	0	+		Przegalinski & Jurkowska, 1987
	Decrease in MHPG	r	0	0 −		Tang et al., 1979
	Sleep	ch	0 −	?		Delbarre & Schmitt, 1971; Ferrari, 1985

Table 6.3 (*cont*)

Drug	Effect	Spec.	AD AC	AD RP	Remarks (mechanism etc.)	Reference
Guanfacine	Hypothermia	r	0	+		Danysz et al., 1987b
Clonidine	Excitation	m	0	+	α-1 receptor dependent	Maj et al., 1980
	Aggression	m	−	+	High dose	Maj et al., 1979
Phenylephrine	Excitation	r	0	+	Intrahippocampally	Plaznik et al., 1984
	Excitation	m	0	0	i.c.	Heal, 1984
Salbutamol	Sedation	r	0	−	β-receptor down-regulation	Przegalinski et al., 1983
Thiopentone	Sleeping time	r	0	−	Reflects β-function	Mason & Angel, 1983

Spec., species; m, mouse; r, rat; g.p., guinea pig; d, dog; c, cat; ch, chicken.
AD, antidepressant column: AC, acutely; RP, repeated. 0 = no effect, − = attenuated, + = potentiated, ? = not known or not used.
TAD = tricyclic AD, ICSS = intracranial self-stimulation.

ing and aggression. However, mianserin is inactive in this test as are serotoninergic agents, whereas blockers of muscarinic and histaminic receptors give false positive results (Sigg & Hill, 1967; Brogden et al., 1978). Responses to centrally administered NA such as rise in blood pressure, hypothermia or NA-induced contraction of the nictitating membrane in cats, are also used for screening noradrenergic properties (see Table 6.3).

Another test introduced early in the history of AD screening was the potentiation of yohimbine toxicity in mice and vascular contraction in dogs (see Malick, 1981, for review). However, as argued by Willner (1984a), the test in mice is rather doubtful ethically, particularly in view of its poor predictive ability since maprotiline, bupropion and some 5-HT reuptake inhibitors are only active in high doses, while electroconvulsive therapy is inactive. Moreover, there are a large number of false positives including atropine, chlorpheniramine, amphetamine, adrenolytics and quipazine (Lapin, 1980; Malick, 1981). In this respect, the test in dogs seems to be more specific, giving negative results with amphetamine, and positive with iprindole (Sanghvi et al., 1969, 1976; Sanghvi & Gershon, 1969).

The potentiation in mice of thyrotropin releasing hormone (TRH)-induced hypothermia was introduced as an AD screening test by Desiles & Rips (1981). This test seems to be more selective than the reserpine tests and to reflect alpha-adrenoceptor stimulation. However, its definite

validity would be established only when data on the effects of atypical ADs have been reported from several independent laboratories.

Other noradrenergic tests are based on chronic or (more usually) subchronic AD administration. Many of them are based on the ability of the ADs to antagonise the actions of the alpha$_2$-adrenoceptor agonist clonidine. Von Voigtlander et al. (1978) first reported that clonidine-induced hypothermia in mice is attenuated after four days of treatment with various ADs, and suggested this as a screening procedure. In addition to hypothermia, other actions of clonidine, such as sedation, EEG-synchronisation, reduction of the startle reflex and reduction of dominant behaviour are decreased following repeated AD treatment (see Table 6.3). NA reuptake inhibitors are particularly active in these tests. In a clonidine-induced sedation test after intrahippocampal injection, and in the socalled social domination test in rats, some 5-HT reuptake inhibitors (zimeldine, alaproclate, citalopram) are also active (Malatynska & Kostowski, 1984; Kostowski et al., 1984a; Danysz et al., 1988). It should be noted, however, that attenuation is not always found: Przegalinski & Jurkowska (1987) reported that ADs potentiated the eating behaviour produced by intrahypothalamic clonidine injection. Although clonidine is selective for alpha$_2$-adrenoceptors at low doses, high doses will interact with alpha$_1$-adrenoceptors, which may explain the potentiation by ADs of the aggression and excitation produced in mice by high doses of clonidine (see Table 6.3).

The function of beta-adrenoceptors may be investigated using agonists such as salbutamol, clenbuterol, etc. It was reported that repeated AD but not an anxiolytic or a neuroleptic treatment attenuated the sedation induced by these agonists (Przegalinski *et al.*, 1983, 1984). Mason & Angel (1983) suggested that attenuation by ADs of thiopentone sleeping time reflects a decreased beta-adrenoceptor function. This latter test, however, awaits verification in other laboratories.

6.4.1.2 Tests related to the dopaminergic system (Table 6.4)

The role of DA in depression has long been neglected (Randrup & Braestrup, 1977; Willner, 1983). The potentiation of the actions of amphetamine, such as excitation, hyperthermia, anorexia etc., was introduced early on as an AD screening procedure (Table 6.4). However, pharmacokinetic interactions between ADs and amphetamine limit the usefulness of this approach (see above). Recent experiments by Dall'Olio *et al.* (1986) indicate that some ADs, scopolamine and the metabolic inhibitor, proadifen, increase the duration of methampethamine excitation (pharmacokinetic interaction). However, all the ADs tested except mianserin enhanced the maximal response (pharmacodynamic interaction). An opposite effect was seen with neuroleptics and there was no change with scopolamine. This evidence suggests the possible usefulness of the test as a screening procedure.

It has been suggested that antagonism of hypothermia produced by a high (16 mg/kg) dose of the direct-acting DA agonist apomorphine may be a useful AD screening test (Puech *et al.*, 1981). However, it is questionable whether dopaminergic mechanisms are solely involved at such a high dose; in fact some results (Pawlowski & Mazela, 1986), though not others (Menon *et al.*, 1984), suggest a noradrenergic involvement in this phenomenon.

Repeated amphetamine treatment may induce (particularly upon withdrawal) a 'depressed' state, which can be used to test for AD actions. Although some ADs acutely antagonised sedation during withdrawal, amitriptyline was ineffective and mianserin only partially effective (Seltzer & Tonge, 1975; Lynch & Leonard, 1978).

Following chronic AD treatment, some of the postsynaptic DA-mediated behaviours, such as amphetamine-induced locomotor stimulation, but not amphetamine-induced stereotypy, are augmented, suggesting perhaps a change in mesolimbic but not nigrostriatal DA receptor function (Spyraki & Fibiger, 1981). The anticholinergic actions of some of the AD compounds may be of crucial importance for the potentiation of amphetamine-induced stimulatory effects (Martin-Iverson *et al.*, 1983). However, other evidence suggests that repeated AD treatment does affect postsynaptic DA receptor function. For example, pimozide antagonised the restoration by imipramine of the decreased sucrose consumption caused in rats by exposure to chronic mild stress (Willner *et al.*, 1987, see also Willner, this vol., Ch. 5).

It has been reported that AD compounds attenuate presynaptic DA receptor function, which may lead to an increased DA release (Serra *et al.*, 1979). However, this result has not been replicated in other laboratories. In addition, no evidence of presynaptic DA receptor subsensitivity was found following 21 days of treatment with desipramine, when presynaptic DA receptor sensitivity was studied by measuring 3-PPP($-$)-induced sedation (S. Ahlenius, W. Danysz & T. Archer, unpublished results). When it occurs, the reduction of presynaptic DA receptor sensitivity appears to be a withdrawal phenomenon rather than a primary action of ADs (Towell *et al.*, 1986).

6.4.1.3 Tests related to the serotonergic system (Table 5)

The involvement of 5-HT in depression was suggested over two decades ago (Coppen, 1967). However, only recently have 5-HT selective uptake inhibitors with AD properties become available (see Table 6.1). These compounds represent an improvement upon the tricyclic ADs with respect to the incidence of some side effects, but at present appear not to be more efficacious.

Several procedures for assessing serotonergic effects of AD and test compounds are available. The effects of some selective 5-HT uptake inhibitors can be studied directly using a two-way active avoidance procedure where disruption of learning is produced by these agents (Archer *et al.*, 1984*a*). The actions of the 5-HT precursor 5-hydroxytryptophan (5-HTP) on several behavioural parameters

Table 6.4. *Biobehavioural assays and screening tests for antidepressant drugs based on interaction with target drugs: dopaminergic mechanisms*

Drug	Effect	Spec.	AD		Remarks (mechanism etc.)	Reference
			AC	RP		
Apomorphine	Hypothermia	m	−	?	Very high dose of apomorphine	Schelkunov, 1977; Puech *et al.*, 1981
	Stereotypy	r	0	0		Spyraki & Fibinger, 1981;
		m	0	?		Maj *et al.*, 1979; Puech *et al.*, 1981
	Sedation	r	0	0 −	Presynaptic DA receptor, low dose of APO	Serra *et al.*, 1979; Spyraki & Fibiger, 1981
	Excitation	m	0	+	Postsynaptic dose	Puech *et al.*, 1981
	Rotation	r	−	−	After substantia nigra lesion AMI, CMI effective; IMI ineffective	Delini-Stula & Vassout, 1979
Amphetamine	Excitation	r	+	+		Spyraki & Fibiger, 1981; Halliwell *et al.*, 1964
	Hyperthermia	r	+	?		Morpurgo & Theobald, 1965
	Anorexia	r	+	?		Claasen & Davies, 1969
	Increase in ICSS	r	+	?		Stein & Seifter, 1961
	Increase in active avoidance	r	+	?		Scheckel & Boff, 1964
	Toxicity	m	−	?	Aggregated mice	Halliwell *et al.*, 1964
	Stereotypy	r	+	0		Halliwell *et al.*, 1964; Spyraki & Fibiger, 1981
	Mydrasis	m	+	?		Halliwell *et al.*, 1964
Amphetamine chronic	Sedation after withdrawal	r	−		Mianserin partly active	Seltzer & Tonge, 1975; Lynch & Leonard, 1978
6-OHDA	DA decrease after substantia nigra lesion	r	0 −	?	Protected by DA uptake inhibitors	Von Voigtlander & Moore, 1973
Haloperidol	Catalepsy	r	− +	?	Antagonised by DA uptake inhibitors	Biziere *et al.*, 1985
(−)-3PPP	Sedation	r		0	Presynaptic DA autoreceptor	Ahlenius *et al.*, unpublished
Dopamine	Excitation	r	0	+	Into n. accumbens	Wedzony & Maj, 1983
Amphetamine	Excitation	r	0	+	ECT and citalopram also active	Plaznik & Kostowski, 1987

Spec., species; m, mouse; r, rat; g.p., guinea pig; d, dog; c, cat; ch, chicken.
AD, antidepressant column: AC, acutely; RP, repeated. 0 = no effect, − = attenuated, + = potentiated, ? = not known or not used.

such as reciprocal forepaw treading and hind limb abduction are potentiated by 5-HT reuptake inhibitors. Some of these methods can also be used to study the effects of repeated AD treatment on 5-HT receptor function (see Table 6.5). Thus, for example, head-twitch responses to 5-HTP seem to be decreased, while the 5-HT syndrome is usually increased following repeated AD treatment (see Willner, 1985a, for review). The 5-HTP-induced behavioural suppression test was advocated by Aprison *et al.* in 1982, but may reflect actions on peripheral rather than central 5-HT receptors (see Willner, this vol., Ch. 5). One interesting method was used by Jones (1980) who found that 5-HT-induced sleep in young chickens was attenuated after acute administration of tricyclic ADs but

Table 6.5. *Biobehavioural assays and screening tests for antidepressant drugs based on interaction with target drugs: 5-HT mechanisms*

Drug	Effect	Spec.	AD		Remarks (mechanism etc.)	Reference
			AC	RP		
5-HTP	Head twitch	m	+	+	Potentiated by MAOI	Corne *et al.*, 1963
		r	+	+ −	5-HT uptake inhibitors selected	Ögren *et al.*, 1982
	5-HT syndrome	m, r	+	+	periventricular n.	Mogilnicka & Klimek, 1979
	Tremor	?	+	?	5-HT uptake inhibitors selected	Lessin, 1959; Carlsson *et al.*, 1969
	Operant performance deficit	r	−	?		Aprison *et al.*, 1982; Nagayama *et al.*, 1981
	Flexor reflex	r	+	?	Potentiated by MAOI	Meek *et al.*, 1970; Tadeschi *et al.*, 1959
PCA	Stereotypy	m	−	?		Depin & Betbeder-Matibet, 1985
	Decrease in 5-HT content	m	−	?	Blockers of 5-HT uptake selected	Von Voigtlander & Losey, 1976
	Sedation					Meek *et al.*, 1971
Fenfluramine	Hyperthermia	r	+	?	High ambient temp. 5-HT uptake inhibitors & 5-HT blockers selected	Sulpizo *et al.*, 1978
Tryptohan (+ MAOI)	Head twitch	m, r	+	?		Unpublished data
	5-HT syndrome	r	+	?		De Montigny *et al.*, 1981
Quipazine	Hyperthermia	r	0 −	?	5-HT-2 receptors	Pawlowski, 1984
5-MeODMT	5-HT syndrome	r	0 −	−		Stolz *et al.*, 1983; Stolz & Marsden, 1982
	Analgesia	r	0 −	− +		Fuxe *et al.*, 1983; Danysz *et al.*, 1986
	Head twitch	r	+	− +	Dependent on dose and treatment duration	Friedman & Dallob, 1979; Fuxe *et al.*, 1983
		m	+	−		Blackshear & Sanders-Bush, 1982
5-HT	Sleep	ch	−	+	5-HT i.v., DMI, IMI & AMI tested	Jones, 1980
RU-24969	Excitation	r	0	0 +		Green *et al.*, 1984
8-OHDPAT	Hypothermia	m	−		Presynaptic 5-HT$_{1A}$ receptor	Goodwin *et al.*, 1985*b*
	Anorexia	r		+		Kennett *et al.*, 1987
	Inhibition of cage leaving	r	+	−	Postsynaptic 5-HT$_{1A}$ receptor	Rényi *et al.*, 1986; Rényi, 1987, unpubl.

Spec., species; m, mouse; r, rat; d, dog; c, cat; ch, chicken.
AD, antidepressant column: AC, acutely; RP, repeated. 0 = no effect, − = attenuated, + = potentiated, ? = not known or not used, MAOI = monoamine oxidase inhibitors.

Table 6.6. *Biobehavioural assays and screening tests for antidepressant drugs: cholinergic and histaminergic mechanisms*

Drug	Effect	Spec.	Reference
Oxotremorine	Tremor	m	Halliwell *et al.*, 1975; Biziere *et al.*, 1985
	Hypothermia	m	Karasawa *et al.*, 19890
	Salivation	m	Karasawa *et al.*, 1980
Physostygmine	Lethality	m	Karasawa *et al.*, 1980; Von Voigtlander *et al.*, 1983
ACh	Contraction of deodenum	g.p.	Karasawa *et al.*, 1980; Boksay *et al.*, 1979
Carbachol	Head shaking	r	Turski *et al.*, 1981
Perphenazine	Catalepsy	r	Boksay *et al.*, 1979
Histamine	Contraction of ileum	g.p.	Karasawa *et al.*, 1980; Depin *et al.*, 1985

Spec., species; m, mouse; r, rat; g.p., guinea pig; d, dog; c, cat; ch, chicken.
All responses attenuated by some antidepressants.

potentiated by repeated administration of tricyclics.

The p-chloroamphetamine (PCA)-induced decrease in 5-HT content can be used to select for 5-HT uptake inhibitors (Von Voigtlander & Losey, 1976). Tryptamine-induced seizures, anorexia, etc., have also been applied to investigate serotonergic actions of AD compounds. MAO inhibitors are particularly effective in these procedures. With regard to MAO inhibitors, the potentiation of tryptophan-induced increases in motor activity has been used as a screening test, but the magnitude of the response varies with different MAO inhibitors Archer *et al.*, 1985).

Use of direct agonists has also been extensively investigated. Thus, for example, hyperthermia induced by the 5-HT agonists quipazine and m-chlorophenylpiperazine (m-CPP) at a high ambient temperature has been used for studying the effects of repeated AD treatment. The same holds true for 5-MeODMT-induced effects, such as head-twitch responses and analgesia, although these results are very variable. In the latter case variability seems to depend on the basal pain threshold level, which can vary following treatment with the different AD compounds (Danysz *et al.*, 1986).

Rényi (for review, see Rényi, 1986) has developed an ingenious battery of biochemical and behavioural tests suitable for new drug screens and tools for the characterisation of 5-HT receptors. For example, Rényi & Ross (1985) studied the effects of MAO inhibitors upon the inhibition

of neuronal accumulation of monoamines *in vitro* and *ex vivo* in rat brain slices, and related this to behavioural measures of 5-HT and NA release. p-Chloroamphetamine and 5-MeODMT-induced ejaculations and inhibition of the cage-leaving response in rats are potentially important methods for analysing 5-HT$_1$-like receptor function (Rényi, 1986; Rényi *et al.*, 1986).

Latterly, new, more selective 5-HT agonists are becoming available. Ru-24969 (a 5-HT$_{1B}$ agonist) increases activity in rats, and this effect is not modified by ADs, but is potentiated by electroconvulsive shock (Green *et al.*, 1984). 8-OH-DPAT (a 5-HT$_{1A}$ agonist) produces hypothermia in mice, and inhibits the cage leaving response in rats. Both actions seem to be attenuated by some tricyclic AD compounds and 5-HT reuptake inhibitors (Goodwin *et al.*, 1985a; L. Rényi, W. Danysz and T. Archer, unpublished results).

6.4.1.4 *Effects on cholinergic and histaminergic systems*

In Table 6.6, some tests used to detect antimuscarinic and antihistaminergic activity are listed. It is questionable whether these effects can be used to predict antidepressant activity of test compounds (see Willner, 1985, Chapter 18, for discussion). However, these tests are useful in predicting side effects related to blockade of these receptors (see above).

6.4.2 Effects of ADs on behaviours resulting from interaction with the environment, surgical lesion of the brain, or stimulation of brain structures

The behavioural tests summarised in the previous section all assume certain biochemical actions of the test compounds; such assumptions are a limiting factor in the search for AD compounds with entirely new mechanisms of action. Theoretically, such compounds can only be developed (excluding clinical serendipity) by the use of screens that do not require particular neurochemical profiles; however, this has not happened so far. The primary purpose of this discussion will be to outline a sequence of objectives both to aid in the development of new antidepressant compounds and to govern the choice of suitable animal models to define the indications for potential AD compounds.

The requirements for an animal model of depression can be summarised as follows:

(a)   The animal model should have the same precipitating factors as depression in humans.
(b)   The animal models should have a similar symptomatology as the disorder in people (a difficulty in itself given the heterogeneity of affective disorders).
(c)   The animal model should respond to the same therapeutic interventions as the disorder itself.

These points are not intended as criteria for a simulation of the depressive state and as such would be inadequate. But they are merely intended to present cornerstones in the search for the 'absolute model'. Thus, only (c) is required in a screening test but (a) and (b) may help to achieve (c).

In fact, the first of these requirements is presumed rather than proven. Since stress is believed to be an important etiological factor in depression, many models use stress to produce the behavioural deficits (for discussion, see Danysz *et al.*, 1987*a*; Willner, this vol., Ch. 5). The responses observed in these tests are usually highly integrated patterns of behaviour, that can be affected by AD treatment (see Table 6.7). The requirement of similar symptoms to human depression entails certain necessary assumptions that include motivational and emotional variables as well as general activity levels and performance of both primary and secondary tasks.

The Porsolt test, in which rats or mice are subjected to an inescapable swim stress has become widely used as a screening procedure for AD action. AD compounds prolong the active behaviour of the rats in this situation (Porsolt *et al.*, 1977; Porsolt, 1978, 1981). This procedure is able to select atypical AD compounds such as iprindole, alaproclate and mianserin, but gives negative results with some 5-HT reuptake inhibitors (fluoxetine, paroxetine, zimeldine, citalopram) administered at reasonably high dose (Gorka *et al.*, 1979, Nomura *et al.*, 1982; Danysz *et al.*, 1988). The Porsolt test may give false positives with antihistaminergic or anticholinergic drugs. This problem may to some extent be eliminated by using a modified version or employing chronic treatment; however chronic treatment reduces considerably the convenience of the test (Nomura *et al.*, 1982; see Willner, this vol., Ch. 5). A tail-suspension test also based on a 'despair-type' reaction has been proposed by Steru *et al.* (1985). Locomotor activity testing should be performed together with this test to reject stimulants. However, analyses utilising a wide range of compounds from several classes of psychoactive drugs are required in order to establish the incidence of false positives and false negatives.

There is an increasing need to show more consideration for the well being as opposed to unnecessary suffering of animals employed in testing (Archer *et al.*, 1988). Some ethical problems not unrelated to scientific considerations may arise from the despair test. Thierry *et al.* (1986) suggest that, in view of the stress-inducing properties and the physiological imbalances such as a decrease in body temperature during forced swimming, it is possible that the pharmacological sensitivity of the test may be compromised. They therefore suggest that the tail suspension test may be preferable on both counts. Ethical problems may also be associated with other tests based on stress, in particular those where muricidal behaviour of rats is scored (see Table 6.7). It should be pointed out that one important component in a discussion of the ethical issues of a given test should be a critical analysis of its reliability and predictive ability as a screening procedure for new AD compounds (see Willner, 1984*a*). Muricide fails dismally on these screening

Table 6.7. *Some behavioural screening methods and models for investigating antidespressants, induced by environmental factors, lesions, or brain stimulation*

Inducing factor	Symptoms	AD (1)	AD (2)	Species	Reference
Stress					
Isolation from mother	Depression	−	ch	monkey	McKinney & Bunney, 1969; McKinney, 1984
	Self-clasping	−	ch	monkey	Suomi *et al.*, 1978
	Vocalisation	−	ch	dog, monkey	Scott *et al.*, 1973; Hrdina *et al.*, 1979
Isolation	Hyperactivity	−	ac	rat	Garzon & Del Rio, 1981; Garzon *et al.*, 1979
	Mouse killing	−	ac	rat	Horowitz *et al.*, 1966; Valzelli & Bernasconi, 1971
	Fighting	+	sc	rat	Willner *et al.*, 1981
Separation of a pair	Weight gain, sedation	?	?	hamster	Crawley, 1983
Different stressors (chronic)	Lack of reactivity	−	ch	rat	Katz, 1981; Katz *et al.*, 1981
	Decreased preferences for saccharin or sucrose	−	ch	rat	Willner *et al.*, 1987
Inescapable foot shock	Decrease in acquisition of escape response	−	ac	rat, dog, mouse	Maier & Seligman, 1976; Anisman & Zacharko, 1982
	Decrease in ICSS	−	ch	mouse	Zacharko & Anisman, 1984
	Aggression in pairs	+	ac	rat	Eichelman & Barchas, 1975; Mogilnicka & Przewlocka, 1981
Restraint	Cortisol response	−	ch	rat	Armario *et al.*, 1985
Tail suspension	Immobility	−	ac	mouse	Steru *et al.*, 1985
Inescapable water swim	Immobility	−	ac ch	mouse, rat	Porsolt *et al.*, 1977; Porsolt, 1981
Differential reinforcement-of-low-rates-of responding (DRL 72)	Low rate of reinforcement	−	ac	rat	Seiden & O'Donnell, 1985
Aggression	Mouse killing	−	ac	rat	Valzelli & Bernasconi, 1971; Onodera *et al.*, 1981; Kostowski *et al.*, 1984b
Surgical					
Olfactory bulbectomy	Mouse killing	−	ac	rat	Sofia, 1969; Shibata *et al.*, 1984
	Passive avoidance deficit	−	ch[a]	rat	Van Riezen *et al.*, 1977; Cairncross *et al.*, 1975; Broekkamp *et al.*, 1980
	Improved active avoidance	0	ac	rat	Archer *et al.*, 1984b
Brain stimulation	Kindling seizures	−	ac	rat	Babington & Wedeking, 1973, 1975

AD(1) − effect of AD treatment: + = potentiated; − = antagonized; ? = not tested yet.
AD(2) − duration of the treatment: ac = acutely; ch = chronically; subc = subchronically.
[a]Sometimes reversed after acute treatment also.

criteria (Horovitz, 1965; Horovitz *et al.*, 1966; Barnett *et al.*, 1969; Dubinsky & Goldberg, 1971; Dubinsky *et al.*, 1973).

The application of uncontrollable stress (mainly shock) is used commonly to produce diverse behavioural changes such as motivational changes including decreased intracranial self-stimulation (ICSS), decrease in preference for saccharine, decreased learning ability and apathy as evidenced by decreased activity in open field and increased 'despair' in the Porsolt test. All these affects are attenuated by AD compounds (see Table 6.7) (Sherman & Petty, 1980; Katz, 1981; Anisman & Zacharko, 1982; Willner, this vol., Ch. 5). Katz and co-workers have developed a procedure utilising chronic unpredictable stresses including electric shocks, immersion in cold water, changes in housing and lighting, tail pinch, etc., which produce changes in open field activity that are antagonised by concurrent administration of ADs (see e.g. Katz & Hersch, 1981; Roth & Katz, 1981; Katz, 1984; Soblosky & Thurmond, 1986). However, these procedures may be of weak practical value since they are clearly time consuming, and may be more fruitful in the further analysis of already selected compounds.

Reactions to the environment may also be studied by other methods such as the startle reflex and shock-induced fighting. Repeated AD treatment decreases startle latency, which may be a result of a decreased function of inhibitory alpha$_2$-adrenoceptors and an increased function of excitatory alpha$_1$-adrenoceptors (Davis, 1982). Shock-induced fighting, which increases after repeated treatment with several AD compounds, may also be a reflection of an alpha$_1$-adrenoceptor supersensitivity (Mogilnicka & Przewlocka, 1981). The social domination test is a complex procedure in which the injection of clonidine (at low doses only) causes the dominant rat to become submissive. This clonidine-induced reversal of domination can be antagonised by many ADs, with very few false positives (Malatynska & Kostowski, 1984).

The above tests represent reactions of animals to unpleasant environmental stimuli, and are thought to reflect the emotional state of the animal. Their usefulness for screening lies in the fact that they do not require a given pharmacological profile for a positive effect (although noradrenergic mechanisms may be implicated in some cases). Of the above-mentioned tests only the Porsolt, tail suspension and domination tests have been evaluated intensively for the primary requirements of screening tests, i.e. for false positives and false negatives. The differential-reinforcement-of-low-rates-of-responding (DRL) 72-s test developed by Seiden *et al.* (O'Donnell & Seiden, 1983; Seiden *et al.*, 1985; Seiden & O'Donnell, 1985) has also been tested extremely stringently for false positives and false negatives (cf. Howard & Pollard, 1984; Pollard & Howard, 1986). (In this test, rats are rewarded only when lever presses are space more than 72 s apart.) Seiden (pers. commun.) seems to have answered most of the problems levelled at the DRL-72 s test but has himself derived results that are intriguing but difficult to understand fully. For example, DA autoreceptor agonists appear to produce an antidepressant-like effect (Seiden, pers. commun.). The Porsolt and tail suspension tests enjoy the advantage of great utility since they are simple and quick to use whereas the DRL-72 s test requires a training period of up to ten weeks before testing.

Olfactory bulbectomy produces various behavioural deficits (see Jesberger & Richardson, 1985, for review), the most robust being a deficit of passive avoidance learning (but not of active avoidance, which may be improved) and increased aggressive behaviour. Both these changes are normalised by AD compounds, but primarily by those affecting the serotonergic system such as 5-HT uptake inhibitors. The validity of the bulbectomy model for general screening may be open to question (Archer *et al.*, 1984*b*), but its selective sensitivity to serotonergic agents should be kept in mind (Cairncross *et al.*, 1978; Broekkamp *et al.*, 1980; Noreika *et al.*, 1981). Kindling induced by repetitive amygdaloid stimulation may be attenuated by AD treatment (Babington & Wedeking, 1973, 1975), although later studies have indicated that this test lacks specificity for AD compounds.

An interesting approach has been the development of tests based on ICSS (Kokkinidis *et al.*, 1980; Fibiger & Phillips, 1981; Cornfeld *et al.*, 1982; Zacharko *et al.*, 1984). This approach relates to the general issue of using a defective reward system to model the anhedonia of endogenous depression (Moore, 1977; Brown & Harris, 1978; Jahoda, 1979; Kanner *et al.*, 1981; Billings *et al.*, 1983; Leibowitz *et al.*, 1985; Willner, 1987).

Finally, it should be mentioned that there are some characteristic effects of acutely administered ADs on human EEG patterns that may allow prediction of antidepressive activity. EEG observations were in fact an important step in introducing compounds such as mianserin and citalopram (Itil *et al.*, 1972, 1984; see also Fink, 1982, for discussion).

6.5 CONCLUSIONS

From the above discussion, it is apparent that none of the tests or models presented here appear to offer an ideal, i.e. specific and predictive, AD screening paradigm. Taking into account the diverse pharmacological actions of AD compounds and the heterogeneity of depressive illness, it seems unlikely that a single ideal screening method will be developed. Generally, most tricyclic antidepressants, and even some atypical ADs, have direct antagonist effects upon serotonergic (Hall & Ögren, 1981), histaminergic (Kanof & Greengard, 1978; Richelson, 1979), muscarinic (Revahi *et al.*, 1977; Snyder & Yamamura, 1977) and alpha-adrenergic receptors (Baumann & Maitre, 1975, 1977; U'Prichard *et al.*, 1978; Svensson & Usdin, 1978) as well as a delayed interaction with beta-adrenoceptors (Vetulani *et al.*, 1976; Banerjee *et al.*, 1977). Combinations of models based on highly integrated behaviour with more elementary screens to exclude unwanted pharmacological actions such as anticholinergic and antihistaminergic effects may lead to the development of novel antidepressant agents that would represent an improvement upon those presently available to the physician. Certain behavioural tests, as for example food/water intake, may have a bearing upon efficacy: for example, patients on zimeldine tended to lose weight whereas those on amitriptyline showed weight gain (Paykel *et al.*, 1973). A valuable discussion on drug effects and side effects is provided by Huskisson & Wojtulewski (1974).

Two main tactics may be used to develop ADs. Derivation from the chemical laboratory requires the synthesis of a new compound similar to AD or potential AD compounds already known (e.g. NA or 5-HT uptake inhibitors, compounds increasing NA transmission, 5-HT_{1A} receptor agonists, 5-HT_2 receptor antagonists, etc.). Those agents should be devoid, or nearly so, of adverse effects like blockade of muscarinic and histaminic receptors, cardiovascular toxicity, etc., and, should have a good oral bioavailability (see for example, Nies *et al.*, 1977; Rodstein & Som Oei, 1979). The second strategy, evaluation of existing compounds for possible AD activity, includes, for example, cases where the compound was developed for such purposes as another illness far removed from 'mental state' or veterinarian therapy.

In each case a similar screening procedure may be followed, involving evaluation of the pharmacological profile in simple biochemical tests and drug screens in parallel with tests for adverse effects, and in the event of survival of the new compound, continuing with a more complex investigation (e.g. chronic treatment, Katz test, DRL-72s).

One such approach is the combination of a procedure with model status (e.g. the chronic mild, unpredictable stress treatment) as the independent variable with procedures relating to affective state (e.g. activity, despair test, saccharin preference) as the dependent variable (e.g. Garcia-Marquez & Armario, 1987; Willner *et al.*, 1987). This recommendation is, of course, modulated by considerations of time-consumption and labour-intensiveness. In these combinations, it would be exciting to see included some measurement of 'pleasure', which appears to be a necessary 'state of mind' missing in endogenous depression. This may be achieved by use of ICSS as the screen (dependent variable) for new antidepressants (Zacharko & Anisman, 1984). Alternatively, the application of the social cooperation measure developed by Berger and associates (Berger *et al.*, 1980; Berger & Schuster, 1982), appears to have great potential (see Willner, this vol., Ch. 5; Willner *et al.*, 1989). Validation of this type of model with respect to accuracy and predictive ability represents a most exciting challenge to behavioural pharmacologists active in the search for new and better antidepressants. It should be borne in mind that the available screening procedures select either catecholaminergic or serotonergic compounds. Thus, the test battery should include procedures sensitive for both, e.g. the Porsolt test and olfactory bulbectomy, respectively. As was indicated earlier, new ADs are often derived from clinical observation rather than direct systematic experimental research. Undoubtedly this trend should ultimately be altered. Such new, possibly 'atypical' ADs are, in their turn, the essential justification of the reliability and predictive value of the existing drug screens.

REFERENCES

Anisman, H. & Zacharko,R.M. (1982). Depression: the predisposing influence of stress. *Behavioral and Brain Sciences* **5**, 89–137.

Aprison, M.M., Hintgen, J.N. & Nagayama, H. (1982). Testing a new theory of depression with an animal model: neurochemical-behavioral evidence for postsynaptic serotonergic receptor involvement. In Langer, S.Z., Takahashi, R., Segawa, T. & Briley, M. (eds.) *New Vistas in Depression*, pp. 171–8. Pergamon Press: New York.

Archer, T., Ögren, S.O., Johansson, G. & Ross, S.B. (1984a). The effect of acute zimeldine and alaproclate administration on the acquisition of two-way avoidance: comparison with other antidepressant agents, test of selectivity and sub-chronic studies. *Psychopharmacology* **84**, 188–95.

Archer, T., Söderberg, U., Ross, S.B. & Jonsson, G. (1984b). Role of olfactory bulbectomy and DSP4 treatment in avoidance learning in the rat. *Behavioral Neuroscience* **98**, 496–505.

Archer, T., Fowler, C.J., Fredricksson, A., Lewander, T., Magnusson, O., Mohringe, B. & Söderberg, U. (1985). Increased total activity in the rat after L-tryptophan plus the monoamine oxidase-A inhibitor amiflamine but not after L-tryptophan plus clorgyline. *British Journal of Pharmacology* **85**, 581–90.

Archer, T., Fredriksson, A., Lewander, T. & Söderberg, U. (1988). Marble burying and spontaneous motor activity in mice: Interactions over days and the effect of diazepam. *Scandinavian Journal of Psychology* **28**, 242–9.

Armario, A., Restrepo, C., Castellanos, J.M. & Balasch, J. (1985). Dissociation between adrenocorticotropin and corticosterone responses to restraint after previous chronic exposure to stress. *Life Science* **36**, 2085–92.

Ask, A.-H., Hellström, W., Norman, S., Ögren, S.-O. & Ross, S.B. (1982). Selective inhibition of the A form of monoamine oxidase by 4-dimethylamino-N-methylphenylalkylamine derivatives in the rat. *Neuropharmacology* **21**, 299–308.

Askew, B.M. (1963). A simple screening procedure for imipramine-like antidepressant agents. *Life Science* **2**, 725–30.

Aulakh, C.S., Cohen, R.M., McLellan, C. & Murphy, D.L. (1983a). Correlation of changes in α-adrenoceptor number and locomotor responses to clonidine following clorgyline discontinuation. *British Journal of Pharmacology* **80**, 10–12.

Aulakh, C.S., Cohen, R.M., Radhan, S.N. & Murphy, D.L. (1983b). Self-stimulation responses are altered following long-term but not short-term treatment with clorgyline. *Brain Research* **270**, 383–6.

Avento, V-P., Koskinen, T., Kylmamaa, T., Lepola, K. & Suominen, J. (1984). A parallel double-blind trial on the selective serotonin uptake inhibitor RO 11-2465 and amitriptyline in depression. *Drug Experimental and Clinical Research* **10**, 127–31.

Babington, R.G. & Wedeking, P.W. (1973). The pharmacology of seizures induced by sensitization with low intensity brain stimulation. *Pharmacology, Biochemistry and Behavior* **1**, 461–76.

Babington, R.G. & Wedeking, P.W. (1975). Blockade of tardive seizures in rat by electroconvulsive shock. *Brain Research* **88**, 141–4.

Ban, T.A., Fujimori, M., Petrie, W.M., Ragheb, M. & Wilson, W.H. (1982). Systematic studies with amoxapine, a new antidepressant. *International Pharmacopsychiatry* **17**, 18–27.

Banerjee, S.P., Kung, L.S., Riggi, S.J. & Chanda, S.K. (1977). Development of β-adrenergic receptor subsensitivity by antidepressants. *Nature* **268**, 455–6.

Barnett, A., Taber, R.I. & Roth, R.E. (1969). Activity of antihistamines in laboratory antidepressant tests. *International Journal of Pharmacology* **8**, 73–9.

Bartholini, G., Scatton, B., Zivkovic, B., Depootere, H. & Lloyd, K.G. (1984). GABAergic mechanisms and antidepressant therapy. *Clinical Neuropharmacology* **7**, Suppl. 1, S167.

Baumann, P.A. & Maître, L. (1975). Blockade of the presynaptic α-receptor in rat cortex by antidepressants. *Experienta* **31**, 726.

Baumann, P.A. & Maître, L. (1977). Blockade of presynaptic α-receptors and of amine uptake in the rat brain by the antidepressant mianserin. *Naunyn-Schmiedeberg's Archives de Pharmacologie* **300**, 31–7.

Berger, B.D. & Schuster, R. (1982). An animal model of social interaction. Implications for the analysis of drug action. In M.Y. Spiegelstein & A. Levy (eds.), *Behavioral Models and the Analysis of Drug Action*, pp. 415–28. Amsterdam: Elsevier.

Berger, B.D., Mesch, D. & Schuster, R. (1980). An animal model of 'cooperation' learning. In R.F. Thompson, L.H. Hicks & V.B. Shvyrkov (eds.), *Neural Mechanisms of Goal-Directed Behavior*, pp. 481–92. New York: Academic Press.

Billings, A., Cronkite, R. & Moos, R. (1983). Social-environmental factors in unipolar depression: comparisons of depressed patients and non-depressed controls. *Journal of Abnormal Psychology* **92**, 119–33.

Biziere, K., Worms, P., Kan, J-P., Mandel, P., Garattini, S. & Roncucci, R. (1985). Minaprine, a new drug with antidepressant properties. *Drug Experimental and Clinical Research* **11**, 831–40.

Blackshear, M.A. & Sanders-Bush, E. (1982). Serotonin receptor sensitivity after acute and chronic treatment with mianserin. *Journal of Pharmacology and Experimental Therapeutics* **221**, 303–8.

Blier, P., DeMontigny, C. & Azzaro, A.J. (1986). Modification of serotonergic and noradrenergic neurotransmissions by repeated administration of monoamine oxidase inhibitors: electrophysiological studies in the rat central nervous system. *Journal of Pharmacology and Experimental Therapeutics* **237**, 987–94.

Boksay, I.J.E., Popendiker, K., Weber, R.O. & Soder, A. (1979). Synthesis and pharmacological activity of befuraline (N-benzo[b]furan-2-ylcarbonyl-N^1N-benzylpiperazine), a new antidepressant compound. *Arzneimittel-Forschung Drug Research* **29**, 193–204.

Bosworth, D.M. (1959). Iproniazid: a brief review of its introduction and clinical use. *Annals of the New York Academy of Sciences* **80**, 809–19.

Brenner, H.D., Alberti, L., Keller, F. & Schaffner, L. (1984). Pharmacotherapy of agitational states in psychiatric gerontology: double-blind study: febarbamat-pipamperon. *Neuropsychobiology* **11**, 187–90.

Brogden, R.N., Heel, R.C., Speight, T.M. & Avery, G.S. (1978). Mianserin: a review of its pharmacological properties and therapeutic efficiency in depressive illness. *Drugs* **16**, 273–301.

Brogden, R.N., Heel, R.C., Speight, T.M. & Avery, G.S. (1979). Nomifensine: a review of its pharmacological properties and therapeutic efficacy in depressive illness. *Drugs* **18**, 1–24.

Broekkamp, C.L., Garrigou, D. & Lloyd, K.G. (1980). Serotonin-mimetic and antidepressant drugs on passive avoidance learning by olfactory bulbectomised rats. *Pharmacology, Biochemistry and Behavior* **13**, 643–6.

Brown, G. & Harris, T. (1978). *Social Origins of Depression*. London: Tavistock.

Cairncross, K.D., King, M.G. & Schofield, S.P.M. (1975). Effect of amitriptyline on avoidance learning in rats following bilateral olfactory ablation. *Pharmacology, Biochemistry and Behavior* **3**, 1063–7.

Cairncross, K.D., Cox, B., Forster, C. & Wren, A.F. (1978). A new model for the detection of antidepressant drugs: Olfactory bulbectomy in the rat compared with existing models. *Journal of Pharmacological Methods* **1**, 131–43.

Campbell, I.C. & McKernan, R.M. (1986). Clorgyline and desipramine alter the sensitivity of (^{3}H)noradrenaline release to calcium but not to clonidine. *Brain Research* **372**, 253–9.

Campbell, I.C., Gallager, D.W., Hamburg, M.A., Tallman, J.F. & Murphy, D.L. (1985a). Electrophysiological and receptor studies in rat brain: effects of clorgyline. *European Journal of Pharmacology* **111**, 355–64.

Campbell, I.C., Durcan, M.J., Cohen, R.M., Pickar, D., Chugani, D. & Murphy, D.L. (1985b). Chronic clorgyline and pargyline increase apomorphine-induced stereotypy in the rat. *Pharmacology, Biochemistry and Behavior* **23**, 921–5.

Carlsson, A., Jonason, J., Lindqvist, M. & Fuxe, K. (1969). Demonstration of extraneuronal 5-hydroxytryptamine accumulation in brain following membrane-pump blockade by chlorimipramine. *Brain Research* **12**, 456–60.

Chen, G. & Bohner, B. (1961). The anti-reserpine effects of certain centrally-acting agents. *Journal of Pharmacology and Experimental Therapeutics* **131**, 179–84.

Claasen, V. (1983). Review of the animal pharmacology and pharmacokinetics of fluvoxamine. *British Journal of Clinical Pharmacology* **15**, 3495–555.

Claasen, V. & Davies, J.E. (1969). Potentiation by tricyclic antidepressasnts of weight loss caused by amphetamine in rats. *Acta Physiologica et Pharmacologica Nederlandica* **15**, 37–40.

Claasen, V., Davies, J.E., Herrting, G. & Placheta, P. (1977). Fluvoxamine, a specific 5-hydroxytryptamine uptake inhibitor. *British Journal of Pharmacology* **60**, 505–16.

Cohen, R.M., Ebstein, R.P., Daly, J.W. & Murphy, D.L. (1982a). Chronic effects of a monoamine-oxidase inhibiting antidepressant: decrease in functional alpha-adrenergic adrenoceptors precedes the decrease in norepinephrine stimulated cyclic AMP systems in rat brain. *Journal of Neuroscience* **2**, 1588–95.

Cohen, R.M., Campbell, I.C., Dauphin, M., Tallman, J.F. & Murphy, D.L. (1982b). Changes in α- and β-receptor densities in rat brain as a result of treatment with monoamine oxidase inhibiting antidepressants. *Neuropharmacology* **21**, 293–8.

Cohen, R.M., Aulakh, C.S. & Murphy, D.L. (1983). Long-term clorgyline treatment antagonizes the eating and motor function responses to m-chlorophenylpiperazine. *European Journal of Pharmacology* **94**, 175–9.

Cohen, R.M., Aulakh, C., McLellan, C. & Murphy, D.L. (1985). 6-Hydroxydopamine pretreatment effects on α- and β-adrenergic receptor adaptation to clorgyline. *Naunyn-Schmiedeberg's Archives de Pharmacologie* **329**, 158–61.

Colpaert, F.C., Lenaerts, F.M., Niemegeers, C.J.E. & Janssen, P.A. (1975). A critical study on RO-4-1284 antagonism in mice. *Archives Internationales de Pharmacodynamie* **215**, 40–90.

Colpaert, F.C., Meert, T.F., Niemegeers, C.J.E. & Janssen, P.A.J. (1985). Behavioral and 5-HT antagonist effects of ritanserin. A pure and selective antagonist of LSD discrimination in rat. *Psychopharmacology* **86**, 45–54.

Coppen, A. (1967). The biochemistry of affective disorders. *British Journal of Psychiatry* **113**, 1237–64.

Coppen, A., RamaRao, V.A., Swade, C. & Wood, K. (1983). Zimeldine: a therapeutic and pharmacokinetic study in depression. *Psychopharmacology* **63**, 199–202.

Cording-Tömmel, C. & Von Zerssen, D. (1982). Mianserin and maprotiline as compared to amitriptyline in severe endogeneous depression. A new methodological approach to the clinical evaluation of the efficacy of antidepressants. *Psychopharmacology* **15**, 197–204.

Corne, S.J., Pickering, R.W. & Warner, B.T. (1963). A method for assessing the effects of the drugs on the central actions of 5-hydroxytryptamine. *British Journal of Pharmacology* **20**, 106–20.

Cornfeldt, M., Fisher, B. & Fielding, S. (1982). Rat internal capsule lesion: A new test for detecting antidepressants. *Federation Proceedings* **41**, 1006.

Crawley, J.N. (1983). Preliminary report of a new rodent separation model of depression. *Psychopharmacology Bulletin* **19**, 537–41.

Cunningham, K.A., Callahan, P.M. & Appel, J.B. (1985). Similarities in the stimulus effects of 8-hydroxy-2-(di-n-propylamino)tetralin (8-OH-DPAT), buspirone, and TVX Q 7821: Implications for understanding the actions of novel anxiolytics. *Society for Neuroscience Abstracts* **11**, 1186.

Dahl, L-E., Lundin, L., LeFevreHonore,P. & Dencker, S.J. (1982). Antidepressant effect of femoxetine and desipramine and relationship to the concentration of amine metabolites in cerebrospinal fluid. *Acta Psychiatrica Scandinavica* **66**, 9–17.

Dall'Olio, R., Vaccheri, A., Gandolfi, O. & Montanaro, N. (1986). Behavioral differentiation between pharmacokinetic and pharmacodynamic components of the interaction of antidepressants or neuroleptics with methamphetamine. *Psychopharmacology* **90**, 18–23.

Danysz, W., Minor, B.G., Post, C. & Archer, T. (1986). Chronic treatment with antidepressant drugs and the analgesia induced by 5-methoxy-N,N-dimethoxytryptamine: attenuation by desipramine. *Acta Pharmacologica et Toxicologica* **59**, 103–12.

Danysz, W., Fowler, C.J. & Archer, T. (1987a). Experimental methods employed in antidepressant research – a critical review. *New Trends in Experimental and Clinical Psychiatry* **III**, 59–88.

Danysz, W., Minor, B.G., Mohammed, A., Pucilowski, P., Plaznik, A., Kostowski, W., Post, C. & Archer, T. (1987b). Chronic treatment with antidepressant drugs and ECT differentially modifies the hypothermic action of clonidine and guanfacine. *Pharmacology and Toxicology* **60**, 305–11.

Danysz, W., Plaznik, A., Kostowski, W., Malatynska, E., Järbe, T.U.C., Hiltunen, A.J. & Archer, T. (1988). Comparison of desipramine, amitriptyline, zimeldine and alaproclate in six animal models used to investigate antidepressant drugs. *Pharmacology and Toxicology* **62**, 42–50.

Da Prada, M., Kettler, R., Burkard, W.P. & Haefely, W.E. (1984). Meclobemide, an antidepressant with short-lasting MAO-A inhibition: brain catecholamines and tyramine pressor effects in rats. In: *Monoamine Oxidase and Disease; Prospects for Therapy with Reversible Inhibitors* (eds. K.F. Tipton, P. Dostert & M. Strolin Benedetti), pp. 137–54. Academic Press: London.

Davis, M. (1982). Agonist-induced changes in behavior as a measure of functional changes in receptor sensitivity following chronic antidepressant treatments. *Psychopharmacology Bulletin* **18**, 137–47.

Davis, M. & Menkes, D.B. (1982). Tricyclic antidepressants vary in decreasing α-adrenoceptor sensitivity with chronic treatment: assessment with clonidine inhibition of acoustic startle. *British Journal of Pharmacology* **77**, 217–22.

Deicken, R.F. (1986). Captopril treatment of depression. *Biological Psychiatry* **21**, 1425–8.

Delbarre, B. & Schmitt, H. (1971). Sedative effects of alpha-sympathomimetic drugs and their antagonism by adrenergic and cholinergic blocking drugs. *European Journal of Pharmacology* **13**, 356–63.

Delini-Stula, A. & Vassout, A. (1979). Modulation of dopamine-mediated behavioral response by antidepressants: effects of single and repeated treatment. *European Journal of Pharmacology* **58**, 443–51.

Delini-Stula, A., Hauser, K., Baumann, P., Olpe, A-R., Waldmeier, P. & Storni, A. (1982). Stereospecificity of behavioural and biochemical responses to oxaprotiline – a new antidepressant. In: E. Costa & G.Racagni (eds.), *Typical and Atypical Antidepressants: Molecular Mechanisms*, pp. 265–75. Raven Press: New York.

De Montigny, C., Tan, A.T. & Caille, G. (1981). Short term lithium enhances 5-HT neurotransmission in rats administered chronic antidepressant treatments. *Society for Neuroscience Abstracts* **7**, 646.

Depin, J.C. & Betbeder-Matibet, J.J. (1985). Pharmacology of lortalamine, a new potent non-tricyclic antidepressant. *Arzneimittel-Forschung Drug Research* **35**, 1655–62.

Desiles, M. & Rips, R. (1981). Use of potentiation of thyrotrophin releasing hormone (TRH)-induced hyperthermia as a test for screening antidepressants which activate the α-adrenoceptor system. *British Journal of Pharmacology* **74**, 81–6.

Dubinsky, B. & Goldberg, M.E. (1971). The effect of imipramine and selected drugs on attack elicited by hypothalamic stimulation in the cat. *Neuropharmacology* **10**, 537–45.

Dubinsky, B., Karpowicz, K. & Goldberg, M. (1973). Effects of tricyclic antidepressants on attack elicited by hypothalamic stimulation: Relation to biogenic amines. *Journal of Pharmacology and Experimental Therapeutics* **18**, 550–2.

Dunner, D., Myers, J., Khan, A., Avery, D., Ishiki, D. & Pyke, R. (1987). Adinazolam – a new antidepressant: findings of a placebo-controlled, double-blind study in outpatients with major depression. *Journal of Clinical Psychopharmacology* **7**, 170–2.

Eichelman, B. & Barchas, J. (1975). Facilitated shock-induced aggression following antidepressive medication in the rat. *Pharmacology, Biochemistry and Behavior* **3**, 601–4.

Eison, A.S. & Vocca, F.D. (1985). Reduction in cortical 5-HT$_2$ receptor sensitivity after continuous gepirone treatment. *European Journal of Pharmacology* **111**, 389–92.

Emrich, H.M., Altmann, H., Dose, M. & VonZerssen, D. (1983). Therapeutic effects of GABA-ergic drugs in affective disorders. A preliminary report. *Pharmacology, Biochemistry and Behavior* **19**, 369–72.

Euvrard, C. & Boissier, J.R. (1980). Biochemical assessment of the central 5-HT agonist activity of RU-24969 (a piperidinyl indole). *European Journal of Pharmacology* **63**, 65–72.

Everett, G.M. (1967). The DOPA response potentiation test and its use in screening for antidepressant drugs. In: S. Garratini & M.N.G. Duke (eds.) *Antidepressant Drugs*, pp. 164–7. Amsterdam: Excerpta Medica.

Ferrari, F. (1985). Behavioural pharmacology of imidazole, a potential antidepressant agent. *Archives Internationales de Pharmacodynamie et Thérapie* **277**, 303–12.

Ferris, R.M. & Beaman, O.J. (1983). Bupropion: a new antidepressant drug, the mechanism of action of which is not associated with down-regulation of postsynaptic β-adrenergic, serotonergic (5-HT$_2$) α_2-adrenergic, imipramine and dopaminergic receptors in brain. *Neuropharmacology* **22**, 1257–67.

Ferris, R.M., White, H.L. & Cooper, B.R. (1981). Some neurochemical properties of a new antidepressant, bupropion hydrochloride (Wellbutrin). *Drug Development Research* **1**, 21–35.

Fibiger, H.C. & Phillips, A. (1981). Increased intracranial self-stimulation in rats after long-term administration of desipramine. *Science* **214**, 683–5.

Fink, M. (1982). Neurophysiologic methods to establish antidepressant activity. *Arzneimittel-Forschung Drug Research* **32**, 857–8.

Finberg, J.P.M. & Tal, A. (1985). Reduced peripheral presynaptic adrenoceptor sensitivity following chronic antidepressant treatment in rats. *British Journal of Pharmacology* **84**, 609–17.

Finberg, J.P.M. & Kopin, I.J. (1986). Chronic clorgyline treatment enhances release of norepinephrine following sympathetic stimulation in the rat. *Naunyn-Schmiedeberg's Archives of Pharmacology* **332**, 236–42.

Florvall, L., Ask, A.-L., Ögren, S.-O. & Ross, S.B. (1978). Selective monoamine oxidase inhibitors. 1. Compounds related to 4-aminophenethylamine. *Journal of Medicinal Chemistry* **21**, 56–63.

Fourtillan, J.B., Bouquet, S., Girault, J., Lefebvre, M.A., Maulet, C. & Courtois, P. (1985). Pharmacokinetics and bioavailability of viqualine a new antidepressant. *European Journal of Drug Metabolism and Pharmacokinetics* **10**, 3–10.

Fowler, C.J. & Ross, S.B. (1984). Selective inhibitors of monoamine oxidase A and B: biochemical, pharmacological and clinical properties. *Medical Research Review* **4**, 323–58.

Friedman, E. & Dallob, A. (1979). Enhanced serotonin receptor activity after chronic treatment with imipramine or amitriptyline. *Communications in Psychopharmacology* **3**, 89–92.

Frost, S.J., Eccleson, D., Marshall, E.F. & Hassanyen, F. (1984). Alaproclate – an open clinical study in depressive illness. *Psychopharmacology* **83**, 285–7.

Fuxe, K., Ögren, S-O., Agnati, L.F., Benfenati, F., Fredholm, B., Anderson, K., Zini, J. & Eneroth, P. (1983). Chronic antidepressant treatment and central 5-HT synapses. *Neuropharmacology* **22**, 389–400.

Garcia-Marquez, C. & Armario, A. (1987). Chronic stress depresses exploratory activity and behavioral performance in the forced swimming test without altering ACTH response to a novel acute stressor. *Physiology and Behavior* **40**, 33–8.

Garratini, S. & Jori, A. (1967). Interaction between imipramine-like drugs and reserpine on body temperature. In: S. Garratini & M.N.G. Dukes (eds.), *Antidepressant Drugs*, pp. 173–93. Amsterdam: Excerpta Medica Foundation.

Garratini, S., Giachetti, A., Pieri, L. & Re, R. (1960). Antagonism of reserpine-induced eyelid ptosis. *Médicine Expérimentale* **3**, 315–20.

Garzon, J. & Del Rio, J. (1981). Hyperactivity induced in rats by long term isolation: further studies on a new animal model for the detection of antidepressants. *European Journal of Pharmacology* **74**, 287–94.

Garzon, J., Fuentes, J.A. & Del Rio, J. (1979). Antidepressants selectively antagonize the hyperactivity induced in rats by long-term isolation. *European Journal of Pharmacology* **59**, 293–6.

Giurgea, C. & Dauby, J. (1966). Sequential screening of antidepressant drugs (a critical procedure). In: P. Mantegazza & F. Piccini (eds.), *Methods in Drug Evaluation*, pp. 251–62. Amsterdam: N. Holland Publishing Company.

Glaser, T. & Traber, J. (1983). Buspirone: action on serotonin receptors in calf hippocampus. *European Journal of Pharmacology* **88**, 137–8.

Glaser, T., Dompert, W.U., Horwath, E., Schuurman, T., Spencer, D.G. & Traber, J. (1987). Tiflucarbine, a potential new antidepressant: biochemical and behavioral pharmacology. Paper presented at First International Congress on the Behavioral Pharmacology of 5-HT, Amsterdam, Abstracts, P22.

Goldstein, F.J., Mojavrian, P., Ossipov, M.H. & Swanson, B.N. (1982). Elevation in analgesic effect and plasma levels of morphine by desipramine in rats. *Pain* **14**, 279–82.

Goodwin, G.M., DeSouza, R.J. & Green, A.R. (1985a). The pharmacology of the hypothermic response in mice to 8-hydroxy-2-(di-n-propylamino)tetralin (8-OH-DPAT): a model of presynaptic 5-HT$_1$ function. *Neuropharmacology* **24**, 1187–94.

Goodwin, G.M., DeSouza, R.J. & Green, A.R. (1985b). Presynaptic serotonin receptor-mediated response in mice attenuated by antidepressant drugs and electroconvulsive shock. *Nature* **317**, 531–3.

Gorka, Z., Wojtasik, E., Kwiatek, H. & Maj, J. (1979). Action of serotoninmimetics in the behavioural despair test in rats. *Communications in Psychopharmacology* **3**, 133–6.

Green, A.R. (1987). Evolving concepts on the interactions between antidepressant treatments and monoamine neurotransmitters. *Neuropharmacology* **26**, 815–22.

Green, A.R., Guy, A.P. & Gardner, C.R. (1984). The behavioral effects of RU 24969, a suggested 5-HT$_1$ receptor agonist in rodents and the effect on the behavior of treatment with antidepressants. *Neuropharmacology* **23**, 655–61.

Haefely, W., Hürlimann, A. & Thoenen, H. (1964). The responses to tyramine of the normal and denervated nictitating membrane of the cat: analysis of the mechanisms and sites of actions. *British Journal of Pharmacology* **22**, 5–21.

Hall, H. & Ögren, S-O. (1981). Effects of antidepressant drugs on different receptors in the brain. *European Journal of Pharmacology* **70**, 393–407.

Hall, H. & Ögren, S-O. (1984). Effects of antidepressant drugs on histamine-H$_1$ receptors in the brain. *Life Science* **34**, 597–605.

Halliwell, G., Quinton, R.M. & Williams, F.E. (1964). A comparison of imipramine, chlorpromazine and related drugs in various tests involving autonomic functions and antagonism of reserpine. *British Journal of Pharmacology* **23**, 330–50.

Haskell, D.S., DiMascio, A. & Prusoff, B. (1975). Rapidity of symptom reduction in depression treated with amitriptyline. *Journal of Nervous and Mental Disorders* **160**, 24–33.

Hauser, K., Olpe, H.R. & Jones, R.S.G. (1985). Trimipramine, a tricyclic antidepressant exerting atypical actions on the central noradrenergic system. *European Journal of Pharmacology* **111**, 23–30.

Heal, D.J. (1984). Phenylephrine-induced activity in mice as a model of central α_1-adrenoceptor function. Effect of acute and repeated administration of antidepressant drugs and electroconvulsive shock. *Neuropharmacology* **23**, 1241–51.

Hjorth, S. & Carlsson, A. (1982). Buspirone: effects on central monoaminergic transmission – possible relevance to animal experimental and clinical findings. *European Journal of Pharmacology* **83**, 299–303.

Horovitz, Z.P. (1965). Selective block of rat mouse killing by antidepressants. *Life Science* **4**, 1909–12.

Horovitz, Z.P., Piala, J.J., High, J.P., Burke, J.C. & Leaf, R.C. (1966). Effects of drugs on the mouse killing (muricide) test and its relationship to amygdaloid function. *International Journal of Neuropharmacology* **5**, 405–11.

Howard, J.L. & Pollard, G.T. (1984). Effects of imipramine, bupropion, chlorpromazine and clozapine on differential-reinforcement-of-low-rate (DRL) > 72-sec and > 36-sec schedules in rat. *Drug Development Research* **4**, 607–17.

Howard, J.L., Soroko, F.E. & Cooper, B.R. (1981). Empirical behavioral models of depression with emphasis on tetrabenazine antagonism. In: S.J. Enna, J.B. Malick & E. Richelson (eds.), *Antidepressants: Neurochemical, Behavioral and Clinical Perspectives*, pp. 107–20. New York: Raven Press.

Hrdina, P.D., Von Kulmiz, P. & Stretch, R. (1979).

Pharmacological modification of experimental depression in infant macaques. *Psychopharmacology* **64**, 89–93.

Huskisson, E.C. & Wojtulewski, J.A. (1974). The measurement of side effects of drugs. *British Medical Journal* **2**, 698–9.

Hyttel, J. (1983). Citalopram: pharmacological profile of a specific serotonin uptake inhibitor with antidepressant activity. *Progress in Neuropsychopharmacology and Biological Psychiatry* **6**, 277–95.

Itil, T.M., Polvan, N. & Hsu, W. (1972). Clinical and EEG effects of GB 94, a 'tetracyclic' antidepressant (EEG model: Discovery of a new psychotropic drug). *Current Therapy Research* **14**, 395–413.

Itil, T.M., Menon, G.N., Bozak, M.M. & Itil, K.Z. (1984). CNS effects of citalopram, a new serotonin inhibitor antidepressant (a quantitative pharmacoelectroencephalography study). *Progress in Neuropsychopharmacology and Biological Psychiatry* **8**, 397–409.

Jahoda, M. (1979). The impact of unemployment in the 1930s and 1970s. *Bulletin of the British Psychological Society* **32**, 309–14.

Jesberger, J.A. & Richardson, S.J. (1985). Animal models of depression: parallels and correlates to severe depression in humans. *Biological Psychiatry* **20**, 764–84.

Johnston, J.P. (1968). Some observations on a new inhibitor of monamine oxidase in brain. *Biochemical Pharmacology* **17**, 1285–97.

Jones, S.G. (1980). Enhancement of 5-hydroxytryptamine-induced behavioral effects following chronic administration of antidepressant drugs. *Psychopharmacology* **69**, 477–86.

Jouvent, R., Lecrubier, Y., Puech, A.J., Frances, H., Simon, P. & Widlocher, D. (1977). De l'étude experimentale d'un stimulant beta-adrenergique à la mise en evidence de son activité antidepressive chez l'homme. *L'Encéphale* **3**, 285–93.

Kafka, M.S., Wirz-Justice, A., Naber, D., Moore, R.Y. & Benedito, M.A. (1983). Circadian rhythms in rat brain neurotransmitter receptors. *Federation Proceedings* **42**, 2796–801.

Kanner, A., Coyne, J., Schaefer, C. & Lazarus, R. (1981). Comparison of two modes of stress measurement: Daily hassles and uplifts versus major life events. *Journal of Behavioral Medicine* **4**, 1–39.

Kanof, P.D. & Greengard, P. (1978). Brain histamine receptors as targets for antidepressant drugs. *Nature* **272**, 329–33.

Karasawa, T., Furukawa, K., Ochi, Y., Ito, T., Yoshida, K. & Shimizu, M. (1980). Antidepressant properties of 2-(4-ethyl-1-piperazinyl)-4-phenylquinoline hydrochloride (AD-1308) and

its mechanism of action as compared with tricyclic antidepressants. *Archives Internationales de Pharmacodynamie et de Thérapie* **245**, 283–302.

Katz, R.J. (1981). Animal models and human depressive disorders. *Neuroscience and Biobehavioral Reviews* **5**, 231–46.

Katz, R.J. (1984). Effects of zometapine, a structurally novel antidepressant, in an animal model of depression. *Pharmacology, Biochemistry and Behavior* **21**, 487–90.

Katz, R.J. & Hersh, S. (1981). Amitriptyline and scopolamine in an animal model of depression. *Neuroscience and Biobehavioral Reviews* **5**, 265–71.

Keller, H.H., Schaffner, R., Carruba, M.O., Burkard, W.P., Pieri, M., Bonetti, E.P., Schershlicht, R., DaPrada, M. & Haefely, W.E. (1982). Diclofensine (RO 8-4650) – a potent inhibitor of monoamine uptake; biochemical and behavioral effects in comparison with nomifensine. In: E. Costa & S. Racagni (eds.), *Typical and Atypical Antidepressants. Molecular Mechanisms*, pp. 287–300. New York: Raven Press.

Kellstein, D.E., Malseed, R.T. & Goldstein, F.J. (1984). Contrasting effects of acute vs. chronic tricyclic antidepressant treatment on central morpine analgesia. *Pain* **20**, 323–34.

Kennett, G.A., Dourish, C.T. & Curzon, G. (1987). Antidepressant and antianorexic properties of 5-HT$_{1A}$ agonists and other drugs in animal models. In: C.T. Dourish, S. Ahlenius & P.H. Hutson (eds.), *Brain 5-HT$_{1A}$ Receptors*, pp. 243–60. Chichester: Ellis Horwood Ltd.

Kinnier, W.J., Tabor, R.D. & Norrell, L.Y. (1984). Neurochemical properties of AHR-9377: A novel inhibitor of norepinephrine reuptake. *Biochemical Pharmacology* **33**, 3001–5.

Koe, B.K., Weissman, A., Welch, W.M. & Browne, R.G. (1983). Sertraline, IS, 45-N-methyl-4-(3,4-dichlorophenyl)-1.2-3.4-tetrahydro-1-naphthylamine, a new uptake inhibitor with selectivity for serotonin. *Journal of Pharmacology and Experimental Therapeutics* **226**, 686–700.

Kokkinidis, L., Zacharko, R. & Predy, P. (1980). Post-amphetamine depression of self-stimulation responding from the substantia nigra. Reversal by tricyclic antidepressants. *Pharmacology, Biochemistry and Behavior* **13**, 379–83.

Kostowski, W. & Dyr, W. (1984). A study of the effect of clonidine on the EEG in rats treated with single and multiple doses of antidepressants. *Psychopharmacology* **84**, 85–90.

Kostowski, W. & Malatynska, E. (1983). Antagonism of behavioral depression produced by clonidine in the mongolian gerbil: a potential screening test for antidepressant drugs. *Psychopharmacology* **79**, 203–8.

Kostowski, W., Plewako, M. & Bidzinski, A. (1984*a*). Brain serotonergic neurones: Their role in a form of dominance-subordination behaviour in rats. *Physiology and Behavior* **33**, 365–71.

Kostowski, W., Valzelli, L., Kozak, W. & Bernasconi, S. (1984*b*). Activity of desipramine, fluoxetine and nomifensine on spontaneous and p-CPA-induced muricidal aggression. *Pharmacological Research Communications* **16**, 265–71.

Kuhn, R. (1970). The imipramine story. In: F.J. Ayd & B. Blackwell (eds.), *Discoveries in Biological Psychiatry*, pp. 199–204. Philadelphia/Toronto: J.B. Lippincott.

Lahti, R.A., Cohn, J.B., Feighner, J.O., Smith, W.T. & Dyke, R.E. (1982). Pharmacological profile and clinical activity of the antidepressant adinazolam, a triazolobenzodiazepine. *Society for Neuroscience Abstracts* **8**, 463.

Lang, W.J. & Gershon, S. (1963). Effects of psychoactive drugs on yohimbine induced responses in conscious dogs. *Archives Internationales de Pharmacodynamie et Thérapie* **124**, 3–4.

Lapin, I.P. (1980). Adrenergic nonspecific potentiation of yohimbine toxicity in mice by antidepressants and related drugs and antiyohimbine action of antiadrenergic and serotonergic drugs. *Psychopharmacology* **46**, 19–25.

Leibowitz, S., Brown, O., Tretter, J. & Kirschgessner, A. (1985). Norepinephrine, clonidine and tricyclic antidepressants selectively stimulate carbohydrate ingestion through noradrenergic system of the paraventricular nucleus. *Pharmacology, Biochemistry and Behavior* **23**, 541–50.

Lessin, A.W. (1959). The pharmacological evaluation of monoamine oxidase inhibitors. *Biochemical Pharmacology* **2**, 290–8.

Leysen, J.E., Gommeren, W., Van Gompel, P., Wynants, J., Janssen, P.A.J. & Laduron, P.M. (1985). Receptor binding properties *in vitro* and *in vivo* of ritanserin. A very potent and long acting S_2 antagonist. *Molecular Pharmacology* **27**, 600–11.

Lynch, M.A. & Leonard, B.E. (1978). Effect of chronic amphetamine administration on the behavior of rats in the open field apparatus: reversal of post-withdrawal depression by two antidepressants. *Journal of Pharmacy and Pharmacology* **30**, 798–9.

Maier, S.F. & Seligman, M.E.O. (1976). Learned helplessness: theory and evidence. *Journal of Experimental Psychology* **105**, 3–46.

Maj, J., Mogilnicka, E. & Klimek, V. (1979). The effect of repeated administration of antidepressant drugs on the responsiveness of rats to catecholamine agonists. *Journal of Neural Transmission* **44**, 221–35.

Maj, J., Mogilnicka, E. & Kordecka-Magiera, A. (1980). Effect of chronic administration of antidepressant drugs on aggressive behavior induced by clonidine in mice. *Pharmacology, Biochemistry and Behavior* **13**, 153–4.

Maj, J., Chojnacka-Wojcik, E., Lewandowska, A., Tatarczynska, E. & Wiczynska, B. (1985). The central action of carbamazepine as a potential antidepressant drug. *Polish Journal of Pharmacology and Pharmacy* **37**, 47–56.

Malatynska, E. & Kostowski, W. (1984). The effect of antidepressant drugs on dominance behavior in rats competing for food. *Polish Journal of Pharmacology and Pharmacy* **35**, 531–40.

Malick, J.B. (1981). Yohimbine potentiation as a predictor of antidepressant action. In: S.J. Enna, J.B. Malick & E. Richelson (eds.), *Antidepressants: Neurochemical, Behavioral and Clinical Perspectives*, pp. 141–55. New York: Raven Press.

Martin-Iverson, M.T., Leclere, J.F. & Fibiger, H.C. (1983). Cholinergic-dopaminergic interactions and the mechanism of action of antidepressants. *European Journal of Pharmacology* **94**, 193–201.

Martorana, P.A. & Nitz, R-E. (1979). The new antidepressant pirlindole. A comparison with imipramine and tranylcypromine. *Arzneimittel-Forschung Drug Research* **29**, 946–9.

Mason, S.T. & Angel, A. (1983). Behavioral evidence that chronic treatment with the antidepressant desipramine causes reduced functioning of brain NA systems. *Psychopharmacology* **81**, 73–7.

McKinney, W.T. (1984). Animal models of depression: An overview. *Psychiatric Developments* **2**, 77–96.

McKinney, W.T. & Bunney, W.E. (1969). Animal models of depression: Review of evidence and implications for research. *Archives of General Psychiatry* **21**, 240–8.

Meek, J., Fuxe, K. & Andén, N-E. (1970). Effects of antidepressant drugs of the imipramine type on central 5-hydroxytryptamine neurotransmission. *European Journal of Pharmacology* **9**, 325–32.

Meek, J.L., Fuxe, K. & Carlsson, A. (1971). Blockade of p-chlorometamphetamine induced 5-hydroxytryptamine depletion by chlorimipramine, chlorpheniramine and mepiridine. *Biochemical Pharmacology* **20**, 707–9.

Menon, M.K., Vivonia, C.A. & Kling, A.S. (1984). Pharmacological studies on the antagonism by antidepressants of the hypothermia induced by apomorphine. *Neuropharmacology* **23**, 121–7.

Mishra, R., Gillespie, D.D., Youdim, M.B.H. & Sulser, F. (1983). Effect of selective monoamine oxidase inhibition by clorgyline and deprenyl on the norepinephrine receptor-coupled adenylate cyclase system in rat cortex. *Psychopharmacology* **81**, 220–3.

Mogilnicka, E. & Klimek, V. (1979). Mianserin, danitracen and amitriptyline withdrawal increases the behavioral responses of rats to L-5-HTP. *Journal of Pharmacy and Pharmacology* **31**, 704–5.

Mogilnicka, E. & Przewlocka, E. (1981). Facilitated shock-induced aggression after chronic treatment with antidepressant drugs in the rat. *Pharmacology, Biochemistry and Behavior* **14**, 129–32.

Mondrup, K., Dupont, E. & Pederson, E. (1983). The effect of the GABA-agonist, progabide, on benign essential tremor. A controlled clinical trial. *Acta Neurologica Scandinavica* **68**, 248–52.

Montanaro, N., Dall'Olio, R., Gandolfi, O. & Vaccheri, A. (1982). Neuroleptic vs antidepressant activity of sulpiride isomers in the rat. In: E. Costa & Racagni (eds.), *Typical and Atypical Antidepressants: Molecular Mechanisms*, pp. 341–6. New York: Raven Press.

Moore, D.C. (1977). Amitriptyline therapy in anorexia nervosa. *American Journal of Psychiatry* **134**, 1303–4.

Morpurgo, C. & Theobald, W. (1965). Influence of imipramine-like compounds and chlorpromazine on reserpine-hypothermia and amphetamine hyperthermia in rats. *Medicine et Pharmacologie Expérimentale* **2**, 226–32.

Morris, J.B. & Beck, A.T. (1974). The efficacy of antidepressant drugs. A review of research (1958 to 1972). *Archives of General Psychiatry* **30**, 667–74.

Nagayama, H., Hingten, J.N. & Aprison, M.H. (1981). Postsynaptic action by four antidepressive drugs in an animal model of depression. *Pharmacology, Biochemistry and Behavior* **15**, 125–30.

Nies, A., Robinson, D.S., Friedman, M.J., Green, R., Cooper, T.B., Ravaris, C.L. & Ives, J.O. (1977). Relationship between age and tricyclic antidepressant plasma levels. *American Journal of Psychiatry* **134**, 790–3.

Nomura, S., Shimizu, J., Kametani, H., Kinjo, M., Watanabe, M. & Nakazawa, T. (1982). Swimming in mice: in search of an animal model for human depression. In: S.Z. Langer, R. Takahashi, T. Segawa & N. Briley (eds.), *New Vistas in Depression*, pp. 203–10. Oxford: Pergamon Press.

Noreika, L., Pastor, G. & Liebman, J. (1981). Delayed emergence of antidepressant efficacy following withdrawal in olfactory bulbectomized rats. *Pharmacology, Biochemistry and Behavior* **15**, 393–8.

O'Donnell, J.M. & Seiden, L.S. (1983). Differential-reinforcement-of-low-rate 72-second schedule: Selective effects of antidepressant drugs. *Journal of Pharmacology and Experimental Therapeutics* **224**, 80–8.

O'Donohue, T.L., Wirz-Justice, A., Kafka, M.S., Naber, D., Campbell, I.C. & Wehr, T.A. (1982). Effects of chronic lithium, clorgyline, imipramine, fluphenazine and constant darkness on the melanotropin content and circadian rhythm in rat brain. *European Journal of Pharmacology* **85**, 1–7.

Ögren, S-O., Fuxe, K., Archer, T., Johansson, G. & Holm, A-C. (1982). Behavioral and biochemical studies on the effects of acute and chronic administration of antidepressant drugs on central serotonergic receptor mechanisms. In: S.Z. Langer, R. Takahashi, T. Segawa & M. Briley (eds.), *New Vistas in Depression*, pp. 11–19. New York: Pergamon Press.

Ögren, S-O., Carlsson, S. & Bartfai, T. (1985). Serotonergic potentiation of muscarine agonist evoked tremor and salivation in rat and mouse. *Psychopharmacology* **86**, 258–64.

Onodera, K., Ogura, Y. & Kisara, K. (1981). Characteristics of muricide induced by thiamine deficiency and its suppression by antidepressants or intraventricular serotonin. *Physiology and Behavior* **27**, 847–53.

Oswald, I. & Adam, K. (1986). Effects of paroxetine on human sleep. *British Journal of Clinical Pharmacology* **22**, 97–9.

Oswald, J., Brezinowa, V. & Dunleavy, D.L.F. (1972). On the slowness of action of tricyclic antidepressant drugs. *British Journal of Psychiatry* **120**, 673–7.

Passarelli, F. & Scotti de Carolis, A. (1983). Effect of chronic treatment with imipramine, trazodon and electro shock on the behavioral and electroencephalographic modifications induced by clonidine in the rat. *Neuropharmacology* **22**, 785–9.

Pawlowski, L. (1984). Amitriptyline and femoxetine, but not clomipramine or citalopram, antagonize hyperthermia induced by directly acting 5-hydroxytryptamine-like drugs in heat adapted rats. *Journal of Pharmacy and Pharmacology* **36**, 197–9.

Pawlowski, L. & Mazela, H. (1986). Effect of antidepressant drugs, selective noradrenaline or 5-hydroxytryptamine uptake inhibitors, on apomorphine-induced hypothermia. *Psychopharmacology* **88**, 18–23.

Paykel, E., Meulter, P. & de la Vergue, P. (1973). Amitriptyline, weight gain and carbohydrate craving: A side effect. *British Journal of Psychiatry* **123**, 501–7.

Plaznik, A. & Kostowski, W. (1987). The effects of antidepressants and electroconvulsive shocks on the functioning of the mesolimbic dopaminergic system: a behavioural study. *European Journal of Pharmacology* **135**, 389–96.

Plaznik, A., Danysz, W. & Kostowski, W. (1984). Behavioral evidence for α_1-adrenoceptor up- and α_2-adrenoceptor downregulation in the rat hippocampus after chronic desipramine treatment. *European Journal of Pharmacology* **101**, 305–6.

Pollard, G.T. & Howard, J.L. (1986). Similar effects of antidepressant and non-antidepressant drugs on behavior under an interresponse-time > 72-s schedule. *Psychopharmacology* **89**, 253–8.

Porsolt, R.D. (1981). Behavioral despair. In: S.J. Enna, J.B. Malick & E. Richelson (eds.), *Antidepressants: Neurochemical, Behavioral and Clinical Perspectives*, pp. 121–39. New York: Raven Press.

Porsolt, R.D., LePichon, M. & Jalfre, M. (1977). Depression; a new animal model sensitive to antidepressant treatments. *Nature* **266**, 730–2.

Porsolt, R.D., Anton, G., Blavet, N. & Jalfre, M. (1978). Behavioral despair in rats, a new model sensitive to antidepressant treatments. *European Journal of Pharmacology* **51**, 291–4.

Post, R.M., Ballenger, J.C., Uhde, T.W., Smith, C., Rubinow, D.R. & Bunney, W.E. (1982). Effect of carbamazepine on cyclic nucleotides in CSF of patients with affective illness. *Biological Psychiatry* **17**, 1037–45.

Post, R.M., Uhde, T.W., Rubinow, D.R., Ballenger, J.C. & Gold, P.W. (1983). Biochemical effects of carbamazepine: Relationship to its mechanism of action in affective illness. *Progress in Neuropsychopharmacology* **7**, 263–71.

Post, R.M., Uhde, T.W. & Ballenger, J.C. (1984). The efficacy of carbamazepine in affective illness. In: E. Usdin, M. Asberg, B. Bertilson & E. Sjöqvist (eds.), *Frontiers in Biochemical and Pharmacological Research in Depression*, pp. 421–37. New York: Raven Press.

Prusoff, B.A., Weissman, M.M. & Charney, J. (1981). Speed of symptoms reduction in depressed outpatients treated with amoxapine and amitriptyline. *Current Therapy Research* **30**, 843–55.

Przegalinski, E. & Bigajska, K. (1983). Antidepressant properties of some phosphodiesterase inhibitors. *Polish Journal of Pharmacology and Pharmacy* **35**, 233–40.

Przegalinski, E. & Jurkowska, T. (1987). Effect of repeated treatment with antidepressant drugs or electroconvulsive shock (ECS) on the increase in food intake induced by clonidine injected into the paraventricular nucleus. *Archives Internationales de Pharmacodynamie et Thérapie* **290**, 257–66.

Przegalinski, E., Baran, L. & Siwanowcz, J. (1983). The effect of chronic treatment with antidepressant drugs on salbutamol-induced hypoactivity. *Psychopharmacology* **80**, 355–9.

Przegalinski, E., Baran, L., Siwanowcz, J. & Bigajska, K. (1984). Repeated treatment with antidepressant drugs prevents salbutamol-induced hypoactivity in rats. *Pharmacology, Biochemistry and Behavior* **21**, 695–8.

Puech, A.J., Chermat, R., Poncelet, M., Doare, L. & Simon, P. (1981). Antagonism of hypothermia and behavioral response to apomorphine: a simple, rapid and discriminant test for screening antidepressants and neuroleptics. *Psychopharmacology* **75**, 84–91.

Quinton, R.M. (1963). The increase in the toxicity of yohimbine induced by imipramine and other drugs in mice. *British Journal of Pharmacology* **21**, 51–66.

Randrup, A. & Braestrup, C. (1977). Uptake inhibition of biogenic amines by newer antidepressant drugs: Relevance to the dopamine hypothesis of depression. *Psychopharmacology* **53**, 309–14.

Rehavi, M., Maayani, S. & Sokolowsky, M. (1977). Tricyclic antidepressants as antimuscarinic drugs: *In vivo* and *in vitro* studies. *Biochemical Pharmacology* **26**, 1559–67.

Remé, C., Wirz-Justice, A., Aeberhard, B. & Rhyner, A. (1984). Chronic clorgyline dampens rat retinal rhythms. *Brain Research* **298**, 99–106.

Rényi, L. (1986). The effects of monoamine oxidase inhibitors on the ejaculatory response induced by 5-methoxy-N,N-dimethyltryptamine in the rat. *British Journal of Pharmacology* **88**, 827–35.

Rényi, L. (1987). The involvement of 5-HT$_1$ and 5-HT$_2$ receptors in different components of the 5-HT syndrome in the rat. *Neuroscience*, Suppl. **22**, 225P.

Rényi, L. & Ross, S.B. (1985). Effects of amiflamine and related compounds on the accumulation of biogenic monoamines in rat brain slices *in vitro* and *ex vivo* in relation to their behavioural effects. *Acta Pharmacologica et Toxicologica* **56**, 416–26.

Rényi, L., Archer, T., Minor, B.G., Tandberg, B., Fredriksson, D. & Ross, S.B. (1986). The inhibition of cage leaving response – a model for studies of the serotonergic neurotransmission in the rat. *Journal of Neural Transmission* **65**, 193–210.

Richelson, E. (1979). Tricyclic antidepressants and histamine H$_1$-receptors. *Mayo Clinic Proceedings* **54**, 669–74.

Rickels, K., Feighner, J.P. & Smith, W.T. (1985). Alprazolam, amitriptyline, doxepin and placebo in the treatment of depression. *Archives of General Psychiatry* **42**, 134–41.

Robertson, M.M. & Trimble, M.R. (1982). Major tranquillizers used as antidepressants. *Journal of Affective Disorders* **4**, 173–93.

Rodstein, M. & Som Oei, L. (1979). Cardiovascular side effects of long-term therapy with tricyclic antidepressants in the aged. *Journal of the American Geriatric Society* **27**, 231–4.

Roth, K.A. & Katz, R.J. (1981). Further studies on a novel animal model of depression: Therapeutic effects of a tricyclic antidepressant. *Neuroscience and Biobehavioral Reviews* **5**, 253–8.

Sanghvi, J. & Gershon, S. (1969). The evaluation of central nervous system stimulants in a new laboratory test for antidepressants. *Life Science* **8**, 449–57.

Sanghvi, J., Bindler, E. & Gershon, S. (1969). The evaluation of a new animal method for the prediction of clinical antidepressant activity. *Life Science* **8**, 99–106.

Sanghvi, J., Geyer, H. & Gershon, S. (1976). Exploration of the antidepressant potential of iprindole. *Life Science* **18**, 569–74.

Savage, D.D., Mendels, J. & Frazer, A. (1980*a*). Decrease in ³H-serotonin binding in rat brain produced by the repeated administration of either monoamine oxidase inhibitors or centrally acting serotonin agonists. *Neuropharmacology* **19**, 1063–70.

Savage, D.D., Mendels, J. & Frazer, A. (1980*b*). Monoamine oxidase inhibitors and serotonin uptake inhibitors: differential effects on ³H-serotonin binding sites in rat brain. *Journal of Pharmacology and Experimental Therapeutics* **212**, 259–63.

Scheckel, C.L. & Boff, E. (1964). Behavioral effects of interacting imipramine and other drugs with d-amphetamine, cocaine and tetrabenazine. *Psychopharmacology* **5**, 198–208.

Schelkunov, E.L. (1977). Efficacy of neuroleptics and antidepressants in the test of apomorphine hypothermia and some data concerning neurochemical mechanisms of the test. *Psychopharmacology* **55**, 87–95.

Schuurman, T., Glaser, T., Spencer, D.G. & Traber, J. (1987). Neurochemical and behavioural effects of the new 5-HT₁A receptor ligand BAY R 1531. Paper presented at First International Congress on the Behavioural Pharmacology of 5-HT, Amsterdam, Abstracts, P4.

Scott, J.P., Stewart, J.M. & De Ghett, V. (1973). Separation in infant dogs: Emotional response and motivational consequences. In: J.P. Scott & E.C. Senay (eds.), *Separation and Depression*, pp. 3–32. Washington: American Association for the Advancement of Science.

Sechter, D., Pirer, M.F. & Loo, H. (1984). Clinical studies with indalpine: a critical review. *Clinical Neuropharmacology* **7**, 870–1.

Seiden, L.S. & O'Donnell, M.J. (1985). Effects of antidepressant drugs on DRL behavior. In: L.S. Seiden & R.L. Balster (eds.), *Behavioral Pharmacology: The Current Status*, pp. 323–38. New York: Alan R. Liss Inc.

Seiden, L.S., Dahms, J.L. & Shaughnessy, R.A. (1985). Behavioral screen for antidepressants. The effects of drugs and electro-convulsive shock on performance under a differential-reinforcement-of-low-rate schedule. *Psychopharmacology* **86**, 55–60.

Seltzer, V. & Tonge, S.R. (1975). Methylamphetamine

withdrawal as a model for the depressive state: antagonism of post-amphetamine depression by imipramine. *Journal of Pharmacy and Pharmacology* **27**, 16–23.

Serra, G., Argiolas, A., Klimek, V., Fadda, F. & Gessa, G.L. (1979). Chronic treatment with antidepressants prevents the inhibitory effect of small doses of apomorphine on dopamine synthesis and motor activity. *Life Science* **25**, 415–24.

Sherman, A.D. & Petty, F. (1980). Neurochemical basis of the action of antidepressants on learned helplessness. *Behavioral and Neural Biology* **30**, 119–34.

Shibata, S., Nakanishi, H., Watanabe, S. & Ueki, S. (1984). Effect of chronic administration of antidepressants on mouse-killing behaviour (muricide) in olfactory bulbectomized rats. *Pharmacology, Biochemistry and Behavior* **21**, 225–30.

Sigg, E.B. & Hill, R.T. (1967). The effect of imipramine on central adrenergic mechanisms. In: O. Brill (ed.), *Neuro-psychopharmacology*, pp. 367–72. Amsterdam: Excerpta Medica.

Sigg, E.G., Gyermek, K. & Hill, R.T. (1965). Antagonism to reserpine induced depression by imipramine, related psychoactive drugs and some autonomic agents. *Psychopharmacology* **7**, 144–9.

Sills, M.A., Heydorn, W.E., Cohen, R.M., Creed, G.J. & Jacobowitz, D.M. (1986). Effect of chronic treatment with clorgyline on the relative concentration of specific proteins in the hippocampus and parietal cortex of the rat. *Neuropharmacology* **25**, 143–50.

Silvestrini, B. (1982). Trazodone – a new type of antidepressant: a discussion of pharmacological data and their clinical implications. *Acta Psychiatrica Scandinavica* **302**, 72–80.

Simon, P., Lecrubier, Y., Jouvent, R., Puech, A., Allilaire, J.F. & Widlocher, D. (1978). Experimental and clinical evidence of the antidepressant effect of a beta-adrenergic stimulant. *Psychological Medicine* **8**, 335–8.

Simon, P., Lecrubier, Y., Jouvent, R., Puech, A. & Widlocher, D. (1984). Beta-receptor stimulation in the treatment of depression. In: E. Usdin, M. Åsberg, B. Bertilson & F. Sjöqvist (eds.), *Frontiers in Biochemical and Pharmacological Research in Depression*, pp. 293–9. New York: Raven Press.

Snyder, S.H. & Yamamura, H.I. (1977). Antidepressants and the muscarinic acetylcholine receptor. *Archives of General Psychiatry* **34**, 236–9.

Soblosky, J.S. & Thurmond, J.B. (1986). Biochemical and behavioral correlates of chronic stress: Effects of tricyclic antidepressants. *Pharmacology, Biochemistry and Behavior* **24**, 1361–8.

Sofia, R.D. (1969). Effect of centrally active drugs on four models of experimentally induced aggression in rodents. *Life Science* **8**, 705–16.

Soroko, F.E., Mehta, N.B., Maxwell, R.A., Ferris, R.M. & Schroeder, D.H. (1977). Bupropion hydrochloride ((±)α-t-butylamino-3-chloropropiophenone HCl): a novel antidepressant agent. *Journal of Pharmacy and Pharmacology* **29**, 767–70.

Spyraki, C. & Fibiger, H.C. (1980). Functional evidence for subsensitivity of noradrenergic α₂-receptors after chronic desipramine treatment. *Life Science* **27**, 1863–7.

Spyraki, C. & Fibiger, H.C. (1981). Behavioral evidence for supersensitivity of postsynaptic dopamine receptors in the mesolimbic system after chronic administration of desipramine. *European Journal of Pharmacology* **74**, 195–206.

Standal, J.E. (1977). Pizotifen as an antidepressant. *Acta Psychiatrica Scandinavica* **56**, 276–9.

Stefanis, C.N., Alevizos, B. & Papadimitrious, G.N. (1984). Controlled study of meclobemide (Ro 11-1163), a new MAO inhibitor, and desipramine in depressive patients. In: K.F. Tipton, P. Dostert & M. Strolin Benedetti, *Monoamine Oxidase and Disease; Prospects for Therapy with Reversible Inhibitors*, pp. 377–92. London: Academic Press.

Stein, L. & Seifter, J. (1961). Possible mode of antidepressive action of imipramine. *Science* **134**, 286–7.

Steru, L., Chermat, R., Thierry, B. & Simon, P. (1985). The tail suspension test: a new method for screening antidepressants in mice. *Psychopharmacology* **85**, 367–70.

Stolz, J.F. & Marsden, C.A. (1982). Withdrawal from chronic treatment with methergoline, dl-propranolol and amitriptyline enhances serotonin receptor mediated behavior in the rat. *European Journal of Pharmacology*, 79, 17–22.

Stolz, J.F., Marsden, C.A. & Middlemiss, D.N. (1983). Effect of chronic antidepressant treatment and subsequent withdrawal on (³H)-5-hydroxytryptamine and (³H)-spiperone binding in rat frontal cortex and serotonin receptor mediated behavior. *Psychopharmacology* **80**, 150–5.

Stratas, N.E. (1984). A double-blind study of the efficacy and safety of dothiepin hydrochloride in the treatment of major depressive disorder. *Journal of Clinical Psychiatry* **45**, 466–9.

Strolin Benedetti, M., Boucher, T., Carlsson, A. & Fowler, C.J. (1983). Intestinal metabolism of tyramine by both forms of monoamine oxidase in the rat. *Biochemical Pharmacology* **32**, 47–52.

Sulpizio, A., Fowler, P.J. & Macko, E. (1978). Antagonism of fenfluramine-induced hyperthermia: a measure of central serotonin inhibition. *Life Science* **22**, 1439–46.

Sulser, F., Watts, J. & Brodie, B. (1962). On the mechanism of antidepressant action of imipramine like drugs. *Annals of the New York Academy of Sciences* **96**, 279–88.

Sulser, F., Owens, M.L. & Dingel, J.V. (1966). On the mechanism of amphetamine potentiation by desipramine. *Life Science* **5**, 2005–10.

Suomi, S.J., Seman, S.F., Lewis, J.K., DeLizio, R.B. & McKinney, W.T. (1978). Effects of imipramine treatment on separation-induced social disorders in rhesus monkeys. *Archives of General Psychiatry* **35**, 321–5.

Svensson, T.H. & Usdin, T. (1978). Feed-back inhibition of brain noradrenaline neurons by tricyclics: α-receptor mediation after acute and chronic treatment. *Science* **202**, 1089–91.

Tachikawa, S., Harada, M. & Maeno, H. (1979). Pharmacological and biochemical studies on a new compound, 2-(7-indeyloxymethyl)morpholine hydrochloride (YM-08054-1), and its derivatives with potential antidepressant properties. *Archives Internationales de Pharmacodynamie et Thérapie* **238**, 81–95.

Tadeschi, D.H., Tadeschi, R.E. & Fellows, E.J. (1959). The effect of tryptamine on the central nervous system, including a pharmacological procedure for the evaluation of iproniazid-like drugs. *Journal of Pharmacology and Experimental Therapeutics* **126**, 223–32.

Tang, S.W., Helmeste, D.M. & Stancer, H.C. (1979). Interaction of antidepressants with clonidine on rat brain total 3-methoxy-4-hydroxyphenylglycol. *Canadian Journal of Physiology and Pharmacology* **57**, 435–7.

Thierry, B., Steru, L., Simon, P. & Porsolt, R.D. (1986). The tail suspension test: ethical considerations. *Psychopharmacology* **90**, 284–5.

Towell, A., Willner, P. & Muscat, R. (1986). Dopamine autoreceptors in the ventral tegmental area show subsensitivity following withdrawal from chronic antidepressant drug treatment. *Psychopharmacology* **90**, 18–23.

Turski, L., Czuczwar, S.J., Turski, W. & Kleinrok, Z. (1981). Effect of antidepressant drugs on carbachol chloride-induced wet dog shake behavior in rats. *Neuropharmacology* **20**, 1193–6.

U'Prichard, D.C., Greenburg, D.A., Sheehan, P.P. & Snyder, S.H. (1978). Tricyclic antidepressants: Therapeutic properties and affinities for α-noradrenergic receptor sites in the brain. *Science* **119**, 197–9.

Valzelli, L. & Bernasconi, S. (1971). Psychoactive drug effects on behavioral changes induced by prolonged socio-environmental deprivation in rats. *Psychological Medicine* **6**, 271–6.

Van Dorth, R.M. (1983). Review of clinical studies with mianserin. *Acta Psychiatrica Scandinavica* **302**, 72–80.

Van Riezen, H. (1972). Differential central effects of the 5-HT antagonists mianserin and cyproheptadine. *Archives Internationales de Pharmacodynamie et Thérapie* **198**, 256–9.

Van Riezen, H., Schneiden, H. & Wren, A. (1977). Olfactory bulb ablation in the rat: Biobehavioural changes and their reversal by antidepressant drugs. *British Journal of Pharmacology* **60**, 521–8.

Vernier, V.G., Alleva, F.R., Hanson, H.M. & Stone, C.A. (1962). Pharmacological action of amitriptyline, nortriptyline and imipramine. *Federation Proceedings* **21**, 419.

Vetulani, J. & Sulser, F. (1975). Action of various antidepressant treatments reduces reactivity of noradrenergic cyclic AMP generating system in limbic forebrain. *Nature* **257**, 495–6.

Vetulani, J., Stawarz, R.J., Dingell, J.V. & Sulser, F. (1976). A possible common mechanism of action of antidepressant treatments. *Naunyn-Schmiedeberg's Archives of Pharmacology* **293**, 109–17.

Von Voigtlander, P.F. & Losey, E.G. (1976). On the use of selective neurotoxic amine analogs to measure the blockade of norepinephrine and 5-hydroxytryptamine uptake systems by antidepressants. *Research Communications in Chemical Pathology and Pharmacology* **13**, 389–99.

Von Voigtlander, P.F. & Losey, E.G. (1978). 6-hydroxydopa depletes both brain epinephrine and norepinephrine: interaction with antidepressants. *Life Science* **23**, 147–50.

Von Voigtlander, P.F. & Moore, K.F. (1973). Turning behavior of mice with unilateral 6-hydroxydopamine lesions in the striatum: effect of apomorphine, L-DOPA, amantadine, amphetamine and other psychomotor stimulants. *Neuropharmacology* **12**, 451–62.

Von Voigtlander, P.F. & Straw, R.N. (1985). Alprazolam: review of pharmacological, pharmacokinetic and clinical data. *Drug Development Research* **6**, 1–12.

Von Voigtlander, P.F., Treiezenberg, H.J. & Losey, E.G. (1978). Interactions between clonidine and antidepressant drugs: a method for identifying antidepressant-like agents. *Neuropharmacology* **17**, 375–81.

Von Voigtlander, P.F., Treiezenberg, H.J., Losey, E.G. & Gray, D.D. (1978). U-48,753E: a potent, structurally novel antidepressant-like agent. *Drug Development Research* **3**, 545–54.

Vossenaar, T. (1980). Introduction. *Current Medical Research and Opinion* **6**, 3–4.

Wedzony, K. & Maj, J. (1983). The effect of repeated treatment with imipramine on locomotor hyperactivity induced in rats by amphetamine administered into the nucleus accumbens. Abstracts of 8th Congress of the Polish Pharmacological Society, Abstract No. 181.

Wehr, T.A. & Wirz-Justice, A. (1982). Circadian rhythm mechanism in affective illness and in antidepressant drug action. *Pharmacopsychiatry* **15**, 31–9.

Willner, P. (1983). Dopamine and depression: A review of recent evidence. *Brain Research Reviews* **6**, 211–46.

Willner, P. (1984*a*). The validity of animal models of depression. *Psychopharmacology* **83**, 1–16.

Willner, P. (1984*b*). The ability of antidepressant drugs to desensitize β-receptors is inversely correlated with their clinical potency. *Journal of Affective Disorders* **7**, 53–8.

Willner, P. (1985). Antidepressants and serotonergic neurotransmission. An integrative review. *Psychopharmacology* **85**, 387–404.

Willner, P. (1987). The anatomy of melancholy: The catecholamine hypothesis of depression revisited. *Reviews in Neuroscience* **1**, 77–99.

Willner, P., Theororou, A. & Montgomery, A. (1981). Subchronic treatment with the tricyclic antidepressant DMI increases isolation-induced fighting in rats. *Pharmacology, Biochemistry and Behavior* **14**, 475–9.

Willner, P., Towell, A., Sampson, D., Sophokleous, S. & Muscat, R. (1987). Reduction of sucrose preference by chronic unpredictable mild stress, and its restoration by a tricyclic antidepressant. *Psychopharmacology* **93**, 358–64.

Willner, P., Sampson, D., Phillips, G., Fichera, R., Foxlow, P. & Muscat, R. (1989). Effects of isolated housing and chronic antidepressant treatment on cooperative social behaviour in cats. *Behavioural Pharmacology* **1**, 85–90.

Youdin, M.B.H. & Finberg, J.P.M. (1983). Implications of MAO-A and MAO-B inhibitors for antidepressant therapy. *Modern Problems in Pharmacopsychiatry* **19**, 63–74.

Zacharko, R.M. & Anisman, H. (1984). Motor, motivational and antinociceptive influences of stress: contribution of neurochemical changes. In: M.D. Tricklebank & G. Curzon (eds.), *Stress-Induced Analgesia*, pp. 33–65. New York: John Wiley.

Zacharko, R., Bowers, Kelley, S. & Anisman, H. (1984). Prevention of stressor-induced disturbances of self-stimulation by desmethylimipramine. *Brain Research* **321**, 1785–9.

Zebrowska-Lupina, I. (1980). Presynaptic alpha-adrenoceptors and the action of tricyclic antidepressant drugs in behavioral despair in rats. *Psychopharmacology* **71**, 169–72.

Zebrowska-Lupina, I. & Kozyrska, C. (1980). The

studies on the role of brain dopamine in the action of antidepressant drugs. *Naunyn-Schmiedeberg's Archives of Pharmacology* **313** (Suppl. 80): R20.

Zeller, E., Stief, H.J., Pflug, B. & Sastre-y-Hernandez, M. (1984). Results of a phase II study of the antidepressant effect of rolipram.

Pharmacopsychiatry **17**, 188–90.

Zis, A.P. & Goodwin, F.K. (1979). Novel antidepressants and the biogenic amine hypothesis of depression. The case for iprindole and mianserin. *Archives of General Psychiatry* **36**, 1097–107.

7

The clinical relevance of animal models of depression
J.F.W. DEAKIN

7.1 INTRODUCTION

Current psychobiological concepts of depression are the result of many years of interplay between discoveries in the basic neurosciences, insights into the neurochemical actions of antidepressant drugs, animal models, and biological and pharmacological studies in patients. Discoveries in the basic neurosciences delimit the areas of interest. In rough chronological order, these have included presynaptic monoamine function, postsynaptic monoamine receptor function, presynaptic receptors, peptides and second messengers. Interest in these areas has been greatly potentiated by discoveries of clinically effective antidepressant drugs and understanding of their pharmacological properties, for example, inhibition of noradrenaline (NA) reuptake. Such findings have led to neurochemical theories of depression (e.g., NA deficiency) and attempts to test them with biological studies in patients (e.g., urinary excretion of the noradrenaline metabolite MHPG). These interacting strands of research encouraged the development of new drugs with more selective actions on systems of interest (e.g., maprotiline, a selective NA uptake inhibitor). Further clinical trials and neurochemical, pharmacological, behavioural and other studies followed. There is no rigid progression in the evolution of psychobiological concepts of depression. Important elements would appear to be: research activity, chance, and a free economy of ideas.

The difficulty in isolating the influence of animal models of depression from the flux of research activity will be apparent. Nevertheless, some real and potential influences of animal models can be discerned.

7.1.1 Detecting new antidepressants

There is no doubt that new antidepressants are needed. Existing drugs have undesirable side-effects which reduce compliance with therapy and significant numbers of depressives (perhaps a third) do not make a satisfactory response. The extent to which animal models can detect drugs with entirely novel mechanisms of action is an important issue addressed by the previous chapters. It seems likely that models which are developed from knowledge of the aetiology (genetics, biochemistry, pathogenesis), symptomatology and treatment of depressive illnesses are the models most likely to detect entirely new ways of treating depression. This chapter begins with an analysis of the fit between animal models and depressive illness in terms of symptoms, aetiology and pathogenesis.

7.1.2 The psychobiology of depression

Animal models are highly relevant to understanding the psychology of depression. Depression could in theory be understood as a neurochemical disorder but an explanation of how this produces the symptoms of depression would be lacking. Such an explanation requires knowledge of the normal behavioural functions of the neurotransmitters concerned so that the effects of disturbances can be understood. Much interest has focussed on the role of monoamines in reinforcement, incentive and aversive mechanisms and the relevance of these ideas to the pathogenesis of depression is discussed.

7.1.3 The mechanism of action of antidepressants

Animal models may become increasingly influential in a second area – in elaborating possible neurochemical mechanisms of action of antidepressants. For example, dopamine (DA) receptor antagonists block the effects of antidepressants in two of the models discussed by Willner (this vol., Ch. 5). Neurochemical studies had given few grounds for suspecting a role for DA in the mechanism of action of antidepressants. Some clincal findings pose difficulties for this view (see Section 7.4.3.1) but they do not weaken the general point: the pharmacology of agents which antagonise antidepressant effects in animal models offers opportunities to falsify theories of how antidepressants work. However, decisive tests of theories of antidepressant action will require similar experiments to be carried out in depressed patients. The animal studies indicate which clinical tests are critical. For example, the ability of 5-hydroxytryptamine (5-HT) antagonists to block antidepressant effects in the bulbectomy model (Broekkamp *et al.*, 1980) greatly encourages the need for similar experiments in patients. Ultimately, it is the clinical tests which are crucial to theories of antidepressant action and the third section of this chapter reviews the clinical efficacy of antidepressants.

7.2 DEPRESSIVE SYMPTOMS AND SYNDROMES

Depression is a syndrome – a collection of symptoms which occur together with a sufficient frequency to constitute a recognisable clinical condition. The symptoms which may be part of the syndrome are listed in Table 7.1. Depressed mood and depressive thought content lie at the heart of the syndrome and it is only in this setting that anorexia, sleep disturbance, apathy, diminished concentration, etc. have significance in indicating that depressed mood is pathological. Such symptoms in themselves are fairly non-specific indicators of illness and occur in many other psychiatric conditions.

The difficulty in assessing the face validity of

Table 7.1. *Symptoms of depression*

Depressed mood

Loss of interest and pleasure

Fatigue

Apathy

Aches and pains

Psychomotor retardation/agitation

Loss of libido

Sleep disturbance

Appetite disturbance and weight loss

Depressive thought content (overvalued ideas – delusions)
    Self is worthless, self deprecation, guilt, deserving punishment
    Future hopeless, disease conviction and phobia, ideas of poverty
    Life pointless, suicidal ideas

Anxiety
    Subjective apprehension and bodily symptoms

animal models is that the core symptoms of depression are subjective and only the illness indicators are behavioural. It is the non-specific illness indicators which are mimicked in the animal models; but what is the animal equivalent of depressed mood, to show that the animals' 'illness' is a 'depressive illness'? One possibility is to regard reversibility by antidepressant treatment in the animal model as a symptom equivalent to depressed mood and this certainly increases the significance of coexisting changes in sleep, appetite and activity. Another solution may lie in formulation of depressed mood as a state of reduced sensitivity to reinforcement (Akiskal & McKinney, 1975) and this can be modelled in animals. In some animal models, reduced sensitivity to reinforcers has been inferred from reductions in intracranial self-stimulation and reduced preference for sucrose solutions. Thus, if an animal model involves antidepressant-reversible anhedonia associated with changes in appetite, sleep, activity, and other functions, it is arguable that this is equivalent to meeting diagnostic criteria for depression in humans. The models that come closest to meeting these criteria are learned helplessness, chronic stress and amphetamine withdrawal (see Willner, this vol., Ch. 5).

Diagnoses in humans are really hypotheses which are falsifiable by clinical course, examination and investigations. It is therefore reasonable to ask what would falsify the hypothesis that, for example, learned helplessness is an animal equivalent of depression? Would the hypothesis be falsified if helpless animals showed reduced activity (reversible by antidepressants) but no change in appetite, sleep or reward processes? Probably not; reversibility by antidepressants remains the major testable prediction. The behavioural analogies are post-hoc, interesting and may suggest further hypotheses but they are not decisive tests of models.

7.2.1 Unipolar vs bipolar affective disorder

It is clear that some depressive illnesses occur as part of a bipolar manic depressive disorder whereas others are part of a recurring pattern of depressive illnesses. The bipolar–unipolar distinction is validated by the genetic distinction; major gene effects operate in the transmission of bipolar illness. In some cases an X-linked gene is implicated (Baron *et al.*, 1987) and amongst the Amish community, the illness segregates with a DNA marker on chromosome 11 (Egeland *et al.*, 1987). In Icelandic families bipolar illness does not segregate with the chromosome 11 marker (Hodgkinson *et al.*, 1987). Therefore there must be at least three distinct genetic types of bipolar affective disorder. Depressive episodes in bipolars are not clinically distinct from other depressive illnesses (Abrams & Taylor, 1980). This is disturbing for attempts to subdivide unipolar depressive illnesses; depressive illnesses which are known to be genetically distinct produce similar clinical pictures.

There are neither behavioural nor genetic animal models of bipolar affective disorder. Nor is the recurrent nature of unipolar depression modelled.

7.2.2 Endogenous vs reactive depression

Numerous classifications of unipolar depressive illnesses have been proposed and some are listed in Table 7.2. Perhaps the greatest area of controversy has been the validity of the endogenous-reactive distinction. The original basis of the distinction

Table 7.2. *Some overlapping concepts of depression*

Endogenous Autonomous Non-situational	Non-precipitated
Endogenomorphic Vital depression Major depression with melancholia (DSMIII)	Biological symptoms
Psychotic Delusional Major depression with psychosis (DSMIII)	Psychotic symptoms
Characterological depression Dysthymic psychopath Depressive personality Melancholic personality Hypomelancholia Dysthymia (DSMIII)	Chronic milder depression
Anxious Atypical Reactive Neurotic Depression spectrum disease	Prominent anxiety and personality difficulties

Table 7.3. *Endogenous vs non-endogenous*

Endogenous	Non-endogenous (reactive, neurotic)
No precipitant	Precipitant
Good premorbid personality	Personality difficulties
Unvarying depression	Fluctuating depression
Mood unresponsive to events	Mood responds to events
Worse a.m.	Worse p.m.
Early morning wakening	Initial insomnia
Retardation	
Delusions	

was that some depressive illnesses are triggered by events in the environment (reactive) whilst others lack precipitants (endogenous). Few psychiatrists now believe that this distinction is valid since the majority of depressive illnesses are preceded by an increased rate of adverse life events. However, the terms endogenous and reactive have come to imply distinct clinical syndromes (Carney *et al.*, 1965) and the two are contrasted in Table 7.3.

Multivariate statistical techniques (factor analysis, principal component analysis, discriminant function analysis, cluster analysis) have been used to investigate whether the symptoms in Table 7.3 tend to associate with one another. This area of research is bedevilled by methodological problems and the reader is referred elsewhere for more lengthy reviews (Kendell, 1976; Nelson & Charney, 1981). Taking an overview of these studies, there is evidence that the symptoms of endogenous depression associate with one another but the remaining depressive illnesses do not form an homogenous group that may be termed reactive depression. The symptoms which are most frequently and strongly associated with the endogenous syndrome are severe depressed mood, loss of reactivity, agitation or retardation, guilt and depressive delusions (Nelson & Charney, 1981). Only the changes in activity and loss of reactivity are potentially observable in animals.

The learned helplessness and bulbectomy models have been advanced as models of endogenous depression (see Willner, this vol., Ch. 5). However, the fact that helpless animals have reduced food intake, sleep and aggression is not strongly suggestive of homology with endogenous depression as these symptoms in humans are non-specific indicators of depressive illness of all forms. If the hypoactivity of helpless animals is a true homologue of psychomotor retardation then this would be more suggestive of a model of endogenous depression. Psychomotor retardation involves subjective slowing of thinking and of the experience of the passage of time – this is not accessible in animals. The behavioural components are reduced verbal fluency, and poverty and slowing of movements. Reduction and slowing of activity in animals might or might not be true homologues of motor retardation in depression – no amount of argument by analogy will settle the issue. The hypoactivity of helpless animals suggests only weakly that retarded and endogenous depression is being modelled. However, hypoactivity in the context of decreased responsiveness to rewards (loss of reactivity) is perhaps more compelling.

The case for homology between hyperactivity in bulbectomised rats with agitation in endogenous depression is particularly weak. The only reason for thinking this model has anything to do with depression is that some of the behaviours are reversible by antidepressants. Antidepressant-reversible anhedonia, possibly a homologue of

depressed mood, has yet to be demonstrated with this model. Thus, arguments from behavioural analogy suggest hypotheses but are weak validators of models.

7.2.2.1 *External validators of the endogenous syndrome*

Two neuroendocrine markers of depression are significantly associated with the endogenous syndrome. The dexamethasone suppression test is more frequently abnormal in endogenous depressives (see Abou-Saleh, 1988) and the growth hormone response to clonidine administration is strikingly attenuated in endogenous depressives compared to non-endogenous ones (Glass *et al.*, 1984; Checkley *et al.*, 1986). The latter has not been studied in animal models of depression. The learned helplessness and bulbectomy models are associated with increased cortisol secretion but this is true of many acute stressors in animals. Dexamethasone resistance and circadian phase shift in cortisol secretion have not been demonstrated in animal models. Reduced latency to REM sleep is a probable marker for endogenous depression and thus a potential validator of animal models of the syndrome.

Patients with the endogenous syndrome are improved by ECT and tricyclic antidepressants. In most animal models both ECT and tricyclics are effective and so survive this test of endogeneity.

7.2.2.2 *Psychotic depression*

Depressive thoughts are sometimes so intense and false as to be delusional; the patient's grasp of reality is impaired and insight is lost. This is psychotic or delusional depression. These terms are sometimes used interchangeably with endogenous depression but many patients with endogenous features do not have delusions. However, the boundary is certainly blurred. It is sometimes difficult to know when a pessimistic view of the future, for example, becomes a delusion of hopelessness. Delusional depression is encompassed within the concept of endogenous depression. However, there is some evidence that delusional epression may be distinct from endogenous and from other forms of depression (Glassman & Roose, 1981). The presence of delusions appears to be a strong predictor of responsiveness to ECT (Clinical Research Centre, 1984). In contrast, there is evidence that delusional depressives are less responsive to tricyclic antidepressants than non-deluded depressives; however, this issue has not been the subject of major clinical trials (Quitkin, 1978; Moradi *et al.*, 1979; Charney & Nelson, 1981; Howarth & Grace, 1985). The differential responsiveness of psychotic depression to ECT rather than tricyclics is not mirrored in any animal model.

7.2.3 The influence of personality

In milder episodes of depressive illness, the clinical picture is often modified by personality and by symptoms which long antedate the onset of depressed mood.

7.2.3.1 Anxiety

Anxiety has a close but perplexing relationship with depression. The difficulty is that anxiety may only occur in some patients during depressive illnesses whereas in others anxiety symptoms antedate and outlive depressive illnesses. Independent cluster analytic studies have tended to produce groupings of depressives with prominent symptoms of anxiety (Overall *et al.*, 1966; Paykel, 1978). Depression with anxiety is part of the concept of neurotic depression and of atypical depression. These concepts arise from cross-sectional studies. Where such patients have been followed up, the group fragments into those who develop more serious depressive illnesses and others who have persisting neurotic syndromes such as obsessional-compulsive disorder, phobic states and generalised anxiety (Akiskal *et al.*, 1978, 1979). Whether depression in the setting of other neurotic diagnoses should be regarded as secondary depression (Akiskal *et al.*, 1978, 1979) or simply as a progression to a more severe level of affective disturbance is difficult to resolve (Foulds & Bedford, 1975). The close relationship between anxiety and depression is emphasised by the recent studies of Goldberg and colleagues (Goldberg *et al.*, 1987). In a large statistical study of symptoms in general practise, separate dimensions of anxiety and depression could be resolved but they were strongly correlated with one other.

The role of anxiety in animal models of depression has been studied little. Animal models of anxiety generally rely on the acute presentation of aversive stimuli (footshock, exposed environments), whereas depression models use chronic or sustained aversive stimuli (repeated footshock, prolonged immersion, etc.). No other general property distinguishes the stimuli used in the models except perhaps that presentation is generally non-contingent in depression models. They are all aversive in the sense that animals will (or would probably) work to reduce them. It seems plausible to regard anxiety as an immediate, normal and protective response to aversive stimuli whereas 'depressive' behaviours result when anxiety-induced behaviours fail to terminate the aversive stimuli.

There is evidence that anxiety symptoms antedate and outlast depressive episodes and depression is regarded by many clinicians as a higher order of affective disturbance than anxiety. Prolonged anxiety in response to stress or by constitution may be involved in the genesis of depression. Such a process may be underway but still reversible in mild depressive states, hence the possible efficacy of benzodiazepines (Rickels *et al.*, 1985). This is also suggested by evidence that anxiolytics prevent the acquisition of learned helplessness in animals but not its expression (Drugan *et al.*, 1985).

Neurotic patients with anxiety or depression or both do not have distinct patterns of responsiveness to drug treatment. There is evidence that tricyclic antidepressants produce a more sustained and complete remission of symptoms than benzodiazepines, irrespective or whether anxiety or depressive symptoms predominate (Johnstone *et al.*, 1980; Ancill *et al.*, 1984; Kahn *et al.*, 1986; Johnston *et al.*, 1988). ECT is probably ineffective in depressive illnesses associated with marked symptoms of anxiety. There are no animal models of depression which are sensitive to antidepressant drugs but not to ECT. If this reasoning is valid then there may be no animal models of common minor depressive illnesses associated with anxiety.

7.2.3.2 Antisocial personality

In an early factor-analytic study of symptoms of depression, Hamilton identified a syndrome he termed 'psychopathic depression' – depressive illnesses in younger people with impulsive and antisocial personality traits (Hamilton & White, 1959). Cluster analytic studies offer some support for such a grouping (Overall *et al.*, 1966; Kendell, 1975; Paykel, 1978). Winokur's classification includes the concept of depression spectrum disease (Winokur, 1979, 1985). These are patients who tend to be younger females with irritable and hostile personalities who lead disorganised lives with frequent problems with relationships. Their male relatives have in increased incidence of antisocial personality disorder and alcoholism and female relatives have higher rates of depression (Merikangas *et al.*, 1985). Winokur suggests there are familial links between alcoholism in males and depression in females.

Some studies suggest that low cerebrospinal fluid (CSF) concentrations of the 5-HT metabolite 5-HIAA are associated with antisocial aggressive and impulsive personality traits (see Traksman-Bendz *et al.*, 1986; Roy *et al.*, 1986). Irritable and impulsive behaviour occur in the olfactory bulbectomy model of depression and Willner (1989*b*; this vol., Ch. 5) points out the possible involvement of reduced 5-HT function in this model. It is therefore plausible to argue that bulbectomised animals model 'psychopathic' depression and this argument seems as strong as the suggestion that they model endogenous depression. This is perhaps an illustration of the limitations of the inferences that can be drawn from analogies between models and depressive illnesses.

7.2.4 Classification of depression and animal models

Symptom patterns, sampled at a single interview, have been used to attempt to subdivide depressive illnesses with the aim of discerning different aetiological mechanisms and distinct prognoses and responsiveness to treatment. The yield has been fairly meagre. The syndromal approach is not encouraged by the fact that similar syndromes of depression can result from at least three different genetic mechanisms in bipolar illnesses and from genetically distinct unipolar processes. Furthermore, longitudinal studies indicate that the different syndromes evolve into one another with the passage of time, for example 'neurotic' illnesses may become 'endogenous' or bipolar. It therefore seems unlikely that seeking behavioural homologues in animal models for depressive syndromes is a promising way of identifying underlying aetiological mechanisms in depression. Furthermore, many of the behavioural abnormalities in depression (loss of appetite, sleep disturbance, apathy) are non-specific indicators of illness rather than of distinct forms of depression. There is evidence that marked agitation or retardation may identify a group of depressives (endogenous) with distinct biological abnormalities. However, the weakness of using changes in activity in animals as a way of identifying animal models of the endogenous process has been pointed out.

Symptoms of anxiety commonly precede and coexist with the syndrome of depression. There has been little evaluation of the relationship between anxiety and depression in the animal literature; indeed, strenuous attempts are made to separate them. This seems misguided in view of clinical association and their similar responsiveness to treatment (see also Lader, Chapter 4 of this volume).

7.3 AETIOLOGY OF DEPRESSION

7.3.1 Genetics

Because no animal model of genetic vulnerability to depression exists, a major contribution to the genesis of depression is absent from existing models. It would be possible to breed strains of laboratory animals selected for high and low resilience or persistence in, for example, a learned helplessness paradigm. This would be analogous to the Maudsley reactive and non-reactive strains of rats which were bred for reactivity in an open-field test and which show genetically determined differences in behavioural measures of anxiety (Broadhurst, 1977). In view of the intimate connections between anxiety and depression, it would be of considerable interest to know how these strains perform in animal models of depression. It is unlikely that animal genetic models will shed light on the genetics of anxiety and depression in humans but they could illuminate the nature of the interaction between genetic, biochemical and environmental factors in the pathogenesis of depression.

7.3.2 Psychosocial factors and the aetiology of depression

7.3.2.1 Life events

It is a commonplace clinical observation that depressive illnesses follow adversity, for example, bereavement. Paykel (1979) found that the number of adverse life events in the six months preceding a depressive illness was increased almost three-fold over the rate of life events in the general population. While increases also occur prior to most other psychiatric illnesses, loss-events are associated with depression whereas no such specificity occurs in the quality of events which precede schizophrenic relapse (Brown & Harris, 1978, 1986). All forms of depressive illnesses are preceded by an increased rate of life events including psychotic or endogenous depressive illnesses (see, e.g., Bebbington *et al.*, 1988).

Loss events include not only interpersonal and financial loss (bereavement, friend moving away, redundancy) but also events which reduce control over circumstances and social status and role. Environmental manipulations in animal models (inescapable shock, immersion) have low face validity as life events and do not capture the element of loss in life events associated with depression. The separation models are an obvious exception.

According to a recent study, the very notion that life events precipitate depression may be an over-simplification. In the first-degree relatives of depressed probands not only was depression more common but they were also more likely to experience life events than controls whether or not they were depressed (McGuffin *et al.*, 1988). Thus the correlation between life events and depression, rather than implying a causal relationship, may in fact be a consequence of an underlying familial cause of both life events and depression. The familial factor could be mediated by genetic or psychosocial influences on personality and lifestyle.

7.3.2.2 Vulnerability

Since life events are common, it seems likely that they are concerned with the timing of depressive episodes and that other factors determine which individuals will develop depressive illnesses in response to life events. Consideration of such vulnerability factors is almost totally lacking in animal models of depression. What is needed is some prior manipulation (e.g., destruction of a monoamine pathway, early separation from mother) which sensitises animals to an acute aversive event which is insufficient by itself to cause depression-like behaviour.

A psychosocial model of vulnerability has been proposed by Brown and colleague on the basis of an extensive series of investigations of large samples of women in the community (Brown & Harris, 1978, 1986). These studies suggest that life events are much more likely to cause depression where there has been the loss of a mother before the age of thirteen, the presence in the household of three or more children under the age of eleven, and the lack of a confiding relationship with a partner. It was proposed that these vulnerability factors give rise to a congitive predisposition of low self esteem or lack of mastery, and that feelings of hopelessness that follow a loss event become generalised in individuals with low self esteem. This generalised feeling of hopelessness may lie at the heart of depressive illness (Brown & Harris, 1978).

The idea that low self esteem mediates vulnerability to depression is difficult to formulate in behavioural terms which might be modelled in animal studies. Low self esteem may be related to Seligman's concept of learned helplessness. A major difference is that learned helplessness is seen as equivalent to depression whereas, in the social models, low self esteem is seen as a predisposition.

The chronic stressor models are perhaps analogous to 'chronic difficulties' (e.g., chronic housing and financial difficulties) which Brown and colleagues have identified as predisposing to depression. The question is whether rats exposed to chronic difficulties are more sensitive (vulnerable) to a new stressor than controls.

7.3.3 Biological factors in the aetiology of depression

Hypersecretion of cortisol and its resistance to dexamethasone suppression are biological markers of the depressed state. The reduced ability of clonidine to increase growth hormone secretion is probably a trait or vulnerability marker (Glass *et al.*, 1984; Checkley *et al.*, 1986), which suggests that reduced function at or beyond alpha-2 adrenergic receptors may predispose individuals to developing depressive illnesses. Low CSF 5-HIAA concentrations may also be a trait marker for depression (Goodwin *et al.*, 1973; Asberg *et al.*, 1976), but other lines of evidence suggest the association is non-specific. Low CSF 5-HIAA has been linked to suicidal tendencies, aggression and impulsiveness (see Roy *et al.*, 1986; Traskman-Bendz *et al.*, 1986; Willner, 1989*b*). Reduced growth hormone and prolactin responses to the 5-HT precursor, tryptophan, occur in depression and may indicate reduced function in some 5-HT systems (Henninger *et al.*, 1984; Deakin & Pennell, 1986; Cowen & Charig, 1987; Deakin, 1989).

Manipulating NA and 5-HT function in experimental animals may be one way of exploring biological vulnerability to depression. Animals with experimentally-induced vulnerability would develop depressive behaviours in response to losses or stressors which are ineffective in controls. This approach could be used to test theories of vulnerability generated from human investigations and to generate hypotheses which could be investigated in human studies.

7.3.4 Animal models and the pathogenesis of depression

Anhedonia and loss of motivation are well recognised but variable features of depression (Fawcett *et al.*, 1983). The idea that these and other symptoms of depression might reflect impaired functioning in brain reward pathways has a long history (Stein, 1962; Crow, 1973; Akiskal & McKinney, 1975; see Deakin & Crow, 1986). It arose from the recognition that reserpine could precipitate depression and the discovery that reserpine powerfully attenuated intracranial self-stimulation, (ICSS) (Olds *et al.*, 1956). These ideas have probably influenced the DSM III requirement that there should be a 'pervasive loss of interests and pleasure' for the diagnosis of major depressive disorder.

7.3.4.1 *Depression, reinforcement and incentive*

Despite the undoubted and considerable appeal of the idea that depressives are insensitive to

reinforcers and incentives, there have been few tests of this hypothesis in patients (Bradshaw & Szabadi, 1978; Szabadi *et al.*, 1981). However, the idea has been tested in animal models of depression using ICSS (decreased in the learned helplessness and amphetamine withdrawal models) and sucrose preference (abolished in the chronic stressor model) (see Willner, this vol., Ch. 5). These findings corroborate earlier formulations of depression which had suggested that aversive stimulation reduces the behavioural impact of reinforcers and incentives (Crow & Deakin, 1979; Deakin, 1983).

The discovery of ICSS suggested the existence of specific neuronal pathways which normally mediate the effects of reinforcers and incentives. DA systems have been very strongly implicated in approach and incentive behaviour (Crow, 1973; Wise *et al.*, 1978; Crow & Deakin, 1979; Beninger, 1983; Everitt, 1982). Suggestions that NA neurones are concerned with the detection of reinforcement (Crow, 1973) are generally discounted but some recent evidence is compatible with this view (Morley *et al.*, 1987*a*, *b*, 1988).

The ability of antidepressants to reverse anhedonia (reduced ICSS or sucrose preference) in some animal models suggests actions on brain reinforcement/incentive systems which may involve catecholamines. Indeed, the ability of antidepressants to restore sucrose preference in the chronic mild stress model is prevented by DA-receptor antagonists (Willner *et al.*, 1988; Willner, this vol., Ch. 5). Furthermore, interference with NA or DA neurotransmission blocks the effects of antidepressants on learned helplessness and behavioural despair (Willner, this vol., Ch. 5). Willner (1989*a*) has summarised the behavioural and biochemical evidence that antidepressants sensitise alpha-adrenergic and mesolimbic dopamine receptor function, both of which have been implicated in reinforcement/incentive mechanisms.

7.3.4.2 *The psychobiology of aversion*

It was pointed out above (see Section 7.2.3.1) that animal models of depression and anxiety both involve the presentation of aversive stimuli. The neurobiology of aversion is therefore of central importance to both sets of models. Some of the acute behavioural effects of aversive stimuli are reduced by 5-HT-receptor antagonists (Geller & Blum, 1970; Wise *et al.*, 1970; Graeff, 1974; Brittain *et al.*, 1987) and by neurotoxin lesions of 5-HT pathways (Tye *et al.*, 1977). These findings gave rise to suggestions that 5-HT neurones mediate the behavioural effects of aversive stimuli (Wise *et al.*, 1970; Deakin, 1983). The clinical prediction is that excessive 5-HT neurotransmission is involved in the aetiology of both anxiety and depression.

5-HT-excess theories of depression stand in opposition to the more prevalent view that depression involves deficient 5-HT neurotransmission. This influential theory is based largely on reports of low CSF 5-HIAA concentrations in depressives and in their post-mortem brains. It led to the development of selective 5-HT reuptake inhibitors as potential antidepressants, which would act to reverse the putative 5-HT deficiency. These drugs are not detected by several animal models of depression but few would take this as strong evidence against 5-HT deficiency theories. It is the clinical trial evidence which is crucial and this is described below (see Section 7.4.3.1).

The olfactory bulbectomy model is sensitive to 5-HT reuptake blockers and Willner (this vol., Ch. 5) argues that the model supports theories of 5-HT deficiency. There is certainly a case that some of the behavioural effects of bulbectomy involve reduced 5-HT function and that the model is sensitive to antidepressants. Furthermore, the action of antidepressants in this model may be blocked by a 5-HT antagonist (Broekkamp *et al.*, 1980). However, aggression, impulsiveness and impaired passive avoidance performance could be homologous to the same difficulties in patients with psychopathic personality disorder (see Section 7.2.3.2); such behaviours are not common in depression. Furthermore, 5-HT excess theories predict that impaired 5-HT function would result in a release of behaviours normally restrained by anxiety (e.g., passive avoidance, aggression). Thus, the bulbectomy findings do not rule decisively in favour of 5-HT deficiency theories of depression and they may have little to do with depression. Despite these uncertainties, the hypothesis that bulbectomy-induced behaviours model human depression is testable; it predicts that bulbectomised rats will be insensitive to reinforcers and incentives and will readily acquire 'helpless' behaviours.

7.3.4.3 *Tolerance to aversion*

The tension between 5-HT excess and deficiency theories of depression has led to various attempts to bring them together, none of which is entirely satisfactory. One suggestion is that 5-HT is involved in the mediation of both acute and chronic responses to aversive stimuli (Deakin & Pennell, 1986; Deakin, 1989). The acute response, which is anxiety, has the protective function of reducing exposure to dangerous environments and there is evidence consistent with the involvement of 5-HT$_2$ receptors (Ceulemans *et al.*, 1985; Deakin, 1988). However, adaptive-protective responses to chronic stress appear to involve 5-HT$_1$ receptors: in a new animal model of depression resilience to repeated immobilisation stress is promoted by antidepressants and associated with increased 5-HT$_1$ receptor function (Kennett *et al.*, 1985, 1987). When 5-HT$_1$ protective mechanisms fail, learned helplessness, or depression in humans, may be the result. This is an example of the heuristic utility of an animal model of depression. It also reinforces the concept of depression as a failure of a protective physiological mechanism rather than as a normal response to adversity.

A proposed resolution of 5-HT excess and deficiency theories suggests that depression is associated with excessive 5-HT$_2$ and deficient 5-HT$_1$ neurotransmission (Deakin & Pennell, 1986; Deakin, 1988). Drugs which enhance 5-HT$_1$ and/or reduce 5-HT$_2$ neurotransmission should be effective antidepressants and detected in animal models. Some of these predictions have been corroborated (see Section 7.4.2.2).

7.4 THE EFFECTIVENESS OF TREATMENT

Monoamine deficiency theories of depression arose from the discovery of the pharmacological properties of early antidepressants – reuptake blockade and monoamine oxidase inhibition. The neuropharmacology of antidepressant drugs will continue to have a pivotal role in pointing to possible neurotransmitter abnormalities in depression. Furthermore, as has been noted, animal models are validated by their sensitivity to antidepressant agents. It is therefore of critical importance that we know which agents are effective antidepressants and which are not. However, establishing the efficacy of a new treatment is no easy task.

At the heart of the difficulty is the high spontaneous remission rate; new drugs have to cause significantly higher remission rates to demonstrate antidepressant efficacy. This necessitates the use of placebos, a sufficiently large sample size to detect statistical significance, double-blind ratings, and randomised entry into treatment groups. Many published trials fall short of these requirements and the weight which becomes attached to the findings is essentially arbitrary. Even where the basic design requirements are fulfilled, there is a lack of certainty. Morris & Beck (1974) reviewed trials of 'established' antidepressants; in one third of them the tricyclic did not cause significantly greater improvement than placebo. A single trial is insufficient to demonstrate or exclude antidepressant efficacy.

Doubts have been raised about the ethics of administering placebo treatment to depressives and this leads some investigators to compare the new agent with an established antidepressant rather than placebo. Such trials are of little scientific value. A common conclusion is that the new drug is as effective as an established antidepressant. The error of this apparently commonsense conclusion can be demonstrated with reference to a trial which compared fluvoxamine with the established antidepressant imipramine in the treatment of a large number of out-patient depressives (Norton *et al.*, 1984). Both treatment groups improved to an equal extent. However, a placebo-treated group was included in the design and these patients improved to the same extent as those treated with active compounds. Results such as these also dispel some of the doubts about the ethics of placebo treatment.

7.4.1 Treatments with non-specific pharmacology

7.4.1.1 Tricyclics

There is no doubt that tricyclic antidepressants alleviate depressive illnesses, but surprisingly little is known about the characteristics of patients who respond. Bielski & Friedel (1976) concluded that certain features of endogenous depression predicted a favourable response to tricyclics, whereas neurotic, hypochondriacal and hysterical traits,

and multiple previous episodes, predicted poorer outcome. However, the same traits, especially the negative predictors, seem likely predictors of response to placebo and non-specific aspects of therapy. Indeed two trials found that neurotic depressive in-patients responded as well as endogenous to tricyclic antidepressants (Wittenborn & Kiremitci, 1975; Simpson *et al.*, 1976). However, non-hospitalised depressives may be less responsive to tricyclics. Schatzberg & Cole (1978) reviewed 20 studies comparing a tricyclic with a benzodiazepine, mainly in neurotic depressives. Three of four trials with in-patients favoured the tricyclic whereas this was observed in only seven of the 16 studies in general practice or out-patients.

In fact most trials find that initial severity is the most powerful predictor of outcome – irrespective of treatment. The exception may be delusional depression, which is a severe illness with little tendency to remission (Glassman & Roose, 1981; Clinical Research Centre, 1984). The evidence that delusional depressives may be resistant to tricyclic antidepressants has been noted above.

Since there is little evidence that particular symptoms differentially predict responsiveness to a tricyclic and not to placebo, it seems likely that tricyclics accelerate a natural tendency to remission. This tendency is least in milder illnesses with disturbances in personality, and perhaps in delusional depression. To the extent that animal models of depression are validated by their responsiveness to tricyclics, they model moderate to severe depressive illnesses but not delusional depression or chronic milder depressive illnesses. Responsiveness to tricyclics is not informative of whether endogenous or neurotic illnesses are being modelled.

7.4.1.2 Monoamine oxidase inhibitors (MAOIs)

MAOIs are less consistently effective antidepressants in placebo-controlled trials than are tricyclics. Nevertheless, there are a convincing number of positive studies. MAOIs have the reputation of being less effective than tricyclics in severer depressive illnesses and more effective in 'atypical' depressive illnesses with marked anxiety, in out-patients (Robinson *et al.*, 1973; Paykel *et al.*, 1983). It is true that a greater proportion of in-patient comparisons with placebo fail to find antidepressant efficacy whereas more studies with out-patients reveal efficacy. Furthermore, in four out of eight in-patient studies, phenelzine was less effective than imipramine (Paykel *et al.*, 1982). However, there is no good evidence that any group of out-patients improve more on an MAOI than a tricyclic. In a definitive study, Paykel *et al.* (1982) showed that amitriptyline and phenelzine were both significantly more effective than placebo in out-patient depressives. There was a striking absence of differential predictors to the drugs. Earlier suggestions that MAOIs possess unique anti-phobic properties are discredited by the trials demonstrating the efficacy of tricyclics in agoraphobia and other anxiety states (see Lader, Chapter 4 of this volume).

It seems probable that MAOIs have a spectrum of activity which is similar to but less extensive than the tricyclics.

7.4.1.3 ECT

Two large trials in the 1960s demonstrated the efficacy of ECT in the treatment of endogenous depression (Greenblatt *et al.*, 1964; Medical Research Council, 1965), and subsequent trials comparing real versus simulated ECT confirm that the electricity does contribute to the clinical effect. ECT is particularly effective in delusional depression (Clinical Research Centre, 1984), and there is evidence for antipsychotic activity in schizophrenia and mania (McCabe, 1976; Taylor & Fleminger, 1980). Efficacy of ECT in non-psychotic endogenous illnesses is suggested by one recent trial (Brandon *et al.*, 1984) but called into question by another (Clinical Research Centre, 1984). ECT is probably ineffective in neurotic depressive illnesses and in patients with personality difficulties (Carney *et al.*, 1965; Brandon *et al.*, 1984).

ECT thus appears to have a spectrum of therapeutic action which differs appreciably from tricyclics and MAOIs. It is somewhat disconcerting that many animal models of depression are sensitive to all these antidepressant treatments.

7.4.2 Treatments with selective actions on 5-HT systems

7.4.2.1 Selective 5-HT reuptake inhibitors

The efficacy of selective 5-HT uptake inhibitors is of critical importance to theories of the role of 5-HT in depression. Some of the trials are summarised in Table 7.4. Assessment of improvement by clinical global impression (CGI, Table 7.4) produces the most positive outcomes and Hamilton depression ratings the least. In some studies (e.g. Cohn & Wilcox, 1985) the outcome depends on whether patients who complete four weeks of therapy are compared (completers, Table 7.4) or whether the last ratings of premature terminators are included (end-point analysis). There is no satisfactory resolution to the problem of how to treat data from patients who do not complete a trial; including or excluding their data both introduce biases. Despite the air of objectivity lent by blind ratings and sophisticated statistics, it can be seen that the evaluation of antidepressant efficacy is in fact highly subjective. Is superiority over placebo in some trials with some measures a sufficient indication of efficacy? In view of these considerations and the bias against publishing negative studies, some caution is required before accepting highly selective 5-HT reuptake blockers as clearly effective.

The questions of whether the 5-HT reuptake blockers are as effective as tricyclics or have a different spectrum of action also do not have a definite answer. In a well designed study, patients treated with citalopram, a highly selective 5-HT reuptake blocker did significantly less well than with clomipramine, a 5-HT reuptake blocker with actions on NA systems (Danish Universities Antidepressant Group, 1986). The disadvantage to citalopram was more marked in endogenous than in neurotic patients. In two studies, clinical observer ratings show equal improvement on a 5-HT reuptake blocker and a tricyclic, but self-ratings show less improvement on the 5-HT reuptake blocker (Chouinard, 1985; Young *et al.*, 1987). One study suggested that a 5-HT reuptake blocker was more effective than imipramine in chronic milder depressions with atypical features (Reimherr *et al.*, 1984). 5-HT reuptake blockers may have a narrower or different spectrum of activity compared

Table 7.4. *Controlled trials of 5-HT uptake inhibitors in depression*

	Improvement ratings		
	HAMD	CGI	Self ratings
Fluvoxamine/placebo			
Itil *et al.*, 1983	NS	> P	NS (Beck)
			> P (Symptom Inventory)
Dominguez *et al.*, 1985	NS	> P	NS
Amin *et al.*, 1984	> P	> P	—
Norton *et al.*, 1984	NS	NS	NS
Fluoxetine/placebo			
Rickels *et al.*, 1986	NS	NS	NS
Fieve *et al.*, 1986	NS	NS	> P
Fabre & Crismon, 1985	> P	> P	NS
Cohn & Wilcox, 1985 end point	> P	> P	> P
completers	NS	NS	> P
Stark & Hardison, 1985	> P	> P	> P
Fluoxetine/tricyclic			
Young *et al.*, 1987	= AMI	—	< AMI
Chouinard, 1985	= AMI	= AMI	< AMI
Levine *et al.*, 1987	< IMI	—	—
Bremner, 1984	> IMI	> IMI	> IMI
Cohn & Wilcox, 1985 end point	> IMI	> IMI	> IMI
complementers	= IMI	= IMI	= IMI
Stark & Hardison, 1985	= IMI	= IMI	= IMI

NS, no significant differences between drug and placebo; > , drug superior to placebo or tricyclic; = , drug and tricyclic equally effective; < , drug less effective than tricyclic. HAMD, Hamilton depression ratings; CGI, clinical global impression.

with established antidepressants but further studies are needed.

All drugs with unequivocal antidepressant efficacy share the ability to down-regulate beta-adrenergic or 5-HT_2 receptor function (Peroutka & Snyder, 1980; Velutani *et al.*, 1981). These actions have been proposed as unified mechanisms of antidepressant action. Some 5-HT reuptake blockers (e.g. citalopram) do not affect beta- or 5-HT_2 receptor function. If they prove to be effective antidepressants, down-regulation theories will be falsified. Thus, whatever the results of studies in animal models, it is the clinical tests which are decisive. There is evidence that citalopram is effective in treatment of depression though less so than clomipramine (see above). If corroborated, this suggests that beta- or 5-HT_2 receptor down-regulation is not necessary for antidepressant efficacy.

It has been suggested that the possible partial efficacy of 5-HT reuptake blockers is due to their ability to potentiate 5-HT_1 receptor stimulation but that this is offset by the concurrent enhancement of 5-HT_2 neurotransmission (Deakin, 1989). Experiments in animal models carried out by Soubrie and colleagues are tests of this hypothesis. They showed that low doses of 5-HT reuptake blockers reversed learned helplessness while high doses were ineffective (Soubrie, 1989). However, high doses became effective when combined with a 5-HT_2 receptor antagonist. Again, results of experiments in an animal model suggest a decisive clinical experiment that is feasible; a comparison of the efficacy of a 5-HT reuptake blocker alone and in combination with a 5-HT_2 receptor antagonist in the treatment of depression.

7.4.2.2 Selective 5-HT receptor agonists and antagonists

Several drugs with selective actions on 5-HT receptor subtypes have been developed. 5-HT_1 agonists (e.g. buspirone, gepirone) and 5-HT_2 antagonists (e.g. ritanserin, ICI169,369) have an anxiolytic profile in some animal models although the effects are inconsistent. The few clincal studies that have been made revealed evidence of anxiolytic activity (Goldberg & Finnerty, 1979; Ceulemans *et al.*, 1985). Some concepts of the role of 5-HT in depression (Deakin, 1989) suggest that these agents should have antidepressant activity. 5-HT$_1$ agonists, for example, not only mimic the effects of 5-HT at 5-HT$_1$ receptors but also induce delayed down-regulation of 5-HT$_2$ receptors (DeSouza *et al.*, 1986; Blackshear *et al.*, 1986). In a well designed study, buspirone and diazepam reduced ratings of anxiety more than placebo in neurotic out-patients but buspirone was superior to diazepam and to placebo in reducing ratings of depression (Goldberg & Finnerty, 1979). There is also a report that the 5-HT$_1$ receptor agonist 8-OH-DPAT reversed learned helplessness (Giral *et al.*, 1988). Thus, theoretical formulations, early clinical results and findings with an animal model strongly suggest the need for further clinical evaluation of 5-HT$_1$ agonists as antidepressants.

7.4.3 Drugs with selective actions on catecholamine mechanisms

Willner (this vol., Ch. 5) points out that the symptoms of stress (learned helplessness, behavioural despair) may be mimicked by catecholamine depletion. In contrast, the symptoms of bulbectomy are mimicked by 5-HT depletion. It is suggested, therefore, that there are two 'relatively independent syndromes' of depression mediated by catecholamine and 5-HT deficiency and modelled respectively by the two paradigms.

There is good evidence that selective NA reuptake blockers (e.g., viloxazine, desmethylimipramine, nortriptyline, maprotiline) and the DA reuptake blocker nomifensine have antidepressant activity. However, these drugs share the ability of other antidepressants to decrease 5-HT$_2$ receptor binding. Hence, their antidepressant efficacy is not decisive evidence for the existence of a group of catecholamine deficiency depressions.

To investigate the possible independence of catecholamine and 5-HT deficiency depressions, clinical trials have been carried out in which a NA and 5-HT uptake inhibitor were compared in the first phase of the trial followed by a cross-over of the treatments (Aberg-Wistedt, 1982; Emrich *et al.*, 1986; Nystrom & Hallstrom, 1987). There is no evidence that patients who fail to respond to one type of drug are particularly responsive to the other. In fact the appropriate design would be very difficult to complete. Those who fail to respond to the first phase drug (about 30% of those entered) would need to be split into groups receiving either the alternative drug or continuing with the first. About 20 patients per group would have to complete the trial to detect statistical significance. This means that roughly 240 patients would have to complete the first phase. Furthermore, the patients would have to be maintained in the trial for almost three months. Such a trial is close to impractical.

In the animal models also, the bulbectomy syndrome does respond to NA uptake inhibitor antidepressants, while, as noted above, learned helplessness may be reversed by 5-HT uptake inhibitors or 5-HT$_1$ agonists. The issue of whether distinct catecholamine and 5-HT deficiency syndromes exist clinically will probably not be resolved by further drug studies.

7.4.3.1 Dopamine

In the stress models, reductions in behavioural output are reversed by catecholamine agonists including L-DOPA. However, the evidence from clinical studies is that L-DOPA is not an antidepressant although it may improve symptoms of retardation (Mendels *et al.*, 1975). This suggests that the ability of DA agonists to reverse reduced behavioural output in chronic stress models is not due to reversal of a state of learned helplessness which is homologous to depression. These findings favour accounts of 'helpless' behaviour as motor deficits. However, the model is not invalidated; the motor deficits due to DA depletion may be downstream to a central motive/neurochemical state of helplessness or depression.

DA receptor antagonists block the effects of antidepressants in some behavioural models but there is little evidence of this in humans. Raskin *et al.* (1970) found chlorpromazine to be superior to placebo in the treatment of endogenous depression and in agitated depressive illnesses. However, the antidepressant activity of neuroleptics must remain uncertain in view of the lack of trial experience. One prospective and four retrospective trials suggest that the combination of a neuroleptic and tricyclic is superior to either alone in the treatment of delusional depression. These results suggest that neuroleptics do not antagonise the therapeutic effects of antidepressants, at least in delusional depression. It therefore seems unlikely that enhanced DA receptor function is of central importance in mechanisms of antidepressant action as suggested by some of the animal models. Again, however, the animal models suggest appropriate clinical studies.

7.4.4 Other drugs

The accompanying chapters (Willner, Danysz *et al.*, this volume, Chapters 5, 6) mention many other drugs with putative antidepressant efficacy (anticholinergics, antihistamines, opioids, GABA agonists, beta-agonists). There are no substantive clinical trials with replications to confirm or refute these suggestions.

7.5 CONCLUSION

There can be little doubt that animal models of depression have contributed to the development of new treatments of depression, suggested possible mechanisms of action of antidepressants, and influenced ideas of the psychobiology of depression. By their very nature, animal models of depression need to draw from clinical knowledge but there is much that is uncertain about the nature of depression. Some of the difficulties in distilling hard clinical facts on which to base animal models have been highlighted in this chapter. Nevertheless, there is room for improving the validity of animal models with attention to what is known about the aetiology and treatment of depression. Several disparities between the models and the illness have been described. They do not seem insurmountable.

The development of animal models would be a sterile exercise if they did not influence clinical thinking. This is perhaps their greatest relevance. This chapter has emphasised that hypotheses about depression are testable in animal models. These tests rest on the assumption that the models are valid. This will always be uncertain and so clinical tests will remain the most decisive. If a hypothesis is not refuted in an animal model, there is still the possibility of refutation in a clinical test. Nevertheless, findings from animal experiments suggest new, clinically testable, ideas and indicate which clinical experiments are important. They have an important influence on the decision to embark on time-consuming, expensive and difficult clinical investigation of new ideas and new drugs.

Original ideas are perhaps the most precious but most unpredictable scientific commodity. They are driven partly by paradoxes in existing ideas. Animal models contribute to the tensions in research on depression and so to the generation of new insights.

REFERENCES

Aberg-Wistedt, A. (1982). A double blind study of zimelidine, a serotonin uptake inhibitor, and desipramine, a noradrenaline uptake inhibitor, in endogenous depression. *Acta Psychiatrica Scandinavica* **66**, 50–65.

Abou-Saleh, T.M. (1988). How useful is a dexamethasone suppression test? *Current Opinion in Psychiatry* **1**, 60–5.

Abrams, R. & Taylor, M.A. (1980). A comparison of unipolar and bipolar depressive illness. *American Journal of Psychiatry* **137**(9), 1084–7.

Akiskal, H.S. & McKinney, Jr. W.T. (1975). Overview of recent research in depression: Integration of ten conceptual models into a comprehensive clinical frame. *Archives of General Psychiatry* **32**, 285–303.

Akiskal, H.S., Bitar, A.H., Puzantian, V.R., Rosenthal, T.L. & Walker, P.W. (1978). The nosological status of neurotic depression. *Archives of General Psychiatry* **35**, 756–66.

Akiskal, H.S., Rosenthal, R.H., Rosenthal, T.L., Kashgarian, M., Khani, M.K. & Puzantian, V.R. (1979). Differentiation of primary affective illness from situational, symptomatic and secondary depression. *Archives of General Psychiatry* **36**, 635–43.

Amin, M.M., Ananth, J.V., Coleman, B.S., Darcourt, G., Parkas, T., Goldstein, B., Lapierre, Y.L., Paykel, E. & Wakelin, J.S. (1984). Fluvoxamine: Antidepressant effects confirmed in a placebo-controlled international study. *Clinical Psychopharmacology* **7**, 580–1.

Ancill, R.I., Poyser, J., Davey, A. & Kennerson, A. (1984). Management of mixed affective symptoms in primary care: a critical experiment. *Acta Psychiatrica Scandinavica* **70**, 463–9.

Asberg, M., Thonen, P., Traskman, L., Bertilsson, L. & Ringberger, V. (1976). 'Serotonin depression' a biochemical subgroup within the Affective Disorders. *Science* **191**, 478–80.

Baron, M., Risch, N., Hamburger, R., Mandel, B., Kushner, S., Newman, M., Drumer, D. & Belmaker, R.H. (1987). Genetic linkage between X-chromosome markers and bipolar affective illness. *Nature* **326**, 289–92.

Bebbington, P.E., Brugha, T., MacCarthy, B., Potter, J., Strut, E., Wykes, T., Katz, R. & McGuffin, P. (1988). The Camberwell collaborative depression study. 1. Depressed probands: Adversity and the form of depression. *British Journal of Psychiatry* **152**, 754–65.

Beninger, R.J. (1983). The role of dopamine in locomotor activity and learning. *Brain Research Reviews* **6**, 173–96.

Bielski, R.J. & Friedel, R.O. (1976). Prediction of tricyclic antidepressant response. *General Psychiatry* **33**, 1479–89.

Blackshear, A.M., Martin, L.L. & Sanders-Bush, E. (1986). Adaptive changes in the 5HT1 binding site after chronic administration of agonists and antagonists. *Neuropharmacology* **25**, 1267–73.

Bradshaw, C.M. & Szabadi, E. (1978). Changes in operant behaviour in a manic-depressive patient. *Behaviour Therapy* **9**, 950–4.

Brandon, S., Cowley, P., McDonald, C., Neville, P., Palmer, B. & Wellstodd-Bason, S. (1984). Electroconvulsive therapy: results in depressive illness from the Leicestershire trial. *British Medical Journal* **228**, 22–5.

Bremner, J.D. (1984). Fluoxetine in depressed patients: A comparison with imipramine. *Journal of Clinical Psychiatry* **45**, 414–19.

Brittain, R.T., Butler, A., Coates, I.H., Fortune, D.H., Hagan, R., Hill, J.M., Ireland, S.J., Jack, D., Jordan, C.C., Oxford, A., Straughan, D.W. & Tyers, M.B. (1987). GR380327, a novel selective 5HT3 receptor antagonist. *British Journal of Pharmacology* **90**, 87P.

Broadhurst, P.L. (1977). Pharmacogenetics. *Handbook of Psychopharmacology*, vol. 7, ed. L.L. Iversen, S.D. Iversen & S.H. Snyder, pp. 265–320. Plenum: New York.

Broekkamp, C.L., Garrigou, D. & Lloyd, K.G. (1980). Serotonin-mimetic and antidepressant drugs on passive avoidance learning by olfactory bulbectomized rats. *Pharmacology, Biochemistry and Behavior* **13**, 643–6.

Brown, G.W. & Harris, T. (1978). *Social Origins of Depression*. Tavistock: London.

Brown, G.W. & Harris, T. (1986). Stressor, vulnerability and depression: a question of replication. *Psychological Medicine* **16**, 739–44.

Carney, M.W.P., Roth, M. & Garside, R.F. (1965). The diagnosis of depressive syndromes and the prediction of ECT response. *British Journal of Psychiatry* **111**, 659–74.

Ceulemans, D.L.S., Hoppenbrowers, M.L.J.A.,

Gelders, Y.G. & Reyntjens, A.J.M. (1985). The influence of ritanserin, a serotonin antagonist, in anxiety disorders: A double blind placebo-controlled study versus lorazepam. *Pharmacopsychiatry* **18**, 303–5.

Charney, B.S. & Nelson, J.C. (1981). Delusional and non-delusional unipolar depression: further evidence of distinct subtypes. *American Journal of Psychiatry* **138**, 328–33.

Checkley, S.A., Corn, T.H., Glass, I.B., Burton, S.W. & Burke, C.A. (1986). The responsiveness of central alpha 2 adrenoceptors in depression. In *The Biology of Depression*, ed. J.F.W. Deakin, pp. 100–20. Gaskell: London.

Chouinard, G. (1985). A double-blind controlled clinical trial of fluoxetine and amitriptyline in the treatment of outpatients with major depressive disorder. *Journal of Clinical Psychiatry* **46**, 32–7.

Clinical Research Centre (1984). The Northwick Park ECT trial: Predictors of response to real and simulated ECT. *British Journal of Psychiatry* **144**, 227–37.

Cohn, T.B. & Wilcox, C. (1985). A comparison of fluoxetine, imipramine and placebo in patients with major depressive disorders. *Journal of Clincal Psychiatry* **46**, 26–31.

Cowen, P.J. & Charig, E.M. (1987). Neuroendocrine resonses to tryptophan in major depression. *Archives of General Psychiatry* **44**, 958–66.

Crow, T.J. (1973). Catecholamine-neurones and self-stimulation. II. A theoretical interpretation and some psychiatric implications. *Psychological Medicine* **3**, 66–73.

Crow, T.J. & Deakin, J.F.W. (1979). Monoamines and the psychoses. In *Chemical Influences on Behaviour*, ed. K. Brown & S.J. Cooper, pp. 503–32. Academic Press: London.

Danish Universities Antidepressant Group. (1986). Citalopram: Clinical effects profile in comparison with clomipramine. A controlled multicenter study. *Psychopharmacology* **90**, 131–8.

DeSouza, R.J., Goodwin, G.M., Green, A.R. & Heal, D.J. (1986). Effect of chronic treatment with 5HT1 agonist (8-OH-DPAT and RU24969) and antagonist (ipsapirone) drugs on the behavioural responses of mice to 5HT1 and 5HT2 agonists. *British Journal of Pharmacology* **89**, 377–84.

Deakin, J.F.W. (1983). Role of serotonergic systems in escape, avoidance and other behaviours. In *Theory in Psychopharmacology*, vol. 12, ed. S.J. Cooper, pp. 149–93. Academic Press: London.

Deakin, J.F.W. & Pennell, I. (1986). 5HT receptor subtypes and depression. *Psychopharmacology* **89**, S24.

Deakin, J.F.W. & Crow, T.J. (1986). Monoamines, rewards and punishment: The anatomy and physiology of the affective disorders. In *The*

Biology of Depression, ed. J.F.W. Deakin, pp. 1–25. Gaskell: London.

Deakin, J.F.W. (1988). 5HT2 receptors, depression and anxiety. *Pharmacology, Biochemistry and Behavior* **29**, 819–20.

Deakin, J.F.W. (1989). Role of 5HT receptor subtypes in depression and anxiety. In *Behavioural Pharmacology of 5HT*, ed. Bevan, A. Cools & T. Archer, pp. 179–204. Lawrence Erlbaum: New York.

Dominguez, R.A., Goldstein, B.J., Jacobson, A.P. & Steinbook, R.M. (1985). A double blind placebo controlled study of fluvoxamine and imipramine in depression. *Journal of Clinical Psychiatry* **46**, 84–7.

Drugan, R.C., Ryan, S.M., Minor, T.R. & Maier, S.F. (1984). Librium prevents the analgesia and shuttlebox escape deficits typically observed following inescapable shock. *Pharmacology, Biochemistry and Behavior* **21**, 749–54.

Drugan, R.C., Maier, S.F., Skolnick, P., Paul, S.M. & Crawley, J.N. (1985). An anxiogenic benzodiazepine receptor ligand induces learned helplessness. *European Journal of Pharmacology* **113**, 453–7.

Egeland, J.A., Gerhard, D.S., Pauls, D.L., Sussex, J.N., Kidd, K.K., Allen, C.R. & Hostetter, A.M. & Housman, D.E. (1987). Bipolar affective disorders linked to DNA markers on chromosome 11. *Nature* **325**, 783–7.

Emrich, H.M., Berger, M. & Von Zerssen, D. (1986). Differential therapy of the depressive syndrome: Results of a therapeutic study with fluvoxamine vs. oxaprotiline. In *Differential Therapy of Depression: Possibilities and Limitations*, H. Hippius & N. Matussek, pp. 57–64. Karger: Basel.

Everitt, B.J. (1982). Functional studies of central catecholamines. *International Review of Neurobiology* **23**, 303–64.

Fabre, L.F. & Crismon, L. (1985). Efficacy of fluoxetine in outpatients with major depression. *Current Therapy Research* **37**, 115–23.

Fawcett, J., Clark, D.C., Scheftner, W.A. & Hedeker, D. (1983). Differences between anhedonic and normally hedonic depressive states. *American Journal of Psychiatry* **140**, 1027–30.

Fieve, R.R., Goodnick, P.J., Peselow, E. & Schlegel, A. (1986). Fluoxetine response: Endpoint vs pattern analysis. *International Clinical Psychopharmacology* **1**, 320–3.

Foulds, G.A. & Bedford, A. (1975). Hierarchy of classes of personal illness. *Psychological Medicine* **5**, 181–92.

Geller, N.E. & Blum, K. (1970). The effects of 5-HTP on parachlorophenylalanine (P-CPA) attenuation of conflict behaviour. *European Journal of Pharmacology* **9**, 319–24.

Giral, P., Martin, P. Soubrie, P. & Simon, P. (1988).

Reversal of helpless behaviour in rats by putative 5HT1A agonists. *Biological Psychiatry* **23**, 237–42.

Glass, I.B., Thompson, C., Corn, T. & Robinson, P. (1984). The GH response to clonidine in endogenous compared to reactive depression. *Psychological Medicine* **14**, 773–7.

Glassman, A.H. & Roose, S.P. (1981). Delusional depression: A distinct clinical entity? *Archives of General Psychiatry* **38**, 424–7.

Goldberg, H.L. & Finnerty, R.J. (1979). The comparative efficacy of buspirone and diazepam in the treatment of anxiety. *American Journal of Psychiatry* **136**, 1184–7.

Goldberg, D.P., Bridges, K., Duncan-Jones, P. & Grayson, D. (1987). Dimensions of neuroses seen in primary care settings. *Psychological Medicine* **17**, 461–70.

Goodwin, F.H., Post, R.M., Dunner, D.L. & Gordon, E.K. (1973). Cerebrospinal fluid amine metabolites in affective illness: The probenecid technique. *American Journal of Psychiatry* **130**, 73–9.

Graeff, F.G. (1974). Tryptamine antagonists and punished behaviour. *Journal of Pharmacology and Experimental Therapeutics* **189**, 344–50.

Greenblatt, M., Grosser, G.H. & Wechsler, H. (1964). Differential response of hospitalized depressed patients to somatic therapy. *American Journal of Psychiatry* **120**, 935–43.

Hamilton, M. & White, J.M. (1959). Clinical syndromes in depressive states. *Journal of Mental Science* **105**, 985–8.

Henninger, G.R., Charney, D.S. & Sternberger, D.E. (1984). Serotonergic function in depression. *Archives of General Psychiatry* **41**, 398–402.

Hodgkinson, S., Sherrington, R., Gurling, H., Marchbanks, R., Reeders, S., Mallet, J., McInnis, M., Petursson, H. & Brynjolfsson, J. (1987). Molecular genetic evidence for heterogeneity in manic depression. *Nature* **325**, 805–8.

Howarth, B.G. & Grace, M.G.A. (1985). Depression, drugs and delusions. *Archives of General Psychiatry* **42**, 1145–7.

Itil, T.M., Shrivastava, R.K., Mukherjeem, S., Coleman, B.S. & Michael, S.T. (1983). A double-blind placebo-controlled study of fluvoxamine and imipramine in out-patients with primary depression. *British Journal of Pharmacology* **15**, 433S–438S.

Johnston, D.G., Troyer, I.E. & Whitsett, S.F. (1988). Clomipramine treatment of agoraphobic women. *Archives of General Psychiatry* **45**, 453–9.

Johnstone, B.C., Cunningham-Owens, D.G., Frith, C.D., McPherson, L., Dowie, C., Riley, G. & Gold, A. (1980). Neurotic illness and its response to anxiolytic and antidepressant treatment. *Psychological Medicine* **10**, 321–8.

Kahn, R.J., McNair, M., Lipman, S.B., Covi, L.,

Rickels, K., Downing, R., Fisher, S. & Frankenthaler, L.M. (1986). Imipramine and chlordiazepoxide in depressive and anxiety disorders. *Archives of General Psychiatry* **43**, 79–84.

Kendell, R.E. (1975). *The Role of Diagnosis in Psychiatry*. Blackwell Scientific Publications: Oxford.

Kendell, R.E. (1976). The classification of depressions: A review of contemporary confusion. *British Journal of Psychiatry* **129**, 15–28.

Kennett, G.A., Dickinson, S. & Curzon, G. (1985). Enhancement of some 5-HT-dependent behavioural responses following repeated immobilization in rats. *Brain Research* **330**, 253–63.

Kennett, G.A., Dourish, C.T. & Curzon, G. (1987). Antidepressant-like action of 5HT1A agonists and conventional antidepressants in an animal model of depression. *European Journal of Pharmacology* **134**, 265–74.

Levine, S., Deo, R. & Mahadeven, K. (1987). A comparative trial of a new antidepressant, fluoxetine. *British Journal of Psychiatry* **150**, 653–5.

McCabe, M.S. (1976). ECT in the treatment of mania: A controlled study. *Archives of General Psychiatry* **133**, 688–90.

McGuffin, P., Katz, R. & Bebbington, P. (1988). The Camberwell collaborative depression study III. Depression and adversity in the relatives of depressed probands. *British Journal of Psychiatry* **152**, 775–82.

Medical Research Council Clinical Psychiatry Committee (1965). Clinical trial of the treatment of depressive illness. *British Medical Journal* **(i)**, 881–6.

Mendels, J., Stinnett, J.L., Burns, D. & Frazer, A. (1975). Amine precursors and depression. *Archives of General Psychiatry* **32**, 22–30.

Moradi, S.R., Muniz, C.E. & Belar, C.D. (1979). Male delusional depressed patients: Response to treatment. *British Journal of Psychiatry* **135**, 136–8.

Morley, M.J., Bradshaw, C. & Szabadi, E. (1987*a*). DSP4 alters the effect of d-amphetamine on variable-interval performance: analysis in terms of Herrnstein's equation. *Psychopharmacology* **92**, 247–53.

Morley, M.J., Bradshaw, C. & Szabadi, E. (1987*b*). Effects of 6-hydroxydopamine-induced lesions of the dorsal noradrenergic bundle on steady state operant behaviour. *Psychopharmacology* **93**, 520–5.

Morley, M.J., Shah, K., Bradshaw, C.M. & Szabadi, E. (1988). DSP4 and Herrnstein's equation: Further evidence for a role of noradrenaline in the maintenance of operant behaviour by positive reinforcement. *Psychopharmacology* **96**, 551–6.

Morris, J.B. & Beck, A.T. (1974). The efficacy of antidepressant drugs. *Archives of General Psychiatry* **30**, 667–74.

Nelson, J.C. & Charney, D.S. (1981). The symptoms of major depressive illness. *American Journal of Psychiatry* **138**, 1–13.

Norton, K.R.W., Sireling, L.I., Bhat, A.V., Rao, B. & Paykel, E.S. (1984). A double-blind comparison of fluvoxamine, imipramine and placebo in depressed patients. *Journal of Affective Disorders* **7**, 297–308.

Nystrom, C. & Hallstrom, T. (1987). Comparison between a serotonin and a noradrenaline reuptake blocker in the treatment of depressed outpatients: a cross over study. *Acta Psychiatrica Scandinavica* **75**, 377–82.

Olds, J., Killam, K.F. & Bach-y-Rita, P. (1956). Self-stimulation of the brain used as a screening method for tranquilizing drugs. *Science* **124**, 265–6.

Overall, J.E., Hollister, L.E., Johnson, M. & Pennington, V. (1966). Nosology of depression and differential response to drugs. *Journal of American Medicine Association* **195**, 946–64.

Paykel, E.S. (1978). Classification of depressed patients: a cluster analysis derived grouping. *British Journal of Psychiatry* **133**, 45–52.

Paykel, E.S. (1979). Recent life events in the development of the depressive disorders. In: *The Psychobiology of the Depressive Disorders: Implications for the Effects of Stress*, ed. R.A. Depue, pp. 245–65. Academic Press: New York.

Paykel, E.S., Rowan, P.R., Parker, R.R. & Bhat, A.V. (1982). Response to phenelzine and amitriptyline in subtypes of outpatient depression. *Archives of General Psychiatry* **39**, 1041–9.

Paykel, E.S., Parker, R.R., Rowan, P.R., Rao, B.M. & Taylor, C.N. (1983). Nosology of atypical depression. *Psychological Medicine* **13**, 131–9.

Peroutka, S.J. & Snyder, S.H. (1980). Long-term antidepressant treatment decreases spiroperidol-labelled serotonin receptor binding. *Science* **210**, 88–90.

Quitkin, F., Rifkin, A. & Klein, D.F. (1978). Imipramine response in deluded depressive patients. *American Journal of Psychiatry* **135**, 786-811.

Raskin, A., Schulterbrandt, J.G., Reatin, H. & McKeon, J.J. (1970). Differential response to chlorpromazine, imipramine, and placebo. *Archives of General Psychiatry* **23**, 164–73.

Reimherr, F.W., Wood, D.R., Byerley, B., Brainard, J. & Grosser, B.I. (1984). Characteristics of responders to fluoxetine. *Psychopharmacology Bulletin* 70–2.

Rickels, K., Feighner, J.P. & Smith, W.T. (1985). Alprazolam, amitriptyline, doxepin, and placebo in the treatment of depression. *Archives of General Psychiatry* **42**, 134–41.

Rickels, K., Amsterdam, J.D. & Avallone, M.F. (1986). Fluoxetine in major depression: A controlled study. *Current Therapy Research* **39**, 559–63.

Robinson, L.S., Nies, A., Ravaris, C.L., Lamborn, K.R. & Burlington, V.T. (1973). The monoamine oxidase inhibitor phenelzine, in the treatment of depressive-anxiety states: A controlled clinical trial. *Archives of General Psychiatry* **29**, 407–13.

Roy, A., Virkkunen, M., Guthrie, S. & Linnoilla, M. (1986). Indices of serotonin and glucose metabolism in violent offenders, arsonists and alcoholics. *Annals of New York Academy of Sciences* **487**, 202–20.

Schatzberg, A.F. & Cole, J.D. (1978). Benzodiazepines in depressive disorders. *Archives of General Psychiatry* **35**, 1359–65.

Simpson, G.M., Lee, J.H., Cuculic, Z. & Kellner, R. (1976). Two dosages of imipramine in hospitalized endogenous and neurotic depressives. *Archives of General Psychiatry* **33**, 1093–101.

Soubrie, Ph. (1989). Role of serotonin-containing neurones and 5HT receptors in animal models of anxiety and depression. In *Behavioural Pharmacology of 5HT*, ed. P. Bevan, A. Cools & T. Archer, pp. 337–52. Lawrence Erlbaum: New York.

Stark, P. & Hardison, C.D. (1985). A review of multicenter controlled studies of fluoxetine vs imipramine and placebo in outpatients with major depressive disorder. *Journal of Clinical Psychiatry* **46**, 53–8.

Stein, L. (1962). Effects and interactions of imipramine, chlorpromazine, reserpine and amphetamine on self-stimulation: possible neurophysiological basis of depression. In *Recent Advances in Biological Psychiatry*, pp. 288–308, vol. 4, ed. J. Wortis. Plenum: New York.

Szabadi, E., Bradshaw, C.M. & Ruddle, H.V. (1981). Reinforcement process in affective illness: Towards a quantitative analysis. In *Quantification of Steady State Operant Behaviour*, ed. C.M. Bradshaw, E. Szabadi & C.F. Lowe, pp. 229–310. Elsevier/North Holland: Amsterdam.

Taylor, P. & Fleminger, J.J. (1980). ECT for schizophrenia. *Lancet* **i**, 1380–3.

Traksman-Bendz, L., Asberg, M. & Schalling, D.

(1986). Serotonergic function and suicidal behaviour in personality disorders. *Annals of New York Academy of Sciences* **487**, 168–74.

Tye, N.C., Everitt, B.J. & Iversen, S.D. (1977). 5-Hydroxytryptamine and punishment. *Nature* **268**, 741–3.

Vetulani, J., Lebrecht, U. & Pilc, A. (1981). Enhancement of responsiveness of the central serotonergic system and serotonin-2 receptor density in rat frontal cortex by electroconvulsive treatment. *European Journal of Pharmacology* **76**, 81–5.

Willner, P. (1989*a*). Sensitization to the actions of antidepressant drugs. In *Psychoactive Drugs: Tolerance and Sensitization*, ed. M.W. Emmett-Oglesby & A.J. Goudie, pp. 275–9. Humana Press: New Jersey.

Willner, P. (1989*b*). Towards a theory of serotonergic dysfunction in depression. In *Behavioural Pharmacology of 5HT*, ed. P. Bevan, A. Cools & T. Archer, pp. 157–78. Lawrence Erlbaum: New York.

Willner, P., Muscat, R., Sampson, D. & Towell, T. (1988). Dopamine and antidepressant drugs: the antagonist challenge strategy. In *Progress in Catecholamine Research, Part C*, vol. 2, ed. H. Belmaker, M. Sandler & A. Dahlstrom, pp. 275–9. Alan Liss: New York.

Winokur, G. (1979). Unipolar depression: Is it divisible into autonomous subtypes. *Archives of General Psychiatry* **36**, 47–52.

Winokur, G. (1985). The validity of neurotic-reactive depression. *Archives of General Psychiatry* **42**, 1116–27.

Wise, C.D., Berger, B.D. & Stein, L. (1970). Serotonin: A possible mediator of behavioural suppression induced by anxiety. *Diseases of the Nervous System*, GWAN Suppl., **31**, 34–7.

Wise, R.A., Spindler, J., De Wit, H. & Gerber, G.J. (1978). Neuroleptics and operant behaviour: the anhedonia hypothesis. *Science* **5**, 39–87.

Wittenborn, J.R. & Kiremitci, N. (1975). A comparison of antidepressant medications in neurotic and psychotic patients. *Archives of General Psychiatry* **32**, 1172–6.

Young, J.P.R., Coleman, A. & Lader, M.H. (1987). A controlled comparison of fluoxetine and amitriptyline in depressed out-patients. *British Journal of Psychiatry* **1512**, 337–40.

PART IV

Models of eating disorders

8

Animal models of eating disorders

ANTHONY M.J. MONTGOMERY

8.1 INTRODUCTION

The relationship between ingestive behaviour and body weight can be described quite simply in terms of energy input and energy expenditure. If energy input matches energy output then a constant body weight should be maintained; a change to one side of the equation, unaccompanied by a compensatory change to the other side, will result in increased or reduced body weight.

In reality this attractively simple account is complicated by a number of factors. Energy output, for most people, will vary from day to day, requiring constant adjustment to energy intake. Meals are usually fairly brief events, terminating before the gut has had time to absorb the ingested calories. This means either that there must be a good short-term mechanism for predicting the caloric consequences of a particular meal, or that the subsequent interval to the next meal, or its size, must be adjusted to compensate for errors made earlier in the day. It seems likely that each of these factors may contribute to maintaining a relatively stable body weight over weeks, months or years. Furthermore, the input and output sides of the energy equation are not independent. The output side of the equation reflects energy used in maintaining metabolic functions together with expenditure incurred through physical activity, the former being much the larger contributor. However, metabolic rate is sensitive to changes in energy intake: metabolic rate declines if energy intake is restricted, and increases with increased energy intake (Bray, 1969, 1970).

Ingestive behaviours are also influenced by cognitive and social factors. For example social events are often associated with eating, usually highly palatable and calorically dense foods, and with drinking calorie-rich alcoholic drinks. Since meal times are usually determined by convention the possibility of adjusting subsequent intake by delaying the next meal is limited, and if the meal size is determined by habit, then the onus of maintaining a steady body weight falls primarily on increasing metabolic rate. If the scope for increasing metabolic rate is limited then the inevitable consequence will be weight gain. It is not intended to suggest that this example explains all instances of weight gain, but rather to point out that if constraints are placed on the adaptability of the factors contributing to the overall equation, then changes in body weight will follow.

The ways in which changes in body weight actually occur must be determined empirically. However, rigorous efforts to specify the relative contribution of factors on either side of the energy equation are lacking and the evidence which does exist is equivocal. It is an unfortunate paradox that there are numerous experimental manipulations that reliably increase or reduce food intake and body weight in animals and yet the success rate of treatments for human conditions is disappointingly low. For example, treatment with fenfluramine reliably reduces both food intake and body weight in rats, but in overweight patients weight loss is less impressive. There are many possible explanations for this difference and a better description of the human disorder would be invaluable to the animal modeller.

8.2 CLINICAL SYMPTOMATOLOGY

8.2.1 Obesity

Obesity is more common among females, the age of onset varying widely from childhood onwards, but having a tendency towards onset during middle-age. Weight gain is associated with overeating and/or underactivity. Studies suggesting that the overweight are underactive have often failed to consider the larger energy requirement of activity in overweight people (Brownwell & Stunkard, 1980) and even if convincing data were available the question of the direction of causality remains open: are they fat because they are inactive or vice-versa? Many studies indicate that the obese do not grossly overeat relative to people with normal body weight (see, e.g. Greene, 1939), although others suggest they do consume more sweet-tasting food (see, e.g. Rodin, 1975). It seems likely that the obese tend to under-report their intake. However, maintaining a steady level of obesity may not require a large energy intake, particularly if accompanied by underactivity. Studies of eating habits during the dynamic phase of weight gain are relatively rare, but there are reports of weight gain starting with a period of enforced bed-rest (Schacter, 1971). Investigations of eating habits in humans with stable obesity reveal a range of eating styles including night eating and bulimia, suggesting that obesity is a heterogeneous disorder. However, a number of generalisations can be made: the obese appear to be metabolically efficient, hyperinsulinaemic and often have low self-esteem (Garfinkel & Kaplan, 1986).

Various attempts have been made to describe a psychological contribution to obesity. According to one formulation (Schachter, 1971) overweight subjects are more under the control of external stimuli (e.g. palatability) and less under the control of internal stimuli (e.g. hunger and satiety signals). However, external responsiveness is not well correlated with level of obesity (Rodin, 1980). In a reformulation of the externality hypothesis, which stresses the interdependence of externality and internality, Rodin (1980) suggests that, while externality may predispose an individual to bouts of overeating, chronic weight gain also requires a failure of the long-term regulatory mechanisms that ordinarily would counteract the effects of over-responsiveness to external stimuli. This reformulation, therefore, suggests that high externality predisposes people towards short-term weight gain; maintenance of this weight gain requires a malfunction of the compensatory influence of unspecified long-term regulatory mechanisms.

A second suggestion is that many of the obese are chronically on (unsuccessful) diets, so that they end a meal before they have reached satiety. This restrained style of eating may paradoxically serve to maintain obesity, because restrained eaters actually overeat in response to stress, alcohol or the perceived failure to maintain restraint (Herman & Polivy, 1980). A third approach suggests that the obese have an impaired ability to identify the physiological correlates of hunger, and interpret changes in arousal, unrelated to energy balance, as hunger signals (Robbins & Fray, 1980). This formulation provides a basis for explaining why some obese humans eat when they are anxious (Ross *et al.*, cited in Schachter, 1971).

Attempts to treat obesity pharmacologically have proven disappointing: weight loss is generally unspectacular and rarely maintained when the drug treatment is withdrawn. The majority of anti-obesity drugs are stimulants (e.g.amphetamine) and associated with adverse side effects such as insomnia, excitement, dizziness and panic attacks. Fenfluramine, on the other hand, is sedative, but adverse side effects include drowsiness and diarrhoea. Problems of tolerance and/or addiction place further restrictions on the suitability of these drugs as acceptable strategies for the treatment of chronic obesity. However, there may be a role for pharmaceuticals as part of a regime combining drug treatment with behaviour therapy and social support (Lasagna, 1987).

8.2.2 Anorexia nervosa (AN)

In AN weight loss is associated with restricted intake, often combined with increased activity. Sufferers exhibit an intense fear of becoming obese, together with a distorted body image. Approximately 90% of anorexics are female, typically with onset occurring in mid- to late teens. Cognitive changes associated with AN include impaired concentration, irritability, anxiety, apathy and social withdrawal (Garfinkel & Kaplan, 1986). Although denial of hunger is common among anorectics, they often confess during remission to having experienced constant hunger at times when intake was being restricted.

Numerous physiological correlates of AN have been noted, including reduced metabolic rate, hypothermia, amenorrhea, reduced plasma levels of noradrenaline (NA) and tryptophan, reduced urinary excretion of catecholamine metabolites (Gross *et al.*, 1979), elevated opiate activity and reduced NA levels in cerebrospinal fluids (CSF) (Kaye *et al.*, 1982, 1984*a*), and changes in hormonal responses (e.g. Vigersky *et al.*, 1976). However, it is unclear whether these changes are causes or effects of starvation.

Dieting is very common among young females, but where weight loss occurs it is rarely maintained. AN usually starts with a diet that is notable for its success – weight loss is pursued and maintained by restricting intake and avoiding highly caloric food (e.g. carbohydrates). In rats, food deprivation biases the underweight animal towards the sensory properties of food and away from its nutritional value (Jacobs & Sharma, 1969) so that the incentive properties of food become prepotent. In people, dietary restraint is often tenuous and the failure of restraint can result in counterregulation: when restraint is perceived to have failed, overeating occurs relative to unrestrained controls. One explanation of counterregulation suggests that the failure to adhere strictly to a diet abolishes the motivation to exert cognitive restraint, allowing internal (hunger) and/or external (incentive) factors to gain control (Herman & Mack, 1975).

Bruch (1974) has suggested that AN is secondary to a personality disorder in which the adolescent female pursues perfection and control, presumably to compensate for low self-esteem. From this perspective AN can be characterised as a maladaptive coping strategy focused on controlling body weight rather than establishing autonomy. Consequently it should be expected that this strategy would be relinquished reluctantly. Consonant with this interpretation is the adoption of vomiting and/or laxative abuse to facilitate weight loss, particularly in anorectics who exhibit occasional bouts of excessive consumption (bingeing). Reports of high incidence of affective disorder in the families of anorectics (Winokur *et al.*, 1980) also point to the possibility that the disordered feeding associated with AN represents a symptom

of an underlying malaise. If this is the case then animal models of AN should be based upon manipulations that produce a suppression of food intake rather than producing weight loss by restricting access to food.

A variety of drugs have been investigated as treatments for AN, but none have yielded clear evidence of enduring efficacy. Neuroleptics (including chlorpromazine, pimozide and sulpiride) have been tested, often in combination with behaviour therapy: overall the results do not support the use of neuroleptic therapy, although patients with severe agitation may benefit (Mitchell, 1987). Despite a possible link between AN and affective disorders treatment with antidepressants does not improve outcome (for review see Mitchell, 1987). Treatment with a high dose of the serotonin (5-HT) antagonist cyproheptadine, in combination with structured inpatient treatment programmes, has been reported to enhance the rate of weight gain in nonbulimic patients, but to impair treatment efficiency in bulimic anorectics (Halmi *et al.*, 1986). This finding of a differential response between bulimic and non-bulimic anorectics suggests that future drug studies should pay close attention to differences in symptomatology between responders and non-responders so as not to overlook potentially valuable treatment strategies in heterogeneous groups of subjects.

8.2.3 Bulimia nervosa (BN)

BN is characterised by bouts of excessive eating that are often followed by vomiting or laxative abuse to prevent weight gain. BN sufferers may have normal body weight, although large fluctuations are common when bulimic episodes alternate with fasting. As with AN, bulimics fear weight gain and the loss of control that accompanies failure to restrict intake. Excessive eating episodes often include consumption of highly caloric carbohydrates or other foods that the sufferer otherwise avoids because of their energy value. It has been suggested that bulimics are unusually anxious (Jimerson *et al.*, 1986) and it may be that bulimic episodes are precipitated by anxiety.

The typical sufferer is a young woman often with a prior history of weight problems and, like anorectics, bulimics often present symptoms of depression and tend to have a high familial incidence of affective disorders (Hudson *et al.*, 1983). Further support for the link between bulimia and depression is provided by the finding that bulimics show incomplete suppression of cortisol by dexamethasone (Gwirtsman *et al.*, 1983).

There is evidence that some normal weight bulimics have previously been anorectic and it has been suggested that dietary restraint (and its failure) may provide a link between the two disorders (Herman & Polivy, 1984). According to this formulation, bulimics may be regarded as anorectics in whom dietary restraint frequently fails, allowing an eating binge to occur; fasting, vomiting or purging are attempts to avoid the consequences.

Abnormalities in the EEG recordings of a large proportion (64%) of BN patients have led to the suggestion that BN might respond to anticonvulsant treatment (Rau & Green, 1975). However, in the only controlled investigation of this hypothesis phenytoin was no more effective than placebo (Wermuth *et al.*, 1977). Three out of five double-blind, placebo-controlled tests of antidepressants have yielded evidence of short-term benefits in normal weight subjects (for review see Walsh, 1987). As Walsh (1987) points out this is a surprising success rate given that antidepressants are associated with weight gain and sugar craving (Paykel *et al.*, 1973). Perhaps even more surprising is the finding that non-depressed bulimics have also benefitted from antidepressant treatment (Walsh *et al.*, 1984; Hughes *et al.*, 1986). It may be that the efficacy of antidepressants is related to their anxiolytic properties rather than their antidepressant effects. Whichever proves to be the case chronic studies are required since withdrawal from short-term treatment is often followed by relapse.

BN has been compared to other self-administrative behaviours, raising the possibility that bulimic episodes might be responsive to treatment with dopamine (DA), opiate and non-opioid peptide antagonists (Halmi *et al.*, 1987). Specific animal models of BN are conspicuous by their absence from the literature, but some models of obesity based on short term enhancements of food intake may be relevant (see Sections 8.4.1.8; 8.4.1.9; 8.4.2.2 and 8.5.1.3).

8.2.4 Implications for animal models

These brief outlines of the clinical sympatomatology of eating disorders place both demands and constraints on putative animal models. Enduring deviations from normal body weight are symptoms of both obesity and AN, so for face validity animal models of these disorders should involve a change in body weight. However, measures of food intake are also required because stable obesity can be maintained in the absence of overeating (relative to those with normal body weight) and AN can be accompanied by intermittent periods of high intake. The need for measures of both body weight and food intake is even more apparent in BN, where the eating disorder may not be associated with stable changes in body weight. Since each of these disorders occurs under circumstances in which a variety of palatable foods is freely available it seems inappropriate to concentrate on paradigms involving limited access to a single bland diet (as is commonly the case in studies of the psychopharmacology of rat feeding behaviour). Finally, the possibility that the disturbed eating behaviour seen in AN and BN are symptoms of personality or affective disorders, questions whether these disorders could be successfully modelled in animals. Certainly if restraint (and its failure) are central to these disorders then the usefulness of animal models may be limited.

Notwithstanding these reservations a good understanding of the physiological mechanisms underlying feeding behaviour is essential. Research on animals might not be able to model all aspects of all human eating disorders accurately, but it might provide useful information on how to treat the symptoms.

8.3 CONTROLS OF FEEDING AND BODY WEIGHT

Like other motivated behaviours, feeding requires the integration of internal and external signals, so that feeding behaviour is initiated, directed towards appropriate sources of food, maintained, and finally terminated when sufficient food has been ingested. An adequate control system would need to be sensitive either to the immediate availability of energy or to its rate of use. Attempts have been made to specify glucostatic, lipostatic, and to a lesser extent aminostatic mechanisms, that monitor blood levels of these three major energy sources, and elicit hunger when levels drop below a critical point. Indeed the discovery that blood transfused from satiated rats to deprived rats inhibits feeding (Davis *et al.*, 1967) provides rudimentary support for this approach.

Injections of insulin, which reduce blood glucose levels, elicit feeding (Booth & Brookover, 1968), suggesting that the onset of feeding may be determined by blood glucose levels. The appealing simplicity of this idea is called into question by the fact that untreated diabetics report hunger despite their high blood glucose levels. However, this problem could be overcome by a glucostatic mechanism sensitive to levels of utilisable glucose (for example, the release of glucose from hepatic glycogen) (Russek, 1976). There is also evidence that mice regulate the amount of energy stored as adipose tissue (Liebelt *et al.*, 1973). The brain may be sensitive to blood-borne free fatty acids since their topical administration increases the firing of cells in the lateral hypothalamus (Oomura, 1976). Evidence for an aminostatic mechanism is less convincing. Protein-rich meals are satiating. However, as diabetics with high blood amino acid levels report hunger, the simple aminostatic theory is untenable. Booth (1972*a*) points out that a wide variety of nutrients suppress feeding in relation to their energy yield (independently of convertibility to glucose). Thus, feeding may occur in order to regulate caloric intake, which might be accomplished by monitoring the rate of energy flow rather than the absolute level of immediate energy supply (Booth, 1976). If this is the case then factors such as the rate of stomach emptying, fat synthesis/catabolism and metabolic expenditure should all be important.

The sensory properties of foods also exert a strong influence on the level of consumption. Animals fitted with an oesophageal fistula (the 'sham feeding' preparation) reliably consume larger quantities of sweet sucrose solutions when the fistula is open (preventing the solution from reaching the stomach) than when it is closed (Janowitz & Grossman, 1949). Furthermore, animals become obese when they are offered a varied palatable diet (Ingle, 1949). Conversely, food-deprived rats reduce their food intake if the food is adulterated with unpalatable flavours such as quinine (Jacobs & Sharma, 1969), demonstrating that palatability can override energy regulation. In people, pleasantness ratings of recently consumed foods are lower than those for foods consumed less recently – a phenomenon known as sensory specific satiety (Rolls *et al.*, 1982). Collectively these findings provide empirical support for a distinction between hunger and appetite. A food-deprived animal given access to lab chow typically eats fairly rapidly for about 20 min by which time it is presumably satiated (i.e., hunger has been dissipated). If the animal is then given access to a sucrose solution feeding will restart, presumably under the control of appetite (i.e. the sucrose is consumed for its sensory properties rather than its caloric value). Obviously ingestion of the sucrose solution does not continue indefinitely so satiety must depend upon what is being eaten as well as how much has been eaten.

The sensory properties of food are central to Le Magnen's (1971) model of the control of free-feeding in rats. He argues that since there is a positive correlation between meal length and time to the start of the next meal (the post-prandial relationship) the initiation of a meal depends upon the level of immediate energy stores (i.e. the size of the preceding meal). However, depletion of energy stores (i.e. the time since the previous meal) does not predict the size of the next meal. Thus, energy levels determine when eating starts, but the sensory properties of food determine how much is eaten. In addition Le Magnen (1971) suggests that the stimulus characteristics of food (e.g. taste) become associated with the long-term metabolic consequences of ingestion: in other words, the organism learns to predict the caloric consequences of eating a particular food. Claims to have demonstrated conditioned satiety and conditioned appetite have been made (Booth, 1972*b*) and criticised (Smith & Gibbs, 1979).

Although digestion of a meal takes something of the order of four hours, meals tend to last less than an hour. This indicates the need for a satiety mechanism that can terminate feeding before the gut has had time to absorb the ingested nutrients. The results of an imaginative series of experiments involving siphoning of food from the stomach led Deutsch (1983) to suggest that meal size is controlled by two types of gastric signal: one measures gastric distension, while the other measures the amount of nutrients in the stomach by means of chemoreceptors sensitive to amino acids and free fatty acids or glycerol. Signals from these chemoreceptors must be calibrated through learning, so that for novel flavours oropharyngeal cues override (uncalibrated) nutrient signals. Plausible as this mechanism may seem it is hard to see why such an exquisitely sensitive system produces obese rats simply as a result of exposure to a varied palatable diet (Ingle, 1949).

8.3.1 Methodology of feeding research in animals

Most of the experimental studies of altered food consumption in animals have investigated the effects of anorectic drugs on the amount of food consumed by rats. Drug-induced enhancements of feeding are less frequently reported. Nevertheless, while pharmacological treatments of obesity are sometimes successful (e.g. Malcolm *et al.*, 1972), they generally yield unimpressive results in terms of weight loss (Scoville, 1976). Why is it that anorectic drugs can be readily identified in the laboratory rat and then prove so disappointing in the clinic? A comparison of the conditions prevailing in the two situations provides a number of plausible reasons for the discrepancy. Firstly the obese human usually has unrestricted access to a wide range of foodstuffs, whereas the laboratory rat is usually maintained on a regime of restricted access to a single, nutritionally sound diet. Secondly, a successful pharmacological treatment of obesity would require chronic treatment, possibly with a gradually increasing dosage to counteract any developing tolerance. The situation in the animal laboratory is markedly different from this: relatively large reductions in food intake are easily obtained using acute treatment with large drug doses and brief test intervals.

Freely-feeding animals eat in bouts, between which are interposed other behaviours. Feeding behaviour, therefore, does not occur in isolation; it is part of a wider behavioural repertoire. Looked at from this perspective any manipulation that interferes with the normal sequencing of behaviours could, in the short term at least, influence the level of food consumption without directly influencing hunger and/or satiety mechanisms. The relative scarcity of drugs that enhance food consumption may, in part, reflect the

fact that there are many non-specific ways in which food consumption can be disrupted (malaise, response competition, motor impairment, sensory impairment, etc.), but relativley few non-specific ways of enhancing food consumption (social facilitation, non-specific arousal and motor facilitation). It follows that understanding the mechanisms underlying changes in food intake requires more information than is provided by a simple measure of how much food is consumed.

In the free-feeding rat, eating is most likely just after the beginning and just before the end of the dark phase (Richter, 1927). Within these meals a number of eating bouts may be distinguished, between which there are gaps in which other behaviours occur. Feeding behaviour has most often been subjected to microstructural analysis following treatment with amphetamine or fenfluramine, drugs that both reduce food intake, but which do so by altering different feeding parameters. Amphetamine has been consistently reported to increase the latency to eat in deprived rats, but to increase the rate of eating within bouts: the net reduction in consumption results from a reduction in the number and duration of eating bouts. By contrast fenfluramine does not influence latency to feed but does reduce both bout length and eating rate (Blundell *et al.*, 1980). There is now some consensus that amphetamine anorexia can be explained to a large extent in terms of response competition from amphetamine-induced motor behaviours (see Lyon & Robbins, 1975) rather than as a specific effect on feeding mechanisms.

Looking beyond the microstructure of feeding bouts, observational studies have revealed that when hungry animals feed to satiety, they exhibit a characteristic sequence of behaviours known as the behavioural satiety sequence: after an initial period of eating they engage in exploratory behaviour and grooming, that then gives way to resting (Antin *et al.*, 1975). Manipulations that enhance satiety advance the onset of resting whereas other manipulations, such as quinine adulteration of the diet, reduce intake (presumably without enhancing satiety) but delay the onset of resting (Antin *et al.*, 1975; Montgomery & Willner, 1990). Thus, investigation of hunger and satiety may be better served by studying the sequencing of behaviours rather than by simply recording feeding episodes or the amount of food consumed.

Finally there have been numerous studies of the effects of various manipulations on the selection of diets that vary in their macronutrient content. For example it has been suggested that fenfluramine selectively suppresses carbohydrate consumption, leaving protein intake unaffected (Wurtman & Wurtman, 1979). Unfortunately, in the vast majority of these studies, changes in macronutrient content have been counfounded with changes in flavour or texture, so it is not clear whether manipulations of the independent variable influence macronutrient selection or, for example, flavour preference.

8.3.2 Central control mechanisms

It has long been known that tumours in the region of the hypothalamus are sometimes associated with gross obesity. The first experimental confirmation of the importance of the hypothalamus in the control of feeding and body weight was reported by Hetherington & Ranson (1939), who showed that lesions of the ventromedial hypothalamus (VMH) in the region of the ventromedial nucleus produced overeating (hyperphagia) and obesity. Destruction of the lateral hypothalamus (LH) on the other hand was later shown to cause aphagia and adipsia (Anand & Brobeck, 1951). These two observations led Stellar (1954) to propose that the LH is an excitatory 'feeding centre' that is inhibited by a 'satiety centre' in the VMH. Subsequent work at first appeared to support this hypothesis. For example, stimulation of the VMH inhibits feeding (Anand & Dua, 1955) and the LH and VMH do have reciprocal neural connections (Arees & Mayer, 1967), so that activity in one nucleus inhibits activity in the other (Oomura *et al.*, 1967).

However, a number of more recent observations have questioned the adequacy of this proposal. Among these is the finding that damage outside the hypothalamus can produce syndromes similar to those seen after damage to the LH or VMH. There are a number of similarities, including aphagia, between LH-lesioned rats and rats with lesions of the nigrostriatal DA pathway. The fact that the nigrostriatal pathway, which appears to be concerned with motor output, is often damaged by LH lesions casts doubt on the assumption that the LH is a hunger centre. Similarly lesions of the ventral

NA bundle (VNAB) (which projects, in part, to the paraventricular nucleus (PVN) of the hypothalamus) mimic many aspects of VMH lesions, including hyperphagia (Kapatos & Gold, 1983), while lesions restricted to the VMH produced neither hyperphagia nor obesity (Gold, 1973). These data effectively refute the 'two-centre' hypothesis of hunger and satiety and indicate that it may be more appropriate to consider the importance of neural pathways and more particularly the areas of hypothalamus to which these pathways project. This tactic has been realised by investigating the effects of administering pharmacological agents to discrete hypothalamic areas. There have been numerous reports of feeding elicited by drugs introduced directly into the hypothalamus. Grossman (1962) induced feeding in satiated rats with hypothalamic administration of 1-NA and Booth (1968) found that this effect was blocked by hypothalamic administration of the alpha-adrenergic antagonist phentolamine. Subsequently Leibowitz (1975) obtained feeding by injecting the alpha-adrenergic agonist, clonidine into the PVN and Martin & Myers (1975) reported large changes in NA turnover and release in the PVN during feeding, thus demonstrating a correlation between feeding behaviour and the release of endogenous hypothalamic NA. Stimulation of beta-adrenoceptors or DA receptors in the perifornical (PFH) region of the LH, on the other hand, reduces food intake and these hypophagic responses can be blocked by the beta-adrenergic antagonist propranolol, or DA antagonists, respectively (Leibowitz, 1973, 1980). In view of these results there can be little doubt that alterations in dopaminergic or noradrenergic transmission in specific hypothalamic regions can influence feeding, but it remains to be demonstrated convincingly that these effects result from manipulations of hunger or satiety mechanisms.

There has also been sustained interest in the possible role of brain 5-HT in feeding behaviour. Early studies failed to find changes in feeding with topical administration to the LH and VMH (see e.g. Wagner & DeGroot, 1963), but Leibowitz & Papadakos (1978) found that 5-HT did suppress feeding when injected into the PVN. Systemic administration of drugs that enhance 5-HT release (e.g. fenfluramine) reliably reduce feeding (for review see Blundell, 1977). However, until recently there were few examples of enhanced feeding responses with treatments that reduce the availability of 5-HT at post-synaptic receptors. Lesions of central 5-HT pathways have occasionally been reported to enhance feeding (Waldbillig *et al.*, 1981), but others have failed to find changes in food intake (Samanin *et al.*, 1972; Carey, 1976; Heym & Gladfelter, 1982). Enhanced feeding following treatment with 5-HT$_{1A}$ agonists (e.g. 8-OH-DPAT) has been reported (see Dourish *et al.*, 1985), but more detailed analysis of 8-OH-DPAT hyperphagia has questioned the suggestion that 8-OH-DPAT acts as a specific orectic agent (Montgomery *et al.*, 1988; Muscat *et al.*, 1989). Overall the picture with 5-HT is rather confusing due to the contradictory results obtained by different workers. Part of the reason for this may be that systemically administered 5-HT drugs influence 5-HT transmission in both the peripheral and the central nervous system and it is now well established that increases in peripheral 5-HT transmission reliably reduce feeding (Pollock & Rowland, 1981; Montgomery *et al.*, 1986). Consequently much of the evidence suggesting a specific role for central 5-HT in feeding behaviour may require re-interpretation.

8.3.3 Peripheral neurochemistry of feeding

In addition to the central mechanisms that must play a role in directing behaviour and integrating information about the nutritional status of the organism, it is evident that there must be signals emanating from the periphery, conveying information about gut distension and the caloric value of stomach contents that have yet to be absorbed. There have been numerous suggestions regarding which neurochemicals might serve as 'satiety signals' and among these 5-HT and various peptides, most notably cholecystokinin (CCK), stand out as strong candidates. Both 5-HT and CCK are released during feeding (Drapanas *et al.*, 1962; Della-Fera *et al.*, 1981), and when administered systemically suppress food intake (Pollock & Rowland, 1981; Antin *et al.*, 1975) and advance the onset of post-prandial resting (Antin *et al.*, 1975; Montgomery & Willner, 1990). However, 5-HT hypophagia is enhanced by vagotomy (Fletcher & Burton, 1985), whereas CCK hypophagia is blocked by vagotomy (Smith *et al.*, 1981); there is also evidence to suggest that CCK hypophagia

may result from non-specific malaise (Deutsch & Hardy, 1977). One possible mechanism for the hypophagic effect of 5-HT is its ability to block insulin secretion (Montgomery & Burton, 1986).

8.4 ANIMAL MODELS OF OBESITY

This section presents many of the well-documented manipulations that have been shown to increase body weight (usually in rats) and assesses their validity as animal models of obesity in the light of the information presented in the preceding sections. Models are assessed in terms of their ability to simulate specific symptoms, the contribution they make to our understanding of the aetiology and mechanisms underlying the clinical condition, and their possible role in the evaluation of pharmacological treatments.

8.4.1 Surgical and environmental manipulations producing weight gain

8.4.1.1 Hypothalamic lesions

The VMH-lesioned rat is probably the most extensively researched animal model of obesity. Lesioning the VMH results in an immediate increase in food intake, and weight gain. The dynamic phase of hyperphagia and weight gain is followed by a static phase in which obesity is maintained by normal levels of food intake. During this static phase VMH rats are finicky; they are reluctant to work for food and will not eat if their food is paired with electric shock administration, or adulterated with unpleasant flavours such as quinine (Ferguson & Keesey, 1975; Franklin & Herberg, 1974). However, if rats are exposed to these types of challenge prior to lesioning, then overeating and obesity occur after lesioning in the absence of finickiness. This indicates that hyperresponsiveness and obesity are dissociable (Grossman, 1979). Similarly if, prior to lesioning, rats are caused to gain weight (by forced feeding) then post-lesion hyperphagia may not be evident – the rats simply eat enough to maintain their elevated body weight (Teitelbaum, 1961). VMH-lesioned rats secrete more insulin, which might explain their hyperphagia and weight gain, particularly as post-lesion vagotomy prevents obesity. However, if vagotomy is performed prior to lesioning then hyperphagia and obesity still occur, suggesting an alteration to the metabolic processes that is independent of altered insulin secretion (Grossman, 1984). The fact that VMH-lesioned rats in the static phase of obesity maintain their increased body weight with normal levels of food consumption indicates that they tend to store nutrients more readily as adipose tissue, and to be less readily able to break down that adipose tissue to meet energy needs.

Lesions of the ventral NA bundle also induce hyperphagia and obesity. However, they do so less reliably than VMH lesions and, unlike VMH-lesion obesity, VNAB-lesion obesity is not associated with finickiness (Hernandez & Hoebel, 1980). The observation that the effects of VNAB and VMH lesions are additive strongly suggests that the underlying mechanisms are different (Ahlskog *et al.*, 1975). The critical VNAB fibres appear to terminate in the PVN (Gold *et al.*, 1977). Electrolytic lesions in the PVN, particularly its caudal region, also cause overeating and weight gain in rats. Again, however, the weight gain following PVN lesions is modest compared to that produced by VMH lesions (Leibowitz *et al.*, 1981). The relevance of these findings to human obesity remains to be determined, though there are reports of low NA levels, possibly indicating increased NA release, in the PVN of genetically obese rats (Cruce *et al.*, 1976).

In terms of magnitude and reliability of the effect, the VMH lesion appears to offer the most useful NA depletion model of obesity. VMH lesion obesity appears to be a good model of obesity in which weight gain results from increased intake, while maintenance of elevated body weight occurs in the absence of continued overeating. From a purely practical point of view the apparently permanent large increase in body weight that results from VMH lesions makes this an ideal model for testing the effects of pharmacological interventions. However, the observation that genetically obese rats have low hypothalamic NA levels indicates that all three types of lesion warrant further investigation.

8.4.1.2 *Depletions of brain serotonin*

While electrolytic lesions of the midbrain 5-HT (raphe) nuclei appear not to alter food intake (Samanin *et al.*, 1972; Carey, 1976; Heym & Gladfelter, 1982), blockade of 5-HT synthesis by intraventricular injections of parachlorophenylalanine (PCPA), has been found to increase the selection of protein rich foods and to cause a daytime hyperphagia, resulting in weight gain (Breisch *et al.*, 1976). However, PCPA-induced hyperphagia can be dissociated from forebrain 5-HT depletion (Coscina *et al.*, 1978; Mackenzie *et al.*, 1979). Nevertheless, the more specific technique of injecting a 5-HT neurotoxin directly into the VMH has revealed a long-lasting hyperphagia in rats consuming a high-fat diet, that was mainly apparent during the daytime (Waldbillig *et al.*, 1981). The effect of this daytime hyperphagia, combined with an increase in nutrient extraction, was to increase adiposity despite the fact that the lesioned animals were more active.

Although changes in diet selection and metabolic efficiency can be seen as positive aspects of these more specific 5-HT depletion models, the shift in the temporal distribution of eating and more particularly the increase in locomotor behaviour are aspects that are not typically associated with obesity in humans. Further investigation of these models, particularly with regard to their effects on metabolic efficiency, should help to determine the extent of their usefulness as models of obesity in humans.

8.4.1.3 *Dietary obesity*

The majority of laboratory rats are maintained on a single, nutritionally sound, composite diet and despite their largely sedentary existence they maintain a relatively stable body weight. Simply giving such animals access to a running wheel increases their energy expenditure and paradoxically reduces their food consumption (Mayer *et al.*, 1955; Berg, 1960), perhaps suggesting that the typical laboratory rat is mildly obese. Nonetheless laboratory rats will put on impressive amounts of weight if they are given access to more palatable calorically dense foods (Sclafani, 1980). Supplementing a chow diet with a sucrose solution for example, reduces chow intake, but insufficiently to compensate for the extra calories consumed (Kanarek & Hirsch, 1978).

Dietary obesity results from a combination of increased caloric intake and an increase in the proportion of energy stored as adipose tissue (Schemmel & Mickelson, 1974). Rats made obese in this way exhibit a number of behavioural similarities to VMH-lesioned rats. They are finicky and unwilling to work for food; further parallels include greater weight gains in older rats (perhaps indicating an age-related increase in both models in responsiveness to the palatability of food), and the failure, in both models, of prior adult experience of being overweight to influence subsequent weight gain (Sclafani & Gorman, 1977). If rats made obese by exposure to a high fat diet are returned to standard laboratory chow they lose weight, but continue to weigh more than rats that have never had access to the high fat food (Rolls *et al.*, 1980). This maintenance of an elevated bodyweight on a bland diet may reflect an increase in the number of fat cells. Forced exercise militates against dietary obesity, but only as long as the exercise is maintained (Pitts & Bull, 1977). Nevertheless, the fact that weight gain through dietary manipulations is largely reversible by returning obese rats to a diet of bland laboratory chow, holds out hope for treatment of the obese human, despite probably permanent increases in the number of adipose cells.

The socalled 'supermarket' or 'cafeteria' diets in which rats are given access to a range of calorically-dense foods, such as peanut butter and chocolate, in addition to laboratory chow also produce impressive weight gain, although it can be reduced by giving access to an activity wheel (Sclafani & Springer, 1976). These procedures more closely resemble the typical human situation. Rats exposed to 'isocafeteria diets', in which the nutritional content of the available food is constant, but variety is obtained by the addition of flavours, odours or changes in texture, also become obese (Louis-Sylvestre *et al.*, 1984). Even rats that have stable elevated body weights as a result of VMH lesions will increase their body weight when given access to a supermarket diet (Sclafani *et al.*, 1979). Taken as a whole these results have led Sclafani (1980) to suggest that both VMH-lesioned rats and dietary obese rats have increased appetites (rather than increased hunger or body weight set point) and that the VMH obesity syndrome can be viewed as an exaggerated form of dietary obesity. However, unlike hypothalamic obesity, dietary obesity is not prevented by vagotomy (Gold *et al.*,

1980) even in VMH-lesioned rats (Sclafani *et al.*, 1979).

The increased cephalic phase of insulin secretion that occurs in response to more palatable foods could explain many of these dietary phenomena. Under the cafeteria regime, meals last longer, but inter-meal intervals are reduced and there is no change in meal frequency (Rogers & Blundell, 1980). Variety enhances insulin secretion which in turn facilitates metabolic storage, increasing adiposity and shortening the interval to the next meal (Louis-Sylvestre *et al.*, 1984). Systemically administered 5-HT, which inhibits insulin secretion, suppresses consumption of a variety of palatable diets (Montgomery & Burton, 1986).

One effect of diet-induced hyperphagia is to elevate the metabolic rate through an increase in heat production by brown adipose tissue (BAT). Beta-adrenergic antagonists (Rothwell & Stock, 1979*a*) and neuroleptics (Rothwell & Stock, 1982) inhibit this effect suggesting that heat production in BAT is stimulated by endogenous NA and DA. Indeed, turnover of both NA and DA in BAT is greatly increased in cafeteria fed rats. Particularly interesting is the fact that systemic administration of DA (which does not cross the blood–brain barrier) stimulates BAT thermogenesis, raising the possibility that a peripheral mechanism could be exploited to increase metabolic rate (this vol., Ch. 9).

A potentially interesting non-pharmacological method of reversing diet-induced hyperphagia and weight gain was reported recently by Geliebter *et al.* (1987). Rats fed a high fat diet showed weight increases of 140 g (23%) over four months, compared to chow-fed controls. Balloons were then passed orally into the stomach and inflated. Food consumption was reduced by approximately 35% and the weight difference between the two groups declined by approximately six grammes per day. These figures are sufficiently striking to suggest that diet-induced obesity might be completely reversible by use of gastric balloons.

The impressive weight gain and the obvious face validity of these dietary manipulations hold out the hope that they should be valuable tools for generating treatments for human obesity. Despite the fact that dietary models may produce irreversible increases in the number of fat cells, the induced obesity does appear to be largely reversible. It may be that manipulations targeted at peripheral mechanisms (e.g. gastric balloons, inhibition of cephalic phase insulin secretion, and enhancement of thermogenesis) provide an avenue for progress that avoids many of the problems associated with centrally acting drug treatments. The marked similarity between dietary obesity and VMH lesion obesity also serves to lend support to the value of the VMH-lesioned rat as an animal model of obesity.

8.4.1.4 Social isolation

Rats raised in social isolation tend to be heavier as adults than rats reared in groups. This increase in body weight is at least partly explained by daytime hyperphagia (Morgan & Einon, 1975). Rats reared in isolation also show greater behavioural activation in response to novelty (Morgan, 1973; Sahakian et al., 1977) and an enhancement of tail pinch-induced eating (Sahakian & Robbins, 1977). A number of lines of evidence indicate that these behavioural changes are associated with elevated catecholamine transmission (Segal et al., 1973; Thoa et al., 1977; Weinstock et al., 1978; Einon & Sahakian, 1979). Cafeteria feeding has been shown to suppress activity in socially reared rats, but to enhance activity in isolates; isolates also exhibited different food preferences to those found in group-reared rats. However, cafeteria feeding did not produce greater weight gain in isolates than in group-reared rats (Sahakian et al., 1982). Thus, isolates appear to be predisposed to overeat in response to mild stress, but the failure of isolation rearing to enhance cafeteria feeding questions the value of this procedure as a widely applicable model of human obesity.

8.4.1.5 Genetic obesity

In both the animal and human literature there are examples of obesity resulting from single genes, e.g. the obese Zucker rat and humans with the Prader-Willi syndrome (Zucker & Zucker, 1961; Clarren & Smith, 1971). Studies of human obesity provide evidence of an appreciable genetic contribution (Biron et al., 1977), but in most cases the heritability of obesity appears most likely to take the form of a polygenic predisposition that interacts with environmental factors such as overfeeding during infancy and overresponsiveness to

external cues associated with food (Foch & McClearn, 1980).

The obese Zucker rat is the most commonly studied form of genetic obesity and has been proposed as an appropriate model of juvenile-onset human obesity. It has near normal body growth associated with excessive fat deposits, resulting from both an increase in fat-cell number and fat-cell size (Keesey, 1980). The obese Zucker rat has been characterised as showing normal defence of an elevated body-weight set point. Like many obese humans, it defends its body weight against caloric dilution and reductions in diet palatability (Cruce et al., 1974). Unlike VMH-lesioned rats, it is prepared to work for food (Greenwood et al., 1974). Obese Zucker rats maintained on a restricted diet (70% of free feeding intake) gradually lose about 6% of their body weight but this is asociated with a decline in resting metabolic rate of almost 15% (Keesey & Corbet, 1984). On the other hand they become hyperphagic in response to a cafeteria diet and put on even more weight (Gale et al., 1981). Obese Zucker rats have lower levels of NA in the PVN (Cruce et al., 1976) and lower levels of DA in the dorsomedial nucleus (DMN) (Levin & Sullivan, 1979), but whether these differences are causes or effects of overeating is not clear. They are also more susceptible to the anorectic action of amphetamine than lean rats (Grinker et al., 1980) and less susceptible to the anorectic effects of CCK (McLaughlin & Baile, 1980).

The various strains of genetically obese rats differ in their susceptibility to particular methods of inducing weight gain. For example the S5B/P3 rat is much less prone to increase its body weight when fed a high fat diet than the Osborne-Mendel rat (Schemmel et al., 1970), although it does exhibit VMH-lesion obesity (Oku et al., 1984). These two strains show similar hypophagic responses to naloxone and amphetamine, while the hyperphagic response to insulin and 2-deoxy-d-glucose is greater in the S5B/P1 rat (Fisler & Bray, 1985). This may be related to the fact that Osborne-Mendel rats are hyperinsulinaemic while S5B/P1 rats are not (Schemmel et al., 1982).

In principle, there is no reason why a particular strain of genetically obese animal should not be an excellent model of a particular type of human obesity, but at present too little is known about the genetics of human obesity to make these judgements.

8.4.1.6 Early overfeeding

Increases in adiposity of normal weight human infants are due to fat-cell enlargement (hypertrophy) during the first year of life, but thereafter to an increase in the number of fat cells (hyperplasia) (Hager *et al.*, 1977). There is now evidence suggesting that moderate obesity is hypertrophic, while more extreme obesity is both hypertrophic and hyperplastic (Sjostrom, 1980). Hyperplastic obesity in adults may have a particularly poor prognosis (Krotkiewski *et al.*, 1977), and so should best be tackled during childhood to prevent further increases in fat cell numbers.

Early overfeeding in rats, for example by reducing litter size, produces higher weaning weights, which persist throughout the rest of life (Miller & Wise, 1975). Early overfeeding may result in a permanent hyperplasia that predisposes the adult rat to obesity; this effect is seen in both obese Zucker rats and their lean littermates (Johnson *et al.*, 1973). Studies in which overnutrition starts post-weaning have failed to find changes in fat-cell numbers (Hirsch & Han, 1969). However, fat cell number may be increased in adult rats by providing them with a highly palatable diet (Faust *et al.*, 1978), or by exposure to low temperatures (Thierriault & Mellin, 1971).

Although there are some differences between species in the effects of nutritional manipulations on adiposity (Bjornstorp *et al.*, 1974), it seems likely that both hypertrophic and hyperplastic obesity can be modelled in animals, thereby providing a means of establishing the best remedy for each type. However, it remains to be demonstrated that fat-cell numbers may be decreased, even when body weight is reduced, so the prevention of further increases in fat-cell numbers may be a more realistic aim.

8.4.1.7 Forced feeding

Rats can be made obese by tube-feeding them a proportion of their normal food intake (e.g. Rothwell & Stock, 1979*b*). As might be expected, tube-fed rats reduce their voluntary intake so as to compensate for the food delivered by tube, but this attempt to regulate intake is unsuccessful (Fabry, 1969). Upon withdrawal of tube-feeding, they eat less than controls, and their body weight returns to normal, mainly as a result of a decrease in body fat (Rothwell & Stock, 1978). This apparent switch from controlling intake to controlling body weight is intriguing and warrants further investigation.

Weight loss in rats after cafeteria feeding results in part from an increase in energy expenditure. However, weight loss after tube-feeding results entirely from hypophagia (Rothwell & Stock, 1979*b*). This difference indicates that it may be important to know why weight gain has occurred in order to propose effective ways of encouraging weight loss.

Tube-feeding in monkeys has also produced obesity (Jen & Hansen, 1984), but in some subjects weight loss after the termination of tube-feeding was incomplete, suggesting that obesity produced by this technique is not always reversible.

8.4.1.8 Electrical stimulation of the brain (ESB)

Repeated electrical stimulation of the lateral hypothalamus can produce both hyperphagia and obesity in rats (Steffens, 1975). Termination of brain stimulation is followed by a period of hypophagia and a gradual return to normal body weight. Blood glucose levels are elevated during the stimulation-induced hyperphagia and below normal during the recovery period. Despite low plasma glucose, plasma insulin is elevated during recovery, as are plasma fat metabolites (Steffens, 1975).

There are many parallels between deprivation-induced and brain stimulation-induced eating. Eating in response to both types of stimulation can be inhibited by quinine adulteration of food, or by intragastric feeding, and enhanced by increasing the palatability of the available food (Tenen & Miller 1964; Hoebel & Thompson, 1969; Smith, 1972). Both ESB and food deprivation will support the learning of operant responses to gain access to food (Mendelson & Chorover, 1965). However, ESB-induced eating can be rigid, and stimulated animals accustomed to eating pellets will sometimes drink water rather than eat powdered food in the absence of pellets (Valenstein *et al.*, 1968), suggesting that the behavioural response to ESB may be partly explained in terms of non-specific arousal. Some of the discrepancies between ESB- and deprivation-induced feeding may reflect the mixed appetitive and aversive effects of ESB: stimulus-bound eating can be facilitated in 'non-eaters' by treatment with anxiolytic drugs (Soper & Wise, 1971), and some of the inappropriate responses seen following ESB can be reproduced by testing food-deprived rats under circumstances designed to make them 'emotional' (Wise & Erdmann, 1973).

Three characteristics of ESB-induced feeding and weight gain suggest that ESB might be more appropriate as a model of a bulimic episode rather than as a model of obesity. Firstly, bulimic episodes can be precipitated by stressful or arousing events. Bulimics might have difficulty in distinguishing between hunger and other forms of arousal, in which case a bulimic episode could be characterised as an inappropriate behavioural response to stress. Secondly, the period of hypophagia that follows termination of ESB parallels the sequence of events in BN where a bulimic episode is followed by strict restriction of further intake. Thirdly, ESB appears to produce both appetitive and aversive effects that may be analogous to the experience of bulimics, who yield to an overpowering urge to eat and then engage in subsequent restriction of intake to reduce their fear of weight gain.

The shortcoming of this account, however, is that it fails to encompass the obsessive need of bulimics to pursue a particular level of body weight and provides no parallel for vomiting as a means of avoiding the consequences of bulimic episodes.

8.4.1.9 Tail pinch

Application of mild pressure to the tail of a rat (tail pinch: TP) reliably elicits eating in sated rats (Antelman & Szechtman, 1975) and repeated TP stimulation causes both hyperphagia and obesity (Rowland & Antelman, 1976). TP-induced eating has been characterised as a response to non-specific activation, having parallels with both ESB- and food deprivation-induced behaviours: responsiveness to external stimuli is increased, such that salient food-related stimuli direct behaviour towards eating and induce cephalic

phase responses that increase both the activation and the eating response (Robbins & Fray, 1980).

The suggestion that many obese people eat in response to stress (Slochower, 1976) is contentious (Abramson & Wunderlich, 1972; Spitzer et al., 1980). However, if this is the case then TP- or ESB-induced feeding may provide valuable animal models that can be used to investigate the efficacy of drug treatment in reducing maladaptive responses to stress. TP-induced behaviours, including eating, are mediated in part by the nigrostriatal DA pathway (as are some ESB-induced behaviours) (see, e.g. Phillips & Fibiger, 1976). This provides a possible starting point for pharmacological investigations, although the involvement of DA in motor output may limit the usefulness of DA manipulations as treatments for feeding disorders. TP-induced behaviours are also influenced by noradrenergic manipulations (Antelman et al., 1975) and benzodiazepines (Robbins et al., 1977).

TP is also associated with the release of endogenous opiates (Morley & Levine, 1980); intraventricular opiate administration enhances food consumption and this effect can be blocked by the opiate antagonist naloxone (Morley & Levine, 1981), as can TP-induced eating (Morley & Levine, 1980). Demonstrations of insensitivity to pain during tail-pinch, and of behaviour similar to that seen during opiate withdrawal after termination of chronic TP, are also compatible with the idea that TP induces eating by releasing endogenous opiates (Morley & Levine, 1980). Naloxone also reduces the feeding responses to intraventricular injection of NA (Morley et al., 1982). This suggests that TP may stimulate the release of endogenous opiates, which modulate hypothalamic NA transmission, and facilitate a feeding response that depends upon nigrostriatal DA for its behavioural expression.

The marked similarities between TP- and ESB-induced behaviours suggest that, as with ESB, TP may be more appropriately regarded as a model of bulimia, rather than as a model of obesity.

8.4.2 Pharmacological models of obesity

8.4.2.1 Antipsychotic drugs

Weight gain is one of the unwanted side-effects of chronic antipsychotic treatment (Doss, 1979).

Neuroleptic drugs are DA-receptor antagonists; low doses of neuroleptics, particularly atypical neuroleptics (e.g. clozapine) dose-dependently increase feeding in rats (Leibowitz, 1980). Since there is a good correlation between the hyperphagic effects of neuroleptics and their ability to inhibit ^{3}H-haloperidol binding to DA receptors it seems that DA-receptor blockade underlies the observed hyperphagia and weight gain (Lawson *et al.*, 1984). Chronic treatment with the DA antagonist sulphiride causes hyperphagia and weight gain, and both effects are blocked by treatment with the DA agonist bromocriptine. In addition, withdrawal from sulpiride is associated with a period of undereating and weight loss (Baptista *et al.*, 1987).

These effects may be mediated by the PFH since direct application of DA antagonists to this hypothalamic region stimulates feeding (McCabe & Leibowitz, 1980). Self-stimulation from the PFH is also blocked by DA antagonists (Wise, 1978), suggesting that blockade of DA receptors in the PFH might increase the drive to eat by depriving the hungry rat of the rewarding effects of food consumption. However, this account of neuroleptic action is controversial (Wise, 1982). In any case, since DA agonists have abuse potential and alter motor output it is unlikely that they could be satisfactory as treatments for eating disorders.

8.4.2.2 8-hydroxy-DPAT (8-OH-DPAT)

While treatments that enhance 5-HT transmission (e.g. fenfluramine) reliably suppress food intake, there are fewer demonstrations that treatments which reduce 5-HT transmission enhance feeding. Recently, however, there have been numerous reports of enhanced feeding following treatment with low doses of 8-OH-DPAT, which reduce 5-HT transmission by stimulating presynaptic 5-HT$_{1A}$ autoreceptors (Dourish *et al.*, 1986). However, the observation that 8-OH-DPAT does not reliably enhance consumption of caloric fluids, but does facilitate gnawing on a wood block (Montgomery *et al.*, 1988), suggests that 8-OH-DPAT hyperphagia, like TP hyperphagia, may be a response to non-specific arousal. 8-OH-DPAT hyperphagia is blocked by DA antagonists, suggesting that 8-OH-DPAT elicits eating by reducing the inhibitory effect of 5-HT

transmission on DA transmission (Muscat *et al.*, 1989).

The similarities between 8-OH-DPAT hyperphagia and TP-induced feeding suggest that 8-OH-DPAT, like TP-induced feeding, may serve better as a model of bulimia rather than obesity, particularly as the hyperphagic response to 8-OH-DPAT shows rapid tolerance (Montgomery & Smer, unpublished finding).

8.4.2.3 Hypothalamic NA

Feeding may be elicited by the application of NA or alpha-adrenergic agonists (e.g. clonidine) directly into the PVN (McCabe *et al.*, 1981). This effect may result from inhibition of the 'VMH satiety function' (Hoebel, 1977) rather than the stimulation of hunger (Leibowitz, 1982): application of NA to the PVN delays the onset of postprandial resting as well as enhancing feeding (Towell *et al.*, 1989). This hyperphagia depends upon corticosterone levels (Leibowitz *et al.*, 1984) and resembles natural feeding in some ways (Leibowitz, 1976), but differs in others (Leibowitz, 1982). As alpha-adrenergic stimulation of the PVN also enhances carbohydrate selection at the expense of protein and fat consumption (Tretter & Leibowitz, 1980) it seems likely that the PVN might play an important role in feeding disorders, particularly those characterised by carbohydrate craving.

8.4.2.4 Anxiolytics

Anxiolytics in general, and benzodiazepines (BZs) in particular, stimulate eating, sometimes quite markedly; the mechanism of these effects is not well understood. BZs have been shown to reinstate eating in satiated rats (Fratta *et al.*, 1976), to potentiate feeding in response to hypothalamic stimulation (Soper & Wise, 1971) and tail-pinch (Robbins *et al.*, 1977), and to increase consumption of novel foods (Poschel, 1971). There has been some debate as to whether the hyperphagic effects of BZs represent a genuine effect on a feeding mechanism or rather an index of their anxiolytic efficacy (e.g. Cooper, 1980). Certainly in a number of paradigms (e.g. electrical stimulation of the hypothalamus, tail-pinch, novel food) it could be argued that BZs act to reduce an aversive aspect of the test situation. Cooper & Estall (1985) have pointed out that midazolam exerts a powerful facilitatory effect on consumption of palatable foods and have suggested that this BZ specifically stimulates appetite. However, chlordiazepoxide also stimulates chewing on a wood block, so at least some of the hyperphagic effects of BZs may reflect a facilitation of chewing (Posadas-Andrews *et al.*, 1983).

BZ-induced eating may be mediated by a hypothalamic mechanism since injections of diazepam into the VMN or flurazepam into the PVN increase food consumption (Anderson-Baker *et al.*, 1979; Kelly & Grossman, 1979). These effects may reflect an enhancement of GABAergic transmission (Costa & Guidotti, 1979); GABA agonists injected into the ventromedial hypothalamus or the PVN also increase feeding (Grandison & Guidotti, 1977; Kelly & Grossman, 1979).

Given the uncertainty about the mechanism through which anxiolytics enhance feeding, the relevance of these effects is unclear. However, recent concern over tolerance and addiction to BZs, together with the probability that inverse agonists acting at the BZ receptor might have anxiogenic properties, appear to limit the likely therapeutic usefulness of this category of drugs.

8.4.2.5 *Opiates and opioid peptides*

The discovery in recent years of endogenous opioid peptides has stimulated interest in the possibility that opiate receptors may be involved in the control of feeding behaviour. There is no shortage of evidence that opiate receptor antagonists reduce food intake (see Section 8.5.3.4), but the hyperphagic effect of opiate receptor agonists such as morphine and RX783030 (a selective agonist at the mu receptor), is limited to a low dose range, perhaps because higher doses induce incompatible behavioural responses (Cooper, 1981). However, injection of morphine into the VMH produces a robust hyperphagia, which appears to depend upon an adrenergic mechanism, since it is blocked by phentolamine (Tepperman *et al.*, 1981). This effect is also influenced by corticosterone manipulations (Bhakthavatsalam & Leibowitz, 1986). Furthermore, both beta-endorphin and the endogenous kappa receptor agonist dynorphin A produce hyperphagia when injected into either the

VMH or PVN (Grandison & Guidotti, 1977; Leibowitz & Hor, 1982; Gosnell *et al.*, 1986), and brain levels of dynorphin have been found to fluctuate with the nutritional state of the rat (Morley *et al.*, 1983). However, butorphanol tartrate (a kappa agonist) stimulates feeding in humans without increasing appetite (as measured by a visual analogue scale) (Morley *et al.*, 1985a).

Two other peptides, neuropeptide Y (NPY) and peptide YY have potent hyperphagic effects particularly when injected directly into the PVN (Clark *et al.*, 1984; Stanley & Leibowitz, 1984). Tolerance does not appear to develop with chronic administration of these peptides (Morley *et al.*, 1985b; Stanley *et al.*, 1985), which can override gut distension satiety signals. NPY appears specifically to increase carbohydrate intake (Morley *et al.*, 1985b), but this effect may reflect a change in flavour preference, rather than an increased requirement for a particular macronutrient.

Elevated pituitary levels of beta-endorphin have been reported in genetically obese mice and rats, although the obesity preceded the changes in endorphin levels (Margules *et al.*, 1978; Bray & York, 1979). Consumption of highly palatable foods is associated with an increase, and food deprivation with a decrease, in the release of beta-endorphin in the hypothalamus (Gambert *et al.*, 1980; Dum *et al.*, 1983). It has been suggested that morphine increases intake either by reducing reward threshold (Dum *et al.*, 1983), or by suppressing an aversive component (Siviy *et al.*, 1982), but the data supporting these interpretations are controversial (e.g. Cooper, 1983). An alternative view arising from observations of the behavioural satiety sequence is that opiate receptor agonists act by abolishing sleep, so that behaviours other than sleeping are necessarily increased (Cooper, 1981).

Opiate antagonists have been tested in a number of animal models of obesity. Obese VMH-lesioned rats reduce their food intake when treated with naloxone (King *et al.*, 1979), but so do VMH-lesioned rats that have been maintained at normal body weight through restricted access to food (Gunion & Peters, 1981). Zucker rats are less sensitive to the hypophagic effects of naloxone after their obesity has developed (McLaughlin & Baile, 1984), but in mature mice, both obese (ob/ob) and lean controls actually over-eat in response to chronic naloxone treatment (Shimomura *et al.*, 1982). More impressively long-acting naloxone zinc tannate reduces intake in cafeteria-fed rats and prevents dietary obesity (Mandenoff *et al.*, 1982). Unfortunately, the results of chronic treatment with naloxone in obese humans have been unimpressive (Levine *et al.*, 1985). The hyperphagic response of obese mice to chronic naloxone treatment, and the disappointing results with obese humans hold out little hope of opiate antagonists proving efficacious in the treatment of human obesity.

8.4.2.6 Antidepressants

Reduced appetite is a common symptom of depressive disorders (Halmi, 1985) and depressed patients maintained on the tricyclic antidepressant amitriptyline frequently show weight gain and carbohydrate craving (Paykel *et al.*, 1973). The neurochemical basis of these effects is poorly understood, but tricyclic antidepressants are known to alter monoamine neurotransmission, and injection of TCAs into the PVN causes a hyperphagia which is mediated via alpha-adrenergic receptors (Leibowitz *et al.*, 1978).

It is tempting to speculate that depression might be associated with a reduction in the incentive value of palatable foods; if so, treatment with antidepressants might reverse this effect to produce carbohydrate craving. However, the observation that weight gain correlates neither with increased appetite nor with recovery from depression (Kupfer *et al.*, 1977) suggests that this simple account will prove inadequate. Indeed, weight gain is seen in recovered patients who are maintained on tricyclic antidepressants (Paykel *et al.*, 1973) and certain antidepressants cause weight loss (Aberg & Holmberg, 1979; Ferguson, 1986).

Chronic systemic antidepressant treatment is typically associated with weight loss in animals (Nobrega & Coscina, 1982; Willner & Montgomery, 1980), so it is unlikely that the antidepressant-treated rat will provide a useful model of obesity.

8.4.2.7 Insulin

Injections of insulin lower blood glucose levels and repeated use of long-acting treatments (e.g. protamine zinc insulin, PZI) increases food consumption by up to 50% and causes weight gain (Hoebel & Teitelbaum, 1966). Upon withdrawal from PZI, fat metabolite levels in the plasma rise, and the rats become hypophagic with normal blood glucose levels, and lose weight (Carpenter & Grossman, 1983). Many obese humans exhibit hyperinsulinaemia as do genetically obese rats, and rats made obese by hypothalamic lesions (Bray & York, 1979). However, the direction of causality implied by this correlation is debatable. Hyperinsulinaemia and obesity can be dissociated: for example, prelesion vagotomy prevents hyperinsulinaemia following VMH lesions, but not the development of obesity (Grossman, 1984). Nonetheless the VMH-lesioned rat whose intake is restricted to that of controls still shows a relative weight gain, suggesting that hyperinsulinaemia does bias the lesioned rat towards fat deposition (Han, 1967).

It seems likely that a treatment that reduces insulin secretion might reduce food consumption and body weight by restricting appetite (by dampening cephalic-phase insulin secretion), increasing inter-meal intervals (by prolonging high blood glucose levels) and by biasing the metabolism against fat deposition. However, the effects of one such treatment (peripherally administered 5-HT, see Section 8.5.3.3) are comparatively modest.

8.4.3 Conclusions

With one exception (forced feeding) the various putative animal models of obesity depend upon manipulations that act to increase food acceptance. This may, in principle, result either from changes in the functioning of mechanisms specifically concerned with the control of feeding (i.e. increased appetite, increased hunger or reduced satiety), or from non-specific effects such as increased arousal or motor facilitiation.

Increasing dietary variety is the least invasive of these techniques. Weight gain associated with variety in the diet depends upon hyperphagia and an increase in the number of fat cells (Faust *et al.*, 1978), together with an increase in the proportion of energy stored as fat (Schemmel & Mickelson, 1974). Isocafeteria diets are also associated with increased metabolic efficiency (Louis-Sylvestre *et al.*, 1984). This pattern of effects might be predicted from the increase in cephalic phase insulin secretion that is associated with exposure to palatable foods (Louis-Sylvestre & Le Magnen, 1980), suggesting that the underlying cause is an increase in appetite. Systemically administered insulin increases food acceptance in rats, and long-acting insulin treatments cause weight gain (Hoebel & Teitelbaum, 1966). Since many obese humans are also hyperinsulinaemic, models involving increased insulin levels obviously have some face validity, but the direction of causality suggested by correlations between obesity and insulin levels remains to be determined. Nonetheless, obesity associated with hyperinsulinaemia might respond to treatments that reduce insulin secretion.

In humans, obesity often begins during middle age, and weight gain produced in rats by dietary manipulations is also more marked in older rats than it is in younger ones (Schemmel & Mickelson, 1974). This may reflect either an age-related malfunction in a satiety mechanism or an increasingly sedentary existence. If the former, then dietary obesity in the rat has parallels with Rodin's (1980) reformulation of the externality hypothesis, in which obesity depends upon increased responsiveness to the sensory properties of food (appetite) and a failure to reduce subsequent intake in response to weight gain.

Early overfeeding in rats produces obesity in adulthood that depends upon an increase in fat-cell number (Johnson *et al.*, 1973). In humans this type of obesity is associated with a particularly poor prognosis in terms of weight loss, so early overfeeding in rats could prove to be a useful model for testing strategies aimed at either weight loss or, more probably, the prevention of further weight gain.

Genetically obese Zucker rats also show increased weight gain when exposed to cafeteria diets (Gale *et al.*, 1981) and like obese humans, they fail to lose substantial amounts of weight even on restricted access to food (Keesey & Corbet, 1984). Obesity in Zucker rats is also associated with reduced hypothalamic catecholamine levels (Cruce *et al.*, 1976) so it seems reasonable to hope that hypothalamic damage might mimic the effects of some types of human genetic obesity. Indeed the VMH-lesioned rat bears a number of similarities to the obese human in terms of increased externality, a reluctance to work for food, and hyperinsulinaemia (Hales & Kennedy, 1964; Ferguson & Keesey, 1975). However, as with the reformulated externality hypothesis, weight gain should not follow from increased responsiveness to the sensory properties of food alone, since compensatory changes (such as an increase in the time to the next meal) should occur, unless the obese rat is also less responsive to interoceptive satiety signals.

Manipulations that deplete brain 5-HT have proved less reliable in inducing weight gain. The most impressive demonstration of weight gain followed 5-HT depletion from the VMH and this was associated with increased locomotor activity, which is not a typical characteristic of human obesity (Stern, 1984). Weight gain in socially isolated rats is also associated with behavioural activation, particularly in cafeteria-fed isolates (Sahakian *et al.*, 1982). However, the enhancement of TP-induced hyperphagia in isolates may shed light on the mechanisms underlying bulimic episodes. ESB- and TP-induced hyperphagia, like social isolation, may be more appropriately regarded as possible models of bulimic episodes. The fact that ESB may produce mixed appetitive and aversive effects holds out the possibility that ESB-induced eating may mimic the ambivalent attitude that the bulimic has towards a bulimic episode, and in addition may shed light on the possible role of stress as a precipitant of bulimic episodes. In a similar vein it may be that TP-induced eating can be characterised as an inappropriate response to stress. Similarities between 8-OH-DPAT- and TP-induced feeding suggest that $5-HT_{1A}$ agonists may elicit feeding through a non-specific increase in arousal, particularly as both effects are blocked by pretreatment with DA antagonists (Muscat *et al.*, 1989). The rapid development of tolerance to 8-OH-DPAT hyperphagia (Montgomery & Smer, unpublished finding) questions the usefulness of this manipulation as an animal model of obesity. However, it remains feasible that $5-HT_{1A}$ autoreceptors may play a role in the development of bulimic episodes.

Neuroleptic-induced hyperphagia appears to depend upon blockade of DA receptors in the PFH, but whether this entails an increase in drive, incentive or reward, or a suppression of satiety remains unclear. Whichever proves to be the case it is unlikely that enhancing DA transmission will prove to be an acceptable treatment strategy for obesity, since amphetamine is no longer regarded as a suitable tool for controlling body weight.

Direct application of alpha-adrenergic agonists to the PVN elicits increases consumption, particularly of carbohydrates, and reduces resting (Tretter & Leibowitz, 1980; Towell *et al.*, 1989). These effects are compatible with a suppression of satiety. Given that genetically obese rats exhibit altered hypothalamic levels of catecholamines (Cruce *et al.*, 1976), it seems that further investigation of this model might yield useful insights into some aspects of human obesity.

The mechanism by which BZs enhance food acceptability is poorly understood, but whatever the mechanism, it is unlikely that inverse agonists at the BZ receptor will provide an acceptable treatment for obesity. Similarly, although opiate antagonists are effective in preventing dietary obesity in rats (Mandenoff *et al.*, 1982) they have yielded disappointing results in obese humans. Some antidepressants induce weight gain in depressed people (Paykel *et al.*, 1973), but it seems most unlikely that antidepressant-treated animals will provide a useful tool for studying the mechanisms underlying obesity.

8.5 ANOREXIA NERVOSA

This section covers a variety of experimental techniques (surgical, pharmacological and behavioural) that have been shown to produce weight loss and/or hypophagia and assesses their value as animal models of AN.

8.5.1 Chronic models of AN

8.5.1.1 Lateral hypothalamus lesions

Lesions of the lateral hypothalamus (LH) render rats aphagic and adipsic; they also exhibit reduced motor output and sensory deficits. However, a gradual and partial recovery is seen if the LH-lesioned rats are maintained by intragastric feeding and then transferred to a palatable wet diet. Eventually, the lesioned rats are able to maintain a reduced body weight on normal dry laboratory food, although they remain finicky (Teitelbaum & Epstein, 1962). It may be that the motor and sensory deficits caused by LH lesions underlie the aphagia, because recovery is facilitated by repeated TP sessions that elicit eating (Antelman *et al.*, 1976) and amphetamine injections produce feeding

in LH-lesioned rats (Wolgin & Teitelbaum, 1976). These findings suggest that LH aphagia results from impaired arousal or attention. On the other hand, this account does not readily explain why food restriction and reduction of body weight prior to LH lesioning facilitates recovery (Powley & Keesey, 1970), or why recovered LH rats show poor responses to challenges such as glucoprivation precipitated by 2-deoxyglucose injection (Stricker *et al.*, 1979). However, recovered LH rats remain hypoinsulinaemic (Opsahl, 1977), suggesting that the cephalic phase of insulin secretion that has been taken as an index of palatability (Le Magnen, 1983) may be suppressed. Recovered LH rats also lose their feeding periodicity and become 'nibblers' rather than 'meal eaters' (Kakolewski *et al.*, 1971); this indicates that they initiate eating bouts frequently but terminate them prematurely – which would not be predicted to result from damage to a 'hunger centre'. It does seem that at least some of the abnormalities of LH rats can be described in terms of a reduced responsiveness to relatively weak stimuli, so that behaviour is initiated only by powerful (arousing?) stimuli and terminated prematurely as such a stimulus (e.g. hunger) begins to wane.

The similarities between LH-lesioned rats and the anorexic patient appear to be limited to hypophagia and loss of body weight. Other aspects of the LH syndrome stand in marked contrast to the symptoms of AN: LH-lesioned rats exhibit reduced motor output as distinct from the excessive exercise regimes exhibited by some anorexic humans, and several aspects of the syndrome suggest a reduction in appetite, whereas the recovered anorexic patient often confesses to having been desperately hungry during periods of dietary restraint.

8.5.1.2 Nigrostriatal DA lesions

Lesions of the nigrostriatal DA pathway, which projects from the substantia nigra to the striatum via the dorsolateral hypothalamus, produce many effects similar to those seen in the LH-lesioned rats. Neurotoxic or electrolytic lesions of the substantia nigra cause aphagia, adipsia, hypokinesia and sensory neglect (Ungerstedt, 1971) and, as with LH lesions, recovery from these deficits is gradual and incomplete (Zigmond & Stricker,

1974). Indeed, there is a good correlation between aphagia induced by LH lesion and that produced by DA depletion (Oltmans & Harvey, 1972), though similar deficits are also seen with lesions that spare DA neurones (Zigmond & Stricker, 1973) or that damage the trigeminal nerve (Zeigler, 1975). As with LH lesions, tail-pinch can facilitate recovery from nigrostriatal DA lesions (Antelman *et al.*, 1976). Overall it seems that there are many similarities between LH- and SN-lesioned rats and that sensory motor and arousal deficits may play an important role in both syndromes. However the finding that kainic acid lesions of the LH (that spare fibres of passage) produce a temporary aphagia suggests that the LH syndrome involves more than DA depletion (Grossman *et al.*, 1978).

The marked similarities between the effects of LH and nigrostriatal lesions suggest that the nigrostriatal-lesioned rat is unlikely to fare any better than the LH-lesioned rat as a model of AN.

8.5.1.3 *Dorsomedial hypothalamic nucleus (DMN) lesions*

DMN lesions produce weight loss, aociated with hypophagia, hypodipsia and hypoactivity, but normal body composition (Bernardis, 1972). Unlike LH- and VMH-lesioned rats those with DMN lesions respond to various challenges so as to defend a reduced bodyweight set point (Bellinger *et al.*, 1979), except that when offered a cafeteria diet DMN-lesioned rats become extremely hyperphagic (Bernardis *et al.*, 1980). So far it seems as though the DMN-lesioned rat is 'homeostatically competent', but 'scaled down' (Bernardis & Bellinger, 1987). The hyperphagic response of DMN-lesioned rats to cafeteria diets indicates that this preparation is unlikely to be a useful model of AN, though it may have some merit as a model of BN.

8.5.1.4 *Olfactory bulbectomy*

Lesioning the olfactory bulbs of rat pups during the first two weeks after birth has a high mortality rate, but survivors exhibit adult weights 20–30% lower than controls. Bulbectomised rats differ from controls on a number of drinking and taste measures (Hill & Almi, 1981); these effects depend upon sex and the age at which bulbectomy was performed. However, the only difference found in feeding tests was that male rats deprived of food for 24 h, then bulbectomised on day 10, subsequently overate relative to controls. There appears to be little to recommend this procedure as a model of AN beyond the superficial similarity in terms of reduced adult body weight.

8.5.1.5 *Early undernutrition*

Relatively brief periods of undernutrition during early life, followed by a long period of ad libitum feeding, result in adult rats having stunted body growth. This is especially the case if rats suffer undernutrition during the period of the brain growth spurt (Smart *et al.*, 1973). Such rats also have a smaller brain and a disproportionately smaller cerebellum (Smart, 1979). Early undernutrition is also associated with hyperactivity, hyperreactivity and faster running in a food-reinforced alleyway test. Since AN is not associated with stunted body growth there is little reason to believe that undernutrition will prove useful as a model of AN. Furthermore, AN is rarely found in circumstances where food availability is restricted.

8.5.1.6 *Vagotomy*

Rats subjected to total subdiaphragmatic vagotomy exhibit long-term reductions in food consumption and body weight (Mordes *et al.*, 1979), and fail to show the expected feeding inhibition in response to a number of challenges, including CCK and glucagon (Martin *et al.*, 1978; Smith *et al.*, 1981). The mechanism by which vagotomy suppresses feeding is poorly understood although the intact vagus may play a role in regulating carbohydrate consumption (Novin & Vander Weele, 1977). For example, using liquid macronutrients, vagotomy has been found selectively to suppress carbohydrate intake but not intake of fat or protein (Fox *et al.*, 1976). However, others report that vagotomy suppressed intake of sucrose solutions but not of sucrose granules (Sclafani & Kramer, 1983); thus, effects on macronutrient selection may reflect a change in flavour or texture preference rather than an alteration in nutrient requirement. Vagotomised humans reduce their consumption of both carbohydrate- and protein-containing foods (Faxen *et al.*, 1979) and report reduced appetite for sweets (Kral & Gortz, 1981).

Although anorectic humans do avoid particular types of food it seems that this is based on caloric value rather than nutrient content, and when restraint fails there is a tendency for the anorexic to seek out the foods that are otherwise avoided. It seems unlikely therefore that vagotomy will prove to be a useful model of AN.

8.5.2 Acute non-pharmacological models of AN

8.5.2.1 *Unconditioned and conditioned emotional responses*

Rats exhibit an unconditioned hypophagia when given access either to a novel food (neophobia) or to a familiar food in a novel environment (hyponeophagia). This suppression of intake has obvious survival value as it permits rats to sample small quantities of potentially dangerous foods, and subsequently to increase their intake (depending upon palatability) if the initial sampling is not followed by aversive consequences. These responses extinguish rapidly in the absence of aversive consequences (Miller & Holzman, 1981). Hypophagic responses can also be conditioned. Food-deprived rats, lever pressing on a variable interval schedule for food pellets, reliably suppress their rate of responding when a conditioned stimulus (CS) that has previously been associated with footshock is delivered (see e.g. Green & Hodges, Chapter 2 of this volume). Under these circumstances the anticipation of an aversive event serves to suppress responding. However, like neophobia, conditioned emotional responses (CERs) extinguish rapidly with the non-delivery of footshock. BZs with anxiolytic properties, which are known to enhance feeding, block the effects of novelty and suppress CERs (Randall *et al.*, 1960; Miczek, 1973; Stephens, 1973).

While these procedures may be useful as screening tests for anxiolytic drugs (see Stephens & Andrews, Chapter 3 of this volume), their relevance to feeding disorders is questionable. Parallels can be drawn between CERs and anorexia nervosa since both involve a suppression of feeding in response to the anticipation of aversive events (i.e. footshock and weight gain respectively). However, the validity of this parallel is under-mined by the ease with which CERs can be extinguished, as compared to the marked resistance to intervention exhibited by the anorexic and by the existence of a subgroup of anorexics who binge and then avoid weight gain by vomiting.

8.5.2.2 *Conditioned taste aversions (CTAs)*

When consumption of a novel food is paired with illness, induced by lithium or radiation poisoning, then subsequent consumption of that food is suppressed (Garcia *et al.*, 1955). This procedure can cause weight loss, but only if food availability is limited to the averted flavour, since the reduction in intake following CTA is specific to the taste paired with illness. There may be some superficial parallels between CTAs and the avoidance of calorically dense foods exhibited by anorexics, but beyond this there is little hope that CTAs will provide useful insights into AN.

8.5.2.3 *Quinine adulteration and caloric dilution*

Free-feeding normal rats adjust their food intake in terms of its caloric value so as to compensate for caloric dilution (e.g. the addition of cellulose to the diet) and changes in palatability. Compensation occurs regardless of whether changes in palatability are achieved by the addition of saccharin or of quinine to the diet (Jacobs & Sharma, 1969). An intact VMH appears to be necessary for this compensation, because VMH-lesioned rats become hypophagic in response to quinine adulteration (Ferguson & Keesey, 1975). However, intact rats maintained on a food-deprivation schedule reduce their consumption of quinine-adulterated foods (Jacobs & Sharma, 1969) just as VMH-lesioned rats do.

Differences in the responsiveness of free-feeding and food-deprived rats (Jacobs & Sharma, 1969) may provide useful insights into AN. For example, the restricting anorectic may be characterised as hungry and particularly sensitive to the palatability of food, rather than to its energy value. These characteristics are comparable to the VMH-lesioned rat during the dynamic phase of weight gain: the lesioned rat is below its body weight set-point, unusually sensitive to the sensory properties of food, and apparently insensitive to the energy value of ingested foods. The obvious difference between the anorectic and the VMH-lesioned rat is the absence, in the rat, of the strict application of cognitive control aimed at maintaining a low body weight. It is precisely this aspect of AN that makes the prospect of useful animal models appear remote.

8.5.2.4 *Food deprivation and exercise*

The most obvious symptom of anorexia nervosa is low body weight. However, it has been recognised for over 20 years that anorectic patients are often restless and overactive (Crisp, 1965; Slade, 1973). Overactivity developed during the course of self-starvation in almost 40% of sufferers in one study (Crisp *et al.*, 1980); with a return to stable eating, activity levels decline (Crisp, 1965). Epling *et al.* (1981, 1983) have tried to incorporate both weight loss and overactivity into an animal model of AN. They allowed 45-day-old rats access to food for one hour each day and free access to a running wheel for the remaining 23 h. Wheel running increased across days, while food consumption and body weight declined, relative to controls with a fixed wheel. Running mainly occurred for six hours prior to and after access to food (Woods, 1969). There are also human data supporting the idea that periods of exercise suppress food consumption (Johnson *et al.*, 1972).

Compared to the other potential models of AN this procedure has more face validity, at least for that subgroup of anorectics who exhibit elevated activity. However, the time course of the effects is reversed in the animal model with increased activity preceding weight loss. Further studies, particularly of the effects of drugs, should help to determine the value of this procedure as an animal model of AN.

8.5.2.5 *Oestrogen and immobilisation stress*

Oestrogen suppresses food intake in a number of species (Wade, 1976). Although it remains unclear whether the hypophagic response to oestrogen depends upon a central or a peripheral mechanism, Young (1975) has suggested that hypothalamic supersensitivity to oestrogen at puberty may be a primary factor in the aetiology of AN. Elaborating upon this idea Donohoe (1984) has suggested that the symptomatology of AN may be explained in terms of a combination of physiological predisposition and psychological stress. Unlike other forms of stress (e.g. TP) bodily immobilisation reduces, rather than enhances, food intake (Perhach & Barry, 1970). In one of the few studies overtly attempting to model AN, Haslam *et al.* (1987) have combined these ideas and investigated the effects of oestradiol and immobilisation stress in ovariectomised rats. Both oestradiol and immobilisation reduced eating and in combination these effects were additive. Oestradiol also reduced body weight, but the effects of immobilisation on body weight were less reliable. Treatment with cyproheptadine (a 5-HT antagonist) reversed the hypophagic effects of immobilisation, but not those of oestradiol. Immobilisation hypophagia was also blocked by low doses of the 5-HT$_{1A}$ agonist 8-OH-DPAT, providing further support for serotonergic involvement in immobilisation hypophagia (Kennett *et al.*, 1987).

This attempt to incorporate both physiological and psychological (stress) factors into an animal model of AN marks an ambitious and imaginative step forward. Clearly the finding that cyproheptadine influences intake in this model, as it does in AN (Halmi *et al.*, 1983) provides a basis for further work, as does the observation that central 5-HT transmission stimulates release of hypothalamic corticotrophin-releasing factor (CRF) (Jones *et al.*, 1976) which has been proposed to mediate stress-induced anorexia (Morley & Levine, 1982). However, the failure of cyproheptadine to reverse oestradiol- or stress-induced loss of body weight questions the value of this model in its present form. Further questions are raised by the existence of AN in males and in prepubertal children.

8.5.3 Pharmacological models of AN

8.5.3.1 *Manipulations of central catecholamines*

Amphetamine stimulates transmission at both NA and DA synapses, and in moderate or high doses suppresses food intake. Amphetamine anorexia appears to be mediated by the stimulation of dopaminergic and beta-adrenergic receptors in the PFH (see Hoebel & Leibowitz, 1981). However, in some circumstances very low doses of amphetamine can stimulate eating (Winn *et al.*, 1982) perhaps by producing a low level of activation similar to that seen with tail-pinch; hyperphagic responses are seen following treatment with low doses of the DA agonist PHNO (Martin-Iverson & Dourish, 1988). Lyon & Robbins (1975) have argued convincingly that anorectic doses of amphetamine may suppress feeding through response competition from locomotor activity and stereotypy. However, diabetic rats exhibit an attenuated anorectic response to amphetamine that can be reinstated by insulin administration (Marshall *et al.*, 1976), so it seems that amphetamine may act in part through a specific feeding mechanism.

Hypothalamic injections of NA and DA into the PFH suppress feeding (Leibowitz & Rossakis, 1979*a*; Leibowitz, 1980) and these hypophagic responses are blocked by beta-adrenergic and DA antagonists respectively (Leibowitz & Rossakis, 1979*b*). Measures of NA release in the PFH have shown that gastrointestinal nutrient loads are associated with an increase in the release of NA (Myers & McCaleb, 1980), indicating that stomach contents can influence events in the PFH. These findings are consistent with the idea that release of NA and/or DA in the PFH acts to potentiate satiety. However, more evidence is needed about the similarities and differences between naturally occurring satiety and PFH hypophagia.

The combination of hypophagia and increased motor activity that is seen with amphetamine treatment has some superficial similarity with AN, but the increased latency to eat following amphetamine administration has been interpreted as a suppression of hunger (Leibowitz *et al.*, 1986), thus questioning the validity of this preparation as a model of AN. The degree to which related manipulations suppress feeding by acting on a specific hunger, appetite or satiety mechanism, rather than through a nonspecific mechanism (e.g. motor competition), remains open. If it transpires that reduced hunger or enhanced satiety do underlie these effects then they are unlikely to provide a valid model of AN.

8.5.3.2 *Manipulations of central 5-HT*

Injection of 5-HT into the PVN blocks eating in response to food deprivation, or (even more effectively) in response to NA injections at the same site (Leibowitz & Papadakos, 1978). This effect may underlie the anorectic properties of peripherally administered fenfluramine, which increases the post-synaptic availability of central 5-HT (Samanin *et al.*, 1980) and reliably suppresses feeding. Fenfluramine anorexia is blocked by pretreatment with 5-HT antagonists such as methysergide (Blundell *et al.*, 1973). However, there has been a suggestion that fenfluramine anorexia may be mediated peripherally (Davies *et al.*, 1983), at least in part.

It has been claimed that fenfluramine suppresses carbohydrate intake while sparing protein intake (Wurtman & Wurtman, 1979), although others have had difficulty in replicating these results (McArthur & Blundell, 1982) and the interpretation of macronutrient selection paradigms is often confounded by parallel changes in taste or texture (see Section 8.3.1). In an exhaustive series of experiments, Even & Nicolaidis (1986) reported that fenfluramine-induced weight loss resulted from a combination of hypophagia, increased energy expenditure, and enhanced mobilisation and utilisation of endogenous fat stores, i.e. the hypophagic rat was also expending more energy and using its energy intake inefficiently. In contrast to this account Booth *et al.*, (1986) suggested that gastric slowing accounts for most of the hypophagic response to fenfluramine, by prolonging postprandial satiety. There is no doubt that fenfluramine reduces food intake, but the mechanism(s) mediating the anorectic action remain(s) to be clarified, and claims that fenfluramine potentiates satiety seem premature. The major effect of fenfluramine on feeding microstructure is to slow the rate of eating and decrease the size of meals (Blundell & Latham, 1978). However, fenfluramine also greatly suppresses post-prandial resting (Montgomery & Willner, 1988), which suggests that fenfluramine does not advance the onset of satiety. If fenfluramine did enhance satiety the case for the fenfluramine-treated rat as an animal model of AN would be weakened, since anorectics restrict their intake despite extreme hunger.

Recently-developed drugs have led to the classification of 5-HT receptors into a number of subtypes (5-HT_{1A}, 5-HT_{1B}, 5-HT_{1C}, 5-HT_2 and 5-HT_3) (Bradley *et al.*, 1986). From studies using specific receptor subtype antagonists it appears that fenfluramine anorexia is mediated by '5-HT_1-like' receptors (Neill & Cooper, 1989); hypophagic responses to m-chlorophenylpiperazine (mCPP) or 1-[3-(trifluoromethyl)phenyl]piperazine (TFMPP) depend upon both 5-HT_{1B} and 5-HT_{1c} receptors, while only 5-HT_{1B} receptors are involved in RU 24969-induced hypophagia (Kennett & Curzon, 1988). Further investigations should clarify the behavioural mechanisms by which these receptor subtype specific drugs influence feeding (see also Sepinwall & Sullivan, Chapter 9 of this volume).

The degree to which manipulations of central 5-HT alter feeding by acting on mechanisms specifically involved in the regulation of appetite, hunger or satiety is still uncertain. However, recent reports that the 5-HT antagonist cyproheptadine had beneficial effects in the treatment of non-bulimic anorexics clearly holds out the hope that 5-HT drugs may provide a useful tool in the treatment of AN. It is also interesting to note that cyproheptadine retarded improvement in a group of anorexics who exhibited bulimia (Halmi *et al.*, 1986), particularly as bulimic anorexics have been reported to have low central 5-HT turnover (Kaye *et al.*, 1984*b*).

8.5.3.3 *Manipulations of peripheral 5-HT*

5-Hydroxytryptophan (5-HTP) is the precursor of 5-HT and readily crosses the blood–brain barrier. Systemically administered 5-HTP reliably suppresses food intake (Joyce & Mrosovsky, 1964). Like fenfluramine, 5-HTP reduces eating rate and it also reduces the number of eating bouts (Blundell & Latham, 1979). Pretreatment with either benserazide (which blocks peripheral conversion of 5-HTP to 5-HT) or xylamidine (a peripherally acting 5-HT antagonist) attenuates 5-HTP anorexia, indicating that at least part of the anorectic action of 5-HTP is mediated peripherally (Fletcher & Burton, 1986*a*; Montgomery & Willner, 1990). Unlike fenfluramine, the effects of 5-HTP on the behavioural satiety sequence are compatible with an enhancement of satiety: 5-HTP slowed eating rate and advanced the onset of resting. However these effects were also reversed by pretreatment with benserazide, again suggesting mediation via peripheral rather than central 5-HT (Montgomery & Willner, 1990).

Systemically administered 5-HT does not cross the blood–brain barrier, but like 5-HTP it does suppress food intake (Pollock & Rowland, 1981;

Kikta *et al.*, 1981). 5-HT anorexia results from a selective reduction in bout size and bout duration (Fletcher & Burton, 1986*b*) and is blocked by the peripheral 5-HT antagonist xylamidine (Fletcher & Burton, 1986*a*). Interestingly, despite being a dipsogen, 5-HT suppresses the consumption of saccharin solutions, and reduces saccharin preference in a two-bottle (saccharin vs water) test (Montgomery & Burton, 1986). One plausible mechanism for these effects is the suppression of insulin secretion by systemic 5-HT (Telib *et al.*, 1968). The taste of saccharin is known to increase plasma insulin levels (Halter *et al.*, 1975) and Le Magnen (1983) has argued that the palatability of food is directly related to this cephalic phase of insulin release. Thus, 5-HT may suppress food intake by reducing appetite for substances possessing food-related cues (e.g. sweetness), rather than by enhancing satiety. Consonant with this interpretation is our finding that although 5-HT advances the onset of resting in rats consuming wet mash, it does not elicit resting in rats consuming saccharin (Montgomery & Willner, 1990). Further support for the role of insulin in 5-HT hypophagia comes from the observation that vagotomy (which prevents hypersecretion of insulin in VMH-lesioned rats) increases 5-HT hypophagia (Fletcher & Burton, 1985).

The hypophagic effects of peripherally administered 5-HTP or 5-HT appear to depend upon an enhancement of satiety and/or a reduction in appetite. These effects may be useful in the treatment of obesity or bulimia, but are unlikely to represent a good model of AN.

8.5.3.4 *Opiate antagonists*

Opiate antagonists (e.g. naloxone) reduce food and water intake in free-feeding rats, deprived rats and obese rats (Margules *et al.*, 1978; Brands *et al.*, 1979; Holzman, 1979), and since these effects are stereospecific they appear to be mediated by opiate receptors (Sanger *et al.*, 1981). Naloxone also advances the onset of post-prandial resting (Cooper, 1981), in contrast to morphine, which has the opposite effect (see Section 8.4.2.5). However, these effects do not simply reflect the sedation because naloxone can facilitate approach to familiar food (Cooper & Posadas-Andrews, 1980) while reducing preference for palatable solutions such as saccharin (Cooper, 1983) and enhancing spontaneous aversion to quinine (Le Magnen *et al.*, 1980).

In human studies naloxone appears to block carbohydrate consumption with an associated increase in fat intake – perhaps indicating a role in the treatment of carbohydrate-craving (Morley *et al.*, 1984*a*). Alternatively, these results may simply reflect a change in flavour or texture preference. However, the observation that anorectics avoid highly caloric foods including those with high fat content argues against the opiate antagonist-treated rat being a good model of AN.

8.5.3.5 *Peptides*

Although psychological stressors can induce overeating (see Section 8.4.1.9), undereating is an equally common response (Willenbring *et al.*, 1986) and appears to involve the corticotropin releasing factor (Morley, 1987). In rats, administration of CRF switches behaviour away from feeding and towards increased grooming (Morley & Levine, 1982). Feeding is also inhibited by intraventricular administration of a range of neuropeptides, including neurotensin, bombesin, thyrotropin-releasing hormone and calcitonin (Morley *et al.*, 1984*b*). Perhaps more interestingly a number of gastrointestinal peptides that suppress feeding have been identified. Most notable among these is CCK.

Hypophagia in response to peripherally administered CCK has been blocked by vagotomy (Smith *et al.*, 1981), supporting the argument that it may constitute a satiety signal carrying information to the brain about the contents of the small intestine (Gibbs *et al.*, 1973). From studies of the behavioural satiety sequence CCK has been claimed to produce 'true satiety' rather than aversion or nausea (Antin *et al.*, 1975). However, Deutsch *et al.* (1978) have argued that CCK abolishes duodenal peristalsis, and suggest drug-induced malaise might be a more appropriate explanation (Deutsch & Hardy, 1977). Further support for this account comes from a study of the effects of CCK on responding for LH stimulation: in some rats, pre-feeding reduced LH self-stimulation, whilst in others satiation had no effect on response rate; however both CCK and lithium chloride reduced responding for LH stimulation in all rats regardless of their response to pre-feeding

(Ettenberg & Koob, 1984). Others have noted that CCK reduces meal size and inter-meal interval in a similar way to diet adulteration (Ettinger *et al.*, 1986). Unlike peripheral 5-HT, CCK also inhibits drinking of NaCl and quinine solutions (Barness & Waldbillig, 1984; Montgomery & Burton, 1986). Also unlike 5-HT, CCK stimulates, rather than inhibits, release of insulin (McLaughlin & Baile, 1981).

It remains unclear whether CCK reduces feeding by functioning as a satiety signal by reducing the palatability of food or by inducing malaise. Whichever proves to be the case it seems unlikely that AN results from any of these malfunctions.

8.5.3.6 *Inverse agonists at the BZ receptor*

Some compounds with high affinity for BZ receptors appear to act as inverse agonists, having pharmacological effects that are the converse of classical agonists (e.g. Braestrup *et al.*, 1983). However, in behavioural tests it is apparent that the categorisation of drugs varies between test situations (File & Pellow, 1986). The inverse agonists FG 7142 and DMCM caused dose-dependent reductions in consumption of palatable foods, which were blocked by the BZ antagonist Ro 15-1788 (Cooper & Estall, 1985). Hypophagia induced by a third inverse agonist, CGS8216, however was not blocked by Ro 15-1788. If the hypophagic effect of these drugs depends upon a reduction in the reward value of palatable foods it seems unlikely that they will provide an entirely accurate model of AN. However, the anxiogenic properties of these compounds may provide a useful model of stress-induced hypophagia.

8.5.4 Conclusions

The majority of the manipulations in the preceding sections have produced reductions in body weight by reducing food acceptability. Suppression of food intake might result either from a specific effect on a feeding mechanism (reduced appetite, reduced hunger or enhanced satiety), or from a non-specific effect such as motor impairment, response competition, malaise or sensory deficits. In the case of AN, however, it seems likely that extreme feeding restraint depends upon a combination of a physiological predisposition and an inappropriate response to psychological stress, such that intake is suppressed in the presence of hunger.

With one exception (early undernutrition) the manipulations that produce chronic weight loss involve surgical intervention, and might at best provide insights into the mechanisms underlying a specific symptom of AN, weight loss. Although the behavioural syndromes produced by lesions of the LH or nigrostriatal DA pathway may not be identical it does seem that, in both cases, a large proportion of the reduction in eating can be explained in terms of impaired arousal and sensory deficits. Since there is no reason to believe that anorexics are insensitive to the incentive properties of food, it is unlikely that either of these lesions accurately model AN. DMN lesions and olfactory bulbectomy also produce body weight loss. However, DMN-lesioned rats appear to exhibit normal defence of a reduced body weight, but are over-responsive to cafeteria diets, whereas OB rats exhibit altered flavour preferences and are over-responsive to 24 h deprivation (Bernardis & Bellinger, 1987). By contrast AN sufferers are usually extremely resistant (although not insensitive) to the incentive properties of food, and strive to minimise their intake in the face of extreme deprivation.

The remaining surgical technique for producing weight loss is vagotomy. Vagotomised rats undereat and exhibit a particular suppression of carbohydrates intake. Vagotomised humans also undereat, and report reduced appetite for sweets (Kral & Gortz, 1981). This latter finding questions the validity of vagotomy as a model of AN: although anorectics typically avoid sweets, those who occasionally binge often do so on sweets; conversely there is no evidence that vagotomised humans suffer distortion of body image, or ambivalence towards food.

The second group of techniques for reducing body weight are acute non-pharmacological models and as such, might, in principle, contribute to our understanding of the aetiology of AN as well as modelling the symptom of weight loss.

Unconditioned and conditioned emotional responses, and conditioned taste aversions, all suppress eating by inducing anticipation of an aversive event. Under these circumstances a fearful, hungry rat refuses food. There may be superficial similarities between these circumstances and the anorec-

tic's refusal to eat for fear of weight gain. However there is clearly more to AN than this because many obese people often fear weight gain and yet do not lose weight.

Studies of how deprived and non-deprived rats respond to alterations in their diet may shed more light on the aetiology of AN, since the disorder has been characterised as a diet that gets out of control (Szmukler, 1987). Like food-deprived rats (Jacobs & Sharma, 1969), the dieter becomes less sensitive to internal signals relating to hunger and satiety and more sensitive to the incentive properties of food, possibly resulting in carbohydrate craving. This account readily explains why so many dieters fail to lose weight and may shed light on the origins of bulimic episodes, but also raises the question of how anorectics succeed in losing weight where so many others fail.

The unusual experiment by Epling *et al.* (1983) found that some effects of food deprivation could be amplified by giving rats access to a running wheel. This pattern of effects parallels the increased activity exhibited by a subgroup of anorexics, except that in AN weight loss precedes increased activity rather than vice versa. In another innovative experiment Haslam *et al.* (1987) used ovariectomised rats to model adolescent hypothalamic supersensitivy to oestrogen: both oestradiol and immobilisation stress caused hypophagia and oestradiol also caused weight loss. Interestingly cyproheptadine, which appears to have therapeutic value in AN, blocked the hypophagic response to stress. It seems that this model or some variant of it may prove to be a useful tool in investigating AN.

The last group of manipulations, those involving pharmacological intervention, may in principle model weight loss and guide the search for effective drug treatments of the clinical condition.

Feeding responses are reduced by increasing hypothalamic availability of catecholamines, either by direct application of NA or DA to the PFH, or by systemic administration of amphetamine (Leibowitz, 1973, 1980). In the case of amphetamine anorexia, the suppression of feeding is largely explained by response competition. It remains possible that direct application to the PFH may act more specifically to suppress hunger or enhance satiety, but even so, there is no evidence that any of these manipulations mimic the ambivalence of the anorexic human, who appears to restrict intake despite severe hunger. Treatments that increase 5-HT transmission in the brain, particularly in the PVN, also suppress food consumption. The mechanisms underlying this effect are, as yet, unresolved. The involvement of central 5-HT in anorexia associated with bulimia is supported by a report that this subgroup of anorexics have low central 5-HT turnover (Kaye *et al.*, 1984*b*). In addition some anorectics report gastric discomfort even after modest meals (Dubois *et al.*, 1979) so fenfluramine-induced gastric slowing may model that aspect of the clinical condition. Increased peripheral 5-HT transmission also suppresses feeding possibly by suppressing appetite. However, this seems unlikely to model AN since anorexics exhibit extreme restraint with regard to eating.

A number of neuropeptides also inhibit feeding. Notable amongst these are CRF and CCK. There is evidence linking hypothalamic CRF release to stress-induced hypophagia (Morley *et al.*, 1984*b*). Elevated CSF levels of CRF have been found in AN, but these return to normal with weight gain (Kaye *et al.*, 1985) and may represent an index, rather than a cause of undereating. There is some dispute as to whether CCK inhibits feeding by enhancing satiety or by inducing malaise; neither of these effects appears to reflect the symptoms of AN.

In rats, opiate antagonists suppress food consumption, reduce preference for palatable foods, and advance the onset of post-prandial resting. In humans the suppression of intake by naloxone is not associated with a change in hunger perception (Trenchard & Silverstone, 1983), but opiate antagonists do appear to shift macronutrient preference away from carbohydrates and towards fats. However, the opposite shift – away from fats is seen in AN (Drewnowski *et al.*, 1984). Furthermore, anorexics exhibit elevated CSF levels of opioid metabolites (Kaye *et al.*, 1982).

Inverse agonists at the BZ receptor have a variety of behavioural effects including inhibition of palatable food intake and anxiogenesis (Cooper & Estall, 1985; Stephens & Kehr, 1985). Antidepressants often produce weight gain that can be dissociated from their antidepressant effects and it has been suggested that the anxiolytic properties of antidepressant treatments may underlie their effects on feeding (Halmi *et al.*, 1987). If this is the case then inverse BZ agonists may provide a model of both the undereating and the anxiety that are features of AN.

8.6 GENERAL CONCLUSIONS

The initiation, maintenance and termination of feeding are under complex physiological control. The hypothalamus seems to play an important role integrating internal and external signals (relating, for example, to energy levels and palatability, respectively). The rat, particularly in the free-feeding situation, appears to regulate energy balance homeostatically by adjusting intake and metabolic efficiency. However, these homeostatic mechanisms can easily be over-ridden by factors including diet palatability and non-specific arousal (e.g. tail-pinch), particularly after food deprivation (Jacobs & Sharma, 1969).

The apparent ease with which feeding behaviour can be influenced by non-nutritional factors serves to indicate that feeding disorders or deviations from 'normal' body weight need not depend upon malfunctioning hunger or satiety mechanisms (although the prolonged maintenance of high or low body weight may depend upon a malfunction in these mechanisms). For example, overeating and body weight gain can result from arousal in the presence of salient food stimuli. If this is the case then the preoccupation of psychopharmacologists with hunger and satiety mechanisms might result in profitable strategies for the treatment of 'eating' disorders being overlooked.

A good understanding of the central mechanisms of hunger and satiety is clearly a valid and important scientific objective, but from the clinical point of view it might prove more effective to manipulate hunger and satiety signals in the periphery where specificity of action is likely to be greater. Central manipulations may have potent effects on feeding but unwanted effects on motor output, arousal and attention can often outweigh the benefits of altered feeding behaviour. Work on gastric balloons, CCK and peripheral 5-HT serves to demonstrate that manipulation of peripheral mechanisms can have appreciable and relatively specific effects on feeding. This is not to say that central mechanisms should be neglected – fenfluramine, although it does not appear to enhance satiety, might well reduce metabolic efficiency and an investigation of fenfluramine in combination with the appetite suppressant effects of peripheral 5-HT could prove fruitful. The perennial problem of tolerance to pharmacological manipulations also points to the importance of behavioural and cognitive strategies.

There are two major obstacles to producing valid animal models of feeding disorders: we lack accurate descriptions of the conditions to be modelled (much of the available information is equivocal and/or incomplete); and although we have many manipulations that alter feeding in rats we do not have effective treatments for eating disorders. This second point suggests that the existing manipulations pick up many false positives. Accounts of AN and BN suggest that these conditions might be described better as affective disorders that are expressed through disordered eating (see also Coscina & Garfinkel, this volume, Chapter 10). It seems that the anorectic is well aware of hunger, so treatments aimed at enhancing hunger might simply substitute BN for AN.

At the present time it seems that genetic models, dietary obesity, tail pinch, social isolation and early overfeeding could provide valuable insights into particular types of obesity. Among the possible models of AN, immobilisation stress and the food deprivation and wheel-running models look encouraging. Our best hope for insights into BN appears to be ESB- and tail pinch-induced eating as these models may provide clues to possible mechanisms involved in the genesis of a bulimic episode.

REFERENCES

Aberg, A. & Holmberg, G. (1979). Preliminary clinical test of zimelidine, a new 5-HT uptake inhibitor. *Acta Psychiatrica Scandinavica* **59**, 45–58.

Abramson, E.E. & Wunderlich, R.A.S. (1972). Anxiety, fear and eating: A test of the psychosomatic concept of obesity. *Journal of Abnormal Psychology* **79**, 317–21.

Ahlskog, J.E., Randall, P.K. & Hoebel, B.G. (1975). Hypothalamic hyperphagia: Dissociation from noradrenergic depletion hyperphagia. *Science* **190**, 399–401.

Anand, B.K. & Brobeck, J.R. (1951). Localization of a feeding centre in the hypothalamus. *Proceedings of the Society for Experimental Biology & Medicine* **77**, 323–4.

Anand, B.K. & Dua, S. (1955). Feeding responses induced by electrical stimulation of the hypothalamus in the cat. *Indian Journal of Medical Research* **43**, 113–22.

Anderson-Baker, W.C., McLaughlin, C.L. & Baile, C.A. (1979). Oral and hypothalamic injections of barbiturates, benzodiazepines and cannabinoids and food intake in rats. *Pharmacology, Biochemistry & Behavior* **11**, 487–91.

Antelman, S.M. & Szechtman, H. (1975). Tail-pinch induces eating in sated rats which appears to be dependent on nigrostriatal dopamine. *Science* **189**, 731–3.

Antelman, S.M., Szechtman, H., Chin, R. & Fisher, A.E. (1975). Tail-pinch induced eating, gnawing and licking behaviour, dependence on the nigrostriatal dopamine system. *Brain Research* **99**, 319–37.

Antelman, S.M., Rowland, N. & Fisher, A.E. (1976). Stress related recovery from lateral hypothalamic aphagia. *Brain Research* **102**, 346–50.

Antin, J., Gibbs, J., Holt, J., Young, R.C. & Smith, G.P. (1975). Cholecystokinin elicits the complete behavioural satiety sequence in rats. *Journal of Comparative & Physiological Psychology* **89**, 784–90.

Arees, E.A. & Mayer, J. (1967). Anatomical connections between medial and lateral regions of the hypothalamus concerned with food intake. *Science* **157**, 1574–5.

Baptista, T., Parada, M. & Hernandez, L. (1987). Long term administration of some antipsychotic drugs increases body weight and feeding in rats. Are D2 dopamine receptors involved? *Pharmacology, Biochemistry & Behavior* **27**, 399–405.

Bartness, T.J. & Waldbillig, R.J. (1984). Cholecystokinin-induced suppression of feeding: An evalution of the generality of gustatory–cholecystokinin interactions. *Physiology & Behavior* **32**, 409–15.

Bellinger, L.L., Bernardis, L.L. & Brooks, S. (1979). The effect of dorsomedial hypothalamic nucleus lesions on body weight regulation. *Neuroscience* **4**, 659–65.

Berg, B.N. (1960). Nutrition and longevity in the rat 1. Food intake in relation to size, health and fertility. *Journal of Nutrition* **71**, 242–54.

Bernardis, L.L. (1972). Hypophagia, hypodipsia and hypoactivity following dorsomedial hypothalamic lesions. *Physiology & Behavior* **8**, 1161–4.

Bernardis, L.L. & Bellinger, L.L. (1987). The dorsomedial hypothalamic nuclei revisited: 1986 update. *Brain Research Reviews* **12**, 321–81.

Bernardis, L.L., Bellinger, L.L., Goldman, J.K. & MacKenzie, R.G. (1980). Somatic and metabolic responses of mature female rats with dietary obesity to dorsomedial hypothalamic lesions: Effect of diet palatability. *Physiology & Behavior* **25**, 911–19.

Bhakthavatsalam, P. & Leibowitz, S.F. (1986). Morphine-elicited feeding, diurnal rhythm, circulating corticosterone and macronutrient selection. *Pharmacology, Biochemistry & Behavior* **24**, 911–17.

Biron, P., Mongeau, J-G. & Bertrand, D. (1977). Familial resemblance of body weight and weight/height in 374 homes with adopted children. *Acta Paediatrica Scandinavica* **65**, 279–87.

Bjornstorp, P., Enzi, G. & Karlsson, K. (1974). The effect of maternal diabetes on adipose tissue cellularity in man and rat. *Diabetologia* **10**, 205–9.

Blundell, J.E. (1977). Is there a role for serotonin in feeding? *International Journal of Obesity* **1**, 15–42.

Blundell, J.E. & Latham, C.J. (1978). Pharmacological manipulation of feeding behaviour: Possible influences of serotonin and dopamine on food intake. In *Central Mechanisms of Anorectic Drugs*, ed. S. Garattini & R. Samanin, pp. 201–54. London: Academic Press.

Blundell, J.E. & Latham, C.J. (1979). Serotonergic influences on food intake: Effect of 5-hydroxytryptophan on parameters of feeding behaviour in deprived and free-feeding rats. *Pharmacology, Biochemistry & Behavior* **11**, 431–7.

Blundell, J.E., Latham, C.J. & Lesham, M.B. (1973). Biphasic action of a 5-hydroxytryptamine inhibitor on fenfluramine-induced anorexia. *Journal of Pharmacy & Pharmacology* **25**, 492–4.

Blundell, J.E., Tombros, E., Rogers, P.J. & Latham, C.J. (1980). Behavioural analysis of feeding: Implications for the pharmacological manipulation of food intake in animals and man. *Progress in Neuro-Psychopharmacology* **4**, 319–26.

Booth, D.A. (1968). Mechanism of action of norepinephrine in eliciting an eating response on injection into the hypothalamus. *Journal of Pharmacology & Experimental Therapeutics* **160**, 336–48.

Booth, D.A. (1972a). Postabsorptively induced suppression of appetite and the energostatic control of feeding. *Physiology & Behavior* **9**, 199–202.

Booth, D. A. (1972b). Conditioned satiety in the rat. *Journal of Comparative & Physiological Psychology* **82**, 457–77.

Booth, D.A. (1976). Approaches to feeding control. In *Appetite and Food Intake*, ed. T. Silverstone, pp. 417–78. Dahlem Konferenzen, Berlin.

Booth, D.A. & Brookover, D. (1968). Hunger elicited by a single injection of bovine crystalline insulin. *Physiology & Behavior* **3**, 439–46.

Booth, D.A., Gibson, E.L. & Baker, B.J. (1986). Gastromotor mechanism of fenfluramine anorexia. *Appetite* **7** [Suppl.] pp. 57–69.

Bradley, P.B., Engel, G., Feniuk, W., Fozard, J., Humphrey, P.P.A., Middlemiss, D.N.,

Mylecharane, E.J., Richardson, B.P. & Saxena, P.R. (1986). Proposals for the classification and nomenclature of functional receptors of 5-hydroxytryptamine. *Neuropharmacology* **97**, 213–18.

Braestrup, C., Nielsen, M., Honore, T., Jensen, L.H. & Petersen, E.N. (1983). Benzodiazepine receptor ligands with positive and negative efficacy. *Neuropharmacology* **22**, 1451–7.

Brands, B., Thornhill, J.A., Hirst, M. & Gowdy, C.W. (1979). Suppression of food intake and body weight gain by naloxone in rats. *Life Science* **24**, 1773–8.

Bray, G.A. (1969). Effect of caloric restriction on energy expenditure in obese patients. *Lancet* **2**, 397–8.

Bray, G.A. (1970). The myth of diet in the management of obesity. *American Journal of Nutrition* **24**, 1482–8.

Bray, G.A. & York, D.A. (1979). Hypothalamic and genetic obesity in experimental animals: an autonomic and endocrine hypothesis. *Physiological Review* **59**, 719–809.

Breisch, S.T., Zemlan, F.P. & Hoebel, B.G. (1976). Hyperphagia and obesity following serotonin depletion with intraventricular PCPA. *Science* **192**, 382–5.

Brownell, K.D. & Stunkard, A.J. (1980). Physical activity in the development and control of obesity. In *Obesity*, ed A.J. Stunkard, pp. 300–24. W.B. Saunders: U.S.A.

Bruch, H. (1974). *Eating Disorders and the Person Within*. London: Routledge & Kegan Paul.

Carey, R.J. (1976). Effects of selective forebrain depletions of NE and 5-HT on the activity and food intake effects of amphetamine and fenfluramine. *Pharmacology, Biochemistry & Behavior* **5**, 519–23.

Carpenter, R.G. & Grossman, S.P. (1983). Reversible obesity and plasma fat metabolites. *Physiology & Behavior* **30**, 51–5.

Clark, J.T., Kalra, P.S., Crowley, W.R. & Kalra, S.P. (1984). Neuropeptide Y and human pancreatic polypeptide stimulate feeding behaviour in rats. *Endocrinology* **115**, 427–9.

Clarren, S.K. & Smith, D.W. (1971). Prader-Willi syndrome: Variable severity and recurrence risk. *American Journal of Diseases of Children* **131**, 798–800.

Cooper, S.J. (1980). Benzodiazepines as appetite-enhancing compounds. *Appetite* **1**, 7–19.

Cooper, S.J. (1981). Prefrontal cortex, benzodiazepines and opiates: Case studies in motivation and behaviour analysis. In *Theory in Psychopharmacology*, vol. **1**, ed. S.J. Cooper, pp. 277–322. Academic Press: New York.

Cooper, S.J. (1983). Effects of opiate agonists and

antagonists on fluid intake and saccharin choice in the rat. *Neuropharmacology* **22**, 323–8.

Cooper, S.J. & Estall, L.B. (1985). Behavioural pharmacology of food, water and salt intake in relation to drug actions at benzodiazepine receptors. *Neuroscience & Biobehavioral Reviews* **9**, 5–19.

Cooper, S.J. & Posadas-Andrews, A. (1980). Familiarity-induced feeding in a food preference test: Effects of chlordiazepoxide and naloxone compared with food deprivation. *British Journal of Pharmacology* **69**, 273–4.

Coscina, D.V., Daniel, J. & Warsh, J.J. (1978). Potential non-serotonergic basis of hyperphagia elicited by intraventricular PCPA. *Pharmacology, Biochemistry & Behavior* **9**, 791–7.

Costa, E. & Guidotti, A. (1979). Molecular mechanisms in the receptor action of benzodiazepines. *Annual Review of Pharmacology & Toxicology* **19**, 531–45.

Crisp, A.H. (1965). Clinical and therapeutic aspects of anorexia nervosa: A study of 30 cases. *Journal of Psychosomatic Research* **9**, 67–78.

Crisp, A.H., Hsu, L.K.G., Harding, B. & Hartshorn, J. (1980). Clinical features of anorexia nervosa: A study of a consecutive series of 102 female patients. *Journal of Psychosomatic Research* **24**, 179–91.

Cruce, J.A.F., Greenwood, M.R.C. & Johnson, P.R. (1974). Genetic versus hypothalamic obesity: Studies of intake and dietary manipulations in rats. *Journal of Comparative & Physiological Psychology* **87**, 295–301.

Cruce, J.A.F., Thoa, N.B. & Jacobowitz, D.M. (1976). Catecholamines in the brains of genetically obese rats. *Brain Research* **101**, 165–70.

Davies, R.F., Rossi, J.R., Panskepp, J., Bean, N.J. & Zolovick, A.J. (1983). Fenfluramine anorexia: A peripheral locus of action. *Physiology & Behavior* **30**, 723–30.

Davis, J.D., Gallagher, R.L. & Ladove, R. (1967). Food intake controlled by a blood factor. *Science* **156**, 1247–8.

Della-Fera, M.A., Baile, C.A., Schneider, B.S. & Grinker, J.A. (1981). Cholecystokinin antibody injected in cerebral ventricles stimulates feeding in sheep. *Science* **212**, 687–9.

Deutsch, J.A. (1983). Dietary control and the stomach. *Progress in Neurobiology* **20**, 313–32.

Deutsch, J.A. & Hardy, W.T. (1977). Cholecystokinin produces bait shyness in rats. *Nature* **266**, 196.

Deutsch, J.A., Thiel, T.R. & Greenberg, L.H. (1978). Duodenal motility after cholecystokinin injection or satiety. *Behavioral Biology* **24**, 393–9.

Donohoe, T.P. (1984). Stress-induced anorexia: Implications for anorexia nervosa. *Life Science* **34**, 203–18.

Doss, F.W.S. (1979). The effects of antipsychotic drugs on body weight: A retrospective review. *Journal of Clinical Psychiatry* **40**, 528–30.

Dourish, C.J., Hutson, P.H. & Curzon, G. (1985). Low doses of the putative serotonin agonist 8-hydroxy-2(di-n-propylamino)tetralin (8-OH-DPAT) elicit feeding in the rat. *Psychopharmacology* **94**, 197–204.

Dourish, C.J., Hutson, P.H., Kennett, G.A. & Curzon, G. (1986). 8-OH-DPAT-induced hyperphagia: Its neural basis and possible therapeutic relevance. *Appetite* **7** (Suppl.), pp. 127–40.

Drapanas, T., McDonald, J.C. & Stewart, J.D. (1962). Serotonin release following instillation of hypertonic glucose into the proximal intestine. *Annals of Surgery* **156**, 528–36.

Drewnowski, A., Greenwood, M.R.C. & Halmi, K.A. (1984). Carbohydrate or fat phobia: Taste responsiveness in anorexia nervosa. *Federation Proceedings* **43**, 475.

Dubois, A., Gross, H.A., Ebert, M.H. & Castell, D.O. (1979). Altered gastric emptying and secretion in primary anorexia nervosa. *Gastroenterology* **77**, 319–23.

Dum, J., Gramsch, C.H. & Herz, A. (1983). Activation of beta endorphin pools by reward induced by highly palatable food. *Pharmacology, Biochemistry & Behavior* **18**, 443–7.

Einon, D. & Sahakian, B.J. (1979). Environmentally-induced differences in susceptibility of rats to CNS stimulants and CNS depressants: evidence against a unitary explanation. *Psychopharmacology* **61**, 299–307.

Epling, W.F., Pierce, W.D. & Stefan, L. (1981). Schedule-induced starvation. In *Quantification of Steady-State Operant Behaviour*, ed. C.M. Bradshaw, E. Szabadi & C.F. Lowe, pp. 393–6. Amsterdam: Elsevier/North Holland Biomedical Press.

Epling, W.F., Pierce, W.D. & Stefan, L. (1983). A theory of activity based anorexia. *International Journal of Eating Disorders* **3**, 27–46.

Ettenberg, A. & Koob, G.F. (1984). Different effects of cholecystokinin and satiety on lateral hypothalamic self-stimulation. *Physiology & Behavior* **32**, 127–30.

Ettinger, R.H., Thompson, S. & Staddon, J.E.R. (1986). Cholecystokinin diet palatability and feeding regulation in rats. *Physiology & Behavior* **36**, 801–9.

Even, P. & Nicolaidis, S. (1986). Metabolic mechanism of the anorectic and leptogenic effects of the serotonin agonist fenfluramine. *Appetite* **7** (Suppl.), pp. 141–63.

Fabry, P. (1969). *Feeding Patterns and Nutritional Adaptations*. London: Butterworth.

Faust, I., Johnson, P. & Stern, J. (1978). Diet-induced

adipocyte number increase in adult rats. *American Journal of Physiology* **235**, E279–E286.

Faxen, A., Rossander, L. & Kewenter, J. (1979). The effect of parietal cell vagotomy with pyloroplasty on body weight and dietary habits. *Scandinavian Journal of Gastroenterology* **14**, 7–11.

Ferguson, J.M. (1986). Fluoxetine induced weight loss in humans. In *Disorders of Eating Behaviour: A Psychoneuroendocrine Approach*, ed. E. Ferrari & F. Brambilla, pp. 313–18. Pergamon Press: Oxford.

Ferguson, N.B.L. & Keesey, R.E. (1975). Effect of quinine adulterated diet upon body weight maintenance in male rats with ventromedial hypothalamic lesions. *Journal of Comparative & Physiological Psychology* **89**, 478–88.

File, S.E. & Pellow, S. (1986). Intrinsic actions of the benzodiazepine antagonist Ro 15-1788: A review. *Psychopharmacology* **88**, 1–11.

Fisler, J.S. & Bray, G.A. (1985). Dietary obesity: Effects of drugs on food intake in S5B/P1 and Osborne-Mendel rats. *Physiology & Behavior* **34**, 225–31.

Fletcher, P.J. & Burton, M.J. (1985). The anorectic effect of peripherally administered serotonin is enhanced by vagotomy. *Physiology & Behavior* **34**, 861–6.

Fletcher, P.J. & Burton, M.J. (1986*a*). Dissociation of the anorectic actions of 5-HTP and fenfluramine. *Psychopharmacology* **89**, 216–20.

Fletcher, P.J. & Burton, M.J. (1986*b*). Microstructural analysis of the anorectic action of peripherally administered 5-HT. *Pharmacology, Biochemistry & Behavior* **24**, 1133–6.

Foch, T.T. & McClearn, G.E. (1980). Genetics, body weight and obesity. In *Obesity*, ed. A.J. Stunkard, pp. 48–71. W.B. Saunders: Philadelphia.

Fox, K.A., Kipp, S.C. & Vander Weele, D.A. (1976). Dietary self-selection following sub-diphragmatic vagotomy in the white rat. *American Journal of Physiology* **231**, 1790–3.

Franklin, B.J. & Herberg, L.J. (1974). Ventromedial syndrome: the rats 'finickiness' results from obesity, not from lesions. *Journal of Comparative & Physiological Psychology* **87**, 410–14.

Fratta, W., Mercu, G., Chessa, P., Paguetti, E. & Gessa, G. (1976). Benzodiazepine-induced voraciousness in cats and inhibition of amphetamine anorexia. *Life Sciences* **18**, 1157–66.

Gale, S.K., Van Itallie, T.B. & Faust, I.M. (1981). Effects of palatable diets on body weight and adipose tissue cellularity in the adult obese female Zucker rat (fa/fa). *Metabolism* **30**, 105–10.

Gambert, S.R., Garthwaite, T.L., Pontzer, C.H. & Hagen, T.C. (1980). Fasting associated with decrease in hypothalamic beta endorphin. *Science* **210**, 1271–2.

Garcia, J., Kimmeldorf, D.J. & Koelling, R.A. (1955). Conditioned aversion to saccharin resulting from exposure to gamma radiation. *Science* **122**, 157–8.

Garfinkel, P.E. & Kaplan, A.S. (1986). Psychoneuroendocrine profiles. In *Disorders of Eating Behavior. A Psychoneuroendocrine Approach*, ed. E. Ferrari & F. Brambilla, pp. 1–8. Pergamon Press: Oxford.

Geliebter, A., Westreich, S., Hashim, S.A. & Gage, D. (1987). Gastric balloon reduces food intake and body weight in obese rats. *Physiology & Behavior* **39**, 399–402.

Gibbs, J., Young, R.C. & Smith, G.P. (1973). Cholecystokinin elicits satiety in rats with open gastric fistulas. *Nature* **245**, 323–5.

Gold, R.M. (1973). Hypothalamic obesity: The myth of the ventromedial nucleus. *Science* **182**, 488–90.

Gold, R.M., Jones, A.P., Sawchenko, P.E. & Kapatos, G. (1977). Paraventricular area: Critical focus of a longitudinal neurocircuitry mediating food intake. *Physiology & Behavior* **18**, 1111–19.

Gold, R.M., Sawchenko, P.E., DeLuca, C., Alexander, J. & Eng, R. (1980). Vagal mediation of hypothalamic obesity but not of supermarket dietary obesity. *American Journal of Physiolgoy* **238**, R447–R453.

Gosnell, B.A., Morley, J.E. & Levine, A.S. (1986). Opioid-induced feeding: Localization of sensitive brain sites. *Brain Research* **369**, 177–84.

Grandison, L. & Guidotti, A. (1977). Stimulation of food intake by muscimol and beta-endorphin. *Neuropharmacology* **16**, 533–6.

Greene, J.A. (1939). Clinical study of the etiology of obesity. *Annals of Internal Medicine* **12**, 1797–803.

Greenwood, M.R.C., Quartermain, D. & Johnson, P.R. (1974). Food motivated behaviour in genetically obese and hypothalamic-hyperphagic rats and mice. *Physiology & Behavior* **13**, 687–92.

Grinker, J.A., Drewnowski, A., Enns, M. & Kissileff, H. (1980). Effects of d-amphetamine and fenfluramine on feeding pattern and activity of obese and lean Zucker rats. *Pharmacology, Biochemistry & Behavior* **12**, 265–75.

Gross, H.A., Lake, C.R., Ebert, M.H., Ziegler, M.G. & Kopin, I.J. (1979). Catecholamine metabolism in primary anorexia nervosa. *Journal of Clinical Endocrinology & Metabolism* **49**, 805–9.

Grossman, S.P. (1962). Direct adrenergic and cholinergic stimulation of hypothalamic mechanisms. *American Journal of Physiology* **202**, 872–882.

Grossman, S.P. (1979). The biology of motivation. *Annual Review of Psychology* **30**, 209–42.

Grossman, S.P. (1984). Contemporary problems concerning our understanding of brain mechanisms that regulate food intake and body weight. In *Eating and its Disorders*, ed. A.J.

Stunkard & E. Stellar, pp. 5–13. New York: Raven Press.

Grossman, S.P., Dacey, D., Halaris, A.E., Collier, T. & Routtenberg, A. (1978). Aphagia and adipsia after preferential destruction of nerve cell bodies in hypothalamus. *Science* **202**, 537–9.

Gunion, M.W. & Peters, R.H. (1981). Pituitary beta endorphin, naloxone and feeding in several experimental obesities. *American Journal of Physiology* **241**, R173–R184.

Gwirtsman, H.E., Roy-Byrne, P., Jager, J. & Gerner, R.H. (1983). Neuroendocrine abnormalities in bulimia. *American Journal of Psychiatry* **140**, 559–63.

Hager, A., Sjostrom, L. & Arvidsson, B. (1977). Body fat and adipose tissue cellularity in infants: a longitudinal study. *Metabolism* **26**, 607–14.

Hales, C.N. & Kennedy, G.C. (1964). Plasma glucose, non-esterified fatty acid and insulin concentrations in hypothalamic hyperphagic rats. *Biochemical Journal* **90**, 620–4.

Halmi, K.A. (1985). Relationship of the eating disorders to depression: Biological similarities and differences. *International Journal of Eating Disorders* **4**, 667–80.

Halmi, K.A., Eckert, E. & Falk, J.R. (1983). Cyproheptadine, an antidepressant and weight inducing drug for anorexia nervosa. *Psychopharmacology Bulletin* **19**, 103–5.

Halmi, K.A., Eckert, E., La Du, T. & Cohen, J. (1986). Anorexia nervosa: Treatment efficacy of cyproheptadine and amitriptyline. *Archives of General Psychiatry* **43**, 177–81.

Halmi, K.A., Ackerman, S., Gibbs, J. & Smith, G. (1987). Basic biological overview of eating disorders. In *Psychopharmacology: The Third Generation of Progress*, ed. H.J. Meltzer, pp. 1255–66. New York: Raven Press.

Halter, J., Kulkosky, P. & Woods, S.C. (1975). Afferent receptors, taste perception and pancreatic endocrine function in man. *Diabetes* **24**, Suppl. 2, 414.

Han, P.W. (1967). Hypothalamic obesity in rats without hyperphagia. *Transactions of the New York Academy of Sciences* **30**, 229–43.

Haslam, C., Stevens, R. & Donohoe, T.P. (1987). The influence of cyproheptadine on immobilisation and oestradiol benzoate induced anorexia in ovariectomised rats. *Psychopharmacology* **93**, 201–6.

Herman, C.P. & Mack, D. (1975). Restrained and unrestrained eating. *Journal of Personality* **43**, 647–60.

Herman, C.P. & Polivy, J. (1980). Restrained eating. In *Obesity*, ed. A.J. Stunkard, pp. 208–25. W.B. Saunders: Philadelphia.

Herman, C.P. (1984). A boundary model for the regulation of eating. In *Eating and its Disorders*, ed. A.J. Stunkard & E. Stellar, pp. 141–56. New York: Raven Press.

Hernandez, L. & Hoebel, B.G. (1980). Basic mechanisms of feeding and weight regulation. In *Obesity*, ed. A.J. Stunkard, pp. 25–47. W.B. Saunders: Philadelphia.

Hetherington, A.W. & Ranson, S.W. (1939). Experimental hypothalamic-hypophyseal obesity in the rat. *Proceedings of the Society for Experimental Biology & Medicine* **41**, 465–6.

Heym, J. & Gladfelter, W.E. (1982). Locomotor activity and ingestive behaviour after damage to ascending serotoninergic systems. *Physiology & Behavior* **29**, 459–67.

Hill, D.L. & Almi, J. (1981). Olfactory bulbectomy in infant rats: Survival, growth and ingestive behaviours. *Physiology & Behavior* **27**, 811–17.

Hirsch, J. & Han, P.W. (1969). Cellularity of rat adipose tissue: Effects of growth, starvation and obesity. *Journal of Lipid Research* **10**, 77–82.

Hoebel, B.G. (1977). Pharmacological control of feeding. *Annual Review of Pharmacology & Toxicology* **17**, 605–21.

Hoebel, B.G. & Leibowitz, S.F. (1981). Brain monoamines in the modulation of self-stimulation, feeding and body weight. In *Brain, Behavior and Bodily Disease*, ed. H. Weiner, M.A. Hofer & A.J. Stunkard, pp. 102–42. New York: Raven Press.

Hoebel, B.G. & Teitelbaum, P. (1966). Weight regulation in normal and hypothalamic rats. *Journal of Comparative & Physiological Psychology* **61**, 189–93.

Hoebel, B.G. & Thompson, R.D. (1969). Aversion to lateral hypothalamic stimulation caused by intragastric feeding or obesity. *Journal of Comparative & Physiological Psychology* **68**, 536–43.

Holzman, S.G. (1979). Suppression of appetitive behaviour in the rat by naloxone: Lack of effect of prior morphine dependence. *Life Science* **24**, 219–26.

Hudson, J.I., Pope, H.G., Jonas, J.M. & Yergelun-Todd, D. (1983). Phenomenologic relationship of eating disorders to major affective disorder. *Psychiatry Research* **9**, 345–54.

Hughes, P.L., Wells, L.A., Cunningham, C.J. & Ilstrup, D.M. (1986). Treating bulimia with desipramine. A double-blind placebo-controlled study. *Archives of General Psychiatry* **43**, 182–6.

Ingle, D.J. (1959). A simple means of producing obesity in the rat. *Proceedings of the Society for Experimental Biology & Medicine* **72**, 604–5.

Jacobs, H.L. & Sharma, K.N. (1969). Taste versus calories: Sensory and metabolic signals in the control of food intake. *Annals of the New York Academy of Science* **157**, 1084–112.

Janowitz, H.D. & Grossman, M.I. (1949). Some factors affecting the food intake of normal dogs and dogs with esophagostomy and gastric fistula. *American Journal of Physiology* **159**, 143–8.

Jen, K-L.C. & Hansen, B.C. (1984). Feeding behavior during experimentally induced obesity in monkeys. *Physiology & Behavior* **33**, 863–9.

Jimerson, D.C., George, D.T., Brewerton, T.D. & Kaye, W.H. (1986). Anxiety in bulimic disorder: Behavioral responses to lactate and isoproterenol infusions. In *Disorders of Eating Behavior. A Psychoneuroendocrine Approach*, ed. E. Ferrari & F. Brambilla, pp. 319–23. Pergamon Press: Oxford.

Johnson, P.R., Stern, J.S. & Greenwood, M.R.C. (1973). Effects of early nutrition on adipose cellularity and pancreatic insulin release in the Zucker rat. *Journal of Nutrition* **103**, 738–43.

Johnson, R.E., Mastropaolo, J.A. & Wharton, M.A. (1972). Exercise, dietary intake and body composition. *Journal of the American Dietetic Association* **61**, 399–403.

Jones, M.T., Hillhouse, E.W. & Burden, J. (1976). Effects of various putative neurotransmitters on the secretion of corticotrophin-releasing hormone from the rat hypothalamus and in vitro: A model of the neurotransmitters involved. *Journal of Endocrinology* **69**, 1–10.

Joyce, D. & Mrosovsky, (1964). Eating, drinking and activity following 5-hydroxytryptophan administration. *Psychopharmacologia* **5**, 417–23.

Kakolewski, J.W., Deaux, E., Christensen, J. & Case, B. (1971). Diurnal patterns in water and food intake and body weight changes in rats with hypothalamic lesions. *American Journal of Physiology* **221**, 711–18.

Kanarek, R.B. & Hirsch, E. (1978). Dietary-induced overeating in experimental animals. *Federation Proceedings* **36**, 154–8.

Kapatos, G. & Gold, R.M. (1973). Evidence for ascending noradrenergic mediation of hypothalamic hyperphagia. *Pharmacology, Biochemistry & Behavior* **1**, 81–7.

Kaye, W.H., Pickar, D., Naber, D. & Ebert, M.H. (1982). Cerebrospinal fluid opioid activity in anorexia nervosa. *American Journal of Psychiatry* **139**, 643–5.

Kaye, W.H., Ebert, M.H., Raleigh, M. & Lake, C.R. (1984*a*). Abnormalities in CNS monoamine metabolism in anorexia nervosa. *Archives of General Psychiatry* **41**, 350–5.

Kaye, W.H., Ebert, M.H., Gwirtsman, H.E. & Weiss, S.R. (1984*b*). Differences in brain serotonergic metabolism between non-bulimic and bulimic patients with anorexia nervosa. *American Journal of Psychiatry* **141**, 1598–601.

Kaye, W.H., Jimerson, D.C., Lake, C.R. & Ebert, M.H. (1985). Altered norepinephrine metabolism following long-term weight recovery in patients with anorexia nervosa. *Psychiatry Research* **14**, 333–42.

Keesey, R.E. (1980). A set-point analysis of the regulation of body weight. In *Obesity*, ed. A.J. Stunkard, pp. 144–65. W.B. Saunders: Philadelphia.

Keesey, R.E. & Corbet, S.W. (1984). Metabolic defense of the body weight set-point. In *Eating and its Disorders*, ed. A.J. Stunkard & E. Stellar, pp. 87–96. New York: Raven Press.

Kelly, J. & Grossman, S.P. (1979). GABA and hypothalamic feeding systems. II. A comparison of GABA, glycine and acetylcholine agonists and their antagonists. *Pharmacology, Biochemistry & Behavior* **11**, 647–52.

Kennett, G.A. & Curzon, G. (1988). Evidence that hypophagia induced by mCPP and TFMPP requires 5-HT$_{1C}$ and 5-HT$_{1B}$ receptors; hypophagia induced by RU24969 only requires 5-HT$_{1B}$ receptors. *Psychopharmacology* **96**, 93–100.

Kennett, G.A., Dourish, C.T. & Curzon, G. (1987). Anti-depressant-like action of 5-HT$_{1A}$ agonists and conventional antidepressants in an animal model of depression. *European Journal of Pharmacology* **134**, 265–74.

Kikta, D.C., Threatte, R.M., Barney, C.C., Fregly, M.J. & Greenleaf, J.E. (1981). Peripheral conversion of 1-5-hydroxytryptophan to 5-HT induces drinking in rats. *Pharmacology, Biochemistry & Behavior* **14**, 889–93.

King, B.M., Castellanos, F.X., Kastin, A.J., Berzas, M.C., Munk, M.D., Olson, G.A. & Olson, R.D. (1979). Naloxone-induced suppression of food intake in normal and hypothalamic obese rats. *Pharmacology, Biochemistry & Behavior* **11**, 729–32.

Kral, J.G. & Gortz, L. (1981). Truncal vagotomy in morbid obesity. *International Journal of Obesity* **5**, 431–5.

Krotkiewski, M., Sjostrom, L. & Bjornstorp, P. (1977). Adipose tissue cellularity in relation to prognosis for weight reduction. *International Journal of Obesity* **1**, 395–416.

Kupfer, D.J., Coble, P.A. & Rubenstein, D. (1979). Changes in weight during treatment for depression. *Psychosomatic Medicine* **41**, 535–44.

Lasagna, L. (1987). The pharmacology of obesity. In *Psychopharmacology: The Third Generation of Progress*, ed. H.J. Meltzer, pp. 1281–4. New York: Raven Press.

Lawson, W.B., Bird, J. & Reed, D. (1984). Effects of neuroleptics on food intake. *Society for Neuroscience Abstracts* **303**, 92–100.

Leibowitz, S.F. (1973). Brain norepinephrine and ingestive behaviour. In *Frontiers of Catecholamine*

Research, ed. E. Usdin & S. Snyder, pp. 711–13. Oxford: Pergamon Press.

Leibowitz, S.F. (1975). Ingestion in the satiated rat: Role of alpha- and beta-receptors in mediating effects of hypothalamic adrenergic stimulation. *Physiology & Behavior* **14**, 745–54.

Leibowitz, S.F. (1976). Brain catecholaminergic mechanisms for control of hunger. In *Hunger: Brain Mechanisms and Clinical Implications*, ed. D. Novin, W. Wyrwicka & G. Bray, pp. 1–18. New York: Raven Press.

Leibowitz, S.F. (1980). Neurochemical systems of the hypothalamus: Control of feeding and drinking behavior and water-electrolyte excretion. In *Handbook of the Hypothalamus*, ed. P.J. Morgane & J. Panksepp, pp. 297–437. Marcel Dekker: New York.

Leibowitz, S.F. (1982). Hypothalamic catecholamine systems in relation to control of eating behavior and mechanisms of reward. In *The Neural Basis of Feeding and Reward*, ed. B.G. Hoebel & D. Novin, pp. 241–58. Haer Institute: Brunswick, Maine.

Leibowitz, S.F. & Hor, L. (1982). Endorphinergic and alpha-noradrenergic systems in the paraventricular nucleus: Effects on eating behavior. *Peptides* **3**, 421–8.

Leibowitz, S.F. & Papadakos, P.J. (1978). Serotonin-norepinephrine interaction in the paraventricular nucleus: Antagonistic effects on feeding behavior in the rat. *Society for Neuroscience Abstracts* **4**, 452.

Leibowitz, S.F. & Rossakis, C. (1979*a*). Mapping study of brain dopamine- and norepinephrine-sensitive sites which cause feeding suppression in the rat. *Brain Research* **172**, 101–13.

Leibowitz, S.F. & Rossakis, C. (1979*b*). Pharmacological characterisation of perifornical hypothalamic dopamine receptors mediating feeding inhibition in the rat. *Brain Research* **172**, 115–30.

Leibowitz, S.F., Arcomano, A. & Hammer, N.J. (1978). Potentiation of eating associated with tricyclic antidepressant drug activation of alpha-adrenergic neurons in the paraventricular hypothalamus. *Progress in Neuropsychopharmacology* **2**, 349–58.

Leibowitz, S.F., Hammer, N.J. & Chang, K. (1981). Hypothalamic paraventricular nucleus lesions produce overeating and obesity in the rat. *Physiology & Behavior* **27**, 1031–40.

Leibowitz, S.F., Roland, C.R., Hor, L. & Squillari, V. (1984). Noradrenergic feeding elicited via the paraventricular nucleus is dependent upon circulating corticosterone. *Physiology & Behavior* **32**, 857–64.

Leibowitz, S.F., Shor-Posner, G., Maclow, C. & Grinker, J.A. (1986). Amphetamine: Effects on meal patterns and macronutrient selection. *Brain Research Bulletin* **17**, 681–9.

Le Magnen, J. (1971). Advances in studies on the physiological control and regulation of food intake. In *Progress in Physiological Psychology*, vol. **4**, ed. J.M. Sprague & E. Stellar, pp. 204–61. New York: Academic Press.

Le Magnen, J. (1983). Body energy balance and food intake. A neuroendocrine regulatory mechanism. *Physiological Review* **63**, 314–86.

Le Magnen, J., Marfaing-Jallet, P., Miceli, D. & Devos, M. (1980). Pain modulating and reward systems: A single brain mechanism? *Pharmacology, Biochemistry & Behavior* **12**, 729–33.

Levin, B.E. & Sullivan, A.C. (1979). Catecholamine levels in discrete brain nuclei of 7 month old genetically obese rats. *Pharmacology, Biochemistry & Behavior* **11**, 77–82.

Levine, A.S., Morley, J.E., Gosnell, B.A., Billington, C.J. & Bartness, T.J. (1985). Opioids and consummatory behaviour. *Brain Research Bulletin* **14**, 662–72.

Liebelt, R.A., Bordelon, C.B. & Liebelt, A.G. (1973). The adipose tissue system and food intake. In *Progress in Physiological Psychology*, vol. **5**, ed. E. Stellar & J.M. Sprague, pp. 211–52. New York: Academic Press.

Louis-Sylvestre, J., Giachetti, I. & Le Magnen, J. (1984). Sensory versus dietary factors in cafeteria-induced overweight. *Physiology & Behavior* **32**, 901–5.

Lyon, M. & Robbins, T.W. (1975). The action of central nervous system drugs: A general theory concerning amphetamine effects. In *Current Developments in Psychopharmacology*, vol. **2**, ed. W.B. Essman & L. Valzelli, pp. 80–163. Spectrum: New York.

Mackenzie, R.G., Hoebel, B.G., Durcet, R.P. & Trulson, M.E. (1979). Hyperphagia following intraventricular PCPA-, leucine-, or tryptophan-methyl esters: lack of correlation with whole brain serotonin levels. *Pharmacology, Biochemistry & Behavior* **10**, 951–6.

Malcolm, A.D., Mace, P.M. & Ontar, K.P. (1972). Experimental evaluation of anorexigenic agents in man: A pilot study. *Proceedings of the Nutrition Society* **31**, 124.

Mandenoff, A., Fumeron, F., Apfelbaum, M. & Margules, D.L. (1982). Endogenous opiates and energy balance. *Science* **215**, 1536–7.

Margules, D.L., Moisset, B., Lewis, M.J., Shibuya, H. & Pert, C.B. (1978). Beta endorphin is associated with overeating in genetically obese mice (ob/ob) and rats (fa/fa). *Science* **202**, 988–91.

Marshall, J.F., Friedman, M.I. & Heffner, T.G. (1976). Reduced anorexia and locomotor stimulant action of d-amphetamine in alloxan-diabetic rats. *Brain Research* **111**, 428–32.

Martin, G.E. & Myers, R.D. (1975). Evoked release of [^{14}C]norepinephrine from the rat hypothalamus during feeding. *American Journal of Physiology* **229**, 1547–55.

Martin, J.R., Novin, D. & Vander Weele, D.A. (1978). Loss of glucagon suppression of feeding after vagotomy in rats. *American Journal of Physiology* **224**, E314–E318.

Martin-Iverson, M.T. & Dourish, C.T. (1988). Role of dopamine D-1 and D-2 receptor subtypes in mediating dopamine agonist effects on food consumption in rats. *Psychopharmacology* **96**, 370–4.

Mayer, J., Marshall, N.B. & Vitale, J.J. (1955). Exercise, food intake and body weight in normal rats and genetically obese adult mice. *American Journal of Physiology* **177**, 544–8.

McArthur, R.A. & Blundell, J.E. (1982). Effects of age and feeding regimen on protein and carbohydrate self-selection. *Appetite* **3**, 153–62.

McCabe, J. & Leibowitz, S.F. (1980). Midbrain catecholamine projections to the perifornical hypothalamus: Their role in the mediation of drug induced anorexia and hyperphagia. *Society for Neuroscience Abstracts* **6**, 784.

McCabe, J., Bitran, D. & Leibowitz, S.F. (1981). The role of the paraventricular nucleus and ascending fibre systems in the mediation of drug induced feeding. Presented at the 52nd Annual Meeting of the Eastern Psychological Association: New York.

McLaughlin, C.L. & Baile, C.A. (1980). Decreased sensitivity of Zucker obese rats to the putative satiety agent cholecystokinin. *Physiology & Behavior* **25**, 543–8.

McLaughlin, C.L. & Baile, C.A. (1981). Serum insulin, glucose and triglyceride responses of Zucker obese and lean rats to cholecystokinin. *Physiology & Behavior* **26**, 995–9.

McLaughlin, C.L. & Baile, C.A. (1984). Increased sensitivity of Zucker obese rats to naloxone is present at weaning. *Physiology & Behavior* **32**, 929–33.

Mendelson, J. & Chorover, S.L. (1965). Lateral hypothalamic stimulation in satiated rats: T-maze learning for food. *Science* **49**, 559–61.

Miczek, K.A. (1973). Effects of scopolamine, amphetamine and benzodiazepines on conditioned suppression. *Pharmacology, Biochemistry & Behavior* **1**, 401–11.

Miller, R.R. & Holzman, A.D. (1981). Neophobia: Generality and function. *Behavioral & Neural Biology* **33**, 17–44.

Miller, D.S. & Wise, A. (1975). Maintenance requirement and adipocyte count of rats from large and small litters, at the same weight. *Proceedings of the Nutrition Society* **34**, 105A.

Mitchell, J.E. (1987). Psychopharmacology of anorexia nervosa. In *Psychopharmacology: The Third Generation of Progress*, ed. H.J. Meltzer, pp. 1273–6. New York: Raven Press.

Montgomery, A.M.J. & Burton, M.J. (1986). Effects of peripheral 5-HT on consumption of flavoured solutions. *Psychopharmacology* **88**, 262–6.

Montgomery, A.M.J., Fletcher, P.J. & Burton, M.J. (1986). Behavioural and pharmacological investigations of 5-HT hypophagia and hyperdipsia. *Pharmacology, Biochemistry & Behaviour* **25**, 23–8.

Montgomery, A.M.J. & Willner, P.J. (1988). Fenfluramine disrupts the behavioural satiety sequence in rats. *Psychopharmacology* **94**, 397–401.

Montgomery, A.M.J. & Willner, P.J. (1990). Peripheral 5-HT and the behavioural satiety sequence (submitted for publication).

Montgomery, A.M.J., Willner, P.J. & Muscat, R. (1988). Behavioural specificity of 8-OH-DPAT-induced feeding. *Psychopharmacology* **94**, 110–14.

Mordes, J.P., El Lozy, M., Herrera, M.G. & Silen, W. (1979). Effects of vagotomy with and without pyloroplasty on weight and food intake in rats. *American Journal of Physiology* **236**, R61–R66.

Morgan, M.J. (1973). Effects of post-weaning environment on learning in the rat. *Animal Behaviour* **21**, 429–42.

Morgan, M.J. & Einon, D. (1975). Incentive motivation and behavioural inhibition in socially-isolated rats. *Physiology & Behavior* **15**, 405–9.

Morley, J.E. (1987). Behavioral pharmacology for eating and drinking. In *Psychopharmacology: The Third Generation of Progress*, ed. H.J. Meltzer, pp. 1267–72. New York: Raven Press.

Morley, J.E. & Levine, A.S. (1980). Stress-induced eating is mediated through endogenous opiates. *Science* **209**, 1259–61.

Morley, J.E. & Levine, A.S. (1981). Dynorphin-(1-13) induces spontaneous feeding in rats. *Life Science* **18**, 1901–3.

Morley, J.E. & Levine, A.S. (1982). Corticotropin releasing factor, grooming and ingestive behavior. *Life Science* **31**, 1459–64.

Morley, J.E., Levine, A.S., Murray, S.S. & Kneip, J. (1982). Peptidergic regulation of norepinephrine induced feeding. *Pharmacology, Biochemistry & Behavior* **16**, 225–8.

Morley, J.E., Elson, M.K., Levine, A.S. & Shafer, R.B. (1983). The effects of stress on central nervous system concentrations of the opioid peptide dynorphin. *Peptides* **3**, 901–6.

Morley, J.E., Levine, A.S., Gosnell, B.A. & Billington,

C.J. (1984*a*). Which opioid receptor mechanism modulates feeding? *Appetite* **5**, 61–8.

Morley, J.E., Gosnell, B.A. & Levine, A.S. (1984*b*). The role of peptides in feeding. *Trends in Pharmacological Sciences*, Nov., pp. 468–71.

Morley, J.E., Parker, S. & Levine, A.S. (1985*a*). Effects of butorphanol tartrate on food and water consumption in humans. *American Journal of Clinical Nutrition* **42**, 1175–8.

Morley, J.E., Levine, A.S., Gosnell, B.A. & Krahn, D.D. (1985*b*). Peptides as central regulators of feeding. *Brain Research Bulletin* **14**, 511–19.

Muscat, R., Montgomery, A.M.J. & Willner, P.J. (1989). Blockade of 8-OH-DPAT-induced feeding by dopamine antagonists. *Psychopharmacology* **99**, 402–8.

Myers, R.D. & McCaleb, M.L. (1980). Feeding: Satiety signal from intestine triggers brain's noradrenergic mechanism. *Science* **209**, 1035–7.

Neill, J.C. & Cooper, S.J. (1989). Evidence that d-fenfluramine anorexia is mediated by 5-HT$_1$ receptors. *Psychopharmacology* **97**, 213–18.

Nobrega, J.N. & Coscina, D.V. (1982). Effects of antidepressant treatment on feeding behaviour in rats. In *The Neural Basis of Feeding and Reward*, ed. B.G. Hoebel & D. Novin, pp. 525–34. Haer Institute: Brunswick, Maine.

Novin, D. & Vander Weele, D.A. (1977). Visceral involvement in feeding: There is more to regulation than the hypothalamus. In *Progress in Psychobiology & Physiological Psychology*, vol. 7, ed. J.M. Sprague & A.N. Epstein, pp. 193–241. New York: Academic Press.

Oku, J., Bray, G.A., Fisler, J.S. & Schemmel, R. (1984). Ventromedial hypothalamic knife-cut lesions in rats resistant to dietary obesity. *American Journal of Physiology* **246**, R943–R948.

Oltmans, G.A. & Harvey, J.A. (1972). Lateral hypothalamic syndrome and brain catecholamine levels after lesions of the nigrostriatal bundle. *Physiology & Behavior* **8**, 69–78.

Oomura, Y. (1976). Significance of glucose, insulin and free fatty acid on the hypothalamic feeding and satiety neurones. In *Hunger: Basic Mechanisms and Clinical Implications*, ed. D. Novin & W. Wyrwicka, pp. 145–57. New York: Raven Press.

Oomura, Y., Ooyama, H., Yamamoto, T. & Naka, F. (1967). Reciprocal relationship of the lateral and ventromedial hypothalamus in the regulation of food intake. *Physiology & Behavior* **2**, 97–115.

Opsahl, C.A. (1977). Sympathetic nervous system involvement in the lateral hypothalamic syndrome. *American Journal of Physiology* **232**, R128–R136.

Paykel, E.S., Mueller, P.S. & De La Vergue, P.M. (1973). Amitriptyline, weight gain and carbohydrate craving. *British Journal of Psychiatry* **123**, 501–7.

Perhach, J.L. & Barry, H. (1970). Stress responses of rats to acute body or neck restraint. *Physiology & Behavior* **5**, 443–8.

Phillips, A.G. & Fibiger, H.C. (1976). Deficits in stimulation-induced feeding and self-stimulation after 6-hydroxydopamine administration in rats. *Behavioral Biology* **16**, 127–43.

Pitts, G.C. & Bull, L.S. (1977). Exercise, dietary obesity and growth in the rat. *American Journal of Physiology* **232**, R38–R44.

Pollock, J.D. & Rowland, N. (1981). Peripherally administered serotonin decreases food intake in rats. *Pharmacology, Biochemistry & Behavior* **15**, 179–83.

Posadas-Andrews, A., Nieti, J. & Burton, M.J. (1983). Chlordiazepoxide induced eating: hunger or voracity? *Proceedings of the Western Pharmacology Society* **26**, 409–12.

Poschel, B.P.H. (1971). A simple and specific screen for benzodiazepine-like drugs. *Psychopharmacologia (Berlin)* **19**, 193–8.

Powley, T.L. & Keesey, R.E. (1970). Relationship of body weight to lateral hypothalamic feeding syndrome. *Journal of Comparative & Physiological Psychology* **70**, 25–36.

Randall, L.O., Schallek, W., Heise, G.A., Keith, E.F. & Bagdon, R.E. (1960). The psychosedative properties of metaminodiazepoxide. *Journal of Pharmacology & Experimental Therapeutics* **129**, 163–97.

Rau, J.H. & Green, R.S. (1975). Compulsive eating: A neuropsychologic approach to certain eating disorders. *Comprehensive Psychiatry* **16**, 223–31.

Richter, C.P. (1927). Animal behaviour and internal drives. *Quarterly Review of Biology* **2**, 307–43.

Robbins, T.W. & Fray, P.J. (1980). Stress-induced eating: Fact, fiction or misunderstanding? *Appetite* **1**, 103–33.

Robbins, T.W., Phillips, A.G. & Sahakian, B.J. (1977). Effects of chlordiazepoxide on tail pinch-induced eating in rats. *Pharmacology, Biochemistry & Behavior* **6**, 297–302.

Rodin, J. (1975). Effects of obesity and set point on taste responsiveness and food intake in humans. *Journal of Comparative & Physiological Psychology* **89**, 1003–9.

Rodin, J. (1980). The externality theory today. In *Obesity*, ed. A.J. Stunkard, pp. 226–39. W.B. Saunders: Philadelphia.

Rogers, P.J. & Blundell, J.E. (1980). Investigation of food selection and meal parameters during development of dietary-induced obesity. *Appetite* **1**, 85–97.

Rolls, B.J., Rowe, E.A. & Turner, R.C. (1980). Persistent obesity in rats following a period of consumption of mixed, high-energy diet. *Journal of Physiology (London)* **298**, 415–27.

Rolls, B.J., Rowe, E.A. & Rolls, E.T. (1982). How sensory properties of food affect human feeding behaviour. *Physiology & Behavior* **29**, 409–17.

Rothwell, N.J. & Stock, M.J. (1978). A paradox in the control of energy intake in the rat. *Nature* **273**, 146–7.

Rothwell, N.J. & Stock, M.J. (1979*a*). A role for brown adipose tissue in diet-induced thermogenesis. *Nature* **281**, 31.

Rothwell, N.J. & Stock, M.J. (1979*b*). Regulation of energy balance in two models of reversible obesity. *Journal of Comparative & Physiological Psychology* **93**, 1024–34.

Rothwell, N.J. & Stock, M.J. (1982). Dopaminergic mechanisms in diet-induced thermogenesis and brown adipose tissue metabolism. *European Journal of Pharmacology* **77**, 45–8.

Rowland, N.E. & Antelman, S.M. (1976). Stress-induced hyperphagia and obesity in rats: a possible model for understanding human obesity. *Science* **191**, 310–12.

Russek, M. (1976). A conceptual equation of intake control. In *Hunger: Basic Mechanisms and Clinical Implications*, ed. D. Novin & W. Wyrwicka, pp. 327–48. New York: Raven Press.

Sahakian, B.J. & Robbins, T.W. (1977). Isolation-rearing enhances tail pinch-induced oral behaviour in the rat. *Physiology & Behavior* **18**, 53–8.

Sahakian, B.J., Robbins, T.W. & Iversen, S.D. (1977). The effects of isolation rearing on exploration in the rat. *Animal Learning & Behaviour* **5**, 193–8.

Sahakian, B.J., Burdess, C., Luckhurst, H. & Trayhurn, P. (1982). Hyperactivity and obesity: The interaction of social isolation and cafeteria feeding. *Physiology & Behavior* **28**, 117–24.

Samanin, R., Ghezzi, D., Valzelli, L. & Garattini, S. (1972). The effects of selective lesioning of brain serotonin or catecholamine containing neurons on the anorectic activity of fenfluramine or amphetamine. *European Journal of Pharmacology* **19**, 318–22.

Samanin, R., Caccia, S., Bendotti, C., Borsini, F., Borroni, E., Invernizzi, R., Pattaccini, R. & Mennini, T. (1980). Further studies on the mechanism of serotonin dependent anorexia in the rat. *Psychopharmacology* **68**, 99–104.

Sanger, D.J., McCarthy, P.S. & Metcalf, G. (1981). The effects of opiate antagonists on food intake are stereospecific. *Neuropharmacology* **20**, 45–7.

Schachter, S. (1971). *Emotion, Obesity and Crime*. Academic Press: New York.

Schemmel, R. & Mickelson, O. (1974). Influence of diet, strain, age and sex on fat depot and body composition of nutritionally obese rat. In *The Regulation of the Adipose Tissue Mass*, ed. J.

Vogue & J. Boyer, pp. 238–253. American Elsevier: New York.

Schemmel, R., Mickelson, O. & Gill, J.L. (1970). Dietary obesity in rats: Body weight and body fat accretion in seven strains of rat. *Journal of Nutrition* **100**, 1941–8.

Schemmel, R., Teague, R.J. & Bray, G.A. (1982). Obesity in Osborne-Mendel and S5B/P1 rats: effects of sucrose solutions, castration and treatment with estradiol and insulin. *American Journal of Physiology* **243**, R347–R353.

Sclafani, A. (1980). Dietary obesity. In *Obesity*, ed. A.J. Stunkard, pp. 166–81. W.B. Saunders: Philadelphia.

Sclafani, A. & Gorman, A.N. (1977). Effects of age, sex and prior body weight on the development of dietary obesity in adult rats. *Physiology & Behavior* **18**, 1021–6.

Sclafani, A. & Kramer, T.H. (1983). Dietary selection in vagotomised rats. *Journal of the Autonomic Nervous System* **9**, 247–58.

Sclafani, A. & Springer, D. (1976). Dietary obesity in adult rats: Similarities to hypothalamic and human obesity syndromes. *Physiology & Behavior* **17**, 461–71.

Sclafani, A., Aravich, P. & Landman, M. (1979). Effects of atropine and vagotomy on appetite in hypothalamic rats. *Society for Neuroscience Abstracts* **5**, 223.

Scoville, B.A. (1976). Review of amphetamine-like drugs by the Food and Drug Administration: Clinical data and value judgements. In *Obesity in Perspective*, ed. G.A. Bray, pp. 441–3. U.S. Government Printing Office: Washington D.C.

Segal, D.S., Knapp, S., Kuczenski, R. & Mandell, A.J. (1973). The effects of environmental isolation on behavior and regional rat brain tyrosine hydroxylase and tryptophan hydroxylase activities. *Behavioral Biology* **8**, 47–53.

Shimomura, Y., Oku, J., Glick, Z. & Bray, G.A. (1982). Opiate receptors, food intake and obesity. *Physiology & Behavior* **28**, 441–5.

Siviy, S.M., Calcagnetti, D.J. & Reid, L.D. (1982). Opioids and palatability. In *The Neural Basis of Feeding and Reward*, ed. B.G. Hoebel & D. Novin, pp. 517–24. Haer Institute: Brunswick, Maine.

Sjostrom, L. (1980). Fat cells and body weight. In *Obesity*, ed. A.J. Stunkard, pp. 72–100. W.B. Saunders: Philadelphia.

Slade, P.D. (1973). A short anorectic behaviour scale. *British Journal of Psychiatry* **122**, 83–5.

Slochower, J. (1976). Emotional labelling and overeating in obese and normal weight individuals. *Psychosomatic Medicine* **38**, 131–9.

Smart, J.L. (1979). Undernutrition and the development of brain and behaviour. In *Chemical*

Influences on Behaviour, ed. K. Brown & S.J. Cooper, pp. 1–33. Academic Press: London.

Smart, J.L., Dobbing, J., Adlard, B.P.F., Lynch, A. & Sands, J. (1973). Vulnerability of developing brain: Relative effects of growth restriction during fetal and suckling periods on behaviour and brain composition of adult rats. *Journal of Nutrition* **103**, 1327–38.

Smith, D.A. (1972). Incentive as a factor in the behaviours of rats given lateral hypothalamic stimulation. *Physiology & Behavior* **8**, 1077–86.

Smith, G.P. & Gibbs, J. (1979). Postprandial satiety. In *Progress in Psychobiology & Physiological Psychology*, vol. **8**, ed. J.M. Sprague & A.N. Epstein, pp. 179–242. Academic Press: New York.

Smith, G.P., Jerome, C., Cushin, B.J., Eterno, R. & Simansky, K.J. (1981). Abdominal vagotomy blocks the satiety effect of cholecystokinin in the rat. *Science* **213**, 1036–7.

Spitzer, L., Marcus, J. & Rodin, J. (1980). Arousal-induced eating: A response to Robbins and Fray. *Appetite* **1**, 343–8.

Soper, W.Y. & Wise, R.A. (1971). Hypothalamically induced eating: eating from 'non-eaters' with diazepam. *TIT Journal of Life Sciences* **1**, 79–84.

Stanley, B.G. & Leibowitz, S.F. (1984). Stimulation of feeding and drinking by injection into the paraventricular nucleus. *Life Science* **35**, 2635–42.

Stanley, B.G., Kyrkouli, S.E. & Lampert, S.F. (1985). Hyperphagia and obesity induced by neuropeptide Y injected chronically into the paraventricular hypothalamus of the rat. *Society for Neuroscience Abstracts* **15**, 16.2.

Steffens, A.B. (1975). Influence of reversible obesity on eating behavior, blood glucose and insulin in the rat. *American Journal of Physiology* **228**, 1738–44.

Stellar, E. (1954). The physiology of motivation. *Psychological Review* **61**, 5–22.

Stephens, R.J. (1973). The influence of mild stress on food consumption in untrained mice and the effect of drugs, *British Journal of Pharmacology* **47**, 146P.

Stephens, D. & Kehr, W. (1985). Beta-carbolines can enhance or antagonize the effects of punishment in mice. *Psychopharmacology* **85**, 143–7.

Stern, J.S. (1984). Is obesity a disease of inactivity? In *Eating and Its Disorders*, ed. A.J. Stunkard & E. Stellar, pp. 131–9. Raven Press: New York.

Stricker, E.M., Cooper, P.H., Marshall, J.F. & Zigmond, M.J. (1979). Acute homeostatic imbalances reinstate sensorimotor dysfunctions in rats with lateral hypothalamic lesions. *Journal of Comparative & Physiological Psychology* **93**, 512–21.

Szmukler, G.I. (1987). Anorexia nervosa: A clinical view. In *Eating Habits: Food, Physiology and Learned Behaviour*, ed. R.A. Boakes, D.A.

Popplewell & M.J. Burton, pp. 25–44. John Wiley & Sons: Chichester.

Teitelbaum, P. (1961). Disturbances in feeding and drinking behaviour after hypothalamic lesions. In *Nebraska Symposium on Motivation*, ed. M.A. Jones, pp. 39–65. University of Nebraska Press: Lincoln: U.S.A.

Teitelbaum, P. & Epstein, A.N. (1962). The lateral hypothalamic syndrome: Recovery of feeding and drinking after lateral hypothalamic lesions. *Psychological Review* **69**, 74–90.

Telib, M., Raptis, S., Schroder, K.E. & Pfeiffer, E.F. (1968). Serotonin and insulin release in vitro. *Diabetologia* **4**, 253–6.

Tepperman, F.S., Hirst, M. & Gowdey, C.W. (1981). A probable role for norepinephrine in feeding after hypothalamic injection of morphine. *Pharmacology, Biochemistry & Behavior* **15**, 555–8.

Tenen, S.S. & Miller, N.E. (1964). Strength of electrical stimulation of lateral hypothalamus, food deprivation and tolerance for quinine in food. *Journal of Comparative Physiological Psychology* **58**, 55–62.

Therriault, D.G. & Mellin, D.B. (1971). Cellularity of adipose tissue in cold-exposed rats and the calorigenic effect of norepinephrine. *Lipids* **6**, 486–91.

Thoa, N.B., Tizabi, Y. & Jacobwitz, D.M. (1977). The effect of isolation on catecholamine concentration and turnover in discrete areas of rat brain. *Brain Research* **131**, 259–69.

Towell, A., Muscat, R. & Willner, P. (1989). Noradrenergic receptor interactions in feeding elicited by stimulation of the paraventricular hypothalamus. *Pharmacology, Biochemistry & Behavior* **32**, 133–9.

Trenchard, E. & Silverstone, T. (1983). Naloxone reduces the food intake of normal human volunteers. *Appetite* **4**, 43–50.

Tretter, J.R. & Leibowitz, S.F. (1980). Specific increase in carbohydrate consumption after NE injections into the paraventricular nucleus. *Society for Neuroscience Abstract* **6**, 532.

Ungerstedt, U. (1971). Adipsia and aphagia after 6-hydroxydopamine induced degeneration of the nigrostriatal dopamine system. *Acta Physiologica Scandinavica*, Suppl. 367, pp. 95–122.

Valenstein, E.S., Cox, V.C. & Kakolewski, J.W. (1968). The motivation underlying eating elicited by lateral hypothalamic stimulation. *Physiology & Behavior* **3**, 969–71.

Vigersky, R., Loriaux, D.L., Andersen, A.E., Mecklenburg, R.S. & Vaitukaitis, J.L. (1976). Delayed pituitary hormone response to LRF and TRF in patients with anorexia nervosa and with secondary amenorrhea associated with simple weight loss. *Journal of Clinical Endocrinology & Metabolism* **43**, 893–900.

Wade, G.N. (1976). Sex hormones, regulatory behaviours and body weight. In *Advances in the Study of Behaviour*, vol. **6**, ed. J.S. Rosenblatt, R.A. Hinde, E. Shaw & C. Beer, pp. 201–79. Academic Press: New York.

Wagner, J.W. & De Groot, J. (1963). Changes in feeding behaviour after intracerebral injections in the rat. *American Journal of Physiology* **204**, 483–7.

Waldbillig, R.J., Bartness, T.J. & Stanley, B.G. (1981). Increased food intake, body weight and adiposity in rats after regional neurochemical depletion of serotonin. *Journal of Comparative & Physiological Psychology* **95**, 391–405.

Walsh, B.T. (1987). Psychopharmacology of bulimia. In *Psychopharmacology: The Third Generation of Progress*, ed H.J. Meltzer, pp. 1277–80. Raven Press: New York.

Walsh, B.T., Stewart, J.W., Roose, S.P., Gladis, S.M. & Glassman, A.H. (1985). Treatment of bulimia with phenelzine: A double-blind placebo-controlled study. *Archives of General Psychiatry* **41**, 1105–9.

Weinstock, M., Speizer, Z. & Ashkenazi, R. (1978). Changes in brain catecholamine turnover and receptor sensitivity by social deprivation in rats. *Psychopharmacology* **56**, 205–9.

Wermuth, B.M., Davis, K.L., Hollister, L.E. & Stunkard, A.J. (1977). Phenytoin treatment of the binge-eating syndrome. *American Journal of Psychiatry* **134**, 1249–53.

Willenbring, M.L., Levine, A.S. & Morley, J.E. (1986). Stress induced eating and food preference in humans: A pilot study. *International Journal of Eating Disorders* **5**, 855–64.

Willner, P. & Montgomery, A.M.J. (1980). Attenuation of amphetamine anorexia in rats following subchronic treatment with a tricyclic antidepressant. *Communications in Psychopharmacology* **4**, 101–6.

Winn, P., Williams, S.F. & Herberg, L.J. (1982). Feeding stimulated by very low doses of amphetamine administered systemically or by microinjection into the striatum. *Psychopharmacology* **78**, 336–41.

Winokur, A., March, V. & Mendels, J. (1980). Primary affective disorder in relatives of patients with anorexia nervosa. *American Journal of Psychiatry* **137**, 695–8.

Wise, R.A. (1978). Catecholamine theories of reward: A critical review. *Brain Research* **152**, 215–47.

Wise, R.A. (1982). Neuroleptics and operant behaviour: The anhedonia hypothesis. *Behavioral and Brain Sciences* **5**, 39–87.

Wise, R.A. & Erdmann, E. (1973). Emotionality, hunger and normal eating: Implications for interpretation of electrically induced behaviour. *Behavioral Biology* **8**, 519–31.

Wolgin, D.L. & Teitelbaum, P. (1978). The role of activation and sensory stimuli in recovery from lateral hypothalamic damage in the rat. *Journal of Comparative & Physiological Psychology* **92**, 474–500.

Woods, D.J. (1969). The effect of age and chlorpromazine on self-starvation in activity wheels. M.A. Thesis, Department of Psychology, Northwestern University.

Wurtman, J.J. & Wurtman, R.J. (1979). Drugs that enhance central serotoninergic transmission diminish elective carbohydrate consumption by rats. *Life Science* **24**, 895–904.

Young, J.K. (1975). A possible neuroendocrine basis of two clinical syndromes: Anorexia nervosa and Kleine-Levin syndrome. *Physiological Psychology* **3**, 322–30.

Ziegler, H.P. (1975). Oral satisfaction and obesity: The sensual feel of food. *Psychology Today* **9**, 62–76.

Zigmond, M.J. & Stricker, E.M. (1973). Recovery of feeding and drinking by rats after intraventricular 6-hydroxydopamine or lateral hypothalamic lesions. *Science* **182**, 717–20.

Zigmond, M.J. & Stricker, E.M. (1974). Ingestive behavior following damage to central dopamine neurons: Implications for homeostasis and recovery of function. *Advances in Biochemical Psychopharmacology* **12**, 385–402.

Zucker, L.M. & Zucker, T.F. (1961). Fatty, a new mutation in the rat. *Journal of Heredity* **52**, 275–8.

9

Screening methods for anorectic, antiobesity and orectic agents

JERRY SEPINWALL and ANN C. SULLIVAN

9.1 INTRODUCTION

Medical, social and commercial interest in controlling body weight is prevalent in many modern societies. A pharmacological means of achieving such control is a frequently sought objective. Therefore, much interest and importance attaches to methods for identifying drugs that can alter body weight, and this chapter describes such methods. For treating obesity, the focus of drug screening programs is to find agents that either lower body weight or prevent weight regain. Accordingly, this search is not confined solely to agents that diminish appetite, i.e., anorectics, but extends to agents that can reduce obesity by any one of a number of mechanisms. Therefore, this chapter will consider methods used to screen for 'antiobesity' agents, including not only anorectics but also thermogenic agents and inhibitors of dietary lipid or carbohydrate absorption. In contrast to the bulk of literature dealing with drugs that reduce appetite, food intake or body weight, the literature covering agents that increase food intake is much smaller and there are relatively few standardized screening methods for orectic compounds.

Within the context of this chapter, the term 'screening methods' typically refers to various experimental procedures involving animals, in which novel compounds are evaluated for their ability to decrease or increase the variable of interest, i.e., body weight, food intake, etc. A frequently used term is 'animal model', which indicates an experimental preparation or procedure that is reliable and valid for identifying compounds with the desired properties. These models can consist of animals with a certain type of neuroanatomical (e.g., ventromedial hypothalamic) or biochemical (e.g., genetic) lesion that influences body weight. Screening methods and animal models for characterizing compounds that alter body weight, appetite or food intake have frequently been reviewed and much useful information is provided by, for example, Blundell (1987), Fuller & Yen (1987), Sullivan (1987), Kanarek & Marks-Kaufman (1988), Triscari et al. (1988), and Montgomery (Chapter 8 of this volume).

There are several critical factors to keep in mind with respect to the issue of developing valid methods in animals for detecting useful antiobesity agents. The first is that there are multiple determinants of body weight regulation, and efforts to control body weight by manipulating one determinant are often futile because of compensation by another. Second, as Montgomery (this vol., Ch. 8) points out, most animal models involve relatively homogeneous rat or mouse strains eating unchanging diets, e.g., lab chow, whereas there are great genetic, environmental and dietary differences among the human population(s) to which the drugs identified in the animal models will be directed. Studying subhuman primates would redress this limitation to some extent, but such studies have been infrequent. Third, there are limitations to the parallels that can be drawn between human feeding disorders and various animal models. For example, 'cafeteria' feeding to induce obesity (Sclafani & Springer, 1976) appears to have much in common with dietary induced obesity in humans, and the cafeteria model does offer rats a choice of foods varying in sensory texture, macro-nutrient composition and postabsorptive consequences. However, Kanarek & Marks-Kaufman (1988) see these variations as limiting the overall usefulness of this model because the differences in individual preferences, in the macronutrient content for each animal, in measuring energy intake accurately, and in other factors, make interpretation of results complex and difficult. Fourth, effective antiobesity activity does not necessarily require a reduction in *food intake*; food intake can actually increase as an individual attempts to compensate for an induced body weight loss. Thus it is not sufficient just to measure food intake. One needs to examine body weight, carcass composition, energy expenditure and whatever else is relevant to the target at which one is aiming.

All of these factors indicate that the task of developing animal models for identifying antiobesity agents and compounds to increase body weight is a most challenging one. Nevertheless, many laboratories have successfully developed screening methods and some examples of these are reviewed in the succeeding sections. To be successful as therapeutic agents, the compounds that are identified by these methods may have to be used in conjunction with one another, e.g., an appetite suppressant together with an agent that enhances energy expenditure. Such a combination might overcome the compensatory responses that occur when only a single process is altered.

9.2 SCREENING METHODS FOR ANTIOBESITY AGENTS

Those engaged in a search for novel compounds to treat obesity need to stratify that search in a manageable way so as to be able to screen a reasonable number of compounds within a given time frame and to have confidence that the tests or models being used will identify active compounds. This stratification typically involves primary, secondary and tertiary levels of screening tests. Pragmatic needs dictate which specific tests or models will be selected to represent these levels.

9.2.1 Primary tests

Above all at the primary level, the investigator is interested in screening procedures that will have high capacity, sensitivity and simplicity. Capacity will be dictated by the number of compounds expected to be available, i.e., to be synthesized by medicinal chemists, to be isolated from microbial broths, or pre-existing in some collection of compounds. At this initial stage, the sensitivity should be such that the procedure being used should be biased toward giving too many positives ('false' positives), rather than too few, in order not to miss any active compounds. The task of secondary and tertiary tests is to distinguish between the true and false positives.

9.2.1.1 CCK-like appetite suppressants

Interest in cholecystokinin (CCK) as an antiobesity agent was stimulated by the initial finding of Gibbs *et al.* (1973*a*) that exogenous administration of the full 33 amino acid form of CCK could apparently induce satiety in sham feeding rats with open gastric fistulas. Their speculations (Smith *et al.*, 1974) about the role of CCK, released endogenously in response to food entering the gut, as a mediator of 'intestinal' satiety led them and others to identify the C-terminal octapeptide (CCK_{26-33}) as the active portion of CCK_{1-33} (Smith & Gibbs, 1984). Because of rapid hydrolysis of CCK-8 by gut and plasma enzymes, however, this compound is impractical to consider as a therapeutic agent. Accordingly, there has been a search for orally active, longer-acting analogs of CCK-8, and chemists in various laboratories have synthesized such compounds. CCK-8 has a number of effects *in vitro* on several target tissues and some of these effects have been used as primary screening measures.

CCK induces contraction of the gall bladder and ileum and stimulates amylase secretion by the pancreas. CCK receptors have also been identified in the pyloric sphincter and the proposal has been advanced that the main satiety actions of CCK are mediated by effects at this site (Smith *et al.*, 1984; Moran & McHugh, 1988). Accordingly, both biochemical and bioassay methods have been established to screen CCK analogs. The biochemical methods typically consist of competition assays in which the affinity of CCK-8 and various analogs or derivatives for CCK receptors is determined by measuring the ability of these compounds to displace radiolabeled CCK from its binding sites in various tissues. In one recent study, IC_{50} values for displacement of radiolabeled CCK from rat pylorus, vagus and brainstem were correlated with functional measures of CCK activity (Moran *et al.*, 1988; Sawyer *et al.*, 1988). It is reasonable to expect that more data of this type will become available in the patent and the scientific literature in the near future, (see e.g., Danho, 1987; Danho *et al.*, 1988, 1989).

The induction of amylase secretion by rat pancreas has been used as a screening procedure by many investigators (Ruiz-Gayo *et al.*, 1985; Mendre *et al.*, 1988; Sawyer *et al.*, 1988; Sugg *et al.*, 1988). However, as is true for all *in vitro* assays, the user must beware of the potential discrepancies between *in vitro* and *in vivo* data. This point has been made by Nagain *et al.* (1987) who compared the potencies of a series of CCK-8 analogs on pancreatic amylase secretion *in vitro* as well as *in vivo* in anesthetized rats. For most compounds, the relative order of potencies was similar *in vitro* and *in vivo*. However, some analogs were anywhere from nine times less potent to two and a half times more potent *in vivo* than *in vitro*. This illustrates the well known contributions made *in vivo* by absorption, metabolism and the presence of endogenous factors, such as other neurotransmitters in the environmental milieu.

Other functional *in vitro* assays used to screen CCK analogs have included measurement of contraction of the guinea pig gall bladder and ileum (Ruiz-Gayo *et al.*, 1985; Blosser *et al.*, 1987; Rosamond *et al.*, 1988; Sugg *et al.*, 1988). Based on a high correlation between efficacy for inducing gall bladder contraction and for inhibiting food intake, one group has concluded that the gall bladder assay is most useful and valid for identifying potential anorectics and that the receptors which mediate the gall bladder and anorectic effects are highly similar in their structural specificity (Rosamond *et al.*, 1988).

Reference was made earlier to the identification of CCK receptors in the pyloric sphincter and to the hypothesis that CCK slows gastric emptying by constricting this sphincter, thereby enhancing satiation (Smith *et al.*, 1984). Accordingly, one might expect measurement of pyloric contraction to be an especially useful screening method. This remains to be established. Sawyer *et al.* (1988) found that the CCK analog U-67827E, was only

two to two and a half times more potent than CCK-8 in the pyloric sphincter assay but ten times more potent at inhibiting food intake. Two other analogs were approximately 40 times weaker than CCK-8 *in vitro* but only seven or eight times weaker *in vivo*. Differences in elimination half-lives *in vivo* or other factors might have been responsible for these differences between tests. Furthermore, the pyloric sphincter procedure might predict relative potencies accurately even if it does not predict absolute potencies. Appraisal will be possible only after additional studies to characterize the pyloric strip (Murphy *et al.*, 1987) and additional screening studies appear.

9.2.1.2 Inhibitors of gastrointestinal lipases (tetrahydrolipstatin)

Controlling body weight by use of anorectic agents typically means attempting to impede meal initiation or accelerate meal termination, so as to reduce the amount of calories ingested. There are other ways, however, to control energy intake including the use of agents that will block the intestinal absorption of either carbohydrates or lipids. Carbohydrate absorption is blocked by α-glucosidase inhibitors, such as acarbose (Pagano *et al.*, 1986; William-Olsson, 1986) and AO-128, a valiolamine derivative (Inoue *et al.*, 1987), and both of these compounds have antiobesity activity. Since fat constitutes as much as 40% of the diet in Western nations and since triglycerides constitute up to 90% of the fat intake, it has been recommended that inhibition of triglyceride absorption would be an efficacious antiobesity therapy. This could be achieved by inhibiting the enzyme pancreatic lipase (PL), which hydrolyzes triglycerides, thereby permitting them to be absorbed across the intestine (Johnson, 1985). To screen compounds for inhibition of hydrolysis, *in vitro* assays can be established, including several to measure the presence or absence of inhibition of other hydrolases, such as α-amylase, phospholipase A_2 and chymotrypsin. This is done to identify selective inhibitors of PL.

Such a screening strategy was adopted by Hadvary *et al.* (1988), using an *in vitro* assay for the hydrolysis of triolein to fatty acids by a readily available PL. This project first identified the compound lipstatin, isolated from *Streptomyces tox-*

ytricini, and later its synthetic derivative, tetrahydrolipstatin (THL). THL exhibited similar dose–response curves and IC_{50} values for inhibition of triolein hydrolysis by purified porcine pancreatic lipase, human duodenal juice and mouse intestinal fluid, but had no effect on the activity of the other hydrolytic pancreatic enzymes. Thus, targeted *in vitro* screening in circumstances where a specific mechanism was being pursued led to identification of a potential antiobesity agent.

In order to characterize the activity of THL *in vivo*, Hogan *et al.* (1987) administered THL orally to mice and squirrel monkeys immediately after a liquid test meal that contained 10% olive oil labeled with [^{3}H]triolein, and measured the fat content of fecal samples collected over the next three days. Vehicle treated animals absorbed more than 90% [^{3}H]triolein whereas THL potently and dose-dependently blocked absorption in both mice and squirrel monkeys.

9.2.1.3 Enhancers of energy expenditure (Thermogenic β-adrenergic agonists)

In addition to decreasing energy intake, increasing energy expenditure, by increasing the production of heat rather than the deposition of calories as fat, has also been advanced as a potential therapeutic approach to obesity (Levin & Sullivan, 1984; Arch *et al.*, 1987). Therefore, compounds have been sought that would enhance thermogenesis. Noradrenaline (NA) is known to act through the sympathetic nervous system as the major endogenous enhancer of thermogenesis and to do so via a β-adrenoceptor. Since NA has various effects at other adrenergic receptor subtypes, however, it has been proposed that more selective β-adrenoceptor agonists might be useful antiobesity agents (Arch *et al.*, 1985).

In rodents, brown adipose tissue (BAT) plays a major role in mediating diet-induced thermogenesis (Rothwell & Stock, 1979). Bukowiecki *et al.* (1980) described an appropriate method for isolating brown adipocytes and studying the effects of adrenergic agonists on rates of oxygen consumption in these cells. The use of such a method to evaluate a novel β-agonist, a bis-phenethylaminobutyl benzamide derivative (Ro 16-8714), was reported by Meier *et al.* (1985). This compound had a bell-shaped dose–response curve and pro-duced a maximum response, at approximately 30 μm, that was 55% of the increase induced by 100 nM NA. These were shown to be β-adrenoceptor-mediated responses, since propranolol blocked the effects of Ro 16-8714 and NA.

Given the difficulty in handling BAT cells for the measurement of respiratory rate, other investigators have used the more reproducible phenomenon of lipolysis in BAT, since this is a step that precedes the increase in oxygen consumption. Thus, Yen *et al.* (1985) showed that a phenethylaminobutyl benzamide derivative (LY104119) dose-dependently enhanced lipolysis in epididymal white adipose tissue from normal a/a mice and yellow obese A^{vy}/a mice. Similarly, Arch *et al.* (1984) compared members of a novel series of phenethylaminopropyl benzoate derivatives for their ability to increase lipolysis in BAT from rats. These compounds, e.g., BRL 28410, differed in potency but all produced the same maximal effect, which was equivalent to that of isoproterenol. Thus they all appeared to be full agonists in this preparation. Based upon differences in the potencies and intrinsic efficacies of the compounds in comparison to known selective β-agonists in various tissue, Arch *et al.* (1984) postulated that several analogs had a selective affinity for BAT and that BAT contains a unique subtype of β-adrenoceptor.

Thermogenic effects of these compounds have been characterized *in vivo*. It was shown by indirect calorimetry that Ro 16-8714 was highly potent and efficacious in increasing oxygen consumption in normal rats and in genetically obese strains of mice (ob/ob) and rats (fa/fa) after a single oral administration. The maximal response in normal rats was a doubling of metabolic rate. Similar efficacy was seen in ob/ob mice but the compound was less potent. In contrast, fa/fa rats were as sensitive as normal rats to low doses of Ro 16-8714 but their maximum response was only about a 60% increase in oxygen consumption. The effect of Ro 16-8714 was mediated directly by β-adrenoceptors since it was blocked by propranolol yet still occurred in animals depleted of endogenous catecholamines by reserpine pretreatment (Meier *et al.*, 1985). Similar effects were seen with LY104119 in obese A^{vy}/a mice: LY104119 enhanced respiratory rate, and propranolol blocked this effect (Yen *et al.*, 1985).

Finally, in pithed female Sprague-Dawley rats,

Arch *et al.* (1987) simultaneously measured the desired therapeutic effect for a thermogenic agent, i.e., an increase in oxygen consumption, and undesired effects, i.e., increases in heart rate and relaxation of the uterus. This experiment combined aspects of primary and secondary screening methods in the sense that it involved only acute administration of a test compound and did not require the animals to be trained, but it did begin to address the crucial issue of 'therapeutic index', which represents the ratio or separation between the dose levels at which desired effects occur versus those at which undesired actions appear. Arch *et al.* (1987) showed that BRL 26830A dose-dependently increased metabolic rate and was more potent than the selective β_1-adrenoceptor agonist denopamine or the selective β_2-agonist salbutamol. Relaxation of the uterus and increase in heart rate also occurred within the same dose range of BRL 26830A. However, BRL 26830A differed from denopamine by producing a smaller and less potent increase of heart rate and from salbutamol by being much less potent at relaxing the uterus. It was concluded that BRL 26830A showed a superior profile to either denopamine or salbutamol because it was more potent than these reference compounds at increasing oxygen consumption but less potent on the heart or uterus. As was the case with their *in vitro* results, Arch *et al.* (1987) postulated that BRL 26830A might be an agonist selective for a novel subtype of the β-adrenoceptor.

9.2.1.4 Feeding tests

Whether or not the compounds to be screened work via a known or postulated mechanism and have been studied previously in the *in vitro* and *in vivo* models discussed above, it is always necessary to measure their effects on food intake. The choice of a feeding test for primary screening will be guided by the desire to have a procedure that requires little, if any, training and little investment of time and resources. Most frequently, investigators have begun their screening sequence by using *lean* animals subjected to some specified period of food deprivation and by measuring inhibition of food intake during a period of several hours after treatment. Many different protocols have been reported in the literature and Blundell

(1987; cf. table 1 in that citation), in the course of discussing different substances that have been observed to reduce food intake, has presented a valuable survey of several of these.

Early papers on fenfluramine and related compounds illustrate procedures commonly used to screen for compounds inhibiting food intake. D,L-fenfluramine and its two enantiomers, D- and L-fenfluramine, were initially compared by Le Douarec *et al.* (1966) in rats trained to have daily access to food for only seven hours. On treatment days, the test compound was administered 60 min before access to food and food intake was measured at two and seven hours; the D-isomer was about twice as potent as the racemate and three to four times as potent as the L-isomer. In studies on D-fenfluramine and related compounds, such as *m*-chlorophenylpiperazine, Garattini and colleagues trained female rats for ten days on a routine of four hours daily access to chow pellets. Subsequently, test compounds were given 30 min before feeding and food intake was measured at the end of the first hour (Samanin *et al.*, 1979; Garattini & Samanin, 1985; Garattini *et al.*, 1987a, b). Similarly, Clineschmidt *et al.* (1974) trained female rats for two to three weeks to eat their daily ration of powdered chow within two hours, and measured the decrease in two-hour food intake induced by test compounds injected intraperitoneally (i.p.) 30 min before access to food. Hill *et al.* (1983) trained male rats for two weeks to consume chow within a five-hour period of daily access; test compounds including a novel series of 1,3-diaryltriazines were administered orally for at least five consecutive days and food intake was determined after one and five hours each day.

One feature that is evident from all of these procedures is the period of training, ranging from two days up to three weeks, invested by the experimenters in these protocols. This is done to surmount problems created by variability in food intake values in untrained rodents and to ensure the stability of the baseline or control food intake readings. Rarely are untrained food-deprived or starved rodents employed.

Earlier (Section 9.2.1.1) procedures for identifying CCK agonists were discussed. In general, similar protocols have been used to study the effects of CCK analogs on feeding behavior as for fenfluramine and related compounds. In one of the earliest papers, Gibbs *et al.* (1973*a*) used 17-h food-deprived rats that had access to liquid or solid diets and received i.p. injections of various forms of CCK. Anika *et al.* (1977) trained rats on a daily 15-h food deprivation schedule. Immediately after i.p. injection of a test compound, the animals received a weighed amount of chow pellets and food intake was measured 30 and 60 min later. Rosamond *et al.* (1988) trained rats on a daily schedule of three-hour access to powdered chow and then measured food intake at 30 min and three hours after i.p. injection of novel CCK analogs. Similarly, Sawyer *et al.* (1988) measured food intake at 75, 135 and 225 min into a daily five-hour period of access to powdered food in animals treated with novel CCK agonists. With CCK itself, which is short acting, the pretreatment times were usually short, e.g., zero or 15 min. However, with synthetic analogs that were made in the hope that they might be long acting, the pretreatment times were often similar to those for other types of antiobesity agents. One approach to determining whether such analogs might indeed be longer acting involved a two-meal feeding test (Danho, 1987; Danho *et al.*, 1989). In this model, after an initial 48-h fast, rats were trained for four days to have access to powdered chow in two separate one-hour periods (0900 to 1000 and 1200 to 1300 h) each day. On the fifth day, test compounds were administered 15 min before the first period of access to food. CCK typically suppressed food intake only during the first access period; in the second, there was a rebound effect. However, certain analogs, e.g., the heptapeptide derivative containing threonine sulphate and N-methylphe-nylalanine instead of aspartate and phenylalanine, respectively, in positions 32 and 33 of CCK, were able to decrease food intake in both access periods.

These examples indicate that there is fairly broad agreement among investigators with respect to primary screening methods that are used for measuring the effects of new compounds on food intake. Although there are differences in strain and sex of rats, type of food, number of days of training, length of the daily access period to food, and times at which readings are taken, the same general strategy is used by many different investigators. Beyond this point, however, things become somewhat more complex.

9.2.2 Secondary and tertiary tests

Whereas most of the primary screening methods involve acute administration of test compounds to lean rodents, secondary screens typically involve a greater variety of models and procedures, including many described by Montgomery (see Chapter 8). Studies are conducted in meal-fed and ad-libitum-fed lean and obese animals. Obesity can be diet-induced, lesion-induced or genetic. Sham feeding preparations with chronically implanted gastric fistulas are sometimes used. Additional species are also included at this stage, such as rabbit, cat, dog and monkey. With compounds expected to work by decreasing food intake, there is some focus upon repeated administration for one to two weeks. For all compounds, some initial attention may be paid to loss of body weight or to reductions in body weight gain. Some initial assessment of effects on meal patterning may also begin in this phase. Finally, duration of action, selectivity (does the compound only affect intake of food and not water, or does it affect only lab chow but not highly palatable foods?) and establishment of a conditioned taste aversion can be determined.

The division between secondary and tertiary levels is to some extent arbitrary. One feature of tertiary tests is duration of treatment: typically this will be one to three months. In addition to the arduous aspects of continuing treatment for such a long period, the studies will often conclude with the labor intensive procedure of carcass analysis. A second feature concerns mechanism of action: in tertiary studies, investigators will often administer the test compound in combination with other agents, e.g., known neurotransmitter antagonists, to identify the relevant receptors at which the drug is acting. A third feature consists of very detailed behavioral analysis of the drug's effects. Finally, at the tertiary level, novel procedures for analyzing antiobesity efficacy may be applied to a compound of interest.

9.2.2.1 CCK-like appetite suppressants

Included in the earliest paper examining the effect of CCK on food intake in rats were two procedures that can be considered secondary screening methods. Gibbs *et al.* (1973*a*) demonstrated that CCK decreased the intake of both solid and liquid diets; this effect was selective since an effective dose of CCK did not suppress drinking in these rats after water deprivation. Gibbs *et al.* (1973*b*) then used sham feeding rats to characterize the effects of CCK after 17 h of food deprivation. When the chronic gastric fistula was open, CCK and synthetic CCK-8 dose-dependently inhibited feeling. Higher doses of CCK or CCK-8 did not affect the initiation of meals but terminated these meals more rapidly. This study thus established the utility of the sham feeding model for providing essential information about the efficacy and point of action of CCK on feeding behavior.

Sham feeding has been employed by others to study CCK-8. For example, Reidelberger & Solomon (1986) compared the effects of intravenous (i.v.) infusions of CCK-8 on normal feeding, sham feeding and pancreatic secretion in rats adapted to overnight food deprivation. Pancreatic amylase output was stimulated maximally at much lower doses than those needed to inhibit food intake, whether the fistulas were closed or open. Among other things, it was concluded that not all of CCK's ability to inhibit food intake is mediated by a mechanism involving gastric distension. However, it is important to bear in mind that sham feeding differs in several significant respects from real feeding. For example, the effects of CCK on real feeding are vagally mediated since they are abolished by vagotomy (Smith *et al.*, 1981), whereas this is not true for sham feeding (Kraly, 1984).

Another area of secondary screening that can be illustrated by reference to the CCK literature concerns assessment of conditioned taste aversions (CTAs). The basic question here is whether a compound that decreases food intake might be doing so by causing malaise. A common way to measure this in rodents is with a 'bait shyness' CTA paradigm which assesses whether administration of a drug following consumption of a novel flavor suppresses consumption of that flavor on a subsequent exposure (Garcia *et al.*, 1974). Unlike lithium chloride (LiCl) and apomorphine, neither CCK nor CCK-8 supported the development of a CTA in rats (Gibbs *et al.*, 1973*a*; Holt *et al.*, 1974). West and coworkers (1984, 1987*a*) argued that the establishment of a CTA does not necessarily mean that the compound in question is reducing intake because of malaise, since agents which do not

cause malaise in humans are capable of establishing aversions in rodents; nevertheless, the absence of a CTA does tend to rule out malaise as the explanation of a decrease in intake. West *et al.* (1987*a*) also compared CCK-8 and LiCl by infusing these compounds shortly after the start of each meal for eight days. LiCl significantly reduced the number of meals per day without affecting meal size or duration. In contrast, CCK-8 significantly reduced meal size while there was a large compensatory increase in the number of meals per day. These differences in patterns of effects between the two compounds were interpreted to mean that infusion of LiCl during a meal led to a CTA whereas infusion of CCK had a true satiety effect. Thus behavioral analysis was able to offer a useful characterization of the basis of CCK's appetite-suppressant effects. Unfortunately, behavioral analysis is not always unambiguous since some studies have demonstrated the establishment of a conditioned aversion to CCK (Deutsch & Hardy, 1977; Swerdlow *et al.*, 1983; Moore & Deutsch, 1985; Verbalis *et al.*, 1986). The main lesson from these studies is that the details of the behavioral methods, i.e., duration of food or water deprivation, attractiveness of the test drinking solution, dose of CCK, one-bottle versus two-bottle choice test, all play essential roles in determining the outcome of a CTA study.

Extension of the assessment of CCK to other species is exemplified by several studies. Houpt *et al.* (1978) adapted rabbits to an overnight fast five days a week. On test days, the rabbits were treated i.v. with CCK, CCK-8, caerulein or vehicle, and food intake was measured for the first 15 min of the daily eight-hour access to chow pellets. Taste aversion tests were also carried out: after several days of saline injections, a novel flavor was introduced into the food for one five-minute test period, after which an i.v. injection of CCK, CCK-8 or caerulein was made. Four days later, the food with the novel flavor was again presented. These taste aversion tests failed to demonstrate a strong aversion of flavors associated with any of the three test compounds, suggesting that the decrease in food intake was not due to such an aversion. Similarly, CCK, CCK-8 and caerulein all dose-dependently decreased food intake in four-hour food-deprived pigs, as did two releasers of endogenous CCK, sodium oleate and protein hydrolysate; as was the case with the rabbits, a CTA could not be formed to any of the test compounds (Anika *et al.*, 1981).

CCK-8 and one analog, Boc-CCK-7, have also been shown to inhibit feeding in cats with chronically implanted gastric fistulas; the analog was somewhat less potent than CCK-8 (Bado *et al.*, 1988). Because there were no large dose–response differences in the effects of CCK on normal versus sham feeding, the authors, like Reidelberger & Solomon (1986), concluded that the suppressive effects of CCK are not mediated solely by gastric distension. In other experiments 24-h intake in nonfasted cats was studied. One of the CCK compounds was infused for 90 min, at the end of which time the cats were given a daily ration of liver. Under control conditions, the cats ate approximately one-third of their daily intake within the first 30 min, and a further 15 to 20% during each of the next two two-hour periods. CCK-8 reduced intake by 40 to 50%, depending on dose, in the first 30 min, and by 45 to 55% during the next two hours. Boc-CCK-7 only decreased food intake during the first 30 min.

The activity of CCK has also been demonstrated in rhesus monkeys (Gibbs *et al.*, 1976). These animals were adapted to three hours of daily access to standard chow. Intravenous infusion of CCK immediately before food access dose-dependently reduced food intake, with most of the efficacy being seen during the first 15 min. Thus, as in other species, CCK was rather short acting. CCK-8 had similar activity. In these animals, large decreases in food intake were also induced by gastric preloads of L-phenylalanine, a potent releaser of endogenous CCK, but not by D-phenylalanine, a weak releaser.

The final area of secondary screening that can be illustrated by reference to studies of CCK agonists is that of initial meal patterning analysis. In one of the earliest meal pattern analyses Gibbs *et al.* (1973*b*) showed in sham feeding rats that CCK terminated meals more rapidly. In a subsequent study Hsiao *et al.* (1979) assessed the effects of CCK-8 in nondeprived rats trained to bar press for food on a fixed ratio (FR) 5 schedule. They analyzed the first and second meals during the dark phase of the light dark cycle, taking advantage of the fact that rats spontaneously eat a large meal shortly after the onset of the dark phase. A meal was defined as the eating of six or more pellets, and

a meal was considered to be terminated when 10 min elapsed without a bar press. The rats were placed into the test chamber five minutes before the dark phase. Administration of CCK-8 10 min after the end of the first meal greatly prolonged the intermeal interval (IMI) without altering the size of the second meal. In another experiment, these investigators showed that three-hour infusions of CCK could likewise increase the length of the IMI dose-dependently in mildly (six-hour) food deprived rats. In a third experiment they administered CCK five minutes before access to the bar press chamber. CCK did not delay the start of the first meal or the rate of eating, but did reduce its size by accelerating its completion. The next question addressed by these investigators (Hsiao & Wang, 1983) was the effect on meal patterns of a continuous two-day infusion of CCK-33 from an implanted osmotic minipump. In this pilot experiment on two rats trained to bar press on the FR 5 schedule, CCK did not influence either meal size (number of pellets per meal) or rate, but did suppress the number of meals and total intake in both the light and dark phases of the daily cycle. Thus over the course of two days, CCK was quite effective at decreasing food intake in these rats.

This meal pattern analysis was extended further by West *et al.* (1984) who administered CCK-8 via implanted i.p. catheters for six days on a meal contingent basis. That is, a computer-controlled infusion of CCK was made over 30 s whenever a rat took two food pellets within 15 s from a food hopper. A meal was defined as the taking of at least five pellets without any pauses longer than five minutes. Infusions of CCK shortly after meal initiation significantly reduced meal size throughout the six-day period. Meal duration, however, was only reduced on days 1 and 2 and had recovered to baseline levels by day 6. Total daily food intake was likewise reduced only on the first three days and then climbed back to baseline. Body weight showed a similar pattern. Perhaps the most significant factor responsible for these latter effects was a reduction in the IMI, leading to a doubling in the number of meals; this occurred on day 1 and was sustained for the full six days. Two main effects were revealed by these results: first, that tolerance did not develop to the ability of CCK to reduce meal size; and, second, that compensation for this effect could occur by means of a greatly increased number of meals, thus mitigating against continued weight loss. In a follow-up study, West *et al.* (1987*b*) attempted to prevent the compensatory increase in number of meals by infusing CCK-8 not only shortly after meal initiation but also periodically during IMI periods. At first, this prevented the decreased IMI and the increased feeding frequency, but by day 6 total daily intake had recovered to baseline levels due to an increased number of meals. In addition to showing the value of behavioral methods for accurately depicting the properties of CCK, these studies highlighted the practical issues, i.e., the behavioral and metabolic adaptations with which those engaged in seeking new antiobesity agents must contend.

The CCK literature provides several examples of tertiary level screening, although these methods are in general better illustrated by the literature on fenfluramine-like anorectics (see section 9.2.2.2 below). However, one method worth considering here concerns the use of pharmacological antagonists to determine whether a drug effect is really associated with a specific biological mechanism. With CCK agonists, this became possible when a nonpeptide compound, L-364,718, was found to be a potent, orally active, long-acting and selective antagonist at peripheral, as opposed to central, CCK receptors (Chang & Lotti, 1986). The availability of this compound as a research tool enabled investigators to establish that the reduction of food intake by CCK-8 in rats is mediated by peripheral receptors (Lotti *et al.*, 1987; Pendleton *et al.*, 1987). An obvious additional step to probe the mechanism of action of CCK was to demonstrate that administration of L-364,718, can increase food intake, presumably by blocking the actions of endogenous CCK-8 (Hansen & Strouse, 1987; Reidelberger *et al.*, 1988). However, in one of these studies, L-364,718 did not increase food intake in an orderly dose-related manner and much higher doses were needed than were required to block the effects of CCK on food intake (Hanson & Strouse, 1987). Furthermore, when the compound was given to rats twice daily for 14 days, there was a gradual loss in efficacy so that by day 7 food intake was no longer elevated above control values (Watson *et al.*, 1988). These results demonstrate some of the benefits and limitations associated with the availability of a specific pharmacological antagonist for tertiary screening of antiobesity compounds.

9.2.2.2 Fenfluramine-like appetite suppressants

Fenfluramine and its dextrorotatory isomer, *d*-fenfluramine, are marketed anorectics which have received much study, some of which was referred to earlier (section 9.2.1.4). Fluoxetine is a newer compound, approved for use as an antidepressant (Benfield *et al.*, 1986), and demonstrated to reduce food intake in animals, that is currently being evaluated for its antiobesity properties in clinical trials. Both fenfluramine and fluoxetine are believed to act by enhancing the effects of endogenous serotonin (5-HT) at certain 5-HT receptors.

One secondary screening strategy illustrated by the literature on fenfluramine and fluoxetine is that of establishing the generality of the antiobesity activity of active compounds. To go beyond the cases where the compounds are evaluated only in lean rats subjected to food deprivation, it is appropriate to examine a variety of models. Rowland *et al.* (1985), for example, compared fenfluramine, fluoxetine and the directly acting 5-HT agonist, quipazine, in three models: 24 h food deprivation; tail pressure induced eating; and 2-deoxy-D-glucose (2-DG) induced eating. The tail pressure model involves inflating a pressure cuff around a rat's tail for five 120 sec trials in rats preselected for responding to this stress by eating chow pellets during the periods of tail cuff occlusion. This is interpreted to represent a model of stress induced eating (Antelman *et al.*, 1976). 2-DG is a glucose antimetabolite that causes cellular glucoprivation, as a consequence of which feeding is rapidly induced (Smith & Epstein, 1969). Rowland *et al.* (1985) found that all three compounds reliably suppressed 24-h deprivation induced eating and 2-DG induced feeding. Fenfluramine was also very effective against tail pressure induced eating; however, fluoxetine and quipazine were less effective in this model.

Another study compared fenfluramine, fluoxetine, quipazine and para-chloroamphetamine (PCA) on 2-DG and insulin-induced feeding in 20-h deprived rats (Carruba *et al.*, 1985). The authors suggested that rats given 2-DG can be considered more like obese diabetic people since

2-DG causes not only hyperphagia but also hyper-glycemia and some relative insulin deficiency or resistance, whereas rats given insulin can be considered more as models of obese non-diabetic people who are hyperphagic and hyperinsulinemic. D-fenfluramine was the most potent compound and fluoxetine the least. The ED_{50} doses of each drug were then administered to ad libitum fed rats immediately before i.p. injection of 2-DG or insulin and food intake was monitored for the next six hours. All four compounds were equieffective in blocking 2-DG and insulin induced eating when given at the 20-h food deprivation ED_{50} dose level.

An example was cited earlier (section 9.2.1.4) of a comparison between the enantiomers of fenfluramine (Le Douarec *et al.*, 1966). A more recent example not only provides a comparison of the enantiomers of fluoxetine, but also includes a comparison across models (Wong *et al.*, 1988). The R- and S-enantiomers of fluoxetine were injected 30 min before either a six-hour period of access to food by rats trained on an 18-h food deprivation schedule, or a four-hour test period in rats fed ad lib and injected with 2-DG 30 min before the test. Both enantiomers were active in both models, with the S-enantiomer being slightly more potent. This parallels the potencies of the two enantiomers as inhibitors of 5-HT uptake *in vivo*.

A secondary screening strategy that was illustrated in the discussion of CCK agonist was the use of species other than the rat. This strategy can be noted briefly in the literature on other appetite suppressants. Hill *et al.* (1983) used adult mongrel dogs and adult squirrel monkeys for their studies on 1,3-diaryltriazenes. The dogs were trained for four to six weeks to eat a daily ration of dry dog food within three hours. A test compound was given orally as a dry powder in a gelatin capsule for five consecutive days and food intake and body weight were monitored throughout. The monkeys were not food deprived; rather, animals were pre-selected whose body weight and drinking were stable. Suspensions of test compounds were administered orally for five days. Only body weight and drinking were monitored since the monkeys scattered their food. The same laboratories also used mongrel dogs and a slightly different protocol to evaluate a halogenated aralkylamine anorectic (SK & F 1-39728-A) (Macko *et al.*, 1972). Dogs were preselected for their ability to consume 50 g

of food every 30 min for six hours after an 18 h fast. ED_{50} values were determined based upon a criterion of refusal of food for two consecutive 30-min periods. Similarly, Brown *et al.* (1986) used exbreeder female dogs trained to have access for only four hours per day to an unlimited amount of meat. Test compounds, e.g., (S)-3-[(benzyloxy)-methyl]morpholine, were administered orally in gelatin capsules one hour before feeding. In each five-day block, all dogs were dosed with placebo for two days and then with test compound on days 3, 4 and 5. Food intake during the first and second hours was the main dependent measure. In all of these experiments only compounds of interest from the primary screening tests were carried on to evaluations in the dogs and monkeys.

The final secondary strategy to be discussed concerns assessment of the development of tolerance to anorectic effects after multiple administrations of a test compound. It seems obvious that loss of efficacy with repeated administration would be an undesirable feature in a potential antiobesity agent. Indeed this is a well known limitation of many anorectics, such as *d*-amphetamine or fenfluramine, not just in animals (Le Douarec & Neveu, 1970; Macko *et al.*, 1972) but also in humans (Nauss-Karol & Sullivan, 1988). Nevertheless, it has been argued (Levitsky *et al.*, 1981) that this effect is not a real pharmacological tolerance, but is secondary to the direct weight-suppressing effects of these drugs. According to this view, the decreasing weight loss and the return to normal appetite reflect an appropriate physiological and behavioral adaptation to a lowered level of body weight. To support this cogent argument, Levitsky *et al.* (1981) manipulated the body weights of rats before subjecting them to repeated daily administration of amphetamine or fenfluramine. Depending upon whether the rats were at ad lib body weights or reduced weights before the start of treatment, they lost or gained weight, respectively, during the regimen of daily drug administration. Similarly, in rats already 'tolerant' to fenfluramine, Levitsky *et al.* were able to re-establish responsiveness to the drug by raising the body weights of the animals through force-feeding a liquid diet after each daily six-hour feeding period. These results are provocative and indeed question the interpretation of apparent tolerance in studies of repeated administration of anorectic agents. Nevertheless, it is desirable to be dealing with a compound to which tolerance does not develop. For this reason, secondary screening for this effect is appropriate.

A few studies have reported instances in which tolerance did not develop. For example, in five-hour deprived rats consuming a high fat diet, tolerance did not develop during 12 days of dosing to the effects on food intake or body weight of either SK & F 1-39728-A or fenfluramine, whereas tolerance did develop to *d*-amphetamine after only three to five days (Macko *et al.*, 1972). In a study in mice maintained under ad lib conditions, tolerance to the effects of fluoxetine, given twice daily for ten days, on food intake and body weight, was slight and only shown by some measures (Morley & Flood, 1987). In six-hour deprived rats, the efficacy of fluoxetine was maintained without tolerance developing over 10 days of administration, as indicated by either one-hour or six-hour food intake readings (Wong & Fuller, 1987). Finally, in rats fed ad lib, either semipurified or purified human 'satietin' was administered by i.p. injection for five days 30 min before the onset of the dark phase. In addition, satietin was delivered for eight days by osmotic minipumps implanted in the peritoneal cavity (Mendel *et al.*, 1988). In all cases, tolerance to satietin developed rapidly; however, food intake in satietin-treated animals remained below the levels of saline controls and there was less tolerance to higher dose levels than to lower doses. All of these results, in addition to illustrating protocols that can be used to evaluate development of tolerance, underline the principle that dose, species and protocol design can all be critical determinants of the outcome of a drug study.

The literature on fenfluramine and related compounds is particularly enriched with examples of tertiary drug evaluation procedures. Among other things, from a mechanistic perspective, fenfluramine, fluoxetine and many other compounds have been linked to a serotonergic mechanism of action. 5-HT agonists, releasers and reuptake inhibitors are all known to reduce food intake and body weight, and behavioral methods have been used extensively to obtain data bearing on this issue. A comprehensive review is beyond the scope of this chapter and excellent reviews are available (e.g., Blundell, 1987; Garattini *et al.*, 1987*b*; Wong & Fuller, 1987; see also Montgomery, Chapter 8 of this volume). Accordingly, only a few selected topics will be considered briefly. First, at the tertiary level, investigators may attempt to obtain a clear picture of what it is that may be unique about a compound of interest as compared to reference compounds or those belonging to other pharmacological classes. Thus Garattini & Samanin (1985) compared and contrasted *d*-amphetamine, fenfluramine and salbutamol, a selective β_2-adrenergic agonist, in six different rat models: starved rats; sated rats eating palatable food; tail pinch induced eating; muscimol induced eating; insulin induced eating; and food rewarded running in a runway. They observed clear distinctions among the compounds: fenfluramine decreased feeding or appetite behavior in all six models whereas *d*-amphetamine decreased only deprivation- and palatable food-induced eating. The profile of salbutamol was intermediate between the other two compounds. Furthermore, the actions of any one of the three drugs could only be blocked by receptor antagonists that were specific for the mechanism through which that particular compound worked. In addition, in rats treated with salbutamol for six days, no cross-tolerance developed to *d*-amphetamine or fenfluramine. All of these experiments served to define distinctions among these three representatives of compounds working through catecholaminergic/dopaminergic, serotonergic or β_2-adrenergic receptors. Another example of this type of characterization is provided by the comparison of four serotonergic drugs and five dopaminergic compounds (Carruba *et al.*, 1985). All of the serotonergic agents were equieffective against 2-DG and insulin induced feeding whereas the dopaminergic compounds were considerably less potent against insulin than against 2-DG. Carruba *et al.* drew the important conclusion that differences in the etiology of the hyperphagia of obese subjects must be taken into consideration when choosing therapy.

Tertiary screening methods that involve analysis of the micro- and macro-structure of feeding behavior have been employed to great advantage for studying fenfluramine and related compounds. Le Magnen (1971, 1981) has been a pioneer in defining the value of a very detailed analysis of meal patterns and of energy regulation both within 24-h periods and over a longer term. Some examples of meal pattern analysis were discussed earlier (section 9.2.2.1). At the tertiary level of drug evaluation, Blundell and his colleagues have been perhaps the most prolific investigators in extending the work of Le Magnen and others (see Blundell, 1987; Montgomery, this vol., Ch. 9). Instead of simply measuring food intake, microstructural analysis consists of comprehensively recording eating and other behaviors (e.g., drinking, grooming, locomotor activity and resting) by either automated ('eatometer') or observational techniques, during short (e.g., one hour) periods of food access. Subsequently, one can derive several parameters that describe an animal's behavior, including total food intake, duration of time spent eating, number of eating bouts, size of bouts (grams of food eaten), duration of bouts, and a local rate of eating.

In a series of studies on serotonergic and dopaminergic compounds, clear differences were observed among these classes. For example, *d*-amphetamine markedly increased the latency to initiate eating and increased the rate of eating whereas fenfluramine, 5-HT reuptake blockers, 5-hydroxytryptophan (5-HTP) and neuroleptics noticeably slowed the rate of eating. With neuroleptics, however, the animals compensated for the slow rate of eating by increasing the time spent feeding, so that the amount eaten did not decrease, whereas such a compensation did not occur after administration of serotonergic compounds. Such behavioral analysis defined a profile shared by various serotonergic drugs that was characterized by a normal latency to begin eating, a slow rate of ingestion, and an early termination of intake. This pattern was interpreted to reflect a facilitation of the process of satiation (Blundell, 1987). Interestingly enough, microstructural analysis has also led to a discrepant conclusion by Montgomery & Willner (1988) who confirmed that *dl-* and *d*-fenfluramine reduced intake of either wet mash or sucrose solution. However, whereas normal satiety was associated with a period of resting behavior at the end of a meal, both forms of fenfluramine dose-dependently decreased resting behavior. Montgomery & Willner speculated that fenfluramine might be acting by slowing gastric emptying (cf. Baker *et al.*, 1988) or by a subtle motor impairment, rather than by enhancing satiety. In contrast to fenfluramine, the normal sequence of post-prandial satiety behavior was observed following the early termination of feeding by fluoxetine (Willner *et al.*, 1990) or other 5-HT uptake inhibitors (Willner, personal communication).

Macroanalysis involves monitoring the patterns of meal taking over days by recording meal size, meal duration, meal frequency, intermeal interval and other measures. This level of analysis gives much the same information as does microanalysis, but it integrates that information over a longer time scale so that it reveals how the animal may be adapting to the action of a particular compound. For example, if the compound delays or reduces feeding during the early part of the dark phase, does the animal compensate by eating more in the late dark phase or during the light phase? Le Magnen (1971) has described several situations in which such regulation occurs in rats by means of a variation in the frequency of meals during a 24-h period. Blundell (1987) presents examples of the changes induced by various drugs (see also Montgomery, this vol., Ch. 8).

The effects of fenfluramine and related compounds upon rats eating cafeteria diets represents a final example of tertiary level studies. It has been postulated that serotonergic agents do not simply reduce food intake but rather selectively reduce carbohydrate ingestion (Wurtman & Wurtman, 1979). Orthen-Gambill & Kanarek (1982) examined the effects of fenfluramine in rats maintained on ground chow or one of two cafeteria regimes, one with a high caloric fat ration and the other with a fat ration isocaloric to the carbohydrate and protein rations. On the diet containing the high caloric fat ration, fenfluramine decreased both protein and fat intake whereas on the diet with the isocaloric fat ration, the drug only decreased fat intake. In neither case did fenfluramine decrease carbohydrate intake. McArthur & Blundell (1986) varied the sensory qualities but not the nutrient composition of pairs of cafeteria diet choices. Not only was the selection of diet choices and consequently protein and carbohydrate intake markedly altered by the sensory qualities, but the effects of fenfluramine were also modified significantly according to the dietary choices available. These two studies indicate not only how differences between compounds can be delineated but also how one must also use a variety of conditions to characterize a compound thoroughly.

9.2.2.3 THL-like inhibitors of gastrointestinal lipases

THL works, as indicated earlier (section 9.2.1.2), by preventing absorption of triglycerides. Therefore, secondary screening methods for such an antiobesity agent focus on whether this action is sustained with repeated dosing and whether body weight is reduced, rather than on food intake per se. Accordingly, Hogan *et al.* (1987) evaluated THL in rats made obese by exposure for 17 weeks to a moderately high fat diet consisting of 47% rat chow, 44% condensed milk and 8% corn oil. Subsequently half of these animals received THL as a dietary admixture for 22 days. Food intake and body weight were measured daily and lipid absorption was determined during two three-day periods. Body weight gain declined steadily throughout the 22 days in the THL group while food intake actually increased significantly on days 10 to 16 and 21 to 22. Overall, there was a significant increase of 8.6% in cumulative food intake. Thus there was a loss of body weight despite development of hyperphagia. Lipid absorption, as determined by fecal analysis, was greater than 90% in the control group and less than 20% in the THL group either midway through the study or at its conclusion. Analogous results were obtained in mice treated with THL (Hadvary *et al.*, 1987), and in rats treated with AO-128, an inhibitor of intestinal carbohydrate absorption (Matsuo *et al.*, 1987).

9.2.2.4 Thermogenic β-agonists

To complete the review of secondary screening methods, a few examples from the literature on thermogenic agents are pertinent because they indicate relevant calorimetry procedures for assessing the energy expenditure enhancing properties of such compounds. In one study, metabolic rate was compared in lean versus obese mice during 18 days of treatment with BRL 26830A (Arch *et al.*, 1985). This β-agonist caused very different changes in the thermogenic response in obese and lean animals. Only in the obese mice was there a large and persistent increase in energy expenditure. Similar results were seen by Arch *et al.* in obese and lean Zucker rats, leading them to conclude that this difference is responsible for the weight reducing efficacy of BRL 26830A in obese, but not lean, animals. Yen *et al.* (1985) likewise similarly demonstrated that the thermogenic response in obese mice was sustained during 28 days of administration of the β agonist LY104119. Meier *et al.* (1985) similarly demonstrated that Ro 16-8714 elevated fat oxidation and, to an even greater extent, carbohydrate oxidation in obese mice, after 15 days of treatment. These effects provided a basis of explaining the antidiabetic properties of β agonists in obese animals.

Since thermogenic activity is likely to be a useful property of any antiobesity compound, it is prudent to assess any potential antiobesity agent, no matter what is its primary mechanism of action, in secondary and tertiary energy expenditure tests. Studies have indeed revealed that several appetite suppressants, including fenfluramine and mazindol, are capable of increasing energy expenditure, an effect that many contribute to their antiobesity efficacy. Various secondary and tertiary methods for defining this type of action are discussed by Bray (1987), Rothwell & Stock (1987) and Wyllie *et al.* (1985). We cite here one example to illustrate some relevant tertiary methods for characterizing thermogenic agents. Meier *et al.* (1985) reported several different experiments in which lean and obese mice or rats were treated with Ro 16-8714 for periods ranging from 42 to 68 days. During and/or at the end of treatment, energy expenditure, cumulative changes in body weight and food intake, carcass composition, weight of BAT, BAT receptor number (determined by GDP binding assays), blood chemistries and glucose tolerance were all determined. Such a combination of measures afforded a comprehensive profile of the efficacy of this β agonist.

9.2.3 Exploratory techniques

It would not be appropriate to conclude a discussion of methods to screen for antiobesity agents without making brief mention of efforts to find novel approaches for identifying efficacious compounds. One way to find such methods is to try to understand more precisely what are the determinants of hunger, appetite and feeding and to translate any insights that are obtained into applied screening tests. Many determinants of food intake have been proposed and discussed: glucose and the glucostatic theory of energy regulation have always been prominent amongst these (Mayer, 1955; Le Magnen, 1971). An example of a current effort to understand the glucostatic theory is furnished by the work of Campfield & Smith (1986; Campfield *et al.*, 1985). Using concurrent behavioral and biochemical recording techniques, including continuous online monitoring of blood glucose levels, they identified a small transient decline in glucose of 6% or more and lasting for at least six minutes before returning to baseline that appeared to be a signal for the initiation of feeding. Not only did the occurrence of such declines predict the onset of feeding, but a critical time window was also defined: meals were initiated within approximately 12 to 24 min after the onset of the transient decline. If the decline was truncated by a computer-controlled injection of a small amount of exogenous glucose (glucose clamp), feeding did not occur. Under normal conditions, if the food cup was covered during the critical 12 to 24 min time period, the animals approached and oriented to the cup, thus demonstrating appropriate food-reinforced appetitive behavior. Interestingly, they did not attempt to ingest the food when the cover was removed after the end of the 'critical window' period. Instead, feeding did not occur until approximately 90 min later when the next transient glucose decline took place. Such results suggest that there is at least one powerful control signal for meal initiation and that suppressing feeding during a critical period associated with its occurrence does not lead to subsequent compensation. Assuming that this peripheral control signal gives rise to feeding by influencing the CNS, potential applications of this phenomenon may derive from efforts to find compounds that affect the coupling between the glucose transients and the CNS neurons that respond to these transients.

9.3 SCREENING METHODS FOR ORECTIC AGENTS

While there are many compounds that decrease food intake or body weight, far fewer have been noted to increase food intake, appetite or body weight. The most consistent reports of orectic activity have been observed for members of certain classes of psychotropic agents: neuroleptics; anti-depressants; benzodiazepines and related anxiolytic agents; antihistaminic agents; 5-HT$_{1A}$ agonists; and kappa opioid agonists (Sepinwall & Cook, 1978; Kalucy, 1980; Cooper *et al.*, 1985a; Dourish *et al.*, 1985; Blundell, 1987; Hewson *et al.*, 1987; Orthen-Gambill, 1988; see also Montgomery, this vol., Ch. 8). Often these observations have been made in patients rather than in animals in laboratory tests. Furthermore, even when such compounds are active in animals, they are not necessarily efficacious in people. Nevertheless, the methods used to identify the activity of compounds belonging to the classes just listed constitute the only currently available guides to screening for orectic activity. During use of these methods, certain investigators have obtained valuable insights about the mechanisms of action of various agents and the reasons for possible discrepancies between animal and human data. Since good reviews are available (Blundell, 1987; Cooper, 1980; Orthen-Gambill, 1988), only a brief discussion of orectic agents will be presented.

In very early testing, Randall *et al.* (1960), recognized that chloradiazepoxide, the prototype of benzodiazepine anxiolytic agents, increased food intake in rats and dogs. Hanson & Stone (1964) set up a screening test simply to measure the increase in body weight (as an estimate of the amount of food eaten) during a two-hour period of feeding by groups of food-deprived immature rats. They regarded this procedure as being extremely useful for identifying potential anxiolytic agents. The hyperphagia included by benzodiazepines is sometimes so pronounced that it has been described by such terms as 'voraciousness' (e.g., Fratta *et al.*, 1976). In the study of Fratta *et al.* (1976), cats were first trained to eat an unlimited amount of food within three hours each day. When given to hungry cats, several benzodiazepines increased total intake over the three-hour test although they varied in their potency and efficacy. In addition, when administered to sated cats offered an additional supply of food for 30 min, all of the compounds studied produced a very robust enhancement of food intake.

Cooper and his colleagues have evaluated many benzodiazepines and other ligands that are believed to act as agonists, antagonists or 'inverse' agonists at the benzodiazepine receptor. In these studies, nondeprived rats were trained over 10 days to consume a highly palatable diet of 50 ml of sweetened condensed milk, 200 ml tap water and 150 ml finely milled food for 30 min each day. During twice weekly drug tests, the animals were given access to the liquid diet for periods ranging from 30 min to three hours. Studies using this technique showed not only that full agonists, including midazolam, triazolam, and other compounds, produced maximal increases in intake, but also that partial agonists could be differentiated into two groups, those that increased intake, such as Ro 16-6028, Ro 23-0364 and ZK 91296, and those that did not, such as CGS 9896 or CGS 9895 (Cooper *et al.*, 1985*a*, 1987; Cooper, 1986). In addition, a more detailed analysis of the effects of one agonist, midazolam, showed that this compound elevated food intake by dose-dependently increasing the duration of feeding bouts without changing their frequency or the rate of eating (Cooper & Yerbury, 1986).

The methods used to study benzodiazepines and related agents, i.e., measurement of increases in food intake above control values by deprived animals or of consumption of a highly palatable diet by nondeprived animals, or a resumption of eating by sated animals, are typical of methods used to identify other classes of orectic agents. For example, cyproheptadine has frequently been listed as a compound that can increase food intake and body weight (Kalucy, 1980). Baxter *et al.* (1970) trained rats to have access to food for only six hours per day and tested them either in groups or individually. In the group setting, all doses of cyproheptadine studied (6.2 to 50 mg/kg s.c.) elevated food intake during the first two hours, but total consumption over six hours was not increased above control on most occasions. In the individual animal test setting, meal analysis showed that cyproheptadine increased the duration of the first meal eaten after food was presented, but eating rate and the number of meals eaten decreased.

Since cyproheptadine has 5-HT antagonist properties and since other 5-HT antagonists, e.g., methysergide, can increase food intake, it has commonly been assumed that the 5-HT receptor is the principal site at which cyproheptadine exerts its orectic action. However, cyproheptadine is also a histamine antagonist, and Orthen-Gambill (1988) tested the hypothesis that histamine blockade stimulates appetite. All animals in the study were fed

ad lib on either ground chow or a palatable liquid diet (vanilla-flavored instant breakfast in milk). In addition to cyproheptadine, two other compounds with antihistaminergic activity were examined: a phenothiazine, promethazine; and the most potent histamine blocker among the tricyclic antidepressants, doxepin. All three compounds produced similar dose-related increases in cumulative food intake over 24 h. Administration of L-histidine, which gives rise to large increases of histamine in brain, produced a clear decrease in food intake. Orthen-Gambill concluded that the antihistaminic potencies of a wide range of antidepressants and antipsychotic agents predict the orectic effects of these compounds. It was also noted that it is easy to miss the orectic effect of a histamine antagonist if one uses too high a dose, at which point the sedative actions of histamine blockers may interfere with feeding. Amitriptyline, a tricyclic antidepressant with strong antihistaminic activity, represents an example of this: most animal experiments have failed to demonstrate the appetite-stimulant effect reported in people because the doses used in animals have apparently been too high.

Compounds with a high affinity for the kappa subtype of the opiate receptor have also been observed to exert orectic actions. Cooper *et al.* (1985*b*) employed the same screening procedure they used to study benzodiazepines to evaluate the specific kappa receptor agonist, U-50,488, and tifluadom, a benzodiazepine. U-50,488, tifluadom and chlordiazepoxide all increased intake of the highly palatable liquid diet by non-deprived rats. Even though tifluadom is a benzodiazepine, its effects were apparently opioid-mediated since they were blocked by the opiate antagonist, naloxone, but not by the specific benzodiazepine antagonist, Ro 15-1788. However, not all kappa agonists are orectic, since bremazocine and ethylketocyclazocine decreased eating, apparently a consequence of a general response suppressant property of these two compounds (Cooper *et al.*, 1985*b*). U-50,488 and another selective kappa agonist, PD117302, have also been evaluated in non-deprived rats eating powdered chow (Hewson *et al.*, 1987; Leighton *et al.*, 1988). In this case, intake was monitored at intervals throughout the 24-h period following dosing. These conditions allowed the investigators to detect a bisphasic effect of the kappa agonists: they induced an initial hyperphagia, but by six hours after dosing this had changed to a reduction in food intake. Accordingly, over a 24-h period total food intake was decreased. These results imply that such compounds might be effective antiobesity agents only if they were given more frequently than once a day.

5-HT$_{1A}$ agonists are the final type of orectic agents that will be discussed here. Whereas most 5-HT agonists and compounds that increase the levels of 5-HT in the synaptic cleft, such as fenfluramine and fluoxetine, decrease food intake, one group of 5-HT agonists increases feeding. These compounds have in common a high affinity for the 5-HT$_{1A}$ receptor subtype. This seemingly paradoxical effect has been attributed to the location of the 5-HT$_{1A}$ receptor, which is believed to be a presynaptic autoreceptor. Thus an agonist at this receptor will inhibit 5-HT release, thereby increasing food intake. This orectic action was first observed for 8-hydroxy-2-(di-n-propylamino)tetralin (8-OH-DPAT) in nondeprived rats (Dourish *et al.*, 1985). The animals were fed laboratory chow ad lib and were then adapted overnight to the chamber in which a single two-hour test was conducted. During this daytime test, a supply of three chow pellets and three wood blocks was placed in the test cage, and a microstructural analysis of feeding and motor behaviors was made. 8-OH-DPAT markedly increased feeding, with a maximum effect at a dose of 500 μg/kg, by increasing the duration of feeding and the number of feeding bouts. Stereotypy and other induced behaviors interfered with feeding at higher dose levels. The action of 8-OH-DPAT was considered to be a specific orectic one since the rats did not increase the time spent gnawing and they consistently chose to eat food pellets rather than gnaw on the wood blocks. 8-OH-DPAT also induced eating without any induction of stereotyped behavior when the compound was microinfused directly into the dorsal or medial raphe nuclei, thus supporting the concept that its effects on food intake could be separated from its postsynaptically mediated behavioral actions (Hutson *et al.*, 1986).

Some other 5-HT$_{1A}$ agonists that have induced an increase in food intake are buspirone, ipsapirone and LY165163 (Dourish *et al.*, 1986; Wong & Fuller, 1987). Screening models used to characterize these effects included not only eating by

nondeprived rats but also the anorexia and loss of body weight induced in rats by immobilization stress. Given all of these results it would seem appropriate to consider 5-HT$_{1A}$ agonists as potentially useful orectic agents. However, the results of studies by Fletcher (1987) and Montgomery *et al.* (1988) suggested that hyperphagia induced by 8-OH-DPAT may be secondary to the elicitation of chewing or gnawing. In deprived or nondeprived rats, 8-OH-DPAT increased only the intake of chow diets (plain or sweetened wet mash) but failed to increase consumption of any of several liquid diets. Furthermore, 8-OH-DPAT increased chewing and gnawing by rats given wood blocks without the concurrent presentation food. These investigators attributed the failure of Dourish *et al.* (1985) to observe an increase in gnawing to the concurrent presence of food pellets, which represented an alternate target for gnawing or chewing behavior. A subsequent study demonstrated that dopaminergic mechanisms are involved in mediating 8-OH-DPAT hyperphagia, which led the investigators to draw a parallel between 8-OH-DPAT and stress-induced eating (Muscat *et al.*, 1989). These results suggested that 8-OH-DPAT may not increase hunger, appetite or motivation to eat in general. Thus careful behavioral analysis, including the use of varied diets, has raised questions about the potential utility of 5-HT$_{1A}$ agonists, as exemplified by 8-OH-DPAT, as orectic agents.

9.4 CONCLUSIONS

Many examples have been presented throughout this chapter of the contributions made by behavioral methods to screening, identifying and understanding the actions of anorectic and orectic compounds and antiobesity agents. Particularly for anorectic and antiobesity agents, a hierarchy of screening strategies (primary, secondary and tertiary) should be employed. This should include not only procedures that measure food intake per se, but also biochemical and physiological tests when these are appropriate to the target mechanism, e.g., thermogenesis or inhibition of intestinal triglyceride absorption. Despite the availability of many methods for identifying active compounds, there remains an urgent need to move back and forth between clinical and animal laboratory settings to validate the animal screening methods. One important question that needs to be addressed concerns the fact that animal tests frequently rely on food intake as the key dependent variable whereas clinical studies, with only a few exceptions, measure body weight and do not measure food intake. This could be one possible reason for the lack of success in finding any truly effective antiobesity agents despite the many active compounds identified in animal tests. Another question concerns the agents believed to work by a mechanism other than appetite suppression. What effect might such compounds have upon food intake, and what impact might this have upon the control of body weight? In animals there are examples of compensatory hyperphagias that occur. Might these counteract the antiobesity action? Such questions deserve careful exploration to provide guidance to those who must devise screening methods for antiobesity or orectic agents.

Our understanding of obesity is in many ways analogous to our comprehension of hypertension or depression, although we are lagging behind advances in treating hypertension or depression by perhaps 10 to 20 years. We are faced with a complex disorder with many different subtypes and multiple determinants. Not only is there a need to study the disorder both in animals and in people from many different perspectives, but it is almost certain that to be successful we shall need to treat many patients by more than one approach. Accordingly, to identify potential new drugs we shall continue to need a variety of approaches, such as those exemplified in this chapter, and a corresponding variety of screening methods.

We wish to express our appreciation to Helen Hipkins, Maggie Johnson, Patty Lotito, Shannon Maloney, Nancy Manion and Lori Zielenski for their excellent work in preparing the manuscript. We are also grateful to Dr Art Campfield for reviewing the manuscript and to Professor Paul Willner for his careful and extensive editorial comments.

REFERENCES

Anika, S.M., Houpt, T.R. & Houpt, K.A. (1977). Satiety elicited by cholecystokinin in intact and vagotomized rats. *Physiology and Behavior* **19**, 761–6.

Anika, S.M., Houpt, T.R. & Houpt, K.A. (1981). Cholecystokinin and satiety in pigs. *American Journal of Physiology* **240**, R310–R318.

Antelman, S.M., Rowland, N.E. & Fisher, A.E. (1976). Stimulation bound ingestive behavior: a view from the tail. *Physiology and Behavior* **17**, 743–8.

Arch, J.R.S., Ainsworth, A.T., Cawthorne, M.A., Piercy, V., Sennitt, M.V., Thody, V.E., Wilson, C. & Wilson, S. (1984). Atypical β-adrenoceptor on brown adipocytes as target for anti-obesity drugs. *Nature* **309**, 163–5.

Arch, J.R.S., Ainsworth, A.T., Ellis, R.D.M., Piercy, V., Thody, V.E., Thurlby, P.L., Wilson, C., Wilson, S. & Young, P. (1985). Treatment of obesity with thermogenic β-adrenoceptor agonists: studies on BRL 26830A in rodents. In *Novel Approaches and Drugs for Obesity*, ed. A.C. Sullivan & S. Garattini, pp. 1–11. New York: John Libbey.

Arch, J.R.S., Piercy, V., Thurlby, P.L., Wilson, C. & Wilson, S. (1987). Thermogenic and lipolytic drugs for the treatment of obesity: old ideas and new possibilities. In *Recent Advances in Obesity Research: V*, ed. E.M. Berry, S.H. Blondheim, H.E. Eliahou & E. Shafrir, pp. 300–11. London: John Libbey.

Bado, A., Rodriguez, M., Lewin, M.J.M., Martinez, J. & Dubrasquet, M. (1988). Cholecystokinin suppresses food intake in cats: structure–activity characterization. *Pharmacology, Biochemistry and Behavior* **31**, 297–303.

Baker, B.J., Duggan, J.P., Barber, D.J. & Booth, D.A. (1988). Effects of dl-fenfluramine and xylamidine on gastric emptying of maintenance diet in freely feeding rats. *European Journal of Pharmacology* **150**, 137–42.

Baxter, M.G., Miller, A.A. & Soroko, F.E. (1970). The effect of cyproheptadine on food consumption in the fasted rat. *British Journal of Pharmacology* **39**, 229P–230P.

Benfield, P., Heel, R.C. & Lewis, S.P. (1986). Fluoxetine – a review of its pharmacodynamic and pharmacokinetic properties, and therapeutic efficacy in depressive illness. *Drugs* **32**, 481–508.

Blosser, J., Augello-Vaisey, S., Barantes, M., Comstock, J., Gawlak, D., Towne, M., Simmons, R. & Rosamond, J. (1987). Comparative effects of CCK-8 and analogs on anorectic activity and *in vitro* gall bladder contraction. *Society for Neuroscience Abstracts* **13**, 879.

Blundell, J.E. (1987). Structure, process, and mechanism: case studies in the psychopharmacology of feeding. In *Handbook of Psychopharmacology*, vol. 19, ed. L.L. Iversen, S.D. Iversen & S.H. Snyder, pp. 123–82. New York: Plenum Press.

Bray, G.A. (1987). Fenfluramine: a thermogenic drug.

In *Recent Advances in Obesity Research: V*, ed. E.M. Berry, S.H. Blondheim, H.E. Eliahou & E. Shafrir, pp. 290–2. London: John Libbey.

Brown, G.R., Foubister, A.J., Foster, G. & Stribling, D. (1986). (S)-3-[(Benzyloxy)methyl]morpholine hydrochloride: a nonstimulant appetite suppressant without conventional neurotransmitter releasing properties. *Journal of Medicinal Chemistry* **29**, 1288–90.

Bukowiecki, L., Folléa, N., Paradis, A. & Collet, A. (1980). Stereospecific stimulation of brown adipocyte respiration by catecholamines via β_1-adrenoreceptors. *American Journal of Physiology* **238**, E552–E563.

Campfield, L.A. & Smith, F.J. (1986). Functional coupling between transient declines in blood glucose and feeding behavior: temporal relationships. *Brain Research Bulletin* **17**, 427–33.

Campfield, L.A., Brandon, P. & Smith, F.J. (1985). On-line continuous measurement of blood glucose and meal pattern in free-feeding rats: the role of glucose in meal initiation. *Brain Research Bulletin* **14**, 605–16.

Carruba, M.O., Ricciardi, S., Spano, P.F. & Mantegazza, P. (1985). Dopaminergic and serotoninergic anorectics differentially antagonize insulin- and 2-DG-induced hyperphagia. *Life Sciences* **36**, 1739–49.

Chang, R.S.L. & Lotti, V.J. (1986). Biochemical and pharmacological characterization of an extremely potent and selective nonpeptide cholecystokinin antagonist. *Proceedings of the National Academy of Sciences USA* **83**, 4923–6.

Clineschmidt, B.V., McGuffin, J.C. & Werner, A.B. (1974). Role of monoamines in the anorexigenic actions of fenfluramine, amphetamine and p-chloromethamphetamine. *European Journal of Pharmacology* **27**, 313–23.

Cooper, S.J. (1980). Benzodiazepines as appetite-enhancing compounds. *Appetite* **1**, 7–19.

Cooper, S.J. (1986). Hyperphagic and anorectic effects of β-carbolines in a palatable food consumption test: comparisons with triazolam and quazepam. *European Journal of Pharmacology* **120**, 257–65.

Cooper, S.J. & Yerbury, R.E. (1986). Midazolam-induced hyperphagia and FG 7142-induced anorexia: behavioural characteristics in the rat. *Pharmacology, Biochemistry and Behavior* **25**, 99–106.

Cooper, S.J., Barber, D.J., Gilbert, D.B. & Moores, W.R. (1985a). Benzodiazepine receptor ligands and the consumption of a highly palatable diet in non-deprived male rats. *Psychopharmacology* **86**, 348–55.

Cooper, S.J., Moores, W.R., Jackson, A. & Barber, D.J. (1985b). Effects of tifluadom on food consumption compared with chlordiazepoxide and

kappa agonists in the rat. *Neuropharmacology* **24**, 877–83.

Cooper, S.J., Yerbury, R.E., Neill, J.C. & Desa, A. (1987). Partial agonists acting at benzodiazepine receptors can be differentiated in tests of ingestional behaviour. *Physiology and Behavior* **41**, 247–55.

Danho, W. (1987). New hepta:peptide(s) and octa:peptide(s) useful for suppressing or reducing food intake and so for reducing body weight. *European Patent Application No.* 87108135.2.

Danho, W., Wagner, R., Tilley, J. & Triscari, J. (1988). The role of the C-terminal carboxyamide in determining the satiety inducing activity of CCK-8. Paper presented at 'CCK-88', Symposium on Cholecystokinin, Cambridge, U.K.

Danho, W., Triscari, J. & Madison, V.S. (1989). Cyclic peptides having appetite regulating activity. United States Patent No. 4,808,701.

Deutsch, J.A. & Hardy, W.T. (1977). Cholecystokinin produces bait shyness in rats. *Nature* **266**, 196.

Dourish, C.T., Hutson, P.H. & Curzon, G. (1985). Low doses of the putative serotonin agonist 8-hydroxy-2-(di-n-propylamino)tetralin (8-OH-DPAT) elicit feeding in the rat. *Psychopharmacology* **86**, 197–204.

Dourish, C.T., Hutson, P.H., Kennett, G.A. & Curzon, G. (1986). 8-Hydroxy-2-di-n-propylamino-tetralin-induced hyperphagia – its neural basis and possible therapeutic relevance. *Appetite* **7**, Suppl., 127–40.

Fletcher, P.J. (1987). 8-OH-DPAT elicits gnawing, and eating of solid but not liquid foods. *Psychopharmacology* **92**, 192–5.

Fratta, W., Mereu, G., Chessa, P., Paglietti, E. & Gessa, G. (1976). Benzodiazepine-induced voraciousness in cats and inhibition of amphetamine-anorexia. *Life Sciences* **18**, 1157–66.

Fuller, R.W. & Yen, T.T. (1987). The place of animal models and animal experimentation in the study of food intake regulation and obesity in humans. *Annals of the New York Academy of Sciences* **499**, 167–78.

Garattini, S. & Samanin, R. (1985). d-fenfluramine and salbutamol: two drugs causing anorexia through different neurochemical mechanisms. In *Novel Approaches and Drugs for Obesity*, ed. A.C. Sullivan & S. Garattini, pp. 151–7. New York: John Libbey.

Garattini, S., Mennini, T. & Samanin, R. (1987a). From fenfluramine racemate to d-fenfluramine: Specificity and potency of the effects on the serotoninergic system and food intake. *Annals of the New York Academy of Sciences* **499**, 156–66.

Garattini, S., Mennini, T. & Samanin, R. (1987b). Advances in understanding the role of serotonin in controlling food intake. In *Recent Advances in*

Obesity Research: V, ed. E.M. Berry, S.H. Blondheim, H.E. Eliahou & E. Shafrir, pp. 272–84. London: John Libbey.

Garcia, J., Hankins, W.G. & Rusiniak, K.W. (1974). Behavioral regulation of the milieu interne in man and rat. *Science* **184**, 824–31.

Gibbs, J., Young, R.C. & Smith, G.P. (1973*a*). Cholecystokinin decreases food intake in rats. *Journal of Comparative and Physiological Psychology* **84**, 488–95.

Gibbs, J., Young, R.C. & Smith, G.P. (1973*b*). Cholecystokinin elicits satiety in rats with open gastric fistulas. *Nature* **245**, 323–4.

Gibbs, J., Falasco, J.D. & McHugh, P.R. (1976). Cholecystokinin-decreased food intake in rhesus monkeys. *American Journal of Physiology* **230**, 15–18.

Hadvary, P., Lengsfeld, H., Barbier, P., Fleury, A., Kupfer, E., Meier, M.K., Schneider, F., Weibel, E. & Widmer, U. (1987). Lipstatin and tetrahydrolipstatin, potent and selective inhibitors of pancreatic lipase. *International Journal of Obesity* **11**, Suppl. 2, 21.

Hadvary, P., Lengsfeld, H. & Wolfer, H. (1988). Inhibition of pancreatic lipase *in vitro* by the covalent inhibitor tetrahydrolipstatin. *Biochemical Journal* **256**, 357–61.

Hanson, H.M. & Stone, C.A. (1964). Animal techniques for evaluating antianxiety drugs. In *Animal and Clinical Pharmacologic Techniques*, ed. J.H. Nodine & P.E. Siegler, pp. 317–24. Chicago: Year Book Medical Publishers.

Hanson, H. & Strouse, J. (1987). Effects of the CCK antagonist, L-364,718, on food intake and on the blockade of feeding produced by exogenous CCK in the rat. *Federation Proceedings* **46**, 1480.

Hewson, G., Hill, R.G., Hughes, J., Leighton, G.E. & Turner, W.D. (1987). The kappa agonists PD117302 and U50488 produce a biphasic effect on 24 hour food intake in the rat. *Neuropharmacology* **26**, 1581–4.

Hill, D.T., Stanley, K.G., Karoglan Williams, J.E., Loev, B., Fowler, P.J., McCafferty, J.P., Macko, E., Berkoff, C.E. & Ladd, C.B. (1983). 1,3-Diaryltriazenes: a new class of anorectic agents. *Journal of Medicinal Chemistry* **26**, 865–9.

Hogan, S., Fleury, A., Hadvary, P., Lengsfeld, H., Meier, M.K., Triscari, J. & Sullivan, A.C. (1987). Studies on the antiobesity activity of tetrahydrolipstatin, a potent and selective inhibitor of pancreatic lipase. *International Journal of Obesity* **11**, Suppl. 3, 35–42.

Holt, J., Antin, J., Gibbs, J., Young, R.C. & Smith, G.P. (1974). Cholecystokinin does not produce bait shyness in rats. *Physiology and Behavior* **12**, 497–8.

Houpt, T.R., Anika, S.M. & Wolff, N.C. (1978). Satiety effects of cholecystokinin and caerulein in rabbits. *American Journal of Physiology* **235**, R23–R28.

Hsiao, S. & Wang, C.H. (1983). Continuous infusion of cholecystokinin and meal pattern in the rat. *Peptides* **4**, 15–17.

Hsiao, S., Wang, C.H. & Schallert, T. (1979). Cholecystokinin, meal pattern, and the intermeal interval: can eating be stopped before it starts? *Physiology and Behavior* **23**, 909–14.

Hutson, P.H., Dourish, C.T. & Curzon, G. (1986). Neurochemical and behavioural evidence for mediation of the hyperphagic action of 8-OH-DPAT by 5-HT cell body autoreceptors. *European Journal of Pharmacology* **129**, 347–52.

Inoue, S., Matasuo, T. & Ikeda, H. (1987). Efficacy of a disaccharidase inhibitor, AO-128 in the treatment of obesity. *International Journal of Obesity* **11**, Suppl. 2, 50.

Johnson, L.R. (1985). *Gastrointestinal Physiology*. St. Louis: C.V. Mosby Company.

Kalucy, R.S. (1980). Drug-induced weight gain. *Drugs* **19**, 268–78.

Kanarek, R.B. & Marks-Kaufman, R. (1988). Animal models of appetitive behavior: interaction of nutritional factors and drug seeking behavior. In *Control of Appetite*, ed. M. Winick, pp. 1–25. New York: John Wiley & Sons.

Kraly, F.S. (1984). Vagotomy does not alter cholecystokinin's inhibition of sham feeding. *American Journal of Physiology* **246**, R829–R831.

Le Douarec, J.C. & Neveu, C. (1970). Pharmacology and biochemistry of fenfluramine. In *Amphetamines and Related Compounds*, ed. E. Costa & S. Garattini, pp. 75–105. New York: Raven Press.

Le Douarec, J.C., Schmitt, H. & Laubie, M. (1966). Étude pharmacologique de la fenfluramine et de ses isomères optiques. *Archives Internationales de Pharmacodynamie et de Thérapie* **161**, 206–32.

Leighton, G.E., Hill, R.G. & Hughes, J. (1988). The effects of the kappa agonist PD-117302 on feeding behaviour in obese and lean Zucker rats. *Pharmacology, Biochemistry and Behavior* **31**, 425–9.

Le Magnen, J. (1971). Advances in studies on the physiological control and regulation of food intake. In *Progress in Physiological Psychology*, ed. E. Stellar & J.M. Sprague, pp. 203–61. London: Academic Press.

Le Magnen, J. (1981). The metabolic basis of dual periodicity of feeding in rats. *The Behavioral and Brain Sciences* **4**, 561–607.

Levin, B.E. & Sullivan, A.C. (1984). Regulation of thermogenesis in obesity. *International Journal of Obesity* **8**, Suppl. 1, 159–80.

Levitsky, D.A., Strupp, B.J. & Lupoli, J. (1981).

Tolerance to anorectic drugs: pharmacological or artifactual. *Pharmacology, Biochemistry and Behavior* **14**, 661–7.

Lotti, V.J., Pendleton, R.G., Gould, R.J., Hanson, H.M., Chang, R.S.L. & Clineschmidt, B.V. (1987). In vivo pharmacology of L-364,718, a new potent nonpeptide peripheral cholecystokinin antagonist. *Journal of Pharmacology and Experimental Therapeutics* **241**, 103–9.

Macko, E., Saunders, H., Heil, G., Fowler, P. & Reichard, G. (1972). Pharmacology of a halogenated aralkylamine with anorectic properties. *Archives Internationales de Pharmacodynamie et de Thérapie* **200**, 102–17.

Matsuo, T., Ikeda, H., Odako, H. & Shino, A. (1987). Antiobesity and antidiabetic actions of a new potent α-glucosidase inhibitor, AO-128, in genetically obese-diabetic mice, KKA^y. *International Journal of Obesity* **11**, Suppl. 2, 21.

Mayer, J. (1955). Regulation of energy intake and body weight: the glucostatic theory and the lipostatic hypothesis. *Annals of the New York Academy of Sciences* **63**, 15–43.

McArthur, R.A. & Blundell, J.E. (1986). Dietary self-selection and intake of protein and energy is altered by the form of the diets. *Physiology and Behavior* **38**, 315–20.

Meier, M.K., Alig, L., Bürgi-Saville, M.E. & Müller, M. (1985). Phenethanolamine derivatives with calorigenic and antidiabetic qualities. In *Novel Approaches and Drugs for Obesity*, ed. A.C. Sullivan & S. Garattini, pp. 215–25. London: John Libbey.

Mendel, V.E., Benitez, R.R. & Tetzke, A. (1988). Human satietin: rapid development of tolerance and its specificity to feeding behavior in rats. *Pharmacology, Biochemistry and Behavior* **31**, 21–6.

Mendre, C., Rodriguez, M., Gueudet, C., Lignon, M.F., Galas, M.C., Laur, J., Worms, P. & Martinez, J. (1988). A pseudopeptide that is a potent cholecystokinin agonist in the peripheral system is able to inhibit the dopamine-like effects of cholecystokinin in the striatum. *Journal of Biological Chemistry* **263**, 10641–5.

Montgomery, A.M.J. & Willner, P. (1988). Fenfluramine disrupts the behavioural satiety sequence in rats. *Psychopharmacology* **94**, 397–401.

Montgomery, A.M.J., Willner, P. & Muscat, R. (1988). Behavioural specificity of 8-OH-DPAT-induced feeding. *Psychopharmacology* **94**, 110–14.

Moore, B.O. & Deutsch, J.A. (1985). An antiemetic is antidotal to the satiety of cholecystokinin. *Nature* **315**, 321–2.

Moran, T.H. & McHugh, P.R. (1988). Gastric and nongastric mechanisms for satiety action of cholecystokinin. *American Journal of Physiology* **254**, R628–R632.

Moran, T.H., Sawyer, T.K., Jensen, R.T., Crosby, R.J. & McHugh, P.R. (1988). Cholecystokinin (CCK) analogs as probes in assessing the site of action of CCK satiety. *Society for Neuroscience Abstracts* **14**, 1108.

Morley, J.E. & Flood, J.F. (1987). An investigation of tolerance to the actions of leptogenic and anorexigenic drugs in mice. *Life Sciences* **41**, 2157–65.

Murphy, R.B., Smith, G.P. & Gibbs, J. (1987). Pharmacological examination of cholecystokinin (CCK-8)-induced contractile activity in the rat isolated pylorus. *Peptides* **8**, 127–34.

Muscat, R., Montgomery, A. & Willner, P. (1989). Blockade of 8-OH-DPAT-induced feeding by dopamine antagonists. *Psychopharmacology* **99**, 402–8.

Nagain, C., Rodriguez, M., Martinez, J. & Roze, C. (1987). *In vivo* activities of peptide and pseudo-peptide analogs of the carboxyl-terminal octapeptide of cholecystokinin on pancreatic secretion in the rat. *Peptides* **8**, 1023–8.

Nauss-Karol, C. & Sullivan, A.C. (1988). Pharmacologic approaches to the treatment of obesity. In *Obesity and Weight Control*, ed. R.T. Frankle & M.-U. Yang, pp. 275–96. Rockville, Maryland: Aspen Publishers, Inc.

Orthen-Gambill, N. (1988). Antihistaminic drugs increase feeding, while histidine suppresses feeding in rats. *Pharmacology, Biochemistry and Behavior* **31**, 81–6.

Orthen-Gambill, N. & Kanarek, R.B. (1982). Differential effects of amphetamine and fenfluramine on dietary self selection in rats. *Pharmacology, Biochemistry and Behavior* **16**, 303–10.

Pagano, G., Cassader, M., Cavallo-Perin, P., Dal Molin, V., Carta, Q., Fasani, R. & Salvini, P. (1986). Comparison of acarbose and phenformin treatment in glibenclamide-treated non-insulin dependent diabetics. *Current Therapeutic Research* **39**, 143–8.

Pendleton, R.G., Bendesky, R.J., Schaffer, L., Nolan, T.E., Gould, R.J. & Clineschmidt, B.V. (1987). Roles of endogenous cholecystokinin in biliary, pancreatic and gastric function: studies with L-364,718, a specific cholecystokinin receptor antagonist. *Journal of Pharmacology and Experimental Therapeutics* **241**, 110–16.

Randall, L.O., Schallek, W., Heise, G.A., Keith, E.F. & Bagdon, R.E. (1960). The psychoactive properties of methaminodiazepoxide. *Journal of Pharmacology and Experimental Therapeutics* **129**, 163–71.

Reidelberger, R.D. & Solomon, T.E. (1986). Comparative effects of CCK-8 on feeding, sham feeding, and exocrine pancreatic secretion in rats. *American Journal of Physiology* **251**, R97–R105.

Reidelberger, R.D., O'Rourke, M.F. & Solomon, T.E. (1988). Comparative effects of the CCK antagonist L-364,718 on food intake and pancreatic exocrine secretion in rats. *Society for Neuroscience Abstracts* **14**, 1196.

Rosamond, J.K., Comstock, J.M., Thomas, N.J., Clark, A.M., Blosser, J.C., Simmons, R.D., Gawlak, D.L., Loss, M.E., Augello-Vaisey, S.J., Spatola, A.F. & Benovitz, D.E. (1988). Structural requirements for the satiety effect of CCK-8. In *Peptides: Chemistry and Biology*, ed. G.R. Marshall, pp. 610–12. Netherlands: ESCOM Science Publishers.

Rothwell, N.J. & Stock, M.J. (1979). A role for brown adipose tissue in diet-induced thermogenesis. *Nature* **281**, 31–5.

Rothwell, N.J. & Stock, M.J. (1987). Effect of diet and fenfluramine on thermogenesis in the rat: possible involvement of serotonergic mechanisms. *International Journal of Obesity* **11**, 319–24.

Rowland, N.E., Antelman, S.M. & Bartness, T.J. (1985). Comparison of the effects of fenfluramine and other anorectic agents in different feeding and drinking paradigms in rats. *Life Sciences* **36**, 2295–300.

Ruiz-Gayo, M., Daugé, V., Menant, I., Bégué, D., Gacel, G. & Roques, B.P. (1985). Synthesis and biological activity of Boc [Nle^{28}, Nle^{31}]CCK$_{27-33}$, a highly potent CCK$_8$ analogue. *Peptides* **6**, 415–20.

Samanin, R., Mennini, T., Ferraris, A., Bendotti, C., Borsini, F. & Garattini, S. (1979). m-Chlorophenylpiperazine: a central serotonin agonist causing powerful anorexia in rats. *Archives of Pharmacology* **308**, 159–63.

Sawyer, T.K., Jensen, R.T., Moran, T., Schreur, P.J.K.D., Staples, D.J., deVaux, A.E. & Hsi, A. (1988). Structure–activity relationships of cholecystokinin-8 analogs: comparison of pancreatic, pyloric sphincter and brainstem CCK receptor activities with in vivo anorexigenic effects. In *Peptides: Chemistry and Biology*, ed. G.R. Marshall, pp. 503–4. Netherlands: ESCOM Science Publishers.

Sclafani, A. & Springer, D. (1976). Dietary obesity in adult rats: similarities to hypothalamic and human obesity syndromes. *Physiology and Behavior* **17**, 461–71.

Sepinwall, J. & Cook, L. (1978). Behavioral pharmacology of antianxiety drugs. In *Handbook of Psychopharmacology*, vol. 13, ed. L.L. Iversen, S.D. Iversen & S.H. Snyder, pp. 345–93. London: Plenum.

Smith, G.P. & Epstein, A.N. (1969). Increased feeding in response to decreased glucose utilization in the rat and monkey. *American Journal of Physiology* **217**, 1083–7.

Smith, G.P. & Gibbs, J. (1984). Gut peptides and postprandial satiety. *Federation Proceedings* **43**, 2889–92.

Smith, G.P., Gibbs, J. & Young, R.C. (1974). Cholecystokinin and intestinal satiety in the rat. *Federation Proceedings* **33**, 1146–9.

Smith, G.P., Jerome, C., Cushin, B.J., Eterno, R. & Simansky, K.J. (1981). Cholecystokinin in the rat. *Science* **213**, 1036–7.

Smith, G.T., Moran, T.H., Coyle, J.T., Kuhar, M.J., O'Donahue, T.L. & McHugh, P.R. (1984). Anatomic localization of cholecycstokinin receptors to the pyloric sphincter. *American Journal of Physiology* **246**, R127–R130.

Sugg, E.E., Serra, M., Shook, J.E., Yamamura, H.I., Burks, T.F., Korc, M. & Hruby, V.J. (1988). Cholecystokinic activity of N$^\alpha$-hydroxysulfonyl-[Nle^{28,31}]CCK$_{26-33}$ analog modified at the C-terminal residue. *International Journal of Peptide and Protein Research* **31**, 514–19.

Sullivan, A.C. (1987). Drug treatment of obesity: a perspective. In *Recent Advances in Obesity Research: V*, ed. E.M. Berry, S.H. Blondheim, H.E. Eliahou & E. Shafrir, pp. 293–9. London: John Libbey.

Swerdlow, N.R., van der Kooy, D., Koob, G.F. & Wenger, J.R. (1983). Cholecystokinin produces conditioned place-aversions, not place-preferences, in food-deprived rats: evidence against involvement in satiety. *Life Sciences* **32**, 2087–93.

Triscari, J., Tilley, J. & Hogan, S. (1988). The pharmacological treatment of obesity. *Annual Reports in Medicinal Chemistry* **23**, 191–200.

Verbalis, J.G., McCann, M.J., McHale, C.M. & Stricker, E.M. (1986). Oxytocin secretion in response to cholecystokinin and food: differentiation of nausea from satiety. *Science* **232**, 1417–19.

Watson, C.A., Schneider, L.H., Corp, E.S., Weatherford, S.C., Shindledecker, R., Murphy, R.B., Smith, G.P. & Gibbs, J. (1988). The effects of chronic and acute treatment with the potent peripheral cholecystokinin antagonist L-364,718 on food and water intake in the rat. *Society for Neuroscience Abstracts* **14**, 1196.

West, D.B., Fey, D. & Woods, S.C. (1984). Cholecystokinin persistently suppresses meal size but not food intake in free-feeding rats. *American Journal of Physiology* **246**, R776–R787.

West, D.B., Greenwood, M.R.C., Marshall, K.A. & Woods, S.C. (1987*a*). Lithium chloride, cholecystokinin and meal patterns: evidence that cholecystokinin suppresses meal size in rats without causing malaise. *Appetite* **8**, 221–8.

West, D.B., Greenwood, M.R.C., Sullivan, A.C., Prescod, L., Marzullo, L.R. & Triscari, J. (1987*b*). Infusion of cholecystokinin between meals into free-feeding rats fails to prolong the intermeal interval. *Physiology and Behavior* **39**, 111–15.

William-Olsson, T. (1986). *α*-Glucosidase inhibition in obesity. *Acta Medica Scandinavica*, Suppl. 706.

Willner, P., McGuirk, J., Phillips, G. & Muscat, R. (1990). A behavioral analysis of the anorectic effects of fenfluramine and fluoxetine. *Psychopharmacology* (in press).

Wong, D.T. & Fuller, R.W. (1987). Serotonergic mechanisms in feeding. *International Journal of Obesity* **11**, Suppl. 3, 125–33.

Wong, D.T., Reid, L.R. & Threlkeld, P.G. (1988). Suppression of food intake in rats by fluoxetine: comparison of enantiomers and effects of serotonin antagonists. *Pharmacology, Biochemistry and Behavior* **31**, 475–9.

Wurtman, J.J. & Wurtman, R.J. (1979). Drugs that enhance central serotonergic transmission diminish elective carbohydrate consumption by rats. *Life Sciences* **24**, 895–904.

Wyllie, M.G., Fletcher, A., Rothwell, N.J. & Stock, M.J. (1985). Thermogenic properties of ciclazindol and mazindol in rodents. In *Novel Approaches and Drugs for Obesity*, ed. A.C. Sullivan & S. Garattini, pp. 85–92. London: John Libbey.

Yen, T.T., McKee, M.M. & Stamm, N.B. (1985). Thermogenesis and weight control. In *Novel Approaches and Drugs for Obesity*, ed. A.C. Sullivan & S. Garattini, pp. 65–78. London: John Libbey.

10

Animal models of eating disorders: a clinical perspective

DONALD V. COSCINA and PAUL E. GARFINKEL

10.1 GENERAL PERSPECTIVE

10.1.1 Overview

The purpose of this chapter is to evaluate how effective current animal models are in addressing clinical issues in patients with eating disorders. This is a daunting task because of the vast number of models that have been generated. As an example of this, the accompanying chapter on this topic (Montgomery, Chapter 8 of this volume) has identified approximately three dozen. That list is by no means exhaustive as additional models have been identified in previous reviews (see e.g. Mrosovsky, 1984; Sclafani, 1984, for additional models of anorexia nervosa and obesity respectively).

The number and diversity of these models immediately raises an important question: Why are there so many? The answer usually put forward is that feeding appears to be controlled by multiple factors. At face value, this conclusion seems most reasonable since a function as important to survival as feeding is likely to be regulated in a fashion that affords many checks and balances. Such an intuitive view has been supported by many basic empirical studies, often in animals, whose specific findings make a unitary explanation of feeding seem unlikely, (e.g., Stricker, 1978; Sclafani, 1984; Blundell, 1987; Keesey, 1986). This same general orientation is compatible with current clinical positions that human eating disorders are 'multidimensional' (e.g., Garfinkel & Garner, 1982). Despite the apparent consensus which supports this position, several perplexing questions arise when one probes this position further. First, as a

basic tenet of scientific investigation, the fact that research has not yet identified a single dimension to explain all aspects of eating behavior does not prove that one does not exist. Given the tools and concepts presently available, we simply may have been unable to isolate it. Secondly, from a clinical perspective, it can easily be argued that interpretations of human feeding controls reflect significant constructs derived from animal studies (e.g. see reviews by Nisbett, 1972; Schachter & Rodin, 1974; Bennett & Gurin, 1982; Garner et al., 1985). Since data on clinical eating abnormalities are every bit as difficult to interpret as those collected from non-human subjects, this information is also too diverse to permit a simple, unitary explanation of causality. Therefore, it is hardly surprising that contemporary clinical positions generally agree with more basic empirical positions in questioning a unitary basis for feeding. Whether or not this view is correct, it is challenging to entertain the possibility that the notion of multiple feeding determinants may represent a default posture. That is, it may have arisen not so much from definitive proofs that multidimensional controls exist, but because we have yet to discover a single global framework under which many lesser factors can be subsumed.

In considering what one construct might account for all feeding controls, the answer most often put forward is *metabolic* (e.g., see recent brief review by Hervey, 1988, as well as the large treatise with accompanying commentaries by Le Magnen, 1981). This orientation derives from the basic fact that the metabolic utilization of foodstuffs is a process fundamental to all living organisms.

Without a continuing supply of nutrients to replace those used to meet bioenergetic demands, life ceases. Indeed, the importance of meeting these vital needs has routinely been invoked as a rationale to explain why diversities and redundancies exist in the control of feeding. However, if we accept this view, why have we been unable to identify such a powerful biological mechanism? This seems all the more bewildering in humans given our success in adapting to a wide variety of environmental conditions. To have been able to do so strongly implies that we have evolved excellent mechanisms to deal with our food requirements (see Brown & Konner, 1987, for support of this view). But if this is so, why are disorders of eating so prevalent, especially in privileged societies? To the extent that they represent a high degree of human evolution, a more reasonable expectation might be that the individuals within such groups would be particularly adept at resisting behavioral disruptions so fundamental to life as those related to feeding.

The purpose of raising these troublesome questions is not to argue for the existence of a unitary explanation for feeding. Rather, they set the tone of this review, which is designed to be probing and reductionistic. We shall attempt to redefine and simplify the multidimensional aspects of human feeding, then provide a general framework under which they can be viewed in relation to animal modeling. Given space limitations, we cannot specifically evaluate in this context the vast number of individual models that have been proposed. Instead, we shall outline what appear to us to be their more global features and comment on their potential relevance to patients with eating disorders.

10.1.2 Simplifying human multidimensional factors

Extensive clinical research has identified at least three major dimensions in the etiology and maintenance of the human eating disorders anorexia nervosa and bulimia nervosa. These distinguishable components, which overlap to some extent, arise from the individual, the family and the culture (see Garfinkel & Garner, 1982). Each of these main categories can be subdivided into a number of constituent elements (*ibid.*, p. 193). While such in-depth categorization is meaningful and useful to

investigations in humans, the constructs that they represent are not immediately transferable to animals. This is so because of divergent evolution, as a result of which the mechanisms that control behaviors, as well as the forms that they take and the functions that they serve, often differ across species. As an example of this problem, it is not possible to use animal data to learn about the attitudinal dimensions, in part related to societal ideals, of the excessive concern with body weight and shape, that are so characteristic of human eating disorders (see Garfinkel & Garner, 1982; Mrosovsky 1984). A second example difficult to model would be the individual's feelings of helplessness due to his or her terror surrounding autonomy. These limitations underscore the major challenge to animal researchers in this field: to translate such human constructs into constructs that are salient in the animal domain.

How can the three major categories described above be viewed in ways that are relevant to animal models? The most common type of reductionism applied to this issue has been to delineate *biological* vs. *environmental* characteristics. Such thinking has had a rich history in science and philosophy, dating back to Rene Descartes' 16th century distinction between body and mind. Over the past century, a variant of this general perspective emerged in the *nature* vs. *nurture* controversy surrounding the ways in which behavior was seen to be controlled. Modern thinking recognizes that an accurate account of such descriptions must include both factors. In many ways, this type of interactive dualism can be seen as the cornerstone of human eating problems, which have traditionally been labeled as *psychosomatic* disorders.

Within this basic context, each of the three identified dimensions can be thought of as containing varying degrees of biological and environmental elements. The simplest biological substrate that comes to mind is *genetic*, which clearly varies from individual to individual as well as across families. At the cultural level, genetic factors are also present. However, their strength varies depending on social groupings. They can be relatively strong if the groups are highly inbred. This can occur for a variety of reasons, ranging from common religious beliefs (e.g., the Amish, Menonites, certain sects of Jews) to geographic or nationality constraints (e.g., island dwellers like the Hawaiians prior to colonization, the Australian aborigines or subcultures in Guam; also, tribally isolated groups within North America such as the Pima Indians). Or they may be relatively weak as in the mixed cultures of modern North America and Western Europe. In the latter cases, racial and subcultural interbreeding across generations has diluted genetic influences, rendering them less powerful in and of themselves. From the environmental standpoint, it takes little imagination to recognize the many external events that impact differently on individuals, families and cultures.

In keeping with this biological/environmental dichotomy, it is interesting to point out that two general spheres of medicine have traditionally dealt with human eating disorders: the 'physiological' branches, such as pharmacology, endocrinology and general internal medicine; and the 'behavioral' branches, such as psychiatry, psychology and, more recently, social work. The primary basis for this appears to be the orientation of treatment modes. However, often embedded within this practical issue are assumptions of etiology; in this case, the cause of the disorders as being biologically (i.e., physiologically) vs. environmentally (i.e., psychosocially or behaviorally) based. In day-to-day practice, this distinction is more apparent than real: since components of each factor overlap, neither can be viewed as mutually exclusive of the other. For example, genetic data derived from twin studies of anorexia nervosa have often been used to infer a biological cause since there is a 50% concordance rate for monozygotic twins but only a 15% rate for dizygotic twins (Holland *et al.*, 1984). However, since twins of either type shared the same environment, these findings could also be invoked to support arguments that a particular genetic makeup *interacts* with certain environmental conditions to promote the disorder. Although either interpretation is viable based on these data, both academics and clinicians have often been trained to try to reduce such issues to unitary explanations. In attempts to do so, attitudes and treatments have evolved which often appear in practice to be of an 'either/or' nature. While recognizing the simplicity of this tendency, it may nevertheless be instructive here to view eating disorders within such an admittedly artificial dichotomy. The purpose of

doing so is to gain perspectives into the relative strengths of the causal vs. sustaining elements of these clinical states, which ultimately must be distinguished in order to generate meaningful animal models.

10.1.3 Disorders of eating vs. weight regulation?

Patients being treated for eating disorders have one very important feature in common: that is, a primary concern about body weight. Whether they are physically overweight (the obese), underweight (restricting anorexics), or normal (many bulimics), they all apparently wish to weigh less than they actually do. In pursuit of that goal, reduction of food intake is often used as a major behavioral strategy. This general issue is a very important one to grasp. It means that eating represents a behavioral *conflict* surrounding weight. Accordingly, successful models should view variations in food intake as *symptoms* or *sustaining factors* rather than as direct *causes* of this problem.

It is possible to view weight conflicts within the biological/environmental dichotomy identified above. One critical antecedent condition surrounding weight seems important in this regard: the presence of excess rather than normative premorbid weight.

In cases where weight has been high before the onset of the eating disorder, for the obese as well as for significant numbers of bulimics (Garfinkel & Garner, 1982), biological and environmental factors can be seen as contributory. In the biological domain, there is evidence for a genetic predisposition towards enhanced body mass by the obese (Stunkard *et al.*, 1986; Bouchard *et al.*, 1988; Sorensen *et al.*, 1989; but see Costanzo & Schiffman, 1989, for an alternative suggestion). Expressions of this state could include general reductions in metabolic rate (e.g., Jequier, 1987; Sims, 1989), specific reductions in thermogenesis associated with diet (Himms-Hagen, 1989), disturbed metabolic rhythms (Bellisle *et al.*, 1988) and/or preferential disposition of nutrients into fat stores (Leibel & Hirsch, 1985; Leibel *et al.*, 1985) leading to nutrient depletion in other body compartments. In the environmental domain, early overfeeding, particularly of high-calorie foods, and/or feeding provided for the child's non-nutritive needs, could lead to learned changes which strengthen the reward value of eating (for broader considerations on this issue, see Rodin & Wing, 1988). In this latter case, secondary metabolic changes might also occur which further sustain abnormal eating habits (see Sahakian, 1982, for discussion of how behavioral and metabolic factors may interact). However, even if one or both conditions exist, the fact that a person may become fat is not a sufficient reason to induce such conflict. The additional factor required is environmental pressure to lose weight. This can take many forms, ranging from subtle advertising to overt messages from peer groups. Regardless of its form, though, one final element is required: that these pressures become internalized by the individual. Once that has occurred, they can act as powerful motivators to modify behavior.

The contextual importance of such internalized weight pressures are perhaps easier to comprehend with an example. Let us imagine an overweight person who lives a solitary life (much like the laboratory animals often studied in model systems). If a variety of foods with high sensory appeal were available at little metabolic or behavioral cost (as is often the case in our privileged Western societies), would there be any reason for that individual to not eat them and/or to shed excess fat? Probably not. The impetus to modify eating patterns or weight arises when biological or individually based learning modes clash with other factors emanating from the environment, leading to conflict. Recent research makes it clear that a host of social and economic variables are powerful determinants of this type (see reviews by Garner *et al.*, 1985; Mustajoki, 1987; Soval & Stunkard, 1989). Whether one chooses to label them as cultural, attitudinal, individual, or familial, they are all learned, hence environmentally bound.

The influence of these learned social variables can be so strong that they can contribute to weight conflict in substantial numbers of people even of normal weight (Whitaker *et al.*, 1989; Laessle *et al.*, 1989; Moses *et al.*, 1989). This fact raises a broader issue: the concept of self as viewed through body image. Hilde Bruch is widely acknowledged for pointing out the importance of this construct in patients with anorexia nervosa (see Bruch, 1974). From her extensive clinical experiences, and as supported by a vast amount of subsequent quantitative work (for examples see summaries in Gar-

finkel & Garner, 1982; Darby *et al.*, 1983; Garner & Garfinkel, 1985), the issue of importance to these people is one of establishing control in their lives and feelings of intrinsic worth. Therefore, conflict surrounding weight often appears to be a cardinal feature in many eating-disordered individuals regardless of their absolute body size. However, the weight itself is not really the issue: instead, it is feelings of self-worth and body size which become operationalized through weight. In this schema, regardless of the problems which the patient has, the thinner they become, the better they believe they will feel about themselves.

The disparity that can exist between one's notion of ideal body weight vs. real weight is reminiscent of Festinger's (1957) concept of *cognitive dissonance*, wherein the larger the distance between ideal vs. real precepts, the greater the conflict. To redress this discrepancy, a variety of strategies are employed. Dieting is the most common. Exercise is another which has become particularly popular in recent years. In addition, the practices of inducing vomiting after eating, the use of diuretics, laxatives, or prescription medications to reduce appetite, or the consumption of common substances like nicotine or caffeine, have all been used in efforts to lose weight and/or maintain weight loss. These diverse approaches have in common the attempt to alter the individual's energy intake vs. output equation in favor of diminishing body mass. However, none of them can be used without incurring additional health risks. The fact that individuals are willing to accept these risks indicates the severity of their conflicts surrounding weight.

In addition to these problems, considerable work has shown that lowering caloric intake is met with metabolic defenses against energy wastage whether people are obese (e.g., Yost & Eckel, 1988; Garrow & Webster, 1989; Segal *et al.*, 1989), anorexic (e.g., Pirke *et al.*, 1985; Krieg *et al.*, 1986; Treasure *et al.*, 1988; Fagher *et al.*, 1989; Jonas *et al.*, 1989), bulimic (Pirke *et al.*, 1985), or normal (Fichter *et al.*, 1986). Furthermore, other evidence suggests that repetitive dieting may itself contribute to enhanced metabolic efficiency (e.g., Brownell *et al.*, 1986; Gray *et al.*, 1988; Steen *et al.*, 1988). Taken together, these factors support the paradoxical conclusion that attempts to lose weight can minimize weight loss or even encourage further weight gain. In addition, dieting alone has been associated with significant emotional problems (e.g., Keys *et al.*, 1950; see also Garner *et al.*, 1985, for discussion). Either or both of these responses make continued diet compliance difficult. When diets are stopped or people see little positive response to them, the conflict surrounding weight is often aggravated further through feelings of failure, hopelessness and ineffectiveness. Since exercise too is often unable to produce weight loss (Bray, 1989), failures using this mode can engender similar feelings of self deprecation (Garfinkel & Coscina, 1989).

Figure 10.1 summarizes diagrammatically the conflictual relationship which can be seen to exist between ideal vs. real weight in patients with eating disorders. It also identifies the broad categories of biological and environmental factors discussed above which contribute to the disparity between the two. In addition, we have listed some of the strategies just described by which individuals attempt to rectify this inequity in order to reduce conflict.

10.2 ASSESSMENT OF EXISTING MODELS

10.2.1 The primary problem

One general problem we have encountered in attempting to assess the clinical relevance of eating-disorder models is the lack of clear statements about the exact clinical dimensions being addressed. In particular, issues of etiology, diagnosis, prognosis or treatment are often poorly delineated, or sometimes overlooked altogether. This may occur because of a lack of clinical knowledge on the part of the modelers. Conversely, available descriptions of the clincial conditions may at times be vague or incomplete, making accurate modeling difficult (see Montgomery, Chapter 8 of this volume). In other cases, the models can be seen as serving a different purpose: that is, as a behavioral assay to probe more fundamental questions about actions or interactions among other variables. A common model of this type can be seen in the many studies where the capacity of neuropharmacological agents to modify short-term feeding has been quantified after direct injection of drugs into the brain (see

reviews by Leibowitz, 1988, and Leibowitz *et al.*, 1988, for examples). While such approaches can have strong heuristic value, they may not directly address the broader clincal dimensions alluded to above.

To assist us in our analysis, we have used Montgomery's chapter as a representative source of models and have identified what appears to us to be the primary etiological feature of each according to the biological vs. environmental dichotomy already described. The results of that assessment are summarized in Tables 10.1 and 10.2 for models of obesity and anorexia, respectively.

Examination of these tables reveals several interesting features. The clear majorities (i.e., two-thirds) in both cases are biologically oriented. However, even this simple nominal breakdown underestimates the true representation of this category. If instead one were to tally the actual number of published studies in the animal literature devoted to each, the preponderance of biological models would far outweigh the environmental ones. In fact, we would venture to say that this would be so if based only on the sheer strength of the ventromedial hypothalamic (VMH) and lateral hypothalamic (LH) lesion models that have been so seminal in our understanding of how feeding processes are controlled. That fact notwithstanding, such biological models alone seem to have little direct utility in helping us understand the causal features of clinical eating disorders. At best, they provide us with a putative stimulus for regulating weight or feeding above or below some statistically defined norm. However, even if functional biological abnormalities of feeding or weight control could be shown to precede all other clinical manifestations, we would still be left with the question as to why the expression of those predispositions evoke such conflict. As indicated above, the answer seems to lie in environmental factors.

If we accept that models which focus on environmental variables may provide opportunities for gaining new insights into eating disorders, what aspects should we be modelling? Returning to the biological/environmental dichotomy, such distinctions, in laboratory settings, are often not readily amenable to simple resolutions. One reason for this is that our basic knowledge of the factors which control behavior are often complicated by elusive constructs such as 'reward', 'need',

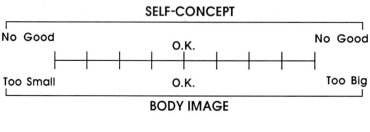

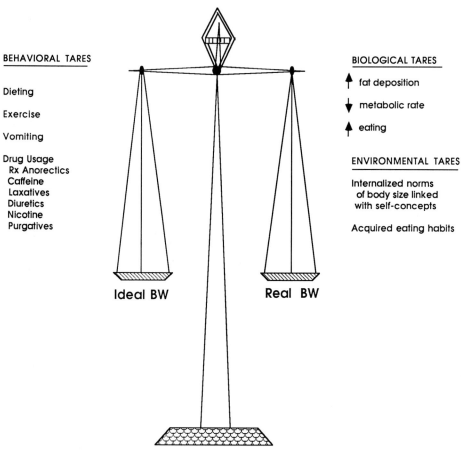

Fig. 10.1. Schematic representation of the conflictual relationship hypothesized to exist between ideal vs. real body weight in patients with eating disorders. The use of a weighing instrument to depict this conflict seems apt as it denotes the objective index, body weight, on which these people focus in their intrapersonal struggles. The factors (i.e. tares) which can add weight to one side or the other of this device are indicated to the sides of the weighing dishes. The rationale for the inclusion of each factor is described more fully in the text. At the same time, this schematized weighing device can be seen as a symbol of the subjective process to which patients may be exposed in their indecisions or conflicts surrounding self-concept as operationalized through perceptions of body image. This has been indicated in the figure by the proposed inter-relationship between these two constructs within the metering scale at the top of the instrument, which has been enlarged above the device to render it readable. Abbreviations used: R_x = prescription; BW = body weight.

Table 10.1. *Animal models of obesity*

Biological etiologies	Acute or chronic[a]
Brain lesions	
Ventromedial hypothalamus	Chronic
Ventral noradrenergic bundle	Chronic
Paraventricular hypothalamus	Chronic
Serotonin neurotoxins	Chronic
Drug-induced	
Serotonin 5-HT$_{1A}$ agonists	Acute
Alpha-2 noradrenergic agonists	Acute***
Anxiolytics (benzodiazepines)	Acute
Opioid agonists	Acute
Antidepressants	Acute
Insulin	Acute***
Antipsychotics/neuroleptics	Acute
Genetic	
Zucker rat	Chronic
ob/ob and db/db mice	Chronic
Environmental etiologies	
Palatable foods	Chronic?
Isolation	Chronic?
Early overfeeding	Chronic
Electrical brain stimulation	Acute***
Tail-pinch	Acute***
Forced feeding	Acute***

[a]In principle, all acute models should be capable of producing long-term changes if stimulus conditions are applied repetitively. Models marked with *** have been shown to do so. Models marked with ? are uncertain as to their chronicity based on inconsistent findings in the literature.

Table 10.2. *Animal models of anorexia*

Biological etiologies	Acute or chronic[a]
Brain lesions	
Lateral hypothalamus	Chronic
Dorsomedial hypothalamus	Chronic
Olfactory bulbectomy	Chronic
Nigrostriatal damage	Chronic
Peripheral lesions	
Vagotomy	Chronic
Drug-induced	
Peripheral serotonin agonists	Acute
Serotonin agonists in the paraventricular hypothalamus	Acute
Amphetamine in the perifornical hypothalamus	Acute
Peripheral opioid antagonists	Acute
Peripheral peptides (e.g., cholecystokinin)	Acute
Peripheral benzodiazepine inverse agonists	Acute
Antipsychotics/neuroleptics	Acute
Environmental etiologies	
Early undernutrition	Chronic
Conditioned emotional response	Acute
Conditioned taste aversions	Acute
Noxious tastes (e.g., quinine)	Acute
Deprivation plus exercise	Acute
Peripheral estrogen plus stress	Acute

[a]In principle, all acute models should be capable of producing long-term changes if stimulus conditions are applied repetitively.

'motivation', 'drive' and 'incentive'. A continuing goal of the behavioral sciences is to improve the definition of these constructs and, where possible, to identify their biological substrates in order to measure and manipulate them more reliably, in the hope of elucidating their interrelationships.

Within the context of eating disorders, the task of separating metabolic from behavioral controls is particularly difficult because of the nature of the response being studied. While environmental factors alone or in combination with biological urgings may be at the root of these disorders, the fact that the conflict takes the form of affecting nutrient intake produces confusion in distinguishing relevant elements. An example of this paradigmatic dilemma has recently been reviewed by Weingarten & Bedard (1989) with regard to studies of *palatability*. The authors discuss the types of methodological and interpretive difficulties that often arise from attempts to separate the hedonic elements of foodstuffs from their metabolic antecedents or consequences. A similar problem is embedded in all basic research on feeding. Perhaps the only way that this necessary sorting of factors can be achieved is to employ technology that permits us simultaneously to assess both metabolic and behavioral events under the same controlled conditions. However, even under those circumstances, the selection of models that operationalize conflict would seem to be an additional ingredient needed to maximize success.

Returning to our contention that conflict surrounding weight regulation is the core issue in patients with eating disorders, how well do existing models address this? To our reading, none do in any comprehensive fashion. Instead, it appears that the vast majority have placed inordinate emphasis on issues of *face* validity in attempts to isolate phenomenological components of eating disorders. Unfortunately, the broader goal of defining underlying *construct* validity often seems lacking. While recognizing that the concept of body weight change is important to incorporate into their models, most investigators have failed to do so in ways which address conflict surrounding weight or its operational representation, food intake. Instead, it appears that the *outcome* of the eating disorder has been targeted for investigation. So, in the case of models attempting to address excess weight or normal-weight bulimia, enhanced feeding has been selected for study; conversely, in the case of anorexic models, depressed feeding has been the focus. In both instances, such approaches seem somewhat ill-conceived. In the case of the anorexic, it has been well documented that these individuals do not lack hunger (see Garfinkel & Garner, 1982). Instead, these patients often feel quite hungry, but refuse to eat because of the conflicts which they perceive surrounding food and fears of losing control. Therefore, studying models in which feeding is primarily depressed contributes little to our understanding of how voluntary restriction of intake occurs despite the existence of hunger (see Montgomery, in this vol., Ch. 8, for further discussion). Furthermore, any model which chooses to study the suppression of feeding is plagued with problems of specificity that render interpretation of the results very difficult (see discussions by Blundell, 1984, as well as Lyon & Robbins, 1976, for prototypic elements of this problem vis-a-vis amphetamine-induced anorexia). Contrasting to this is the case of obese or bulimic patients in which modeling excess intake mirrors only the behavioral pressure of one component in elevating their real weights (see Fig. 10.1). But this too does nothing to address the issues of conflict that arise in associ-ation with exaggerated intake or the frenzied eating patterns characteristic of episodic binging. And what of the evidence that substantial subsets of the overweight do not actually eat more than those of normal weight (e.g., Westerterp *et al.*, 1988) yet maintain elevated weight or continue to gain it (see Garner *et al.*, 1985, for discussion)? Of those who seek treatment for their 'eating disorder', the apparent conflict between ideal and real weight may be all the stronger since they cannot identify excess intake as a factor that can be modified successfully to achieve weight-loss goals.

10.2.2 Additional considerations

The foregoing comments have largely focused around issues of biological predispositions to gain or maintain weight and/or environmental pressures to lose it. There is one additional issue which has not been addressed that seems quite germane to human eating disorders: that is, the clinical psychopathology of eating-disordered patients. With regard to the obese, there is now strong evidence against the historical notion that they are emotionally disturbed in any general sense that is causal to their problem; however, social stigmatization and/or the consequences of dieting may evoke significant psychological distress (reviewed by Mustajoki, 1987; Wadden & Stunkard, 1987). Therefore, from a modeling stand-point, this again suggests that conflict surrounding weight, which emanates from the environment, is the fundamental element for most obese people. Nevertheless, there may be a smaller subset who demonstrate bulimia and may be more disturbed (see Hudson *et al.*, 1988). In that respect, they may resemble some normal-weight bulimics or a subset of anorexics in displaying higher than normal prevalence of depression, anxiety, obsessive-compulsive disorders, substance abuse, or personality disorders (see Garner & Garfinkel, 1988, for several reviews; also, Pyle *et al.*, 1981; Williamson *et al.*, 1985). In some instances these conditions may be consequences of the unresolved conflicts related to their reasons for abnormal eating. More recently it has been suggested that portions of such psychopathologies may be evoked by the chaotic dietary patterns characteristic of these conflicts (Garner *et al.*, 1989). In either case, these psychiatric features would be seen as *state dependent*.

Contrary to this, other work suggests that underlying processes exist in a significant portion of these patients that are more causal to their problems, i.e., *trait dependent*. Impulsivity – the incapacity to withhold responding – appears to be one such trait as it has recently been reported to be a poor prognostic indicator for both anorexics and bulimics (see Sohlberg *et al.*, 1989). In keeping with this contention, poor impulse control has generally been associated with depressed serotonergic function in humans (e.g., Coccaro *et al.*, 1989; Fishbein *et al.*, 1989; Roy & Linnoila, 1988; Van Kammen, 1987; Virkkunen *et al.*, 1989*a, b*). Therefore, it is not surprising that a major form of current treatment for all of the above-named conditions, including eating disorders, has been the use of pharmacological agents whose primary effect is to enhance serotonergic neurotransmission (see Halmi *et al.*, 1986; Freeman & Hampson, 1987; Garfinkel & Garner, 1987; Byerley *et al.*, 1988; Coppen & Doogan, 1988; Fernstrom & Kupfer, 1988; Lopez-Ibor, 1988). Of course, the latter fact does not mean that a fundamental flaw necessarily exists in the 5-HT systems of these patients. Rather, altered 5-HT function may simply be a means of expressing impulsivity. The more general conclusion is that unabated impulse control, whether constitutional or learned, may be a general feature in significant numbers of eating-disordered patients, but that this means of expressing or dealing with conflict is only one of several plastic forms that it may take.

10.3 FUTURE MODELS

10.3.1 Specific suggestions

The issues raised above lead to important points which should be considered in research involving animal models. This first is that feeding *per se* may only be a behavioral conduit for the underlying events which characterize eating disorders. Of the existing models, we concur with Montgomery (Chapter 8) that the arousal-based models based on brain stimulation or tail pinch appear the closest to addressing this element. The general strength of these models is that they incorporate non-specific arousal and the undifferentiated drive associated with it as an important motivator of behavior. This mechanism can be seen as providing the impetus for plastic expressions of behavior that can be shaped by learning through the availability of specific stimulus objects with which the subject can interact to channel the arousal. Secondly, to embrace more fully the notions raised above, future work might benefit by extending these models into situations where behavioral conflict has been integrated into the stimulus situations as either a pre-existing or co-varying condition. Recent work by Soubrie (1986, 1988) suggests this general point in the context of 5-HT dysfunction in relation to models of depression and anxiety. Given the clear clinical interrelationships which exist between eating disorders and the other psychiatric entities already discussed, it seems prudent to incorporate aspects of these other, more general, arousal models into new or existing models of eating disorders. Thirdly, portions of the foregoing discussion have indicated that an important element in significant numbers of eating-disordered patients may be a predisposition towards obesity. In such instances, that tendency may be a critical dimension which interacts with environmentally-mediated events to evoke or sustain conflicts surrounding feeding. Therefore, it seems appropriate to accommodate this likelihood in future attempts at modeling. One example of how this might be achieved is to determine how genetically prone models of overweight (e.g., Zucker fatty rat) or animals made obese by exposure to palatable foods (i.e., dietary obese) respond to stress or conflict surrounding feeding. Fourthly, an important dietary dimension to keep in mind for any future modeling should be the induction of voluntary feeding suppression, despite the availability of food, as a choice made by the *animal*. This more accurately mirrors the circumstances which surround patients in which they initially play an *active* behavioral role in their dieting. Such an approach stands in marked contrast to the majority of current models wherein food deprivation occurs as a forced consequence controlled by the *experimenter*, thereby provoking a *passive* stance on the part of the animal. Parenthetically, it may be the very use of this tactic which accounts for the need of animals to fall back on to metabolic mechanisms as a primary coping strategy, as opposed to a more naturalistic strategy in which metabolic and behavioral factors interact, as seems to occur in patients.

It should be clear from our remarks that we

favor the pursuit of models which incorporate stress or conflict. This position is not new. The essence of it has been discussed by Mrosovsky (1984) in an earlier review which examined models of anorexia nervosa, as well as by Montgomery (this vol. Ch.8). Not only do we agree with this position, but would see it as so fundamental to all eating disorders that future investigations of this type should be extended to models of bulimia and regulation of excess weight.

These points notwithstanding, the institution of conflict surrounding food intake does not suffice to generate a comprehensive model of eating disorders. A second stage of investigation seems necessary in which separate series of tests, both behavioral and physiological, should be employed to confirm that the conflict was effective in producing changes which mirror clinical conditions. The advantages of this stage are two-fold: it provides independent measurements which confirm that the conflict has been produced; and it permits a means of assessing whether subsequent environmental or pharmacological treatments designed to alleviate the conflict are effective. This second stage, then, can be seen as providing the means of establishing both construct and predictive validity for the first stage.

10.3.2 A cautionary note

While the conceptual dimensions outlined above may give rise to models that more comprehensively address features of human eating disorders, a practical problem remains. As in all research with non-human subjects, pursuit of these issues will be forced to use behavioral output as an index of underlying function. Therefore, it is inevitable that studies along the lines suggested will engender alterations of motoric control as important causal or sustaining elements. This means that factors which influence motoric control must be manipulated and measured with great care. The larger goal of defining more exactly the mechanism(s) through which feeding behavior is globally controlled further requires that we learn how reward and sensory information are integrated to support motoric acts. It is in this integrative sphere that the task will undoubtedly be most difficult. At what point can we comfortably conclude that the model constructed has accurately delineated the correct

balance of these factors? In the end, the ultimate challenge will be to distinguish the sensory, motor, and reward components of conflicts, which must inevitably combine to create and sustain them.

10.4 CONCLUDING REMARKS

In the Overview we raised several rhetorical questions which served to identify issues that have been developed in this paper. Having done so, we will conclude by attempting to address them more directly.

As regards the scientific possibility that one unitary dimension may exist in the control of food intake and its disorders, the foregoing discussion has tried to argue that this issue seems largely reducible to two broad general categories: biological and environmental. However, from a philosophical perspective which may be more in keeping with the humanistic goals of clinical medicine, such an operational dichotomoy may appear artificial, if not practically irrelevant. If one accepts that the responsibility of the practicing physician must inevitably be the well-being of each individual patient, then organismic variables must be seen as paramount. This can be interpreted to mean that environmental factors should be viewed as biological entities, for their functional representation depends on how experience is encoded within the constitutional makeup of each individual. If thought about in this way, we are forced to see that any unitarian theory would be biological. However, even if we accept this position, it does not necessarily follow that the cause and sustaining elements of eating disorders are wholly rooted in *aberrant* biological processes. Rather, these statistically defined patterns of abnormality may be seen as appropriate, to the extent that they are compensatory or adaptive within the limitations of that individual's biological make-up and environmental constraints. The problem with such a position is that it leads to a logical dilemma: a person's behavior can only be judged as abnormal if it is made in comparison to others under equally comparable conditions. From our understanding of the subtleties and diversities of the human condition, it seems difficult to imagine this ever being possible given the unlikelihood that the same combinations of internally- and externally-mediated events can be wholly represented in any two people. Therefore, out of practical necessity, we search for larger processes that seem able to account for significant aspects of behavior, in the hope that they are sufficient to interact with or override other factors that are more uniquely individual. This epistemological position provides a conceptual basis for modelers to reduce the complex functional states seen in individual eating disorders to more broad biological ones that are objective to work with.

The premise developed in this paper has been that conflict surrounding weight regulation is a cardinal feature of human eating disorders, and that food intake represents only one, albeit major, behavioral representation of this problem. However, since the more fundamental feature of conflict surrounds concepts of self, its form may be plastic, depending on the environmental pressures which intercede. Within this perspective, the fact that eating disorders are over-represented in privileged societies does not really represent a challenge to general biological concepts surrounding survival. In fact, at its inception, these patterns of behavior can be seen as almost sensible given the relatively high availability of food in our societies, hence the limited possibility of real starvation based on factors *external* to the individual. Instead, such behavior represents an attempt to adapt, i.e., to exert *internal* control over a perceived situation which would otherwise be deemed uncontrollable. In this sense, the patient has displayed an *active* role in his or her life based on the original volition required to precipitate the eating disorder. Unfortunately, given the many health risks which follow, this strategy is unwise. Furthermore, while originally volitional, many of the ensuing patterns become ritualized and difficult to change. At this stage, even the faulty illusion of voluntary control may be lost, plunging the individual back into the perceptions of passivity which gave rise to the original conflict. The only way out of this spiraling cycle is to introduce new volitional controls through additional learning. This, in essence, represents the basis for all forms of psychotherapeutic intervention. For the therapist and animal modeler alike, this view of eating disorders supports the suggestion that conflicts surrounding feeding are relative. When faced with more salient or rewarding environmental pressures, either the patient in therapy or the animal tested in a model paradigm should relinquish the abnormal feeding pattern in favor of a more normal one. For basic researchers to assist clinicians, it is important to unravel the specific conditions under which such behavioral shifts can be induced, as their identification could provide new insights into the causes and perpetuating elements of human eating disorders.

The authors thank Dr P.J. Fletcher and Ms E. de Rooy for helpful comments and criticisms on earlier drafts of this paper. Preparation of this manuscript was supported by a grant to the first author from the Natural Sciences and Engineering Research Council of Canada as well as funds from the Clarke Institute of Psychiatry.

REFERENCES

Bellisle, F., Rolland-Cachera, M.-F., Deheeger, M. & Guilloud-Bataille, M. (1988). Obesity and food intake in children: evidence for a role of metabolic and/or behavioral daily rhythms. *Appetite* **11**, 111–18.

Bennett, E. & Gurin, J. (1982). *The Dieter's Dilemma: Eating Less and Weighing More.* New York: Basic Books.

Blundell, J.E. (1984). Systems and interactions: an approach to the pharmacology of eating and hunger. In: *Eating and Its Disorders*, eds. A. Stunkard & E. Stellar, pp. 39–64. New York: Raven Press.

Blundell, J. E. (1987). Nutritional manipulations for altering food intake. Towards a causal model of experimental obesity. *Annals of the New York Academy of Sciences* **499**, 144–55.

Bouchard, C., Perusse, L., Leblanc, C., Tremblay, A. & Theriault, G. (1988). Inheritance of the amount and distribution of human body fat. *International Journal of Obesity* **12**, 205–15.

Bray, G.A. (1989). Exercise and obesity. In: *Proceedings of the International Conference on Fitness and Health*, ed. C. Bouchard. New York: Humana Press (in press).

Brown, P.J. & Konner, M. (1987). An anthropological perspective on obesity. *Annals of the New York Academy of Sciences* **499**, 29–46.

Brownell, K.D., Greenwood, M.R.C., Stellar, E. & Shrager, E.E. (1986). The effects of repeated cycles of weight loss and regain in rats. *Physiology and Behavior* **38**, 459–64.

Bruch, H. (1973). *Eating Disorders: Obesity, Anorexia Nervosa and the Person Within.* New York: Basic Books.

Byerley, W.F., Reimherr, F.W., Wood, D.R. & Grosser, B.I. (1988). Fluoxetine, a selective serotonin uptake inhibitor, for treatment of out patients with major depression. *Journal of Clinical Psychopharmacology* **8**, 112–15.

Coccaro, E.F., Siever, L.J., Klar, H.M., Maurer, G., Cochrane, K., Cooper, T.B., Mohs, R.C. & Davis, K.L. (1989). Serotonergic studies in patients with affective and personality disorders. Correlates with suicidal and impulsive aggressive behavior. *Archives of General Psychiatry* **46**, 587–99.

Coppen, A.J. & Doogan, D.P., eds. (1988). *Serotonin in Behavioral Disorders. Journal of Clinical Psychiatry* **49**, Suppl. 8, 1–73.

Costanzo, P.R. & Schiffman, S.S. (1989). Thinness – not obesity – has a genetic component. *Neuroscience and Biobehavioral Reviews* **13**, 55–8.

Darby, P.L., Garfinkel, P.E., Garner, D.M. & Coscina, D.V., eds. (1983). *Anorexia Nervosa. Recent Developments in Research.* New York: Alan Liss.

Fagher, B., Monti, M. & Theander, S. (1989). Microcalorimetric study of muscle and platelet thermogenesis in anorexia nervosa and bulimia. *American Journal of Clinical Nutrition* **49**, 476–81.

Fernstrom, M.H. & Kupfer, D.J. (1988). Imipramine treatment and preference for sweets. *Appetite* **10**, 149–55.

Festinger, L. (1957). *A Theory of Cognitive Dissonance.* Evanston, Illinois: Row, Peterson.

Fichter, M.W., Pirke, K.-M. & Holsboer, F. (1986). Weight loss causes neuroendocrine disturbances: Experimental study in healthy starving subjects. *Psychiatry Research* **17**, 61–72.

Fishbein, D.H., Lozovsky, D. & Jaffe, J.H. (1989). Impulsivity, aggression, and neuroendocrine responses to serotonergic stimulation in substance abusers. *Biological Psychiatry* **25**, 1049–66.

Freeman, C.P.L. & Hampson, M. (1987). Fluoxetine as a treatment for bulimia nervosa. *International Journal of Obesity* **11**, Suppl. 3, 171–7.

Garfinkel, P.E. & Coscina, D.V. (1989). Exercise and fitness: obesity, anorexia nervosa and bulimia nervosa. In: *Proceedings of the International Conference on Fitness and Health* (in press), ed. C. Bouchard. New York: Humana Press.

Garfinkel, P.E. & Garner, D.M. (1982). *Anorexia Nervosa: A Multidimensional Perspective.* New York: Brunner/Mazel.

Garfinkel, P.E. & Garner, D.M. (1987). *The Role of Drug Treatments for Eating Disorders.* New York: Brunner/Mazel.

Garner, D.M. & Garfinkel, P.E., eds. (1985). *Handbook of Psychotherapy for Anorexia Nervosa and Bulimia.* New York: Guilford Press.

Garner, D.M. & Garfinkel, P.E. (1988). *Diagnostic Issues in Anorexia Nervosa and Bulimia Nervosa.* New York: Brunner/Mazel.

Garner, D.M., Rockert, W., Olmsted, M.P., Johnson, C. & Coscina, D.V. (1985). Psychoeducational principles in the treatment of bulimia and anorexia nervosa. *Handbook of Psychotherapy for Anorexia Nervosa and Bulimia,* eds. D.M. Garner & P.E. Garfinkel, pp. 513–72. New York: Guilford Press.

Garner, D.M., Olmsted, M.P., Davis, R., Rockert, W., Goldbloom, D. & Eagle, M. (1989). The association between bulimic symptoms and reported psychopathology. *International Journal of Eating Disorders* **8**, (in press).

Garrow, J.S. & Webster, J.D. (1989). Effects on weight and metabolic rate of obese women of a 3.4 mj (800 kcal) diet. *Lancet,* June 24, 1429–31.

Gray, D.S., Fisler, J.S. & Bray, G.A. (1988). Effects of repeated weight loss and regain on body composition in obese rats. *American Journal of Clinical Nutrition* **47**, 393–9.

Halmi, K.A., Eckert, E., Ladu, T.J. & Cohen, J. (1986). Anorexia nervosa: treatment efficacy of cyproheptadine and amitriptyline. *Archives of General Psychiatry* **43**, 177–81.

Hervey, G.R. (1988). Physiological factors involved in long-term control of food intake. *International Journal of Vitamin and Nutrition Research* **58**, 477–90.

Himms-Hagen, J. (1989). Brown adipose tissue thermogenesis and obesity. *Progress in Lipid Research* **28**, 67–115.

Holland, A.J., Hall, A., Murray, R., Russell, G.F.M. & Crisp, A.H. (1984). A study of 34 twin pairs and one set of triplets. *British Journal of Psychiatry* **145**, 414–19.

Hudson, J.I., Pope, H.G., Wurtman, J., Yurgelun-Todd, D., Mark, S. & Rosenthal, N.E. (1988). Bulimia in obese individuals. Relationship to normal-weight bulimia. *Journal of Nervous and Mental Disease* **176**, 144–52.

Jequier, E. (1987). Energy utilization in human obesity. *Annals of the New York Academy of Sciences* **499**, 73–83.

Jonas, J.M., Ehrenkranz, J. & Gold, M.S. (1989). Urinary basal body temperature in anorexia nervosa. *Biological Psychiatry* **26**, 289–96.

Keesey, R.E. (1986). The body-weight set point. What can you tell your patients? *Postgraduate Medicine* **83**, 114–27.

Keys, A., Brozek, J., Henschel, A., Mickelsen, O. & Taylor, H.L. (1950). *The Biology of Human Starvation.* Minneapolis: University of Minnesota Press.

Krieg, J.-C., Backmund, H. & Pirke, K.-M. (1986). Endocrine, metabolic, and brain morphological abnormalities in patients with eating disorders. *International Journal of Eating Disorders* **5**, 999–1005.

Laessle, R.G., Tuschl, R.J., Kotthus, B.C. & Pirke, K.M. (1989). Behavioral and biological correlates of dietary restraint in normal lilfe. *Appetite* **12**, 83–94.

Leibel, R.L. & Hirsch, J. (1985). Metabolic characterization of obesity. *Annals of Internal Medicine* **103**, 1000–2.

Leibel, R.L., Hirsch, J., Berry, E.M. & Gruen, R.K. (1985). Alterations in adipocyte free fatty acid re-esterification associated with obesity and weight reduction in man. *American Journal of Clinical Nutrition* **42**, 198–206.

Leibowitz, S.F. (1988). Hypothalamic paraventricular nucleus: interaction between alpha2-noradrenergic system and circulating hormones and nutrients in relation to energy balance. *Neuroscience and Biobehavioral Reviews* **12**, 101–9.

Leibowitz, S.F., Weiss, G.F. & Shor-Posner, G. (1988). Hypothalamic serotonin: pharmacological, biochemical, and behavioral analyses of its feeding-suppressive action. *Clinical Neuropharmacology* **11**, Suppl. 1, 51–71.

Le Magnen, J. (1981). The metabolic basis of dual periodicity of feeding in rats. *Behavioral and Brain Sciences* **4**, 561–607.

Lopez-Ibor, J.J. (1988). The involvement of serotonin in psychiatric disorders and behavior. *British Journal of Psychiatry* **153**, Suppl. 3, 26–39.

Lyon, M. & Robbins, T.W. (1976). The action of central nervous system stimulant drugs: A general theory concerning amphetamine effects. In: *Current Developments in Psychopharmacology*, vol. 2, eds. M.V. Essman & L. Valzelli, pp. 81–162. New York: Spectrum Publications Inc.

Moses, N., Banilivy, M.-M. & Lifshitz, F. (1989). Fear of obesity among adolescent girls. *Pediatrics* **83**, 393–8.

Mrosovsky, N. (1984). Animal models: anorexia yes, nervosa no. In: *The Psychobiology of Anorexia Nervosa*, eds. K.-M. Pirke & D. Ploog, pp. 22–34. Berlin: Springer-Verlag.

Mustajoki, P. (1987). Pychosocial factors in obesity. *Annals of Clinical Research* **19**, 143–6.

Nisbett, R.E. (1972). Eating behavior and obesity in men and animals. *Advances in Psychosomatic Medicine* **7**, 173–93.

Pirke, K.-M., Pahl, J., Schweiger, U. & Warnhoff, M. (1985). Metabolic and endocrine indices of starvation in bulimia: A comparison with anorexia nervosa. *Psychiatry Research* **15**, 33–9.

Pyle, R.L., Mitchell, J.E. & Eckert, E.D. (1981). Bulimia: a report of 34 cases. *Journal of Clinical Psychiatry* **42**, 60–4.

Rodin, J. & Wing, R.R. (1988). Behavioral factors in obesity. *Diabetes/Metabolism Reviews* **4**, 701–25.

Roy, A. & Linnoila, M. (1988). Suicidal behavior, impulsiveness and serotonin. *Acta Psychiatrica Scandinavica* **78**, 529–35.

Sahakian, B.J. (1982). The interaction of psychological and metabolic factors in the control of eating and obesity. *Human Nutrition: Applied Nutrition* **36A**, 262–71.

Schachter, S. & Rodin, J. (1974). *Obese Humans and Rats*. Potomac, Maryland: Lawrence Erlbaum.

Sclafani, A. (1984). Animal models of obesity: classification and characterization. *International Journal of Obesity* **8**, 491–508.

Segal, K.R., Lacayanga, I., Dunaif, A., Gutin, B. & Pi-Sunyer, F.X. (1989). Impact of body fat mass and percent fat on metabolic rate and thermogenesis in men. *American Journal of Physiology* **256**, E573–E579.

Sims, E.A.H. (1989). Storage and expenditure of energy in obesity and their implications for management. *Medical Clinics of North America* **73**, 97–110.

Sobal, J. & Stunkard, J. (1989). Socioeconomic status and obesity: a review of the literature. *Psychological Bulletin* **105**, 260–75.

Sohlberg, S., Norring, C., Holmgren, S. & Rosmark, B. (1989). Impulsivity and long-term prognosis of psychiatric patients with anorexia nervosa/bulimia nervosa. *Journal of Nervous and Mental Disease* **177**, 249–58.

Sorensen, T.I.A., Price, R.A., Stunkard, A.J. & Schulsinger, F. (1989). Genetics of obesity in adult adoptees and their biological siblings. *British Medical Journal* **298**, 87–90.

Soubrie, P. (1986). Reconciling the role of central serotonin neurons in human and animal behavior. *Behavioral and Brain Sciences* **9**, 319–64.

Soubrie, P. (1988). Serotonin and behavior, with special regard to animal models of anxiety, depression and waiting ability. In: *Neuronal Serotonin*, eds. N.N. Osborne & M. Hamon, pp. 255–70. New York: John Wiley.

Steen, S.N., Oppliger, R.A. & Brownell, K.D. (1988). Metabolic effects of repeated weight loss and regain in adolescent wrestlers. *Journal of the American Medical Association* **260**, 47–50.

Stricker, E.M. (1978). Hyperphagia. *New England Journal of Medicine* **298**, 1010–13.

Stunkard, A.J., Sorensen, T.I.A., Hanis, C., Teasdale, T.W., Chakraborty, R., Schull, W.J. & Schulsinger, F. (1986). An adoption study of human obesity. *New England Journal of Medicine* **314**, 193–8.

Treasure, J.L., Wheeler, M., King, E.A., Gordon, A.L. & Russell, G.F.M. (1988). Weight gain and reproductive function: ultrasonographic and endocrine features in anorexia nervosa. *Clinical Endocrinology* **29**, 607–16.

Van Kammen, D.P. (1987). 5-HT, a neurotransmitter for all seasons? *Biological Psychiatry* **22**, 1–3.

Virkkunen, M., De Jong, J., Bartko, J., Goodwin, F.K. & Linnoila, M. (1989a). Relationship of psychobiological variables to recidivism in violent offenders and impulse fire setters. A follow-up study. *Archives of General Psychiatry* **46**, 600–3.

Virkkunen, M., De Jong, J., Bartko & Linnoila, M. (1989*b*). Psychological concomitants of history of suicide attempts among violent offenders and impulse fire setters. *Archives of General Psychiatry* **46**, 604–6.

Wadden, T.A. & Stunkard, A.J. (1987). Psychopathology and obesity. *Annals of the New York Academy of Sciences* **499**, 55–65.

Weingarten, H.P. & Bedard, M. (1989). Diet palatability: Its definition, measurement, and experimental analysis. In: *The Neuropharmacology of Appetite*, eds. S.J. Cooper & J.M. Liebman (in press). London: Oxford University Press.

Westerterp, K.R., Nicolson, N.A., Boots, J.M.J., Mordant, A. & Westerterp, M.S. (1988). Obesity, restrained eating and the cumulative intake curve. *Appetite* **11**, 119–28.

Whitaker, A., Davies, M., Shaffer, D., Johnson, J., Abrams, S., Walsh, T.B. & Kalikow, K. (1989). The struggle to be thin: a survey of anorexic and bulimic symptoms in a non-referred adolescent population. *Psychological Medicine* **19**, 143–63.

Williamson, D.A., Kelly, M.L., Davis, C.J., Ruggiero, L. & Blouin, D.C. (1985). Psychopathology of eating disorders: a controlled comparison of bulimic, obese and normal subjects. *Journal of Consulting and Clinical Psychology* **53**, 161–6.

Yost, T.J. & Eckel, R.H. (1988). Fat calories may be preferentially stored in reduced-obese women: a permissive pathway for resumption of the obese state. *Journal of Clinical Endocrinology and Metabolism* **67**, 259–64.

PART V

Models of mania and schizophrenia

Animal models of mania and schizophrenia

MELVIN LYON

11.1 INTRODUCTION

11.1.1 Earlier reviews

While there are a number of excellent recent reviews of animal models for mania and schizophrenia (Kumar, 1976; Robbins & Sahakian, 1980; Segal *et al.*, 1981; McKinney & Moran, 1981; Iversen, 1986, 1987), the present review is designed to be more behaviorally descriptive, with specific attention to symptom modeling characteristics. In this connection, it may be useful to list some leading attributes of the earlier reviews, so as to place the present one in perspective.

Kumar (1976) sounded the keynote of many criticisms of animal models up to that time. He noted that animal models were often limited to particular neuropharmacological aspects, and did not reflect the complexity of the human syndrome. Narrowness is still a criticism which can be leveled at many extant animal models of psychosis (Mailman *et al.*, 1981). Giving a more extreme view, Kornetsky (1977; Kornetsky & Markowitz, 1978) cast doubt on many descriptions of animal behavior, including those derived from amphetamine and LSD/mescaline experiments, as true models of human psychosis. As he put it,

The difficulty with all these models is that they are all trying to model a human state that may be so uniquely human that no global model can possibly be adequate . . . No matter how often you ask the animal he will not be able to tell you that other animals or people are plotting against him. (Kornetsky, 1977, p. 4).

This sort of criticism is not particularly relevant. To argue that animals do not talk is *not* to argue convincingly that they do not communicate clearly by their actions. As will be documented below, there are many examples from animal models which demonstrate extreme sensitivity to the actions, and nearness, of conspecifics and even unprovoked attack upon them. In cats and monkeys, attacks have been directed at points in space where no object exists (Ellinwood & Kilbey, 1975; Sudilovsky, 1975; Lyon & Nielsen, 1979; Nielsen *et al.*, 1983; Iversen 1986), and Segal and his coworkers (Segal & Janowsky, 1978; Segal *et al.*, 1981) have argued convincingly for a parallel relationship between the development of 'searching and examining' behavioral stereotypies in animals, and the development of paranoid ideation in human thought. Randrup *et al.* (1981) provided further support for this view, noting close parallels between the fragmentation of animal behavior under stimulant drugs and the development of what closely resembles persecutory behavior with apparent suspiciousness or anxiety. Their brief article has an extended list of references comparing animal and human symptomatology.

McKinney & Moran (1981) agree that many seeming criteria for schizophrenia are based upon drug related speculations, rather than upon the behavioral symptomatology of schizophrenia. They conclude that a closer ethological analysis ought to be made of *human* behavior during psychosis, so that more direct comparisons are possible with the available animal measures. In a sense, this is also the object of the present chapter. Unfortunately, in the end, they revert to the psychiatric view that 'schizophrenia is a thought disorder', thus again implying that it is uniquely human. Beyond questioning the implication that animals do not think, one might also ask whether thought disorder is not simply one additional symptom type, among many others, rather than the basic fault in schizophrenia.

Matthysse (1983) suggested that animal models should be used not only to seek for evidence of motor stereotypies, but possibly for 'motivational stereotypies' as evidenced by the loss of natural switching behavior between response alternatives in rats with nucleus accumbens lesions, or by the persistence in irrational behaviors (from a motivational viewpoint) in animals with hippocampal lesions. Most recently, Iversen (1986, 1987) has written two excellent reviews, centred upon the central stimulant drug models, and summarizing many of the cogent neuropharmacological and neuroanatomical arguments supporting these models and others. These particular reviews deserve special commendation for their clarity and overview.

Taken together, the previous reviews make unnecessary any extensive anatomical and neuropharmacological discussions, and the major weight will therefore be placed upon the efficacy of the various models in producing the behavioral symptoms which resemble those in schizophrenia.

11.1.2 Mania and schizophrenia

It seems appropriate to discuss mania and schizophrenia in the same context since the diagnostic signs for the two syndromes partially overlap at the behavioral level (Zigler & Glick, 1988), and there is additional evidence implying that they may be related disorders (Crow, 1986). It thus remains possible that affective illness, at least in some subgroups of patients, is a symptomatic variant produced by the same basic physical abnormality which also underlies schizophrenia (Tissot, 1975; Randrup *et al.*, 1981).

Several factors have clouded the view of parallels between mania and schizophrenia, or even between subgroups within these categories. Principal among these is the growing evidence that, although schizophrenia depends upon a genetic predisposition, many environmental factors, occurring both before and after birth, can lead to different behavioral syndromes in schizophrenia (Crow, 1982). This has led to a proposed subdivi-

sion of symptomatology into *positive* (active, florid) vs. *negative* (passive, cataleptic) categories. This suggests that it has been misleading to use symptomatic evidence from mixed groups of schizophrenics to test clinical hypotheses, or to serve as the basis for animal behavioral models.

Nevertheless, there remain many obvious behavioral similarities between some common symptoms of mania and of schizophrenia (e.g. increased activity and arousal, stereotyped forms of responding, inappropriately placed or excessive emotional reactions, etc.). Even the depressive phases in affective illness parallel certain states in schizophrenia. Also, it is striking how well the same drug treatments work in both instances. Neuroleptic drugs are beneficial in many cases of mania, where they are used in conjunction with lithium to aid initial calming (see Cookson, this volume, Chapter 13), and lithium treatment has also been found to improve some symptoms of schizophrenia (Zemlan *et al.*, 1984).

This makes it pertinent to examine the potentially close behavioral parallels that also are apparent in animal models of mania and schizophrenia, notwithstanding that the simplification of symptomatological description, which is often necessary in animal studies (see below), may tend to blur differences that appear obvious to the clinician evaluating human symptoms (see Cookson, this vol., Ch. 13). However, in view of the boundary problems in classification and diagnosis, a flexible attitude will be taken to the relatively fixed categorizations of symptoms that permeate psychiatric diagnosis (i.e. DSM-III classifications – American Psychiatric Association, 1980). It will be assumed here that one of the primary goals of experimentation with animals is not merely to find models for the known diagnostic categories, but also to suggest whether or not some of these categories should be simplified or totally replaced.

From an experimentalist's point of view, psychiatric diagnosis has all too frequently centered merely upon the '*content*' of the behaviors seen to be aberrant, and not sufficiently on the similarity of '*structure*' which is to be found in many of the purportedly different symptoms. For instance, diagnoses for thought disorder, verbal aberrations and stereotyped behavior are typically listed as separate symptoms of schizophrenia (e.g.

DSM-III). It will be argued here that the basic structure of behavior available to the schizophrenic is limited, just as it is in animal models, but that it is precisely the *same* basic structural limitation of the behavior which predisposes to all three of the above symptoms. Furthermore, some of these structural differences in schizophrenic behavior overlap heavily with those found in both manic and depressive phases of affective illness. Thus the present chapter is not only a review of the available animal models for mania and schizophrenia, but also a direct attempt to show how contributions from animal models have powerful implications for the direction that psychiatric diagnosis might take to escape the present morass of symptomatology (see also, Brockington, 1986).

11.2 VALIDATING CRITERIA FOR ANIMAL MODELS OF MANIA AND SCHIZOPHRENIA

In an introductory chapter, Willner (this volume) has aptly suggested dividing problems of validity into *face validity*, *predictive validity* and *construct validity*. Each of these will be considered separately.

11.2.1 Assessment of face validity

Face validity is understood as the outward similarity of appearance of symptoms and behavioral changes in the animal and human cases being compared.

11.2.1.1 Mania

According to DSM-III (1980), with slight modifications from Murphy (1977), mania is characterized by the following symptoms:

(1) A generally labile affective state, characterized by elation, but also containing periods of depression and irritability.
(2) Hyperactivity, which includes increased social contact and increased aggressive and sexual behavior.
(3) Pressure of speech.
(4) Flight of ideas.
(5) Inflated self-esteem.
(6) Decreased need for sleep.
(7) Distractibility.

(8) Excessive involvement in acts leading to painful consequences with no recognition of this relationship. (This particular symptom is better described by Murphy, 1977, as 'impaired judgment and impulse control.')

It should be noted that all of these symptoms, except for the 'periods of depression' under (1), are active *positive* symptoms. Even the brief periods of depression could be conceived of as parts of an active, but 'fragmented' emotional response which finally becomes stereotyped as a 'compulsive', and continuous, depressive activity (see Randrup *et al.*, 1981).

At first glance, it seems difficult to identify reasonable parallels with animal behaviors. However, it is evident that there is much overlap between the symptoms, and Robbins & Sahakian (1980) have suggested that, for the purpose of studying basic processes in animal models, the list could be shortened to *hyperactivity*, *elation* and *irritability*.

Hyperactivity and *irritability* yield to rather straightforward analyses (see below). *Elation* is a more difficult problem, but some aspects may be measurable in animals. The significant feature of elation in humans is a perceptual experience, which is difficult, but not impossible, to measure in animals. Since reactions to primary reinforcing stimuli may vary greatly in normal individuals, elation in animals may be more reliably detected by a generalized sharp increase in the positive secondary reinforcing aspects of stimuli, including those not previously associated with specific rewards. Since it would be difficult to separate the element of elation, in the sense of a gladness about everything, from increased behavior elicited for other reasons, the measure of reinforcing value should preferably include estimations not based on simple response rate, to avoid confounding with other rate-changing conditions. This matter is especially critical in studies of intracranial self-stimulation (ICSS), as pointed out by Atrens *et al.* (1983).

Furthermore, as in human cases, a single one of these behavioral features – hyperactivity, irritability or elation – cannot function by itself as a critical symptom, but requires corroboration by some other signs such as decreased sleep activity, and increased frequency of shifting between periods of irritability and depression. Thus, what distinguishes mania from normal mood changes is not elation, depressive episodes, or irritability, but rather the intensity of these changes and the rapidity of shifting between these states. In fact, many of the human symptoms can be related to rapid changes in responding (labile emotionality, ranging from elation to irritability to depression; increased social contact, aggression and sexual behavior; pressure of speech; flight of ideas; distractibility; impaired impulse control). It seems obvious that this *rapid switching of ideas and responses* is one main structural feature of manic behavior (see also Lyon, 1986; Lyon & Gerlach, 1988).

In fact, the early manic phase is the period of greatest creativity in writers and artists who suffer from manic/depressive illness (Andreasen & Powers, 1975), and this in turn is suggestive of increased switching and a greater variety of response patterns during this period. Interestingly, the latter signs are also documented features of animal behavior following amphetamine treatment (Evenden & Robbins, 1983*a, b*; Moerschbaecher *et al.*, 1979), as well as of human schizophrenics (Armitage *et al.*, 1964; Lyon *et al.*, 1986). Seemingly opposed to increased switching, and flexibility of responding, is a feature not so evident in the symptom list, but which relates to the perseverative, and even stereotyped, tendencies also present in manic individuals as they approach the peak of the manic phase and begin the downward trend to depression. The 'irritability' noted above usually follows directly upon the frustration induced by not being allowed, or not being physically able, to continue rapidly enough upon the chain of behavior (including thought) which is in progress. Andreasen (personal communication) has described how creative writers suffering from mania can sense when they are about to 'peak out' on a creative manic phase, because their activity becomes too frenetic to be useful (distractibility, flight of ideas) and they become increasingly irritable, with a tendency to get 'hung up' on minor details in their work. This perseveration and stereotyped occupation with the same problem signals the onset of a depressive period. These behavioral features are very suggestive of the perseveration and stereotypy that are characteristic of amphetamine psychosis, and of certain phases in schizophrenia, as will be seen below. It may even

be true that the depressive phase represents an intense perseveration on one area of thought: the all-pervading sense of personal worthlessness and loss of positive sensation (Bleuler, 1950; Randrup *et al.*, 1981).

If this reasoning is at all correct, there may be valid grounds for a re-evaluation of the symptoms of mania, and their grouping as given above, as well as for a second look at the tacitly accepted separation of mania and schizophrenia. However, in mania, unlike schizophrenia, the sequences of behavior, though rapidly executed, and perhaps partially stereotyped, are still essentially normal in many respects. What distinguishes them from the everyday normal state is the forced hurriedness of their execution, and it is only when that execution reaches the limits of speed and induces a tendency to repeat the same activity (stereotypy), that the behavior loses its normal sequence.

11.2.1.2 Schizophrenia

According to DSM-III (American Psychiatric Association, 1980), the 'Initial' phase of schizophrenia is characterized by the following four 'Positive' symptoms (1–4) associated with overtly active behavior, and three 'Negative' symptoms (5–7) with a passive behavioral quality:

Positive symptoms
(1) Peculiar or bizarre behaviors.
(2) Digressive speech, or overelaboration, unnecessary repetitions or excessive punning.
(3) Odd ideation, especially as bizarre or extremely unusual associations.
(4) Unusual perceptual experiences.

Negative symptoms
(5) Lack of personal hygiene or grooming.
(6) Flat or inappropriate affect.
(7) Social isolation.

In the 'Final' phase of the developing illness a further two Positive (9, 10) and two Negative (11, 12) symptoms are described:

Positive symptoms
(9) Delusional thinking.
(10) Hallucinations, especially of the auditory type.

Negative symptoms
(11) Incoherence or poverty of speech.
(12) Flat, or inappropriate affect.

Two additional symptom categories are proposed in the present account (see 8, 13 below), which belong to the Initial and Final phases respectively and are largely derived from Bleuler's (1950) description of schizophrenic symptoms (see also Astrachan *et al.*, 1972). These are assumed here to be *cardinal* symptoms, which will always appear, even though their behavioral content may vary extremely, including both Positive and Negative symptom dimensions:

(8) Excessive response alternation or switching between several response alternatives. (This symptom may occur independently or in relation to any of the Positive symptoms (1–4). It can also be a contributing factor to each of the Negative symptoms (5–7), when the switching behavior begins to dominate all other behavior patterns.)

(13) Frequently occurring repetition or stereotypy of the same response type, or constant (stereotyped) shifting of response modes. (This may occur in all response modes (speech, thought, writing, bodily movement) typically without regard for the consequences of the action. It may apply to Positive symptoms (9, 10) as well as contributing to Negative symptoms (11, 12)).

It may be added that the DSM-III listing for schizophrenia, which was based on the supposition of independence between symptomatologies of mania, schizophrenia, and a number of other disorders, was purposely constructed to omit symptoms which would overlap with those found in affective illness, schizoaffective disorders and cycloid psychoses. While this has the advantage of narrowly defining the schizophrenic syndrome, it also ignores these two further aspects of schizophrenia.

Another feature omitted from the above list, which may distinguish schizophrenics, but only when considered in combination with other symptoms, is an *excessively high or excessively low level of autonomic activity*. High levels of autonomic activity can be seen even in chronic patients who are extremely withdrawn and catatonic (Venables & Wing, 1962). More recently, Venables (1978) has suggested that there may be two groups of schizophrenics, one unduly hyper-responsive autonomically, the other hypo-responsive (see also Dawson

& Neuchterlein, 1984). Excessively high, or low, autonomic arousal can be found in both the early and the late phases of the illness, and in both positive and negative types of symptom states, particularly in the subgroup of schizophrenics characterized by hypothalamic damage and ventricular widening (Andreasen *et al.*, 1982).

All in all, what characterizes the DSM-III (1980) description of the schizophrenic state is not, as in mania, the obvious acceleration of normal behaviors and thought processes, but the apparent breakdown of speech, thought, perception and bodily movement into abnormal, non-adaptive and bizarre components, that no longer can achieve their normal functions. There is, however, a gradual change in severity of symptoms. In the 'Initial' phase, the symptoms develop from those that are bizarre or unusual, but still socially interactive, to symptoms indicating a greater and greater loss of contact with, or reference to, the actions of others, culminating in social isolation. In the 'Final' phase no semblance of contact with reality, or with other persons, may remain, and all of the symptoms are essentially negative with respect to social contact, even though the more active symptoms are labelled positive since they are productive of delusions, hallucinations, and stereotyped behavior.

Despite the problems engendered by their highly varied content, all thirteen of the above symptoms can be grouped structurally under just four major types of behavioral change: *response switching, focusing, fragmentation,* and *stereotypy*. Furthermore, these four types are listed in their usual order of appearance as the illness progresses. As will be seen below, the progressive nature of the behavioral changes seems to follow a pattern that is common to both mania and schizophrenia.

It is probably no coincidence that evidence of increased response switching is prominent in the symptoms that are suggested to occur during the *initial* phases of schizophrenia, such as bizarre behaviors or verbal associations, digressive speech and overelaboration. Such switching mirrors similar changes noted earlier to represent the creative early phase of mania. In both illnesses, this highly creative phase is accompanied by a lack of concern for personal appearance, and an increasing concentration on the task at hand, to the exclusion of interest in, or attention to, the social responses of others. This leads, in turn, to the flat or inappropriate (too rapidly switching?) affect, and the final stage of social isolation. This type of sequence, leading from hyperactivity with normal behaviors, to increasing concentration on single behaviors with poor grooming and inappropriate affect, and finally ending with a period of social isolation has been seen both in human amphetamine psychosis (Connell, 1958; Bell, 1965) and in animal models of psychosis (Ellison, 1979; Ellison & Eison, 1983).

As the schizophrenic episode deepens, the response switching increases rapidly, ultimately becoming too frenetic to resemble normal responding. Individuals have less and less time to react to the responses of others and, almost by default, become increasingly focused upon what is happening to themselves and their thoughts. This apparent 'withdrawal', and self-preoccupation, is thus probably best viewed, not as a choice, but as a necessity in the face of increasingly rapid shifts in the stream of attention-demanding thoughts. The dominant delusional thinking, hallucinations and paranoid thoughts are very likely the rationalization of this forced introspection.

Parallels to this focusing of attention by individuals upon themselves have also been reported in non-human primate models of psychosis (Goosen, 1981; Nielsen *et al.*, 1983). However, another type of focusing, which frequently goes unrecognized, is the focusing of responses into a limited number of categories, such that the variety of responding becomes reduced and inflexible. This can occur as repetition of thoughts or words, surfacing as thought disorder, unnecessary repetitions of words or incoherency. It can also be seen in a wide variety of repetitive motor activities engaged in by schizophrenics. Such repetitive activity produces 'response competition' by interfering directly with other responses, or by reducing the time available to make them correctly (Grilly, 1977; Robbins, 1984). It has been hypothesized that this repetitional activity, as it increases in frequency, and competes for the available response time, results in the failure to complete normal sequences of behavior, and thus in the *fragmentation* of response strategies and other sequential activities (Randrup *et al.*, 1981; Lyon *et al.*, 1986; Lyon & Gerlach, 1988). This explanation could apply to such varying symptoms as inappropriate affect, lack of

attention to personal hygiene, and bizarre ideation. Here also, there are animal models showing inappropriate maternal behavior (Schiørring, 1977; Wegener, 1986) and abortive body grooming (Ellinwood *et al.*, 1972) apparently due to an increased tendency to initiate responses more quickly, causing a fragmentary response that does not achieve its full purpose.

When the fragmentation reaches its maximum, there are almost no behaviors of any great sequential length which can be completed before they are fragmented. The result is that a very few simple behavioral patterns are retained which are then used more and more frequently. Excessive repetition, and finally *stereotypy of thought and action* are the end result of this breakdown.

11.2.1.3 The definitional symptoms as related to animal models

Tables 11.1 and 11.2 show the relationship between the human symptoms and the presumed animal behavior correlates for mania (Table 11.1) and for schizophrenia (Table 11.2). Table 11.2 owes much to the listings and descriptions of: Ellinwood & Kilbey (1977); Sudilovsky (1975); Haber *et al.* (1977); and Nielsen *et al.* (1983). These tables will form the basis of the subsequent animal/human comparisons and it is hoped that they will provide a clear guideline for further experiments. It is important to note that *both* human and animal symptom descriptions include activities or states which are sometimes *directly* measurable (such as 'hyperactivity'), and sometimes only *indirectly* measurable (such symptoms as 'elation'). The implication of 'indirect' measurement is that it requires a judgmental interpretation by the observer of, for instance, a verbal statement or a body movement or gesture (non-verbal communication).

The supposedly great advantage of verbal experiental report in humans, as opposed to animals, is taken with some skepticism, since inaccuracy, poor use of speech, and delusional errors abound in the psychoses. In animal studies, this source of information is all too often totally ignored, despite the known complexity of vocal communication, from the ultrasonic mating calls of rats to the variety of sounds used by monkeys. On the other hand, unquestioning transfer of results from conditioned behaviors in animals to all manner of human situations is unwarranted. Therefore, an attempt is made in the tables to specify the exact conditions for data collection. What turns out to be striking is the similarity of available data sources, when viewed in this manner.

From the descriptions in Tables 11.1 and 11.2, it is obvious that the *face validity* of the animal models for human psychosis is potentially much better than might be assumed from the loosely defined listings of human symptoms. The specific intention of the present view of animal models for schizophrenia and mania is to show that these models encourage us to look at simpler, yet more fundamental, behavioral changes which very possibly underlie most of the extremely varied human symptom expressions. *What is needed most is attention to the structure of the behavior, and not to its idiosyncratic communicative content.* Useful and accurate predictive information can be gained from animal models by emphasizing the structural principles involved in the behavioral change, rather than the uniquely human expression of these changes (e.g. as in speech).

11.2.2 Assessment of predictive validity

Among important predictive factors for mania and schizophrenia are the following: (a) *drug treatment specificity*; (b) *sex*; (c) *etiological factors*; (d) *temporal course*; and (e) *laterality of brain function*. Since the degree to which a given model has been subjected to investigation on these dimensions will determine, in large measure, how accurate its predictive validity can be at this time, those models which have been investigated on several dimensions deserve more attention than those which model a single symptom or were tested along only one dimension.

11.2.2.1 Drug treatment specificity

There is a tendency to view schizophrenia as a condition specifically sensitive to the typical and atypical neuroleptic drugs. However, the close relationship between the antipsychotic effects and dopamine (DA) receptor blocking functions of neuroleptics has led to a relatively narrow view of drug specificity for schizophrenia. Still, up to this point, except for the atypical neuroleptics which

only indirectly affect the dopamine balance, no other antipsychotic drugs strongly benefit the majority of schizophrenics. The ability of neuroleptics to block effects in animal models of schizophrenia is therefore of prime importance for their predictive validity.

Mania is thought to be more specifically sensitive to lithium treatment, which is the treatment of choice once the immediate effects of the mania have been dampened by neuroleptics. However, this connection is less clear, since although lithium does well in the treatment of human mania, it has only mixed results in tests against animal models of this condition. This may result from a faulty choice of models, or from the non-specificity of lithium, but, at this time, the failure of lithium to block apparently manic behavior in an animal model cannot be considered sufficient evidence to reject the model.

11.2.2.2 Sex

It is worth noting that the majority of schizophrenics are men while the majority of depressives are women, and that in both cases the ratio is approximately 3 : 2. Furthermore, schizophrenia in females develops later and is less intense than in males, and there is a sharp increase in the incidence of female schizophrenia after menopause. These facts suggest that the female sex hormones, which are generally anti-dopaminergic in function, may serve to protect females against the illness. There is direct evidence that the activity of the nigrostriatal DA system is altered during the normal estrus cycle of the female rat (Hruska & Silbergeld, 1980; Falardeau *et al.*, 1987), while chronic estrogen treatment produces a 20% increase in DA receptors of both D1 and D2 types (Levesque & Di Paolo, 1987).

In a two-lever situation with food reinforcement, testosterone increased response perseveration on the non-reinforced lever in both male and female rats. However, the females tended to persevere on the lever which had delivered food on the previous trial, while males showed a more fixed preference for one lever, regardless of its previous association with reinforcement (van Hest *et al.*, 1987). Given such findings, it is astonishing to see how many animal studies still rely exclusively upon male animals.

Table 11.1. *Comparative human and animal symptoms of mania*

DIRECTLY AND INDIRECTLY MEASURABLE HUMAN SYMPTOMS	DIRECTLY AND INDIRECTLY MEASURABLE ANIMAL CORRELATES

Elation

Indirect: Verbal report.   *Indirect:* Difficult to evaluate.
    Direct: Sharp increase in secondary reinforcing value, independent of response rate changes.

Irritability

Indirect: Overly strong, vocal bodily response to stimuli.
    Direct: Decreased threshold for startle and for provocation of aggressive response.

Depression

Indirect: Verbal report.   *Indirect:* Vocal response minimal, humped-over posture, non-responsive to social contact, fetal position, self-directed aggression.
    Direct: Sharp decrease in secondary reinforcing value, independent of response rate.

Increased mood switching (elation, irritability, depression)

Indirect: Verbal report and non-verbal communication.   *Indirect:* Changing vocal response, posture, and responses to stimuli.
    Direct: Abrupt changes in secondary reinforcement functions, and in thresholds for startle and aggression.

Hyperactivity (general)

Direct: Rapid, too frequent vocalizing, increased rates of responding and of locomotion.

Increased general social contact

Indirect: Verbal, non-verbal communication directed at attracting others.   *Indirect:* Social or localizing calls, eliciting approach or grooming.
Direct: Increased bodily contacts.   *Direct:* Increased touching, hugging, huddling, grooming with others.

Increased aggressive acts

Indirect: Verbal and non-verbal threatening behavior.   *Indirect:* Vocal or posturing threats, not necessarily leading to attack.

Direct: Increased verbal or physical attacks.   *Direct:* Attack-oriented snarling and growling, assault and biting.

Increased sexual contacts

Indirect: Verbal and non-verbal elicitations.   *Indirect:* Mating calls, presenting posture, sexual eliciting stimuli.
Direct: Increased frequency of direct sexual investigation, foreplay, and copulation. Indexed in animals as mounting, intromission, and ejaculation.

Increased switching of social, aggressive, sexual contacts

Direct: Abrupt and inappropriate switches, and increased rate of changing between these types of activity.

Pressure of speech

Direct: Overly rapid speech or non-verbal communication.   *Direct:* Increased rate of any sequential activity of a social type, including rates of vocalization, maternal behavior, grooming, sexual and social contact.

Table 11.1. (*cont*)

DIRECTLY AND INDIRECTLY MEASURABLE HUMAN SYMPTOMS	DIRECTLY AND INDIRECTLY MEASURABLE ANIMAL CORRELATES

Flight of ideas

Direct: Rapid switching of thoughts (in speech, writing, overt acts).

Direct: Rapid switching of activities (as above under social, aggressive and sexual acts).

Inflated self-esteem

Indirect: No respect for others, even direct superiors.

Indirect: Ignores dominance hierarchy, and attempts dominant role.

Direct: Grandiose statements, extreme self-compliments.

Direct: Not measurable.

Decreased sleeping time

Direct: Decreased total sleep, and shorter sleep periods.

Distractability

Direct: Disruption of stimulus control by irrelevant stimuli.

Impaired judgment and loss of impulse control

Direct: Poor choice behavior where consequences are aversive. (Must be controlled for memory.)

Direct: Loss of avoidance without loss of escape response or of memory capacity (test with DRL!).

Perseveration or stereotypy of responding

Direct: Repetitious patterns, or single response perseveration.

11.2.2.3 Etiological factors

The weight of evidence now suggests that schizophrenics inherit some predispositional factor which interacts with an environmental influence such as emotional stress, viral attack, or X-irradiation, to produce schizophrenic behavior (Gottesman & Shields, 1972; Torrey *et al.*, 1983; Mednick *et al.*, 1988). The nature of the inherited weakness is not known with certainty, but it now appears likely that it affects the fetal development of brain structure, with especially noticeable effects in the pallidum, hippocampal formation, and vermis of the cerebellum (Heath *et al.*, 1979; Kovelman & Scheibel, 1984; Bogerts *et al.*, 1985). There are at this time no genetically based animal models designed to test for schizophrenia or mania *per se*, but Nowakowski (1987) has reported finding an autosomal dominant mutation in Hld mice with hippocampal defects closely paralleling those found in human schizophrenics, including some

who have not received neuroleptic drugs (B. Bogerts, personal communication). Other possibilities for genetically based animal models are related to the demonstration that certain amphetamine-related behaviors seem to have a specific genetic basis (Moisset, 1977).

Prenatal (maternal hormonal) stress factors (Huttunen & Niskanen, 1978), and viral infections (Mednick *et al.*, 1988), have been suggested as important possible agents in the etiology of schizophrenia. While there are studies showing cross-generational effects of DA overstimulation in the offspring of rat mothers treated with drugs during pregnancy (Gauron & Rowley, 1974, 1976), no animal studies have yet used prenatal techniques in an attempt to model schizophrenia or mania.

The most important perinatal factors with possible implications for schizophrenia are those related to hemorrhage and/or hypoxia occurring because of birth complications. The brain damage following hypoxia has been carefully documented

Table 11.2. *Comparison of the human and animal symptoms related to schizophrenia*

DIRECTLY AND INDIRECTLY MEASURABLE HUMAN SYMPTOMS	DIRECTLY AND INDIRECTLY MEASURABLE ANIMAL CORRELATES

INITIAL PHASE: POSITIVE SYMPTOMS

Digressive speech, overelaboration, repetitions, many puns

Direct: Spoken or written language.	*Direct:* Not measurable. *Indirect:* Inappropriate switching or increased repetition of certain responses with exclusion of others, interfering with communicative, sexual or social behaviors. Excessively repeated vocalization.

Odd ideation, bizarre or extremely unusual associations

Direct: As revealed in speech, writing, and action.	*Direct:* Not measurable. *Indirect:* Peculiar or bizarre behavior towards others. Incorrect vocalizations, or social behaviors. Danger calls in absence of danger. Close contact with others despite being attacked or bitten. Self-destructive and self-punitive behaviors.

Peculiar or bizarre behavior

Direct: Observation of inappropriate, awkward, fragmented, or dissociated responses.	*Direct:* Head or paw shakes, ataxia, twitches, obstinate progression (persistent pushing against an immovable resistance), abortive grooming, postural imbalance and awkwardness, hyperextension and hyperflexion of limbs, akathisia.

Unusual perceptual experiences

Indirect: Verbal report.	*Indirect:* Extremely concentrated examination of details, especially near objects, or body and hands. Hyperstartle, or sudden recoil to humanly undetectable stimuli. Crouching, hiding, and fearful responses to innocuous stimuli, or familiar individuals.

INITIAL PHASE: NEGATIVE SYMPTOMS

Lack of personal hygiene or grooming

Direct: Lacks normal grooming, hair in disorder, does not wash or clean body. Attentive to specific points on body, to exclusion of all others. In humans, clothes unclean and in disorder.

Flat or inappropriate affect

Indirect: Verbal report.	*Indirect:* Not measurable.

Direct: Flat, or inappropriate affect to social contact or provocation. Staring into space, no emotional response to others, inappropriate sexual or aggressive actions.

Social isolation

Direct: Loss of bodily contact, failure to respond to others, failure to make eye contact, hiding, and avoiding others. In animals, added loss of social grooming.

Table 11.2. (*cont.*)

DIRECTLY AND INDIRECTLY MEASURABLE HUMAN SYMPTOMS	DIRECTLY AND INDIRECTLY MEASURABLE ANIMAL CORRELATES

INITIAL PHASE: POSITIVE/NEGATIVE SYMPTOMS

Excessive response switching or alternations

Direct: Excessive response alternation, or switching. Initially visible in positive symptoms such as akathisia, but contributes to negative symptoms when behavioral stereotypy occurs. Behavior becomes extremely limited and focused on the individual, no locomotion, little response to environmental stimuli. Catalepsy and waxy flexibility of limbs can occur in end-stage.

FINAL PHASE: POSITIVE SYMPTOMS

Delusional thinking

Indirect: Verbal report.	*Indirect:* Not reliably measurable.
Direct: Verbal and non-verbal insistence on reacting as if in unreal life situations.	*Direct:* May relate to failures to observe social hierarchy, including that toward dominant male animal.

Hallucinations

Indirect: Verbal report of auditory, visual, olfactory, or somatosensory stimulation in the absence of any real detectable stimulus source. Includes hearing voices, seeing objects or persons, smelling body odors or noxious fumes, and sensing presence of parasites on the body (hallucinatory parasitosis).	*Indirect:* As behavior directed at invisible objects, and having a sequential nature, with attention remaining concentrated on one spot during the activity. 'Fly-catching' with coordinated eye, tongue, and mouth movements (cats). Attack, flight, or feeding behaviors associated with invisible stimuli (monkeys). Excessive parasitotic-like grooming and biting at specific points on the skin (monkeys). Not readily measurable in rodents.

FINAL PHASE: NEGATIVE SYMPTOMS

Incoherence or poverty of speech

Direct: Mumbling, fragmented, or inadequate vocalization.	*Direct:* Atypical, fragmented calls or noises, muteness.

Flat or inappropriate affect

Direct: Flat, or inappropriate affect to social contact or provocation. Staring and lack of eye contact. Emotionally unresponsive. Remains isolated and passive.

FINAL PHASE: POSITIVE/NEGATIVE SYMPTOMS

Response stereotypy

Direct: Frequently occurring repetition or stereotypy of one response type, or constant (stereotyped) shifting of response mode. Single or double alternation of responses, ending in fixed stereotypy of one response type.

in non-human primate models for this condition (Ranck & Windle, 1959; Faro & Windle, 1969). While enlarged brain ventricles (i.e. reduced brain volume) in humans are sometimes the result of such birth complications, the enlargement can also occur for other reasons. In any case, a significantly higher proportion of schizophrenic individuals, especially those with negative symptoms, have enlarged ventricles (Andreasen *et al.*, 1982). No animal studies were found with a direct relationship to perinatal factors and schizophrenic-like symptoms. However, a model for schizophrenia involving periventricular damage in the adult rat following intraventricular perfusion with lyso-phosphatidyl choline has recently been suggested (Kline & Reid, 1985, see section 11.4.11.3).

11.2.2.4 Temporal course

The terms 'acute' and 'chronic' are frequently used in the clinical descriptions of psychosis; 'acute' usually referring to the initial, and often most florid, phase of the illness, while the word 'chronic' is usually reserved for those with extensive hospitalization. This state of affairs should be compared with the typical animal study, which involves either a single acute drug treatment, or perhaps a few well-spaced injections with intervening tests. In some fewer cases, a so-called chronic treatment regimen is used, but this commonly involves daily drug injections, for a period of, at best, six to twelve days. In still fewer cases, there is a more truly chronic drug treatment by means of implanted capsules or minipumps, but even here the resulting doses have a rapidly attained peaking effect early in treatment, with a drop-off at the end of the drug period (Nielsen & Lyon, 1982).

Some cyclic changes also deserve recognition. While mania is typically regarded as one phase within a cyclic disturbance, Robbins & Sahakian (1980) noted that manic phases are also frequently interrupted by brief periods of depression. These periods could be caused by a relatively stable underlying depression, which occasionally is brought to overt expression by external events (Carlson, 1986, p. 699), or they could be the result of a cyclic behavioral rhythmicity with relatively brief duration. In the rat, there is also a periodicity of several days between the peaks in 'spontaneous' hyperactivity produced by constant infusion of dopamine into the nucleus accumbens (Costall *et al.*, 1984*a, b*; see section 11.4.1.3).

In schizophrenia, there is also growing recognition of phasic changes, and of periodicity in the severity of symptoms. There is, therefore, a need for greater awareness of cyclical behavioral changes and corresponding measurement of behavior at more than one time prior to and after treatment.

11.2.2.5 Laterality of brain function

There is increasing evidence that left hemispheric dominance is reduced in schizophrenia (Gruzelier & Flor-Henry, 1979). Handedness and laterality have also been demonstrated in monkeys, cats and rats, and there is a corresponding laterality in the dopamine systems of the rat brain (Ungerstedt, 1971; Jerussi & Glick, 1976). Hemispheric influences in animal models of schizophrenia and mania form an important future line of research (see section 11.4.9).

11.2.3 Assessment of construct validity

From a behavioral aspect, there are at least two major theoretical viewpoints on mania and schizophrenia which have direct relevance to studies of animal behavior.

The '*defective stimulus filter theory*' (Payne *et al.*, 1964; Frith, 1979; see also Braff *et al.*, 1977) is based mainly upon clinical observations, but is amenable to test by animal models (Matthysse, 1983). This theory suggests that the basic fault in schizophrenia is not in the response per se, but in the decisional process regarding the choice of which stimuli should receive attention (Norman & Shallice, 1980), and be allowed to elicit a response. It is assumed that the normal filtering of irrelevant stimuli does not occur, or is weakened, so that the stimuli chosen for response initiation may be relevant or irrelevant to the ongoing activity. This theory also appears to have some relationship to 'attentional deficit' hypotheses, and possibly also to the non-human primate model of cognitive deficit (thought disorder) proposed by Goldman-Rakic (1987). Models of mania and schizophrenia that emphasize stimulus effects and attentional factors can be compared within this theoretical framework.

The '*response acceleration/competition theory*' of Lyon & Robbins (1975) is based directly upon an amphetamine animal model, and has already been tested clinically (Lyon *et al.*, 1986; Lyon & Gerlach, 1988). This theory, as already noted, suggests that the psychosis acts very much as a central stimulant drug overdose would do, producing increasingly rapid activity, which by its nature must be relegated to responses of briefer and briefer duration. Initially, the accelerating response rate leads to increased response switching, and more complex response patterns, but the end result is complete stereotypy. A partially related concept is found in the '*reaction potential response ceiling theory*' of Broen & Storms (1977), which states that as the schizophrenic behavior gains in strength, so do the response initiation probabilities for all responses. The result is that the differences between the response probabilities decrease as they all approach the response probability ceiling, which leads to increased interference from irrelevant responses for the task in progress. However, the Lyon & Robbins' theory additionally specifies how certain responses then come to dominate behavior, and predicts which of them will survive the process. This theory can be applied best to models of mania or schizophrenia that emphasize changes in response patterns or timing, where basic stimulus functions are assumed to be intact. However, changes in stimulus functions are expected to occur secondarily, as many internal brain functions are also presumed to be accelerated.

11.3 ANIMAL MODELS OF MANIA

Animal models of mania can be placed into two broad categories. One group consists of *Neuropharmacological models* related to specific neurotransmitters such as dopamine (DA), noradrenaline (NA), and serotonin (5-HT). This category also includes models using lesions produced by transmitter-related substances such as 6-hydroxydopamine (6-OHDA) and 5, 7 dihydroxytryptamine (5, 7-DHT). The second group may be characterized as *Treatment effect models* and includes models based on drug treatment effects in reducing mania (lithium), or on the induction of manic changes in behavior and perception produced by specific non-pharmacological treatments (kindling, intracranial self-stimulation (ICSS), brain lesions not limited to neurotransmitter systems, social isolation etc.).

11.3.1 Catecholamine (CA) related models

Most of the following models were either designed to support a catecholaminergic (DA or NA) factor as a causal event, or to examine how lithium or other agents might antagonize such activity.

11.3.1.1 Amphetamine models

The amphetamine models of mania are principally related to the hyperactivity, hypersensitivity, and elation which are common to the central stimulant drug effects and to mania. Murphy (1977), Robbins & Sahakian (1980), and Iversen (1986) have all reviewed these models at some length, so only a summary is presented here.

Amphetamine produces a noticeable hyperactivity, which is not dependent upon obvious reinforcing events. As noted above, it is necessary to separate the effects of hyperactivity from those of elation by using response measures not dependent upon rate of responding (see for example, Fantie & Nakajima, 1987). However, Hill (1970), and Robbins (1975, 1976), have shown that amphetamine, and other central stimulant drugs (pipradrol and methylphenidate in particular), can increase the strength of responding to conditioned reinforcers. These changes are not totally dependent upon a simple stimulation of response rate, since they are specific to the lever producing the conditioned reinforcer (Robbins *et al.*, 1983). Therefore, both hyperactivity and elation can be said to be modeled by amphetamine, although the latter effect is apparently even stronger with other central stimulants.

In addition, there are signs of increased irritability after low doses of amphetamine. These changes appear as increased startle reactions (Davis *et al.*, 1975) and increased aggression in social situations with conspecifics (Miczek, 1974). Amphetamine is also known to interfere with sexual activity in rats by increasing the initiation of such activity, but disturbing the normal pattern of behaviors (Gessa & Tagliamonte, 1975; Michanek & Meyerson, 1975). As in mania, insomnia is also a symptom of amphetamine use in both animals and people.

Given the above multiplicity of symptoms paralleling those found in mania, it would appear that amphetamine stimulation provides an excellent model for that illness. To this may be added that the classical neuroleptic drugs, such as haloperidol, are effective in controlling the manic symptoms, as well as being effective amphetamine antagonists. The CA synthesis inhibitor alpha-methyl-para-tyrosine also tends to counteract the manic-like effects of amphetamine in rats (Davies *et al.*, 1974), and may have some anti-manic properties in humans (Brodie *et al.*, 1971).

Despite the apparent success of the amphetamine model in producing symptoms similar to those of mania, there are several problems with the predictive validity of the model. Fessler *et al.* (1982) showed that pretreatment of animals for fourteen days with lithium did not affect amphetamine-induced hyperactivity, and even potentiated that produced by phencyclidine (PCP), another widely abused stimulant drug which has been suggested to mimic schizophrenia.

Interestingly, the DA reuptake inhibitor cocaine produces hyperactivity closely resembling that seen with amphetamine, or in mania, but lithium pretreatment does prevent this action of cocaine (Mandell & Knapp, 1976). Lithium also appears to counteract amphetamine-induced locomotor activity (Robbins & Sahakian, 1980), which suggests the potential separation of amphetamine effects into two categories: activity level increases and stereotyped responding. The nigrostriatal and mesolimbic dopaminergic systems appear to subserve stereotypy and activity level respectively (Kelly *et al.*, 1975), suggesting that lithium mainly counteracts mesolimbic activity. This suggests that mania may be more closely related to the mesolimbic system than to the nigrostriatal system.

11.3.1.2 6-OHDA nucleus accumbens lesion model

Selective destruction of DA neurons in the nucleus accumbens of the rat with 6-OHDA leads to a *decrease* in d-amphetamine induced hyperactivity, but to marked *increases* in locomotor activity following treatment with the direct DA agonist, apomorphine (Kelly *et al.*, 1975; Kelly, 1977). This, together with two other reports linking DA receptor sensitivity with hyperactivity (Shaywitz *et al.*, 1976; Sahakian *et al.*, 1976), suggests that 'denervation supersensitivity' following the 6-OHDA treatment might be the important factor in the increased activity (Robbins & Sahakian, 1980). Antelman & Chiodo (1981), on the other hand, suggested that autoreceptor (presynaptic) *subsensitization*, rather than postsynaptic supersensitivity, could be the cause of amphetamine psychosis.

Interpretation of the apparent receptor super-

sensitivity at a behavioral level is further complicated by the fact that the hyper-reactive behavioral syndrome, earlier called 'septal rage' (Brady & Nauta, 1953), but incorrectly associated with the septal nuclei themselves as shown by Harrison & Lyon (1957), probably depends instead on damage to the nucleus accumbens. Animals with such lesions:

... appear to exhibit many of the 'affective' signs of mania, or the 'elation' and 'irritability' of our own designation, without a persistent accompanying hyperkinesia. (Robbins & Sahakian, 1980, p. 55).

However, it should be noted that most of the symptoms of an affective nature following lesions involving the nucleus accumbens and the basal septal region, do not last beyond the period necessary for recovery from the immediate lesion effects (fourteen days). Presurgical handling and petting of the animals leads to a great reduction, and sometimes the complete disappearance, of the post-lesion affective syndrome. Nevertheless, certain forms of sensitivity to stimulus change are retained. It is quite apparent in attempting to handle such animals that they are hypersensitive to novel stimuli, and that attempts to handle them, or other means of sensory stimulation, lead more easily to aggressive reactions. This suggests that the basic fault following electrolytic lesions is not increased affectivity per se, but *hyperesthesia*, which leads to the rapid learning of protective and defensive responses.

However, Evetts *et al.* (1970) found that the more specific dopaminergic lesions induced by intraventricular 6-OHDA did not increase motor activity, but that this treatment did increase irritability, aggressive activity, hyper-reactivity and vocalization (Nakamura & Thoenen, 1972; Coscina *et al.*, 1973). In agreement with the above evaluation of the 'septal' (accumbens) syndrome, Coscina *et al.* (1975) found that intense handling of 6-OHDA treated animals could largely prevent the hyper-reactive and aggressive responding. Petty & Sherman (1981) therefore proposed that intraventricular injections of 6-OHDA provide an excellent animal model for mania. Although the 6-OHDA treated animals were not initially hyperactive prior to shock training, the 6-OHDA treatment could induce a relatively long-lasting hypersensitivity to a 1.0 mA electric foot-shock. Especially interesting, in this connection, was that this hypersensitivity increased in strength over days one to six for rats treated only with 6-OHDA. Lithium counteracted the effect, and there was partial reduction by chlorpromazine. The tricyclic antidepressant imipramine increased reactivity, while chronic ECS partially reduced it. These findings are in good agreement with what has been reported earlier with respect to other animal models of mania, and Petty & Sherman (1981) suggest that such increasing sensitization may be a causal factor in the development of insomnia, distractibility, and even the flight of ideas. Neuroanatomically, it might be mentioned, intraventricular injections in the rat will be bathing a large surface of the medial corpus striatum, and the nucleus accumbens. Thus both of these structures, as well as the lateral septal nuclei, are among the brain regions most immediately affected by intraventricular neurotoxic agents.

11.3.1.3 Ventral tegmental lesion model

Lesions of the ventral tegmental area (VTA), whether by electrolytic methods or by microinjections of 6-OHDA, result in hyperactivity, which has been shown to be correlated with the reduction in level of frontal cortical dopamine (Le Moal *et al.*, 1976, 1977). In addition, lesions of the VTA seem to lead to fractionation of behavior and more rapid shifting from one behavior to another (Stinus *et al.*, 1978). These signs appear similar to those seen with increasing DA release, but injection of low doses of the DA agonist apomorphine, instead of exacerbating these symptoms, *reduces* locomotor activity. This is consistent with the observation of Post *et al.* (1976), that low doses of piribedil (ET-495), a DA receptor agonist, can have a beneficial effect in some cases of mania. However, the dose level is important, since Gerner *et al.* (1976) found that piribedil could evoke manic episodes in one patient, after apparent sensitization to dopaminergic agents following treatment with amphetamine.

The fact that a DA agonist could reduce the hyperactivity following VTA lesions seems to resemble the 'paradoxical' effect of amphetamines in hyperactive children, and Stinus *et al.* (1978) suggested this might provide an animal model for hyperactivity or mania. However, since the VTA contains the cell bodies of the mesolimbic system

which project to the nucleus accumbens, it is curious that after 6-OHDA lesions of the accumbens, apomorphine elicited great *increases* in locomotor activity (Kelly *et al.*, 1975).

11.3.1.4 Other catecholaminergic models

A number of other pharmacological methods induce changes in catecholaminergic functions, and also specifically increase hyperactivity in animals. However, since this is the only important symptom that has been studied in these models (Robbins & Sahakian, 1980), they are listed here only briefly:

(i) The benzodiazepine tranquilizer chlordiazepoxide potentiates and prolongs the hyperactivity induced by amphetamine in mice and rats tested on a holeboard or in a Y-maze (U'Prichard & Steinberg, 1972; Davies *et al.*, 1974). The effect of the drug mixture, but not of amphetamine alone, could be reduced by lithium (Cox *et al.*, 1971). However, in certain instances, chlordiazepoxide alone induced hyperactivity of a sort (Davies *et al.*, 1974), which lithium also reduced.

(ii) The hyperactivity induced by the combination of an NA reuptake inhibitor (desmethylimipramine) and a reserpine-like drug (tetrabenazine) has been likened to that produced in mania, but the major available evidence of similarity is based on the reversal of the hyperactivity by pretreatment with lithium (Matussek & Linsmayer, 1968; see also Murphy, 1977). Although this pretreatment was more effective if continued chronically, which is similar to the effect of lithium in mania, the causes of the behavioral changes remain unclear. Robbins & Sahakian (1980) have pointed out that pretreatment with low doses of amphetamine also reduced the hyperactivity to some extent.

(iii) The combination of a reserpine-like drug (tetrabenazine) with nialamide, a monoamine oxidase inhibitor (MAOI), also induced hyperactivity in mice. However, this behavior was *not* effectively reduced by lithium (Furukawa *et al.*, 1975).

(iv) A combination of the dopamine precursor

L-dopa, with a peripheral decarboxylase inhibitor (benserazide) produces hyper-activity which can be effectively blocked by lithium pretreatment (Smith, 1976). If other symptoms could be tested with this model, it might prove useful.

(v) Daily post-natal injection of 1-tri-iodothyro-nine for thirty days induced significant increases in DA release and behavioral hyperactivity (Rastogi & Singhal, 1977). Lithium reduced both the hyperactivity and some of the major catecholaminergic effects. There is known to be a close association between hyperactivity and hyperthyroidism, but clinical cases of mania do not typically involve thyroid hyperfunction, even though *hypo*thyroidism is frequently linked with depression. The fact that lithium reduces this hyperactivity may indicate only that lithium also counteracts some forms of hyperactivity which are not necessarily related to mania (Robbins & Sahakian, 1980; Fessler *et al.*, 1982). The validity of this model for mania remains questionable.

11.3.2 Indoleamine related models

11.3.2.1 Depletion of 5-HT by parachlorophenylalanine (PCPA)

Treatment with the 5-HT synthesis inhibitor para-chlorophenylalanine (PCPA) produces an interest-ing combination of hyperactivity (Fibiger & Cam-pbell, 1971), which includes increased frequency of social and aggressive acts, increased irritability or hypersensitivity to new stimuli (Ellison & Bresler, 1974; Marsden & Curzon, 1976), and disturbance of circadian rest–activity rhythms (Marsden & Curzon, 1976). These changes make this one of the more complete models for the symptoms of mania. While PCPA affects other neurotransmitter systems (Miller *et al.*, 1970), the behavioral effects seem to be dependent on 5-HT depletion since tryptophan and 5-hydroxytryptophan treatment can reverse them. There is also evidence of possible elational effects in that intracranial self-stimulation in the lateral hypothalamus is increased by PCPA treatment (Poschel & Ninte-man, 1971; Phillips *et al.*, 1976; van der Kooy *et al.*,

1977); however, ICSS in the hippocampus and caudate nucleus is concomitantly *decreased* by PCPA treatment (Phillips *et al.*, 1976). All this points to the need for better rate-free measures (Atrens *et al.*, 1983), though it may also be asking too much to expect that self-stimulation should increase in all brain areas during elational effects, especially since, unlike the lateral hypothalamic region, the caudate and hippocampus are not thought to be as specifically related to positive reward.

Pimozide, a specific dopamine receptor blocker, does not readily affect the PCPA hyperactivity in doses which did block amphetamine-induced hyperactivity (Marsden & Curzon, 1976). More importantly, lithium also fails to reduce activity in PCPA pretreated animals (Murphy, 1977), which casts some doubt on this model, even though the increased aggression following PCPA treatment can be blocked by lithium (Sheard, 1970).

11.3.2.2 Destruction of serotonergic neurons

The neurotoxins 5,6 and 5,7-dihydroxytryptamine (5,6 and 5,7-DHT) cause a relatively specific destruction of 5-HT neurons, when injected into the brain (Baumgarten & Schlossberger, 1973). Diaz *et al.* (1974) found that repeated intraventri-cular injections of 5,6-DHT resulted in different types of activity change dependent upon the size of the induced lesion. With smaller lesions, rats showed an increase in bodily locomotion, but a reduction in rearing on the hind legs. With larger lesions, there was decreased locomotion with increased rearing. The effect after the larger lesions is similar to that seen after PCPA. Poschel *et al.* (1974) reported increased ICSS in the lateral hypo-thalamus following 5,6-DHT lesions which might suggest elational effects, but once more, these results do not generalize to self-stimulation of the hippocampus (van der Kooy *et al.*, 1977), nor do we have a rate-free measurement.

11.3.2.3 5-HT precursor + MAOI models

Grahame-Smith (1971) found that a mixture of the 5-HT precursor 1-tryptophan, together with a MAOI, tranylcypromine, induced a very strong behavioral activation, with increased locomotion, accompanied initially by many other physiological signs such as increased body temperature, piloerection, and penile erection. However, ICSS rates were apparently reduced by treatment with l-tryptophan + the MAOI pargyline (Herberg & Franklin, 1976), suggesting that any elation effect is lessened, rather than increased.

The MAOI tranylcypromine did not produce hyperactivity when given alone (Grahame-Smith, 1971), but there are reports of strongly increased activity in animals treated with MAOI nialamide (Modigh & Svensson, 1972). However, lithium *potentiated* the activity increase, which suggests that it is not relevant to mania.

11.3.2.4 *Lysergic acid diethylamide (LSD) models*

Jacobs *et al.* (1976) described some symptoms in the cat following treatment with LSD, which seemed to mimic reactions seen after amphetamine. These included a type of fractionated behavior described as 'abortive grooming' (which has been frequently reported in rats, cats, and monkeys following treatment with central stimulant drugs – Schiørring, 1971; Sudilovsky, 1975; Ellinwood & Kilbey, 1977), and more or less continuous switching between forms of activity (suggested as an important component of amphetamine induced hyperactivity by Evenden & Robbins, 1983*a*). It will be noted that these symptoms are perhaps more characteristic of schizophrenia than mania. Furthermore, the attribution of these behavioral effects to 5-HT systems alone is suspect, since LSD may also act directly as a DA agonist in certain mesolimbic structures (Kelly & Iversen, 1975).

11.3.3 Models related to exogenous and endogenous opioids

11.3.3.1 *Morphine-induced hyperactivity model*

Morphine induces a clear hyperactivity in many animal species, and such morphine effects were suggested as an animal model for mania by Carroll & Sharp (1971). Although the mouse shows an early and excessive behavioral activation to relatively low doses of morphine, in early experiments rats seemed to be sedated, rather than activated,

while cats were thought to show only 'feline mania', without other typical early morphine effects. However, both of these supposed species differences have mostly to do with the choice of dosage, and the rapidity with which the behavioral changes occur (Ayhan & Randrup, 1973; Villablanca *et al.*, 1984). In the rat, low doses (1.5–5.0 mg/kg) of morphine are especially effective, while higher doses produced hyperactivity after a period of inactivity (sedation). In drug-tolerant rats, hyperactivity may be interspersed with periods of stereotyped sniffing, licking, and biting, which are similar to those seen under higher doses of amphetamine. However, morphine tends to induce a more rapid shifting ('distractibility') from one extremely different mode of activity to another – even from sedation to bursts of locomotion (E. Schiørring & A. Hecht, personal communication). As in mania, social interaction is increased in rats by morphine (Schiørring & Hecht, op. cit.), as are hindleg grooming, eating and drinking (Ayhan & Randrup, 1973). These effects differ from those produced by higher doses of amphetamine in that they do not seem to follow the simple rule of increased activity within categories of response requiring increasingly shorter response chains (Lyon & Robbins, 1975). Rather, the morphine induced hyperactivity seems to provoke rapid shifting between widely divergent categories of responding, which more closely resembles the increased response shifting found in amphetamine treated rats by Evenden & Robbins (1983*a*), and in manic humans by Lyon *et al.* (1986). On the other hand, behaviors with characteristically longer time sequences, such as hind-leg grooming and sedation, are stimulated only at extremely low levels of amphetamine dosage, yet under morphine, these behaviors alternate with briefer activities in an apparently meaningless sequence. This makes the morphine hyperactivity a particularly appealing parallel to mania where similarly divergent behaviors may all be stimulated.

Furthermore, lithium reduces this hyperactivity, as do neuroleptics. So too, however, do PCPA depletion of 5-HT, a number of opiate antagonists, and certain NA antagonists (Carroll & Sharp, 1972; Ayhan & Randrup, 1973). The effect of the neuroleptics supports the contention that DA stimulation may play an important role in the manic-like behavior produced by morphine (Iversen & Joyce, 1978; Joyce & Iversen, 1979; Robbins & Sahakian, 1980). Perhaps opiate antagonists may be expected to counteract morphine effects via opiate receptors, and NA antagonists do tend to counteract hyperactivity, but the effect of PCPA is more difficult to explain, since PCPA itself seems to provide an excellent model for mania (see above).

11.3.3.2 *Enkephalin related model in mice*

Katz and his colleagues (Katz, 1982) have demonstrated that d-Ala2-substituted amides of the endogenous opioids Leu and Met enkephalin, produced identical stereotyped running to that seen with morphine. They also found that the enkephalinase inhibitor kyotorphin, which potentiates endogenous enkephalin activity, also produced behavioral activation. In this case, low doses of kyotorphin elicited general increases in several behaviors, including rearing on the hind legs, while higher doses resulted in purely stereotyped running, which in extreme cases ended in convulsions and death. The obvious behavioral relationship between this opioid model of mania and the DA-related models needs further clarification. It appears that the opioid influence may produce a more general increase in switching behavior, while direct DA stimulation has a greater influence upon perseveration and repetition of *specific* responses. It is worth noting that both influences peak out in a pure stereotypy of behavior.

11.3.3.3 *Opioid administration to the VTA*

The above findings suggested that some endogenous opioids might play an important role in mania, but the locus of this action is not yet clear. Morphine increased locomotion after infusions into the ventral tegmental area (VTA), but not into the substantia nigra (Joyce & Iversen, 1979; Broekkamp *et al.*, 1979), thus suggesting that the mesolimbic (but not the nigrostriatal) system might be involved. Liebman & Segal (1976) found that lithium antagonized the facilitation by morphine of ICSS in the mesencephalon, but not the facilitation by amphetamine in the same location. These data suggest that the control of elational effects by midbrain structures may be under greater influence from opioid than DA systems.

11.3.4 Lithium treatment effect models

Since lithium is the treatment of choice in mania, sometimes with the combined use of a neuroleptic in the initial stages, it is possible to consider lithium as a behavioral antagonist for manic responding. The ability or failure of lithium to counteract certain forms of hyperactivity etc. is frequently used as a criterion for accepting or rejecting a model for mania. For instance, the fact that lithium does not reduce the spontaneous activity of animals pretreated with the MAOI pargyline, and actually *increases* the hyperactivity induced by other MAOIs (see section 11.3.1.2) has led almost directly to the rejection of MAOI induced hyperactivity as a viable model for mania.

As noted above, lithium antagonizes many of the social, aggressive, and irritability-related changes induced by the catecholamine, indoleamine, and opioid models of mania. However, lithium is ineffective in reducing general, or spontaneous, activity in PCPA-treated animals and it does not affect, or may even increase, the hyperactivity produced by d-amphetamine.

In animals not pretreated with any other drug, lithium does reduce certain behaviors, although not uniformly across species (reviewed by Murphy, 1977). Territorial aggression and foot shock-induced fighting behavior are reduced, even though under the latter condition, flinch and jumping responses were not altered. This suggests that it is irritability that is diminished, rather than general muscular response. As Murphy (1977) suggests, the counteracting of the aggressive, irritable behavior is one of the essential features for an anti-manic drug.

In all of the three major models for mania so far mentioned (catecholamine, indoleamine and opioid), rearing on the hind legs is dose-dependently increased, but this effect is not uniformly counteracted by lithium treatment. However, Johnson & Wormington (1972) did find a reduction in rearing in normal rats treated with lithium. Since the height of rearing, when it did occur, was unaltered, they concluded that it was the exploratory aspects of rearing which were suppressed. Other studies have also indicated a lessened reaction to environmental stimuli, including social contacts, in lithium treated normal animals (Sheard, 1970; Syme & Syme, 1973; Smith & Smith, 1973).

On the basis of such evidence, Johnson (1972) suggested that lithium's principal action might be to raise the threshold of salience for external stimuli to be processed by the central nervous system. This view was tested in rats and goldfish by assessing their response to pairs of stimuli which were either clearly discriminable, or only slightly discriminable; the response to those stimuli which were least different, was most affected by lithium (Johnson, 1975, 1979). Johnson suggested that in mania such small differences would be given exaggerated attention. This theoretical account is reminiscent of the stimulus generalization theory of schizophrenia (Mednick, 1958), and the overgeneralization effects resulting from the defective stimulus filter model of the same illness (Frith, 1979).

11.3.5 Brain lesion models from toxic treatment effects

Environmental agents, such as lead poisoning, may be responsible for the abnormal hyperactivity syndrome frequently seen in children. A key factor in differentiating this type of hyperactivity has been the fact that central stimulant drugs, especially methylphenidate and amphetamine, appear to *reduce* the activity, rather than increasing it as might be expected. However, the changes induced may be due to an increased perseveration and a decreased flexibility of response, rather than to a simple reduction in hyperactivity. Accordingly, this apparently 'paradoxical' effect of the stimulant drugs may be due to an overstimulation of DA systems with a corresponding change in activity from the high locomotion phase to the more restricted activity preceding the induction of stereotypy (Robbins & Sahakian, 1979; Mailman *et al.*, 1981).

A subgroup of hyperactive children are not sensitive to stimulant drugs. This finding is paralleled by findings of amphetamine responsive and non-responsive dogs in an extremely hyperactive hybrid breed of telomian-beagle origin (Bareggi *et al.*, 1979). Responsive dogs were found to have a higher blood level of an active metabolite of amphetamine, which would support the view that the 'paradoxical' effect may be due to a higher dose

effect, leading to greater perseveration of responding, but to less extensive motor activity.

Mice exposed to lead acetate before and after weaning, were more aggressive, three times more hyperactive than untreated mice, and exhibited certain forms of stereotyped self-grooming. Most importantly, d-amphetamine and methylphenidate reduced this hyperactivity in the same, seemingly paradoxical, manner as these drugs appear to 'calm' hyperactive children (Silbergeld & Goldberg, 1974). Similar effects have been reported for another heavy metal, cadmium, in rats (Rastogi *et al.*, 1977), and for the alkaline earth metal, rubidium, in monkeys (Meltzer *et al.*, 1969). Rubidium appears to potentiate morphine induced hyperactivity in mice in a manner similar to that of tricyclic antidepressants and MAOIs, which suggests some relationship of the rubidium effects to increased availability of the catecholamines (Stolk *et al.*, 1970). Some evidence suggests that heavy metals within the brain accumulate most readily in the hippocampus (Fjerdingstad *et al.*, 1974), and therefore any potential behavioral effects may be due to malfunctioning in the limbic system.

Norton *et al.*(1976) examined the behavior of animals that had received brain lesions due to X-irradiation of the fetus at gestational days 14–15, carbon monoxide exposure on post-natal Day 5, or bilateral electrolytic lesions of the globus pallidus. Of particular interest here, are the two first procedures, which resemble the prenatal and perinatal factors mentioned earlier as possible causes of manic/depressive or schizophrenic symptoms. By photographing the animals each second and then subjecting the individual frames to analysis into fifteen different behavioral categories, Norton *et al.* were able to specify that 'exploratory acts' were increased in frequency, but of shortened duration, compared with control rats. Activities requiring longer periods of time such as grooming or sitting in an attentive posture, were decreased in frequency and duration. There was also an apparent decrease in the normal sequential appearance of activities, which suggested a more frequent, and more random, switching between activities. These results are highly suggestive of the type of hyperactivity seen after amphetamine stimulation, while the emphasis on the increased randomness of behavioral acts seems to mirror the effects reported under morphine stimulation.

11.3.6 ICSS as a model for elation

An important suggestion by Robbins & Sahakian (1980) was that the 'elation' symptom in mania might be measured by the reaction to intracranial self-stimulation (ICSS) reinforcement in animal models (Panksepp & Trowill, 1969, 1970; Koob, 1977). Increases in ICSS rates following stimulant drug treatment might well be related to those increases in 'elation, flight of thoughts, rapid speech and increased sexual thoughts and overtures' found with ICSS in humans (for reviews see: Delgado, 1976; Sem-Jacobsen, 1976).

Robbins & Sahakian also noted that brain stimulation frequently seems to elicit a more generalized activating state, such that the same brain stimulation, in the same location, can produce varying behaviors dependent upon the relevance of these behaviors to the external environment (Valenstein, 1976). When food was available the animal would eat, when water was available it might drink, and so on; similar effects are brought about if DA systems are activated by tail-pinch stimulation (Antelman & Szechtman, 1975). A cardinal feature of both ICSS and tail-pinch induced activity is that when food, or other objects which can be orally explored or gnawed at, are not present, then the animal shows the effects of the stimulation by increased bodily hyperactivity, including locomotion. These effects are increased by amphetamine and other central stimulants, and decreased by neuroleptics (Gallistel & Karras, 1984).

However, it is necessary to separate responses due to increasing motor activity alone from those directly associated with a change in positive drive state. Gallistel & Karras (1984) have shown that amphetamine tends to increase the reinforcing efficacy of ICSS, while pimozide reduces it, which they suggested was the result of antagonistic effects on the same set of DA receptors. It has also been suggested that positive behavioral contrast is closely related to elation, and low doses of pimozide have been shown to reduce, selectively, positive contrast effects of ICSS (Phillips & LePiane, 1986). It appears that the major pathway involved is a descending system of fibers in the medial forebrain bundle that terminates in the VTA and has excitatory connections to the ascending mesolimbic DA projection to the nucleus accumbens

(Gallistel *et al.*, 1981; Gallistel, 1986). If we suppose that overactivation of these regions represents a model for elation in mania, then what might cause a shift to the depressive phase? One possible explanation is illustrated by some suggestive evidence that uncontrollable stress appears to reduce responding for brain stimulation, and that these reductions are found in the medial forebrain bundle and nucleus accumbens of the mesolimbic system, but not in the substantia nigra (Zacharko *et al.*, 1983).

Using a different type of ICSS model, Leith & Barrett (1980) showed that *chronic* amphetamine and reserpine, in rats, led to a reduction in ICSS rates, but that an initial phase of stimulation was seen with amphetamine. They suggested that the effects of amphetamine provided a model for the close relationship between manic and depressive phases frequently seen in the clinic. This model, unlike many others, takes the time course of the development of the symptoms into account, and it also makes excellent use of the known fact that extended amphetamine stimulation is followed by a downward-going, depressive phase which closely resembles clinical depression (Murphy, 1977; see also Willner, this vol., Ch. 5).

In brief, ICSS models seem to provide some of the best hope for defining behavioral categories that are closely related to 'elational' effects.

11.3.7 Sensitization and kindling treatment effect models

The electrical activity of the cortex is changed over time, by low, repeated, subthreshold electrical stimulation, until finally the induced electrical changes convert to a form of convulsive activity. This activity may then demonstrate a relatively permanent existence, independent of the initial 'kindling' stimulus (Goddard *et al.*, 1969). It has also been frequently observed that low-dose repeated injections of central stimulant drugs such as amphetamine or cocaine, may cause sensitization to the behavioral effects of these drugs, and Post (1977) suggested that the interval between drug treatments may have the same importance as the interstimulus interval in the induction of cortical 'kindling'. The effects of drug sensitization are not commonly thought to share the relatively long-lasting property of the 'kindling' procedure, yet apparently spontaneous 'flashbacks' have been reported by chronic drug abusers. Furthermore, vervet monkeys subjected to long-term continuous low-dose treatment with d-amphetamine via implanted capsules, may show recurring behavioral symptoms, first seen under the drug, for as long as four to six months after the capsule is removed (unpublished personal observations).

Post (1977) reported that repeated low-dose treatments with cocaine in the rhesus monkey, over periods as long as six months, led to an initial phase of behavioral activation, which was followed by:

an inhibitory syndrome, characterized by motor inhibition, catalepsy, abnormal visual searching, and staring behvaior, and in four of 13 animals, prominent oral–buccal–lingual dyskinesias. This inhibitory syndrome, as well as the dyskinesias, appeared to develop with increasing severity over time. (p. 201)

Such effects have also been reported following long-term treatment of vervet monkeys with low doses of amphetamine from implanted capsules (Nielsen & Lyon, 1982; Nielsen *et al.*, 1983). Repeated dosing of rhesus monkeys with cocaine at subconvulsive levels also led to an increasing frequency of drug-induced convulsive episodes (Post *et al.*, 1981). The progressive sensitization produced by this central stimulant drug in monkeys was paralleled in rats by the progressive development of seizures induced by the repeated cortical application of the local anesthetic lidocaine, or by amygdaloid stimulation. In the latter case, periodic stimulation of the amygdala led with increasing rapidity to prolonged after-discharge effects and to convulsions, in agreement with the drug-induced effects (Post, 1977; Post *et al.*, 1977*a, b*). Furthermore, about 40% of the animals developed a rhythmic alternation in sensitivity to this stimulation, showing sharply increased or decreased after-discharge durations on a definite cycle varying from one to several days in length. The peak duration values of the afterdischarge varied by a factor of three or four times from the lowest values of the cycle. Very similar results have been observed during kindling in the frontal cortex of the cat (Wake & Wada, 1975).

This strong cyclic variation is more rapid than the cyclic manifestations of manic and depressive periods in humans, but it might be a similar type of phenomenon. Therefore, Post *et al.* (1984) suggested that carbamazepine, an anti-convulsant drug, might also be effective as an anti-manic agent. This proved to be the case, and carbamazepine may also be effective in patients not responding to lithium or neuroleptics. This animal model is thus one of the few that has resulted directly in the development of a new means of treatment for mania (Rubin & Zorumski, 1985). It is also the only model presented in this review which attempts to provide a realistic account of the cyclic factors in manic/depressive illness. The apparently similar relationship between induction of the seizures by a DA agonist (cocaine), and by means of amygdaloid stimulation, or the local anesthetic lidocaine, suggests that there may be a single brain function which can be disturbed by several processes, only some of them being dopaminergic. Thus, the sensitization/kindling model may be useful in testing the involvement of other neurotransmitter systems in the genesis of mania.

11.3.8 Social isolation models

Robbins & Sahakian (1980) have summarized a number of points suggesting some parallels between the effects of social isolation and mania. Several of these points are related to the increased sensitivity to novel environments in isolated animals. There are also reported increases in open-field activity, and an increase in visits to a novel object (though without a corresponding increase in the time used to 'explore' the object).

Weinstock *et al.* (1977) have reported that alpha-methyl-para-tyrosine, which inhibits the synthesis of catecholamines, selectively reduced the activity of isolates in the open-field. While these authors have attempted to link this effect with a reduction in NA levels, none of the test methods used (i.e. intraventricular 6-OHDA, chronic amphetamine treatment) are free from decided effects upon other neurotransmitters, such as DA.

In further support of the possibility that DA systems might be involved, Guisado *et al.* (1980) reported that 12–15 months of isolation in rats significantly increased the DA receptor binding in the corpus striatum. While this factor could be important in the increased motor activity shown by isolates in the open-field, and for their apparent supersensitivity to the effects of DA stimulating drugs (Sahakian *et al.*, 1975), there are conflicting reports on the effects of neuroleptics and amphetamine on isolation-induced hyperactivity (Garzon *et al.*, 1979; Garzon & Del Rio, 1981 – see also Willner this vol., Ch. 5).

5-HT balance may also be extremely affected by social isolation. Dominant male vervet monkeys showed an approximately 50% drop in 5-HT levels during isolation, which, however, quickly recovered when they rejoined the social group and re-established their social dominance (Raleigh *et al.*, 1984). This isolation came during adulthood, where it produces much less, although not a negligible, effect on NA systems in the rat (Stolk *et al.*, 1974). There is also considerable evidence for decreased 5-HT levels in the isolated mouse (Garattini *et al.*, 1969; Valzelli & Garattini, 1972), and both mice and rats exhibit more frequent aggression upon release from isolation. The reasons for this aggression are not clear.

Beyond increased activity and social aggression, there is little evidence to suggest that isolated animals provide a good model for mania. They do, however, show changes in neurotransmitter influences (increased DA; decreased 5-HT) which are reminiscent of those observed in other models.

11.3.9 General summary of animal models of mania

Catecholaminergic models for mania are those with the widest coverage of symptomatology at this time, with evidence for hyperactivity, irritability, increased startle and social aggression, sleep disturbances, forced behavioral patterns, and strong evidence of elational effects as revealed by increases in the rewarding properties of stimuli. On the other hand, most of the evidence is related to hyperactivity itself, and its consequences, which may not duplicate successfully all of the human symptoms. Thus, of the catecholaminergic models, only the amphetamine model seems to have a broadly supported overlap with the symptoms of mania. Most of the other models have a close relationship to hyperactivity, but not to very much else. In many cases, however, this is due to lack of testing rather than to negative results. The failure of lithium to reverse the amphetamine hyperactivity remains a problem.

Of the indoleamine models mentioned here, the PCPA model seems especially relevant behaviorally, since it documents several symptoms clearly related to mania. However, lithium does not appear to reduce PCPA hyperactivity, although there is still too little evidence available on the effects of lithium under severe 5-HT imbalance. There does seem to be a clear difference between the PCPA model and the amphetamine-based model in their respective reactions to neuroleptic drugs. Models based on selective destruction of 5-HT neurons provide supportive evidence for a possible role of 5-HT imbalance in mania, but the behavioral evidence is still weak.

As implied above, the opioid models are among the most interesting of those proposed, since like the 5-HT depletion and amphetamine models, they actually mirror not just one, but many symptomatic aspects of mania. Morhpine, and opioid models in general, have some different characteristics from the catecholaminergic models, in that they induce a more marked tendency for switching of behaviors to occur. This switching may include a diverse mixture of stimulated behaviors, such as resting, running, and eating, which is seen only fleetingly with DA stimulant drugs at very low dose levels, but which is perhaps more characteristic of the early distractibility in mania. Social interaction is also increased within a broader dose range than with amphetamine. It would appear that a morphine activation model, based on the dopaminergic connections from VTA through the hypothalamic region to the nucleus accumbens is the most likely anatomical substrate for these effects. Furthermore, since there are clear inter-relationships between DA activation and 5-HT depletion, it is perhaps not surprising that changes in all of these neurotransmitter systems can cause similar behavioral syndromes. Certainly the opioid models are prime candidates for further study, since they provide such an excellent picture of manic symptomatology.

Among the neurotoxic lesion models, it appears that the 6-OHDA lesions induced by intraventricular infusion provide the best parallel to the symptoms of mania. Such lesion models, especially with involvement of the nucleus accumbens, striatum, and hippocampus as possible sites of action, seem particularly inviting for further study, since more and more evidence links the mesolimbic structures to psychosis. On the other hand, the widespread lesions produced by intraventricular infusion do not ease the problems of attempting to localize the pertinent effects in the brain. It may be important to look at some of these lesion effects for much longer periods than the six days used by Petty & Sherman (1981), since the post-lesion recovery of the brain tissue will not be complete for at least eighteen days.

Each of the remaining models presents some strong points, but the significance for a general model of mania is still lacking. Although attempts have been made to associate lithium effects with a general model for mania, this remains highly speculative, since lithium itself can produce behavioral effects, and it potentiates some manic-like effects of amphetamine and other drugs. It is probably too early to neglect models of mania solely on the grounds that they are not affected by lithium.

ICSS methods seem to provide some of the most direct evidence on elation, but in many cases there is no clear differentiation between motor stimulation effects and some independent measure of reward value. However, where this has been attempted, the results do suggest that changes in rewarding value are also involved. The only model type mimicking the rhythmic changes of the bipolar syndrome, is the amygdalar kindling model, which suggests that the effectiveness of mesolimbic system lesions in producing manic-like symptoms is not entirely adventitious. The possible relationship of kindling phenomena to long term potentiation of memory is also suggestive of how inherited brain weaknesses may dispose to the development of long term psychoses. Finally, isolation models alter 5-HT, NE, and DA levels, but, aside from general activity changes, their relationship to mania is unclear at this time.

All in all, the presently available animal models of mania provide an impressive set of symptomatic parallels, and a number of excellent possibilities for further research.

11.4 ANIMAL MODELS OF SCHIZOPHRENIA

Animal models of schizophrenia can be subdivided roughly into the same categories used previously. *Neuropharmacological* models include those related to: DA (including brain asymmetry of DA),

NA, phenylethylamine, indoleamines, opioids (including endogenous types), gamma-aminobutyric acid (GABA), and glutamate. *Treatment effect models* are related to: neuroleptic effects on reward functions (anhedonia), kindling, brain lesions, and social isolation. The major weight in discussion of both types of models will be laid upon their efficacy in producing the behavioral symptoms which were listed in Table 11.2 above.

11.4.1 DA related models

The DA related animal models of schizophrenia have enjoyed the greatest popularity, and are, at present, the best documented (Iversen, 1986, 1987). This arises primarily from the fact that DA blocking drugs are extremely effective anti-psychotic medicines in direct relation to their DA receptor binding capacity (Creese *et al.*, 1976; see also Ahlenius, Cookson, this vol., Ch. 12, 13). Furthermore, the ability of the amphetamine, and other DA stimulating drugs, to induce a psychosis extremely similar to schizophrenia has given a high face validity to the DA models. In addition, virtually all of the following types of DA related behavioral changes induced by systemic injection are directly antagonized by the standard neuroleptics in a dose-related manner. In fact, this predictive validity is so consistent that corroborative evidence for neuroleptic blocking of DA-related drugs is not given for each individual experiment, but is noted occasionally for drugs with non-DA effects.

11.4.1.1 Amphetamine models using repeated injections

The initial work with amphetamine-produced abnormal behavior (see Randrup & Munkvad, 1968) was frequently characterized by single massive injections, compared with lower doses. Kjellberg & Randrup (1972), however, studied the effects of repeated injections in monkeys and showed how their unconditioned cage behavior became increasingly fragmented and stereotyped. A diversity of models is now available using the multiple injection method, and three representative examples using rats, cats, and monkeys, respectively, are given below.

Chronic injection of amphetamine for 65 days in

Table 11.3. *Behavioral symptoms in cats during the end stages of chronic amphetamine intoxication (Ellinwood et al., 1972; Ellinwood & Sudilovsky, 1973)*

MAJOR SYMPTOMS	MODES OF EXPRESSION
Hyper-reactivity	Excessive startle. Increase in head and paw shakes. 'Hallucinatory-like' behaviors. 'Fearful hyper-reactivity' (crouching).
Dysjunctive behavior	Abortive or fragmented behaviors.
Dystonia	Awkward posture or movement. Hyperextension or hyperflexion.
Dyskinesias and twitches	Tongue torsion. Blepharospasm (eye twitches). Facial twitches.
Obstinate progression	'Persistent propulsion; pushing against an immovable obstacle'.
Akathisia	Restless behavior. Repeated shifting from side to side.
Ataxia	Poor control of posture and movement. Staggering or swaying during movements.

rats was suggested as a model for paranoid schizophrenia by Eichler *et al.* (1980) who found that sniffing was increasingly sensitized with repeated drug treatments. Licking, on the other hand, first showed some tolerance, but after about three weeks there was also an increasing sensitization. The increasing sensitization to certain effects of the stimulant drug is similar to the residual after-effects of such drugs in human chronic drug abusers (Bell, 1973; Utena, 1974). In such cases, even a small dose of the drug at a later point may induce the full-blown psychotic reaction.

Ellinwood and his coworkers (Ellinwood *et al.*, 1972; Ellinwood & Sudilovsky, 1973; Ellinwood & Kilbey, 1977, 1980) were among the first to emphasize the richness of the behavioral symptoms in cats and monkeys under amphetamine, as compared to the description of human symptoms of psychosis. Even the more complicated forms of human disturbance such as thought disorder and hallucinatory activity, appear to find certain parallels in the behavior of amphetamine-treated animals. Table 11.3 defines more specifically the most important symptomatic characteristics seen in these animals during the 'end-stage' of chronic amphetamine treatment. The major difference between this listing of symptoms and that given earlier in Table 11.2 lies in the large number of motor disturbances such as restless shifting, awkward postures, and muscular twitches, which are also significantly increased both in adult schizophrenics and in children genetically at high risk for schizophrenia. These symptoms appear in the end-stage of the drug treatment model, which in the closely related cocaine treatment experiments was three to four months after inception (Ellinwood & Kilbey, 1977). The treatment periods in the latter study are thus considerably longer than any of the other extant studies of animal models using drug treatment, which may explain the frequency of these neurological symptoms. As a result of this extensive testing, it is clear that a multiple injection regime in cats offers some excellent parallels to human schizophrenic symptoms.

Another important emphasis in Table 11.3 is upon the fragmentation of behaviors, with abortive sequences, and situationally irrelevant responses. Added to the inappropriate social behavior sequences, the frequent appearance of perseverative behaviors, and the strong evidence for continuous shifting of responses, these features all point directly to severe 'thought disorder'.

Haber *et al.* (1977) made observations on colonies of five rhesus macaques, who were treated daily with amphetamine for three weeks. After amphetamine there were sharp decreases in sitting idly or in huddling/sleeping activities, while sitting tensely and orienting and indulging in agonistic behavior increased markedly. It appeared that dominant animals asserted their dominance more aggressively under amphetamine, while submissive animals became more submissive to the animals directly above them in the social hierarchy. In group situations, the monkeys tended to show reduced social contact with other animals, usually restricting (focusing) their relationship to one other animal which had also previously been their most common object of affiliation.

Both the restriction of social contact to a single, previously preferred, animal, and the increase in submissiveness in previously submissive animals seem to follow the rule that amphetamine will tend most obviously to increase those behaviors which already have a high operant strength (Evenden & Robbins, 1983a). Haber *et al.* (1977) suggested that the increased aggressivity of the dominant males in an environment that was objectively no more threatening to them than before, represents a close parallel to the development of human paranoia. The fact that the submissive animals became even more submissive may also be consistent with this view, if they became, correspondingly, 'overvigilant' and 'hyper-reactive' to the other animals, as has been reported in amphetamine-treated submissive rats (Gambill & Kornetsky, 1976).

11.4.1.2 *Continuous amphetamine intoxication via implanted pellets*

Since the development of schizophrenia presumably depends upon a gradual, continuous progress involving some internal factors, it seems likely that a continuous treatment procedure with amphetamine might better resemble the progress of the human illness. The development of the implanted slow-release amphetamine pellet (Huberman *et al.*, 1977) made it possible to study the almost continuous effect of amphetamine over periods from 7–14 days in rats and monkeys. There was some general agreement with the results of multiple

daily injections mentioned above. For instance, experiments with singly caged vervet monkeys disclosed a behavioral sequence following pellet implantation which began with hyperactivity and lack of sleep, and then led to increasing stereotypy, with fragmented or abortive responses, sudden orienting, visual tracking of a moving non-existent 'object' in space (described by Ellinwood & Kilbey, 1980, as 'fly-catching'), hallucinatory activity, and finally to passive sitting with a total lack of social contact and an excessive 'empty' staring or visual fixation (Lyon & Nielsen, 1979; Nielsen *et al.*, 1983).

Ellison & Eison (1983) suggested that, in rats, these changes follow three general stages. In the first, animals show hyperactivity, leading into more and more frequently occurring motor stereotypies. The second stage is a relative inactivity and social isolation, in which the animals often sit staring into space and seemingly oblivious of other individuals. The third stage is characterized by the development of motor abnormalities such as 'wet-dog shakes', and by the development of excessive and stereotyped parasitotic-like grooming, and apparent hallucinations.

In comparing both the multiple injection model and the continuous slow-release model with other types of animal models of schizophrenia, it is especially important to note the detailed observations of fragmented or abortive behaviors (Ellinwood & Kilbey, 1977, 1980; Lyon & Nielsen, 1979; Ellison & Eison, 1983) and of apparent hallucinatory behavior (Nielsen *et al.*, 1983). Four different types of hallucinatory behavior were described, resulting in flight, attack, eating behavior, or other behaviors such as threat responses repeatedly oriented to a particular spot on the cage wall. To be classified as hallucinatory, each of these behaviors had to meet stringent requirements:

> ... it must not be simply a single reaction, but rather sequences of well defined behaviors that could apparently only be accounted for by the reaction of the animal to non-existent stimuli. Special care was taken to avoid classifying any behavior as 'hallucinatory' if observers had the least doubt. (Nielsen *et al.*, 1983, p. 225)

These criteria meant that a number of suspiciously similar activities, such as sudden startle to no obvious stimulus, visual tracking of a moving point in mid-air where no object was present ('fly-catching'), and fixed intense staring at a single point on the wall or floor where no obvious tiny object or unusual detail could be seen by a human observer, *were all excluded* from the definition of genuine hallucinatory behavior.

Nielsen & Lyon (1982) also used a family group of three monkeys to show that implantation of one member (adult male) led to a severe loss of social interaction, and prolonged periods of fixed staring in the amphetamine treated monkey. Furthermore, in other experiments, this type of staring behavior, and tendency toward social isolation, was found to return during periods as much as six months after removal of the pellet (unpublished observations). Even allowing for a possible role of learning, these are strong residual after-effects following the pellet treatment, which resemble those previously reported for multiple injections over long periods of cocaine or amphetamine (Ellinwood & Kilbey, 1977; Utena, 1974).

All in all, this sort of evidence is the most compelling yet obtained that long-term treatment with amphetamine by implanted pellet, or by multiple daily injections (see also Segal *et al.*, 1980; Owen *et al.*, 1981) provides an excellent parallel even for the most 'human' of schizophrenic symptoms, hallucinations and apparently related thought disorder. Furthermore, treatment with the neuroleptic haloperidol caused almost immediate cessation of the hallucinatory behaviors, as well as removing motor stereotypies and other amphetamine-induced dysfunctions (personal observations).

11.4.1.3 Dopamine effects within the nucleus accumbens

Injections of dopamine into the nucleus accumbens in rats yield hyperactivity (Pijnenburg & Van Rossum, 1973; Jackson *et al.*, 1975) but this symptom alone is not an adequate indicator for schizophrenia. Costall *et al.* (1984a, b) have introduced a more elaborate model for intra-accumbens infusion of DA by testing this effect in subgroups of rats which showed low or high activity respectively to systemic injections of ($-$)N-n-propylnorapomorphine [($-$)NPA]. They demonstrated that the DA infusion could induce a sensitivity to ($-$)NPA challenge which had a cyclic variation over days despite continuous DA infusion,

and also showed a relatively permanent ($>$one year) tendency to return under drug challenge, even though spontaneous activity appeared normal (Costall *et al.*, 1984a, b). Perhaps most importantly, Costall *et al.* showed that these DA infusion effects were changed in a complex manner by systemic injections of haloperidol and sulpiride. Most particularly, the neuroleptics appeared to change the periodicity of the cyclic activity changes induced by the DA infusion. This model is potentially important because of its demonstrated long-term cyclic effect of DA treatment, and the interesting relationship to neuroleptic action. Nevertheless, the lack of any sophisticated behavioral measurements cannot support any direct extrapolation to schizophrenia: perhaps the cyclic interactions between DA treatment and neuroleptic challenge are more closely related to the course of affective disorders. However, as with continuous amphetamine intoxication and kindling procedures, these studies further demonstrate the value of repeated or continuous treatment measures.

A more complete symptomatic picture is provided by squirrel monkeys following DA injections into the ventral aspect of the nucleus accumbens, which yielded principally hyperactive responses, but with the addition of intense self-grooming, exaggerated startle responses and greatly increased exploratory behavior (Dill *et al.*, 1979; Jones *et al.*, 1981). These signs appear to be related to some of the intermediate stages described for systemic amphetamine treatment. However, Dubach & Bowden (1983) found that in long-tailed macaques (*M. fascicularis*), the major changes induced by DA injections into the dorsal nucleus accumbens were losses in social grooming and self-grooming, and increases in scanning the environment, or looking intensely at close objects. These symptoms are more similar to those from the *earlier* stages of amphetamine intoxication, where grooming begins to weaken, and watchfulness and close examination of details begin to predominate. Dubach & Bowden attempted to relate the loss of social grooming to the flat affect seen in schizophrenia, and argued, for this reason, that the nucleus accumbens model is superior to many others. However, a reduction in social contact has been widely documented in other models which also show a wider range of schizophrenic symptoms.

Probably any model using such a precise delivery spot in the brain will not be able to mimic more than part of the full schizophrenic syndrome.

11.4.1.4 Methylphenidate effects on genetically nervous dogs

The exacerbation of psychotic signs in schizophrenics produced by injection of methylphenidate, a CA releasing psychostimulant, is accompanied by prolonged heart rate and blood pressure increases (Janowsky *et al.*, 1973). Newton *et al.* (1976) investigated the effects of methylphenidate on blood pressure and heart rate in dogs bred selectively to be 'pathologically nervous'. This model is interesting because it is one of the few with any genetic input. However, it is complicated, in practice, by the fact that methylphenidate causes a deceleration in heart rate in dogs, rather than an increase, as in humans: both the control and genetically nervous dogs showed a quiet, drowsy state after injection, rather than the hyperactivity found in rats.

After the drug treatment, the genetically nervous animals showed a delayed return of heart rate to normal, and a significantly greater number of extrasystoles and atrioventricular blocks in their heart rhythms. There was also a marked delay in blood pressure increase following methylphenidate in the genetically nervous dogs. All of these signs correlated well with measures of increased nervousness. Combined with behavioral measures, these signs of stress reaction may provide a concise model. Significant correlations have also been found between changes in skin resistance and the width of the third ventricle in persons at risk for schizophrenia (Cannon *et al.*, 1988). The major question is whether or not the 'pathological inertia' (Pavlov, 1927) in physiological reactions, which seems to be observed in genetically nervous dogs, is typical of a larger sample of unselected schizophrenics.

11.4.1.5 Acoustic startle response model

The eye-blink startle response habituates more slowly in schizophrenics than in normal controls, and there is an increased peak startle latency in these subjects. Furthermore, if a preliminary auditory signal is applied, control subjects show a reduced startle response ('prepulse inhibition') but schizophrenics do not (Geyer & Braff, 1982).

Similar effects on the startle response were seen in 6-OHDA treated rats that also received a systemic injection of the direct DA agonist apomorphine (Swerdlow *et al.*, 1986). The effect was very clear in rats with nucleus accumbens injections, and to a lesser degree in rats with injections into the caudate nucleus, but it was not found following 6-OHDA injections into the dorsomedial frontal cortex. These results suggest that the nucleus accumbens may also be supersensitive to DA in schizophrenia. Since these results with the acoustic startle response also fit with those from many other studies implicating the mesolimbic system in schizophrenia, this model deserves further investigation. It remains to be seen whether the effect of the denervation supersensitivity ascribed to the 6-OHDA injections can be separated from the specific neurotoxic effects of 6-OHDA. There is increasing evidence that neuronal loss during development may play a role in schizophrenia (S.A. Mednick & T. Cannon, personal communication).

11.4.1.6 Dopamine autoreceptor subsensitivity model

Amphetamine psychosis is frequently associated with changes in postsynaptic DA receptor effects, but Antelman & Chiodo (1981) suggested that reduction of presynaptic 'autoreceptor' sensitivity might be more specifically related to the psychosis, based on the observation that repeated treatment with DA agonists reduced the responsiveness to apomorphine of substantia nigra pars compacta DA neurons. Apomorphine, at the proper dose level, is thought to stimulate presynaptic DA autoreceptors selectively (Skirboll *et al.*, 1979). Similar effects were also observed following electroconvulsive shock (ECS) (Chiodo & Antelman, 1980). However the behavioral effects of apomorphine appear to be *increased* following ECS (White & Barrett, 1981) or repeated daily administration of amphetamine or cocaine (Kilbey & Ellinwood, 1977; Hitzemann *et al.*, 1977) which suggests that the increase in postsynaptic receptor sensitivity is the predominant effect. The potential importance of the subsensitivity model is that it would suggest different therapeutic strategies to those presently in use (see Ahlenius, this vol., Ch. 12).

11.4.2 Other catecholamine models

11.4.2.1 *Phenylethylamine (PEA) models*

PEA-like substances have attracted particular attention in models of schizophrenia, since PEA is an endogenous amphetamine-like substance found in the human brain, particularly in limbic system structures (Borison *et al.*, 1977). Subsequently, it was shown in rat brain that the most significant amounts of PEA were found in the caudate nucleus and the hypothalamus (Karoum *et al.*, 1981), and that there are specific PEA binding sites in these two structures (Hauger *et al.*, 1982).

Sandler & Reynolds (1976) and Wyatt *et al.* (1977) suggested that over-production of PEA might be the cause of schizophrenia. Supportive evidence is found in the increased PEA excretion in the urine of paranoid (though not non-paranoid) schizophrenics (Potkin *et al.*, 1979). Chronic PEA treatment in rats selectively increased NA content in the hypothalamus and nucleus accumbens, and increased hypothalamic 3-methoxy-4-hydroxy-phenylglycol (MHPG), a metabolite of NA (Karoum *et al.*, 1982); a similar elevation of CSF NA has been reported in the brains of paranoid schizophrenics (Sternberg *et al.*, 1981). As PEA is broken down rapidly by monoamine oxidase, experimental models frequently employ an MAOI along with the PEA treatment. This also suggests that a deficiency in MAO formation might underlie an abnormal increase in endogenous PEA; the evidence for this is still meager, although Demisch *et al.* (1977) have reported that, using PEA as a substrate, MAO activity was significantly reduced in the blood platelets of paranoid schizophrenics.

However this may be, PEA is known to produce extremely hyperactive and stereotyped sniffing and head movements in rats (Braestrup & Randrup, 1978). Dourish (1982) found in mice, that PEA:

> . . . produced a syndrome consisting of three distinct phases. The brief initial phase (0–5 min after injection), which consisted of forward walking, sniffing and headweaving, was succeeded by a locomotor depressant phase (5–20 min after injection) which consisted of abortive grooming, headweaving, splayed hindlimbs, forepaw padding, sniffing and hyper-reactivity, and a late locomotor stimulant phase (20–35 min after injection), which was characterized by forward walking, sniffing, hyperreactivity, rearing and licking. (p. 129)

The sequence of motor changes is apparently not the same in mice as that following administration of d-amphetamine in rats, cats or monkeys, where there is steadily decreasing locomotion, rather than a late, extremely active, phase. However, Tinklenberg *et al.* (1978) found that the effects of single doses of PEA in rhesus monkeys were very similar to those of single doses of amphetamine. By using various antagonists, Dourish was able to show that some of the behavioral symptoms were closely related to 5-HT functions (headweaving, splayed hindlimbs), while others were more closely related to DA stimulation (rearing and licking), but inhibited by 5-HT or NA stimulation. Abortive grooming was described as the 'dominant' behavior after PEA injection, but it was prevented by all of the antagonists for DA, 5-HT and NA, which suggests that all three neurotransmitters may be involved in the expression of this symptom. However, selective depletion of NA antagonizes PEA stereotypy greatly, but has little effect upon stereotypies induced by d-amphetamine, or cocaine (Borison & Diamond, 1978).

Perhaps the most compelling evidence in favor of a role for PEA in psychosis is that the hyperactivity response to this drug in inbred mice seems to be highly heritable (82%), and may be localized to a single major gene locus (Jeste *et al.*, 1984). Moisset (1977) similarly demonstrated that amphetamine-induced reductions in open-field rearing are dependent upon a single autosomal dominant gene. The specificity of the genetic effect in both cases suggests that the behaviors observed are not dependent upon multiple causative factors, and are not simply vague determinants of general behavior, but rather represent basic behavioral traits. The same reasoning is gradually emerging for the causative factors in schizophrenia (Gottesman & Shields, 1972; S.A. Mednick & T. Cannon, personal communication).

In summary, the PEA model is interesting and provocative because it involves an endogenous compound, but there remain some problems in explaining how PEA could reach abnormally high levels in the brain. In addition, the behavioral aspects of schizophrenia that are modeled by PEA are still somewhat more limited than those produced by other drugs which principally affect the DA systems. For instance, PEA treatment has apparently not yet produced, in animal models, parasitosis-like behaviors, or sign-tracking ('fly-catching') visual behavior, or apparent hallucinations, all of which are present during the later stages of chronic amphetamine intoxication (Nielsen *et al.*, 1983).

11.4.2.2 Noradrenaline models

Considering the closeness of the synthetic pathways for DA and NA, it is surprising that very few animal models of schizophrenia have focused on NA effects. While Sudilovsky (1975) has shown that the amphetamine model of psychosis in the cat is not adversely affected by blocking NA synthesis, there is suggestive evidence of NA changes from human studies of postmortem brain tissue from schizophrenics (Farley *et al.*, 1978; Bird *et al.*, 1979). Furthermore, CSF NA concentrations are higher in schizophrenic patients not under medication, but during treatment with the neuroleptic pimozide, which is primarily a DA receptor-blocker, there can also be a significant decrease in NA in the CSF, simultaneously with the decrease in the psychotic symptoms (Sternberg *et al.*, 1981).

The above points suggest a need for animal models to differentiate the influences of NA and DA. Kokkinidis & Anisman (1981) have suggested a combined DA/NA model. Briefly, they hypothesize that DA overactivity is responsible not only for increased locomotor activity, but also for the subsequent response stereotypy. Following this initial DA effect, they assume that tolerance develops to the locomotor effects, which partially accounts for the changes after long-term amphetamine treatment. However, they attribute behavioural perseveration in the early phases of drug treatment to NA activity, with a resultant high degree of attention to most recently presented stimuli. Later on, as the continuously raised DA activity results in reduced NA levels, there is purported to be a loss of perseveration so great that the end result is an attentional deficit. Finally, Kokkinidis & Anisman suggest that continued DA overactivity causes an imbalance in 5-HT systems, with resulting hallucinations. The main fault in this reasoning is that the measures used for attentional deficit are the same as those which might be used to support an increasing response perseveration. For instance, Kokkinidis & Anisman mention the increased resistance to extinction following dorsal

NA bundle lesions as evidence for selective attentional deficit. Yet one could argue with equal persuasion, and supported by the same data, that the resistance to extinction is a perseveration of the original response! (See also Rebec & Bashore, 1982.)

Nevertheless, NA deficiency may lie behind deficits in autonomic functioning such as electrodermal response (Cannon *et al.*, 1988), and the dysfunction of mesolimbically controlled eye movements in schizophrenics (Stevens, 1978). Clarification of these NA/DA inter-relationships will require further study of long-term treatments with catecholaminergic agents.

11.4.3 Overall summary of central stimulant drug models

Taken as a group, the DA models are the most complete animal models of all the major symptoms of schizophrenia which have yet been found. The evidence that thought disorder, hallucinations, motor disorders, stereotypies of speech and action and many other specific symptoms of schizophrenia find close parallels, especially within the multiple injection or continuous slow-release treatment methods with amphetamine, indicates that increases in DA activity are central to the psychotic state.

Sudilovsky (1975), using cats, has provided perhaps the best evidence of this from an animal model, by using a dopamine beta-hydroxylase inhibitor (disulfiram) to prevent the synthesis of NA, while treating the animals with amphetamine to produce the psychotic symptoms. The striking parallels thus produced to most of the major symptoms of schizophrenia clearly support the preeminence of DA related symptoms as the central features of the illness. A variety of monkey models has simply confirmed that what is seen in the cat closely resembles the more complicated symptom forms seen in primates, including humans.

Table 11.4 provides an impressive (and extensive) list of symptoms typically observed in animal models using stimulant drugs that are parallel to those found in human schizophrenics (with selected typical references to the best descriptions), and we have seen that there are both experimental evidence and plausible pharmacological reasoning behind this assumption of similarity. Furthermore,

Table 11.4. *Selected symptoms from catecholaminergic animal models of schizophrenia*

POSITIVE SYMPTOMS

Response sequence disturbances

Peculiar or bizarre behavior (Ellinwood & Sudilovsky, 1973).

Inappropriate switching (Evenden & Robbins, 1983*a, b*).

Excessive switching (two-choice) (Evenden & Robbins, 1983*a, b*).

Abortive grooming and fragmentary movements (Sudilovsky, 1975).

Response stereotypy (Randrup & Munkrad, 1967).

Visual disturbances

Excessive visual scanning (Tinklenberg *et al.*, 1978).

Eye movement (saccadic) (Stevens, 1978; Tinklenberg *et al.*, 1978).

Fixed staring at conspecifics (Utena, 1974).

Fixed staring at wall or floor points (Lyon & Nielsen, 1979).

Perceptual and hallucinatory disturbances

Sudden orienting and startle reaction (Ellinwood & Kilbey, 1977).

Visual sign-tracking (no object visible) (Nielsen *et al.*, 1983).

Hallucinations (Nielsen *et al.*, 1983).

Inappropriate affect disturbances

Inappropriate and repetitive vocal warnings (Lyon & Nielsen, 1979).

Inappropriate affect to provocation (Ellison *et al.*, 1978).

Increased aggression toward conspecifics (Haber *et al.*, 1977).

Ignores dominance rules (Haber *et al.*, 1977; Ellison *et al.*, 1978).

NEGATIVE SYMPTOMS

Lack of normal grooming (Schiørring, 1971).

Flat or inappropriate affect (Ellison, 1977; Sudilovsky, 1975).

Loss of normal social contact/response (Nielsen & Lyon, 1982).

Social isolation (Haber *et al.*, 1977).

OTHER SYMPTOMS

Neurological soft signs (Sudilovsky, 1975; Haber *et al.*, 1977).

Motor disturbances (dyskinesias) (Ellinwood & Kilbey, 1977, 1980).

although the human symptom of 'thought disorder' is not specifically named, it seems obvious from many of the other symptoms that it must also occur in these animals (see section 11.4.1.1). It now appears that even the genetic predisposition that lies behind the development of the human illness finds parallels in the genetic specificity of certain behavioral signs in the catecholaminergic animal models (hyperactivity; open-field rearing). As will be seen in the remainder of this review, no other animal model even approaches the CA based models in terms of broadness of symptomatology, and closeness of relationship to known neurotransmitter interactions in the human nervous system. Most of the other models mentioned either indirectly involve the CA systems, or they model only a single symptom of schizophrenia. Some models are directed at neurotransmitters other than DA, but so far, the relationship to the CA systems appears to be the principal, if not the only, relevant factor.

11.4.4 Indoleamine models

11.4.4.1 Lysergic acid diethylamide (LSD) model

Following the above statements concerning the catecholaminergic models of schizophrenia, it is most fitting to reconsider the model once thought to be the most closely related to that illness. LSD is a powerful synthetic hallucinogen in people and seemed to allow non-psychotic individuals to experience powerful and insightful visions which had an unreal perceptual background, which seemed extremely realistic from a phenomenological viewpoint. Of more specific interest here is the fact that approximately one of every three schizophrenic patients is partially or wholly insensitive to neuroleptic medications (Domino, 1975). However, more recently it has become apparent that many patients not amenable to treatment with typical neuroleptics have suffered brain damage (Crow, 1980; Andreasen *et al.*, 1982).

Since LSD is a synthetic compound, Domino (1975) attempted to find evidence of endogenously produced substances, which like PEA, could be suspected of causing schizophrenia. He suggested the indole alkyl amines as a particular group of such naturally occurring substances, with tryptamine (T), dimethyltryptamine (DMT), and 5-

methoxy-dimethyltryptamine (5-MeODMT) as possible specific agents with known hallucinogenic properties in people. However, it has not been demonstrated that these agents are actually found in excessive quantities in schizophrenics.

Jacobs *et al.* (1977) examined the reactions of cats to several drugs including LSD, d-amphetamine, the non-hallucinogenic methysergide (which is structurally similar to LSD), and caffeine. They found that LSD tended to increase limb flicks (the motion made by a cat to flick off some foreign substance from its paws), abortive grooming, and in some animals but not in others, either excessive hallucinatory-like behaviors or excessive investigatory or play-like behaviors.

Hallucinatory-like behavior was scored when the animal 'pounced on, batted at, attacked, or seemed to be visually tracking something not apparent to the experimenter.' Play-like behavior consisted of 'pawing or sniffing at objects or in corners, chasing the tail, or batting at pieces of food, feces, etc.' Other behaviors seen in many cats under LSD were:

standing or sitting in bizarre positions, e.g., with hindleg extended in space; kitten-like behavior, e.g., chasing their tails and pawing the air while lying on their sides or backs; leaping about the cage (often falling from their perch); sitting on the perch and staring down and back (they also frequently appeared to be responding to their own reflection in the stainless steel walls of their cages); continual scanning of environment by moving the head about; compulsive scratching in the litter pan; biting wood perch or metal of cage; pawing in water. (Jacobs *et al.*, 1977, p. 307)

The specificity to LSD of most of these reactions is countered by the descriptions of precisely the same behaviors in cats under the influence of chronic amphetamine or cocaine (Ellinwood & Kilbey, 1977; Sudilovsky, 1975), and in monkeys under continuous d-amphetamine (Haber *et al.*, 1977; Lyon & Nielsen, 1979; Nielsen *et al.*, 1983) (see section 11.4.1.1).

However, the most constant sign associated with LSD treatment, even in low to moderate doses, was an increase in 'limb flicks'. This sign is far less frequent under amphetamine in cats and monkeys. Trulson & Jacobs (1977) suggested, therefore, that limb flicks might be used as a specific test for drugs having hallucinogenic action. But even their own results with the non-hallucinogenic LSD-like sub-

stance methysergide (Jacobs *et al.*, 1977), cast some doubt on this conclusion, and subsequently the DA receptor agonists lisuride and apomorphine have also been shown to induce limb flicks in cats (Marini & Sheard, 1981; White *et al.*, 1981).

Geyer *et al.* (1979; Geyer & Light, 1979) showed that LSD, and a number of other hallucinogenic compounds, produced a change in the temporal sequence of exploratory behaviour on a holeboard, during three successive 8-min observation periods. Hallucinogen treated rats gave fewer responses than controls during the first eight minutes, but more than controls during the last eight minutes. The effect seemed to be fairly specific to hallucinogens, with only the DA agonist apomorphine producing a similar effect. Braff & Geyer (1980) also demonstrated that the habituation of the startle response to air-puff was slower under LSD, but as with the other tests, the use of a very specific, single response as an animal model for schizophrenia poses many questions. As noted earlier, the lack of inhibition of the acoustic startle response in schizophrenics can also be mimicked in rats by making DA receptors in the mesolimbic system of the brain supersensitive (Swerdlow *et al.*, 1986).

11.4.4.2 5-HT behavioral syndrome model

Griffiths *et al.* (1975) found that a high dose of fenfluramine, which releases 5-HT, resulted in psychotic disturbances with accompanying hallucinations, in human volunteers. Green & Grahame-Smith (1974) and Jacobs (1976) showed that a selective increase in 5-HT release in the rodent brain, by p-chloroamphetamine, fenfluramine, or administration of tryptophan after pretreatment with a MAOI, leads to a syndrome which includes 'increased locomotion, hindlimb adduction, head weaving, forepaw treading, tremor, "wet dog" shakes, and straub tail.' (Iversen, 1986). Very high doses of d-amphetamine (> 15 mg/kg) in the rat also cause forepaw treading, splayed hindlimbs and side-to-side head movements or head tremor. These symptoms appear to result from 5-HT receptor stimulation, rather than by DA overstimulation. It was therefore suggested that, while the DA is important in the preliminary symptoms of amphetamine intoxication, the induction of paranoid psychosis by the amphetamines is due mainly to the high dose effects upon 5-HT receptors (Slo-viter *et al.*, 1980).

However, in some human studies amphetamine psychosis could be induced by low doses of the drug (Gold & Bowers, 1978), if a predisposition existed, and the animal models using continuous amphetamine release (see section 11.4.1.1) suggest that it may be the continuity of the low dose effect, rather than the very high dose range which is important. It is also important to note that almost all of the symptoms in the amphetamine models (see Tables 11.3 and 11.4) are produced by dosages of *less than* 10 mg/kg d-amphetamine. In fact, in most behavioral studies, 10 mg/kg is considered the upper limit of treatment because of the direct neurological effects produced by higher doses (Sudilovsky, 1975; Ellinwood & Kilbey, 1980). This implies that many, if not most, of the important parallel symptoms in animal models can be produced *without* the high dose range necessary to overstimulate the 5-HT receptors in the manner suggested by Sloviter *et al.* (1980).

The main question is whether or not the syndrome described for 5-HT effects is an essential part of the behavioral changes induced by DA stimulation. As at least two of the symptoms, hindlimb splaying and head tremor are relatively severe neurological signs which find no obvious parallel in many human schizophrenics, and especially not in paranoid individuals (who show a minimum of neurological signs), it seems doubtful that 5-HT imbalance can be the principal feature of schizophrenia.

11.4.4.3 Summary of indoleamine models

There are still no convincing animal models of schizophrenia that can be shown to rely specifically upon disruption of the indoleamines. Claridge (1978) has criticized the available models as lacking clear relevance to complex human symptomatology, while pointing instead to other similarities between schizophrenia and LSD effects. Thus, both in schizophrenics, and in normal individuals after LSD, low levels of autonomic arousal (as measured by the electrodermal response) are accompanied by a 'paradoxically excessively high' perceptual sensitivity (as measured by the two-flash threshold), whereas these two measures tend to be positively correlated in normal, non-drugged individuals.

Despite some closeness of the LSD hallucinatory effect in humans with the schizophrenic experience, there is still little to recommend the indoleamine models in preference to the CA models. Many of the known behavioral effects of chronic amphetamine treatment are also reproduced with acute LSD treatment, but only at dose levels that affect the DA systems as well. The only signs that may be somewhat, although not exclusively, specific to the indoleamine treatments are those related to the 'hard' neurological signs, which are not common in schizophrenics who do not have other brain damage, and are, in any case, certainly not limited to that disorder. The doses of amphetamine required to produce these signs are far higher than those necessary to produce the schizophrenic-like symptoms, and in fact are not typical of the doses that have been used in many humans to induce psychotic-like states. At the very most, the 5-HT effects partially modulate the CA effects, but without the latter, 5-HT cannot be shown to produce any viable model of schizophrenia. PEA, which has, relative to its DA action, the strongest NE and 5-HT stimulating function of the central stimulant drugs, is also unable to duplicate the entire variety of schizophrenic symptoms as they are found in the amphetamine model, without at the same time reaching dosages that produce the same unnecessary neurological 'hard' signs.

11.4.5 Opioid models

In low doses, morphine appears to induce arousal as measured electrophysiologically, but the additional stimulation of tail-pinch, applied with an 'alligator clip', induces a state of immobilization, which is accompanied by a marked loss of EEG arousal (Stille & Sayers, 1975). These investigators attempted to draw a parallel between this immobile, non-reactive state and the reduction in the span of attention in catatonic schizophrenics (Venables, 1964). Buchsbaum *et al.* (1982) subsequently showed that the N120 component of the cortical evoked potential, which is closely related to attention selectively given to new stimuli, is decreased by morphine, and that the spread of evoked potentials to the parietal somatosensory cortex is strongly diminished in catatonic schizophrenia.

The catatonic state described by Stille & Sayers

(1975) in the rat was not relieved by treatment with 'typical' neuroleptics such as haloperidol or fluphenazine, or by benzodiazepines. The only effective drug to counteract the catatonic effect was the 'atypical' neuroleptic clozapine, which is also one of the few antipsychotic drugs that can induce some improvement in the impassivity accompanying the negative symptoms of schizophrenia.

de Wied (1979) has speculated that inherited disturbances in the degradation of the endogenous opioid beta-endorphin (beta-LPH 61-91) might lead to various symptoms of schizophrenia. Of the varying beta-endorphin fractions that might thus be produced, gamma-endorphin (beta-LPH 61-77) has a similar effect to neuroleptic drugs, in that it facilitates extinction of active (pole-climbing) avoidance, and weakens passive (one-trial step through) avoidance in rats. Des-tyrosine-gamma-endorphin has even more powerful neuroleptic-like effects, including decreases in ICSS rates with mesolimbic electrodes, and direct antagonistic effects on apomorphine in the nucleus accumbens (van Ree *et al.*, 1982). Apparently gamma-endorphins specifically modulate some types of DA receptors, particularly in the mesolimbic system. On the other hand, alpha-endorphin (beta-LPH 61-76), beta-endorphin (beta-LPH 61-69), and metenkephalin (beta-LPH 61-65) delayed the extinction of active avoidance, and facilitated passive avoidance. These particular effects of the beta-endorphin fragments were not dependent upon opioid receptors, since they persisted after pretreatment with the specific opiate antagonist naltrexone (van Ree *et al.*, 1982). This is an interesting point since the opiate antagonists naltrexone and naloxone have not had much success in reducing schizophrenic symptoms during clinical treatment; the after effects of naloxone can apparently influence some major symptoms such as hallucinations, but this may not be attributable to the blocking of opiate receptors (Barchas *et al.*, 1985). Naloxone was also ineffective in blocking amphetamine or PEA-induced stereotyped behaviors (Diamond & Borison, 1982).

Since some fragments of beta-endorphin can *increase* avoidance and sensitivity to noxious stimuli, such disturbances might lead to positive symptoms of schizophrenia, and perhaps some form of paranoia (de Wied, 1979). The more passive response to noxious stimuli and a general failure to respond rapidly, which can result in a catatonic state, have already been produced in an animal model by using excessive amounts of morphine (see above) or beta-endorphin (Bloom *et al.*, 1976). Thus the various positive and negative symptoms in schizophrenia could be related to only very slightly different changes in the breakdown of beta-endorphin.

Despite this generally hopeful picture for a role of the endorphins, clinical trials with beta-endorphins in schizophrenics have not produced any strong evidence of their efficacy (for an overview see Verebey, 1982), but there are still many methodological problems. It is possible that the extremely simple pole-climbing avoidance and step-through tasks, as well as the relationships to yawning and grooming, which have predominated in the study of the endorphins, are not adequate predictors of antischizophrenic activity in humans (see Ahlenius, this vol., Ch, 12).

There seems to be a definite relationship between many neuropeptide fragments and DA functions, and this may well extend to other neurotransmitters and to other (non-endorphin) neuropeptides. (For example, intra-accumbens injections of cholecystokin octapeptide (CCK-8) have been found to antagonize amphetamine-induced hyperactivity, but to potentiate the stereotyped behavior seen at higher doses (Weiss *et al.*, 1988)) What is not known, is whether the neuropeptide action is preliminary to, or even necessary for, the CA actions which seem to form the major basis for schizophrenic systems. Attentional dysfunction is an area not yet covered adequately by the DA hypothesis in which the neuropeptides may be of great importance.

11.4.6 Gamma-aminobutyric acid (GABA) related models

11.4.6.1 GABA-controlled DA model

In the early 1970s, Scheel-Kruger pointed out the necessity of extending the DA model to include the other obvious neuropharmacological steps which must precede, or run parallel to, the DA activation. In particular, he focused on the interactions between the DA and GABA systems in the striatum, substantia nigra, and mesolimbic structures.

As an animal model of schizophrenia, the GABA-controlled DA model does not distinguish itself *behaviorally* from the symptoms of the DA model described earlier. The difference lies in the explanation for the development of these symptoms as the GABA system applies its influence. This has led to a number of different attempts to control schizophrenic behavior by medicating with GABA active agents. But due to the complexity of the GABA system, plus the fact that sometimes GABA exacerbates, and sometimes reduces, DA activation, no successful GABA-related treatment has yet evolved.

The most important point made by Scheel-Kruger and his colleagues, is that the traditional view of GABA as simply acting to inhibit DA is not accurate. The combination of GABA and DA agonists can be shown to elicit a large number of widely varying symptoms, ranging from sedation and catalepsy to excitatory, stereotyped, and even aggressive behavior. The first two of these broaden the symptomatology produced by the DA activation, to include a number of more 'passive', or negative, symptoms, and thus make it even more expressive of the total range of schizophrenic symptoms.

Stereotyped gnawing was strongly facilitated by GABA agonists, such as muscimol, while other amphetamine stereotypies such as sniffing, head movements and locomotion, were reduced by this drug combination (Scheel-Kruger *et al.*, 1978). Similar effects were also found in marmosets by Ridley *et al.* (1979). In this connection, it may be important to note that systemic muscimol, or local injections in the substantia nigra pars reticularis (SNR), increased the synthesis and release of striatal DA (Scheel-Kruger *et al.*, 1979). There are large amounts of GABA in the corpus striatum and globus pallidus, but GABA agonists were effective in producing *catalepsy* only in the ventral region of the striatum, and only at an intermediate rostro-caudal level (Scheel-Kruger *et al.*, 1980). In the globus pallidus, only the medial portion of the rostral and intermediate levels of that structure were sensitive to the cataleptic effects of GABA agonists.

GABA systems can actually control the elicitation of stereotyped behavior and circling asymmetries independently of the DA systems. Bilateral injections of muscimol into the SNR induced a stereotyped behavior in the rat consisting of sniffing, licking, biting (at the cage bars), and self-mutilation (licking and biting at specific points on the limbs until tissue damage occurred), similar to that seen after high doses of amphetamine or apomorphine. Furthermore, the catalepsy following large systemic doses of DA antagonists was immediately removed by exceedingly small (nanogram) doses of muscimol in the SNR (Scheel-Kruger *et al.*, 1980).

These data suggest that there must be other pathways from the SNR, in addition to the nigrostriatal DA pathways, that are important to these abnormal behaviors. The SNR can be subdivided into a rostral region which has a much closer relationship to DA functions and striatal influence, while the caudal region seems to be the source of the GABA dependent, non-DA influences (Scheel-Kruger *et al.*, 1979, 1980). The relevance of these latter pathways to schizophrenia is as yet unclear. Connections to the ventromedial thalamus may be particularly important for the control of postural asymmetry (circling behavior) and possibly cataleptic reactions.

Other symptoms of importance for schizophrenia have been suggested to arise from GABAergic influences upon the ventral tegmental area (VTA). GABA stimulation of the VTA in the cat leads to fear-like reactions, with many of the abnormal eye and head movements seen in schizophrenics (Stevens, 1979). In rats, GABA agonists injected into the rostral portion of the VTA induced a decrease in general activity which progressed to sedation, with some signs of muscular rigidity; there were no signs of aggression or outward emotional display. However, in the caudal part of the VTA, GABA produced a dose-dependent increase in 'non-explorative' locomotion, a strong biting attack on anything touching the animal, and immediate fighting if another rat were present. There were no stereotypies of sniffing, head movements, licking or gnawing, and there was no turning after unilateral injection, thus separating these behaviors clearly from those typically produced by GABA stimulation in the SNR (Arnt & Scheel-Kruger, 1979). The affective attack and fighting are probably closely related to the extremely similar hyperesthesia ('septal') syndrome reported to follow lesions which usually include the nucleus accumbens to which the VTA projects (see section 11.3.1.2). In this case, GABA appears to modulate the DA systems, rather than working independently of them. Injection of GABA agonists directly into the nucleus accumbens generally inhibits the locomotor stimulant effects of morphine, amphetamine or apomorphine, but *increases* apomorphine stereotypy (Scheel-Kruger *et al.*, 1977).

It is very possible that GABA agonists do not fare well in clinical trials with schizophrenics because the decrease of DA activity in the mesolimbic system is countered by an increase in striatal DA activity produced by the same agents. That this may be the case is supported by the fact that the GABA agonists muscimol and progabide will decrease mesolimbic DA activity only at doses ten times higher than those which begin to increase DA activity in the striatal system (Bartholini, 1985). The ingenious and thorough experiments of Scheel-Kruger and his colleagues illustrate the importance of establishing the balances between different neurotransmitter effects in animal models before leaning on simple theoretical assumptions based upon a single neurotransmitter.

However, our main interest is in how the more complex model of Scheel-Kruger, involving as it does, GABA, DA, NA, ACh, and 5-HT, might provide some behavioral evidence which could not be gained from a less complicated model. Unfortunately, almost all of the knowledge we have about the behavioral influences of the elements in this system is based on extremely simple behavioral observations and ratings. There is a need for more technically sound tests of the animal equivalents for selective attention, learning, problem-solving, stimulus discrimination and generalization, and memory. Since many neurotransmitters will be active simultaneously during testing, there is also a need for multifactorial tests of the many sources of individual and interactive variance.

11.4.6.2 Tetanus-toxin induced stereotyped behavior

Among the agents known to have an effect on GABA functions are tetanus toxin, which is thought to block the release of GABA from presynaptic terminals, thus hindering its inhibitory action (Curtis *et al.*, 1973; Davis & Tongroach,

1977), and penicillin and picrotoxin, which also function as GABA agonists.

Following bilateral injections of these agents into the rostral part of the caudate nucleus in rats, Kryzhanovsky & Aliev (1981) reported the induction of a stereotyped behavior which was an exact replica of that described for DA activation in the caudate (Fog *et al.*, 1967). This syndrome could be blocked by the DA antagonist haloperidol injected systemically, or by the microinjection of GABA into the rostral caudate. Lithium, and the benzodiazepine diazepam, also suppressed the syndrome, and a combination of minimal doses of haloperidol, lithium chloride, and diazepam was particularly effective (Kryzhanovsky & Aliev, 1981). However, neither lithium nor diazepam are treatments of choice in schizophrenia (although the combination of lithium chloride and haloperidol has been used with benefit in the treatment of mania, and the benzodiazepines have been used as ancillary drugs in psychosis with accompanying severe anxiety). Furthermore, diazepam has also been reported to facilitate the effect of GABA on the gnawing stereotypy produced in mice by DA agonists (Scheel-Kruger *et al.*, 1980).

These disagreements suggest that the behavioral stereotypy produced by the tetanus toxin may have sources other than the interruption of GABA functions alone, and that, as in other models, the activational aspects of the stereotyped behavior need to be considered within a broader symptom picture which includes aspects such as sedation, catalepsy, and neurological signs of motor and eye movement disturbance.

11.4.7 Glutamate antagonist model

Glutamate (GLU) has been implicated as an important excitatory neurotransmitter in widespread regions of the brain. There are glutamatergic pathways from the medial prefrontal cortex to the rostral corpus striatum, nucleus accumbens and substantia nigra, all of which are important dopaminergic regions (Fonnum, 1984; Kornhuber *et al.*, 1984; Christie *et al.*, 1986). Of particular interest in relation to schizophrenia, GLU appears to be a neurotransmitter, or modulator, for several cell types within the hippocampus. This is also a region where prenatally developed, possibly genetically determined, aberrant dendritic orientation and cellular migration have been reported in the brains of schizophrenics (Kovelman & Scheibel, 1984; Bogerts *et al.*, 1985). A reduction of GLU levels in the CSF of schizophrenic patients has been reported (Kim *et al.*, 1980; Kim & Kornhuber, 1982).

On the basis of such evidence, Kornhuber & Fischer (1982) injected a relatively specific GLU antagonist, glutamic acid diethyl ester (GDEE), into the lateral ventricles of rats. At a low dose, there was a great increase in catalepsy (measured on a simple step-down task), while at a higher dose there was an initial catalepsy, which was masked about an hour later by a large increase in locomotion. (At the highest dose, some animals showed tachypnea and died.) There did *not* seem to be any significant changes in wakefulness, response to auditory stimuli, general equilibrium, or muscle tone. The latter negative findings contrast with the observation that in schizophrenics, the most withdrawn and catatonic states are frequently associated with extremely high arousal levels (Venables, 1964), and with obvious muscular and postural derangements.

Evaluation of GLU antagonism as a model of negative symptoms is difficult, partly because of the intraventricular method of administration, which favors effects upon those structures with large surfaces facing the ventricles, and partly because the cataleptic symptoms, and their later change to excessive locomotion at the higher dose of GDEE parallels exactly the biphasic behavioral effects caused by high-dose GABA stimulation of the ventromedial striatum (Scheel-Kruger *et al.*, 1980). The ventromedial striatum borders on the base of the lateral ventricle, in a region that might easily be affected directly by the GDEE injections into the ventricle. This raises the question of the neurotransmitter basis for the biphasic catalepsy/ locomotor-excitation effect. It should be noted that intraventricular injection of the demyelinating agent lysophosphatidyl choline also induces a very strong cataleptic reaction (Kline & Reid, 1985; see section 11.4.10.3 below), along with other signs that have been likened to the negative symptom complex in schizophrenia. This indicates a third way by which the cataleptic reaction might be induced, and undermines the specificity of this symptom alone for a given neurotransmitter action.

11.4.8 Motivational change treatment effect models

The original model of this type was based on the supposition that basic reward processes are dependent upon DA (Wise, 1981), and that neuroleptic drugs, by blocking DA function, produce 'anhedonia' – the lack of positive enjoyment of rewards. If one begins with this premise, one might ask how the hedonic function is related to schizophrenia in the *absence* of neuroleptic treatment. But this would seem to suggest that there must be an overactive positive rewarding element in schizophrenia, which has to be reduced by the neuroleptic treatment. This seems unlikely; however, there is a persistent element of hedonic elation in the almost compulsive behavior of mania (Szabadi *et al.*, 1981), and a similar, misplaced, positive hedonic effect is often seen in schizophrenia as 'inappropriate affect'. These somewhat perseverative mood changes, in the face of a quite different reality, could be interpreted as the result of DA overactivity, and as we have seen above, there are some animal data to support such an interpretation. In general, however, it seems unlikely that schizophrenic behaviour results from an overactive reward system, which has to be normalized by neuroleptics.

Is anhedonia the primary effect of neuroleptics? Wise *et al.* (1978) first formulated this hypothesis, on the basis of observations that the neuroleptic pimozide, which is a highly specific DA receptor antagonist, suppressed behaviors rewarded by drug-taking, intracranial self-stimulation, and basic food reinforcement. However, the purported anhedonic effect does not run strictly parallel to the antipsychotic effect of the neuroleptics, since the reduction in performance of the rewarded behavior can be shown to occur immediately (i.e. even before the first reward is delivered on a partial reinforcement schedule) (Phillips & Fibiger, 1979; Gray & Wise, 1980), while the antipsychotic effects can take several weeks to become evident. Wise suggested that the anhedonic effect is not specific to the primary rewarding events alone, but to the secondary rewarding attributes of the situation in which reward has been delivered.

However, other interpretations of the neuroleptic effects on rewarded behavior have been suggested. Ettenberg *et al.* (1981) showed that, if lever pressing for reward were replaced by nose-poking, animals did *not* show decreased responding after neuroleptic treatment. They suggested therefore that the neuroleptic effect was more closely related to the motor requirements of the task than to any reward properties, an interpretation also supported by many others (Beninger *et al.*, 1980a, b; Asin & Fibiger, 1984; Blackburn *et al.*, 1987). In general, these results appear to demonstrate that it is the initiation of the response, rather than the reward properties of the situation, which is altered by the neuroleptic. However, this view is more tenuous in the light of recent experiments by Ettenberg & Horvitz (1987), who have shown that haloperidol treatment during relearning of an extinguished response, may reduce the effectiveness of such relearning on a later test *without* drug. Similarly, Duvauchelle & Ettenberg (1987) showed that place preference induced by lateral hypothalamic self-stimulation was decreased on later trials, if the training had taken place under haloperidol. Furthermore, in rats responding on variable interval schedules, it is clear that pimozide impairs incentive motivation over and above any motor impairment that may be present (Willner *et al.*, 1987). A recent study by Fantie & Nakajima (1987) also suggests that there is a reduction in the rewarding value of brain stimulation under pimozide. Rats were conditioned to make *either* a 3-sec bar-holding response, *or* to produce a 3-sec burst of hippocampal theta wave activity, for a brain stimulation reward in the lateral hypothalamus. The theta wave response does not require bodily movement, or active motor responses, as demonstrated by testing the rats under the influence of a muscle relaxant. Pimozide reduced both bar-holding and also theta wave production, although control measurements showed that the rats were still capable of producing theta wave bursts at the same rate as preinjection. This test seems to confirm that although decreased motor response initiation may be involved in the effect of neuroleptics, there is also some kind of reduced reward effect, at least in so far as lateral hypothalamic brain stimulation is concerned. However, the relationship of hippocampal theta waves to the orienting responses associated with novel stimuli (Adey, 1966) may indicate that theta waves are not the best choice for bias-free response parameters.

A different way in which anhedonia may be important is in relation to its possible role as one of the cardinal *negative* symptoms of schizophrenia (Andreasen *et al.*, 1982), which suggests that in these negative symptom cases, there may be a decrease in DA activity. With this in mind, it is interesting to note that neuroleptics do not substantially benefit a large subgroup of patients with negative symptoms.

In experimental animals, low doses of the specific DA agonists apomorphine or 3-PPP (n-n-propyl-3(3-hydroxyphenyl)-piperidine) selectively decrease DA activity by stimulating inhibitory presynaptic autoreceptors. This effect is mirrored in reduced motor activity, yawning, and increased grooming. Carnoy *et al.* (1986) tested the effect of low doses of apomorphine, or 3-PPP, in rats shifted from a continuous reinforcement schedule to a fixed ratio of four responses per food reinforcement (FR4). Apomorphine selectively reduced responding on the more demanding schedule, although there was no evidence of an inability to fulfill the response requirements, and during extinction, the conditioned reward properties of the feeder noise were diminished by the drug treatment. Carnoy *et al.* suggested that these effects could serve as a model for at least one of the important negative symptoms of schizophrenia. They also found that low to moderate doses of some neuroleptics (including pimozide and the atypical neuroleptic sulpiride) would counteract the low-dose anhedonic effect of apomorphine, while other typical neuroleptics (haloperidol, fluphenazine, chlorpromazine) not only did not counteract the effect, but actually enhanced it.

The most important problem for these models is to resolve the question of whether or not 'anhedonia' is the proper name for the effects produced. There is a certain face validity in relation to the reports of anhedonia in schizophrenics following neuroleptic treatment, but the relationship between initiation of activity and its potential rewarding properties remains unclear. The concept of incentive motivation may offer a more powerful alternative.

11.4.9 Cerebral asymmetry models

There is growing evidence that cerebral laterality, or perhaps the loss of bilateral interaction of the hemispheres, is an important factor in schizophrenia (Gruzelier & Flor-Henry, 1979). Without reviewing the literature here, several generalities can be stated:

(a) the dominant (usually left) hemisphere tends to be overactive in schizophrenia (Gur, 1978);

(b) interhemispheric interactions through the corpus callosum are disturbed in schizophrenia (Beaumont & Dimond, 1973);

(c) cerebral injury, epileptic foci, reduced blood circulation and oxygen content, and reduced size of the cortical mass, are all significantly more often associated with the left cerebral hemisphere in schizophrenics (Gruzelier & Flor-Henry, 1979).

The problem with much of this evidence is that it is highly statistical, with the prediction of individual outcome dependent on many unknown factors. The potential importance of animal models has therefore been very much increased by the demonstration that the animal brain also shows interhemispheric differences, both in neurotransmitter content, and in response to drug treatment.

Two very important animal models have developed in relation to cerebral asymmetry. The first is the rotational behavior model for dopaminergic effects in the striatum and substantia nigra (Anden *et al.*, 1966; Ungerstedt, 1971; see also Ahlenius this volume), in particular those related to the effects of amphetamine and other DA stimulant drugs (Iversen, 1986); and the second is the observation that rats have an asymmetry of DA content on the two sides of the brain (Glick *et al.*, 1977, 1982).

Haloperidol in rats tends to increase DA turnover in the side opposite the animals' normal direction of rotation (Jerussi & Taylor, 1982). Since amphetamine tends to induce increased rotation to the same side as that preferred by rats without drug (Glick & Cox, 1978), this suggests that haloperidol tends selectively to block DA receptors on the side of the brain that is dominant for the normal side preference. Neuroleptic-induced dyskinesias tend to be greater on the right side in right-handed schizophrenics (Waziri, 1980).

Whether or not the rotational model is strictly related to handedness in animals or humans is not yet entirely clear, since there is a tendency for more frequent left-hand preference in schizophrenia, yet the left hemisphere of schizophrenics appears to be more easily aroused, with a higher power spectrum in the left temporal EEG (Flor-Henry, 1976), which would seem to favor right-handed preference. Perhaps the induced overactivity in the left hemisphere tends to give an imperfect control over the right hand (Shimizu *et al.*, 1985). Vervet monkeys solving a complicated mechanical puzzle showed a significant change from the preferred left hand to the right hand under methylphenidate (Lyon & Magnusson, 1982).

Glick *et al.* (1981) have demonstrated that rats with ICSS electrodes implanted bilaterally in the lateral hypothalamus also show a distinct side preference in terms of sensitivity to ICSS, and that this preferred side is contralateral to the direction of bodily rotation. Both effects are strengthened by amphetamine, suggesting that the preference is related to a lateralization of DA reward systems.

Further tests are needed of the relationship between handedness, hemispheric dominance, and schizophrenia in humans, and also of the relationship of these measures to paw preference, hemispheric dominance, and direction of rotation in animal models of schizophrenia, in order to establish reliable animal/human parallels for this model. However, the fact that asymmetry also exists for intracranial self-stimulation, brings the model into the forefront among those which relate not only to motor aspects of responding, but also to motivational/emotional characteristics. This is clearly an important aspect of the DA asymmetry; the possibility of similar asymmetries in other transmitter systems should also be investigated with animal models.

11.4.10 Kindling treatment models

Early experiments (Stevens *et al.*, 1974) had shown that injection of the GABA blocking agent bicuculline into the ventral tegmental area in cats produced a psychotic-like state characterized by 'intense arousal, staring, fear, withdrawal, waxy flexibility, statuesque posture, looking, searching, sniffing, and hiding behavior'. The same agent injected into the substantia nigra, and thus affecting the neostriatal system, produced only 'ipsilateral head-turning, circling, and grooming.' (see section 11.4.6.1, and Scheel-Kruger *et al.*, 1980; Scheel-Kruger, 1986). After finding the sensitive area in the VTA (probably the caudal region according to Scheel-Kruger *et al.*, 1980), by injecting bicuculline, Stevens & Livermore (1978) then stimulated the chosen point electrically to produce gradual kindling over a period of days. Several cats developed a very strong syndrome similar to that induced by bicuculline, and some became highly aggressive and difficult to handle as well. These effects lasted in two of the animals for a period of weeks after stimulation had been stopped. This is a characteristic that has also been described after continuous amphetamine treatment, or less commonly after multiple high dose injections. Rapid recovery is, in fact, what usually distinguishes the clinical recovery of amphetamine psychosis from that of a true schizophrenic episode (Kalant, 1973).

The relationship between kindling and amphetamine-induced abnormal behavior was addressed directly by Sato (1983). Methamphetamine-induced stereotyped behavior was enhanced in animals that had received amygdaloid kindling treatments, and this hypersensitivity to the DA effects was still present at least ten days after the last kindled convulsion. In addition, the usual autonomic responses to methamphetamine were all augmented significantly until at least three days after the last kindled convulsion. Since intraventricular injection of 6-OHDA facilitates amygdaloid kindling, and there is evidence for decreased DA and NA activity in the amygdaloid-kindled brain, then the kindling may be the source of a DA supersensitivity, which increases in step with the kindling stimulus presentations (Engel & Ackermann, 1980). Interestingly, Sato (1983) also found that during methamphetamine-induced stereotyped behavior, kindled convulsions were suppressed, and furthermore, that this suppression was removed by pimozide. The fact that stereotyped behavior is correlated with a reduction in abnormal reactivity produced by limbic system (amygdaloid) stimulation may be related to the suggestion of Sørensen (1987) that stereotypy may, in part, develop in animals as a protection against severe emotional stress.

11.4.11 Brain lesion models

There is evidence that increases in the width of the third ventricle in the region of the anterior hypothalamus are significantly associated with a specific loss of autonomic function, as measured by the electrodermal response in persons at high risk for schizophrenia (Cannon *et al.*, 1988). Disoriented neuronal dendrites have been discovered in the hippocampus of schizophrenics (Kovelman & Scheibel, 1984), and reductions in total volume have also been reported (Bogerts *et al.*, 1985). Widening of the fourth ventricle and poor development of cerebellar vermis have recently been found in schizophrenics who also fulfill the criteria for a genetically related prenatal structural deficit (S.A. Mednick, personal communication).

Most of the structures making up the mesolimbic DA system, as well as the striatal part of the nigrostriatal system, border on the ventricles of the brain. It should not be surprising that lesions or drug treatment introduced via, or near, the ventricles should provide some important animal models for schizophrenia. The following account is based on three basic methods involving hippocampal lesions, VTA lesions, and intraventricular injection of neurotoxins.

11.4.11.1 Hippocampal lesion models

Devenport *et al.* (1981) found that food-deprived rats with hippocampal lesions showed a highly significant rise over days in their locomotor activity and their stereotyped behavior ratings during an observation period of one hour prior to their daily meal. After approximately two weeks of this schedule, the hippocampal animals were given an injection of the DA receptor antagonist haloperidol, while control animals, with neocortical lesions, received an injection of d-amphetamine. This resulted in an almost perfect reversal of the behaviors for the two groups. The authors concluded that removal of the hippocampus leads to the same condition of DA overstimulation as follows injection with d-amphetamine.

Schmajuk (1987) has recently reviewed all of the evidence supporting a hippocampal lesion model, and concludes that there are, indeed, many similarities between amphetamine overstimulation and hippocampal lesion effects on behavior. An important feature of the lesion effects on behavior is that learning *per se* is not prevented, but the ability to use this learning is interfered with by rapid shifts in attention (Matthysse, 1981). Schmajuk also points to the known sensitivity of the hippocampus to hypoxia and other perinatal disturbances (see also Gilles *et al.*, 1983; Brann, 1985).

Since incomplete migration of cells during fetal development, cumulative concentration of heavy metals such as lead, and high vulnerability to perinatal hypoxia and hemorrhage, all specifically implicate the hippocampus (Fjerdingstad *et al.*, 1974; Kovelman & Scheibel, 1984; Bogerts *et al.*, 1985; Brann, 1985; Nowakowski, 1987), *it seems extremely likely that this model of schizophrenia is one of the most important yet discovered.* From these studies, it also appears likely that the most important relationships of the hippocampus needing further investigation are those involving the nucleus accumbens, the globus pallidus, and the amygdala.

11.4.11.2 Ventral tegmental area (VTA) lesion models

The delayed alternation task, in which the animal must learn to alternate between the two arms of a T-maze, even though there is a delay between trials, has often been used as a test for cognitive deficits following prefrontal cortex lesions; lesions of the rostral striatum produce even greater performance deficits. So, too, did 6-OHDA induced lesions in the VTA; control experiments showed that motivation and discriminative abilities were intact (Simon *et al.*, 1980). However, since VTA lesions increase exploratory behavior (Scheel-Kruger *et al.*, 1980), which would tend to interfere with performance of the delayed-alternation task, these results are difficult to interpret.

Clearer data were obtained by Oades (1982) using a hole-board as a foraging area. The apparatus had sixteen holes in the floor, four of which, the same on each trial, were baited with food pellets which the food-deprived animal had to find. VTA lesions resulted in more visits to empty holes, no improvement in the ratio of food-holes visited vs. non-food holes visited, and no significant increase in the repetition of any preferred sequential strategy of hole visits during each session. The

VTA animals also changed their strategy more often between sessions than did the control animals. All of this indicates that the VTA animals could still search, but that they were severely impaired when it came to the more complex aspects, such as a preferred strategy of search. Oades suggests that this may provide a simple animal model for thought disorder, and if so, this would be one of the few animal models directly testing this attribute.

Unfortunately, the method of making VTA lesions which Oades chose, did not match the ingenuity of the behavioral measure. A mechanical method was used to produce a large lesion of the VTA, but it also produced extensive damage to the regions dorsal to it, including the region of the red nucleus. Even though this was controlled for by a group with cannulae implanted just above the VTA, there is still ground for doubt, since it has long been known that lesions of the region around the red nucleus can impair feeding behavior (Blatt & Lyon, 1968; Lyon *et al.*, 1968).

It is of further interest that the same hole-board search task, in animals with hippocampal lesions, also produced an increase in errors, and fewer overall strategies. Haloperidol corrected for the errors to some extent, but did not normalize the strategy sequences of the hippocampal lesioned animals (Oades & Isaacson, 1978). This suggests that neuroleptic treatment has only a partial effect on behavioral choice tasks, which may be the reason for persistent deficits in medicated schizophrenics on active and passive avoidance (Gruzelier *et al.*, 1980), and on simple two-choice positively rewarded tasks (Frith & Done, 1983; Lyon *et al.*, 1986).

The fact that GABA stimulation in the VTA appears to separate rostrocaudally into influences upon the striatal system and the mesolimbic system respectively (Scheel-Kruger, 1986; Stevens & Livermore, 1978; see section 11.4.6), suggests that a more refined VTA lesion model should be able to differentiate the motor vs. the emotional aspects of DA system damage. It may be that the problem of attention deficits in schizophrenia will show itself to be related to these two basic functions, rather than to a separate quality of perceptual deficit. The VTA lesion model, in conjunction with behavioral tests that are not confounded with the ability to initiate motor responding, may be able to solve this important issue.

11.4.11.3 Periventricular lesion model

There is a great deal of evidence that brain ventricular enlargement is related to schizophrenia, particularly in those cases with more negative symptoms (Andreasen *et al.*, 1982). Kline & Reid (1985) have recently proposed what appears to be an excellent model for periventricular damage, and enlargement of the ventricles, by injecting the demyelinating agent lysophosphatidyl choline into the lateral ventricles of rats. The treatment produced a transient weight loss, and subsequently the animals begin to show a number of 'negative' symptoms among which were decreased emotionality (measured in the open-field), extreme postural indifference (catalepsy), inappropriate aggressive responses, and impaired grooming. Accompanying these symptoms was an enlargement of the cerebral ventricles, and periventricular damage to cell bodies and to myelin sheaths of nerve fibers. Some animals became extremely aggressive during handling; the same animals were the most cataleptic of the group, and the ones most likely to be poorly groomed. Unfortunately, this model was observed for only fourteen days immediately post-treatment, and thus needs to be tested for longer survival times, to allow full recovery of the nervous tissue and the ventricular flow, to insure the sealing off of the punctures inflicted into the ventricles during the injection procedure, and to allow recovery from the early post-operative hyperesthesia known to follow lesions of the basal periventricular region (Brady & Nauta, 1953; Harrison & Lyon, 1957). Hyperesthesia itself would not be considered a typical symptom of schizophrenia, but it may lead to other behaviors, such as explosive reactions or aggressive attack, that may be considered symptomatic of psychosis.

At present, there is no clear knowledge about whether the physical deficit introduced by the genetic loading in schizophrenia is principally due to a neurochemical (or neurotransmitter) defect, or whether it is mainly a developmentally incurred, structural defect in the anatomical connections of the brain. It is therefore essential at this point to develop models which embrace both points of view – the neurochemical and the neuroanatomical.

11.4.12 Social isolation models

Since the early observations of social withdrawal as a symptom of schizophrenia, it has frequently been suggested that disturbances in social communication and contact might be central to this illness. Among the earliest animal models of psychosis attempting to use isolation as a causative parameter are the experiments of Harlow and his associates (Suomi *et al.*, 1973), which were concerned with the effects of social isolation on infant monkeys. The resulting syndrome resembled depression, rather than schizophrenia (see Willner, this vol., Ch. 5), but the important point was made, that once subjected to early social deprivation for a long period, it was extremely difficult to eradicate the behavioral consequences in the adult animal. In the monkey, as in rats, it is the early social deprivation of 'isolation rearing' which produces the social effects, since adult animals can often be isolated for similar periods without noticeable effect (see however Raleigh *et al.*, 1984).

Of greater relevance here, in the light of the DA hypothesis of schizophrenia, is the realization that isolation rearing in rats leads to increased DA activity in the striatum. This was first revealed by the over-reaction of isolation reared animals to amphetamine and other DA agonists, including apomorphine. Recently, Jones *et al.* (1987) have shown increased DA activity in the striatum by *in vivo* dialysis in isolated rats, thus providing direct confirmation. Since there is also evidence (Einon & Morgan, 1977; Uchimura *et al.*, 1982) that the DA overactivity induced by early social isolation leaves relatively permanent changes in DA sensitivity in the adult animal, this model also meets one of the more difficult criteria for an animal model of schizophrenia – that the effect should outlast the specific inducing treatment.

Only two specific models using social isolation will be examined in detail here, but both are especially interesting because they involve primates.

11.4.12.1 Social deprivation syndrome model in rhesus monkeys

Goosen (1981) has described a 'social deprivation syndrome' in isolated female rhesus monkeys, which consists of two basic categories of misdirected normal behaviors:

(1) *Stereotyped locomotion and gross rhythmic stereotypies*

These include fixed patterns of locomotion, and other rhythmic rocking, bouncing, and somersaulting movements, as in autistic children, accompanied by a clear attempt to avoid eye contact, and to break off social contact with other animals, if they should interfere, and the specific loss of ability to interpret and respond to facial expressions of another monkey (Mirsky, 1968). The excessive appearance of any, or all, of these signs is exceedingly familiar in the clinical picture of autism or schizophrenia (Hutt & Hutt, 1970). In addition to the above, they may show severe disturbances in sexual behavior and mothering of young, so that mothers not only fail to care for young, but may attack and kill them.

(2) *Self-directed and bizarre behaviors*

These are movements which seem to be part of normal behavioral patterns, but which are either inappropriately expressed acts, 'bizarre' response forms, or resemble 'infantile' behavioral patterns which an adult animal typically does not show. These activities are either self-oriented, or oriented as if to a non-existent partner, and include: aggressive, or warding-off, responses; sexual self-stimulation, or presenting; grooming and caressing movements; hugging and holding behaviors; and three behaviors closely related to infant response patterns (face-pressing, sucking digits, crouching in a prenatal position). Most responses in the isolated animals were related to grooming and caressing, and to hugging and holding; all animals also showed at least one aspect of infantile behavior.

The percentage of time spent in behaviors related to grooming, and close body contact was almost identical to that of feral animals in the normal group context. Goosen takes the ethologically sound viewpoint that the source of these abnormal-appearing movements is a *persistent searching behavior for the social contact forms that are missing* from the animal's environment which,

in the monkey, would consist of huddling, holding, grooming, and aspects of aggressive and sexual behavior. Sørensen (1987) has suggested a similar cause for stereotyped activity in the caged bank vole as a means of surviving in a highly stressful environment. In the context of schizophrenia, such responses would be interpreted as attempts to ward off the worst effects of an environment radically altered in appearance, and effect, by responding in an abnormal (psychotic) manner. However, even if the monkey behaviors are initially coping responses, they persist even if the animals are returned to a social context. They are thus not situation-specific, and like schizophrenic symptoms, they can outlast the stimulus conditions which initially caused them to appear.

Goosen notes the close similarity of stereotyped and apparently aimless activity in the monkeys to the same symptoms in psychotic humans, and one may also point to the noticeably increased self-oriented behavior seen in vervet monkeys under continuous amphetamine intoxication (Nielsen *et al.*, 1983). Finally, one may note that in both the isolated monkeys and in human psychotics, there is a massive reduction in social communication, such that social isolation is ensured, even in situations where other individuals are present.

It can readily be seen that Goosen's social deprivation syndrome model, and chronic amphetamine treatment models, have a great deal in common, including the presumed overactivation of the DA systems.

11.4.12.2 *Amphetamine challenge in previously isolated rhesus monkeys as a model for high risk in schizophrenia*

Early social deprivation can induce lasting internal changes in the organism, so that even the apparent normalization of behavior of social experience does not remove the risk of a return of the abnormal behaviors seen during isolation (Mason, 1968). Even monkeys isolated during the second six months of life tend to be extremely aggressive when returned to a social group.

Although the careful use of social interactions with other young monkeys can be therapeutic for many of the social deficits in isolated monkeys (Novak & Harlow, 1975), severe stress will reinstate withdrawal or aggression, and stereotypic behaviors. This suggests some permanent change in brain function, but the only structural defect found so far in formerly isolated monkeys is a decrease in prefrontal cortical dendritic branching (Struble & Riesen, 1978). However, as described above, much evidence in rodents suggests that social isolation increases DA activation. These increases continue to be present in adult animals, long after the social isolation itself has terminated.

Accordingly, Kraemer *et al.* (1983) studied the effects of amphetamine challenge on eight rhesus monkeys of which four were socially isolated for one month after having spent six months with their mothers, and four were isolated for eleven months beginning one month after birth. Both sets of isolates were rehabilitated to social grouping for approximately two years after the end of their isolation period.

At the end of the rehabilitation period, no differences in social or other behaviors were noticeable in the group setting, but whereas non-isolated control animals responded to amphetamine by showing greatly reduced social contact with other animals and intense stereotyped explorative activity, the previously isolated animals showed, instead, a large increase in two, nearly diametrically opposed, behaviors, 'contact cling' (close body contact, hugging) and 'submit' (fear grimace, screeching, withdrawal). The result of this highly agitated and socially maladaptive behavior was that the animals were so severely attacked and bitten that the experiment had to be terminated. In addition to these problems, five of the previously isolated monkeys, including all of those isolated for eleven months, showed severe toxic reactions (postural collapse and convulsions) to the higher doses of amphetamine. A significantly increased stereotypy following amphetamine was subsequently demonstrated in the remaining three one-month-isolated monkeys despite their relatively brief exposure to social isolation. Measurements of CSF-NA showed significantly higher values for the previously isolated monkeys, which is consistent with the findings of increased NA in the CSF of paranoid schizophrenics (Lake *et al.*, 1980; Sternberg *et al.*, 1981).

This model is extremely valuable since it highlights the way in which susceptibility to psychotic behavior can be hidden from casual observation, only to appear at full force when a stressful challenge is applied. The model of Kramer *et al.* (1983) thus provides a tool for investigating the relationship between an underlying neurochemical deficit, and the stress factors that provoke a schizophrenic episode (cf. Gold & Bowers, 1978).

11.4.13 General summary of animal models of schizophrenia

There is a close relationship betwen the DA agonist-induced models and two other model types, PEA and social isolation, both of which are dependent upon increased DA activity. The PEA model is the best, presently known, instance of a potential endogenous neurotransmitter agent, and the social isolation model relies upon an increase in DA activity produced by a non-pharmacological behavioral treatment. Of the many models considered, the DA-related models are presently the best in terms of both face and predictive validity. The close parallels with human symptoms, seen in Tables 11.3 and 11.4, provide an excellent face validity for schizophrenic symptoms and, in each of the DA-related models, neuroleptic drugs are highly effective in reducing the symptomatic behaviors, so providing good predictive validity.

The indoleamine related models, including LSD, have been disappointing in terms of mimicking the full range of schizophrenic symptomatology. Specific symptoms, such as limb-flick in cats, have not proven to be reliable indicators of hallucinogenic action, and it appears that the most important 5-HT related effect is that of an imbalance with the more important DA systems.

There is also apparently some relationship between three other, seemingly disparate, model types: (1) lesions of the VTA, hippocampus, or periventricular structures; (2) kindling in the VTA or amygdala; (3) glutamate related models. The common factor is that all three types have significant relationships to basic mesolimbic structures and to the hippocampus itself: glutamate as the potential neurotransmitter which may be affected within the hippocampus; kindling as a procedure which may affect long-term potentiation in the hippocampus and elsewhere; lesions of the limbic and periventricular structures as possible correlates of pre- or perinatal damage predisposing to schizophrenia. These factors together also underline the importance of the hippocampal lesion model of schizophrenia.

Of the remaining models, the opioid and GABA related models seem most aptly related to a neuromodulatory function, which interacts with the DA systems to produce specific symptoms. Neither of these models has yet been shown to replicate the entire range of schizophrenic symptoms, yet individual symptoms are extremely well modeled in many cases.

Cerebral asymmetry of DA functions is doubtless important, but the exact relevance to schizophrenia is not yet clear. The same applies to the anhedonia models, which may turn out to be better characterized as incentive motivation models. These motivational models may actually provide a better parallel to mania, since in that case the untreated patients do show excessive levels of positive motivation, which are reduced by neuroleptics.

Many of the more complex models suffer from the fact that they introduce a multiplicity of new factors, both biochemically and behaviorally. To test such systems adequately, a wide behavioral 'net' is necessary, yet this condition is rarely met. In fact, the complexity and depth of the behavioral analysis tend to be *negatively correlated* with the complexity of the neuropharmacological treatments. Until this situation is corrected, most of these models will be intriguing, but not compelling. It might also be pointed out that animal models relying on many neurotransmitters, or neuromodulators, are not necessarily better models simply because they are more complex. Emphasis always has a cognitive advantage over completeness, and attempting to emphasize at an early stage in thinking permits a much clearer stipulation of relevant hypotheses.

11.5 GENERAL CONCLUSIONS

We may conclude that animal models of mania and schizophrenia are a great deal more robust, in terms of their direct parallels to human symptomatology, than is commonly admitted. There has been too great an emphasis on the DA hypothesis of schizophrenia in much of the work reviewed here, but it is becoming equally obvious that all of the major model types have extremely close relationships to the DA systems.

There remains the possibility that the entire search for a neurotransmitter based systemic defect in schizophrenia and mania, may be confounded by the failure to recognize that anatomically localized, developmentally caused, brain defects could be affecting the neural systems. This might, for instance, be the cause of the difficulty, so far, in finding very clear postmortem evidence of neurotransmitter anomalies in the brains of schizophrenics. However, there is now an increasing amount of data from positron-emission tomography (PET) studies of the living brain in schizophrenics, and these data document changes in DA D2 receptors within localized brain regions (Wong *et al.*, 1986; Andreasen, 1988). The exact localization of relevant changes in DA systems remains to be determined.

The conclusion regarding the future of animal models is therefore that, while the search for neurotransmitter deficits continues, models related to anatomical localization and structural differences should not be neglected. To this should be added the possibility that the vast families of neuropeptides, with their hormonal, neurotransmitter, and neuromodulating influences, may be the most crucial of all potential sources for a fundamental imbalance in brain function.

The least investigated aspects in most animal models are in the areas of genetic and developmental factors. With a very few worthy exceptions, most of the supposed causal factors for human schizophrenia, such as genetic disposition, viruses, stress, and even social environmental influences are virtually absent from the literature in any properly detailed form. Given the excellent behavioral modelling which the animals can provide, as documented above, it seems obvious that these areas should be developed as soon as possible. The problem is especially acute since animals provide almost the only non-medicated group of 'patients'.

The relationships drawn between the various theories of mania and schizophrenia and the actual models have been limited, mostly because the animal models do not frequently encompass sufficiently precise measures of psychological variables. Without more extensive tests, over much longer periods of time, of cognitive deficits, memory, thought and problem-solving abilities, and perceptual changes, the exact nature of the schizophrenic deficits will not be discovered. Animal models are particularly well-suited for such extensive and long-term investigations, which seldom can be performed on human subjects. The truly long-term effects of both neuroleptic and other types of drug are still a very much neglected task.

In summary, there is every reason to use animal models for the study of schizophrenia and mania, and the future promise is great.

This work was supported in part by grant no. 15-5958 from the Danish Research Council for the Humanities. Grateful appreciation is also expressed to the Social Science Research Institute, University of Southern California, including director Ward Edwards and staff, for the use of facilities and personal encouragement during preparation of this review.

REFERENCES

Adey, W.R. (1966). Neurophysiological correlates of information transaction and storage in brain tissue. In *Progress in Physiological Psychology*, eds. E. Stellar & J.M. Sprague, pp. 17–39. New York: Academic Press.

American Psychiatric Association (1980). *DSM-III: Diagnostic and Statistical Manual of Mental Disorders*, 3rd edn. Washington: American Psychiatric Association.

Anden, N.E., Dahlstrom, A., Fuxe, K. & Larsson, K. (1966). Functional role of the nigro-striatal dopamine neurons. *Acta Pharmacologica et Toxicologia* **24**, 263–74.

Andreasen, N.C. (1988). Brain imaging: Applications in psychiatry. *Science* **239**, 1381–8.

Andreasen, N.C. & Powers, P.S. (1975). Creativity and psychosis. An examination of conceptual style. *Archives of General Psychiatry* **32**, 70–3.

Andreasen, N.C., Olsen, S.A., Dennert, J.W. & Smith, M.R. (1982). Ventricular enlargement in schizophrenia: Relationship to positive and negative symptoms. *American Journal of Psychiatry* **139**, 297–302.

Antelman, S.M. & Chiodo, L.A. (1981). Dopamine autoreceptor subsensitivity: A mechanism common to the treatment of depression and the induction of amphetamine psychosis? *Biological Psychiatry* **16**, 717–27.

Antelman, S.M. & Szechtman, H. (1975). Tail pinch induces eating in sated rats which appears to depend on nigrostriatal dopamine. *Science* **189**, 731–3.

Armitage, S.G., Brown, C.R. & Denny, M.R. (1964). Stereotypy of response in schizophrenics. *Journal of Clinical Psychology* **20**, 225–30.

Arnt, J. & Scheel-Kruger, J. (1979). GABA in the ventral tegmental area: Differential regional effects on locomotion, aggression and food intake after microinjection of GABA agonists and antagonists. *Life Sciences* **25**, 1351–60.

Asin, K.E. & Fibiger, H.C. (1984). Force requirements in lever-pressing and responding after haloperidol. *Pharmacology, Biochemistry and Behavior* **20**, 323–6.

Astrachan, B.M., Harrow, M., Adler, D., Brauer, L., Schwartz, A., Schwartz, C. & Tucker, G. (1972). A checklist for the diagnosis of schizophrenia. *British Journal of Psychiatry* **121**, 529–39.

Atrens, D.M., Sinden, J.D. & Hunt, G.E. (1983). Dissociating the determinants of self-stimulation. *Physiology and Behavior* **31**, 787–99.

Ayhan, I. & Randrup, A. (1973). Behavioural and pharmacological studies on morphine-induced excitation of rats. Possible relation to brain catecholamines. *Psychopharmacologia (Berl.)* **29**, 317–28.

Barchas, J.D., Evans, C., Elliott, G.R. & Berger, P.A. (1985). Peptide neuroregulators: The opioid system as a model. *Yale Journal of Biology and Medicine* **58**, 579–96.

Bareggi, S.R., Becker, R.E., Ginsburg, B. & Genovese, E. (1979). Paradoxical effect of amphetamine in an endogenous model of the hyperkinetic syndrome in a hybrid dog: Correlation with amphetamine and p-hydroxyamphetamine blood levels. *Psychopharmacology* **62**, 217–24.

Bartholini, G. (1985). GABA receptor agonists: Pharmacological spectrum and therapeutic actions. *Medicinal Research Reviews* **5**, 55–75.

Baumgarten, H.G. & Schlossberger, H.G. (1973). Effects of 5,6-dihydroxytryptamine on brain monoamine neurons in the rat. In *Serotonin and Behavior*, eds. J.D. Barchas & E. Usdin, pp. 209–24. New York: Academic Press.

Beaumont, J.G. & Dimond, S.J. (1973). Brain disconnection and schizophrenia. *British Journal of Psychiatry* **123**, 661–2.

Bell, D.S. (1965). A comparison of amphetamine psychosis and schizophrenia. *British Journal of Psychiatry* **3**, 701–6.

Bell, D.S. (1973). The experimental reproduction of amphetamine psychosis. *Archives of General Psychiatry* **29**, 35–40.

Beninger, R.J., Mason, S.T., Phillips, A.G. & Fibiger, H.C. (1980*a*). The use of conditioned suppression to evaluate the nature of neuroleptic-induced avoidance deficits. *Journal of Pharmacology and Experimental Therapeutics* **213**, 623–7.

Beninger, R.J., MacLennan, A.J. & Pinel, J.P. (1980*b*). The use of conditioned defensive burying to test the effects of pimozide on associative learning. *Pharmacology, Biochemistry and Behavior* **12**, 445–8.

Bird, E.D., Barnes, J., Iversen, L., Spokes, E.G., Mackary, A.V.P. & Shepherd, M. (1977). Increased brain dopamine and reduced glutamic acid decarboxylase and choline acetyl transferase activity in schizophrenia and related psychoses. *Lancet* **II**, 1157–9.

Bird, E.D., Spokes, E.G. & Iversen, L.L. (1979). Brain norepinephrine and dopamine in schizophrenia. *Science* **204**, 93–4.

Blackburn, J.R., Phillips, A.G. & Fibiger, H.C. (1987). Dopamine and preparatory behavior: I. Effects of pimozide. *Behavioral Neuroscience* **101**, 352–60.

Blatt, B. & Lyon, M. (1968). The interrelationship of forebrain and midbrain structures involved in feeding behavior. *Acta Neurologica Scandinavica* **44**, 576–95.

Bleuler, E. (1950). *Dementia Praecox*. New York: International University Press.

Bloom, F., Segal, D., Ling, N. & Guillemin, R. (1976). Endorphins: Profound behavioral effects in rats suggest new etiological factors in mental illness. *Science* **194**, 630–2.

Bogerts, B., Meertz, E. & Schonfeldt-Bausch, R. (1985). Basal ganglia and limbic system pathology in schizophrenia. *Archives of General Psychiatry* **42**, 784–91.

Borison, R.L. & Diamond, B.I. (1978). A new animal model for schizophrenia: Interactions with adrenergic mechanisms. *Biological Psychiatry* **13**, 217–25.

Borison, R.L., Havdala, H.S. & Diamond, B.I. (1977). Chronic phenylethylamine stereotypy in rats: A new animal model for schizophrenia? *Life Sciences* **21**, 117–22.

Brady, J.V. & Nauta, W.J.H. (1953). Subcortical mechanisms in emotional behavior: Affective changes following septal forebrain lesions in the albino rat. *Journal of Comparative and Physiological Psychology* **46**, 339–46.

Braestrup, C. & Randrup, A. (1978). Stereotyped behavior in rats induced by phenylethylamine, dependence on dopamine and noradrenaline, and possible relation to psychoses? In *Noncatecholic Phenylethylamines, Part 1, Phenylethylamine: Biological Mechanisms and Clinical Aspects*, eds. A. Mosnaim & M.E. Wolf, pp. 245–69. New York: Marcel Dekker.

Braff, D.L. & Geyer, M.A. (1980). Acute and chronic LSD effects on rat startle: Data supporting an LSD-rat model of schizophrenia. *Biological Psychiatry* **15**, 909–16.

Braff, D.L., Callaway, E. & Naylor, H. (1977). Very short-term memory dysfunction in schizophrenia. *Archives of General Psychiatry* **34**, 25–30.

Brann, A.W., Jr. (1985). Factors during neonatal life that influence brain disorders. In *Prenatal and Perinatal Factors Associated with Brain Disorders*, ed. J.M. Freeman, pp. 263–358. NIH Publication No. 85-1149.

Brockington, I. (1986). Diagnosis of schizophrenia and schizoaffective psychoses. In *The Psychopharmacology and Treatment of Schizophrenia*, eds. P.B. Bradley & S.R. Hirsch, pp. 166–99. Oxford: Oxford University Press.

Brodie, H.K.H., Murphy, D.L., Goodwin, F.K. & Bunney, W.E., Jr. (1971). Catecholamines and mania: The effect of alpha-methyl-para-tyrosine on manic behavior and catecholamine metabolism *Clinical Pharmacology and Therapy* **12**, 218.

Broekkamp, C.L.E., Phillips, A.G. & Cools, A.R. (1979). Stimulant effects of enkephalin microinjection into the dopamine A-10 area. *Nature* **278**, 560.

Broen, W.E. & Storms, L.H. (1977). A theory of response interference in schizophrenia. In *Contributions to the Psychopathology of Schizophrenia*, ed. B. Maher, pp. 267–318. New York: Academic Press.

Buchsbaum, M.S., Reus, V.I., Davis, G.C., Holcomb, H.H., Cappelletti, J. & Silberman, E. (1982). Role of opioid peptides in disorders of attention in psychopathology. In *Opioids in Mental Illness*, vol. 298, *Annals New York Academy of Sciences*, ed. K. Verebey, pp. 352–65. New York: The New York Academy of Sciences.

Carlson, N.R. (1986). *Physiology of Behavior*, 3rd edn. Boston: Allyn and Bacon, Inc.

Cannon, T.D., Fuhrmann, M., Mednick, S.A., Machon, R.A., Parnas, J. & Schulsinger, F. (1988). Third ventricle enlargement and reduced electrodermal responsiveness. *Psychophysiology* **25**, 153–7.

Carnoy, P., Soubrie, P., Puech, A.J. & Simon, P. (1986). Performance deficit induced by low doses of dopamine agonists in rats: Toward a model for approaching the neurobiology of negative schizophrenic symptomatology. *Biological Psychiatry* **21**, 11–22.

Carroll, B.J. & Sharp, P.T. (1971). Rubidium and lithium: Opposite effects on amine-mediated excitement. *Science* **172**, 1355–7.

Carroll, B.J. & Sharp, P.T. (1972). Monoamine mediation of the morphine-induced activation of mice. *British Journal of Pharmacy* **46**, 124–39.

Chiodo, L.A. & Antelman, S.M. (1980). Electroconvulsive shock: Progressive dopamine autoreceptor subsensitivity independent of repeated treatment. *Science* **210**, 799–801.

Christie, M.J., James, L.B. & Beart, P.M. (1986). An excitatory amino acid projection from rat prefrontal cortex to periaqueductal gray. *Brain Research Bulletin* **16**, 127–9.

Claridge, G. (1978). Animal models of schizophrenia:

The case for LSD-25. *Schizophrenia Bulletin* **4**, 186–209.

Connell, P. (1958). *Amphetamine Psychosis*. Maudsley Monograph 5. London: Oxford University Press.

Coscina, D.V., Seggie, J., Godse, D.D. & Stancer, H.C. (1973). Induction of rage in rats by central injection of 6-hydroxydopamine. *Pharmacology Biochemistry and Behavior* **1**, 1–6.

Coscina, D.V., Goodman, J., Godse, D.D. & Stancer, H.C. (1975). Taming effects of handling on 6-hydroxydopamine induced rage. *Pharmacology Biochemistry and Behavior* **3**, 525–8.

Costall, B., Domeney, A.M. & Naylor, R.J. (1984*a*). Locomotor hyperactivity caused by dopamine infusion into the nucleus accumbens of rat brain: specificity of action. *Psychopharmacology* **82**, 174–80.

Costall, B., Domeney, A.M. & Naylor, R.J. (1984*b*). Long-term consequences of antagonism by neuroleptics of behavioural events occurring during mesolimbic dopamine infusion. *Neuropharmacology* **23**, 287–94.

Cox, C., Harrison-Read, P.E., Steinberg, H. & Tomkiewicz, M. (1971). Lithium attenuates drug-induced hyperactivity in rats. *Nature* **232**, 336–8.

Creese, I.N.R., Burt, D.R. & Snyder, S.H. (1976). Dopamine receptor binding predicts clinical and pharmacological potencies of antischizophrenic drugs. *Science* **192**, 481–3.

Crow, T. (1980). Molecular pathology of schizophrenia. *British Medical Journal* **280**, 66–8.

Crow, T. (1982). Two dimensions of pathology in schizophrenia: Dopaminergic and non-dopaminergic. *Psychopharmacology Bulletin* **18**, 22–9.

Crow, T. (1986). The continuum of psychosis and its implication for the structure of the gene. *British Journal of Psychiatry* **149**, 419–29.

Curtis, D.R., Felix, D., Game, C.J.A. & McCulloch, R.M. (1973). Tetanus toxin and the synaptic release of GABA. *Brain Research* **51**, 358–62.

Davies, C., Sanger, D.J., Steinberg, H., Tomkiewicz, M. & U'Prichard, D.C. (1974). Lithium and alpha-methyl-p-tyrosine prevent 'manic' activity in rodents. *Psychopharmacologia (Berl.)* **36**, 263–74.

Davis, J. & Tongroach, P. (1977). Antagonism of synaptic inhibition in the rat substantia nigra by tetanus toxin. *British Journal of Pharmacy* **59**, 489–90.

Davis, M., Svennson, T.H. & Aghajanian, G.K. (1975). Effects of d- and l-amphetamine on habituation and sensitization of the acoustic startle response in rats. *Psychopharmacologia (Berl.)* **43**, 1–11.

Dawson, M.E. & Neuchterlein, K.H. (1984). Psychophysiological dysfunctions in the developmental course of schizophrenic disorders. *Schizophrenia Bulletin* **10**, 204–32.

Delgado, J.M.R. (1976). New orientations in brain stimulation in man. In *Brain-Stimulation Reward*, ed. A. Wauquier & E.T. Rolls, pp. 481–503. New York: North Holland/American Elsevier.

Demisch, L., von der Muhlen, H., Bocknik, H.J. & Seiler, N. (1977). Substrate-typic changes of platelet monamine oxidase activity in subtypes of schizophrenia. *Archiv fur Psychiatrie und Nervenkrankheiten* **224**, 319–29.

Devenport, L.D., Devenport, J. & Holloway, F.A. (1981). Reward-induced stereotypy: Modulation by the hippocampus. *Science* **212**, 1288–9.

de Wied, (1979). Schizophrenia as an inborn error in the degradation of beta-endorphin – a hypothesis. *Trends in Neurosciences* **2**, 79–82.

Diamond, B.I. & Borison, R.L. (1982). Regulatory peptides in animal paradigms of neuropsychiatric illness. In *Regulatory Peptides: From Molecular Biology to Function*, eds. E. Costa & M. Trabucchi, pp. 541–7. New York: Raven Press.

Diaz, J., Ellison, G. & Matsuoka, D. (1974). Opposed behavioral syndromes in rats with partial and more complete central serotonergic lesions made with 5,6-dihydroxytrypamine. *Psychopharmacologia (Berl.)* **37**, 67–79.

Dill, R.E., Jones D.L., Gillin, C. & Murphy, G. (1979). Comparison of behavioral effects of systemic L-DOPA and intracranial dopamine in mesolimbic forebrain of nonhuman primates. *Pharmacology, Biochemistry and Behavior* **10**, 711–16.

Domino, E.F. (1975). Indole alkyl amines as psychotogen precursors: possible neurotransmitter imbalance. In *Neurotransmitter Balances Regulating Bheavior*, eds. E.F. Domino & J.M. Davis, pp. 185–228. Ann Arbor: E.F. Domino & J.M. Davis.

Dourish, C.T. (1982). A pharmacological analysis of the hyperactivity syndrome induced by beta-phenylethylamine in the mouse. *British Journal of Pharmacy* **77**, 129–39.

Dubach, M.F. & Bowden, D.M. (1983). Response to intracerebral dopamine injection as a model of schizophrenic symptomatology. In *Ethopharmacology: Primate Models of Neuropsychiatric Disorders*, ed. K.A. Miczek, pp. 157–84. New York: Alan R. Liss.

Duvauchelle, C.L. & Ettenberg, A. (1987). Haloperidol blocks conditioned place preferences induced by rewarding lateral hypothalamic stimulation in rats. *Society for Neuroscience Abstracts* **13**, 1323.

Eichler, A.J., Antelman, S.M. & Black, C. (1980). Amphetamine stereotypy is not a homogeneous phenomenon: Sniffing and licking show distinct profiles of sensitization and tolerance. *Psychopharmacology* **68**, 287–90.

Einon, D.F. & Morgan, M.J. (1977). A critical period for social isolation in the rat. *Developmental Psychobiology* **10**, 123–32.

Ellinwood, E.H., Jr. & Kilbey, M.M. (1975). Amphetamine stereotypy: the influence of environmental factors and prepotent behavioural patterns on its topography and development. *Biological Psychiatry* **10**, 3–16.

Ellinwood, E.H., Jr. & Kilbey, M.M. (1977). Chronic stimulant intoxication models of psychosis. In *Animal Models in Psychiatry and Neurology*, eds. I. Hanin & E. Usdin, pp. 61–74. Oxford: Pergamon Press.

Ellinwood, E.H., Jr. & Kilbey, M.M. (1980). Fundamental mechanisms underlying altered behavior following chronic administration of psychomotor stimulants. *Biological Psychiatry* **15**, 749–57.

Ellinwood, E.H., Jr. & Sudilovsky, A. (1973). Chronic amphetamine intoxication: Behavioral model of psychoses. In *Psychopathology and Psychopharmacology*, eds. J.O. Cole, A.M. Freedman & A.J. Friedhoff, pp. 51–70. Baltimore: Johns Hopkins University Press.

Ellinwood, E.H., Jr., Sudilovsky, A. & Nelson, L.M. (1972). Behavioral analysis of chronic amphetamine intoxication. *Biological Psychiatry* **4**, 215–25.

Ellison, G.D. (1977). Animal models of psychopathology: The low-norepinephrine and low-serotonin rat. *American Psychologist*, December, 1036–45.

Ellison, G.D. (1979). Animal models of psychopathology: Studies in naturalistic colony environments. In *Psychopathology in Animals: Research and Clinical Applications*, ed. J.D. Keehn, pp. 81–101. New York: Academic Press.

Ellison, G. & Bresler, O. (1974). Tests of emotional behavior in rats following depletion of norepinephrine, of serotonin, or of both. *Psychopharmacologia (Berl.)* **34**, 275–88.

Ellison, G. & Eison, M.S. (1983). Continuous amphetamine intoxication: An animal model of the acute psychotic episode. *Psychological Medicine* **13**, 751–61.

Ellison, G., Eison, M.S. & Huberman, H.S. (1978). Stages of constant amphetamine intoxication: Delayed appearance of paranoid-like behaviours in rat colonies. *Psychopharmacology* **56**, 293–9.

Engel, J. & Ackermann, R.F. (1980). Interictal EEG spikes correlate with decreased, rather than increased, epileptogenicity in amygdaloid kindled rats. *Brain Research* **190**, 543–8.

Ettenberg, A. & Horvitz, J.C. (1987). Haloperidol blocks the incentive motivational properties of food reinforcement. *Society for Neuroscience Abstracts* **13**, 219.

Ettenberg, A., Koob, G.F. & Bloom, F.E. (1981). Response artifact in the measurement of neuroleptic-induced anhedonia. *Science* **213**, 357–9.

Evenden, J.L. & Robbins, T.W. (1983a). Increased response switching, perseveration and perseverative switching following d-amphetamine in the rat. *Psychopharmacology* **80**, 67–73.

Evenden, J.L. & Robbins, T.W. (1983b). Dissociable effects of d-amphetamine, chlordiazepoxide and alpha-flupenthixol on choice and rate measures of reinforcement in the rat. *Psychopharmacology* **79**, 180–6.

Evetts, K.D., Uretsky, N.J., Iversen, L.L. & Iversen, S.D. (1970). Effects of 6-hydroxydopamine on CNS catecholamines, spontaneous motor activity and amphetamine-induced hyperactivity in rats. *Nature* **225**, 961–2.

Falardeau, P., Morrissette, M. & Di Paolo, T. (1987). The striatal agonist binding site of the D-2 dopamine receptor fluctuates during the rat oestrous cycle. *Society for Neuroscience Abstracts* **13**, 1344.

Fantie, B.D. & Nakajima, S. (1987). Operant conditioning of hippocampal theta: Dissociating reward from performance deficits. *Behavioral Neuroscience* **101**, 626–33.

Farley, I.F., Price, K.S., McCullough, E., Deck, J.H.N., Hordynski, W. & Hornykiewicz, O. (1978). Norepinephrine in chronic paranoid schizophrenia: Above-normal levels in limbic forebrain. *Science* **200**, 456–8.

Faro, M.D. & Windle, W.F. (1969). Transneuronal degeneration in brains of monkeys asphyxiated at birth. *Experimental Neurology* **24**, 38–53.

Fessler, R.G., Sturgeon, R.D., London, S.F. & Meltzer, H.Y. (1982). Effects of lithium on behaviour induced by phencyclidine and amphetamine in rats. *Psychopharmacology* **78**, 373–6.

Fibiger, H.C. & Campbell, B.A. (1971). The effect of parachlorophenylalanine on spontaneous locomotor activity in the rat. *Neuropharmacology* **10**, 25–32.

Fjerdingstad, E., Danscher, G. & Fjerdingstad, E.J. (1974). Hippocampus: selective concentration of lead in the normal rat brain. *Brain Research* **80**, 350–4.

Flor-Henry, P. (1976). Lateralized temporal-limbic dysfunction and psychopathology. *Annals New York Academy Science* **280**, 777–95.

Fog, R., Randrup, A. & Pakkenberg, H. (1967). Aminergic mechanisms in corpus striatum and amphetamine induced stereotyped behaviour. *Psychopharmacologia (Berl.)* **11**, 179–83.

Fonnum, F. (1984). Glutamate: A neurotransmitter in mammalian brain. *Journal of Neurochemistry* **42**, 1–11.

Frith, C.D. (1979). Consciousness, information processing and schizophrenia. *British Journal of Psychiatry* **134**, 225–35.

Frith, C.D. & Done, D.J. (1983). Stereotyped responding by schizophrenic patients on a two-choice guessing task. *Psychological Medicine* **13**, 779–86.

Furukawa, T., Ushizima, I. & Ono, N. (1975). Modification by lithium of behavioral responses to methamphetamine and tetrabenazine. *Psychopharmacologia (Berl.)* **42**, 243–8.

Gallistel, C.R. (1986). The role of the dopaminergic projections in MFB self-stimulation. *Behavioral Brain Research* **20**, 313–21.

Gallistel, C.R. & Karras, D. (1984). Pimozide and amphetamine have opposing effects on the reward summation function. *Pharmacology, Biochemistry & Behavior* **20**, 73–7.

Gallistel, C.R., Shizgal, P. & Yeomans, J.S. (1981). A portrait of the substrate for self-stimulation. *Psychological Review* **88**, 228–73.

Gambill, J.D. & Kornetsky, C. (1976). Effects of chronic d-amphetamine on social behavior of the rat: Implications for an animal model of paranoid schizophrenia. *Psychopharmacology* **50**, 215–23.

Garattini, S., Gialcalone, E. & Valzelli, L. (1969). Biochemical changes during isolation-induced aggressiveness in mice. In *Aggressive Behaviour*, eds. S. Garattini & E.B. Sigg, pp. 179–87. Amsterdam: Excerpta Medica.

Garzon, J. & Del Rio, J. (1981). Hyperactivity induced in rats by long-term isolation: further studies on a new animal model for the detection of antidepressants. *European Journal of Pharmacology* **74**, 287–94.

Garzon, J., Fuentes, X. & Del Rio, J. (1979). Antidepressants selectively antagonize the hyperactivity induced in rats by long-term isolation. *European Journal of Pharmacology* **59**, 293–6.

Gauron, E.F. & Rowley, V.N. (1974). Effects of chronic methylphenidate administration on learning and offspring behavior. *Journal of General Psychology* **91**, 157–8.

Gauron, E.F. & Rowley, V.N. (1976). Chronic methylphenidate effects on learning in an F2 generation. *Journal of General Psychology* **95**, 71–6.

Gerner, R.H., Post, R.M. & Bunney, W.E., Jr. (1976). A dopaminergic mechanism of mania. *American Journal of Psychiatry* **133**, 1177–9.

Gessa, G.L. & Tagliamonte, A. (1975). Role of brain serotonin and dopamine in male sexual behavior. In *Sexual Behavior: Pharmacology and Biochemistry*, eds. M. Sandler & G.L. Gessa, pp. 117–28. New York: Raven Press.

Geyer, M.A. & Braff, D.L. (1982). Habituation of the blink reflex in normals and schizophrenic patients. *Psychophysiology* **19**, 1–6.

Geyer, M.A. & Light, R.K. (1979). LSD-induced alterations of investigatory responding in rats. *Psychopharmacology* **65**, 41–7.

Geyer, M.A., Light, R.K., Rose, G.J., Petersen, L.R., Horwitt, D.D., Adams, L.M. & Hawkins, R.L. (1979). A characteristic effect of hallucinogens on investigatory responding in rats. *Psychopharmacology* **65**, 35–40.

Gilles, F.H., Leviton, A. & Dooling, E.C. (1983). *The Developing Human Brain. Growth and Epidemiological Neuropathology.* Boston: Wright PSG.

Glick, S.D., Zimmerberg, B. & Jerussi, T.P. (1977). Adaptive significance of laterality in the rodent. *Annals New York Academy of Science* **299**, 180–5.

Glick, S.D. & Cox, R.D. (1978). Nocturnal rotation in normal rats: Correlation with amphetamine-induced orientation and effects of nigrostriatal lesions. *Brain Research* **150**, 149–61.

Glick, S.D., Weaver, L.M. & Meibach, R.C. (1981). Amphetamine enhancement of reward asymmetry. *Psychopharmacology* **73**, 323–7.

Glick, S.D., Ross, D.A. & Hough, L.B. (1982). Lateral asymmetry of neurotransmitters in human brain. *Brain Research* **234**, 53–63.

Goddard, G.V., McIntyre, D.C. & Leech, C.K. (1969). A permanent change in brain function resulting from daily electrical stimulation. *Experimental Neurology* **25**, 295–301.

Gold, M.S. & Bowers, M.B. (1978). Neurobiological vulnerability to low-dose amphetamine psychosis. *American Journal of Psychiatry* **135**, 1546–8.

Goldman-Rakic, P.S. (1987). Development of cortical circuitry and cognitive function. *Child Development* **58**, 601–22.

Goosen, C. (1981). Abnormal behavior patterns in rhesus monkeys: Symptoms of mental disease? *Biological Psychiatry* **16**, 697–716.

Gottesman, I.I. & Shields, J. (1972). *Schizophrenia and Genetics: A Twin Study Vantage Point.* New York: Academic Press.

Grahame-Smith, D.G. (1971). Studies in vivo on the relationships between brain tryptophan, brain 5-HT synthesis, and hyperactivity in rats treated with a MAO inhibitor and l-tryptophan. *Journal of Neurochemistry* **18**, 1053–66.

Gray, T. & Wise, R.A. (1980). Effects of pimozide on lever pressing behavior maintained on an intermittent reinforcement schedule. *Pharmacology, Biochemistry & Behavior* **12**. 931–5.

Green, A.R. & Grahame-Smith, D.G. (1974). The role of brain dopamine in the hyperactivity syndrome produced by increased 5-hydroxytryptamine synthesis in rats. *Neuropharmacology* **13**, 949–59.

Griffiths, J.D., Nutt, J.G. & Jasinski, D.R. (1975). A

comparison of fenfluramine and amphetamine in man. *Clinical Pharmacology and Therapy* **18**, 563–70.

Grilly, D.M. (1977). Rate-dependent effects of amphetamine resulting from behavioural competition. *Biobehavioral Reviews* **1**, 87–93.

Gruzelier, J.H. & Flor-Henry, P. (1979). *Hemisphere Asymmetries and Psychopathology*. Amsterdam: Elsevier.

Gruzelier, J.H., Thornton, S., Staniforth, D., Zaki, S. & Yorkston, N. (1980). Active and passive avoidance learning in controls and schizophrenic patients on racemic propranolol and neuroleptics. *British Journal of Psychiatry* **137**, 131.

Guisado, E., Fernandez-Tome, P., Garzon, J. & Del Rio, J. (1980). Increased dopamine receptor binding in the striatum of rats after long-term isolation. *European Journal of Pharmacology* **65**, 462–4.

Gur, R.E. (1978). Left hemisphere dysfunction and left hemisphere overactivation in schizophrenia. *Journal of Abnormal Psychology* **87**, 226–38.

Haber, S., Barchas, P.R. & Barchas, J.D. (1977). Effects of amphetamine on social behaviors of rhesus macaques: An animal model of paranoia. In *Animal Models in Psychiatry and Neurology*, eds. I. Hanin & E. Usdin, pp. 107–14. New York: Pergamon Press.

Harrison, J.M. & Lyon, M. (1957). The role of the septal nuclei and components of the fornix in the behavior of the rat. *Journal of Comparative Neurology* **108**, 121–37.

Hauger, R.L., Skolnick, P. & Paul, S.M. (1982). (3H)beta-phenylethylamine binding sites in rat brain. *European Journal of Pharmacology* **83**, 147–8.

Heath, R.G., Franklin, D.E. & Shraberg, D. (1979). Gross pathology of the cerebellum in patients diagnosed and treated as functional psychiatric disorders. *Journal of Nervous and Mental Diseases* **167**, 585–92.

Herberg, L.J. & Franklin, K.B.J. (1976). The 'stimulant' action of tryptophan-monoamine oxidase inhibitor combinations: Suppression of self-stimulation. *Neuropharmacology* **15**, 349–51.

Hill, R.T. (1970). Facilitation of conditioned reinforcement as a mechanism for psychomotor stimulation. In *Amphetamines and Related Compounds*, eds. E. Costa & S. Garattini, pp. 791–5. New York: Raven Press.

Hitzemann, R.J., Tseng, L.F., Hitzemann, B.A., Sampath-Khanna, S. & Loh, H.H. (1977). Effects of withdrawal from chronic amphetamine intoxication on exploratory and stereotyped behaviors in the rat. *Psychopharmacology* **54**, 295–302.

Hruska, R.E. & Silbergeld, E.K. (1980). Estrogen

treatment enhances dopamine receptor sensitivity in the rat striatum. *European Journal of Pharmacology* **61**, 397–400.

Huberman, H.S., Eison, M.S., Bryan, K. & Ellison, G. (1977). A slow-release pellet for chronic amphetamine administration. *European Journal of Pharmacology* **45**, 237–40.

Hutt, C. & Hutt, S.J. (1970). Stereotypies and their relation to arousal. A study of autistic children. In *Behaviour Studies in Psychiatry*, eds. S.J. Hutt & C. Hutt, pp. 175–200. Oxford: Pergamon Press.

Huttunen, M.O. & Niskanen, P. (1978). Prenatal loss of father and psychiatric disorders. *Archives of General Psychiatry* **35**, 429–31.

Iversen, S.D. (1986). Animal models of schizophrenia. In *The Psychopharmacology and Treatment of Schizophrenia*, eds. P.B. Bradley & S.R. Hirsch. Oxford: Oxford University Press.

Iversen S.D. (1987). Is it possible to model psychotic states in animals? *Journal of Psychopharmacology* **1**, 154–76.

Iversen, S.D. & Joyce, E.M. (1978). Effect in the rat of chronic morphine treatment on the behavioural response to apomorphine. *British Journal of Pharmacology* **62**, 390P–391P.

Jackson, D.M., Anden, N. & Dahlstrom, A. (1975). A functional effect of dopamine in the nucleus accumbens and in some other dopamine-rich parts of the brain. *Psychopharmacologia (Berl.)* **45**, 139–49.

Jacobs, B.L. (1976). An animal behaviour model for studying central serotonergic synapses. *Life Sciences* **19**, 777–86.

Jacobs, B.L., Trulson, M.E. & Stern, W.C. (1976). An animal behavior model for studying the actions of LSD and related hallucinogens. *Science* **194**, 741–3.

Jacobs, B.L., Trulson, M.E. & Stern, W.C. (1977). Behavioral effects of LSD in the cat: Proposal of an animal behavior model for studying the actions of hallucinogenic drugs. *Brain Research* **132**, 301–14.

Janowsky, D.S., El-Yousef, M.K., Davis, J.M. & Sekerke, H.J. (1973). Provocation of schizophrenic symptoms by intravenous administration of methylphenidate. *Archives of General Psychiatry* **28**, 185–91.

Jerussi, T.P. & Glick, S.D. (1976). Drug-induced rotation in rats without lesions: Behavioral and neurochemical indices of a normal asymmetry in nigrostriatal function. *Psychopharmacology* **47**, 249–60.

Jerussi, T.P. & Taylor, C.A. (1982). Bilateral asymmetry in striatal dopamine metabolism: Implications for pharmacotherapy of schizophrenia. *Brain Research* **246**, 71–5.

Jeste, D.V., Stoff, D.M., Rawlings, R. & Wyatt, R.J. (1984). Pharmacogenetics of phenylethylamine:

Determination of heritability and genetic transmission of locomotor effects in recombinant inbred strains of mice. *Psychopharmacology* **84**, 537–40.

Johnson, F.N. (1972). Chlorpromazine and lithium: Effects on stimulus significance. *Diseases of the Nervous System* **33**, 235–41.

Johnson, F.N. (1975). Behavioural and cognitive effects of lithium: Observations and experiments. In *Lithium Research and Therapy*, ed. F.N. Johnson, pp. 315–27. London: Academic Press.

Johnson, F.N. (1979). The effects of lithium chloride on response to salient and nonsalient stimuli in *Carassius auratus. International Journal of Neuroscience* **9**, 185–90.

Johnson, F.N. & Wormington, S. (1972). Effects of lithium on rearing activity in rats. *Nature* **235**, 159.

Jones, D.L., Berg, S.L., Dorris, R.L. & Dill, R.E. (1981). Biphasic locomotor response to intra-accumbens dopamine in a non-human primate. *Pharmacology, Biochemistry & Behavior* **15**, 243–6.

Jones, G.H., Hernandez, T.D. & Robbins, T.W. (1987). Isolation-rearing impairs the acquisition of schedule-induced polydipsia. *Society for Neuroscience Abstracts* **13**, 405.

Joyce, E.M. & Iversen, S.D. (1979). The effect of morphine applied locally to mesencephalic dopamine cell bodies on spontaneous motor activity in the rat. *Neuroscience Letters* **14**, 207–12.

Kalant, O.J. (1973). *The Amphetamines. Toxicity and Addiction.* Brookside Monograph of the Addiction Research Foundation, No. 5. Toronto: University of Toronto Press.

Karoum, F., Chuang, L-W. & Wyatt, R.J. (1981). Presence and distribution of phenylethylamine in the rat spinal cord. *Brain Research* **225**, 442–5.

Karoum, F., Speciale, S.G. Jr., Chuang, L-W. & Wyatt, R.JU. (1982). Selective effects of phenylethylamine on central catecholamines: A comparative study with amphetamine. *Journal of Pharmacology and Experimental Therapeutics* **223**, 432–9.

Katz, R.J. (1982). Morphine- and endorphin-induced behavioral activation in the mouse: Implications for mania and some recent pharmacogenetic studies. In *Opioids in Mental Illness*, vol. 398, *Annals New York Academy of Science*, ed. K. Verebey, pp. 291–300.

Kelly, P.H. (1977). Drug-induced motor behavior. In *Handbook of Psychopharmacology*, vol. 8, eds. L.L. Iversen, S.D. Iversen & S.H. Snyder, pp. 295–331. New York: Plenum Press.

Kelly, P.H. & Iversen, L.L. (1975). LSD as an agonist at mesolimbic dopamine receptors. *Psychopharmacologia (Berl.)* **45**, 221–4.

Kelly, P.H., Seviour, P.W. & Iversen, S.D. (1975).

Amphetamine and apomorphine response in the rat following 6-OHDA lesions of the nucleus accumbens septi and corpus striatum. *Brain Research* **94**, 507–22.

Kilbey, M.M. & Ellinwood, E.H., Jr (1977). Reverse tolerance to stimulant-induced abnormal behavior. *Life Sciences* **20**, 1063–76.

Kim, J.S. & Kornhuber, H.H. (1982). The glutamate theory in schizophrenia: Clinical and experimental evidence. In *Psychobiology of Schizophrenia*,eds. M. Namba & H. Kaiya, pp. 221–34. Oxford: Pergamon Press.

Kim, J.S., Kornhuber, H.H., Schmidburgh, W. & Holzmuller, B. (1980). Low cerebrospinal fluid glutamate in schizophrenic patients and a new hypothesis of schizophrenia. *Neuroscience Letters* **20**, 379–82.

Kjellberg, B. & Randrup, A. (1972). Stereotypy with selective stimulation of certain items of behavior observed in amphetamine-treated monkeys (*Cercopithecus*). *Pharmakopsychiatrie* **5**, 1–12.

Kline, J., Jr. & Reid, K.H. (1985). The acute periventricular injury syndrome: A possible animal model for psychotic disease. *Psychopharmacology* **87**, 292–7.

Kokkinidis, L. & Anisman, H. (1980). Amphetamine models of paranoid psychosis: An overview and elaboration of animal experimentation. *Psychological Bulletin* **88**, 551–79.

Kokkinidis, L. & Anisman, H. (1981). Amphetamine psychosis and schizophrenia: A dual model. *Neuroscience and Biobehavioral Reviews* **5**, 449–61.

Koob, G.F. (1977). Incentive shifts in intracranial self-stimulation produced by different series of stimulus intensity presentations. *Physiology and Behavior* **18**, 131–5.

Kornetsky, C. (1977). Animal models: Promises and problems. In *Animal Models in Psychiatry and Neurology*, eds. I. Hanin & E. Usdin, pp. 1–8. New York: Pergamon Press.

Kornetsky, C. & Markowitz, R. (1978). Animal models of schizophrenia. In *Psychopharmacology: A Generation of Progress*, eds. M.A. Lipton, A. Dimascio & K.F. Killam, pp. 583–93. New York: Raven Press.

Kornhuber, J. & Fischer, E.G. (1982). Glutamic acid diethyl ester induces catalepsy in rats. A new model for schizophrenia? *Neuroscience Letters* **34**, 325–9.

Kornhuber, J., Kim, J., Kornhuber, M.E. & Kornhuber, H.H. (1984). The cortico-nigral projection: reduced glutamate content in the substantia nigra following frontal cortex ablation in the rat. *Brain Research* **322**, 124–6.

Kovelman, J.A. & Scheibel, A.B. (1984). A neurohistological correlate of schizophrenia. *Biological Psychiatry* **19**, 1601–21.

Kraemer, G.W., Ebert, M.H., Lake, C.R. & McKinney, W.T. (1983). Amphetamine challenge: Effects in previously isolated rhesus monkeys and implications for animal models of schizophrenia. In *Ethopharmacology: Primate Models of Neuropsychiatric Disorders*, ed. K. Miczek, pp. 199–218. New York: Alan R. Liss, Inc.

Kryzhanovsky, G.N. & Aliev, M.N. (1981). The stereotyped behavior syndrome: A new model and proposed therapy. *Pharmacology, Biochemistry and Behavior* **14**, 273–81.

Kumar, R. (1976). Animal models for screening new agents: A behavioural view. *British Journal of Clinical Pharmacy*, Suppl., 13–17.

Lake, C.R., Sternberg, D.E., van Kammen, D.P., Balenger, J.C., Ziegler, M.G., Post, R.M., Kopin, I.J. & Bunney, W.E. (1980). Schizophrenia: Elevated cerebrospinal fluid norepinephrine. *Science* **207**, 331–3.

Leith, N.J. & Barrett, R.J. (1980). Effects of chronic amphetamine or reserpine on self-stimulation responding: Animal model of depression? *Psychopharmacology* **72**, 9–15.

Le Moal, M., Stinus, L. & Galey, D. (1976). Radiofrequency lesion of the ventral mesencephalic tegmentum: Neurological and behavioral considerations. *Experimental Neurology* **50**, 521–35.

Le Moal, M., Stinus, L., Simon, H., Tassin, J.P., Thierry, A.M., Blanc, G., Glowinski, J. & Cardo, B. (1977). Behavioral effects of a lesion in the ventral mesencephalic tegmentum: Evidence for involvement of A10 dopaminergic neurons. In *Advances in Biochemical Psychopharmacology*, vol. 16, eds. E. Costa & G.L. Gessa, pp. 237–45. New York: Raven Press.

Levesque, D. & Di Paolo, T. (1987). Physiological and pharmacological hormonal modulation of the rat striatal D1 dopamine receptor. *Society for Neuroscience Abstracts* **13**, 1344.

Liebman, J.M. & Segal, D.S. (1976). Lithium differentially antagonizes self-stimulation facilitated by morphine and (+)-amphetamine. *Nature* **260**, 161.

Lyon, M. (1986). Temporal problems in the definition of behavioral acts. In *Delhed og Helhed: Teoretiske og Metodiske Studier over Komplicerede Psykobiologiske Faenomener*, eds. I.D. Petersen & A.F. Petersen, pp. 75–99. Copenhagen: Forlaget Politiske Studier.

Lyon, M. & Magnusson, M.S. (1982). Central stimulant drugs and the learning of abnormal behavioral sequences. In *Behavioral Models and the Analysis of Drug Action*, eds. M.Y. Spiegelstein & A. Levy, pp. 135–53. Amsterdam: Elsevier Scientific Publishing Company.

Lyon, M. & Nielsen, E.B. (1979). Psychosis and drug induced stereotypies. In *Psychopathology in Animals: Research and Clinical Implications*, ed. J.D. Keehn, pp. 103–42. New York: Academic Press.

Lyon, M. & Robbins, T.W. (1975). The action of central nervous system stimulant drugs: A general theory concerning amphetamine effects. In *Current Developments in Psychopharmacology*, vol. 2, eds. W. Essman & L. Valzelli, pp. 79–163. New York: Spectrum Publications Inc.

Lyon, M., Halpern, M. & Mintz, E. (1968). The significance of the mesencephalon for coordinated feeding behavior. *Acta Neurologica Scandinavica* **44**, 323–46.

Lyon, N. & Gerlach, J. (1988). Perseverative structuring of responses by schizophrenic and affective disorder patients. *Journal of Psychiatric Research* **23**, 261–77.

Lyon, N., Mejsholm, B. & Lyon, M. (1986). Stereotyped responding in schizophrenic outpatients: Cross-cultural confirmation of perseverative switching on a two-choice task. *Journal of Psychiatric Research* **20**, 137–50.

Mailman, R.B., Lewis, M.H. & Kilts, C.D. (1981). Animal models related to developmental disorders: Theoretical and pharmacological analyses. *Applied Research in Mental Retardation* **2**, 1–12.

Mandell, A.J. & Knapp, S. (1976). A neurobiological model for the symmetrical prophylactic action of lithium in bipolar affective disorder. *Pharmakopsychiatrie* **9**, 116–26.

Marini, J.L. & Sheard, M.H. (1981). On the specificity of a cat behavior model for the study of hallucinogens. *European Journal of Pharmacology* **70**, 479–87.

Marsden, C.A. & Curzon, G. (1976). Studies on the behavioural effects of tryptophan and p-chlorophenylalanine. *Neuropharmacology* **15**, 165–71.

Mason, W.A. (1968). Early social deprivation in the nonhuman primates: Implications for human behavior. In *Environmental Influences*, ed. D.C. Glass, pp. 70–100. New York: Rockefeller University Press and Russell Sage Foundation.

Matthysse, S. (1981). Nucleus accumbens and schizophrenia. In *The Neurobiology of the Nucleus Accumbens*, eds. R.B. Chronister & J.F. DeFrance, pp. 351–9. Brunswick, Maine: Haer Institute.

Matthysse, S. (1983). Making animal models relevant to psychiatry. *Annals New York Academy of Science* **406**, 133–9.

McKinney, W.T. & Moran, E.C. (1981). Animal models of schizophrenia. *American Journal of Psychiatry* **138**, 478–83.

Matussek, N. & Linsmayer, M. (1968). The effect of lithium and amphetamine on desmethyl-imipramine – RO 4-1284 – induced motor hyperactivity. *Life Sciences* **7**, 371–5.

Mednick, S.A. (1958). A learning theory approach to research in schizophrenia. *Psychological Bulletin* **55**, 316–27.

Mednick, S.A., Machon, R.A., Huttunen, M.O. & Bonett, D. (1988). Adult schizophrenia following prenatal exposure to an influenza epidemic. *Archives of General Psychiatry* **45**, 189–92.

Meltzer, H.L., Taylor, R.M., Platman, S.R. & Fieve, R.R. (1969). Rubidium: A potential modifier of affect and behaviour. *Nature* **223**, 321–2.

Michanek, A. & Meyerson, B.J. (1975). Copulatory behavior in the female rat after amphetamine and amphetamine derivatives. In *Sexual Behavior: Pharmacology and Biochemistry*, eds. M. Sandler & G.L. Gessa, pp. 51–7. New York: Raven Press.

Miczek, K.A. (1974). Intraspecies aggression in rats: effects of d-amphetamine and chlordiazepoxide. *Psychopharmacologia (Berl.)* **39**, 275–301.

Miller, F.P., Cox, R.H., Snodgrass, W.R. & Maikel, R.P. (1970). Comparative effects of p-chlorophenylalanine, p-chloroamphetamine, and p-chloro-N methylamphetamine on rat brain norepinephrine, serotonin, and 5-hydroxyindole-3-acetic acid. *Biochemical Pharmacology* **19**, 435–42.

Mirsky, I.A. (1968). Communication of affects in monkeys. In *Environmental Influences*, ed. D.C. Glass, pp. 129–37. New York: Rockefeller University Press and Russell Sage Foundation.

Modigh, K. & Svensson, T.H. (1972). On the role of central nervous system catecholamines and 5-hydroxytryptamine in the nialamide-induced behavioral syndrome. *British Journal of Pharmacology* **46**, 32–45.

Moerschbaecher, J.M., Thompson, D.M. & Thomas, J.F. (1979). Effects of methamphetamine and scopolamine on variability of response location. *Journal of Experimental Analysis of Behavior* **32**, 255–63.

Moisset, B. (1977). Genetic analysis of the behavioral response to d-amphetamine in mice. *Psychopharmacology* **53**, 263–7.

Murphy, D.L. (1977). Animal models for mania. In *Animal Models in Psychiatry and Neurology*, eds. I. Hanin & E. Usdin, pp. 211–22. New York: Pergamon Press.

Nakamura, K. & Thoenen, H. (1972). Increased irritability – A permanent behavior change induced in the rat by intraventricular administration of 6-hydroxydopamine. *Psychopharmacologia (Berl.)* **24**, 359–72.

Nielsen, E.B. & Lyon, M. (1982). Behavioral alterations during prolonged low level continuous amphetamine administration in a monkey family

group (*Cercopithecus aethiops*). *Biological Psychiatry* **17**, 423–4.

Nielsen, E.B., Lyon, M. & Ellison, G. (1983). Apparent hallucinations in monkeys during around-the-clock amphetamine for seven to fourteen days. Possible relevance to amphetamine psychosis. *Journal of Nervous and Mental Disease* **171**, 222–33.

Newton, J.E.O., Chapin, J.L. & Murphree, O.D. (1976). Correlations of normality and nervousness with cardiovascular functions in pointer dogs. *Pavlovian Journal of Biological Science* **11**, 105–20.

Norman, D.A. & Shallice, T. (1980). *Attention to Action: Willed and Automatic Control of Behavior*. Office of Naval Research Report No. 8006. San Diego: University of California Press.

Norton, S., Mullenix, P. & Culver, B. (1976). Comparison of the structure of hyperactive behavior in rats after brain damage from X-irradiation, carbon monoxide and pallidal lesions. *Brain Research* **116**, 49–47.

Novak, M.A. & Harlow, H.F. (1975). Social recovery of monkeys isolated for the first year of life: 1. Rehabilitation and therapy. *Developmental Psychology* **11**, 453–65.

Nowakowski, R.S. (1987). Basic concepts of CNS development. *Child Development* **58**, 568–95.

Oades, R.D. (1982). Search strategies on a hole-board are impaired in rats with ventral tegmental damage: Animal model for tests of thought disorder. *Biological Psychiatry* **17**, 243–58.

Oades, R.D. & Isaacson, R.L. (1978). The development of food search behavior by rats: The effects of hippocampal damage and haloperidol treatment. *Behavioral Biology* **24**, 327.

Owen, F., Baker, H.F., Ridley, R.M., Cross, A.J. & Crow, T.J. (1981). Effect of chronic amphetamine administration on central dopaminergic mechanisms in the vervet. *Psychopharmacology* **74**, 213–16.

Panksepp, J. & Trowill, J.A. (1969). Positive and negative contrast effects with hypothalamic reward. *Physiology & Behavior* **4**, 173–5.

Panksepp, J. & Trowill, J.A. (1970). Positive incentive contrast with rewarding electrical stimulation of the brain. *Journal of Comparative & Physiological Psychology* **70**, 358–63.

Pavlov, I.P. (1927). *Conditioned Reflexes*. New York: Oxford University Press.

Payne, R.W., Caird, W.K. & Laverty, S.G. (1964). Overinclusive thinking and delusions in schizophrenic patients. *Journal of Abnormal and Social Psychology* **68**, 561–6.

Petty, F. & Sherman, A.D. (1981). A pharmacologically pertinent animal model of mania. *Journal of Affective Disorders* **3**, 381–7.

Phillips, A.G. & Fibiger, H.C. (1979). Decreased

resistance to extinction after haloperidol: Implications for the role of dopamine in reinforcement. *Pharmacology Biochemistry and Behavior* **10**, 751–60.

Phillips, A.G. & LePiane, F.G. (1986). Effects of pimozide on positive and negative incentive contrast with rewarding brain stimulation. *Pharmacology Biochemistry and Behavior* **24**, 1577–82

Phillips, A.G., Carter, D.A. & Fibiger, H.C. (1976). Dopaminergic substrates of intracranial self-stimulation in the caudate-putamen. *Brain Research* **104**, 221–32.

Pijnenburg, A.J.J. & van Rossum, J.M. (1973). Stimulation of locomotor activity following injection of dopamine into the nucleus accumbens. *Journal of Pharmacy and Pharmacology* **25**, 1003–4.

Poschel, B.P.H. & Ninteman, F.W. (1971). Intracranial reward and the forebrain's serotonergic mechanism: Studies employing para-chlorophenylalanine and para-chloroamphetamine. *Physiology and Behavior* **7**, 39–66.

Poschel, B.P.H., Ninteman, F.W., McLean, J.R. & Potaczak, D. (1974). Intracranial reward after 5,6-dihydroxytryptamine: Further evidence for serotonin's inhibitory role. *Life Sciences* **15**, 1515–22.

Post, R.M. (1977). Approaches to rapidly cycling manic-depressive illness. In *Animal Models in Psychiatry and Neurology*, eds. I. Hanin & E. Usdin, pp. 201–10. New York: Pergamon Press.

Post, R.M., Gerner, R.H., Carmen, J.S. & Bunney, W.E., Jr. (1976). Effects of low doses of a dopamine receptor stimulator in mania. *Lancet* **I**, 203–4.

Post, R.M., Squillace, K.M. & Part, A. (1977*a*). Kindling and oscillation in amygdala excitability. In *Proceedings of the Annual Meeting of the Society of Biological Psychiatry*, Toronto, May 1977.

Post, R.M., Squillace, K.M., Sass, W. & Pert, A. (1977*b*). Drug sensitization and electrical kindling. In *Abstracts of the Society for Neuroscience*, Anaheim, California.

Post, R.M., Squillace, K.M., Sass, W. & Pert, A. (1981). The effect of amygdaloid kindling on spontaneous and cocaine-induced motor activity and lidocaine seizures. *Psychopharmacology* **72**, 189–96.

Post, R.M., Rubinow, D. & Ballenger, J. (1984). Conditioning, sensitization, and kindling: Implications for the course of affective illness. In *Neurobiology of Mood Disorders*, eds. R.M. Post & J. Ballenger, pp. 432–66. Baltimore: Williams and Wilkins Co.

Potkin, S.G., Karoum, F., Chuang, L-W.,

Cannon-Spoor, H.E., Phillips, I. & Wyatt, R.J. (1979). Phenylethylamine in paranoid schizophrenia. *Science* **206**, 470–1.

Raleigh, M.J., McGuire, M.T., Brammer, G.L. & Yuwiler, A. (1984). Social and environmental influences on blood serotonin concentrations in monkeys. *Archives of General Psychiatry* **41**, 405–10.

Ranck, J.B., Jr. & Windle, W.F. (1959). Brain damage in the monkey, *Macaca mulatta*, by asphyxia neonatorum. *Experimental Neurology* **1**, 130–54.

Randrup, A. & Munkvad, I. (1967). Stereotyped activities produced by amphetamine in several species and man. *Psychopharmacologia (Berl.)*, **11**, 300–10.

Randrup, A. & Munkvad, I. (1968). Behavioural stereotypies induced by pharmacological agents. *Pharmakopsychiatri und Neuropsychopharmakologi* **1**, 18–26.

Randrup, A., Munkvad, I. & Fog, R. (1981). Mental and behavioural stereotypies elicited by stimulant drugs. Relation to the dopamine hypothesis of schizophrenia, mania, and depression. In *Recent Advances in Neuropsychopharmacology*, eds. B. Angrist, G.D. Burrows, M. Lader, O. Lingjaerde, G. Sedvall & D. Wheatley, pp. 63–74. Oxford: Pergamon Press.

Rastogi, R.M. & Singhal, R.L. (1977). Lithium suppresses elevated behavioural activity and brain catecholamines in developing hyperthyroid rats. *Canadian Journal of Physiology and Pharmacology* **55**, 490–5.

Rastogi, R.B., Merali, Z. & Singhal, R.L. (1977). Cadmium alters behaviour and biosynthetic capacity for catecholamine and serotonin in neonatal rat brain. *Journal of Neurochemistry* **28**, 789–94.

Rebec, G.V. & Bashore, T.R. (1982). Comments on 'Amphetamine models of paranoid schizophrenia': A precautionary note. *Psychological Bulletin* **92**, 403–9.

Ridley, R.M., Scraggs, P.R. & Baker, H.F. (1979). Modification of the behavioural effects of amphetamine by a GABA agonist in a primate species. *Psychopharmacology* **64**, 197–200.

Robbins, T.W. (1975). The potentiation of conditioned reinforcement by psychomotor stimulant drugs: A test of Hill's hypothesis. *Psychopharmacologia (Berl.)* **45**, 103–12.

Robbins, T.W. (1976). Relationship between reward-enhancing and stereotypical effects of psychomotor stimulant drugs. *Nature* **254**, 57–9.

Robbins, T.W. (1984). Cortical noradrenaline, attention and arousal. *Psychological Medicine* **14**, 13–21.

Robbins, T.W. & Sahakian, B.J. (1979). 'Paradoxical' effects of psychomotor stimulant drugs in hyperactive children from the standpoint of behavioural pharmacology. *Neuropharmacology* **18**, 931–50.

Robbins, T.W. & Sahakian, B.J. (1980). Animal models of mania. In *Mania: An Evolving Concept*, eds. R. Belmaker & H. van Praag, pp. 143–216. New York: Spectrum Publishing Co.

Robbins, T.W., Watson, B.A., Gaskin, M. & Ennis, C. (1983). Contrasting interactions of pipradrol, d-amphetamine, cocaine, cocaine analogues, apomorphine and other drugs with conditioned reinforcement. *Psychopharmacology* **80**, 113–19.

Rubin, E.H. & Zorumski, C.F. (1985). Limbic seizures, kindling and psychosis: A link between neurobiology and clinical psychiatry. *Comprehensive Therapy* **1**, 54–8.

Sahakian, B.J., Robbins, T.W., Morgan, M.J. & Iversen, S.D. (1975). The effects of psychomotor stimulants on stereotypy and locomotor activity in socially deprived and control rats. *Brain Research* **8**, 195–205.

Sahakian, B.J., Robbins, T.W. & Iversen, S.D. (1976). Alpha-flupentixol-induced hyperactivity by chronic dosing in rats. *European Journal of Pharmacology* **37**, 169–78.

Sandler, M. & Reynolds. G.P. (1976). Does phenylethylamine cause schizophrenia? *Lancet* **i**, 70–1.

Sato, M. (1983). Long-lasting hypersensitivity to methamphetamine following amygdaloid kindling in cats: the relationship between limbic epilepsy and the psychotic state. *Biological Psychiatry* **18**, 525–36.

Scheel-Kruger, J. (1986). Dopamine-GABA interactions: Evidence that GABA transmits, modulates and mediates dopaminergic functions in the basal ganglia and the limbic system. *Acta Neurologica Scandinavica (Suppl.)* **107**, 1–54.

Scheel-Kruger, J., Cools, A.R. & van Wel, P.M. (1977). Muscimol a GABA-agonist injected into the nucleus accumbens increases apomorphine stereotypy and decreases the motility. *Life Sciences* **21**, 1697–702.

Scheel-Kruger, J., Christensen, A.V. & Arnt, J. (1978). Muscimol differentially facilitates stereotypy but antagonizes motility induced by dopaminergic drugs: A complex GABA–dopamine interaction. *Life Sciences* **22**, 75–84.

Scheel-Kruger, J., Arnt, J., Braestrup, C., Christensen, A.V. & Magelund, G. (1979). Development of new animal models for GABA-ergic actions using muscimol as a tool. In *GABA-Neurotransmitters*, eds. P. Krogsgaard-Larsen, J. Scheel-Kruger & H. Kofod, pp. 447–64. New York: Academic Press.

Scheel-Kruger, J., Arnt, J., Magelund, G., Olianas, M., Przewlocka, B. & Christensen, A.V. (1980).

Behavioural functions of GABA in basal ganglia and limbic system. *Brain Research Bulletin* **5**: Suppl. 2, 261–7.

Schiørring, E. (1971). Amphetamine induced selective stimulation of certain behaviour items with concurrent inhibition of others in an open-field test with rats. *Behaviour* **39**, 1–17.

Schiørring, E. (1977). Changes in individual and social behavior induced by amphetamine and related compounds in monkeys and man. In *Cocaine and Other Stimulants*, eds. E.H. Ellinwood & M. Kilbey, pp. 481–522. New York: Plenum Press.

Schmajuk, N.A. (1987). Animal models for schizophrenia: The hippocampally lesioned animal. *Schizophrenia Bulletin* **13**, 317–27.

Schwartz, J.M., Ksir, C., Koob, G.F. & Bloom, F.E. (1982). Changes in locomotor response to beta-endorphin microinfusion during and after opiate abstinence syndrome – A proposal for a model of the onset of mania. *Psychiatry Research* **7**, 153–61.

Segal, D.S. & Janowsky, D.S. (1978). Psychostimulant-induced behavioral effects: Possible models of schizophrenia. In *Psychopharmacology: A Generation of Progress*, eds. M.A. Lipton, A. Dimascio & K.F. Killam, pp. 1113–23. New York: Raven Press.

Segal, D.S., Weinberger, S.B., Cahill, J. & McCunney, S.J. (1980). Multiple daily amphetamine administrations: Behavioral and neurochemical alterations. *Science* **207**, 904–7.

Segal, D.S., Geyer, M.A. & Schuckit, M.A. (1981). Stimulant-induced psychosis: An evaluation of animal models. *Essays in Neurochemistry and Neuropharmacology* **5**, 95–129.

Seligman, M.E., Weiss, J., Weinraub, M. & Schulman, A. (1980). Copying behavior: Learned helplessness, physiological change and learned inactivity. *Behavior Research and Therapy* **18**, 459–512.

Sem-Jacobsen, C.W. (1976). Electrical stimulation and self-stimulation in man with chronic implanted electrodes. Interpretation and pitfalls of results. In *Brain-Stimulation Reward*, eds. A. Wauquier & E.T. Rolls, pp. 508–20. New York: North Holland/American Elsevier.

Shaywitz, B.A., Yager, R.D. & Klopper, J.H. (1976). Selective brain dopamine depletion in developing rats: An experimental model of minimal brain dysfunction. *Science* **191**, 305–7.

Sheard, M. (1970). Effect of lithium on foot shock aggression in rats. *Nature* **228**, 284.

Shimizu, A., Endo, M., Yamaguchi, N., Torii, H. & Isaki, K. (1985). Hand preference in schizophrenics and handedness conversion in their childhood. *Acta Psychiatrica Scandinavica* **72**, 259–65.

Silbergeld, E.K. & Goldberg, A.M. (1974). Lead-induced behavioral dysfunction: An animal model of hyperactivity. *Experimental Neurology* **42**, 146–57.

Simon, H., Scatton, B. & Le Moal, M. (1980). Dopaminergic A10 neurones are involved in cognitive functions. *Nature* **286**, 150–1.

Skirboll, L.R., Grace, A.A. & Bunney, B.S. (1979). Dopamine auto- and postsynaptic receptors: Electrophysiological evidence for differential sensitivity to dopamine agonists. *Science* **206**, 80–2.

Sloviter, R.S., Damiano, B.P. & Connor, J.D. (1980). Relative potency of amphetamine isomers in causing the serotonin behavioral syndrome in rats. *Biological Psychiatry* **15**, 789–96.

Smith, D.F. (1976). Antagonistic effect of lithium chloride on L-dopa induced locomotor activity in rats. *Pharmacological Research Communications* **8**, 575–9.

Smith, D.F. & Smith, H.B. (1973). The effect of prolonged lithium administration on activity, reactivity, and endurance in the rat. *Psychopharmacologia (Berl.)* **30**, 83.

Sørensen, G. (1987). Animal experiments indicating behavioural pathologies as high cost strategy of survival. In *Problems of Constancy and Change: The Complementarity of Systems Approaches to Complexity*, eds. P. Checkland & I. Kiss, pp. 1059–63. Hungary: International Society for General Systems Research.

Sternberg, D.E., van Kammen, D.P., Lake, C.R., Ballenger, J.C., Marder, S.R. & Bunney, W.E., Jr. (1981). The effect of pimozide on CSF norepinephrine in schizophrenia. *American Journal of Psychiatry* **138**, 1045–51.

Stevens, J.R. (1978). Disturbances of ocular movements and blinking in schizophrenia. *Journal of Neurology, Neurosurgery and Psychiatry* **41**, 1024–30.

Stevens, J.R. (1979). Schizophrenia and dopamine regulation in the mesolimbic system. *Trends in Neuroscience* **2**, 102–5.

Stevens, J.R. & Livermore, A.L., Jr. (1978). Kindling of the mesolimbic dopamine system: Animal model of psychosis. *Neurology* **28**, 36–46.

Stevens, J.R., Wilson, K. & Foote, W. (1974). GABA blockade, dopamine and schizophrenia: Experimental studies in the cat. *Psychopharmacologia (Berl.)* **39**, 105–19.

Stille, G. & Sayers, A. (1975). Die Immobilisationsreaktion der Ratte als tierexperimentelles Modell fur die Katatonie. *Pharmakopsychiatrie* **8**, 105–14.

Stinus, L., Gaffori, O., Simon, H. & Le Moal, M. (1978). Disappearance of hoarding and disorganisation of eating behavior after ventral mesencephalic tegmentum lesion in rats. *Journal of Comparative & Physiological Psychology* **92**, 289–96.

Stolk, J.M., Nowack, W.J., Barchas, J.D. & Platman, S.R. (1970). Brain-norepinephrine enhanced turnover after rubidium treatment. *Science* **168**, 501–3.

Stolk, J.M., Conner, R.L. & Barchas, J.D. (1974). Social environment and brain biogenic amine metabolism in rats. *Journal of Comparative and Physiological Psychology* **87**, 203–7.

Struble, R.G. & Riesen, A.H. (1978). Changes in cortical dendritic branching subsequent to partial social isolations in stumptailed monkeys. *Developmental Psychobiology* **11**, 479–86.

Sudilovsky, A. (1975). Effects of disulfiram on the amphetamine-induced behavioral syndrome in the cat as a model of psychosis. *National Institute on Drug Abuse research, Monograph Series 3.*, pp. 109–35.

Suomi, S.J., Collins, M.L. & Harlow, H.F. (1973). Effects of permanent separation from mother on infant monkeys. *Developmental Psychology* **9**, 376–84.

Swerdlow, N.R., Braff, D.L., Geyer, M.A. & Koob, G.F. (1986). Central dopamine hyperactivity in rats mimics abnormal acoustic startle response in schizophrenics. *Biological Psychiatry* **21**, 23–33.

Syme, L.A. & Syme, G.J. (1973). Effects of lithium chloride on the activity of rats tested alone or in pairs. *Psychopharmacologia (Berl.)* **29**, 85–7.

Szabadi, E., Bradshaw, C.M. & Ruddle, H.V. (1981). Reinforcement processes in affective illness: Towards a quantitative analysis. In *Quantification of Steady-State Operant Behaviour*, eds. C.M. Bradshaw, E. Szabadi & C.F. Lowe, pp. 299–310. Amsterdam: Elsevier/North Holland Biomedical Press.

Tinklenberg, J.R., Gillin, J.C., Murphy, G.M. Jr., Staub, R. & Wyatt, R.J. (1978). The effects of phenylethylamine in rhesus monkeys. *American Journal of Psychiatry* **135**, 576–8.

Tissot, R. (1976). Un concept hypothétique: l'unité physiopathologique des psychoses monoaminergiques. *L'Encephale* **1:4**, 289–339.

Torrey, E.F., Yolken, R.H. & Albrecht, P. (1983). Cytomegalovirus as a possible etiological agent in schizophrenia. In *Research on the Viral Hypothesis of Mental Disorders*, vol. 12, ed. P.V. Morozov, pp. 150–60. Basel, Switzerland: S. Karger AG.

Trulson, M.E. & Jacobs, B.L. (1977). Usefulness of an animal behavior model in studying the duration of action of LSD and the onset and duration of tolerance to LSD in the cat. *Brain Research* **132**, 315–26.

Uchimura, H., Matsumoto, T., Hirano, M., Kim, J.S.

& Nakahara, T. (1982). Effects of isolation induced behavioral abnormalities and haloperidol on homovanillic acid levels in individual dopaminergic neuron systems of rat brain. In *Advances in Dopamine Research*, eds. M. Kohsaka, T. Shohmori, Y. Tsukada & G.N. Woodruff, pp. 95–106. New York: Pergamon Press.

Ungerstedt, U. (1971). Striatal dopamine release after amphetamine or nerve degeneration revealed by rotational behavior. *Acta Physiological Scandinavica*, Suppl. 367 49–68.

U-Prichard, D.C. & Steinberg, H. (1972). Selective effects of lithium on two forms of spontaneous activity. *British Journal of Pharmacology* **44**, 349–50.

Utena, H. (1974). On relapse-reliability: Schizophrenia, amphetamine psychosis and animal model. In *Biological Mechanisms of Schizophrenia and Schizophrenia-Like Psychoses*, eds. H. Mitsuda & T. Fukuda, pp. 285–90. Tokyo: Igaku Shoin.

Valenstein, E. (1976). The interpretation of behavior evoked by brain stimulation. In *Brain-Stimulation Reward*, eds. A. Wauquier & E.T. Rolls, pp. 557–75. New York: North Holland/American Elsevier.

Valzelli, L. & Garattini, S. (1972). Biochemical and behavioral changes induced by isolation in rats. *Neuropharmacology* **11**, 17–22.

van der Kooy, D., Fibiger, H.C. & Phillips, A.G. (1977). Monoamine involvement in hippocampal self-stimulation. *Brain Research* **136**, 119–30.

van Hest, A., van Haaren, F. & van de Poll, N.E. (1987). Perseverative responding in male and female rats: Effects of testosterone. *Society for Neuroscience Abstracts* **13**, 1315.

van Ree, J.M., Verhoeven, W.M. A., de Wied, D. & van Praag, H.M. (1982). The use of the synthetic peptides gamma-type endorphins in mentally ill patients. In *Opioids in Mental Illness*, vol. 398, *Annals New York Academy of Science*, ed. K. Verebey, pp. 478–95. New York: The New York Academy of Sciences.

Venables, P.H. (1964). Attentional deficits in chronic schizophrenics. In *Progress in Experimental Personality Research*, ed. B. Maher, pp. 1–42. New York: Academic Press.

Venables, P.H. (1978). Cognitive disorder. In *Schizophrenia – Towards a New Synthesis*, ed. J.K. Wing, pp. 117–37. London: Academic Press.

Venables, P.H. & Wing, J.K. (1962). Level of arousal and the subclassification of schizophrenia. *Archives of General Psychiatry* **7**, 114–19.

Verebey, K. (ed.) (1982). *Opioids in Mental Illness: Theories, Clinical Observations, and Treatment Possibilities. Annals New York Academy of Science*, vol. **398**.

Villablanca, J.R., Harris, C.M., Burgess, J.W. & de Andres, I. (1984). Reassessing morphine effects in

cats: I. Specific behavioral responses in intact and unilaterally brain-lesioned animals. *Pharmacology, Biochemistry & Behavior* **21**, 913–21.

Wake, A. & Wada, J. (1975). Frontal cortical kindling in cats. *Canadian Journal of Neurological Science* **2**, 493–6.

Waziri, R. (1980). Lateralization of neuroleptic-induced dyskinesia indicates pharmacologic asymmetry in the brain. *Psychopharmacology* **68**, 51–3.

Wegener, S. (1986). Dopaminerge Kontrolle des Eintrageverhaltens bei laktierenden Hausmäusen (*Mus musculus*). Diplomarbeit, Biologisches Institut der Universität Stuttgart, Abteilung Tierphysiologie.

Weinstock, M., Speiser, Z. & Ashkenazi, R. (1977). Biochemical and pharmacological studies on an animal model of hyperactivity states. In *The Impact of Biology on Modern Psychiatry*, eds. E.S. Gershon, R.H. Belmaker, S.S. Kety & M. Rosenbaum, pp. 149–61. New York: Plenum Press.

Weiss, F., Tanzer, D.J. & Ettenberg, A. (1988). Opposite actions of CCK-8 on amphetamine-induced hyperlocomotion and stereotypy following intracerebroventricular and intra-accumbens injections in rats. *Pharmacology, Biochemistry & Behavior* **30**, 309–17.

White, D.K. & Barrett, R.J. (1981). The effects of electroconvulsive shock on the discriminative stimulus properties of d-amphetamine and apomorphine: Evidence for dopamine receptor alteration subsequent to ECS. *Psychopharmacology* **73**, 211–14.

White, F.J., Holohean, A.M. & Appel, J.B. (1981). Lack of specificity of an animal behavior model for hallucinogenic drug action. *Pharmacology, Biochemistry & Behavior* **14**, 339–43.

Willner, P., Towell, A. & Muscat, R. (1987). Effects of amphetamine and pimozide on reinforcement and motor parameters in variable interval performance. *Journal of Psychopharmacology* **1**, 140–53.

Wise, R.A. (1981). Brain dopamine and reward. In *Theory in Psychopharmacology* vol. 1, ed. S.J. Cooper, pp. 103–22. New York: Academic Press.

Wise, R.A., Spindler, J., de Wit, H. & Gerber, G.J. (1978). Neuroleptic-induced 'anhedonia' in rats: Pimozide blocks the reward quality of food. *Science* **201**, 262–4.

Wong, D.F., Wagner, H.N., Jr., Tune, L.E., Dannals, R.F., Pearlson, G.D., Links, J.M., Tamminga, C.A., Broussole, E.P., Ravert, H.T., Wilson, A.A., Toung, J.K.T., Malat, J., Williams, J.A., O'Tuama, L.A., Snyder, S.H., Kuhar, M.J. & Gjedde, A. (1986). Positron emission tomography reveals elevated D2 dopamine receptors in drug-naive schizophrenics. *Science* **234**, 1558–63.

Wyatt, R.J., Gillin, J.C. & Stoff, D.M. (1977). Beta-phenylethylamine (PEA) and the neuropsychiatric disturbances. In *Neuroregulators and Hypotheses of Psychiatric Disturbances*, eds. J.D. Barchas & E. Usdin, pp. 31–45. London: Oxford University Press.

Zacharko, R.M., Bowers, W.J., Kokkinidis, L. & Anisman, H. (1983). Region-specific reductions of intracranial self-stimulation after uncontrollable stress: possible effects on reward processes. *Behavioral Brain Research* **9**, 129–41.

Zemlan, F., Hirschowitz, J., Sautter, F. & Garver, D. (1984). Impact of lithium on core psychotic symptoms of schizophrenia. *British Journal of Psychiatry* **144**, 64–9.

Zigler, E. & Glick, M. (1988). Is paranoid schizophrenia really camouflaged depression? *American Psychologist* **43**, 284–90.

12

Pharmacological evaluation of new antipsychotic drugs
SVEN AHLENIUS

12.1 INTRODUCTION

12.1.1 Definitions

Antipsychotics are compounds used primarily in the treatment of schizophrenia, and in the present context an antipsychotic will be synonymous with an antischizophrenic agent. However, there are other forms of functional and organic psychoses, as well as other conditions, where this group of agents is used, e.g. the manic phase of manic-depressive psychosis, Gilles de la Tourette's syndrome and Huntington's chorea (see Baldessarini, 1985). The use of antipsychotics on indications other than schizophrenia is empirical, however, and is seldom considered in the preclinical evaluation of this group of agents. The reason for this is the necessity to focus on biological mechanisms rather than behavioral symptoms since, in contrast to the evaluation of e.g. antidiuretics, antibiotics or antihypertensive drugs, we have no analogue functions in animals corresponding to schizophrenia. Thus, fundamentally the evaluation is based on a clinical hint, brought into the laboratory and further studied biochemically and by use of homologue behavioral functions. For this reason, however, these latter functions are of little value in predicting effects outside the initial hypothesis. This important distinction, and other possible routes of investigation, are discussed further below (Section 12.2), and by Lyon (this vol., Ch. 11).

12.1.2 Schizophrenia – a brain disease?

The nature of schizophrenia is vital to any search for drug therapies, or other somatic remedies. Is schizophrenia really an organic brain disease? Can the disturbance be localized and defined morphologically or biochemically? The evidence for an organic component is circumstantial, but unmistakable. Firstly, it has been shown that there is a genetic component in schizophrenia (see Rosenthal & Kety, 1968; Kendler & Robinette, 1983; Kendler, 1987), and there have been recent attempts to localize the chromosomal aberration responsible for the illness (Sherrington et al., 1988; Kennedy et al., 1988). In this connection, it should be noted that the persistence of the illness throughout history (Jeste et al., 1985) has been puzzling, and it has been suggested that, in an evolutionary perspective, the symptoms of schizophrenia may be associated with traits of survival value like creativity (see Allen & Sarich, 1988). Secondly, taking national differences in diagnostic criteria into account, epidemiological data indicate a fairly constant incidence and prevalence of the disease (see Babigian, 1980; Eaton, 1985). Anthropological studies have shown that schizophrenic symptoms can be identified in widely different societies, like the Inuites on Greenland, and the Yorubas in Nigeria, with about the same prevalence as in western societies like Canada and Sweden (Murphy, 1976).

The cognitive and affective behavior of an individual is strongly influenced during ontogeny, and is formed by a complex interaction between organic maturational processes, and the environmental context. This implies that any disease of cognitive or affective behavior will have a strong environmental component. Incidentally, it is interesting to note that the implication of an interaction is seldom understood. Most mental health therapy policies, as expressed by health authorities, seem to swing on a pendulum between the extremes of the nature–nurture concept.

Missing a specific unambiguous and measurable organic disturbance, the field is open for speculation, and the various therapies used to treat schizophrenia would be the topic of a separate chapter. These therapies range from general somatic treatments like insulin therapy, brain surgery, and hemodialysis (see e.g. Kalinowsky & Hippius, 1969; Wagemaker & Cade, 1977) to drug therapies (see Barchas *et al.*, 1977) and psychotherapy (see e.g. Gelder *et al.*, 1983; Schooler & Hogarty, 1987). There has even been questioning of the disease concept of schizophrenia, suggesting radical changes in mental health care (cf. Szasz, 1974; Laing, 1976). In the field of neurotransmitters and drug therapy alone, several suggestions have appeared, including changes in brain GABA (van Kammen, 1977), dopamine-β-hydroxylase (Wise & Stein, 1973), endorphins (Terenius *et al.*, 1976) and, not the least, dopamine (DA), as will be discussed in detail below.

It was the accidental discovery of the effectiveness of chlorpromazine and reserpine in the treatment of schizophrenia about 30 years ago, that started a new era in psychopharmacology. The title of a book from that period, *The New Chemotherapy of Mental Illness* (Gordon, 1958), is perhaps significant, and suggested a similarity, at least in principle, to other forms of chemotherapy in somatic medicine. Indeed, the marked discharge of resident patients from mental hospitals, and the increase in ambulant treatment, that followed the introduction of these new alternatives in the drug treatment of schizophrenia, were by themselves no less than a revolution (e.g. Kline, 1968). The effect was new, and of a distinct antipsychotic quality, which initially led to the labeling of these compounds as major tranquilizers, as distinct from the minor tranquilizers represented by barbiturates and similar generally sedative compounds (Jacobsen, 1958). It is not surprising that laboratories in many countries tried to pinpoint the mechanism of action of these drugs in the hope of finding clues to the etiology of schizophrenia. After a short period of intensive research, the finding that chlorpromazine and haloperidol increased brain catecholamine metabolism provided the crucial break-

through, and it was suggested that the increased metabolism was due to compensatory mechanisms in response to the receptor blockade by these compounds (Carlsson & Lindqvist, 1963). It was already known that reserpine depletes brain catecholamine stores (see Carlsson, 1965), and thus both types of treatment probably worked by inhibiting brain catecholamine neurotransmission. Continued laboratory experiments and clinical observations have focussed on DA as primarily involved in the therapeutic effects of antipsychotics (see Carlsson, 1978).

Conversely, an overdose of amphetamine may in susceptible individuals produce a clinical picture close to, or identical with, acute psychotic behavior (Kalant, 1966; Randrup & Munkvad, 1967; Segal & Janowsky, 1978). This amphetamine-induced psychosis can be antagonized by treatment with antipsychotics (Angrist *et al.*, 1974). Furthermore, schizophrenic symptoms are exacerbated by the administration of *d*-amphetamine or L-DOPA (e.g. Yaryura-Tobias *et al.*, 1970; Angrist *et al.*, 1973; Angrist & Gershon, 1977). Since *d*-amphetamine has been shown to enhance brain DA release (see Carlsson, 1970), these observations provide further support for a link between brain DA neurotransmission and schizophrenia.

Disturbingly enough, these findings and observations have remained just clues over the years. In spite of the fact that both chlorpromazine and reserpine inhibit brain DA neurotransmission, it has not been possible to demonstrate consistently any endogenous change in DA metabolism associated with schizophrenia. This is a difficult task, and can only be addressed either by making CSF measurements of DA metabolites or by making biochemical or morphological examinations of post-mortem tissues. No clear picture has as yet emerged from this approach (e.g. Bunney *et al.*, 1979; Kleinman *et al.*, 1979; Shelton & Weinberger, 1987; Losonczy *et al.*, 1987). As an example of continued efforts to this end, it was recently reported that DA D_1 receptor binding was decreased in postmortem examination of brains from schizophrenia patients, concomitant with an increase in DA D_2 receptor binding, resulting in a highly significant difference in the ratio of D_2/D_1 receptor binding in schizophrenics as compared to controls (Creese, 1988). This finding may very well prove to be important in the evaluation of new and more selective antipsychotics. It is important to consider the enormous potential in diagnosis, treatment and prophylaxis to understand the efforts that have been made, and continue to be made, to identify somatic changes associated with schizophrenia. At the moment, the fact is that we have a useful treatment for schizophrenia, and that this is in all probability linked to an inhibition of brain DA neurotransmission. We cannot say, however, that brain DA necessarily is of importance in the etiology of schizophrenia.

Finally, mention should be made of results from brain imaging techniques like computerized tomography (CT), positron emission tomography (PET), magnetic resonance imaging (MRI), and techniques for studying regional changes in brain blood flow and glucose metabolism (see Andreasen, 1988). These techniques can all be applied to the living human brain. CT and MRI scans have demonstrated ventricular enlargement (lateral and 3rd ventricle) in schizophrenic patients (see Shelton & Weinberger, 1987), whereas monitoring of regional blood flow and glucose utilization, has focussed interest on the role of the frontal lobes in schizophrenia (see Buchsbaum, 1987; Weinberger, 1988). These changes, however, have not been shown to be specific for schizophrenia, nor do they discriminate this syndrome from other psychiatric or neurological disorders; thus they cannot be employed for diagnostic purposes. The potential of these techniques is not yet fully explored, however, and they will continue to provide details of morphological and functional changes associated with schizophrenia, and in the study of drug-induced changes of brain function. Perhaps the most suggestive contribution, so far, has been the interesting link found between ventricular enlargement in schizophrenia and negative symptomatology (Johnstone *et al.*, 1976; see Crow, 1980). The general distinction between negative and positive symptoms dates back to the writings of Hughlings-Jackson (1931), and denotes the difference between loss of function, as compared to exaggeration or distortion of normal functions. It is interesting to note that the positive symptoms of schizophrenia, such as hallucinations, are sensitive to presently used antipsychotic drugs, whereas negative symptoms are not (see Lyon, this vol., Ch. 11, and Section 12.4 below).

12.1.3 Drugs used in the treatment of schizophrenia

Compounds presently used in the treatment of schizophrenia can be subdivided into five major groups: (1) *phenothiazines*, represented by chlorpromazine and thioridazine; (2) *thioxanthenes*, represented by chlorprothixene; (3) *butyrophenones* and *diphenylbutylpiperidines*, represented by haloperidol and pimozide; (4) substituted *benzamides*, represented by sulpiride and, most recently, remoxipride and raclopride; and finally (5) *dibenzodiazepines* represented by clozapine (Fig. 12.1). The latter two groups have sometimes been called atypical antipsychotics, because of their lesser liability to produce severe extrapyramidal side effects, possibly due to a selective action on the mesolimbic DA system (see e.g. Tamminga & Gerlach, 1987; Bunney *et al.*, 1987).

There is increasing awareness of differences in the receptor affinity profile of antipsychotic compounds, both within the DA domain, and at other types of receptors, and the implications of this for future drug development will be discussed briefly below (section 12.4). There have also been great efforts to understand structure–activity relationships of these, and other compounds, as regards antipsychotic efficacy (e.g. Chretien *et al.*, 1985; Buydens *et al.*, 1986). Suffice to say that the structural requirements for a selective antipsychotic compound are not known, and the evaluation depends heavily on effects of potential compounds in biological test systems, the topic of the present chapter. All antipsychotics block brain DA neurotransmission, and their close relationship is illustrated by the fact that the prototype compound chlorpromazine, after more than 30 years of intensive drug development, is still in use.

Chlorpromazine, in common with other compounds used initially, not only inhibits brain DA receptors, but also possesses anticholinergic, antiadrenergic and antihistamine properties (see e.g. Sack, 1977). However, all these latter effects have been excluded in later developments (e.g. pimozide or remoxipride), with retained antipsychotic efficacy. Furthermore, receptor binding studies have shown a very good correlation between clinical potency and potency to bind to DA D_2 receptors (see Seeman, 1980). Patient heterogeneity, however, precludes clear distinctions, and the

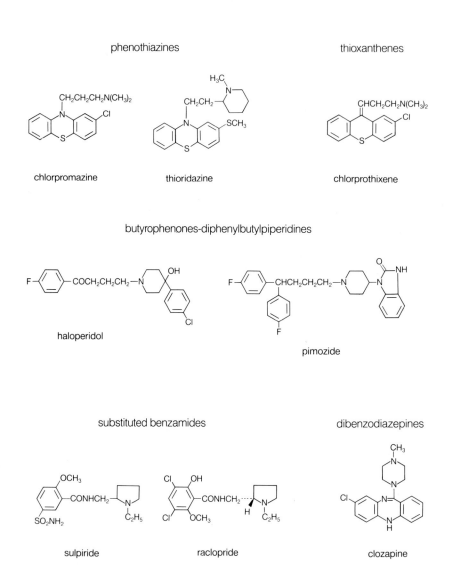

Fig. 12.1. Major groups of antipsychotic drugs with some typical examples.

antiadrenergic component in particular, as for instance in thioridazine, appears to be of advantage in some patients (see Carlsson, 1978).

There have been several contestants to the DA receptor blocking agents, but none have gained general acceptance. These alternatives include the α_2-receptor agonist clonidine and the β-receptor blocker propranolol (see van Kammen & Gelernter, 1987). Treatment with lithium has also been

suggested, especially as an adjunct to traditional antipsychotics – a finding that also applies to propranolol. To this list can be added benzodiazepines and methadone (see Rifkin & Siris, 1987). Although some of these drugs may provide leads to more specific treatment of schizophrenia, it is notable that little progress has been made, although all these compounds have been available for some 20 years. In fact, the DA hypothesis has so far withstood all assaults from outside forces. At this point, we have to conclude that a blockade of brain DA receptors is symptomatically effective in the treatment of schizophrenia, and all drug evaluation so far has been based on this assumption. We shall return to how this assumption affects the interpretation of preclinical experiments. Few, if any, alternatives to this approach are available, and those that have been proposed have seen little use in screening programs. For discussion of efforts in this direction, see Lyon (this vol., Ch. 11).

12.2 VALIDITY OF LABORATORY OBSERVATIONS

12.2.1 Predictive validity

On the assumption of the involvement of brain DA in schizophrenia, a number of critical tests are available with good predictive validity. These critical tests are the topic of the present chapter, and will be presented below (section 12.3). Test models that can select the DA receptor blocking properties of new compounds reliably predict antipsychotic activity, and if this was the only requirement the task would not be too difficult. In addition, however, test models predicting side effects must be included, and pharmacokinetics and toxicity must be considered. These additional considerations, however, do not change the possibility of making a reasonable prediction of the clinical utility of a new compound. As will be discussed below, a number of tests on major side effects are available, with good predictive validity. This applies particularly to side effects related to the DA receptor blocking properties of a new compound. In fact, in relation to these properties of a new compound, the test situations for minimizing side effects are more accurate and reliable than tests for the main effect. This may even lead to the conclusion that what we have done, up to this point, is to reproduce chlorpromazine, stripped of side effects.

Generally, the predictive validity of a test situation can be based on empirical observations, i.e. the pharmacological profile of a particular drug, say chlorpromazine, can be defined in whole animals or in tissue preparations, and this profile, if sufficiently detailed, can be used to produce new compounds of the same kind (see e.g. Boren, 1966;

Kelleher & Morse, 1968). This approach is somewhat outdated, however, and today predictive validity is primarily based on *mechanisms*, i.e. the supposed mechanism of action *in vivo* is defined, and is used to select new compounds. The best-known example in the present context is perhaps the antagonism by antipsychotics of amphetamine-induced stereotypies or hyperlocomotion in animals. Since amphetamine has been shown to enhance the release of catecholamines, and DA in particular (see Carlsson, 1970), such tests can be used to study antidopaminergic effects of putative antipsychotic compounds. Unfortunately, the interpretation of this test is clouded by the early observations that amphetamine can produce symptoms of acute schizophrenia in people (cf. section 12.1.2 above), and it should be noted that it primarily provides information on possible antidopaminergic actions of the compound under investigation. At the same time, however, this will also aid the understanding of behavioral effects maintained by this mechanism, and ties behavioral and biochemical information together in the evaluation program. The link between selective changes in the synaptic machinery of DA and other neurotransmitters produced by specific chemical tools, together with detailed behavioral observations, defines a very viable field of psychopharmacology of great interest in the present context (see e.g. Seiden & Dykstra, 1977; Iversen & Iversen, 1981; Mason, 1984). As will be discussed below, this approach has given new insights into the machinery of DA synapses and has opened entirely new possibilities, within the boundaries of the DA hypothesis of schizophrenia.

12.2.2 Face validity

Ambitious and promising attempts to achieve face validity have been made (see Lyon, this vol., Ch. 11), but they are most likely to succeed if we limit ourselves to particular symptoms of schizophrenia (cf. Henn & McKinney, 1987). It is important to be clear about the fact that, at present, no animal models of schizophrenia are widely accepted at this level of validity (see also Iversen, 1987). This brings us to a different but related question, the question of resolution in observations of animal behavior. In order to make predictions from animal behavior to humans we need a more thorough knowledge of *the neurology of behavior*, i.e. the animal (usually the Norwegian rat) as a sensory-motor organism, since all derived measures of assumed cognitive behavior of an animal are dependent on knowledge of effects on sensory mechanisms, as well as on motor mechanisms needed for the expression of behavior. When this information is available we can make behavioral observations of animals based on sound ethological knowledge of the needs and motivation of the particular species (see Maier & Schneirla, 1964; Lorenz, 1974).

It may, at this point, be appropriate to outline how we can make any predictions at all from observations of animal behavior. On the evolutionary scale we are close relatives even to a rat. We have the same basic sensory-motor build-up, the same brain structures, although greatly distorted (some parts expand others diminish in relative importance). Given these general similarities, the question, as regards the DA hypothesis of schizophrenia, becomes that of establishing where DA is located within the rat brain, and what functions these areas subserve: we have not yet arrived at this point. These *homologue* functions can be used to predict something about human behavior. To be specific, on the basis of human clinical and laboratory findings we assume that the limbic forebrain and the prefrontal cortical functions are disturbed in schizophrenia (see e.g. Mesulam & Geschwind, 1978; Sulkowski, 1983; Weinberger, 1988). Let us find out what functions DA in limbic and cortical areas subserve in the rat. Such functions may be something quite 'infrahuman' in character like 'grooming', 'hoarding behavior' or the like. The importance of this approach, however, is only now being fully realized, and it is only too obvious that in most cases *analogy* of function has been the guiding principle. For further discussions on animal models of schizophrenia, and the semantics of the issue, see McKinney & Moran (1981), Kornetsky & Markowitz (1978), and Lyon (this vol., Ch. 11).

12.2.3 Construct validity

At this level of construct validity we immediately come to the question of the nosology of schizophrenia. The possibility and limitations of animal observations have been outlined above. Thus, it is clear that the present knowledge of animal behav-ior does not provide a foundation for a test situation with construct validity. However, limitations in the diagnosis of schizophrenia alone are limitation enough (see e.g. Andreasen, 1987). This sounds like discouraging any attempt to produce a theory of schizophrenia for this purpose. It is not, but we have to realize that such attempts have not met with success and, as explained above, it is not possible to give the DA hypothesis of schizophrenia status at this level of validity.

Needless to say, validity cannot be claimed or proven by a theoretical analysis of animal behavior. In the present context, the concept implies that a direct comparison is made with the ability to predict clinical success. In this sense 'predictive validity' is of proven value. However, we can claim face or construct validity if, and only if, by use of such models we can predict clinical efficacy of treatments not related to presently known mechanisms. We have not yet reached this point, and I would again like to stress the importance of considering the distinction between homologue and analogue functions in continued work towards this end.

12.3 GENERAL PROCEDURES IN THE LABORATORY EVALUATION OF NEW DOPAMINERGIC ANTIPSYCHOTIC DRUGS

At this point, after all limitations and qualifications raised above, we may ask whether there really is any need for further drug development in this area: we have produced 'super-chlorproma-zines', but after more than 30 years the original is still in use. The answer is clearly yes. Presently used compounds have limited efficacy, particularly when it comes to negative symptoms. Side effects alone are a reason: acute and late extrapyramidal side effects are severe, and may be troublesome to treat. There are also endocrine disturbances linked to the dopaminergic effects of antipsychotic compounds. It may be well to remember that Albert's classic textbook of pharmacology is called *Selective Toxicity* (Albert, 1965): the clinical effect is usually bought at a price. Presently, however, this price is increasingly being considered too high, and control of side effects would be no less than a revolution in this group of frequently used drugs. Finally, we cannot exclude the exciting possibility that there may be drugs acting in the CNS at levels other than DA, though unfortunately these drugs would be picked up by few, if any, of the screening procedures outlined here. It is important to keep in mind that most drugs originally used in the treatment of mental disorders, such as chlorpromazine, imipramine and iproniazide, were detected serendipitously by skilled clinical observations, followed by systematic laboratory investigations and drug development (see e.g. Parnham & Bruinvels, 1983).

12.3.1 Initial screening

For the reasons outlined above, the initial screening for antipsychotic activity aims to detect a blockade of brain DA receptors. This can be tested by the antagonism of apomorphine-induced emesis in the dog, or antagonism of apomorphine-induced stereotypies in rodents (see section 12.3.4.2). A biochemical test on brain DA synthesis, which is receptor regulated, can in most cases more readily provide the same information (see Anden, 1980). Structure–activity relationships are of great importance at this stage, and we have to decide on critical structures or group of compounds. We need great capacity at the expense of resolution of the possible behavioral and biochemical actions of the drug under evaluation. A basic requirement is that there must be no false negatives. Occasional false positives are a minor problem since they will be detected easily in the continued evaluation.

12.3.2 Pharmacological characterization of mechanism of action

To characterize the mechanism of action may sound like a contradiction. From all that has been said above it is evident that we have not only adopted a DA hypothesis of schizophrenia, but also put on colored glasses affecting our initial screening. Furthermore, it has been said that other drugs probably will not be detected! The fact is that the initial screening as regards mechanism of action is only the beginning. The initial screening procedures are of great capacity, but have poor resolution. The next step is to improve the resolution as regards mechanism of action on brain DA neurotransmission.

In the initial screening, we may have looked at

50 compounds, four of which clearly have the defined profile. It is now of paramount importance to know the dose–effect curve on some relevant measures, in comparison with relevant reference compounds. This will allow us to make an estimate of the *potency* of the compound, usually given as an ED_{50} value, and to assess maximal effect i.e. *intrinsic activity*. It has been mentioned in passing above that brain DA synthesis is receptor regulated. Blockade of forebrain DA receptors gives rise to a marked compensatory increase in synthesis whereas stimulation of these receptors decreases synthesis (see Anden, 1980). This observation can now be used in the further evaluation, for example by measuring the accumulation of DOPA after inhibition of aromatic amino acid decarboxylase (e.g. by the hydrazine compound NSD-1015), which gives a good estimate of tyrosine hydroxylase activity (Carlsson *et al.*, 1972) – the rate-limiting step in the catecholamine synthesis (Udenfriend, 1966). Behavioral measures like antagonism of apomorphine-induced emesis in the dog, or antagonism of amphetamine-induced stereotypies in rodents can also be used for this purpose (Niemegeers, 1971; Niemegeers & Janssen, 1979). It is important to keep in mind at this stage that the use of whole animals does not necessarily imply anything beyond the level of a bioassay. Biochemical indices will give us an estimate of effects on DA nerve impulse flow (disappearance of catecholamines after inhibition of syntheses) (Anden *et al.*, 1969), turnover (monoamines plus metabolites) and release (relation between different metabolic pathways intra- and extraneuronally). This part of the evaluation program, straightforward but never simple, is further complicated by the existence of presynaptic autoreceptors on the DA neuron, as well as subtypes of postsynaptic DA receptors (see Carlsson, 1975; Bunney *et al.*, 1987; Creese, 1987). Depending on the type of compound under study it may be considered more or less important to study its relative effects pre- and post-synaptically. Furthermore, the distinction between the D_1 and D_2 DA receptor subtypes is recognized as increasingly important. The D_1, but not the D_2, receptor is linked to adenylate cyclase (see Kebabian & Calne, 1979). However, most behavioral effects are linked to the D_2 receptor subtype, and the clinical implications of this distinction are not yet clear. This also applies to the pre- and postsynaptic distinc-

tion. These problems, however, are now intensively investigated, and this may result in new and more selective agents than presently used. This perspective applies at least as much to pharmacotherapy in Parkinson's disease as the schizophrenia (Carlsson, 1983; Svensson *et al.*, 1986). An example of this is the Boehringer compound B-HT 920. Originally characterized as an autoreceptor selective DA agonist (Anden *et al.*, 1982), continued studies with this drug have shown it to have selective agonist properties at postsynaptic D_2 receptors (Anden & Grabowska-Anden, 1987; Pifl & Hornykiewicz, 1988). Thus, this compound theoretically is a potential antipsychotic and antiparkinsonian drug at the same time. Clinical research will show us whether these new distinctions of brain DA receptors will lead to more precise pharmacological treatments in psychiatry and neurology.

The aim of the evaluation program, at this stage, is to characterize the mechanism of action at the present state of knowledge of brain DA neurotransmission: this is a basic requirement, necessary for a sound understanding of biological actions and interactions in animals and humans. The menu of tests may be slightly modified depending on type of compound(s) under study. In addition to understanding its mechanism of action on brain DA mechanisms, some positive controls are needed, i.e. little or no effect on other monoamines like noradrenaline (NA) or serotonin (5-HT). It is also important that there is reasonable confidence in the mechanism of action since further activities, as detailed below (sections 12.3.3, 12.3.4), are a much greater investment in time and money. Furthermore, the continued evaluation is not necessarily decisive as regards mechanisms of action.

Needless to say, if a new type of compound is under study, it is highly probable that there will also be effects in a biological system, other than those on DA mechanisms. Continued studies of chlorpromazine, haloperidol, and other antipsychotics revealed new biological actions long after they had an established role in psychiatry.

12.3.3 Safety and pharmacokinetics

Before we enter into further examination of the pharmacological profile, it is important to have a rough estimate of possible toxicity. This can include an Ames test for carcinogenicity (see Bartsch & Malaveille, 1987), and considerations of chemical structure, if the toxicity of similar compounds is known. Gross observations of animals for obvious neurological side effects are desirable, and an LD_{50} test may be carried out at this stage. It is important to note that, although such determinations should not be done routinely, as they may have been in the past, information on the safety of the compound, including lethal doses, is literally of vital importance. We also need to know its half-life, its bioavailability, and possible active metabolites. With the type of compounds under investigation, there will be effects on the circulation, which will require elucidation. Poor properties in any of these respects do not necessarily stop the evaluation, but have to be considered seriously in relation to results in further test situations. Among other things, we shall have to assess a 'therapeutic ratio', by comparing the potency of the compound in various biological tests with its safety and pharmacokinetics.

12.3.4 Pharmacodynamics

What follows is a mixture of tests, some of which were at least at some point considered to possess face validity. As discussed above this can now be seriously questioned. However, the frequent use of some of these standard laboratory tests makes them, at least empirically, useful (see also Costall & Naylor, 1980). As far as test situations regarding neurological side effects are concerned, the connections to human behavior are more obvious, as will be described below. Most of the tests presented justify themselves because they were used in the evaluation of presently available antipsychotic drugs. Thus, negative results are also of interest, and the menu presented below can best be considered as a matrix or fingerprint of classic antipsychotics. This also implies that new and entirely different antipsychotics would at present have to pass at least portions of this 5,000 m hurdles race.

12.3.4.1 Conditioned avoidance behavior

The early observation that antipsychotic compounds selectively suppressed avoidance behavior was taken as an index of their special sedative properties (Courvoisier, 1956). Typically a rat is trained to avoid an electric shock on the grid floor of a shuttle-box. In this situation chlorpromazine was found to dissociate the avoidance and escape behavior – the former was lost and the latter remained intact; barbiturates or benzodiazepines affected both components similarly (Cook & Weidley, 1957; Verhave *et al.*, 1958). These initial observations were made before the DA receptor blocking properties of antipsychotic compounds were known. Extensive experimentation on the role of brain DA in animal behavior has linked the loss of avoidance to a sensory-motor deficit, rather than a cognitive deficit. Thus, the animals know what to do, but cannot initiate the behavior until pushed by strong stimulation like an electric shock (Fibiger *et al.*, 1975; see Ahlenius & Archer, 1985). The comparison with Parkinson's disease is not farfetched, and several studies suggest that this test situation is more a test for extrapyramidal side effects than anything else (see Ahlenius, 1979). For example, the antipsychotic-induced avoidance deficit can be antagonized by anticholinergics (Morpurgo & Theobald, 1964; Hanson *et al.*, 1970) and cholinomimetics produce a loss of avoidance behavior (Goldberg *et al.*, 1965). Parkinson's disease can be treated with anticholinergics, and extrapyramidal side effects produced by antipsychotics are sensitive to anticholinergic treatment without loss of antipsychotic efficacy (Brune *et al.*, 1962; McEvoy, 1983). This latter observation is of particular interest since in laboratory experiments it has been shown that the effects of DA-receptor blocking antipsychotics on neostriatal DA turnover, but not limbic DA turnover, are antagonized by anticholinergics (Anden, 1972). Thus, anticholinergic agents may dissociate neostriatal extrapyramidal functions from limbic brain functions. As regards side effects, more simple and direct tests are available (see below), and the interesting possibility that the conditioned avoidance test can be used to predict the antipsychotic potential of new compounds remains to be clarified.

12.3.4.2 Antagonism of apomorphine- or amphetamine-induced behavior in rodents

Both apomorphine and *d*-amphetamine produce a stimulation, albeit by different mechanisms, of brain DA receptors (e.g. Anden, 1970; Carlsson, 1970), leading to an increase in locomotion and stereotyped behavior in rats and mice. These effects are easily observed by use of activity meters or by scoring of degree of stereotypy (e.g. Niemegeers & Janssen, 1979). Initially, the association of psychotic-like behavior produced by amphetamine, and effects on animal behavior, motivated the use of this test in the evaluation of antipsychotic compounds. The clinical correlation is unclear, and has faded over the years, but the test remains a simple and good estimate of antidopaminergic effects.

12.3.4.3 Spontaneous locomotor activity

Observations of exploratory locomotor activity in an open field have often been used in the evaluation of potential antipsychotic drugs. Available antipsychotic compounds suppress the exploratory locomotor activity (e.g. Marriott & Spencer, 1965; Watzman *et al.*, 1967), and this suppression, unlike suppression of treadmill locomotion, is not antagonized by treatment with an anticholinergic agent (Ahlenius & Hillegaart, 1986; Hillegaart & Ahlenius, 1987). We have already speculated above that limbic forebrain mechanisms may be involved in the development of schizophrenia. The suppression of exploratory behavior may represent a relevant function in the rat, since the limbic forebrain has been shown to be an important site for this effect (e.g. Ahlenius *et al.*, 1987). It should be noted, however, that the open field or other activity meters provide neither specific nor selective tests in the evaluation of antipsychotic compounds. However, these tests are of great value together with other information.

Various locomotor activity boxes have been constructed to provide detailed information on the spontaneous locomotor activity (see e.g. Ljungberg & Ungerstedt, 1978). Two behavior items often recorded are horizontal ('locomotion') and vertical ('rearing') activity. Interestingly, in support of this empirical distinction, these two types of activity, horizontal and vertical, were identified by principal component analysis of the effects of apomorphine on ten different items of behavior in an open field test (Ståhle & Ungerstedt, 1986).

12.3.4.4 Temperature regulation

Administration of DA agonists produces hypothermia. Paradoxically, the same effect can be seen after administration of antipsychotic drugs. This latter effect, however, is dependent on ambient temperature, and is due to a poikothermic effect of these agents (Clark & Lipton, 1985; Lee *et al.*, 1985). This is clearly not a screening test, but requires some consideration, since dopaminergic effects on temperature regulation may produce serious side effects (Clark & Lipton, 1984).

12.3.4.5 *Catalepsy and treadmill locomotion*

Most, or all, antipsychotics produce catalepsy and loss of treadmill locomotion at high doses, and these effects are clearly related to their propensity to produce extrapyramidal side effects (e.g. Morpurgo, 1965). This is a pivotal point in the evaluation of new DA receptor blocking antipsychotics, at which they must demonstrate an edge over existing drug therapies in order to justify further investigation. The extrapyramidal side effects of antipsychotics are disabling, not least the late-developing tardive dyskinesias (Baldessarini, 1980). It was mentioned above that anticholinergics can antagonize antipsychotic-induced Parkinsonism. Catalepsy and loss of treadmill locomotion, which provide an animal model for Parkinson's disease, are antagonized by anticholinergics (e.g. Morpurgo, 1965; Hillegaart & Ahlenius, 1987) and, if the extrapyramidal effects were induced by reserpine, by the administration of DA agonists or the precursor L-DOPA (Carlsson, 1959). In further support of a neostriatal localization of extrapyramidal effects, it has been shown that amphetamine-induced stereotypies are abolished by neostriatal 6-OHDA lesions, whereas the increase in locomotion produced by *d*-amphetamine was abolished by lesions in the nucleus accumbens (Kelly *et al.*, 1975).

There are almost as many different types of apparatus for testing catalepsy and treadmill locomotion as there are laboratories. This will inevitably lead to different results being recorded by different groups, and underscores the importance of including proper reference controls (cf. Krsiak & Janku, 1971). Since the purpose is not only to observe type and severity of the effect, but also to calculate ED_{50} values, *reliability*, rather than validity, is of primary concern. This is an example of the difference between biological and physical measurements. We can build instruments

for estimating and comparing temperature and air pressure in Dunedin and Irkutsk, provided the instruments have been constructed according to specification. This is not so with the measurements under consideration here. Theoretically, all variables of importance, if known, could be controlled, like species, strain, sex, age, light-dark conditions, environmental stress (sound, smell, illumination etc.) but this is not practicable. As but one example of the methodological pitfalls, we have shown recently that repeated testing of animals affects later treadmill performance or degree of catalepsy (Hillegaart *et al.*, 1987, 1988). This finding alone has strong implications for the interpretation of time–effect relationships, and drug interactions. Consequently, the apparatus and procedures must be clearly specified, and proper controls must be included.

12.3.4.6 *Seizure thresholds*

It is well known that brain DA can influence seizure thresholds (see Meynert *et al.*, 1975). It appears that stimulation of brain DA receptors provides some protection, whereas a blockade of these receptors may produce a proconvulsive effect (e.g. Criborn *et al.*, 1986, 1987, 1988). This was evident from the very beginning of the use of DA-receptor blocking antipsychotics, as convulsions were often mentioned as a side effect, particularly with the phenothiazines (see Kalinowsky & Hippius, 1969). Many antipsychotics produce EEG discharge patterns associated with epileptic seizures (see Itil, 1978), and these effects may interact both with individual predisposition and with concomitant drug treatments. Thus, it is important to investigate possible effects of a new drug on seizure thresholds in a suitable model (e.g. Meynert *et al.*, 1975).

12.3.4.7 *Prolactin secretion*

Effects on prolactin secretion are both a useful index of antidopaminergic activity as well as of an endocrine side effect. DA in the tuberoinfundibular system appears to have an important endocrine role as the prolactin inhibitory factor (PIF) (Fuxe & Hökfelt, 1969). Thus, a blockade of brain DA receptors, or depletion of its intraneuronal stores by reserpine or *α*-methyltyrosine, produces a marked increase in plasma prolactin levels (Meltzer *et al.*, 1978). Side effects related to this include amenorrhea, galactorrhea and gynecomasty. At the same time prolactin secretion has been used as an index of DA activity after treatment with antipsychotics in schizophrenia (see Rubin, 1987; Meltzer & Busch, 1983). The measurement of effects on plasma prolactin levels is an indispensible part of the evaluation program for antipsychotic drugs.

12.3.4.8 Emesis

DA agonists produce emesis in a number of species, including man, and DA receptor blocking agents are used as antiemetics. The area postrema in the floor of the 4th ventricle has been suggested as a possible site for this effect (Hatcher & Weiss, 1923; Shapiro & Miselin, 1985). Antagonism of apomorphine-induced emesis in dogs has also been used as a screening procedure in the evaluation of new antipsychotics (see Niemegeers & Janssen, 1979).

12.3.4.9 Intracranial selfstimulation (ICSS)

Rats with electrodes implanted in the medial forebrain bundle can learn to bar-press with a weak electric stimulation via the electrode as reward, and this behavior has been shown to depend on intact DA functions for its motor execution (Breese & Cooper, 1975; Wauquier & Niemegeers, 1972; Mason *et al.*, 1980). Interestingly, there is also the possibility that DA pathways mediate the reward properties, and various experimental observations have led to the formulation of an anhedonia hypothesis of antipsychotic action (Wise *et al.*, 1978; Wise, 1982; Beninger, 1983). In the present context, however, we have entered a gray-zone as regards utility in the evaluation of antipsychotic drugs. As a test for DA receptor blockade, although useful, ICSS is clearly too complicated a phenomenon, and the validity of the test depends on theoretical speculation, which is still in need of validation (see also Lyon, this vol., Ch. 11, for further discussion).

12.3.4.10 Receptor binding studies

In vitro and *in vivo* receptor binding techniques have gained a prominent place in classifying DA receptors, and in antipsychotic drug evaluation (Seeman, 1980; Boyson *et al.*, 1986; Creese, 1987). The value of these techniques is greater, however, in a late phase of the evaluation because intrinsic activity and biological effects are not clearly related to binding behavior of drugs. *In vitro* binding is most useful when we have decided on a particular compound in our initial evaluation (cf. 12.3.1 above), and now need to know with which other kinds of receptors the compound interacts. Intrinsic activity is not a primary concern at this stage, but will be considered in later experiments depending on the particular pattern of results obtained. *In vivo* autoradiography gives a good illustration of the distribution of the compound to different receptor populations; this technique is qualitative, and the poor relationship between function and receptor binding in animal studies limits the value of this approach (Staunton *et al.*, 1981). Maybe the most exciting application of *in vivo* binding is the visualization of DA receptors in the living human brain by positron emission tomography (see Farde *et al.*, 1986).

12.3.4.11 Single unit activity

Electrophysiological techniques allow the recording of firing rate and patterns of firing rates from DA cell bodies in the mesencephalon (Bunney & Aghajanian, 1978). Changes in activity can be monitored after systemic drug administration, and it is also possible to make direct iontophoretic application of a compound onto cell bodies. This technique is not commonly used in the early evaluation of antipsychotic drugs, but provides unique information of DA neurotransmission, not easily estimated by other means. Thus, interesting details on the mesolimbic selectivity of some of the socalled atypical antipsychotics have been disclosed. Whereas all antipsychotics appear to activate both nigrostriatal and mesolimbic DA turnover and firing rate on acute administration to rats (Westerink, 1978; Chiodo & Bunney, 1983), on chronic administration the atypical antipsychotic clozapine was found selectively to produce a depolarization inactivation of mesolimbic (ventral tegmental area) neurons only (Chiodo & Bunney, 1983; White & Wang, 1983; see Bunney *et al.*, 1987). It is interesting to note that this particular profile of clozapine may be due to its anticholinergic and antiadrenergic properties, since haloperidol in combination with either the anticholinergic trihexphenidyl or the α-adrenergic antagonist prazosin, produced a clozapine-like profile in similar experiments (Chiodo & Bunney, 1985). It is generally assumed that mesolimbic selectivity is a desirable goal and electrophysiological experiments may provide important information in this regard.

12.3.5 Concluding remarks

The pharmacodynamics of existing antipsychotic compounds have been thoroughly investigated in a large variety of tests, and in particular those described under sections 12.3.4.1 to 12.3.4.8. Thus, it becomes important to include these tests in the evaluation program, if for no other purpose than to allow comparisons between candidate compounds and established drugs. The purpose of testing so far is mechanistic and descriptive – to understand the mechanism of action, and to describe the pharmacological profile regardless of mechanism of action. All this is examined after acute administration of the test compound at suitable doses. We also need to study behavioral and biochemical effects after subchronic or chronic treatment (Gianutsos & Moore, 1977; Clow *et al.*, 1980). Usually an easily observable behavioral response, like spontaneous locomotion, and some biochemical measures of brain DA metabolites are selected. The aim is to assess changes in DA *receptor sensitivity* after long-term antipsychotic treatment. Up-regulation of brain DA receptors following administration of the neurotoxin 6-OHDA is well known (Ungerstedt, 1971; Moore & Goodale, 1975), and provides a model for studying the behavior of brain DA receptors after prolonged inactivation of DA neurotransmission. Usually animals are given a standard dose of apomorphine, and the effects obtained are compared with the behavioral response in animals withdrawn from long-term drug treatment. Conversely, continuous infusion of DA into the nucleus accumbens of the rat produces persistent changes in the response to a challenge dose of a DA agonist, and this model has been used to study effects of putative antipsychotics (Costall *et al.*, 1983, 1984). Interesting, but not easily explained, are the phasic effects on locomotor activity pro-

duced by continuous intra-accumbens infusion with peaks around days 4 and 11 of a 14-day infusion schedule, due perhaps to oscillations in receptor sensitivity and feedback mechanisms. Possibly, this model could be of value in studying the mechanisms behind time dependent phenomena in the onset of clinical effects and their persistence for a period of time upon withdrawal.

In parallel with the program outlined in this section, an extensive program on the general pharmacology and pharmacokinetics of the test compound would have been initiated. As mentioned above, poor performance in these respects does not immediately stop the evaluation. Gradually, however, a picture emerges from which we can make estimates of effects possibly related to clinical use – side effects, cardiovascular effects, metabolism, distribution and toxicity – which allow us to pass judgement on the clinical potential of our new compound. The toxicity and safety evaluation is a distinct procedure which usually starts at the end of the program outlined above. It is also physically separated, in the sense that safety evaluation is performed in a separate laboratory under strict control, as required by health authorities.

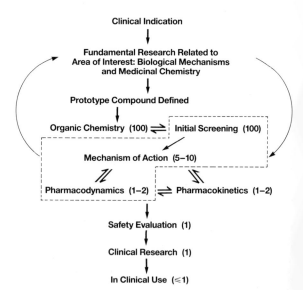

Fig. 12.2. A general scheme of drug development. The numbers within parentheses indicate approx. the relative numbers of compounds examined at each stage. The scope of the present chapter is enclosed by the Broken line.

The investment, in terms of time and money, of this process only underscores the importance of a distinct and decisive pharmacological evaluation. The general procedure of drug evaluation is schematically shown in Figure 12.2, which also indicates the scope of the present chapter.

12.4 LIMITATIONS AND POSSIBILITIES

The strength and limitation of the program outlined above is its dependence on the DA hypothesis of schizophrenia. Let us start by considering the strength and possibilities of this approach. In the first place, it allows us to formulate an evaluation program with reasonable predictive validity. Secondly, new evidence on the dynamics of brain DA neurotransmission offers new possibilities of pharmacological intervention. These possibilities are not merely cosmetic, and promise new avenues in the treatment of schizophrenia. Two areas are currently in focus: (1) the distinction between D_1 and D_2 brain DA receptors (Kebabian & Calne, 1979; Stoof & Kebabian, 1984); and (2) the existence of presynaptic DA autoreceptors (Carlsson, 1975). It is only recently that selective tools in these respects have been developed. Thus, SK&F 38,393 and SCH 23,390 are examples of a typical agonist (Setler *et al.*, 1978) and antagonist (Iorio *et al.*, 1983), respectively, at the D_1 receptor. The corresponding agonists and antagonist at the D_2 receptor are quinpirole (Titus *et al.*, 1983) and (−)sulpiride (Spano *et al.*, 1978). Furthermore, agonists and antagonists with preference for the autoreceptor, have been developed, represented by (−)3-PPP (Hjorth *et al.*, 1983) and (+)AJ-76 (Svensson *et al.*, 1986), respectively. Needless to say, selective in this context simply means selectivity as regards the particular mechanism under consideration, i.e., a selective D_1 DA agonist should have no detectable effects at the D_2 receptor. This is well to keep in mind, since there may be unexpected effects of any of these compounds on other neurochemical systems. The therapeutic potential of these new distinctions is only now being explored. The hope is that new compounds, with more selective actions on brain DA neurotransmission, will be more selective in their functional effects. In particular, a better separation is sought between antipsychotic effects and extrapyramidal side effects. Further-

more, there is also the interesting possibility that new drugs will be found with effects also on negative symptoms in schizophrenia. Although outside the scope of the present chapter, it should be mentioned that new avenues of pharmacological intervention in brain DA neurotransmission may have strong implications for drug treatments also in other conditions, like Parkinson's disease, Huntington's chorea and Gilles de la Tourette's syndrome. Some of the compounds mentioned above are now in early clinical trials.

The first task, in the present context, is to describe how classic antipsychotic compounds behave in these regards. The second is to evaluate whether more selective agents are more promising in terms of relationships between indices of antipsychotic effect and of side effects. At the moment several compounds acting as agonists at inhibitory DA autoreceptors are being explored preclinically; these include (−)3-PPP (Hjorth *et al.*, 1983), terguride (Kehr, 1984), and B-HT 920 (Anden *et al.*, 1982). Activation of inhibitory autoreceptors would be expected to result in a decrease in DA cell firing and DA release, leading to an inhibition of DA neurotransmission, i.e. a net effect similar to that obtained by blockade of postsynaptic DA receptors. This difference in synaptic site of action is perhaps most interesting because compensatory feedback mechanisms operating at the synaptic level would be affected differently. A postsynaptic receptor blockade is compensated for by an increase in DA synthesis and release (to a large extent caused by a concurrent blockade of presynaptic receptors), whereas this compensation would not be expected following an activation of presynaptic DA autoreceptors. Inhibition of tyrosine hydroxylase, which blocks the compensatory increase in DA synthesis, markedly enhances the behavioral effects of haloperidol and other receptor blocking antipsychotics in animals and people (Ahlenius & Engel, 1971, 1977; Wålinder *et al.*, 1976), suggesting that absence of compensatory feedback effects following autoreceptor stimulation may have important functional consequences.

The limitations of the DA-based approach are maybe too obvious to need further discussion. There is no doubt that DA receptor blockade is beneficial in schizophrenia, but DA is not necessarily critically involved in the etiology of the disease process. However, to find biological mechanisms outside the DA hypothesis we need animal models of a different kind (see section 12.2, above, and Lyon, this vol., Ch. 11). In order to make comparisons of homologue functions in people and animals, as regards brain areas, and neurotransmitters, involved in the development of schizophrenia, we also need more information on nosology and biological markers of the disease in man. If we are looking for analogous functions, a personal guess would be to use simple sensory–motor mechanisms, like attention and reaction time, where differences have been found between schizophrenic patients and normal controls (Kietzman, 1985; Geyer & Braff, 1987). This is possibly an area where we can meet our animal relatives on common ground.

12.5 CLINICAL EVALUATION

Following a preclinical pharmacological evaluation as outlined above, and a detailed toxicological investigation in different animal species, the drug is investigated in people in three distinct phases. *Phase I* is carried out in healthy volunteers under continuous observation. Information is primarily obtained on human pharmacokinetics, on prolactin and on cardiovascular functions. There is also the interesting possibility of obtaining verbal information on the subjective experience of drug effects. In fact, skilled observations at this stage can provide information not easy, or even impossible, to obtain in laboratory studies. *Phase II* consists of limited open clinical trials, followed by double-blind trials in *Phase III*. These investigations, which are usually carried out by independent medical practitioners, are performed in close contact with the clinical research staff at the pharmaceutical company. It is particularly important to evaluate the clinical profile of the compound, and to establish how this differs from other compounds on the same indication. It is equally important to be alerted to possible side effects, expected as well as unexpected, and not least to have an open mind on other indications for the new compound. If it really is novel in terms of chemical structure and preclinical profile, its potential in people will often not be fully realized until clinical trials have been initiated.

It is beyond the scope of the present chapter to

go into further details of a prolonged and careful clinical evaluation. For other aspects of the drug development see, e.g., Smith (1985). The focus of the present chapter, in this broader perspective, is schematically shown in Fig. 12.2.

12.6 CONCLUSIONS

As described here the development of a new antipsychotic compound presents a fairly rigid schedule of events. Paradoxically, the adherence to such a program is the more necessary since the possibility and necessity of making detours are numerous at each stage, and unexpected findings will demand consideration throughout the preclinical evaluation. The program must also be flexible enough to permit deviations from the determined plan and, in a great tradition, to suggest new and unexpected indications for the compound under evaluation.

For fruitful discussions and constructive criticism on an earlier draft of this manuscript I should like to thank Dr Tommy Lewander, Astra Research Centre, and the editor, Professor Paul Willner who, in a creative mixture of question and praise gave me new views on old antipsychotics! The figures were prepared by Ms Birgitta Pålsson-Stråe, Trosa, Sweden.

REFERENCES

Ahlenius, S. (1979). An analysis of behavioural effects produced by drug-induced changes of dopaminergic neurotransmission in the brain. *Scandinavian Journal of Psychology* **20**, 59–64.

Ahlenius, S. & Archer, T. (1985). Normal and abnormal animal behaviour: Importance of catecholamine release by nerve impulses for the maintenance of normal behaviour. In *Neurobiology*, eds. R. Gilles & J. Balthazart, pp. 329–43. Springer-Verlag: Berlin–Heidelberg.

Ahlenius, S. & Engel, J. (1971). Behavioral effects of haloperidol after tyrosine hydroxylase inhibition. *European Journal of Pharmacology* **15**, 187–92.

Ahlenius, S. & Engel, J. (1977). Potentiation by α-methyltyrosine of the suppression of food-reinforced lever-pressing behaviour induced by antipsychotic drugs. *Acta Pharmacologica et Toxicologica* **40**, 115–25.

Ahlenius, S. & Hillegaart, V. (1986). Involvement of extrapyramidal motor mechanisms in the suppression of locomotor activity by antipsychotic drugs: A comparison between the effects produced by pre- and post-synaptic inhibition of dopaminergic neurotransmission. *Pharmacology, Biochemistry and Behavior* **24**, 1409–15.

Ahlenius, S., Hillegaart, V., Thorell, G., Magnusson, O. & Fowler, C.J. (1987). Suppression of exploratory locomotor activity and increase in dopamine turnover following the local application of cis-flupenthixol into limbic projection areas of the rat striatum. *Brain Research* **402**, 131–8.

Albert, A. (1965). *Selective Toxicity*. The Broadwater Press Ltd: Welwyn Garden City.

Allen, J.S. & Sarich, V.M. (1988). Schizophrenia in an evolutionary perspective. *Perspectives in Biology and Medicine* **32**, 132–53.

Anden, N.-E. (1970). Effects of amphetamine and some other drugs on central catecholamine mechanisms. In *Amphetamines and Related Compounds*, eds. E. Costa & S. Garattini, pp. 447–62. Raven Press: New York.

Anden, N.-E. (1972). Dopamine turnover in the corpus striatum and the limbic system after treatment with neuroleptic and anti-acetylcholine drugs. *Journal of Pharmacy and Pharmacology* **24**, 905–6.

Anden, N.-E. (1980). Regulation of monoamine synthesis and utilization by receptors. In *Handbook of Experimental Pharmacology*, ed. L. Szekeres, pp. 429–62. Springer-Verlag: Berlin–Heidelberg.

Anden, N.-E. & Grabowska-Anden, M. (1987). Increased motor activity following combined stimulation of B-HT 920-sensitive and D-1 dopamine receptors. *Acta Physiologica Scandinavica* **131**, 157–8.

Anden, N.E., Corrodi, H. & Fuxe, K. (1969). Turnover studies using synthesis inhibition. In *Metabolism of Amines in the Brain*, ed. G. Hooper, pp. 38–47. MacMillan: London.

Anden, N.E., Golembiowska-Nikitin, K. & Thornstrom, U. (1982). Selective stimulation of dopamine and noradrenaline autoreceptors by B-HT 920 and B-HT 933, respectively. *Naunyn-Schmiedeberg's Archives of Pharmacology* **321**, 100–4.

Andreasen, N.C. (1987). Schizophrenia: diagnosis and assessment. In *Psychopharmacology: The Third Generation of Progress*, ed. H.Y. Meltzer, pp. 1087–94. Raven Press: New York.

Andreasen, N.C. (1988). Brain imaging: applications in psychiatry. *Science* **239**, 1381–8.

Angrist, B. & Gershon, S. (1977). A clinical response to several dopamine agonists in schizophrenic and nonschizophrenic subjects. *Advances in Biochemical Psychopharmacology* **16**, 677–80.

Angrist, B., Sathananthan, G.S. & Gershon, S. (1973). Behavioral effects of L-dopa in schizophrenic patients. *Psychopharmacology* **31**, 1–12.

Angrist, B., Lee, H.K. & Gershon, S. (1974). The antagonism of amphetamine-induced

symptomatology by a neuroleptic. *American Journal of Psychiatry* **131**, 817–19.

Babigian, H.R. (1980). Schizophrenia: epidemiology. In *Comprehensive Textbook of Psychiatry*, eds. H.I. Kaplan, A.M. Freedman & B.J. Saduk, pp. 1113–21. Williams & Wilkins: Baltimore.

Baldessarini, R.J. (1980). Dopamine and the pathophysiology of dyskinesias induced by antipsychotic drugs. *Annual Review of Neuroscience* **3**, 23–41.

Baldessarini, R.J. (1985). Drugs and the treatment of psychiatric disorders. In *Goodman and Gilman's The Pharmacological Basis of Therapeutics*, eds. A.G. Gilman, L.S. Goodman, T.W. Rall & F. Murad, pp. 387–445. MacMillan Publishing Company: New York.

Barchas, J.D., Berger, P.A., Ciaranello, R.D. & Elliott, G.R. (eds.) (1977). *Psychopharmacology: From Theory to Practice*. Oxford University Press: New York.

Bartsch, H. & Malaveille, C. (1987). The Ames test. *ISI Atlas of Science & Pharmacology* **1**, 1–4.

Beninger, R.J. (1983). The role of dopamine in locomotor activity and learning. *Brain Research Reviews* **6**, 173–96.

Boren, J.J. (1966). The study of drugs with operant techniques. In *Operant Behavior: Areas of Research and Application*, ed. W.K. Honig, pp. 531–64. Appleton-Century-Crofts: New York.

Boyson, S.J., McGonigle, P. & Molinoff, P.B. (1986). Quantitative autoradiographic localization of the D1 and D2 subtypes of dopamine receptors in rat brain. *Journal of Neuroscience* **6**, 3177–88.

Breese, G.R. & Cooper, B.R. (1975). Effects of catecholamine-depleting drugs and amphetamine on self-stimulation obtained from lateral hypothalamus, substantia nigra and locus coeruleus. In *Brain-Stimulation Reward*, eds. A. Wauquier & E.T. Rolls, pp. 190–3. North-Holland: Amsterdam.

Brune, G., Morpurgo, C., Bielkus, A., Kobayashi, T., Tourlentes, T.T. & Himwich, H.E. (1962). Relevance of drug-induced extrapyramidal reactions to behavioral changes during neuroleptic treatment. I. Treatment with trifluoperazine singly and in combination with trihexyphenidyl. *Comprehensive Psychiatry* **3**, 227–34.

Buchsbaum, M.S. (1987). Positron emission tomography in schizophrenia. In *Psychopharmacology: The Third Generation of Progress*, ed. H.Y. Meltzer, pp. 783–92. Raven Press: New York.

Bunney, B.S. & Aghajanian, G.K. (1978). Mesolimbic and mesocortical dopaminergic systems: physiology and pharmacology. In *Psychopharmacology: A Generation of Progress*, eds. M.A. Lipton, A. DiMascio & K.F. Killam, pp. 159–69. Raven Press: New York.

Bunney, B.S., Sesack, S.R. & Silva, N.L. (1987). Midbrain dopaminergic systems: neurophysiology and electrophysiological pharmacology. In *Psychopharmacology: The Third Generation of Progress*, ed. H.Y. Meltzer, pp. 113–26. Raven Press: New York.

Bunney Jr, W.E., van Kammen, D.P., Post, R.M. & Garland, B.L. (1979). A possible role for dopamine in schizophrenia and manic-depressive illness (a review of the evidence). In *Catecholamines: Basic and Clinical Frontiers*, eds. E. Usdin, I.J. Kopin & J. Barchas, pp. 1807–19. Pergamon Press: New York.

Buydens, L., Massart, D.L. & Geerlings, P. (1986). Pharmacological activity of neuroleptic drugs and physiochemical, topological and quantum chemically calculated parameters: a QSAR study. *European Journal of Medicinal Chemistry* **21**, 34–43.

Carlsson, A. (1959). The occurrence, distribution and physiological role of catecholamines in the nervous system. *Pharmacological Reviews* **11**, 490–3.

Carlsson, A. (1965). Drugs which block the storage of 5-hydroxytryptamine and related amines. In *Handbook of Experimental Pharmacology*, eds. O. Eichler & A. Farah, pp. 529–92. Springer-Verlag: Berlin.

Carlsson, A. (1970). Amphetamine and brain catecholamines. In *Amphetamines and Related Compounds*, eds. E. Costa & S. Garattini, pp. 289–300. Raven Press: New York.

Carlsson, A. (1975). Dopaminergic autoreceptors. In *Chemical Tools in Catecholamine Research*, eds. O. Almgren, A. Carlsson & J. Engel, pp. 219–25. North-Holland Publishing Company: Amsterdam.

Carlsson, A. (1978). Antipsychotic drugs, neurotransmitters, and schizophrenia. *American Journal of Psychiatry* **135**, 164–73.

Carlsson, A. (1983). Dopamine receptor agonists: intrinsic activity vs. state of receptor. *Journal of Neural Transmission* **57**, 309–15.

Carlsson, A. & Lindqvist, M. (1963). Effect of chlorpromazine or haloperidol on formation of 3-methoxytyramine and normetanephrine in mouse brain. *Acta Pharmacologica et Toxicologica* **20**, 140–4.

Carlsson, A., Kehr, W., Lindqvist, M., Magnusson, T. & Atack, C.V. (1972). Regulation of monoamine metabolism in the central nervous system. *Pharmacological Reviews* **24**, 371–84.

Chiodo, L.A. & Bunney, B.S. (1983). Typical and atypical neuroleptics: differential effects of chronic administration on the activity of A9 and A10 midbrain dopaminergic neurons. *Journal of Neuroscience* **3**, 1607–19.

Chiodo, L.A. & Bunney, B.S. (1985). Possible mechanisms by which repeated clozapine administration differentially affects the activity of two subpopulations of midbrain dopamine neurons. *Journal of Neuroscience* **5**, 2539–44.

Chretien, J.R., Szymoniuk, J., Dubois, J.E., Poirier, M.F., Garreau, M. & Deniker, P. (1985). Structure–activity relation studies and pharmacological data bank: a DARC method of CAD applied to neuroleptic compounds. *European Journal of Medicinal Chemistry* **20**, 315–25.

Clark, W.G. & Lipton, J.M. (1984). Drug-related heatstroke. *Pharmacology and Therapy* **26**, 345–88.

Clark, W.G. & Lipton, J.M. (1985). Changes in body temperature after administration of amino acids, peptides, dopamine, neuroleptics and related agents: II. *Neuroscience and Biobehavioral Reviews* **9**, 299–371.

Clow, A., Theodorou, A.E., Jenner, P. & Marsden, C.D. (1980). Changes in rat striatal dopamine turnover and receptor activity during one years neuroleptic administration. *European Journal of Pharmacology* **63**, 135–44.

Cook, L. & Weidley, E. (1957). Behavioral effects of some psychopharmacological agents. *Annals of the New York Academy of Sciences* **66**, 740–52.

Costall, B. & Naylor, R. (1980). Assessment of the test procedures used to analyse neuroleptic action. *Reviews of Pure and Applied Pharmacological Sciences* **1**, 3–83.

Costall, B., Domeney, A.M. & Naylor, R.J. (1983). A comparison of the behavioral consequences of chronic stimulation of dopamine receptors in the nucleus accumbens of rat brain effected by a continuous infusion or by single daily injections. *Naunyn-Schmiedeberg's Archives of Pharmacology* **324**, 27–33.

Costall, B., Domeney, A.M. & Naylor, R.J. (1984). Locomotor hyperactivity caused by dopamine infusion into the nucleus accumbens of rat brain: specificity of action. *Psychopharmacology* **82**, 174–80.

Courvoisier, S. (1956). Pharmacodynamic basis for the use of chlorpromazine in psychiatry. *Journal of Clinical and Experimental Psychopathology* **17**, 25–37.

Creese, I. (1987). Biochemical properties of CNS dopamine receptors. In *Psychopharmacology: The Third Generation of Progress*, ed. H.Y. Meltzer, pp. 257–64. Raven Press: New York.

Creese, I. (1988). Dopamine receptor subtypes: differential regulatory characteristics and role in schizophrenia. *Neurochemistry International* **13**(S1), 24.

Criborn, C.-O., Clemedson, C.J. & Henriksson, C. (1986). Amphetamine as a protective agent against oxygen-induced convulsions in mice. *Aviation, Space and Environmental Medicine* **57**, 777–81.

Criborn, C.-O., Henriksson, Ch., Ahlenius, S. & Hillegaart, V. (1987). Partial protection against hyperbaric oxygen induced convulsions by dopaminergic agents in mice: possible involvement of autoreceptors? *Journal of Neural Transmission* **63**, 277–86.

Criborn, C.-O., Muren, A., Ahlenius, S. & Hillegaart, V. (1988). Effects of lisuride and quinpirole on convulsions induced by hyperbaric oxygen in the mouse. *Aviation, Space and Environmental Medicine* **59**, 723–7.

Crow, T.J. (1980). Molecular pathology of schizophrenia: more than one disease process? *British Medical Journal* **280**, 66–8.

Eaton, W.W. (1985). Epidemiology of schizophrenia. *Epidemiology Reviews* **7**, 105–26.

Farde, L., Hall, H., Ehrin, E. & Sedvall, G. (1986). Quantitative analysis of D2 dopamine receptor binding in the living human brain by PET. *Science* **231**, 258–61.

Fibiger, H.C., Zis, A.P. & Phillips, A.G. (1975). Haloperidol-induced disruption of conditioned avoidance responding: attenuation by prior training or by anticholinergic drugs. *European Journal of Pharmacology* **30**, 309–14.

Fuxe, K. & Hökfelt, T. (1969). Catecholamines in the hypothalamus and the pituitary gland. In *Frontiers in Neuroendocrinology*, eds. W.F. Ganong & L. Martini, pp. 47–96. Oxford University Press: New York.

Gelder, M., Gath, D. & Mayou, R. (1983). *Oxford Textbook of Psychiatry*. Oxford University Press: Oxford.

Geyer, M.A. & Braff, D.L. (1987). Startle habituation and sensorimotor gating in schizophrenia and related animal models. *Schizophrenia Bulletin* **13**, 643–68.

Gianutsos, G. & Moore, K.E. (1977). Dopaminergic supersensitivity in striatum and olfactory tubercle following chronic administration of haloperidol or clozapine. *Life Sciences* **20**, 1585–92.

Goldberg, M.E., Johnson, H.E. & Knaak, J.B. (1965). Inhibition of discrete avoidance behavior by three anticholinesterase agents. *Psychopharmacologia* **7**, 72–6.

Gordon, H.L. (1958). *The New Chemotherapy of Mental Illness*. Philosophical Library, Inc.: New York.

Hanson, H.M., Stone, C.A. & Witoslawski, J.J. (1970). Antagonism of the antiavoidance effects of various agents by anticholinergic drugs. *Journal of Pharmacology and Experimental Therapeutics* **173**, 117–24.

Hatcher, R.A. & Weiss, S. (1923). Studies on vomiting. *Journal of Pharmacology and Experimental Therapeutics* **22**, 139–93.

Henn, F.A. & McKinney, W.T. (1987). Animal models in psychiatry. In *Psychopharmacology: The Third Generation of Progress*, ed. H.Y. Meltzer, pp. 687–95. Raven Press: New York.

Hillegaart, V. & Ahlenius, S. (1987). Effect of raclopride on exploratory locomotor activity, treadmill locomotion, conditioned avoidance behaviour and catalepsy in rats: behavioural profile comparisons between raclopride, haloperidol and preclamol. *Pharmacology and Toxicology* **60**, 350–4.

Hillegaart, V., Ahlenius, S., Magnusson, O. & Fowler, C.J. (1987). Repeated testing of rats markedly enhances the duration of effects induced by haloperidol on treadmill locomotion, catalepsy, and a conditioned avoidance response. *Pharmacology, Biochemistry and Behavior* **27**, 159–64.

Hillegaart, V., Ahlenius, S., Fowler, C.J. & Magnusson, O. (1988). Increased duration of dopamine receptor antagonist-induced effects on both behavior and striatal dopamine turnover by repeated testing in rats. *Pharmacology and Toxicology* **63**, 114–17.

Hjorth, A., Carlsson, A., Clark, D., Svensson, K., Lindberg, M., Wikström, H., Sanchez, D. & Lindberg, P. (1983). Central dopamine receptor agonist and antagonist actions of the enantiomers of 3-PPP. *Psycopharmacology* **81**, 89–99.

Hughlings-Jackson, J. (1931). *Selected Writings*. Hodder & Stoughton: London.

Iorio, L.C., Barnett, A., Leitz, F.H., Houser, V.P. & Korduba, C.A. (1983). SCH 23390, a potential benzazepine antipsychotic with unique interactions on dopaminergic systems. *Journal of Pharmacology and Experimental Therapeutics* **226**, 462–8.

Itil, T.M. (1978). Effects of psychotropic drugs on qualitatively and quantitatively analyzed human EEG. In *Principles of Psychopharmacology*, eds. W.G. Clark & J. del Guidice, pp. 261–77. Academic Press: New York.

Iversen, S.D. (1987). Is it possible to model psychotic states in animals. *Journal of Psychopharmacology* **1**, 154–76.

Iversen, S.D. & Iversen, L.L. (1981). *Behavioural Pharmacology*. Oxford University Press: Oxford.

Jacobsen, E. (1958). The pharmacological classification of central nervous depressants. *Journal of Pharmacy and Pharmacology* **10**, 273–94.

Jeste, D., Del Carmen, R., Lohr, J. & Wyatt, R. (1985). Did schizophrenia exist before the 18th century? *Comprehensive Psychiatry* **26**, 493–503.

Johnstone, E.C., Crow, T.J., Frith, C.D., Husband, J. & Kreel, L. (1976). Cerebral ventricular size and cognitive impairment in chronic schizophrenia. *Lancet* **2**, 924–6.

Kalant, O.J. (1966). *The Amphetamines.* University of Toronto Press: Toronto.

Kalinowsky, L.B. & Hippius, H. (1969). *Pharmacological, Convulsive and Other Somatic Treatments in Psychiatry.* Grune & Stratton Inc.: New York.

Kebabian, J.W. & Calne, D.B. (1979). Multiple receptors for dopamine. *Nature* **277**, 93–6.

Kehr, W. (1984). Transdihydrolisuride, a partial dopamine receptor antagonist: effects on monoamine metabolism. *European Journal of Pharmacology* **97**, 111–19.

Kelleher, R.T. & Morse, W.H. (1968). Determinants of the specificity of behavioural effects of drugs. *Reviews of Physiology* **60**, 1–56.

Kelly, P.H., Seviour, P.W. & Iversen, S.D. (1975). Amphetamine and apomorphine responses in the rat following 6-OHDA lesions of the nucleus accumbens septi and corpus striatum. *Brain Research* **94**, 507–22.

Kendler, K.S. (1987). The genetics of schizophrenia. In *Psychopharmacology: The Third Generation of Progress*, ed. H.Y. Meltzer, pp. 705–13. Raven Press: New York.

Kendler, K.S. & Robinette, C.D. (1983). Schizophrenia in the National Academy of Sciences National Research Council Twin Registry: A 16-year update. *American Journal of Psychiatry* **140**, 1551–63.

Kennedy, J.L., Giuffra, L.A., Moises, H.W., Cavalli-Sforza, L.L., Pakstis, A.J., Kidd, J.R., Castiglione, C.M., Sjögren, B., Wetterberg, L. & Kidd, K.K. (1988). Evidence against linkage of schizophrenia to markers on chromosome 5 in a northern Swedish pedigree. *Nature* **336**, 167–70.

Kietzman, M.L. (1985). Sensory and perceptual behavioral markers of schizophrenia. *Psychopharmacology Bulletin* **21**, 518–22.

Kleinman, J.E., Bridge, P., Karoum, F., Speciale, S., Staub, R., Zalcman, S., Gillin, J.C. & Wyatt, R.J. (1979). Catecholamines and metabolites in the brains of psychotics and normals: post-mortem studies. In *Catecholamines: Basic and Clinical Frontiers*, eds. E. Usdin, I.J. Kopin & J. Barchas, pp. 1845–7. Pergamon Press: New York.

Kline, N.S. (1968). Presidential address. In *Psychopharmacology: A Review of Progress 1957–1967*, eds. D.H. Efron, J.O. Cole, J. Levine & J.R. Wittenborn, pp. 1–3. US Government Printing Office: Washington DC.

Kornetsky, C. & Markowitz, R. (1978). Animal models in schizophrenia. In *Psychopharmacology: A Generation of Progress*, ed. M.A. Lipton, A. DiMascio & K.F. Killam, pp. 583–93. Raven Press: New York.

Krsiak, M. & Janku, I. (1971). Measurement of pharmacological depression of exploratory activity in mice: a contribution to the problem of time-economy and sensitivity. *Psychopharmacology* **21**, 118–30.

Laing, R.D. (1976). *The Facts of Life.* Pantheon: New York.

Lee, T.F., Mora, F. & Myers, R.D. (1985). Dopamine and thermoregulation: an evaluation with special reference to dopaminergic pathways. *Neuroscience and Biobehavioral Reviews* **9**, 589–98.

Ljungberg, T. & Ungerstedt, U. (1978). A method for simultaneous recording of eight behavioral parameters related to monoamine neurotransmission. *Pharmacology, Biochemistry and Behavior* **8**, 483–9.

Lorenz, K. (1974). Analogy as a source of knowledge. *Science* **185**, 229–34.

Losonczy, M.F., Davidson, M. & Davis, K.L. (1987). The dopamine hypothesis of schizophrenia. In *Psychopharmacology: The Third Generation of Progress*, ed. H.Y. Meltzer, pp. 715–26. Raven Press: New York.

Maier, N.R.F. & Schneirla, T.C. (1964). *Principles of Animal Psychology.* Dover Publications Inc.: New York.

Marriott, A.S. & Spencer, P.S.J. (1965). Effects of centrally acting drugs on exploratory behavior in rats. *British Journal of Pharmacology* **25**, 432–41.

Mason, S.T. (1984). *Catecholamines and Behaviour.* Cambridge University Press: Cambridge.

Mason, S.T., Beninger, R.J., Fibiger, H.C. & Phillips, A.G. (1980). Pimozide-induced suppression of responding: evidence against a block of food reward. *Pharmacology, Biochemistry and Behavior* **12**, 917–23.

McEvoy, J.P. (1983). The clinical use of anticholinergic drugs as treatment for extrapyramidal side effects of neuroleptic drugs. *Journal of Clinical Psychopharmacology* **3**, 288–302.

McKinney, W.T. & Moran, E.C. (1981). Animal models of schizophrenia. *American Journal of Psychiatry* **138**, 478–83.

Meltzer, H.Y. & Busch, D. (1983). Serum prolactin response to chlorpromazine and psychopathology in schizophrenics: implications for the dopamine hypothesis. *Psychiatric Research* **9**, 285–99.

Meltzer, H.Y., Goode, D.J. & Fang, V.S. (1978). The effect of psychotropic drugs on endocrine function. I. Neuroleptics, precursors, and agonists. In *Psychopharmacology: A Generation of Progress*, eds. M.A. Lipton, A. DiMascio & K.F. Killam, pp. 509–29. Raven Press: New York.

Mesulam, M.-M. & Geschwind, N. (1978). On the possible role of neocortex and its limbic connections in the process of attention and schizophrenia: clinical cases of inattention in man and experimental anatomy in monkey. *Journal of Psychiatric Research* **14**, 249–59.

Meynert, E.W., Maraczynski, T.J. & Browning, R.A. (1975). The role of neurotransmitters in the epilepsies. *Advances in Neurology* **13**, 79–147.

Moore, K.E. & Goodale, D.B. (1975). 6-Hydroxydopamine-induced dopaminergic receptor supersensitivity: studies with D- and L-DOPA. In *Chemical Tools in Catecholamine Research*, eds. G. Jonsson, T. Malmfors & Ch. Sachs, pp. 343–7. North-Holland Publishing Company: Amsterdam.

Morpurgo, C. (1965). Antiparkinson drugs and neuroleptics. *Progress in Brain Research* **16**, 121–34.

Morpurgo, C. & Theobald, W. (1964). Influence of antiparkinson drugs and amphetamine on some pharmacological effects of phenothiazine derivatives used as neuroleptics. *Psychopharmacologia* **6**, 178–91.

Murphy, J. (1976). Psychiatric labeling in cross-cultural perspective. *Science* **191**, 1019–28.

Niemegeers, C.J.E. (1971). The apomorphine antagonism test in dogs. *Pharmacology* **6**, 353–64.

Niemegeers, C.J.E. & Janssen, P.A.J. (1979). A systematic study of the pharmacological activities of dopamine antagonists. *Life Sciences* **24**, 2201–16.

Parnham, M.J. & Bruinvels, J. (eds.) (1983). *Discoveries in Pharmacology: Psycho- and Neuro-Pharmacology*. Elsevier: Amsterdam.

Pifl, C. & Hornykiewicz, O. (1988). Postsynaptic dopamine agonist properties of B-HT 920 as revealed by concomitant D-1 receptor stimulation. *European Journal of Pharmacology* **146**, 189–91.

Randrup, A. & Munkvad, I. (1967). Stereotyped activities produced by amphetamine in several species and man. *Psychopharmacologia* **1**, 300–10.

Rifkin, A. & Siris, S. (1987). Drug treatment of acute schizophrenia. In *Psychopharmacology: The Third Generation of Progress*, ed. H.Y. Meltzer, pp. 1095–101. Raven Press: New York.

Rosenthal, D. & Kety, S.S. (1968). *The Transmission of Schizophrenia*. Pergamon Press: Oxford.

Rubin, R.T. (1987). Prolactin and schizophrenia. In *Psychopharmacology: The Third Generation of Progress*, ed. H.Y. Meltzer, pp. 803–8. Raven Press: New York.

Sack, R.L. (1977). Side effects of and adverse reactions to psychotropic medications. In *Psychopharmacology: From Theory to Practice*, eds. J.D. Barchas, P.A. Berger, R.D. Ciaranello & G.R. Elliott, pp. 276–89. Oxford University Press: New York.

Schooler, N.R. & Hogarty, G.E. (1987). Medication and psychosocial strategies in the treatment of schizophrenia. In *Psychopharmacology: The Third Generation of Progress*, ed. H.Y. Meltzer, pp. 1111–19. Raven Press: New York.

Seeman, P. (1980). Brain dopamine receptors. *Pharmacological Reviews* **32**, 229–313.

Segal, D.S. & Janowsky, D.S. (1978). Psychostimulant-induced behavioral effects: possible models of schizophrenia. In *Psychopharmacology: A Generation of Progress*, eds. M.A. Lipton, A. DiMascio & K.F. Killam, pp. 1113–23. Raven Press: New York.

Seiden, L.S. & Dykstra, L.A. (1977). *Psychopharmacology: A Biochemical and Behavioral Approach*. Van Nostrand Reinhold Company: New York.

Setler, P.E., Sarau, H.M., Zirkle, C.L. & Saunders, H.L. (1978). The central effects of a novel dopamine agonist. *European Journal of Pharmacology* **50**, 419–30.

Shapiro, R.E. & Miselin, R.R. (1985). The central neural connections of the area postrema of the rat. *Journal of Comparative Neurology* **234**, 344–64.

Shelton, R.C. & Weinberger, D.R. (1987). Brain morphology in schizophrenia. In *Psychopharmacology: The Third Generation of Progress*, ed. H.Y. Meltzer, pp. 773–81. Raven Press: New York.

Sherrington, R., Brynjolfsson, J., Petursson, H., Potter, M., Dudleston, K., Barraclough, B., Wasmuth, J., Dobbs, M. & Gurling, H. (1988). Localization of a susceptibility locus for schizophrenia on chromosome 5. *Nature* **336**, 164–7.

Smith, R.B. (1985). *The Development of a Medicine*. The Macmillan Press Ltd: Houndmills.

Spano, P.F., Govoni, S. & Trabucchi, M. (1978). Studies on the pharmacological properties of dopamine receptors in various areas of the central nervous system. In *Advances in Biochemical Psychopharmacology*, eds. P.J. Roberts, G.N. Woodruff & L.L. Iversen, pp. 155–65. Raven Press: New York.

Ståhle, L. & Ungerstedt, U. (1986). Different behavioural patterns induced by the dopamine agonist apomorphine analysed by multivariate statistics. *Pharmacology, Biochemistry and Behavior* **24**, 291–8.

Staunton, D.A., Wolfe, B.B., Groves, P.M. & Molinoff, P.B. (1981). Dopamine receptor changes following destruction of the nigrostriatal pathway: Lack of a relationship to rotational behavior. *Brain Research* **211**, 315–27.

Stoof, J.C. & Kebabian, J.W. (1984). Two dopamine receptors: biochemistry, physiology and pharmacology. *Life Sciences* **35**, 2281–6.

Sulkowski, A. (1983). Psychobiology of schizophrenia: A neo-Jacksonian detour. *Perspectives in Biology and Medicine* **26**, 205–18.

Svensson, K., Johansson, A.M., Magnusson, T. & Carlsson, A. (1986). (+)-AJ 76 and (+)-UH 232: central stimulants acting as preferential dopamine autoreceptor antagonists. *Naunyn-Schmiedeberg's Archives of Pharmacology* **334**, 234–45.

Szasz, T.S. (1974). *The Myth of Mental Illness*. Harper & Row: New York.

Tamminga, C.A. & Gerlach, J. (1987). New neuroleptics and experimental antipsychotics in schizophrenia. In *Psychopharmacology: The Third Generation of Progress*, ed. H.Y. Meltzer, pp. 1129–40. Raven Press: New York.

Terenius, L., Wahlström, A., Lindström, L. & Widerlov, E. (1976). Increased levels of endorphins in chronic psychoses. *Neuroscience Letters* **3**, 157–62.

Titus, R.D., Kornfeld, E.C., Jones, N.D., Clemens, J.A., Smalstig, E.B., Fuller, R.W., Hahn, R.A. & Hynes, M.D. (1983). Resolution and absolute configuration of an ergoline-related dopamine agonist, trans-4,4a,5,6,7,8,8a,9-octohydro-5-propyl-1H (or 2H)-pyrazolo(3,4,-g)quinoline. *Journal of Medicinal Chemistry* **26**, 1112–16.

Udenfriend, S. (1966). Tyrosine hydroxylase. *Pharmacological Reviews* **18**, 43–51.

Ungerstedt, U. (1971). Postsynaptic supersensitivity after 6-hydroxydopamine induced degeneration of the nigro-striatal dopamine system. *Acta Physiologica Scandinavica* **S367**, 69–93.

van Kammen, D.P. (1977). γ-Aminobutyric acid and the dopamine hypothesis of schizophrenia. *American Journal of Psychiatry* **134**, 138–43.

van Kammen, D.P. & Gelernter, J. (1987). Biochemical instability in schizophrenia I: The norepinephrine system. In *Psychopharmacology: The Third Generation of Progress*, ed. H.Y. Meltzer, pp. 745–51. Raven Press: New York.

Verhave, T., Owen Jr, J.E. & Robbins, E.B. (1958). Effects of chlorpromazine and secobarbital on avoidance and escape behaviour. *Archives Internationales de Pharmacodynamie et de Thérapie* **116**, 45–53.

Wagemaker, H. & Cade, R. (1977). The use of hemodialysis in chronic schizophrenia. *American Journal of Psychiatry* **134**, 684–5.

Wålinder, J., Skott, A., Carlsson, A. & Roos, B.-E. (1976). Potentiation by metyrosine of thioridazine effects in chronic schizophrenics. *Archives of General Psychiatry* **33**, 501–5.

Watzman, N., Barry III, H., Kinnard Jr, W.J. & Buckley, J.P. (1967). Influence of chlorpromazine on spontaneous activity of mice under various parametric conditions. *Archives Internationales de Pharmacodynamie et de Thérapie* **165**, 352–68.

Wauquier, A. & Niemegeers, C.J.E. (1972). Intracranial self-stimulation in rats as a function of various stimulus parameters. II. Influence of haloperidol, pimozide and pipamperone on medial forebrain bundle stimulation with monopolar electrodes. *Psychopharmacology* **27**, 191–202.

Weinberger, D.R. (1988). Schizophrenia and the frontal lobe. *Trends in Pharmacological Sciences* **11**, 367–70.

Westerink, B.H.C. (1978). Effect of centrally acting drugs on regional dopamine metabolism. *Advances in Biochemical Psychopharmacology* **19**, 255–66.

White, F.J. & Wang, R.Y. (1983). Comparison of the effects of chronic haloperidol treatment on A9 and A10 dopamine neurons in the rat. *Life Sciences* **32**, 983–93.

Wise, C.D. & Stein, L. (1973). Dopamine-β-hydroxylase deficits in the brains of schizophrenic patients. *Science* **181**, 344–7.

Wise, R.A. (1982). Neuroleptics and operant behavior: the anhedonia hypothesis. *Behavioral and Brain Sciences* **5**, 39–87.

Wise, R.A., Spindler, J., deWit, H. & Gerber, G.J. (1978). Neuroleptic-induced 'anhedonia' in rats: Pimozide blocks reward quality of food. *Science* **201**, 262–4.

Yaryura-Tobias, J., Diamond, B. & Merlis, S. (1970). The action of L-DOPA on schizophrenic patients. *Current Therapeutic Research* **12**, 528–31.

13

Animal models of schizophrenia and mania: clinical perspectives

JOHN COOKSON

Anti-psychotic drugs, of which the phenothiazines were the first, are effective in both mania and schizophrenia. This suggests either that there are similarities in the pathogensis of the two conditions, or that these drugs work in a non-specific manner reducing behavioural disturbance or 'psychotic' symptoms irrespective of the aetiology or diagnosis. There is much clinical evidence, to be reviewed below, that mania and schizophrenia are best regarded as separate medical conditions (Kraepelin, 1921), and the family and genetic evidence point strongly in that direction. There is also some pharmacological evidence that particular drugs are effective in mania but not in schizophrenia. If animal models were able to distinguish between mania and schizophrenia, this would open up possibilities to clarify the neuropharmacology and pathogenesis of the two conditions.

This chapter reviews the clinical features of mania and schizophrenia relevant to animal models, with particular emphasis on distinctions that may be drawn between the two disorders.

13.1 DIAGNOSIS

13.1.1 Diagnostic criteria for mania

The essential criterion in the diagnosis of mania is elevation of mood in the form of elation or irritability (Cookson, 1987a). There are also objective behavioural elements and vegetative symptoms such as insomnia and increased appetite. The insomnia is experienced as a decreased need for sleep, in contrast to the exhaustion of the depressed patient. The increased appetite is not necessarily accompanied by increased food intake or weight gain because the patients may be too distractible to complete their meal (Robinson et al., 1975). Likewise there is an increased interest in sexual activity leading to new or promiscuous relationships, but the patient may be too distracted to consummate a relationship. The distractibility means that any minor sensory stimulation catches the attention and leads to new responses in speech or motor activity. There is a feeling of increased energy and a general increase in motor activity and in communicative gestures. Although the motor activity is apparently purposeful, in the sense that it has understandable goals, these goals are often left unachieved, because new activities intervene and the actions are hurried and clumsy. In severe mania, the activity may become stereotyped and limited (Carlsson & Goodwin, 1973). For instance, patients may spend hours repeatedly sorting through their belongings, and explain this by saying that someone may steal them.

Speech is increased in volume and speed. The association of thoughts proceeds in a fast and lively but usually understandable way, and puns and other sorts of word-play are common (flight of ideas). The thought content reflects the mood and displays increased self-esteem and ambitiousness. Mania is the most insightless form of mental illness and patients not only believe but behave in accordance with their increased estimation of their ability, wealth or status, spending too much money, or causing offence to people in authority. The grandiosity may become delusional, when they hold with conviction ideas such that they have great wealth or are of royal descent, or have some special relationship or identity with a deity. There may be auditory hallucinations but these are usually limited to voices addressing the patient with a reassuring or exciting content, e.g. the comforting voices of dead relatives, or God's voice encouraging them to religious acts. Manic patients lose normal social inhibitions and become over-familiar or even abusive or assaultative. Often there is an infectious quality to their good humour which cheers other people and leads to laughter. However, their condition is wearing on those around them. Any attempts to challenge their plans are viewed with resentment and suspicion leading to the formation of ideas of persecution. Tearful swings of mood occur briefly, especially when the patients are confronted with their personal problems by an interviewer. Likewise, there may be swings of anger in response to frustration. In some patients the irritability masks the elation and the picture is of hostile-paranoid mania, as opposed to the more classical elated-grandiose mania (Beigel *et al.*, 1971). Without treatment, mania usually lasts three to nine months and is complicated by physical exhaustion and intercurrent infection.

13.1.2 Diagnostic criteria for schizophrenia

The clinical picture of schizophrenia is not so uniform. Indeed different diagnostic systems classify different patients as schizophrenic (Brockington *et al.*, 1978).

It is useful to distinguish the positive and negative symptoms of schizophrenia. The negative symptoms are flattening of affect and poverty of speech reflecting a general loss of drive and goal-orientated behaviour; it is difficult to motivate the patient into activity (Andreasen, 1982). The negative symptoms make a large contribution to the overall disability of chronic schizophrenic patients, tending to persist when the positive symptoms have subsided. They then constitute the so-called 'deficit syndrome' (Carpenter *et al.*, 1988) and result in self-neglect and social withdrawal.

The positive symptoms are formal thought disorder, affective incongruity (laughing, crying or anger that is inappropriate to the context), delusional ideas, hallucinations, and phenomena such as experiences of passivity and thought alienation. Formal thought disorder shows itself as loosening of associations, very illogical thinking, the use of neologisms (newly constructed words), or unusual phrases that seem to convey special or obscure meanings. The effect on the listener is that patients seem not be able to communicate their ideas, but to be abnormally vague or obscure.

13.1.2.1 Delusions

A delusion is a belief, inconsistent with the available information and not shared by others of the same cultural group, which is firmly held and cannot be dispelled by argument or proof to the contrary. Delusions are false beliefs; they may coincidentally be true but it is the logical basis from which the belief is derived that is at fault. Delusions that arise suddenly with no understandable connection to what has gone before, are called primary delusions, whereas secondary delusions develop out of an underlying abnormal mental state (abnormal mood or hallucinations). Secondary delusions have some understandable connections between their content and the abnormal state from which they arose, for instance, mood-congruent grandiose delusions in mania. Primary delusions are usually followed by a secondary delusional system.

Primary delusional experiences (delusional mood, sudden delusional ideas and delusional perception), are almost pathognomonic of the onset of schizophrenia. A delusional mood is a conviction that something is going on around the person which concerns him but he does not know what it is. This may continue for some time before the patient develops a sudden delusional idea or a delusional perception. Sudden delusional ideas occur fully formed in the patient's mind. In a delusional perception the patient attaches new meaning to a normal perception and the delusional content cannot be understood as arising from the previous affective state or attitudes. By contrast, delusional misinterpretation occurs when a delusional meaning arises from the patient's pre-existing disorder.

13.1.2.2 Hallucinations

Schizophrenics experience many abnormal perceptions of which hallucinations are the clearest examples. A hallucination is experienced as a true percept but arises in the absence of an external sensory stimulus. In schizophrenia, hallucinations frequently occur in the auditory modality. Less commonly, somatic, tactile or olfactory hallucinations occur. Visual hallucinations are unusual in schizophrenia; their occurrence usually signifies an organic brain disease or drug-induced psychosis. The frequency of visual hallucinations in animal models may signify a short-coming of the models.

Certain types of hallucination are considered pathognomonic of schizophrenia if they occur in clear consciousness and in the absence of brain disease. Hearing one's own thoughts spoken aloud, hearing voices speaking to each other about one in a hostile way, or hearing a voice commenting on one's actions are characteristic (Schneider, 1959). The occurrence of hallucinations can sometimes only be recognized from the behaviour of the patient who appears to listen and reply to non-existent voices. The acutely ill patient may also display fear and anger at the alien presence.

'Passivity phenomena' are also diagnostic of schizophrenia and consist of the experience that one's feelings, impulses or actions are imposed on one, and not one's own. Similarly, thoughts may be experienced as alien (thought insertion) or taken away (thought withdrawal).

13.1.2.3 Different types of schizophrenia

It is traditional to distinguish paranoid, hebephrenic, catatonic and simple forms of schizophrenia. In paranoid schizophrenia persecutory delusions predominate together with hallucinations, but formal thought disorder is limited to some loosening of associations or over-inclusiveness; the mood is one of suspicion and hostility but many areas of personality are unaffected. In hebephrenic (or disorganized) schizophrenia, there is severe formal thought disorder and affective incongruity, as well as hallucinations, but without systematised delusions. The personality as a whole tends to deteriorate. Catatonic symptoms, described below, are less common now in schizophrenia, perhaps because of modern treatment. Simple schizophrenia presents as a gradual social withdrawal and vagueness, without delusions or hallucinations: in other words, the negative symptoms predominate.

More recent formulations have emphasised

distinctions between three different components of symptomatology, psychomotor poverty (negative symptoms), disorganisation (formal thought disorder, and incongruous affect), and reality distortion (delusions and hallucinations) (Liddle, 1987).

13.1.3 Diagnostic problems

13.1.3.1 Schizoaffective and mixed affective states

Mania and depression involve changes of mood, activity and thinking, and these three areas can vary independently to produce 'mixed states'. The best recognised are depression with flight of ideas, and manic stupor, in which the patient has extreme motor retardation but appears elated and may later describe racing thoughts (Himmelhoch *et al.*, 1976). The term 'dysphoric mania' has been used to encompass patients in whom classical manic symptoms are accompanied by marked anxiety, depression or anger (Post *et al.*, 1989).

Some patients show a mixture of manic and schizophrenic symptoms in the same episode and are regarded as having schizoaffective disorder according to Research Diagnostic Criteria (Spitzer *et al.*, 1978) or mania with mood-incongruent psychotic features according to DSM-III (American Psychiatric Association, 1980).

13.1.3.2 Catatonic symptoms

Catatonia is a syndrome rather than a diagnosis (Cookson, 1987*b*). The term refers to a variety of abnormalities of posture, movement and speech. The commonest abnormalities of movement are stereotypies, mannerisms, grimacing, catalepsy, waxy flexibility, echopraxia, negativism and automatic obedience. These symptoms, which seem to indicate abnormalities of volitional control of the neuromuscular apparatus, can occur in mania (Abrams *et al.*, 1979) or in association with lesions in the region of the basal ganglia, temporal lobes or third ventricle, as well as in schizophrenia (Gelenberg, 1976). Stress-induced (hysterical) catatonia has also been described. Nowadays, florid catatonic states are less common and are more likely to occur in patients with a foreign language or culture.

Stereotypies and catalepsy have obvious correlates in animal models, though speech disorders such as the use of neologisms, and echolalia (the verbatim repetition of what is heard), are less capable of modelling in animals. Stereotypies are readily induced in animals by dopamine (DA) agonists, with a presumed site of action in the neostriatum (see Ahlenius, this vol., Ch. 12). This might suggest that stereotypies occurring in psychotic patients correspond to the development of excessive DA transmission in the nigro-striatal pathway. On the other hand, catalepsy in animals can be induced by DA antagonists and reversed by anti-cholinergic drugs; a similar state (waxy flexibility) occurs in acute catatonia, but is not reversed by anti-cholinergics.

In catatonic schizophrenia, a state of catatonic excitement can intervene. This bears some resemblance to mania, in that there is great overactivity and aggression, but the activity in this case has an incomprehensible, unpredictable, purposeless and often extreme quality, and the affect has a peculiar dead-pan quality (Kahlbaum, 1874).

13.1.4 Neuroendocrine findings

The neuroendocrinology of mania (Cookson, 1985), of schizophrenia (Cleghorn & Brown, 1988), and of schizoaffective disorder (Cookson, 1989) have been reviewed. Mania is associated with increased nocturnal cortisol levels and, in severe cases, with increased daytime cortisol levels (Cookson *et al.*, 1983). TSH responses to TRH are often blunted in both mania and schizophrenia, perhaps more often in the former. Growth hormone responses to clonidine are blunted in mania as in depression. Growth hormone responses to the DA agonist apomorphine may be increased in association with psychotic symptoms in mania, or with positive symptoms in schizophrenia.

13.1.5 Animal behaviour and clinical symptoms

Although schizophrenia and mania may result in superficially similar symptoms, for instance persecutory delusions, incomprehensible speech, or aggressive behavior, and despite the possibility of mixed states, closer analysis reveals obvious and important differences. The symptoms of mania should be understandable in relation to the elevated or irritable mood, whereas most of those in schizophrenia arise from abnormalities in perception, thinking or volition. Moreover, mania is invariably accompanied by alterations in sleep and appetitive behaviour and by overactivity, which do not usually occur in schizophrenia.

Lyon (this vol., ch. 11) has listed the individual symptoms and corresponding animal behaviour to be sought in models of mania. His analysis of manic symptoms and particularly the rapid changes in responding and eventual emergence of stereotypies carry some conviction with regard to clinical presentation. However, from a clinical perspective this account places too much emphasis on change in affect between elation, depression and irritability; manic patients can be divided into the predominantly elated-grandiose and the predominantly paranoid-destructive (Beigel *et al.*, 1971). Shortening the list of basic symptoms to elation, hyperactivity and irritability, also overlooks distinctions between disinhibition, elation or overactivity as the source of new patterns of behaviour. The shorter list could also lead to important biological correlates, such as insomnia and changes in eating behaviour, being overlooked.

Lyon has also listed schizophrenic symptoms and their counterparts, to be sought in animal behavioural models. It is clear that most schizophrenia-like symptoms could have animal counterparts with the possible exception of formal thought disorder. Obviously, the range of delusional phenomena that can be recognised in animals is limited. One persisting problem is the relative importance of visual as opposed to auditory hallucinations in animal models.

13.2 ETIOLOGY

In addition to the different clinical presentation of the two conditions, there are differences as well as similarities in their natural history and genetics.

13.2.1 Natural history

Both schizophrenia and mania often start in late adolescence (Joyce, 1984; Schur, 1988). The fact that mania is virtually unknown before puberty, distinguishes it from schizophrenia and may have some hormonal correlate.

Schizophrenia is a chronic illness liable to relapses and usually with incomplete recovery of the personality and persistent negative and sometimes positive symptoms after each episode. Occasionally a brief schizophreniform episode may occur, perhaps as a result of extreme stress, without ensuing chronicity (Fenton & McGlashan, 1987). Mania is usually part of a bipolar manic-depressive illness; between episodes there is complete remission, and patients recover their normal personalities. There is, on average, a gradual shortening of the interval between the first four episodes, although there is much individual variation (Goodwin & Jamison, 1984). Some patients enter a period when mania and depression alternate almost continuously; if there are four or more episodes in a single year they are said to have 'rapid cycling' (Roy-Byrne *et al.*, 1984). In rare cases there is rhythmic cycling with a 48-h pattern (Jenner *et al.*, 1968).

The phenomenon of kindling (see Lyon, this vol., Ch. 11) may be relevant to the initial increase in frequency of episodes and the occasional development of rapid cycling. Post & Weiss (1987) have distinguished between the kindling of behavioural stimulation following injections of cocaine and the kindling of seizure activity following electrical stimulation of the amygdala or repeated systemic injections of local anaesthetics. Behavioural kindling by cocaine or amphetamine is inhibited by lithium but not by carbamazepine. However, electrical kindling of seizures is antagonised by carbamazepine. This led Ballenger & Post (1980) to investigate whether carbamazepine might be useful in bipolar affective illness (see Section 13.4.2.3 below). Subsequently the effectiveness of carbamazepine in rapid cycling has suggested that this condition develops through an analogous process of kindling in limbic brain areas. However, those bipolar patients who experience more temporal lobe epilepsy-like psychosensory phenomena appear to respond better to lithium than patients who experience fewer of such phenomena (Silberman *et al.*, 1985).

There is some seasonal variation in the incidence of mania, with more cases developing in the hot summer months, possibly related to greater exposure to sunlight (Walter, 1977). Some patients with a milder form of bipolar illness show repeated episodes of winter depression and summer elation,

so-called 'seasonal affective disorder' (Rosenthal *et al.*, 1984). The biological basis of the effects of season and light cycles upon mood is unclear and this is an area where animal work could be helpful (Mason, 1989). It is known that there is a seasonal variation in 5-hydroxytryptamine (5-HT) levels in the brain, with higher levels in winter than in summer (Carlsson *et al.*, 1980).

13.2.1.1 The occurrence of depression in mania and schizophrenia

Depression occurs as an integral part of bipolar manic-depressive illness and often either immediately precedes or succeeds an episode of mania (Kotin & Goodwin, 1971; Morgan, 1972). The switch between mania and depression may occur quite suddenly, for instance, overnight, but usually more gradually with an intervening mixed state or a period of normality. Bipolar disorder may involve three separate biological abnormalities, one predisposing the individual to affective disorder, and the others producing the symptoms of mania or of depression; it has been hypothesised that 5-HT deficiency may be part of the predisposition (Prange *et al.*, 1974; Mendels & Frazer, 1975) and models based on reduction of 5-HT function, for example by the synthesis inhibitor PCPA, are relevant to this (but see below, Section 13.3.1.5).

Depression can occur in schizophrenia in the prodromal phase, as an indicator of imminent relapse, during the acute episode, or after the acute episode has subsided (post-psychotic depression) (Knights & Hirsch, 1981; Johnson, 1988). Depressive equivalents may occur in animal models of mania and schizophrenia; for instance, amphetamine withdrawal can produce severe clinical depression, which may be relevant to post-psychotic or post-manic depressive symptoms.

13.2.2 Family and genetic studies

In first-degree relatives of patients with a history of mania, there is a greatly increased risk of mania and of unipolar depression as well as personality types with prominent mood swings (cyclothymia); there is no excess of schizophrenia in this group (Weissman *et al.*, 1984). By contrast, the first-degree relatives of patients with schizophrenia have an increased risk of schizophrenia and of abnormal personalities of the schizotypal (highly sensitive or suspicious) kind (Kendler *et al.*, 1985).

Among monozygotic (MZ) (identical) twin pairs, the concordance rate for schizophrenia, when one twin is affected, is 45%; there must therefore be environmental as well as genetic factors governing the development of schizophrenia. A familial/ sporadic distinction has been used as a research strategy and has indicated an increased ventricular brain ratio (VBR) in sporadic cases (i.e. those without a known family history) (Lewis *et al.*, 1987). In particular, the affected member of a discordant MZ pair was found to have such an increase (Reveley *et al.*, 1984). This increased VBR is evidence of long-standing brain damage (Winberger, 1988) due perhaps to previous obstetric complications (Owen *et al.*, 1988) or childhood head injury or intracranial infection (De Lisi *et al.*, 1986). In some cases there are both familial factors and evidence of birth complications suggesting an effect of brain damage upon a genetically predisposed individual (De Lisi *et al.*, 1987).

The situation in manic-depressive illness has not been explored in the same way using CAT scans of twin pairs, but there is evidence of ventricular enlargement in some cases of bipolar illness (Jeste *et al.*, 1988). The concordance rate for major affective illness in MZ twins, one of whom has a bipolar disorder, is 69% (Bertelsen *et al.*, 1977).

13.2.3 Epidemiology of mania and schizophrenia

In contrast to mania, schizophrenia is more common in people of lower socio-economic class, a consequence in large part of drift down the social scale in predisposed individuals (Goldberg & Morrison, 1963) and a drift into inner city areas where they find anonymity and cheap lodgings (Hare, 1956).

Schizophrenia occurs equally in men and women although the age of onset may be slightly earlier in males (Loranger, 1984; Shur, 1988). The problems posed in rehabilitation are often greater in men as they have greater difficulty in forming new relationships. There is a slightly greater incidence of mania in women (Winokur *et al.*, 1969). Females with bipolar disorder are also more likely than males to develop rapid-cycling phases in their illness (Roy-Byrne *et al.*, 1984). There is no appar-

ent difference in the age of onset between men and women (Loranger & Levine, 1978). Since sex differences in the two disorders are relatively insignificant, the omission of this factor from animal models is probably unimportant. However, a minority of cases of mania may be genetically X-linked (Baron *et al.*, 1981).

There is a slight excess (5–10%) of winter births in schizophrenia (Hare & Walter, 1978), the causes of which are speculated to be viral (Crow, 1984; Watson *et al.*, 1984), or obstetric (McNeil & Kaij, 1978).

13.2.4 Psychological stress and social environment

Episodes of schizophrenia and mania may develop following a major life event such as bereavement, personal separation, work-related problems and loss of role (Brown & Birley, 1968; Kennedy *et al.*, 1983; Ambelas, 1987; Aronson & Shukla, 1987). The life event seems to act as a trigger in a predisposed individual. In mania, it has been suggested that the first episode is more likely than subsequent ones to have been triggered by life events (Astrup *et al.*, 1959; Sclare & Creed, 1987).

Psychological stress may also arise from the emotional environment in which the patient lives (see Hirsch, 1986). The concept has arisen of families showing high 'expressed emotion' (Brown *et al.*, 1972); this consists of either emotional over-involvement, or hostility and criticism directed to the patients and their symptoms. Patients with schizophrenia who return to 'high expressed emotion' homes, are more likely to relapse (Vaughn & Leff, 1976). The same appears also to be true to a moderate extent for bipolar patients (Mikowitz *et al.*, 1985).

There is evidence that lack of social stimulation may exacerbate the development of negative symptoms in schizophrenia (Wing & Brown, 1961). However, in some patients there is evidence of high autonomic arousal (Sturgeon *et al.*, 1984); further stimulation by excessive social contact can exacerbate their withdrawn negative state. In primates, social deprivation can lead to later stereotyped locomotion, gross rhythmic stereotypies and reduced social interaction (see Lyon, this vol., Ch. 11). Social isolation leads to biochemical alterations including increased DA receptors and altered levels of 5-HT and noradrenaline (NA) (see Lyon, this vol., Ch. 11).

13.3 SECONDARY MANIA AND SCHIZOPHRENIA

A manic or acute schizophrenia-like state may develop following a physical disturbance by drugs or by disease. When this occurs in a patient without a past or family history of the disorder, the term 'secondary' mania or schizophrenia is used. These states occur in clear consciousness (patients are fully orientated in their surroundings) and this distinguishes them from acute toxic-confusional states. Table 13.1 shows the recognised causes of secondary mania (Krauthammer & Klerman, 1978). Table 13.2 shows the causes of secondary schizophreniform states (Davison & Bagley, 1969). Secondary states could provide important clues to the neurobiological mechanisms of the conditions.

13.3.1 Drug induced secondary mania

13.3.1.1 Psychostimulants

Low doses of amphetamines in normal subjects will produce a state resembling mania (Jacobs & Silverstone, 1986), involving euphoria, irritability, increased talkativeness and racing thoughts, social disinhibition, distractibility, increased activity, and insomnia. Unlike mania, however, hunger and appetite (as well as food intake) are reduced. It is thought that the anorectic effect of amphetamine in humans is caused by release of NA, whereas stimulant effects are due to release of DA (Silverstone *et al.*, 1981; Goodall *et al.*, 1987); in animal models however, the evidence points to both DA and beta-adrenergic receptors mediating the anorectic effects of amphetamine (see Montgomery, this volume, Chapter 8).

The neuroendocrine effects of low doses of amphetamine in humans include an increase in circulating growth hormones and cortisol levels. The latter is particularly marked when the drug is given in the evening (Besser *et al.*, 1969), and is thought to be mediated by released NA acting on alpha-1 receptors in the hypothalamus (Rees *et al.*, 1979). However, animals differ from humans in the pharmacology of the cortisol response to amphetamine. The changes in cortisol levels in mania are

Table 13.1. *Common causes of secondary mania*

Psychostimulants
Amphetamines
Cocaine

Recreational drugs
Cannabis
Alcohol

Medication
Dopamine agonists (bromocriptine, etc.)
L-dopa
Anti-cholinergics (procyclidine, benzhexol)
Antidepressants
Thyroid hormones
Corticosteroids

Metabolic conditions
Thyrotoxicosis
Childbirth: puerperal psychosis

Organic brain disorders
Cerebral tumour
Cerebrovascular disease
Head injury
Dementia
Epilepsy

Table 13.2. *Common organic causes of schizophreniform states*

Psychostimulants
Amphetamines
Cocaine

Psychotomimetics
Phencyclidine
Cannabis

Dopamine agonists
Bromocriptine, etc.

Organic brain disorders
Temporal lobe epilepsy
Cerebral tumour
Head injury
Dementia

in keeping with these findings, but circulating growth hormone levels are not elevated in mania (Cookson *et al.*, 1982).

Amphetamine can model most of the symptoms

of mania in animals (see Lyon, this vol., Ch. 11), although the hyperactivity is apparently less evident in non-human primates than in other species, and the anorectic effects of amphetamine are anomalous. Thus, in both animals and people, low doses of amphetamine produce a syndrome resembling mania but with anomalous effects on appetite and hormonal levels.

Directly-acting DA agonists such as bromocriptine, used either in the treatment of pituitary tumours or to suppress lactation, can lead to psychotic states with predominantly manic or mixed manic and schizophrenic symptoms (Turner *et al.*, 1984). In lower doses these drugs do not have psychostimulant properties like amphetamine, but are somewhat sedative, perhaps because they stimulate presynaptic DA autoreceptors (Zetterström *et al.*, 1986). The DA precursor L-dopa can lead to manic symptoms but this appears far more likely to occur in individuals with a bipolar predisposition (Murphy, 1972). Likewise, a manic episode may be triggered by amphetamine (Post *et al.*, 1976) or by antidepressant drugs (Bunney, 1978; Wehr & Goodwin, 1987) in predisposed individuals.

These findings seem to indicate a role for excessive DA and possibly NA transmission in the pathogenesis of mania. Direct biochemical evidence from the measurement of transmitters and metabolites in cerebrospinal fluid (CSF) is somewhat disappointing, perhaps because lumber CSF is remote from the relevant sites of abnormal transmission in the brain. However, levels of the DA metabolite HVA are higher in mania than in depression or the normal (euthymic) state (Post, 1980; Post *et al.*, 1989). The more striking abnormality is an increase in CSF levels of NA, but levels of both NA and its metabolite MHPG correlate with the severity of anxiety, anger and depression rather than with mania scores (Swann *et al.*, 1987; Post *et al.*, 1989).

13.3.1.2 *Cannabis, opioids and alcohol*

Cannabis in large doses or in susceptible individuals, can produce a state resembling mania (Tunving, 1982; Rottanburg *et al.*, 1982). However, manic symptoms are not characteristically associated with opiate use, despite the parallels between mania and the behavioural effects of

morphine and enkephalins in rodents (see Lyon, this vol., Ch. 11).

Many patients consume excessive amounts of alcohol during mania, and it cannot be ruled out that alcohol may have triggered the manic episode in some cases (Reich *et al.*, 1974).

13.3.1.3 *Endocrine causes of secondary mania*

Corticosteroids in high doses can produce elated psychotic states resembling mania (Hall *et al.*, 1979; Ling *et al.*, 1981), although Cushing's disease (with high circulating levels of pituitary and adrenocortical hormones) is associated with depression in some 40% of cases, and with mania in only 2% (Jeffcoate *et al.*, 1979; Cohen, 1980; Kelly *et al.*, 1980). The animal literature indicates that corticosteroids can modulate NA receptor-coupled adenylate cyclase, which in turn could affect receptor sensitivity (Mobley & Sulser, 1980). Of relevance to the role of DA in mania, corticosteroids may also increase DA receptor numbers (Crocker *et al.*, 1984). It is of interest to know the effect of corticosteroids in the biogenic-amine-related animal models of mania.

Thyrotoxicosis is often accompanied by hyperactivity and irritability, but the main diagnostic similarity is with an anxiety state, rather than mania. However, thyrotoxicosis can precipitate mania in predisposed individuals (Corn & Checkley, 1983). The commencement of treatment of myxoedema (hypothyroidism) with thyroid hormones leads, in some cases, to a worsening of pre-existing (myxoedematous) psychotic symptoms and to the emergence of a state resembling mania (Josephson & McKenzie, 1980). The number of beta-adrenergic receptors is decreased in hypothyroid animals and number and affinity for agonists are increased in hyperthyroidism (Stiles & Lefkowicz, 1981; Minneman, 1981). The clinical and animal evidence has been integrated in relation to affective illness in a hypothesis of 'thyroid-catecholamine-receptor interaction' (Whybrow & Prange, 1981); it is suggested that manic states can be triggered by increased beta-adrenergic receptor activity. However, thyroid hormone also has indirect effects on DA turnover, through an interaction with corticosteroids (Crocker, 1988).

The effects in animals of pre-natal exposure to

high levels of thyroid hormone (see Lyon, this vol., Ch. 11) are relevant to the development of hyperactive behaviour in the children of mothers who had raised levels of thyroid-stimulating antibodies during pregnancy; the relevance to bipolar disorder is not established.

13.3.1.4 Puerperal psychosis

During pregnancy, the risk of psychosis is reduced, but for two weeks after childbirth, there is a very much greater risk of psychosis developing in mothers with a predisposition to bipolar disorder. This 'puerperal psychosis' usually takes the form of mania or schizo-affective schizomania (Dean & Kendell, 1982). Animal work suggests that the mechanism for the effect of pregnancy and childbirth may lie in the interaction of oestrogens with DA receptors. It has been pointed out that in both the pituitary and the striatum, the observed effects of oestrogens are to reduce responses to DA (Cookson, 1981). Also, chronic oestradiol treatment leads to an increase in putative DA receptors in the striatum (Di Paolo *et al.*, 1981). Puerperal mania might therefore be explained by the fall in oestrogen levels after childbirth exposing supersensitive DA receptors in the mesolimbic system (Cookson, 1982). This is an area open to investigation using the DA-related animal models of mania and schizophrenia.

13.3.1.5 Para-chlorophenylalanine (PCPA)

PCPA inhibits tryptophan hydroxylase and leads to depletion of 5-HT. In rodents this results in increased activity, aggression and sexual behaviour and altered circadian rhythms (see Lyon, this vol., Ch. 11). PCPA is used clinically for the treatment of carcinoid syndrome (Cremata & Koe, 1966), but there have been no reports of PCPA precipitating mania. Levels of 5-HIAA, the metabolite of 5-HT, are not significantly altered in mania (Post *et al.*, 1989).

13.3.2 Drug-induced secondary schizophrenia

A psychosis closely resembling paranoid schizophrenia may be produced by amphetamine-like drugs (Connell, 1958; Ellinwood, 1967), or DA agonists (Turner *et al.*, 1984). Volunteer addicts,

taking large doses of amphetamines, may experience symptoms such as thought insertion and passivity phenomena, but being aware of having taken a drug will not necessarily have secondary delusional ideas of influence from an outside agency (Angrist & Gershon, 1970; Bell, 1973). Cocaine addiction is frequently associated with a similar psychosis, and large intravenous doses or continuous infusions in volunteer addicts, can induce suspiciousness or psychosis (Sherer *et al.*, 1988). Cocaine psychosis is particularly associated with 'formication', a sensation as of insects in the skin with secondary delusional parasitosis. This symptom can occur alone as a form of monosymptomatic hypochondriacal psychosis, which often responds favourably to pimozide (Reilly, 1988). A similar condition can occur with DA agonists (Turner *et al.*, 1984).

The use of cannabis can exacerbate schizophrenic symptoms and consumption of large amounts can induce schizophreniform symptoms (Tunving, 1982). Phencyclidine (PCP) produces a psychotic state in humans resembling paranoid schizophrenia but with more cognitive impairment, visual hallucinations and hyperactivity. The response to antipsychotic drugs such as chlorpromazine is weak. The PCP model is particularly interesting because of the evident involvement of transmitters other than DA. PCP blocks potassium channels and prolongs the action potential leading to more transmitter release; it also blocks the NMDA-type receptors for glutamate (Zukin *et al.*, 1988). Increased glutamate binding sites have been reported in the orbital cortex of schizophrenics at post-mortem (Deakin *et al.*, 1988).

There is substantial evidence (see Lyon, this vol., Ch. 11) that long-term treatment with amphetamine by implanted pellet or multiple daily injections, can produce a behavioural syndrome in animals that closely resembles paranoid schizophrenia in humans. The picture seems to result from DA release, possibly accompanied by modified receptor sensitivity induced by repeated or prolonged stimulation.

13.3.3 Lesion-induced secondary mania

Affective disorders are frequently seen in patients with brain damage due to stroke, tumour or head injury; in these patients, depression is more commonly associated with lesions in the left hemisphere, and mania with lesions in the right (Robinson *et al.*, 1984). In one series of 17 brain-injured patients presenting with mania, 12 had right hemisphere lesions, and all but one of the remainder had bilateral lesions (Robinson *et al.*, 1988). Some 25 cases have been reported of mania following head injury (Bracken, 1987; Riess *et al.*, 1987; Nizamie *et al.*, 1988), though only nine of these fulfill the criteria for secondary mania (Krauthammer & Klerman, 1978). Left hemisphere lesions in frontal and temporal regions have also been associated with secondary mania (Yatham, 1987).

Although frontal and temporal lobe lesions are particularly involved, a male patient has been described who developed a propensity for manic attacks in association with an infarct limited to the right parietal area (Cohen & Niska, 1980). Frontal lobe damage will commonly produce fatuous euphoria and disinhibition, but not the full manic picture (Lishman, 1978). Right hemispherectomy has also been described as leading to bipolar disorder (Forrest, 1982). A case of secondary seasonal affective disorder has been described in association with a deep right hemisphere tumour (Hunt & Silverstone, 1990). On the basis of the EEG power spectral analysis, Flor-Henry (1969) postulated that mania is generated in the right hemisphere leading to activation of the left frontal and temporal regions by reduced transcallosal inhibition.

Epilepsy can lead to secondary manic states, but no particular lateralisation or lobar focus has been reliably associated (Perez & Trimble, 1980). Shulman & Post (1980) described several elderly patients presenting for the first time with mania in whom no precipitating factor was found although underlying cerebral pathology was suspected. Berrios (1986) has reviewed the concept of 'presbyophrenia' in which memory impairment suggestive of organic brain disease is associated with features of mania such as excitement and overactivity.

Whether deeper 'limbic' areas are affected by these lesions is unclear and one might hope for help from animal models. Experiments in rats have demonstrated hyperactivity resulting from right cerebral hemisphere lesions, particularly in the vicinity of the frontal pole (Pearlson *et al.*, 1982). The authors considered the effects to be linked to interruption of ascending NA pathways and the hyperactivity was prevented by concurrent administration of a tricyclic antidepressant (Robinson & Bloom, 1977); the latter finding suggests this model may relate to agitated depression rather than mania.

13.3.4 Lesion-induced schizophrenia

Illnesses virtually indistinguishable from schizophrenia may be associated with brain lesions, a subject comprehensively reviewed by Davison & Bagley (1969). In spite of the difficulties in interpretation, a causal association emerges with lesions in the temporal lobes and diencephalon. The association with the temporal lobes applies with tumours, head injury and epilepsy, and that with the diencephalon applies to tumours of the pituitary which lead to compression of adjacent structures such as the hypothalamus. Late-onset schizophrenia may be significantly associated with cerebrovascular disease or cerebral atrophy, but insufficient information is available on the localisation of lesions. In dementia, paranoid ideation may arise in response to forgetfulness or to other psychological deficits, or to the changing attitudes of those around the patient. Sensory deprivation, for instance deafness or blindness, and social isolation, have also long been recognised as contributing to paranoid developments and hence the greater frequency of these in the elderly.

In addition there is a statistical association between lesions of the hemisphere dominant for language (usually the left) and schizophrenia. This applies especially with temporal lobe epilepsy (TLE) but also with non-epileptogenic lesions. Slater *et al.* (1963) described 69 patients in whom a long-term schizophrenic-like illness had developed following epilepsy, after a mean duration of epilepsy of 14 years. The majority of patients had TLE, mostly with a focal lesion in one or both temporal lobes. The association with TLE is interesting in relation to the evidence that amygdaloid kindling in animals may lead to DA supersensitivity (Lyon, this vol., Ch. 11). However, the development of psychosis was not related to the severity or frequency of fits. There is a strong association between foci in the left temporal lobe and the development of schizophrenia (Flor-Henry, 1969). A separate mechanism probably underlies the acute transient schizophrenia-like states that can occur during a period of reduced epileptic activity, including the commencement of anticonvulsant medication in some cases (Landolt, 1958).

EEG abnormalities occur in up to 25% of young acute schizophrenics, especially in catatonic cases (Hill, 1957). There is CAT scan evidence of left-sided pathology (hypodensity) in schizophrenia (Reveley *et al.*, 1987) and the higher resolution of nuclear magnetic resonance (NMR) has revealed temporal lobe shrinkage, although not specifically left-sided (Suddath *et al.*, 1989). Post-mortem studies have sometimes found more left-sided neuropathology in schizophrenia (Brown *et al.*, 1986). Cerebral blood flow studies have indicated greater left hemisphere blood flow in acute schizophrenia (Gur *et al.*, 1987), and reduced frontal lobe blood flow in chronic schizophrenics with negative symptoms (Weinberger *et al.*, 1986). Abnormal interhemispheric signals via the corpus callosum have been postulated, but a patient with schizophrenia has been reported with NMR evidence of aplasia of the callosum (Lewis *et al.*, 1988). Finally, there is evidence of asymmetrical levels of DA and higher concentrations on the left side, in the amygdala of schizophrenics (Reynolds, 1983). This is distinct from the asymmetry of neurotransmitter that is found in the normal human brain (Glick *et al.*, 1982; Flor-Henry, 1986).

Animal models based on hippocampal lesions (see Lyon, this vol., Ch. 11) suggest that bilateral lesions of this area of the temporal lobes produce a behavioural state resembling that of DA overstimulation. As yet, however, it is unclear whether this provides a model of schizophrenia rather than mania, for locomotor activity is increased as well as stereotyped behaviour. The neuropharmacological relationships between the hippocampus and amygdala, and the nucleus accumbens and basal ganglia, appear to require much further animal work, which should include the possibility of lateralised differences.

13.4 DRUG TREATMENT OF MANIA AND SCHIZOPHRENIA

13.4.1 Antipsychotic drug treatment

13.4.1.1 *Schizophrenia*

Antipsychotic drugs have in common the ability to block DA receptors (Carlsson & Lindqvist, 1963).

Many of these drugs have additional pharmacological actions but the potency of their antipsychotic effect correlates closely with their potency as DA antagonists (Creese *et al.*, 1976). However, the selective DA receptor-blocking drugs such as pimozide and sulpiride, are less useful in the initial stages of treatment of acute schizophrenia, especially when the patient is behaviourally disturbed, as many patients with paranoid schizophrenia are. It seems that drugs which have additional actions, such as blockade of histamine and alpha-1-adrenergic receptors, have an advantage at the start of treatment. This may correspond to their sedative property. There is evidence that NA release may be activated in acute paranoid schizophrenia; for instance CSF levels of NA and its metabolite MHPG are raised in some patients (Lake *et al.*, 1980; Glazer *et al.*, 1987).

When schizophrenic symptoms are measured over several weeks of treatment under double-blind placebo-controlled conditions, three facts become apparent (Cole *et al.*, 1964; Johnstone *et al.*, 1978). Firstly, the time-course of drug-related improvement in positive psychotic symptoms is slow, being only slight during the first week and showing greater change during the second and third weeks of treatment, followed by further gradual improvement which may continue for several months if the illness has been untreated for a long time. Secondly, different antipsychotic drugs, used in adequate doses, show no difference in the time-course or extent of their effects on schizophrenic symptoms (Cole *et al.*, 1964). This applies also to the newer benzamide remoxipride compared to haloperidol (Lewander, 1989). The differences that occur are in the incidence of side-effects. The third observation is that whereas positive psychotic symptoms generally improve – although to varying extents in different patients – there is a tendency for some of the negative symptoms to persist; that is, antipsychotic drugs seem to make little difference to the residual schizophrenic defect state (Carpenter *et al.*, 1988). However, some negative symptoms that are apparent at the time of the acute episode may resolve and should be regarded as secondary (Goldberg, 1985; Andreasen, 1985; Breier *et al.*, 1987). Even positive symptoms show a poor response in some patients; these cases have not been found to differ from cases showing a good response, in terms of family history, obstetric history or CAT scans, but do tend to have an earlier age of onset (Nimgaonkar *et al.*, 1988).

13.4.1.2 Antipsychotic drugs in mania

Manic symptoms also improve in response to antipsychotic drugs. In clinical practice, haloperidol is preferred by many psychiatrists for the treatment of mania. Chlorpromazine is effective, but is more sedative, and manic patients often resent being made to feel drowsy by medication. The selective DA antagonist, pimozide, is also effective in mania (Cookson & Silverstone, 1976; Post *et al.*, 1976). In comparison with chlorpromazine, pimozide produces less immediate improvement in behavioural disturbance, being less sedative, but over the course of two weeks it produces at least as much improvement in manic symptoms (Post *et al.*, 1980; Cookson *et al.*, 1981). The time-course of improvement with these drugs is notably faster in mania than in schizophrenia, improvement being seen within 24 h, increasing for three days, then more gradually thereafter, for two weeks (Silverstone & Cookson, 1983).

Raised cortisol levels in mania become normalised more rapidly during treatment with haloperidol than with pimozide, perhaps because haloperidol is an alpha-1 adrenergic blocker as well as a DA antagonist (Cookson *et al.*, 1985).

13.4.1.3 Antipsychotic drugs in animal models

Antipsychotic drugs, including haloperidol and pimozide (Marsden & Curzon, 1976), antagonise amphetamine effects in animals, and in people pimozide reduces amphetamine-induced arousal (Silverstone *et al.*, 1981). However, it is not clear whether there are differences in the time-course of effects of antipsychotic drugs upon the animal models of mania and schizophrenia. It would seem important to establish the time-course of these effects in animal models.

In the PCPA model of mania, pimozide antagonises hyperactivity but only at doses higher than those effective in the amphetamine model (Marsden & Curzon, 1976).

13.4.1.4 Atypical antipsychotic drugs

Drugs that block behaviour mediated by mesolimbic DA pathways, but not behaviour associated with the nigrostriatal DA pathway, can be expected to have antipsychotic properties without producing Parkinsonism. Such drugs are sometimes referred to as 'atypical antipsychotics'. They include the substituted benzamides (sulpiride, remoxipride, raclopride), thioridazine, clozapine (a dibenzazepine), and zetidoline. Thioridazine is probably a typical antipsychotic with potent intrinsic anti-cholinergic activity. Likewise, zetidoline produces fewer extrapyramidal side effects than haloperidol (Silverstone *et al.*, 1984), but has anti-cholinergic activity (Szabadi *et al.*, 1980). Clozapine is also a potent acetylcholine (ACh) receptor-blocker and 5-HT$_2$ receptor-blocker. However, the benzamide drugs appear to be more specific for the D2 subtype of DA receptors.

Clozapine and thioridazine have anti-manic as well as anti-schizophrenic effects. Sulpiride has also shown anti-manic properties (Christie *et al.*, 1988) but there is some doubt whether another benzamide drug remoxipride has anti-manic effects, at doses that are effective in schizophrenia (Cookson & Mitchell, 1985; Chouinard & Steiner, 1986).

The precise mechanisms whereby these drugs spare the basal ganglia or avoid producing Parkinsonism, are unclear. It is known, for instance, that thioridazine (Reynolds *et al.*, 1982), the benzamides and clozapine (Farde *et al.*, 1988), bind to striatal D2 receptors producing more than 70% occupancy at clinically effective doses. What appears to distinguish the benzamide drugs, is that they lack activity at D1 receptors (Kebabian & Calne, 1979). The clinical relevance of D1 receptors and their blockade, is unknown. It is also possible that the benzamides block only a subpopulation of D2 receptors, although this remains to be clarified (Hamblin *et al.*, 1984; Ögren *et al.*, 1986). These questions can be addressed in animal models.

The basis for the atypical nature of clozapine is also unclear. Recent work has drawn attention to the role of 5-HT at different receptor subtypes that may be blocked by antipsychotic drugs. The specific 5-HT$_2$ receptor antagonist ritanserin, has been found to counteract Parkinsonian side-affects

(Gelders *et al.*, 1985). It may therefore transpire that intrinsic 5-HT$_2$ receptor blocking properties, as well as anti-muscarinic properties, can endow an antipsychotic drug with typical properties. Clozapine is ten times more potent in blocking 5-HT$_2$ receptors and five times more potent as an anticholinergic, than in blocking D2 receptors (Leysen & Niemegeers, 1985).

Another drug, rimcazole, which was effective in open studies in schizophrenia, has an action on sigma opioid receptors but does not block DA receptors (Ferris *et al.*, 1982).

13.4.1.5 *The role of acetylcholine*

Anticholinergic drugs are commonly used in conjunction with antipsychotic drugs to counteract extrapyramidal side effects such as dystonia, Parkinsonism, and to some extent, akathisia. There is a tendency for positive psychotic symptoms (and insomnia) to worsen slightly when the anticholinergic drug procyclidine, is given to relieve extrapyramidal symptoms during the initial treatment of acute schizophrenia (Johnstone *et al.*, 1983; Singh *et al.*, 1987) although this does not happen during the longer term treatment of chronic schizophrenia (Bamrah *et al.*, 1986). There are also suggestions that anticholinergic drugs may counteract negative symptoms (Fisch, 1987). However, procyclidine had no influence on negative symptoms in a placebo-controlled study (Johnstone *et al.*, 1983).

In the case of mania, there is stronger evidence for a cholinergic–dopaminergic balance. Thus, manic symptoms can be improved by drugs which potentiate ACh such as the cholinesterase inhibitor physostigmine (Janowsky *et al.*, 1973*b*) and the ACh precursor lecithin (Cohen *et al.*, 1982). However, anticholinergics usually produce only a slight reversal of the effects of anti-manic drugs (Johnstone *et al.*, 1988). The interpretation of this is difficult since some anticholinergics, particularly benzhexol, have DA reuptake-inhibiting properties (Horn *et al.*, 1971). It is important to determine the effect of cholinesterase inhibitors in animal models of mania.

In animals, anticholinergic drugs counteract certain DA induced behaviours such as stereotypies and catalepsy but have little effect upon locomotor stimulation by DA agonists.

13.4.1.6 *Negative symptoms and 'activating' drugs*

Some negative symptoms such as mutism and emotional withdrawal can be secondary to positive psychotic phenomena, e.g., when the patient is preoccupied with delusions and hallucinations, or when depression accompanies the acute psychotic phase. These secondary negative symptoms tend to improve during treatment with antipsychotic drugs. However, the negative symptoms that persist (the schizophrenic defect state) can lead to great disabilities. At least two drugs in use (pimozide and sulpiride) are claimed to improve this state in low doses by 'activating' the patient (Falloon *et al.*, 1978; Petit *et al.*, 1987). Animal models have suggested that an activating effect on locomotor behaviour can arise through preferential blockade of presynaptic DA autoreceptors (Zetterström *et al.*, 1986). Clinically, the activating effect is not specific to schizophrenic patients but may also be useful in other patients who are profoundly lacking in drive. This activating effect occurs with small doses, but with larger doses, the postsynaptic dopamine-blocking effect becomes dominant (Petit *et al.*, 1987); thus, improvement in negative symptoms is not seen in those chronic schizophrenics who need higher doses to control positive symptoms (Gerlach *et al.*, 1985).

A beneficial effect on negative symptoms has also been claimed for the 5-HT$_2$ antagonist ritanserin (Gelders *et al.*, 1985), and for the MAOI tranylcypromine (Bucci, 1987), both of which have mood-improving properties. Further studies are needed in animal models of negative symptoms, to explore such drug effects.

13.4.1.7 *DA agonists*

Indirect DA agonists such as amphetamine or L-dopa, tend usually to worsen the positive symptoms of schizophrenia (Janowsky *et al.*, 1973*a*; Angrist *et al.*, 1973, 1980). Paradoxically, a proportion of schizophrenics (Van Kammen *et al.*, 1982) and of manics (Garvey *et al.*, 1987) are improved by single small doses of amphetamine. Likewise, although postsynaptic D2 agonists such as bromocriptine can precipitate psychosis, they can sometimes be given to schizophrenic patients without exacerbating the psychosis. Negative symptoms and depressed mood improve in some schizophrenics given intravenous d-amphetamine in placebo-controlled tests (Angrist *et al.*, 1982; Van Kammen & Boronow, 1988).

Preferential stimulation of inhibitory presynaptic autoreceptors is a possible strategy for new antipsychotic drugs which might, however, be expected to worsen negative symptoms (Carlsson, 1988). Improvement of manic symptoms has been reported with the presynaptic DA agonist, piribedil (Post *et al.*, 1976). Early open studies with single doses of apomorphine also suggested some anti-manic and anti-schizophrenic effects (Corsini *et al.*, 1977) but placebo-controlled studies in schizophrenia indicated that this may be only a non-specific sedation (Levy *et al.*, 1984). Moreover, no improvement in schizophrenic symptoms occurred using repeated oral doses of N-propyl-norapomorphine for three weeks (Tamminga *et al.*, 1986). Such negative findings with DA agonists may arise either through the development of tolerance to the presynaptic effect, or because the DA neurones of the mesolimbic system do not all contain autoreceptors; it has been suggested that the neurones of the mesocortical DA cells may not (Bannon & Roth, 1983).

Animal models of schizophrenia and mania may be used to investigate whether individual DA agonists have predominantly postsynaptic or presynaptic effects. Some drugs are partial agonists and the findings in animal models suggest they may be more useful than apomorphine in the treatment of acute psychosis (see Ahlenius, this vol., Ch. 12).

13.4.2 Other drugs in the treatment of schizophrenia and mania

Some drugs are more useful in the treatment of mania than of schizophrenia. These present useful tools for exploring the validity of animal models of the two conditions.

13.4.2.1 Lithium

Lithium has therapeutic effects in acute mania (Cade, 1949). It is not sedative; the anti-manic effect is not usually evident until after three days or more of treatment, and develops gradually over two weeks. This makes it unsuitable alone for the treatment of moderate or severe mania (Prien *et al.*, 1972). Only 60–70% of patients with acute

mania can be expected to show significant improvement on lithium (Tyrer, 1985). Patients with a family history of bipolar illness and with predominantly elated-grandiose mania without psychotic features, are more likely to benefit (Prien *et al.*, 1972). Patients in whom mania precedes depression may be more likely to respond to lithium than those in whom depression precedes mania (Kukopulos & Reginaldi, 1980; Kukopulos *et al.*, 1980). A comprehensive review of the use of lithium in schizophrenia led to the conclusion that although about one third of patients appeared to have derived some benefit while on lithium, virtually none made a full recovery (Delva & Letemendia, 1982). A recent placebo-controlled study of acute psychotic patients showed no effect of lithium on positive schizophrenic symptoms, but a trend for manic symptoms to improve in all diagnostic groups (Johnstone *et al.*, 1988).

13.4.2.2 Effects of lithium in models

Lithium attenuates but does not abolish amphetamine-induced euphoria in people (Van Kammen & Murphy, 1975; Angrist & Gershon, 1979). In schizophrenia, lithium attenuates amphetamine-induced activation-euphoria but not the increase in psychosis (Van Kammen *et al.*, 1985). The much shorter plasma half-life of lithium in rodents, compared to people, imposes difficulties in reproducing similar plasma levels in animals (Wood *et al.*, 1986). In animals, lithium is reported to reduce amphetamine-induced hyperactivity, with an effect on locomotor stimulation but not on stereotypies (see Lyon, this vol., Ch. 11). Lithium has no effect on stereotypy in response to apomorphine (Pittman *et al.*, 1984), but does antagonise the ability of systemic apomorphine to reduce locomotor activity (Staunton *et al.*, 1982) and does reduce the hyperactivity induced by infusions of DA into the nucleus accumbens (Costall *et al.*, 1985). It seems, therefore, that the models of mania based on DA agonists show responses to lithium similar to, but less impressive than, those of manic patients. Lithium is known to antagonise receptors linked to adenylate cyclase (Hendler, 1978), and is therefore thought to reduce DA responses mediated by D-1 but not D-2 receptors.

It is possible that the anti-manic effect of lithium is exerted partly on neurotransmitters other than

DA. There is growing evidence that lithium potentiates 5-HT responses, particularly at 5-HT$_1$ receptors (Wood & Goodwin, 1987). In the PCPA model of mania, lithium reduces the aggression but not the hyperactivity (see Lyon, this vol., Ch. 11). This model might therefore correspond to a lithium-resistant type of mania. However, lithium is known to have beneficial effects on aggressive behaviour in situations other than mania (Tyrer, 1988). The interaction of antagonists specific for different subtypes of 5-HT receptors with the PCPA model will be of interest. Likewise, the ability of lithium to potentiate ACh responses, perhaps through the phospho-inositol second messenger system, suggests that the interaction of lithium with anti-cholinergic drugs in animal models would be informative.

Lithium effects upon aspects of normal animal behaviour, for instance hind-leg rearing, a feature of exploratory behaviour in rats (see Lyon, this vol., Ch. 11), have counterparts in the subjective effects of lithium in euthymic bipolar patients (Johnson, 1984).

13.4.2.3 Carbamazepine

The use of carbamazepine in acute mania is increasingly well established following the work of Okuma *et al.* (1973) and Ballanger & Post (1980). A review of controlled trials indicated that 60% of cases of mania show a good response to carbamazepine alone (Cookson, 1988). Another anticonvulsant, valproate, may also be effective (Emrich *et al.*, 1985).

Manic patients who improve on carbamazepine are not necessarily the same as those who improve on lithium. Indeed, 61% of lithium non-responders improved on carbamazepine (Post *et al.*, 1987). Okuma *et al.* (1979) found no association between the degree of improvement on carbamazepine and the severity of mania, number of previous manic episodes, type of illness, age, sex, or age at initial onset of illness. But in another series, patients who improved on carbamazepine had higher ratings of dysphoria (anxiety, depression and anger) and of psychosis, than those who did not; also, patients with a family history of bipolar disorder were less likely to improve on carbamazepine than manic patients with no such family history (Post *et al.*, 1987). If rapid cyclers were more likely than other

manic patients to respond to carbamazepine, this would support the role of 'kindling' in the development of bipolar disorder. The proportion of rapid cyclers reported to respond, in the literature (which may however be biassed against publication of negative findings) is 74% (Emrich, 1987), compared to the proportion (60%) of responders in other controlled studies (Cookson, 1988). Also, Post *et al.* (1987) found that responders to carbamazepine had more frequent episodes in the year prior to treatment than non-responders. However, Joyce (1988) found carbamazepine effective in the long-term treatment of only a minority of rapid cyclers.

The evidence so far does not suggest that carbamazepine has much anti-schizophrenic action, although it may be useful as an adjunct to antipsychotic drugs in behaviourally disturbed or aggressive patients, including schizophrenics (Hakola & Laulumaa, 1982; Neppe, 1982; Luchins, 1983; Mattes, 1984; Klein *et al.*, 1984). However, valproate has been reported to increase symptoms in schizophrenics withdrawn from neuropleptic medication, although this might be coincidental (Lautin *et al.*, 1980).

Carbamazepine does not antagonise methylphenidate-induced euphoria in people (Meyendorff *et al.*, 1985). The effect of carbamazepine on kindling is described above (Section 13.2.1); further studies of the effects of carbamazepine in animal models of mania and schizophrenia are required.

13.4.2.4 Calcium antagonists

There are several reports of the efficacy of the calcium antagonist, verapamil, in acute mania (Dubovsky *et al.*, 1982; Giannini *et al.*, 1984; Dinan *et al.*, 1988). Verapamil is not effective in schizophrenia (Pickar *et al.*, 1987). Other classes of calcium antagonist with more specific actions remain to be investigated.

13.4.2.5 Drugs interacting with serotonin (5-HT)

Non-specific 5-HT antagonists were at first thought to be beneficial in mania, but a placebo-controlled trial of methysergide found that it worsened the condition (Coppen *et al.*, 1969). PCPA has no therapeutic effect in manic patients (Goodwin & Ebert, 1973). Trials of antagonists with greater specificity for particular subtypes of

5-HT receptor would be of interest. These agents seem not to benefit positive schizophrenic symptoms although beneficial effects on Parkinsonism and on negative symptoms, have been claimed for the 5-HT$_2$ antagonist, ritanserin (Gelders *et al.*, 1985).

L-tryptophan has been reported to ameliorate manic symptoms in mild cases (Prange *et al.*, 1974), but this was not confirmed (Chambers & Naylor, 1978). d,l-Fenfluramine has non-specific sedative affects in mania (Cookson & Silverstone, 1976). No effects on clinical ratings were found in a placebo-controlled trial of d-fenfluramine, in overweight chronic schizophrenics on antipsychotic medication (Goodall *et al.*, 1988). A non-significant improvement in negative symptoms was observed in schizophrenics on d,l-fenfluramine (Stahl *et al.*, 1985); the racemic mixture is less specific than the d-isomer in releasing presynaptic 5-HT.

13.4.2.6 Ineffective drugs

Important negative findings on drug treatment include the absence of any consistent improvement in either mania (Davis *et al.*, 1980) or schizophrenia (Mueser & Dysken, 1983), using the opiate antagonist naloxone given regularly under placebo-controlled double-blind conditions. Likewise, in spite of encouraging reports based on animal models measuring passive avoidance and the extinction of active avoidance (De Wied *et al.*, 1978; see Lyon, this vol., Ch. 11), des-tyr-gamma-endorphin was ineffective in placebo-controlled double-blind trials in schizophrenia (Manchanda & Hirsch, 1981). Another peptide, CCK, has similar properties in these screening models, but this may also be a 'false-positive' (see Lyon, Ahlenius, this vol., Ch. 11, 12); CCK has not been found clinically effective, although the trials so far can be criticised (Montgomery & Green, 1988). It is claimed that the use of buprenorphine or methadone as an adjunct to neuroleptics leads to symptomatic improvement in therapy-resistant chronic schizophrenics (Brizer *et al.*, 1985; Schmauss *et al.*, 1987), but this may be a non-specific effect.

13.4.3 Prophylactic drug treatment

The mainstay of prophylactic treatment for schizophrenia is antipsychotic medication, which can be most satisfactorily delivered by injections of depot preparations given at intervals of one to six weeks by Community Psychiatric Nurses. Placebo-controlled studies have shown that without prophylactic antipsychotic medication, there is a one-year relapse rate of 60–70% and a two-year relapse rate of 80%. This compares with a one-year relapse rate of about 10% in patients maintained on medication (Davis, 1975; Kane, 1984). The doses required are quite small compared to those used in acute treatment. The possibility that abrupt withdrawal of antipsychotic medication may increase the risk of relapse has been mooted but not proved.

The first line of prophylaxis for mania and bipolar disorder, is lithium treatment. However, this is effective in only 50–70% of cases and patients with rapid cycling are less likely to benefit (Dunner & Fieve, 1975). Furthermore, many patients refuse to take lithium because of perceived side effects. In practice, only a small proportion of those at risk of further manic episodes are actually taking prophylactic lithium (McCreadie & Morrison, 1985). The advent of lithium treatment has not dramatically altered the statistics for hospital admissions of manic patients (Dickson & Kendell, 1986). One possible reason for this is that sudden cessation of lithium may lead to recurrence of mania within two weeks in as many as 50% of cases (Mander & Loudon, 1988). Thus, lithium withdrawal effects upon animal behaviour warrant study as models of mania.

Clearly, alternatives to lithium are needed. Carbamazepine is often effective in patients who have not benefitted from lithium. The usefulness of verapamil remains to be established. For intractable manic-depressives who have not benefitted from lithium or carbamazepine, or a combination of these two drugs, depot antipsychotic medication is sometimes effective in stabilising their condition, although the side effects, particularly sedation and tardive dyskinesia, seem more troublesome than in schizophrenic patients (Ahlfors *et al.*, 1981; Waddington & Youssef, 1988).

The profound effect of abrupt withdrawal of lithium in triggering manic relapses can be elucidated in animal models. Does lithium withdrawal render animals more sensitive to the stimulant effects of amphetamine or to the effects of PCPA? It is known that following antipsychotic drug treatment, a number of behavioural responses are potentiated, but the relevance of this to withdrawal psychosis needs further investigation using animal models.

13.5 ANIMAL MODELS AND THE NEUROBIOLOGY OF SCHIZOPHRENIA AND MANIA

The available biochemical and psychopharmacological evidence reviewed above can be interpreted as providing a tentative basis for neurobiological models of schizophrenia and mania. In mania, there is compelling evidence for an important role for DA pathways (Cookson & Silverstone, 1976; Gerner *et al.*, 1976). The localisation of the relevant pathways is uncertain. The fact that the mesolimbic DA pathway will support electrical intracranial self-stimulation suggests that this pathway may be involved (see Lyon, this vol., Ch. 11; Crow; 1973; Redgrave & Dean, 1981). The projection to the nucleus accumbens is of particular interest since DA infusions locally will produce locomotor stimulation and other features of the manic syndrome (see Lyon, this vol., Ch. 11); infusion of DA to the amygdala also produces hyperactivity (see above).

Hippocampal lesions appear to activate the mesolimbic DA pathway (see Lyon, this vol., Ch. 11). Mogenson (1984) has provided evidence that DA afferents to the nucleus accumbens modulate its function as a gating mechanism for signals from the hippocampus and amygdala requiring exploratory motor behaviour. Much remains to be learned of the function of DA pathways and their integration with other neurotransmitters and limbic structures. Non-DA pathways, which are sensitive to lithium, must also be involved in mania. There is evidence of a reciprocal relationship between DA and ACh pathways; also, 5-HT may play an inhibitory role and NA pathways may be activated in mania (Silverstone & Cookson, 1982).

In schizophrenia there is pharmacological evidence for a possible role of DA pathways in the pathogensis of positive symptoms, with a small reciprocal interaction with ACh; there is little evidence about the basis of negative symptoms, but inactivity of DA pathways has been postulated (Crow, 1980). It has also been postulated that a developmental abnormality associated with diminished prefrontal cortical function underlies some deficit symptoms and leads to compensatory overactivity of the mesolimbic DA pathway (Weinberger, 1987).

Post-mortem studies in schizophrenia have revealed few replicable abnormalities (Reynolds, 1988). There is no evidence of damage to DA neurones, so any DA underactivity in negative states must be functional. Evidence for overactivity of DA pathways is limited to the increase in DA in the left amygdala (see above) and to increased numbers of D2 receptors in the brain. The latter could arise through previous drug treatment; this could be clarified by *in vivo* visualisation of receptors in drug-naive schizophrenics, using position emission tomographic (PET) scans, but the findings as yet are difficult to reconcile (Waddington, 1989). Negative symptoms have been associated at post mortem with reduced CCK and somatostatin levels in temporal lobe areas (Ferrier *et al.*, 1983).

For the future, the techniques of molecular biology and genetics and of functional brain imaging, may provide insights into the neurobiology of these illnesses, and animal models will remain important in understanding their neurochemistry and in the development of new treatments.

REFERENCES

Abrams, R., Taylor, M.A. & Coleman Solurow, A. (1979). Catatonia and mania: patterns of cerebral dysfunction. *Biological Psychiatry* **14**, 111–17.

Ahlfors, U.G., Baastrup, P.C., Dencker, S.J. & Elgin, K. (1981). Flupenthixol decanoate in recurrent manic depressive illness. *Acta Psychiatrica Scandinavica* **64**, 226–37.

Ambelas, A. (1987). Life events and mania: a special relationship? *British Journal of Psychiatry* **150**, 235–40.

American Psychiatric Association (1980). *Diagnostics and Statistical Manual of Mental Disorders*, 3rd edn. Washington: American Psychiatric Association.

Andreasen, N.C. (1982). Negative symptoms in schizophrenia: definition and reliability. *Archives of General Psychiatry* **39**, 784–8.

Andreasen, N.C. (1985). Positive vs negative schizophrenia: a critical evaluation. *Schizophrenia Bulletin* **11**, 380–9.

Angrist, B.M. & Gershon, S. (1970). The phenomenology of experimentally induced amphetamine psychosis – preliminary observations. *Biological Psychiatry* **2**, 95–107.

Angrist, B.M. & Gershon, S. (1979). Variable attenuation of amphetamine effects by lithium. *American Journal of Psychiatry* **136**, 806–10.

Angrist, B.M., Sathananthan, G. & Gershon, S. (1973). Behavioral effects of L-Dopa in schizophrenic patients. *Psychopharmacology* **31**, 1–12.

Angrist, B.M., Rotrosen, J. & Gershon, S. (1980). Differential effects of amphetamine on negative vs positive symptoms in schizophrenia. *Psychopharmacology* **11**, 1–3.

Angrist, B.M., Pesselow, E., Rubinstein, M., Corwin, J. & Rotrosen, J. (1982). Partial improvement in negative schizophrenic symptoms after amphetamine. *Psychopharmacology* **78**, 128–30.

Aronson, T.A. & Shukla, S. (1987). Life events and relapse in bipolar disorder: the impact of a catastrophic event. *Acta Psychiatrica Scandinavica* **75**, 571–6.

Astrup, C., Fossum, A. & Holmboe, R. (1959). A follow-up study of 270 patients with acute affective psychoses. *Acta Psychiatrica Scandinavica* **34**, Suppl. 135.

Ballenger, J. & Post, R.M. (1980). Carbamazepine (Tegretol) in manic-depressive illness: a new treatment. *American Journal of Psychiatry* **137**, 782–9.

Bamrah, J.S., Kumar, V., Krska, S.J. & Soni, S.D. (1986). Interactions between procyclidine and neuroleptic drugs. Some pharmacological and clinical aspects. *British Journal of Psychiatry* **149**, 726–33.

Bannon, M.J. & Roth, R.H. (1983). Pharmacology of mesocortical dopamine neurones. *Pharmacology Reviews* **35**, 53–68.

Baron, M., Rainer, J.D. & Risch, N. (1981). X-linkage in bipolar affective illness. *Journal of Affective Disorders* **3**, 141–57.

Beigel, A., Murphy, D.C. & Bunney, W.E. (1971). The manic-state rating scale. *American Journal of Psychiatry* **25**, 256–62.

Bell, D.S. (1973). Comparison of amphetamine psychosis and schizophrenia. *British Journal of Psychiatry* **111**, 701–7.

Berrios, G.E. (1986). Presbyophrenia: the rise and fall of a concept. *Psychological Medicine* **16**, 267–75.

Bertelsen, A., Harvala, B. & Hauge, M. (1977). A Danish twin study of manic-depressive disorders. *British Journal of Psychiatry* **130**, 330–51.

Besser, G.M., Butler, P.W.P., Landon, J. & Rees, L. (1969). Influences of amphetamine on plasma corticosteroid and growth hormone levels in man. *British Medical Journal* **iv**, 528–30.

Bracken, P. (1987). Mania following head injury. *British Journal of Psychiatry* **150**, 690–2.

Breier, A., Wolkowitz, O.M., Doran, A.R., Roy, A., Boronow, J., Hommer, D.W. & Pickar, D. (1987). Neuroleptic responsivity of negative and positive symptoms in schizophrenia. *American Journal of Psychiatry* **144**, 1549–55.

Brizer, D.A., Hartmann, N., Sweeney, J. & Millman, R.B. (1985). Effect of methadone plus neuroleptics on treatment-resistant chronic paranoid schizophrenia. *American Journal of Psychiatry* **142**, 1106–7.

Brockington, I.F., Kendell, R.E. & Leff, J.P. (1978). Definitions of schizophrenia: concordance and prediction of outcome. *Psychological Medicine* **8**, 387–98.

Brown, G.W. & Birley, J.T. (1968). Crises and life changes and the onset of schizophrenia. *Journal of Health and Social Behaviour* **9**, 203.

Brown, G.W., Birley, J.T. & Wing, J.K. (1972). Influence of family life on the course of schizophrenic disorders: a replication. *British Journal of Psychiatry* **121**, 241–58.

Brown, R., Colter, N., Corsellis, J.A.N., Crow, T.J., Frith, C.D., Jagoe, R., Johnstone, E.C. & Marsh, L. (1986). Post-mortem evidence of structural brain changes in schizophrenia. *Archives of General Psychiatry* **43**, 36–42.

Bucci, L. (1987). The negative symptoms of schizophrenia and the monoamine oxidase inhibitors. *Psychopharmacology* **91**, 104–8.

Bunney, W.E. (1978). Psychopharmacology of the switch process in affective disorders. In: *Psychopharmacology: A Generation of Progress*, eds. M.A. Lipton, A. DiMassio & K.F. Killam, pp. 1249–59. New York: Raven Press.

Cade, J.F.J. (1949). Lithium salts in the treatment of psychotic excitement. *Medical Journal of Australia* **36**, 349–52.

Carlsson, A. (1988). Dopamine-autoreceptors and schizophrenia. In: *Receptors and Ligands in Psychiatry*, eds. A.K. Sen & T. Lee, pp. 1–10. Cambridge University Press: Cambridge.

Carlsson, A. & Lindqvist, M. (1963). Effect of chlorpromazine and haloperidol on formation of 3-methoxytyramine and normetanephrine in mouse brain. *Acta Pharmacologica Toxicologica* **20**, 140–4.

Carlsson, A., Svennerholm, L. & Winblad, B. (1980). Seasonal and circadian monoamine variations in human brain examined post-mortem. *Acta Psychiatrica Scandinavica* **61**, Suppl. 280.

Carlsson, G.A. & Goodwin, F.K. (1973). The stages of mania: A longitudinal analysis of the manic episode. *Archives of General Psychiatry* **28**, 221–8.

Carpenter, W.T., Heinrichs, D.W. & Wagman, A.M.I. (1988). Deficit and non-deficit of schizophrenia: The concept. *American Journal of Psychiatry* **145**, 578–83.

Chambers, A.C. & Naylor, G.J. (1978). A controlled trial of L-tryptophan in mania. *British Journal of Psychiatry* **132**, 555–9.

Chouinard, G. & Steiner, W. (1986). Remoxipride in the treatment of acute mania. *Biological Psychiatry* **21**, 1429–33.

Christie, J.E., Whalley, L., Hunter, R., Bennie, J. & Fink, G. (1988). Sulpiride treatment of acute mania with a comparison of effects on plasma hormone concentrations of lithium and sulpiride treatment. *Journal of Affective Disorders* **16**, 115–20.

Cleghorn, J.M. & Brown, G.M. (1988). Neuroendocrine studies in schizophrenia. In: *Receptors and Ligands in Psychiatry*, eds. A.K. Sen & T. Lee, pp. 127–46. Cambridge University Press: Cambridge.

Cohen, S.I. (1980). Cushings syndrome: A psychiatric study of 29 patients. *British Journal of Psychiatry* **136**, 120–4.

Cohen, B.M., Lipinski, J.F. & Altesman, R.I. (1982). Lecithin in the treatment of mania. *American Journal of Psychiatry* **139**, 1162–4.

Cohen, M.R. & Niska, R.W. (1980). Localised right cerebral hemisphere dysfunction and recurrent mania. *American Journal of Psychiatry* **137**, 847–8.

Cole, J.O., Klerman, G.L., Goldberg, S.C. *et al.* (1964). Phenothiazine treatment in acute schizophrenia. *Archives of General Psychiatry* **10**, 246–61.

Connell, P.H. (1958). *Amphetamine Psychosis*, Maudsley Monograph No. 5. Oxford University Press: Oxford.

Cookson, J.C. (1981). Oestrogens, dopamine and mood (Letter). *British Journal of Psychiatry* **139**, 365–6.

Cookson, J.C. (1982). Post-partum mania, dopamine and oestrogens. *Lancet* **ii**, 672.

Cookson, J.C. (1985). The neuroendocrinology of mania. *Journal of Affective Disorders* **8**, 233–41.

Cookson, J.C. (1987*a*). Excitement states. In: *Psychiatric Differential Diagnosis*, eds. J. Pfeffer & G. Waldron, pp. 127–33. Churchill Livingstone: Edinburgh.

Cookson, J.C. (1987*b*). Funny movements. In: *Psychiatric Differential Diagnosis*, eds. J. Pfeffer & G. Waldron, pp. 15–22 Churchill Livingstone: Edinburgh.

Cookson, J.C. (1988). Carbamazepine in acute mania: a practical review. *International Clinical Psychopharmacology* (Suppl.), pp. 11–22.

Cookson, J.C. (1989). The neuroendocrinology of schizo-affective disorders. In: *New Directions in Affective Disorders*, eds. B. Lerer & E. Gershon. Springer Verlag: New York (in press).

Cookson, J.C. & Mitchell, M.J. (1985). Clinical outcome and plasma prolactin in manic patients treated with a novel dopamine antagonist. Abstract Number 421.3 IVth World Congress of Biological Psychiatry. Philadelphia.

Cookson, J.C. & Silverstone, T. (1976). 5-Hydroxytryptamine and dopamine pathways in mania: a pilot study of fenfluramine and pimozide. *British Journal of Clinical Pharmacology* **3**, 942–3.

Cookson, J.C., Silverstone, T. & Wells, B. (1981). A double-blind comparative clinical trial of pimozide and chlorpromazine in mania: a test of the dopamine hypothesis. *Acta Psychiatrica Scandinavica* **64**, 381–97.

Cookson, J.C., Silverstone, T. & Rees, L.H. (1982). Plasma prolactin and growth hormone levels in manic patients treated with pimozide. *British Journal of Psychiatry* **140**, 274–9.

Cookson, J.C., Moult, P., Wiles, D. & Besser, G.M. (1983). The relationship between prolactin levels and clinical ratings in manic patients treated with oral and intravenous 'test' doses of haloperidol. *Psychological Medicine* **13**, 279–85.

Cookson, J.C., Silverstone, T., Williams, S. & Besser, G.M. (1985). Plasma cortisol levels in mania: Associated clinical ratings and changes during treatment with haloperidol. *British Journal of Psychiatry* **146**, 498–502.

Coppen, A., Prange, A.J., Whybrow, P.C., Noguera, R. & Paez, J.M. (1969). Methysergide in mania: A controlled trial. *Lancet* **ii**, 338–40.

Corn, T.H. & Checkley, S.A. (1983). A case of recurrent mania with recurrent hyperthyroidism. *British Journal of Psychiatry* **143**, 74–6.

Corsini, G.U., Del Zompo, M., Marconi, S., Cianchetti, C., Mangoni, A. & Gessa, G.L. (1977). Sedative, hypnotic and antipsychotic effects of low doses of apormorphine in man. *Advances in Biochemical Psychopharmacology* **16**, 645–9.

Costall, B., Domeney, A.M. & Naylor, R.J. (1985). Dopamine and schizophrenia: a changing perspective. In: *Schizophrenia: New Pharmacological and Clinical Developments*, Royal Society of Medicine Monograph, Number 94, ed. A.A. Schiff, M. Roth & H. Freeman, pp. 65–77. Royal Society of Medicine Services Limited: London.

Creese, I., Burt, D.R. & Snyder, S. (1976). Dopamine receptor binding predicts clinical and pharmacological potencies of antischizophrenic drugs. *Science* **192**, 481–3.

Cremata, V.Y. & Koe, B.K. (1966). Clinical–pharmacological evaluations of p-chlorophenylalanine: A new serotonin-depleting agent. *Clinical Pharmacology and Therapeutics* **7**, 768–76.

Crocker, A.D. (1988). Interactions between thyroid and adrenocortical hormones on responses elicited by dopamine receptor agonists. *Psychopharmacology* **96**, Suppl. 249.

Crocker, A.D., Crocker, J.M., Wright, A. & Overstreet, D.H. (1984). The effect of adrenocortical hormones on dopamine receptor sensitivity. Paper presented at 14th CINP Congress. Abstract Number F64.

Crow, T.J. (1973). Catecholamine-containing neurones and electrical self-stimulation: 2. A theoretical interpretation and some psychiatric implications. *Psychological Medicine* **3**, 66–73.

Crow, T.J. (1980). Molecular pathology of schizophrenia: more than one disease process? *British Medical Journal* **280**, 66–86.

Crow, T.J. (1984). A re-evaluation of the viral hypothesis: Is psychosis the result of retroviral integration at a site close to the cerebral dominance gene? *British Journal of Psychiatry* **145**, 243–53.

Davis, J. (1975). Overview: Maintenance therapy in psychiatry: 1, Schizophrenia. *American Journal of Psychiatry* **132**, 1237–45.

Davis, G.C., Extein, R.I., Reus, V.I., Hamilton, W., Post, R.M., Goodwin, F.K. & Bunney, W.E. (1980). Failure of naloxone to reduce manic symptoms. *Archives of General Psychiatry* **137**, 1583–5.

Davison, K. & Bagley, C.R. (1969). Schizophrenia-like psychoses associated with organic disorders of the central nervous system, in R.N. Harrington, (ED): *Current Problems in Neuropsychiatry*, ed. O. Harrington, Royal College of Psychiatrists Special Publication No. 4, pp. 113–84. Ashford, England: Headley Bros. Ltd.

De Lisi, L.E., Goldin, L.R., Hamovit, V.R., Maxwell, M.E., Kutz, D. & Gershon, E.S. (1986). A family study of the association of increased ventricular size with schizophrenia. *Archives of General Psychiatry* **43**, 148–53.

De Lisi, L.E., Goldin, L.R., Maxwell, E., Kazuba, D.M. & Gershon, E.S. (1987). Clinical features of illness in siblings with schizophrenia or schizo-affective disorder. *Archives of General Psychiatry* **44**, 891–6.

De Wied, D., Bohus, J.M., van Ree, J.M. & Urban, F. (1978). Behavioural and electrophysiological effects of peptides, related to lipotropin. *Journal of Pharmacology and Experimental Therapeutics* **204**, 570–80.

Deakin, J.F.W., Simpson, M.D.C., Gilchrist, A.C., Skan, W.J. & Slater, P. (1988). Cortical glutamate dysfunction in schizophrenia. XVIth CINP Congress. *Psychopharmacology*, **96**, Suppl. 08.

Dean, C. & Kendell, R.E. (1982). The symptomatology of puerperal illness. *British Journal of Psychiatry* **139**, 128–33.

Delva, N.J. & Letemendia, F.J.J. (1982). Lithium treatment in schizophrenia and schizo-affective disorders. *British Journal of Psychiatry* **141**, 387–400.

Di Paolo, T., Poyet, P. & Labrie, F. (1981). Effect of chronic oestradiol and haloperidol treatment on striatal dopamine receptors. *European Journal of Pharmacology* **73**, 105–6.

Dickson, W.E. & Kendell, R.E. (1986). Does maintenance lithium prevent recurrences of mania under ordinary clinical conditions? *Psychological Medicine* **16**, 521–30.

Dinan, T., Silverstone, T. & Cookson, J.C. (1988). Cortisol, prolactin and growth hormone levels with clinical ratings in manic patients treated with verapamil. *International Clinical Psychopharmacology* **3**, 151–6.

Dubovsky, S.L., Franks, R.D., Lifschitz, M. & Coen, P. (1982). Effectiveness of verapamil in the treatment of a manic patient. *American Journal of Psychiatry* **139**, 502–4.

Dunner, D.L. & Fieve, R.R. (1975). Clinical factors in lithium carbonate prophylaxis failure. *Archives of General Psychiatry* **30**, 229–33.

Ellinwood, E.M. (1967). Amphetamine psychosis: Description of the individuals and process. *Journal of Nervous and Mental Disease* **144**, 273–83.

Emrich, H.M. (1987). The psychopharmacology of rapid-cycling. Presented to British Association for Psychopharmacology, Cambridge.

Emrich, H.M., Dose, M. & Von Zerssen, D. (1985). The use of sodium valproate, carbamazepine and oxcarbamazepine in patients with affective disorders. *Journal of Affective Disorders* **8**, 243–50.

Falloon, I., Watt, D.C. & Shepherd, M. (1978). The social outcome of patients in a trial of long-term continuation therapy in schizophrenia: pimozide vs fluphenazine. *Psychological Medicine* **8**, 265–74.

Farde, L., Wiesel, F.A., Halldin, C. & Sedvall, G. (1988). Central D2-dopamine receptor occupancy in schizophrenic patients treated with anti-psychotic drugs. *Archives of General Psychiatry* **45**, 71–6.

Fenton, W.S. & McGlashan, T.H. (1987). Sustained remission in drug-free schizophrenic patients. *American Journal of Psychiatry* **144**, 1306–9.

Ferrier, I.N., Roberts, G.W., Crow, T.J., Johnstone, E.C., Owens, D.G.C., Lee, Y.C., O'Shaughnessy, D., Adrian, T.E., Polack, J.R. & Bloom, S.R. (1983). Reduced cholecystokinin-like and somatostatin-like immunoreactivity in limbic lobe is associated with negative symptoms in schizophrenia. *Life Sciences* **33**, 475–82.

Ferris, R.M., Harfenist, M., McKenzie, G.M., Cooper, B., Soroko, F.E. & Maxwell, R.A. (1982). BW 234U, a novel antipsychotic agent. *Journal of Pharmaceutical Pharmacology* **34**, 388–90.

Fisch, R.Z. (1987). Trihexyphenidyl abuse: therapeutic implications for negative symptoms of schizophrenia. *Acta Psychiatrica Scandinavica* **75**, 91–4.

Flor-Henry, P. (1969). Psychosis in temporal lobe

epilepsy: a controlled investigation. *Epilepsia* **10**, 363–95.

Flor-Henry, P. (1986). Observations, reflections and speculations on the cerebral determinants of mood and on the bilaterally asymmetric distributions of the major neurotransmitter systems. *Acta Neurologica Scandinavica* **74**, Suppl. 109, 75–89.

Forrest, D.V. (1982). Bipolar illness after right hemispherectomy. *Archives of General Psychiatry* **39**, 817–19.

Garvey, M.J., Hwang, S., Teubner-Rhodes, D., Zander, J. & Rhem, C. (1987). Dextroamphetamine treatment of mania. *Journal of Clinical Psychiatry* **48**, 412–13.

Gelenberg, A.J. (1976). The catatonic syndrome. *Lancet* i, 1339–41.

Gelders, Y., Ceulemans, D., Reyntjens, A., Hoppenbrouwers, M.L. & Ferauge, M. (1985). The influence of selective serotonin antagonism on conventional neuroleptic therapy. Abstract Number 532.7. IVth World Congress of Biological Psychiatry, Philadelphia.

Gerlach, J., Behnke, K., Heltberg, J., Munk-Anderson, E. & Neilsen, H. (1985). Sulpiride and haloperidol in schizophrenia: a double-blind cross-over study of therapeutic effect, side effects and plasma concentrations. *British Journal of Psychiatry* **147**, 283–8.

Gerner, R.H., Post, R.M. & Bunney, W.E. (1976). A dopaminergic mechanism in mania. *American Journal of Psychiatry* **133**, 1177–80.

Giannini, A.J., Houser, W.L., Loiselle, R.H., Giannini, M.C. & Price, W.A. (1984). Antimanic effects of verapamil. *American Journal of Psychiatry* **141**, 1602–3.

Glazer, W.M., Charney, D.S. & Henninger, G.R. (1987). Noradrenergic function in schizophrenia. *Archives of General Psychiatry* **44**, 898–904.

Glick, D.S., Ross, D.A. & Hough, L.B. (1982). Lateral asymmetry of neurotransmitters in the human brain. *Brain Research* **234**, 53–63.

Goldberg, E. & Morrison, S.L. (1982). Schizophrenia and social class. *British Journal of Psychiatry* **109**, 785–802.

Goldberg, S.C. (1985). Negative and deficit symptoms in schizophrenia do respond to neuroleptics. *Schizophrenia Bulletin* **11**, 453–6.

Goodall, E., Trenchard, E. & Silverstone, T. (1987). Receptor blocking drugs and amphetamine anorexia in human subjects. *Psychopharmacology* **97**, 484–90.

Goodall, E., Oxtoby, C., Richards, R., Watkinson, G., Brown, D. & Silverstone, T. (1988). A clinical trial of the efficacy and acceptability of d-fenfluramine in the treatment of neuroleptic-induced obesity. *British Journal of Psychiatry* **153**, 208–13.

Goodwin, F.K. & Ebert, M.H. (1973). Lithium in mania: Clinical trials and controlled studies. In: *Lithium: Its Role in Psychiatric Research and Treatment*, S. Gershon & B. Shopsin, pp. 237–52. Oxford University Press: New York.

Goodwin, F.K. & Jamison, K.R. (1984). The natural course of manic-depressive illness. In: *Neurobiology of Mood Disorders*, eds. R.M. Post & J.C. Ballenger, pp. 20–38. Baltimore: Williams and Wilkins.

Gur, R.E., Resnick, S.M. & Gur, R.C. (1987). Regional brain function in schizophrenia. II. Repeated evaluation with position emission tomography. *Archives of General Psychiatry* **44**, 126–9.

Hakola, H.P.A. & Laulumaa, V.A. (1982). Carbamazepine in treatment of violent schizophrenics. *Lancet* i, 1358.

Hall, R., Popkin, M., Stickney, S. & Gardner, E. (1979). Presentation of the steroid psychoses. *Journal of Nervous and Mental Disorders* **167**, 229–36.

Hamblin, M.W., Leff, S.E. & Creese, I. (1984). Interactions of agonists with D-2 dopamine receptors: evidence for a single receptor population existing in multiple agonist affinity-states in rat striatal membranes. *Biochemical Psychopharmacology* **33**, 877–87.

Hare, E.H. (1956). Family setting and the urban distribution of schizophrenia. *Journal of Mental Sciences* **102**, 753–60.

Hare, E.H. & Walter, S.D. (1978). Seasonal variation in admissions of psychiatric patients and its relation to seasonal variation in their birth. *Journal of Epidemiology and Community Health* **32**, 47–52.

Hendler, N.H. (1978). Lithium pharmacology and physiology. In: *Handbook of Psychopharmacology*, vol. 14, eds. L.L. Iversen, S.D. Iversen & S.H. Snyder, pp. 233–72. Plenum Press: New York.

Hill, D. (1957). Electro-encephalogram in schizophrenia. In: *Schizophrenia: Somatic Aspects*, ed. D. Richter. Pergamon: London.

Himmelhoch, J.M., Mulla, D., Neil, J.F., Detre, T.P. & Kupfer, D. (1976). Incidence and significance of mixed affective states in a bipolar population. *Archives of General Psychiatry* **33**, 1062–6.

Hirsch, S.R. (1986). Influence of social experience and environment on the course of schizophrenia. In: *The Psychopharmacology and Treatment of Schizophrenia*, ed. P.B. Bradley & S.R. Hirsch, British Association for Psychopharmacology Monograph No. 8, pp. 200–11. Oxford University Press: Oxford.

Horn, A.S., Coyle, J.T. & Snyder, S.H. (1971). Catecholamine uptake by synaptosomes from rat brain: structure activity relationships of drugs with differential effects on dopamine and norepinephrine neurones. *Molecular Pharmacology* **7**, 66–80.

Hunt, N. & Silverstone, T. (1990). Secondary seasonal affective disorder: a case following brain injury. *British Journal of Psychiatry*, in press.

Jacobs & Silverstone, T. (1986). Dextroamphetamine-induced arousal in human subjects as a model for mania. *Psychological Medicine* **16**, 323–9.

Janowsky, D.S., El-Yousef, M.K. & Davis, J.M. (1973a). Parasympathetic suppression of manic symptoms by physostigmine. *Archives of General Psychiatry* **28**, 542–7.

Janowsky, D.S., El-Yousef, M.K., Davis, J.M. & Sekerke, H.S. (1973b). Provocation of schizophrenic symptoms by intravenous administration of methylphemidate. *Archives of General Psychiatry* **28**, 185–91.

Jeffcoate, W.J., Silverstone, J.T., Edwards, C.R.W. & Besser, G.M. (1979). Psychiatric manifestations of Cushings syndrome: response to lowering of plasma cortisol. *Quarterly Journal of Medicine* **48**, 465–72.

Jenner, F.A., Goodwin, J.C., Sheridan, M., Tauber, E.J. & Loban, M.C. (1968). The effect of an altered time regime on biological rhythms in a 48 hour periodic psychosis. *British Journal of Psychiatry* **114**, 215–24.

Jeste, D.V., Lohr, J.B. & Goodwin, F.K. (1988). Neuroanatomical studies of major affective disorder. *British Journal of Psychiatry* **153**, 444–59.

Johnson, D. (1988). The significance of depression in the prediction of relapse in chronic schizophrenia. *British Journal of Psychiatry* **152**, 320–3.

Johnson, F.N. (1984). *The Psychopharmacology of Lithium*. MacMillan: London.

Johnstone, E.C., Crow, T.J., Frith, C.D., Carney, M.M.P. & Price, J.S. (1978). Mechanism of the antipsychotic effect in the treatment of acute schizophrenia. *Lancet* **i**, 848–51.

Johnstone, E.C., Crow, T.J., Ferrier, I.N., Owens, D.C., Bourne, R. & Gamble, S. (1983). Adverse effects of anticholinergic medication on positive schizophrenic symptoms. *Psychological Medicine* **13**, 513–27.

Johnstone, E.C., Crow, T.J., C.D. & Owens, D.G.C. (1988). The Northwick Park 'functional' psychosis study: diagnosis and treatment response. *Lancet* **ii**, 119–25.

Josephson, M.A. & MacKenzie, T.B. (1980). Thyroid-induced mania in hypothyroid patients. *British Journal of Psychiatry* **137**, 222–8.

Joyce, P.R. (1984). Age of onset in bipolar affective disorder and misdiagnosis as schizophrenia. *Psychological Medicine* **14**, 145–9.

Joyce, P.R. (1988). Carbamazepine in rapid cycling

bipolar affective disorder. *International Clinical Psychopharmacology* **3**, 123–9.

Kahlbaum, K.L. (1874). *Catatonia*. (Translated from the German). Johns Hopkins University Press: Baltimore, Maryland (1974).

Kane, J.M. (1984). The use of depot neuroleptics: clinical experience in the United States. *Journal of Clinical Psychiatry* **45**[5, Sec. 2], 5–12.

Kebabian, J.W. & Calne, D.B. (1979). Multiple receptors for dopamine. *Nature* **277**, 93–6.

Kelly, W.F., Checkley, S.A. & Bender, D.A. (1980). Cushings syndrome, tryptophan and depression. *British Journal of Psychiatry* **136**, 125–32.

Kendler, K.S., Grunberg, A.M. & Tsuang, M.T. (1985). Psychiatric illness in first degree relatives of schizophrenic and surgical control patients. *Archives of General Psychiatry* **42**, 770–9.

Kennedy, S., Thompson, R., Stancer, H.C., Roy, A. & Persad, E. (1983). Life events precipitating mania. *British Journal of Psychiatry* **142**, 358–403.

Klein, E., Bental, E., Lerer, B. & Belmaker, R.H. (1984). Carbamazepine and haloperidol vs placebo and haloperidol in excited psychoses. *Archives of General Psychiatry* **41**, 165–70.

Knights, A. & Hirsch, S. (1981). Revealed depression and drug treatment of schizophrenia. *Archives of General Psychiatry* **38**, 28–33.

Kotin, J. & Goodwin, F.K. (1971). Depression during mania: Clincal observations and theoretical implications. *American Journal of Psychiatry* **129**, 687–92.

Kraepelin, E. (1921). *Manic-depressive Insanity and Paranoia*, ed. G.M. Robertson (transl. R.M. Barclay). E & S Livingstone: Edinburgh.

Krauthammer, C. & Klerman, G.A. (1978). Secondary mania. *Archives of General Psychiatry* **35**, 1333–9.

Kukopulos, A. & Reginaldi, D. (1980). Recurrences of manic-depressive episodes during lithium treatment. *Handbook of Lithium Therapy*, ed. F. Neil Johnson, pp. 109–17. Lancaster, MTP.

Kukopulos, A., Reginaldi, D., Laddomada, P., Floris, G., Serra, G. & Tonso, L. (1980). Course of manic-depressive cycle and changes caused by treatment. *Pharmakopsychiatrie* **13**, 156–7.

Lake, C.R., Sternberg, D.E., van Kammen, D.P., Ballanger, J.C., Ziegler, M.G., Post, S.M., Kopin, J.J. & Bunney, W.E. (1980). Schizophrenia: elevated cerebrospinal fluid norepinephrine. *Science* **207**, 331–3.

Landolt (1958). Serial electroencephalographic investigations during psychotic episodes in epileptic patients and during schizophrenic attacks. In: *Lectures on Epilepsy*, A.M. Lorentz de Hass. Elsevier: Amsterdam.

Lautin, A., Angrist, B., Stanley, M., Gershon, S., Heckl, K. & Karobath, M. (1980). Sodium valproate in schizophrenia: some biochemical correlates. *British Journal of Psychiatry* **137**, 240–4.

Levy, M.I., Davis, B.M., Mohs, R.C., Kendler, K.S., Mathé, A.A., Trigos, G., Horvath, T.B. & Davis, K.L. (1984). Apomorphine and schizophrenia. *Archives of General Psychiatry* **41**, 520–4.

Lewander, T. (1989). Multicentre comparative trial of remoxipride and haloperidol in schizophrenia. *Acta Psychiatrica Scandinavica*, Supplement (In press).

Lewis, S.W., Reveley, A.M., Reveley, M.A., Chitkara, B. & Murray, R.M. (1987). The familial-sporadic distinction as a strategy in schizophrenia research. *British Journal of Psychiatry* **15**, 306–13.

Lewis, S.W., Reveley, M.A., David, A.S. & Ron, M. (1988). Agenesis of the corpus callosum and schizophrenia: a case report. *Psychological Medicine* **18**, 341–8.

Leysen, J. & Niemegeers, C.J.E. (1985). In: *Handbook of Neurochemistry*, vol. 9, ed. A. Lajtha, pp. 331–61. Plenum Press: New York.

Liddle, P.F. (1987). The symptoms of chronic schizophrenia: a re-examination of the positive-negative dichotomy. *British Journal of Psychiatry* **151**, 145–51.

Ling, M.H.M., Perry, P.J. & Tsuang, M.T. (1981). Side-effects of corticosteroid therapy. *Archives of General Psychiatry* **38**, 471–7.

Lishman, O. (1978). *Organic Psychiatry. The Psychological Consequences of Organic Disorder.* Blackwell Scientific Publications: Oxford.

Loranger, A.W. (1984). Sex difference in age at onset of schizophrenia. *Archives of General Psychiatry* **41**, 157–61.

Loranger, A.W. & Levine, J. (1978). Age at onset of bipolar affective illness. *Archives of General Psychiatry* **35**, 1345–8.

Luchins, D.J. (1983). Carbamazepine for the violent psychiatric patient. *Lancet* **i**, 766.

Manchanda, R. & Hirsch, S.R. (1981). (Des-tyr)-gamma-endorphin in the treatment of schizophrenia. *Psychological Medicine* **11**, 401–4.

Mander, A.J. & Loudon, J.B. (1988). Rapid recurrence of mania following abrupt discontinuation of lithium. *Lancet* **ii**, 15–17.

Marsden, C.A. & Curzon, G. (1976). Studies on the behavioural effects of tryptophan and p-chlorophenylalanine. *Neuropharmacology* **15**, 165–71.

Mason, R. (1989). The effect of light on CNS noradrenaline and serotonin receptor function. In: *Seasonal Affective Disorders*, eds. C. Thompson & T. Silverstone. CNS Publications: London.

Mattes, J.A. (1984). Carbamazepine for uncontrolled rage outbursts. *Lancet* **ii**, 1164–5.

McCreadie, R.D. & Morrison, D.P. (1985). The impact

of lithium in South West Scotland. *British Journal of Psychiatry* **146**, 70–80.

McNeil, T.F. & Kaij, L. (1978). Obstetric factors in the development of schizophrenia. Complications in the births of preschizophrenics and in reproduction by schizophrenic parents. In: *High Risk and Premorbid Development*, eds. C. Wynne, R.L. Cromwell & S. Matthysse, pp. 407–29. Wiley: New York.

Mendels, J. & Frazer, A. (1975). Reduced central serotonergic activity in mania: implications for the relationship between depression and mania. *British Journal of Psychiatry* **126**, 241–8.

Meyendorff, E., Lerer, B., Moore, N.C., Bow, J. & Gershon, S. (1985). Methylphenidate infusion in euthymic bipolars: effects of carbamazepine pretreatment. *Psychiatry Research* **16**, 303–8.

Miklowitz, D.J., Goldstein, M.J. & Neuch Terlein, K.H. (1985). Expressed emotion, lithium compliance and relapse in early onset mania. Presented at American College of Neuropsychopharmacology, Maui.

Minneman, K.P. (1981). Adrenergic receptor molecules. In: *Neurotransmitter Receptors, Part 2, Biogenic Amines*, eds. H.I. Yamamura & S.J. Enna, pp. 185–268. Chapman and Hall: London.

Mobley, P.L. & Sulser, F. (1980). Adrenal corticoids regulate sensitivity of noradrenaline receptor-coupled adenylate cyclase in brain. *Nature* **286**, 608–9.

Mogenson, G.J. (1984). Limbic-motor integration – with emphasis on initiation of exploratory and goal-directed locomotion. In: *Modulation of Sensorimotor Activity During Alterations in Behavioural States*, ed. R. Bandler, pp. 121–37. Alan R. Liss, Inc.: New York.

Montgomery, S.A. & Green, M.D. (1988). The use of cholecystokinin in schizophrenia: a review. *Psychological Medicine* **18**, 593–604.

Morgan, H.G. (1972). The incidence of depressive symptoms during recovery from hypomania. *British Journal of Psychiatry* **120**, 537–9.

Mueser, K. & Dysken, M. (1983). Narcotic antagonists in schizophrenia: a methodological review. *Schizophrenia Bulletin* **9**, 213–25.

Murphy, D.C. (1972). L-dopa, behavioural activation and psychopathology. *Research Publications of the Association for Research into Nervous and Mental Diseases* **50**, 472–93.

Neppe, V.M. (1982). Carbamazepine in the psychiatric patient. *Lancet* **ii**, 334.

Nimgaonkar, V.L., Wessely, S., Tune, L.E. & Murray, R.M. (1988). Response to drugs in schizophrenia: the influence of family history, obstetric complications and ventricular enlargement. *Psychological Medicine* **18**, 583–92.

Nizamie, S.M., Nizamie, A., Borde, M. & Sharma,

A.S. (1988). Mania following head injury: case reports and neuropsychological findings. *Acta Psychiatrica Scandinavica* **77**, 637–9.

Ögren, S.O., Hall, H., Köhler, C., Magnusson, O. & Sjostrand, S.E. (1986). The selective dopamine D2 receptor antagonist raclopride discriminates between dopamine-mediated motor functions. *Psychopharmacology* **90**, 287–94.

Okuma, T., Inanaga, K., Otsuki, S., Sarai, K., Takahashi, R., Hazama, H., Mori, A. & Watanabe, M. (1979). Comparison of the antimanic efficacy of carbamazepine and chlorpromazine: a double-blind controlled study. *Psychopharmacology* **66**, 211–17.

Okuma, T., Kishimoto, A., Inoue, K., Matsumoto, H., Ogura, A., Matsushita, T., Naklao, T. & Ogura, C. (1973). Antimanic and prophylactic effects of carbamazepine on manic-depressive psychosis. *Folia Psychiatrica et Neurologica Japonica* **27**, 283–97.

Owen, M.J., Lewis, S.W. & Murray, R.M. (1988). Obstetric complications and abnormalities in schizophrenia. *Psychological Medicine* **18**, 331–40.

Pearlson, G.D., Kubos, K.L. & Robinson, R.G. (1982). Novel cortical localization of a lateralized behaviour in the rat: relationship to noradrenergic innervation. *Society for Neurosciences Abstracts* **8**, 894.

Perez, M.M. & Trimble, M. (1980). Epileptic psychosis – diagnostic comparison with process schizophrenia. *British Journal of Psychiatry* **137**, 245–49.

Petit, M., Zann, M., Lesieur, P. & Colonna, L. (1987). The effect of sulpiride on negative symptoms of schizophrenia. *British Journal of Psychiatry* **150**, 270–1.

Pickar, D., Wolkowitz, O.M., Doran, A.R., Labarca, R., Roy, A., Breier, A. & Narang, P.K. (1987). Clinical and biochemical effects of verapamil administration to schizophrenic patients. *Archives of General Psychiatry* **44**, 113–18.

Pittman, K.J., Jakubovic, A. & Fibiger, H.C. (1984). The effects of chronic lithium treatment on behavioural and biochemical indices of dopamine receptor supersensitivity in the rat. *Psychopharmacology* **82**, 371–7.

Post, R.M. (1980). Biochemical theories of mania. In: *Mania*, eds. R.H. Belmaker & H.M. Van Praag, pp. 217–65. MTP: Lancaster.

Post, R.M. & Weiss, S.R.B. (1987). Behavioural pharmacology of carbamazepine: differential effects on kindling. *International Clinical Psychopharmacology*, (Suppl.) 51–73.

Post, R.M., Gerner, R.H., Carman, J.S. & Bunney, W.E. (1976). Effects of low doses of a dopamine-receptor stimulator in mania. *Lancet* **i**, 203–4.

Post, R.M., Jimerson, D.C., Bunney, W.E. & Goodwin, F.K. (1980). Dopamine and mania: behavioural and biochemical effects of the dopamine receptor blocker pimozide. *Psychopharmacology* **67**, 297–305.

Post, R.M., Uhde, T.W., Roy-Byrne, P.P. & Joffe, R. (1987). Correlates of anti-manic response to carbamazeine. *Psychiatry Research* **21**, 71–83.

Post, R.M., Rubinow, D.R., Uhde, T.W., Roy-Byrne, P.P., Linnoila, M., Rosoff, A. & Cowdry, R. (1989). Dysphoric mania: Clinical and biological correlates. *Archives of General Psychiatry* **46**, 353–8.

Prange, A.J., Wilson, I.C. & Lynn, C.W. (1974). L-Tryptophan in mania. *Archives of General Psychiatry* **30**, 56–62.

Prien, R.F., Caffey, E.M. & Klett, C.J. (1972). Comparison of lithium carbonate and chlorpromazine in the treatment of mania. *Archives of General Psychiatry* **26**, 146–53.

Redgrave, P. & Dean, P. (1981). Intracranial self-stimulation. *British Medical Bulletin* **37**, 141–6.

Rees, L., Butler, P.W.P., Gosling, C. & Besser, G.W. (1979). Adrenergic blockade and the corticosteroid and growth hormone response to methylamphetamine. *Nature* **228**, 565–6.

Reich, L.H., Davies, R.K. & Himmelhoch, J.M. (1974). Excessive alochol use in manic-depressive illness. *American Journal of Psychiatry* **131**, 83–6.

Reilly, T. (1988). Delusional infestations. *British Journal of Psychiatry* **153** (Suppl. 2), 44–6.

Reveley, A.M., Reveley, M.A. & Murray, R.M. (1984). Cerebral ventricular enlargement in nongenetic schizophrenia: a controlled twin study. *British Journal of Psychiatry* **144**, 89–93.

Reveley, M.A., Reveley, A.M. & Baldy, R. (1987). Left cerebral hemisphere hypodensity in discordant schizophrenic twins. *Archives of General Psychiatry* **44**, 625–32.

Reynolds, G.P. (1983). Increased concentrations and lateral asymmetry of amygdala dopamine in schizophrenia. *Nature* **305**, 527–9.

Reynolds, G.P. (1988). Post-mortem neurochemistry of schizophrenia. *Psychological Medicine* **18**, 793–7.

Reynolds, G.P., Cowey, L., Rosser, M.N. & Iversen, L.L. (1982). Thioridazine is not specific for limbic dopamine receptors. *Lancet* **ii**, 499–500.

Riess, H., Schwartz, C.E. & Klerman, G.L. (1987). Manic syndrome following head injury: another form of secondary mania. *Journal of Clinical Psychiatry* **48**, 29–30.

Robinson, R.G. & Bloom, F.E. (1977). Pharmacological treatment following experimental cerebral infarction: implications for understanding psychological symptoms of human stroke. *Biological Psychiatry* **12**, 669–80.

Robinson, R.G., Kubos, K.L., Starr, L.B., Rao, K. &

Price, T.R. (1984). Mood disorders in stroke patients, importance of location of lesion. *Brain* **107**, 81–93.

Robinson, R.G., McHugh, P.R. & Folstein, M.F. (1975). Measurement of appetite disturbances in psychiatric disorders. *Journal of Psychiatric Research* **12**, 59–68.

Robinson, R.G., Boston, J.D., Starkstein, S.E. & Price, T.R. (1988). Comparison of mania and depression after brain injury: causal factors. *American Journal of Psychiatry* **145**, 172–8.

Rosenthal, N.E., Salk, D.A., Gillin, J.C., Levy, A.J., Goodwin, F.K., Davenport, Y., Mueller, P.S., Newson, D.A. & Wehr, T.A. (1984). Seasonal affective disorder. *Archives of General Psychiatry* **4**, 72–80.

Rottanburg, D., Robins, A.H., Ben-Arie, O., Teggin, A. & Elk, R. (1982). Cannabis-associated psychosis with manic features. *Lancet* **ii**, 1364–6.

Roy-Byrne, P.P., Joffe, R.T., Uhde, T.W. & Post, R.M. (1984). Approaches to the evaluation and treatment of rapid-cycling affective illness. *British Journal of Psychiatry* **145**, 543–50.

Schur, E. (1988). The epidemiology of schizophrenia. *British Journal of Hospital Medicine* **39**, 38–45.

Schmauss, C., Yassouridis, A. & Emrich, H.M. (1987). Antipsychotic effect of buprenorphine in schizophrenia. *American Journal of Psychiatry* **144**, 1340–2.

Schneider, K. (1959). *Clinical Psychopathology* (transl. M.W. Hamilton). Grune and Stratton: New York.

Sclare, P. & Creed, F. (1987). Life events and mania. *British Journal of Psychiatry* **150**, 875.

Sherer, M.A., Kumar, K.M., Cone, E.J. & Joffe, J.H. (1988). Suspiciousness induced by 4-hour intravenous infusions of cocaine, preliminary findings. *Archives of General Psychiatry* **45**, 673–7.

Shulman, D. & Post, F. (1980). Bipolar affective disorder in old age. *British Journal of Psychiatry* **136**, 26–32.

Silberman, E.K., Post, R.M., Nurnberger, J.L., Theodore, W. & Boulanger, J-P. (1985). Transient sensory, cognitive and affective phenomena in affective illness: a comparison with complex partial epilepsy. *British Journal of Psychiatry* **146**, 81–9.

Silverstone, T. & Cookson, J.C. (1982). The biology of mania. In: K. Granville-Grossman (ed.) *Recent Advances in Clinical Psychiatry*, vol. 4, pp. 201–41. Churchill Livingstone: Edinburgh.

Silverstone, T. & Cookson, J.C. (1983). Examining the dopamine hypothesis of schizophrenia and of mania using the prolactin response to antipsychotic drugs. *Neuropharmacology* **22**, 539–41.

Silverstone, T., Fincham, T., Wells, B. & Kyriakides, M. (1981). The effect of the dopamine receptor blocking drug pimozide on the stimulant and anorectic actions of dextroamphetamine. *Neuropharmacology* **19**, 1235–7.

Silverstone, T., Levine, S., Freeman, H. & Dubini, A. (1984). Zetidoline, a new antipsychotic: first controlled trial in acute schizophrenia. *British Journal of Psychiatry* **145**, 294–9.

Singh, M.M., Kay, S.R. & Opler, L.A. (1987). Anticholinergic–neuroleptic antagonism in terms of positive and negative symptoms of schizophrenia: implications for psychobiological subtyping. *Psychological Medicine* **17**, 39–48.

Slater, E., Beard, A.W. & Clithero, E. (1963). The schizophrenia-like psychoses of epilepsy. *British Journal of Psychiatry* **109**, 95–150.

Spitzer, R., Endicott, J. & Robbins, E. (1978). Research diagnostic criteria: rationale and reliability. *Archives of General Psychiatry* **35**, 773–82.

Stahl, S.M., Uhr, S.B. & Berger, P.A. (1985). Pilot study on the effects of fenfluramine on negative symptoms in twelve schizophrenic in patients. *Biological Psychiatry* **20**, 1098–102.

Staunton, D.A., Magistretti, P.J., Shoemaker, W.J. & Bloom, F.E. (1982). Effects of chronic lithium treatment on dopamine receptors in rat corpus callosum. I. Locomotor activity and behavioural supersensitivity. *Brain Research* **232**, 401–12.

Stiles, G.L. & Lefkowicz, R.J. (1981). Thyroid hormone modulation of agonist-beta-adrenergic receptor interactions in the rat heart. *Life Sciences* **28**, 2529–36.

Sturgeon, D., Turpin, D., Kuipers, L., Berkowitz, R. & Leff, J. (1984). Psychophysiological responses of schizophrenic patients to high and low expressed emotion relatives: a follow-up study. *American Journal of Psychiatry* **145**, 62–9.

Suddath, R.L., Casanova, M.F., Goldberg, T.E., Daniel, D.G., Kelsoe, J.R. & Weinberger, D.R. (1989). Temporal lobe pathology in schizophrenia: a quantitative magnetic resonance imaging study. *American Journal of Psychiatry* **146**, 464–72.

Swann, A.C., Koslow, S.H., Katz, M.M., Maas, J.W., Javaid, J., Secunda, S.K. & Robins, E. (1987). Lithium carbonate treatment of mania: cerebro-spinal fluid and urinary monoamine metabolite and treatment outcome. *Archives of General Psychiatry* **44**, 345–54.

Szabadi, E., Bradshaw, C.M. & Glaszner, P.B. (1980). The comparison of the effects of DL-308, a potential new neuroleptic agent, and thioridazine on some psychological and physiological functions in healthy volunteers. *Psychopharmacology* **68**, 125–34.

Tamminga, C.A., Gotts, M.O., Thaker, G.K., Alphs, L. & Foster, N.L. (1986). Dopamine antagonist treatment of schizophrenia with

N-propylnorapomorphine. *Archives of General Psychiatry* **43**, 398–402.

Tunving, K. (1982). Psychiatric effects of cannabis. *Acta Psychiatrica Scandinavica* **72**, 209–17.

Turner, T., Cookson, J.C., Wass, J.A.H., Drury, P.C., Price, P.A. & Besser, G.M. (1984). Psychotic reactions during treatment of pituitary tumours with dopamine agonists. *British Medical Journal* **289**, 1101–3.

Tyrer, S.P. (1985). Lithium in the treatment of mania. *Journal of Affective Disorders* **8**, 251–7.

Tyrer, S.P. (1988). Lithium in aggression. In: *Lithium: Inorganic Pharmacology and Psychiatric Use*, ed. M.J. Birch, pp. 39–42. IRL Press: Oxford.

Van Kammen, D.P. & Boronow, J.J. (1988). Dextro-amphetamine diminishes negative symptoms in schizophrenia. *International Clinical Psychopharmacology* **3**, 111–21.

Van Kammen, D.P. & Murphy, D.L. (1975). Attenuation of the euphoriant and activatory effects of d- and l-amphetamine by lithium carbonate treatment. *Psychopharmacology* **44**, 215–24.

Van Kammen, D.P., Bunney, W.E., Docherty, J.P., Marder, S.R., Ebert, M.H., Rosenblatt, J.E. & Rayner, J.N. (1982). d-Amphetamine-induced heterogeneous changes in psychotic behaviour in schizophrenia. *American Journal of Psychiatry* **139**, 991–7.

Van Kammen, D.P., Docherty, J.P., Marder, S.R., Rosenblatt, J.E. & Bunney, W.E. (1985). Lithium attenuates the activation-euphoria but not the psychosis induced by d-amphetamine in schizophrenia. *Psychopharmacology* **87**, 11–15.

Vaughn, C. & Leff, J. (1976). The influences of family and social factors on the course of psychiatric illness. *British Journal of Psychiatry* **129**, 125–37.

Waddington, J.L. (1989). Sight and insight: Brain dopamine-receptor occupancy by neuroleptics visualized in living schizophrenic patients by position emission tomography. *British Journal of Psychiatry* **154**, 433–6.

Waddington, J.L. & Youssef, M.A. (1988). Tardive dyskinesia in bipolar affective disorder: aging, cognitive dysfunction, course of illness, and exposure to neuroleptics and lithium. *American Journal of Psychiatry* **145**, 613–16.

Walter, S.D. (1977). Seasonality of mania: a reappraisal. *British Journal of Psychiatry* **131**, 345–50.

Watson, G.C., Tilleskjor, C. & Jacobs, L. (1984).

Schizophrenic birth seasonality in relation to the incidence of infectious diseases and temperature extremes. *Archives of General Psychiatry* **41**, 85–90.

Wehr, T.A. & Goodwin, F.K. (1987). Can antidepressants cause mania and worsen the course of affective illness? *American Journal of Psychiatry* **144**, 1403–11.

Weinberger, D.R. (1987). Implications of normal brain development for the pathogenesis of schizophrenia. *Archives of General Psychiatry* **44**, 660–9.

Weinberger, D. R. (1988). Premorbid neuropathology in schizophrenia. *Lancet* **ii**, 445.

Weinberger, D.R., Berman, K.F. & Zec, R.F. (1986). Physiologic dysfunction of dorsolateral prefrontal cortex in schizophrenia. *Archives of General Psychiatry* **43**, 114–25.

Weissman, M.M., Gershon, E.S., Kidd, K.K., Prusoff, B.A., Leckman, J.F., Dibble, E., Hamovit, J., Thompson, W.D., Pauls, D.L. & Guroff, J.J. (1984). Psychiatric disorders in the relatives of probands with affective disorders. *Archives of General Psychiatry* **41**, 13–21.

Whybrow, P.C. & Prange, A.J. (1981). A hypothesis of thyroid–catecholamine receptor interaction. *Archives of General Psychiatry* **38**, 106–13.

Wing, J.K. & Brown, G.W. (1961). Social treatment of chronic schizophrenia: a comparative survey in three hospitals. *Journal of Mental Science* **107**, 847.

Winokur, G., Clayton, P.A. & Reich, T. (1969). *Manic Depressive Illness*. Mosby: Illinois.

Wood, A.J. & Goodwin, G.M. (1987). A review of the biochemical and neuropharmacological actions of lithium. *Psychological Medicine* **17**, 579–600.

Wood, A.J., Goodwin, G.M., De Souza, R.J. & Green, A.R. (1986). The pharmacokinetic profile of lithium in the rat and mouse: an important factor in the psychopharmacological investigation of the drug. *Neuropharmacology* **25**, 1285–8.

Yatham, L.N. (1987). Mania following head injury. *British Journal of Psychiatry* **151**, 558.

Zetterström, T., Sharp, T. & Ungerstedt, U. (1986). Effect of dopamine D-1 and D-2 selective drugs on dopamine release and metabolism in rat striatum in vivo. *Naunyn-Schmiedebergs Archives of Pharmacology* **334**, 117–24.

Zukin, R.S., Kushner, L., Lerma, J. & Bennett, M.V.L. (1988). Co-expression of N-Methyl-D-Aspartate and phenylcyclidine receptors in *Xenopus* oocytes injected with rat brain mRNA. *Journal of Psychopharmacology* **2** (2, Abstr.).

PART VI

Models of dementia

Animal models of Alzheimer's disease and dementia (with an emphasis on cortical cholinergic systems)

STEPHEN B. DUNNETT and TIMOTHY M. BARTH

The proportion of elderly in the general population in Western countries is progressively rising, in association with improvements in the standards of general health care (Isaacs, 1983), so that age-related disorders are providing an increasing pressure on the health services. As a consequence, increasing clinical, industrial and research resources are being directed towards the understanding and treatment of disorders associated with old age, the most prevalent being the intellectual impairments associated with dementia. The present chapter reviews a variety of approaches to modelling human dementia in experimental animals. This enterprise has the tripartite purpose of providing a better understanding of the nature of the human disorders themselves, an assessment of the primary and secondary features of the neuropathology, and an experimental tool for the development and screening of alternative strategies for treatment. Our focus in the present chapter will be on the second neurobiological aspect: namely that of attempting to reproduce in experimental animals particular features of the pathological changes observed in humans, and establishing their functional consequences. We will adopt a particular, but not exclusive, bias to considering one aspect in greater detail, *viz.* the involvement of subcortical-cortical cholinergic systems in the pathogenesis of dementia. The subsequent chapters (Salamone, Altman *et al.*, this vol., Chapters 15, 16) will consider the implications such models have for contemporary clinical practice and the development of improved pharmacological and therapeutic strategies for treatment.

14.1 AGEING, DEMENTIA AND ALZHEIMER'S DISEASE

Normal ageing is associated with a decline in many functions, in physical, sensory and motor as well as psychological performance. Within this context, dementia is a cluster of conditions characterized by a widespread, progressive decline in intellectual functions, developing in middle or later life to the point where the ability to continue normal lifestyle is seriously disrupted, and usually associated with distinctive neuropathological changes in the brain (Corsellis, 1976; Gurland & Birkett, 1983). The dementia is considered primary when it occurs in relative isolation or precedes other significant physical or psychiatric impairment. Dementia is usually associated with developing brain atrophy (Roth *et al.*, 1967), and a variety of specific neuropathological and biochemical changes, described in more detail below. Whereas some forms of dementia are unequivocally attributable to genetic disorder or infectious disease (e.g. Huntington's disease, or Creutzfeld-Jacob's disease, respectively), it remains contentious whether all dementias result from a disease process or whether some represent exaggerated normal ageing of the nervous system. It has often been argued that mild dementia is extremely common within the elderly population, perhaps encompassing 5–20% of all people over the age of 65 (Kay *et al.*, 1972; Henderson & Huppert, 1984), although in the majority of such cases the problems are contained within the family and community rather than coming to medical attention.

14.1.1 Neuropathology

At the turn of the century, Alois Alzheimer described a range of neuropathological changes in the *post mortem* brains of a 51-year-old woman and a 50-year-old man with dementia. He provided the first clear description of the disease, characterized by cerebral atrophy and microscopic neuritic plaques and neurofibrillary tangles, that has come to bear his name. 'Alzheimer's disease' properly refers to presenile cases of dementia with this particular profile of neuropathological degeneration. However, qualitatively similar changes are associated with many cases of dementia with senile (older than 65) onset – strictly 'senile dementia of the Alzheimer type' (SDAT). In line with widespread practice, we will consider these presenile and senile types of dementia as representing a continuous spectrum in Alzheimer's disease, which can only be diagnosed with certainty on the basis of the *post mortem* neuropathology (Corsellis, 1976).

Between 50% and 70% of all demented patients manifest the distinctive pathology of Alzheimer's disease (Tomlinson *et al.*, 1970; Terry & Katzman, 1983). The other common form, multi-infarct (or arteriosclerotic) dementia, appears to have a primary vascular origin involving multiple small foci of ischaemic brain destruction and tissue softening in subcortical as well as cortical loci (Corsellis, 1976). Less common forms are associated with a variety of other disorders including Parkinson's disease, Huntington's chorea, Pick's disease, Creutzfeldt-Jacob's disease, Wilson's disease, infection, hydrocephalus, trauma and metabolic deficiency (Mulder, 1975; Corsellis, 1976; Rossor, 1982; Gurland & Birkett, 1983), each with distinctive pathological features and in most cases unknown cause.

The two primary neuropathologic features of Alzheimer's disease are neuritic (senile) plaques and neurofibrillary tangles. Neuritic plaques are disorganized extracellular aggregations of granular and filamentous fragments, often surrounding a central amyloid core. There have been recent rapid advances in the determination of the molecular composition of the amyloid core of neuritic plaques (Delabar *et al.*, 1987; Kang *et al.*, 1987), with the hope that this may hold the key to understanding the aetiology of the neural degener-

ation underlying Alzheimer's disease, although this goal has yet to be realized. Neurofibrillary tangles are dense intracellular accumulations of fibrils, with highest incidence in the hippocampus but also occurring frequently throughout the cortex (Tomlinson *et al.*, 1970; Hyman *et al.*, 1984). In the electron microscope, neurofibrillary tangles are seen to be comprised of accumulations of paired helical filaments 10 nm in diameter and with a complete twist every 160 nm (Kidd, 1963; Wisniewski *et al.*, 1976). Their molecular structure has not yet been fully identified, and there is disagreement about the degree to which they involve a disorganization of normal cytoskeletal components, including neurofilaments or microtubules (Wischik & Crowther, 1986), as opposed to the formation of novel proteins found only in the pathological brain. Thus, whereas antibodies raised against the major protein constituents of normal neurofilaments do label neurofibrillary tangles *in situ* (Grundke-Iqbal *et al.*, 1979; Anderton *et al.*, 1982; Dahl *et al.*, 1982), they do not cross-react with filaments from Alzheimer tangles when the latter are purified *in vitro* (Ihara *et al.*, 1983). Conversely, some antibodies to purified Alzheimer tangle filaments do not recognize normal brain proteins (Ihara *et al.*, 1983).

Other less well-characterized neuropathological features of Alzheimer's disease include granulovacuolar degeneration, particularly in the pyramidal cell layers of the hippocampus (Tomlinson & Kitchener, 1972; Corsellis, 1976), and Hirano bodies throughout the CNS but particularly in the hippocampus (Tomonaga, 1974; Gibson & Tomlinson, 1977).

The co-occurrence of these neuropathological hallmarks are diagnostic of Alzheimer's disease in presenile cases of dementia (McMenemy, 1963). However they are all found at a low level in the aged brain of cases in which there is no evidence of dementia (Tomlinson *et al.* 1968; Corsellis, 1976), and they may also feature predominantly in other distinct disease processes associated with dementia, such as Creutzfeld-Jacob's disease. Therefore these signs cannot be considered a unique feature of Alzheimer's disease. Nevertheless, in an extended series of brains, derived from demented elderly patients, the coincidence of high levels of senile plaques, neurofibrillary tangles and granulovacuolar degeneration was present in 25 of the 50 patients (Tomlinson *et al.*, 1970), confirming the diagnosis of SDAT and demonstrating highly significant quantitative if not qualitative differences from the brains of age-matched non-demented elderly. In the majority of the remaining cases, other pathological features were clearly identified, such as extensive areas of brain softening in the cortex and basal ganglia indicative of dementia of arteriosclerotic origin, or a mixed combination of senile and arteriosclerotic indices.

Atrophy of the brain, apparent as a general shrinkage of both the grey and white matter, is a mild feature of ageing *per se* (Tomlinson *et al.*, 1968). However, the degree of generalized atrophy is also significantly greater in Alzheimer's disease (Tomlinson *et al.*, 1970; Miller *et al.*, 1980), which is both more marked and less regionally specific in cases where the onset of dementia is relatively early (Hubbard & Anderson, 1981). The decrease in total and regional brain volume in Alzheimer's disease appears to be attributable primarily to neuronal cell loss, which is most marked in the frontal and temporal cortices (Terry *et al.*, 1981; Mountjoy *et al.*, 1983). The association between cerebral atrophy and *post mortem* determination of Alzheimer's neuropathology, has been employed to validate the development of discriminant functions based on measures from computed tomography of enlarged ventricular spaces and increased interhemispheric fissures, in order to aid and refine the clinical diagnosis of dementia (Damasio *et al.*, 1983). Additionally, clear differences are apparent in cellular metabolic activity determined by positron emission tomography, not only between Alzheimer's patients and age-matched controls, but also as a basis for distinguishing between senile and multi-infarct dementia in the living patient (Benson *et al.*, 1983).

Not only do the *post mortem* neuropathological features of Alzheimer's disease serve to confirm the diagnosis in the senile as well as presenile cases of dementia, but also they appear to correlate with simple indices of the severity of dementia determined prior to death. Thus, Blessed *et al.* (1968) assessed patients from general, geriatric and psychiatric wards in terms of a dementia scale (involving capacity to perform everyday tasks, changes in general habits, personality, interest and drive) and a series of simple cognitive (information, memory and concentration) tests. In the group as a whole

they found highly significant correlations between mean cortical plaque counts in the *post mortem* brains and the previous dementia and cognitive performance scores (although the correlation declined substantially within the small group of severely demented patients). Subsequent studies have demonstrated a similar relationship between dementia scales and the incidence of neurofibrillary tangles, in particular in the hippocampus (Wilcock & Esiri, 1982). Indeed, Wilcock & Esiri (1982) suggested that the regional incidence of tangles provides a better correlation with dementia ratings than plaque counts, but their four-point dementia rating scale based on clinical notes was so crude that direct comparison with the earlier study (Blessed *et al.*, 1968) is extremely difficult.

14.1.2 Neurochemical pathology

In parallel with the analysis of classical neuropathology changes associated with ageing and dementia, the last ten years have seen a dramatic rise in a second line of neuropathological investigation focussing on changes in identified neurotransmitter systems in the brain.

In 1976 and 1977, three groups independently identified a significant decrease in choline acetyltransferase (ChAT) activity – an efficient biochemical marker for ACh-containing neurons – in the *post mortem* brains from demented patients (Bowen *et al.*, 1976; Davies & Maloney, 1976; Perry *et al.*, 1977). The decline in ChAT activity is much more marked in younger (< 79 y) than older (> 79-y-old) patients (Rosser *et al.*, 1981; Mountjoy *et al.*, 1984) and has been found to correlate both with classical neuropathological signs and with the scores on tests of cognitive function obtained prior to death (Perry *et al.*, 1978; Wilcock *et al.*, 1982).

A loss of cortical ChAT activity in Alzheimer's disease suggests a decline in the integrity of the cholinergic nerve terminals in the neocortex, which derive primarily from a subcortical innervation originating from nuclei in the basal forebrain. Whitehouse *et al.* (1981, 1982) first demonstrated cell loss in the nucleus basalis of Meynert (NBM), the basal forebrain site where the magnocellular cortically-projecting cholinergic neurons are found. Subsequent studies have confirmed the cholinergic nature of the cell loss in the NBM (Rosser *et al.*, 1982*b*; Henke & Lang, 1983; Nagai *et al.*, 1983; Jacobs & Butcher, 1986). Although there has been a suggestion that the reduction in number of magnocellular cholinergic neurons in the NBM is attributable to atrophy rather than cell loss *per se* (Pearson *et al.*, 1983*b*), several recent studies have indicated substantial declines in number of ChAT immunoreactive cells in the NBM, particularly in the younger onset patient groups (Etienne *et al.*, 1986). Moreover, loss of the cholinergic innervation is associated with changes in cholinergic receptor density in the cortex and hippocampus of the aged brain (Rinne *et al.*, 1977; Reisine *et al.*, 1978; Norberg & Winblad, 1981; Palacios, 1982; Reinikainen *et al.*, 1987).

Cell loss in the NBM has been shown to correlate both with cortical plaque counts in the Alzheimer brain (Arendt *et al.*, 1985; Mann *et al.*, 1985*a*) and with the severity of dementia manifested by the patients prior to death (Perry *et al.*, 1978; Wilcock *et al.*, 1982). The magnocellular cholinergic cells of the NBM are also suceptible to the accumulation of neuritic plaques and neurofibrillary tangles (Candy *et al.*, 1983; Saper *et al.*, 1985), and degenerating cholinergic terminals are found in association with senile plaque formation in the neocortex (Kitt *et al.*, 1984). These observations indicate an extension of the classic cortical neuropathology of Alzheimer's disease to other subcortical sites implicated in the disease neuropathology, although there continues to be a major dispute over the extent to which the subcortical changes are secondary to the initial degenerative changes originating in neo- and allocortical areas (See Section 14.6.4.2).

The degeneration in forebrain cholinergic systems has captured much attention in the last decade, in particular response to the proposal that the cholinergic degeneration underlies the memory impairments associated with dementia (Bartus *et al.*, 1982; Coyle *et al.*, 1983), but it is necessary to note that neurochemical and immunohistochemical changes have been recorded in many other neurotransmitter systems in the aged brain, and in particular in association with the occurrence of senile and presenile dementia. Thus, in cortical areas, the brains from demented patients show a decline in classical amino acid neurotransmitters, including both the primary excitatory amino acid glutamate (Arai *et al.*, 1984; Hardy *et al.*, 1987*b*),

and the primary inhibitory transmitter GABA (Mountjoy *et al.*, 1984; Rosser *et al.*, 1984; Hardy *et al.*, 1987*a*). The peptides somatostatin, corticotrophin-releasing factor and neuropeptide Y also decline in concentration (Davies *et al.*, 1980; Rosser *et al.*, 1982*a*; De Souza *et al.*, 1986; Beal *et al.*, 1986; Chan-Palay *et al.*, 1986; Dawbarn *et al.*, 1986; Chan-Palay, 1987), and neurofibrillary tangle formation in the neocortex is particularly associated with somatostatin immunoreactive cells (Morrison *et al.*, 1985). These changes do not reflect a non-specific degeneration of all cortical neurons since the number of other neuropeptidergic cells, such as cholecystokinin, vasopressin or vasoactive intestinal polypeptide, do not appear to be reduced in Alzheimer's disease (Perry *et al.*, 1981; Rosser *et al.*, 1982*a*).

In addition to the involvement of the cholinergic innervation of subcortical origin, many other diffusely projecting systems of the forebrain also show an age-related decline, which is particularly marked in patients manifesting dementia. Benton *et al.* (1982) have characterized these systems as comprising a regulatory 'isodendritic core' of the brain, with cell bodies located through pontine, mesencephalic and basal forebrain sites associated with the classical reticular formation, and providing an extensive, diffuse and highly-branched innervation of multiple cortical, limbic and basal ganglia target areas in the forebrain. Age-related changes were first identified biochemically, with the demonstration of a decline in the concentration and other indices of metabolic activity, of dopamine (DA), noradrenaline (NA) and serotonin (5-HT) in the aged brain (Adolfsson *et al.*, 1979; Cross *et al.*, 1981; Benton *et al.*, 1982; Carlsson & Winblad, 1976), which in the case of NA and 5-HT tends to be more marked in the brains from demented patients (Bowen *et al.*, 1983; D'Amato *et al.*, 1987; Palmer *et al.*, 1987). These changes have been associated with cell loss and other neuropathological changes in the locus coeruleus (Bondareff *et al.*, 1981; Tomlinson *et al.*, 1981; Marcyniuk *et al.*, 1986) and raphe nuclei (Terry & Davies, 1983), the respective brain stem nuclei in which these forebrain monoamine systems originate.

In summary, whereas the classical neuropathological features of Alzheimer's disease provide the keystone for *post mortem* diagnosis, they have been associated with changes in the functional activity of a wide variety of identified cortical and subcortical neurotransmitter systems. In the context of this plethora of covarying neuropathological changes, it has proved more difficult to ascertain which components are primary in the aetiology and development of the disease, or which might be responsible for different features of the psychopathological changes of the dementia.

14.1.3 Psychopathology

In parallel with the neuropathological and neurochemical studies, recent years have seen substantial advances in the refinement of the psychological characteristization of the cognitive impairments associated with dementia. Of the many definitions that have been proposed, one of the most succinct characterizes dementia as:

a global impairment of higher cortical functions, including memory, the capacity to solve the problems of everyday living, the performance of learned perceptual motor skills and the correct use of social skills and control of emotional reactions, in the absence of gross clouding of consciousness (Royal College of Physicians, 1981).

Thus, dementia as widely understood involves multifaceted impairments and, as such, the psychological study and assessment of dementia can be undertaken for several different reasons, can take many different forms, and can focus on either a generalized assessment of impairment or on specific features of the disorder. For example, the behavoioural analysis of deficits associated with dementia has been conducted with several different purposes in mind: (i) to refine the diagnosis and early detection of dementia (e.g. Henderson & Huppert, 1984); (ii) to assess the efficacy of treatment strategies (e.g. Corkin *et al.*, 1982); (iii) to identify and characterize the nature of age- and dementia-related impairments in performance; and (iv) as one approach to the study of normal cognitive abilities by considering the consequences of their disruption (e.g. Craik, 1984; Baddeley *et al.*, 1986).

In the present context, that of establishing the validity of animal models of dementia, our primary concern is with purpose (iii): what is the precise nature of the behavioural deficits in dementia, in particular in patients with Alzheimer's disease, against which the behavioural changes in animals

can be compared? The various definitions of dementia all indicate that an impairment in memory and new learning is a necessary diagnostic feature appearing early in the time course of dementia, and mnemonic abilities have been the most extensively studied components of dementia in the laboratory. However, dementia is characterized by a more widespread disruption of performance both in specific psychological tests of cognitive and perceptual abilities, and more generally in the 'activities of daily living'.

14.1.3.1 Perceptual and motor impairments

Visual perceptual performance in normal ageing and Alzheimer's patients has been elegantly studied by Schlottere *et al.*, (1983). In comparison to young controls, both Alzheimer's patients and normal elderly showed similar mild impairments on perceptual tasks that are dependent on relatively peripheral elements of the visual system, such as visual acuity, contrast sensitivity of gratings and backward masking by a homogenous flash of light. By contrast, perceptual tests that were dependent on higher-order visual processing revealed more substantial impairments selectively in the Alzheimer's patient group. These tests included not only time to name letters, but also the duration of the disruptive effects of patterned backward masks. From these observations, Schlotterer and colleagues conclude that both the peripheral visual system and central cortical processing systems undergo some mild age-related decline, but only the higher-order central mechanisms are further disrupted in Alzheimer's disease. Moscovitch (1982) notes that other visual functions that are processed anterior to the striate cortex, such as colour perception, are also impaired to a greater extent in demented patients than in normal elderly subjects.

Other studies have indicated that ageing in general and Alzheimer's disease in particular is associated with a decline in the speed and coordination of motor and sensorimotor performance in a wide variety of tests (Corkin *et al.*, 1981; Perret & Birri, 1982; Semple *et al.*, 1982). Consequently, the design of studies of intellectual and mnemonic performance in demented patients must take account of the extent to which impairments might be attributable to perceptual or performance factors rather than central cognitive factors.

14.1.3.2 *Intellectual and language impairments*

There is a long history of use of intelligence tests in the assessment of dementia. Thus, in the pioneering studies by Roth and colleagues (1967; Blessed *et al.*, 1968), *post mortem* indices of neuropathology were compared with the patients' performance on standardized batteries of tests administered before death. Intelligence tests such as the Weschler Adult Intelligence Scale (WAIS) are frequently employed in the initial screening and selection of patients, with the common observation that Performance scales tend to reveal greater impairments than Verbal scales, both in normal ageing (e.g. Drachman & Leavitt, 1974) and in Alzheimer's patients (e.g. Alexander, 1973; Weingartner *et al.*, 1981).

Although scores on verbal tests may be relatively preserved in demented patients, they nevertheless manifest distinct and substantial impairments in language function. For example, Rochford (1971) compared dysphasic and demented patients in their ability to name drawings of common objects and to indicate parts of the body. Whereas the dysphasic group had much reduced verbal output, the demented patients made many errors by misnaming the objects or body parts. Indeed, naming ability provides a very simple test of language ability in dementia and such items are included in many diagnostic screening tests. Bayles (1982) found that demented elderly subjects were impaired on a variety of verbal tests, but the greatest impairments were found on tests sensitive to semantic content of language, such as the inability to recognize or identify any problem with a sentence such as 'I lost John's temper'. Typically, speech production is not clearly affected in dementia. Rather, fluent speech is found to be relatively disorganized or devoid of content.

Impairments of linguistic ability cannot be readily represented in animal models of dementia. However, since many cognitive tests in human patients are dependent upon verbal instructions and verbal responses, account must be taken of the extent that linguistic impairments might contribute to performance deficits of patients, when developing analogues of clinical tasks for use with experimental animals.

14.1.3.3 *Memory impairments*

Disorders of memory and new learning have always provided a cardinal symptom of dementia, and often provide the earliest signs of its insidious onset (Sjogren, 1952). Normal human memory has been the subject of very extensive experimental psychological research over several decades, and as a consequence has been subject to a variety of overlapping divisions and theoretical formulations (see Baddeley, 1976). The decline of memory in dementia has also been reviewed on many occasions; in the present summary we follow the recent overview provided by Morris & Kopelman (1986).

A distinction can be made between short-term primary or working memory for the temporary storage of material undergoing active processing, and the longer-term storage of processed information for recall on a later occasion. Working memory deficits are apparent in Alzheimer's patients in both verbal and non-verbal tests (Corkin, 1982; Morris, 1984; Kopelman, 1985). Moreover, the degree of impairment appears to be proportional to the severity of dementia assessed by qualitative disruption of independent function in daily life (Corkin, 1982).

More precise data have been obtained with the recent use of non-verbal delayed matching to sample (DMTS) tasks in aged and demented humans, based on tests originally designed for primates (Flicker *et al.*, 1983; Sahakian *et al.*, 1988). These tests use computer-controlled screens or panels, on which various stimuli can be presented and responses recorded. In the DMTS task, a sample stimulus is presented (to which the subject must respond), and then after a variable delay a choice is presented between two stimuli, only one of which has been seen before. The correct response is to respond to the previously presented sample stimulus. On this task, Alzheimer's patients had no problem at very short delays, when the memory load was minimal, indicating that they could detect the stimuli, make the relevant decisions, and had learned the task rule. However as the memory load was increased by lengthening the delay intervals, the patients showed progressively greater impairments in comparison with age-matched controls. These studies provide the least ambiguous and best controlled demonstration of specific and quantifiable working memory impairments associated with dementia.

An even greater disruption is seen of long-term or secondary memory for newly learned information. Thus for example, Corkin (1982) found that demented patients were severely impaired in paired-associate learning with both verbal and non-verbal material, which was again proportional to the degree of dementia. However, several studies have found that if the demented subjects are given an increased number of trials or longer exposure to the information, so that their retention over a short time period of ten minutes is matched to the level seen in control subjects, then the memory is as good one to seven days later as that of the controls (Kopelman, 1985; Corkin, 1982). It therefore appears that the patients have particular difficulty at the initial stage of encoding to to-be-remembered information, rather than with the storage and subsequent retrieval of information.

In the early stages of dementia, the memory of recently acquired information may be more disrupted than remote memories, in accord with the difficulty lying more in the registration than the retrieval of information. For example, Moscovitch (1982) has reported that patients in the early stages of Alzheimer's disease were impaired at recognizing the faces of famous people from recent years, but were as good as normal elderly subjects at recognizing faces of famous people from earlier years. By contrast, several other studies have reported that Alzheimer's patients do have difficulty with recall of information from long ago, both in the famous faces test (Wilson et al., 1981), and in memory for both personal and public events (Corkin et al., 1982, 1987).

Although the primary/secondary distinction in memory was originally derived from models involving temporal stages in the processing of information to be remembered, Tulving (1972) has drawn an overlapping distinction (which is relevant to the animal models to be discussed below) between episodic and semantic memory. The primary memory digit span and Brown-Peterson tests involve memory for discrete elements of information within a spatial and temporal context (episodic memory), whereas the secondary memory tests probe previously established knowledge and the learning of new semantic associations (semantic memory).

14.1.3.4 Cholinergic involvement

Drachman & Leavitt (1974) first described a parallel between deficits induced by the anticholinergic drug scopolamine and the cognitive impairments of aged subjects. The association with Alzheimer's disease has been considered even stronger, leading to the hypothesis that age-related memory impairments generally and the more severe impairments of Alzheimer's dementia in particular may be attributable to a decline in the functional activity of central cholinergic systems (Bartus et al., 1982; Coyle et al., 1983; Kopelman, 1986). As a consequence there have been a number of trials attempting cholinergic replacement therapy in Alzheimer's disease, by treatment with either an excess of precursors to ACh (such as choline or lecithin), cholinesterase inhibitors to inhibit metabolism (such as physostigmine), or direct cholinergic receptor agonists (such as arecoline). Early studies generally found very little effect (Bartus et al., 1982; Kopelman, 1986), although several recent double-blind trials have found that physostigmine provided a small but significant benefit (Davis et al., 1979; Christie et al., 1981; Beller et al., 1985; Sahakian et al., 1987; see also Salamone, this vol., Ch. 15).

In the original studies by Drachman & Leavitt (1974), the class of function most seriously disrupted in drug-treated young subjects and elderly subjects alike was memory storage, although both groups also showed substantial impairments in non-mnemonic cognitive functions such as WAIS Performance IQ. Similarly, physostigmine has been reported to produce improvements in visuo-spatial and constructional abilities of demented patients in addition to the improvements reported in mnemonic performance (Muramato et al., 1984; Sahakian et al., 1987). This suggests that the involvement of cholinergic systems in geriatric psychopathology is not restricted to memory functions, as might be inferred from an initial reading of early versions of the 'cholinergic hypothesis'.

14.1.4 Heterogeneity of dementia

Recent refinements in neuropathological and psychopathological analysis of Alzheimer's disease reveal that it is not a uniform, homogeneous disorder. Although the risk of dementia increases with age, as indicated above, the pattern of neuropathological and neurochemical changes may be distinctive in patients with early- and late-onset of the disease (Rossor *et al.*, 1984; Roth, 1986). Alzheimer's disease of late-onset (Type I) is qualitatively similar to an exacerbation of the decline associated with normal ageing, with little significant neuronal cell loss or changes in catecholamine, amino acid or peptide transmitter systems, moderate decline in central cholinergic markers, and a substantial increase in cortical and hippocampal plaques and neurofibrillary tangles over that observed in the intact eldery brain. By contrast, the onset of dementia at the presenile stage (Type II), is associated with similar cellular neuropathological signs, but additional neuronal loss, atrophy and enlarged ventricles. Neurochemically, the Type II syndrome shows even more marked cholinergic decline, and additional involvement of noradrenergic, GABAergic and some peptidergic systems. Such dissociation within the dementias of the Alzheimer-type accounts for the paradoxical observation that whereas GABA and ChAT concentrations show a significant decline with increasing age in the normal elderly, the concentrations of these markers show a significant positive correlation with age in diagnosed Alzheimer's cases (Rossor *et al.*, 1984; Rossor & Iversen, 1986).

While memory disturbance and impairments in new learning are equally impaired in all subtypes of Alzheimer's dementia (Roth, 1986), a higher prevalence of speech disorder, and possibly also spatial disorientation and visuospatial dysfunction, has been found in early onset cases, whereas increased muscle tone and abnormality of gait appears to be higher in late onset cases (Selzer & Sherwin, 1983; Roth, 1986).

Thus, even within the population of cases in which the diagnosis of Alzheimer's disease is sustained there exists a clear diversity of clinical, psychopathological, neuropathological and neurochemical changes. When account is also taken of other diseases (such as multi-infarct dementia, Parkinsonism, Creutzfeld-Jacob's disease of Huntington's chorea) as well as the decline associated with normal ageing, the complexity of structural changes underlying overlapping profiles of functional deficit has rendered the task of disentangling the structural–functional associations and

the potential design of rational therapies extremely difficult.

14.2 WHY HAVE ANIMAL MODELS?

The preceding overview of the pathology of dementia in humans, in particular as it relates to Alzheimer's disease, has highlighted a number of critical issues that remain unresolved. To what extent is the pathology of Alzheimer's disease an exacerbation of natural ageing as opposed to representing a distinct disease process? What is the relative importance of the cortical and subcortical changes in the aetiology and symptomatology of dementia? More specifically, might it be possible to identify particular elements of neuronal or neurochemical pathology with particular subsets of functional deficit? What therapeutic strategies have the greatest prospect of halting or reversing the insidious cognitive decline associated with dementia? A major purpose of animal experimentation is to develop models of the disease process that permit the systematic manipulation of variables necessary to provide controlled experimental investigation of such issues.

14.2.1 Theoretical issues

A major problem of experimental and clinical studies with human patients is the lack of experimental control of the neurobiological variables underlying the disorder. Thus, for example, there are generally large individual differences in intelligence, education, health and experience that complicate the selection of appropriate age-matched elderly control groups for comparison with the patient groups manifesting dementia. In general, only relatively crude pre-morbid indices are available to provide the basis for the selection of controls. As a consequence large sample sizes are generally necessary to overcome the intrinsic high variability of clinical studies when investigating issues such as whether demented patients manifest differences in a particular neurotransmitter system, unless the difference is itself very large (for example the cholinergic decline in early onset Alzheimer's patients).

A more serious difficulty is to determine the relevance of individual neuropathological deficits to distinct features of the functional deficit. Thus, for example, there continues to be a major disagreement about whether either classic neuropathological indices or cholinergic deficiency are necessary or sufficient to account for mnemonic or other cognitive impairments in Alzheimer's disease, i.e. which feature is fundamental to the occurrence of dementia or its individual components. Evidence for either hypothesis is drawn from the observation that memory impairments correlate with both the cholinergic deficit and the incidence of classic neuropathological indices, but since these two indices of pathological change covary with one another it is difficult to demonstrate unequivocally the causative role of either one. A possible resolution of this discrepancy might arise out of the demonstration that the induction of cholinergic decline in the absence of cortical neuropathological changes, or vice versa, is sufficient to reproduce learning and memory impairments akin to dementia in experimental animals, whilst the converse manipulation does not. Alternatively, if it is found that the experimental induction of cholinergic dysfunction in animals leads to degeneration and neurofibrillary pathology in postsynaptic neurons, or conversely that the induction of cortical neurofibrillary pathology leads to secondary retrograde degeneration of subcortical cholinergic neurons, then again experimental corroboraton would be obtained for the primacy of one feature of the neuropathology over the other.

14.2.2 Therapeutic consequences

In addition to the pursuit of such overall theoretical issues, animal studies have been oriented to understanding the neural basis of complex cognitive, learning and memory functions. In particular, as will be outlined in more detail below, discrete neuronal and neurochemical systems have been implicated in different components of task performance. Not surprisingly, cortical and subcortical systems are generally found to function in close interaction. Nevertheless, cholinergic, adrenergic and amino acid neurotransmitter systems sustain distinctive contributions to overall task performance, and the animal studies suggest particular indices of performance indicative of the functional integrity of the individual components. To the extent that the neural circuitry underlying task

performance is resolved, advances will be made in understanding the course of neuropathology and the nature of psychopathology in patients broadly categorized as manifesting 'dementia', with dual therapeutic relevance.

The first therapeutic consequence of an improved theoretical understanding of the psychopathology and neuropathology of Alzheimer's and related dementias is the provision of a rational basis for the design of novel pharmaceutical and other treatment strategies. Animal models provide a powerful tool in this research, but the programmes can only be as good as the models on which they are based. For example, there exists considerable contemporary research into novel cholinergic drugs which are effective in ameliorating cognitive deficits in animals that have been subjected to various cholinergic-depleting treatments (see below). This provides an efficient approach to dementia only to the extent that the cholinergic deficit is found to be primary or at least fundamentally implicated in the human condition, rather than simply a secondary consequence of, e.g., neocortical pathology that is independently responsible for the cognitive impairments.

A second aim is to provide refined diagnostic procedures for the rational application of treatment appropriate to the individual case. Since the dementias are heterogeneous in the pattern of neurological decline and the related profile of cognitive deficit, understanding the relationship between structure and function should enable the precise diagnostic inference of neurological decline from the cognitive and psychological capacities of the individual patient, in combination with non-invasive neurological and neurochemical tests. The combination of these two therapeutic strategies holds the ultimate prospect of the design of appropriate therapeutic regimens rationally tailored to the individual patient's profile of dysfunction.

14.2.3 Classes of animal model

A widely accepted premise in the development of animal models of dementia is that the more accurately a particular model mimics the aetiology and symptomatology of ageing in the human brain and changes in behaviour, the greater will be its validity and predictive value. In advocating this premise, Bartus *et al.* (1983) outline five criteria against which alternative behavioural models can be evaluated: (i) the behaviour measured should display natural age-related deficits in the species used; (ii) the behaviour should have conceptual and operational similarities with relevant symptoms in humans; (iii) the species used should show age-related neurochemical changes similar to those observed in humans, particularly those that correlate with the behaviours measured; (iv) CNS manipulations to induce behavioural deficits in younger subjects should involve changes in CNS function that mimic those known to exist in the aged subject; and (v) some of the drugs or other treatments that are effective in clinical trials in geriatric patients should also produce positive effects in the animal model.

Bartus *et al.* (1983) designate models which are oriented towards reproducing in experimental animals the rich spectrum of neuropathological, neurochemical and functional changes that are seen in human ageing and dementia as 'Class A' models. The clearest examples of such models are the studies of controlled memory tests in aged primates (Bartus *et al.*, 1983), and these are ideal for purposes such as the screening of novel drugs of different classes. However, other types of model have been developed for the study of drug actions: 'Class B' models involve the artificial induction of neural and behavioural deficits in young animals, and 'Class C' models measure changes of particular parameters of interest *in vivo* or *in vitro* following drug administration. Whereas these latter types of model may be less efficient for screening purposes, they can be considerably more powerful in disentangling the neurobiological variables implicated in a particular aspect of the deficit in the ageing brain and its manifestation in dementia.

Thus, the adequacy of any particular animal model of dementia does not stand in isolation but must be considered in the context of the particular experimental questions to be addressed. For example, if the issue of concern is the relative contribution of cortical and subcortical pathology to deficits in a particular aspect of cognitive performance, then a Class B model involving induction of specific aspects of cortical pathology alone in some animals and of subcortical pathology alone in others is more powerful than attempting to disentangle these components in aged animals,

which develop multifaceted cortical and subcortical pathological changes. Conversely, if the experimental question is to assess the possible efficacy of a novel nootropic drug X, then its administration in a Class A model, such as to elderly primates, will provide a more direct assessment of its potential benefit in elderly patients, whereas its effects in a variety of Class B models are more likely to be informative as to the mechanism of drug action at a functional level. At a third level of analysis, precise information of the site and molecular mechanism of drug action might best be ascertained in an *in vitro* Class C model.

To summarize, the validity of any particular animal model of dementia is not necessarily determined by its similarity to human syndromes associated with ageing, dementia or Alzheimer's disease in all their richness and complexity, although for some purposes (such as exploratory drug screening) this may be the primary criterion. For other experimental questions, in particular neurobiological issues of the relationship between particular features of neurological, neurochemical and behavioural pathology, the validity of any particular model is determined in large part by the specificity of the particular manipulations and their appropriateness to the particular hypotheses under investigation.

14.3 THE AGED ANIMAL

The aged animal provides the most valid animal model of human ageing. However, its validity as a model of dementia is directly related to the extent to which Alzheimer's disease represents an exacerbation or extreme form of normal ageing as opposed to a distinct neuropathological disease process (see Section 14.1). As summarized by Landfield (1983), the basic argument for the aged animal brain as a suitable model for Alzheimer's disease is based on the qualitative (but not quantitative) similarities between the two conditions, whereas the major counterargument is represented by the observation that only a minority of humans develop the specific Alzheimer profile of neuropathology and dementia, although this minority becomes substantial at very advanced ages.

In this Section we will therefore summarize the primary neuropathological, biochemical and function changes that have been characterized in aged animals in the laboratory. In subsequent sections we consider the experimental manipulations that have been considered to model the specific features of pathological ageing more explicitly.

14.3.1 Neuropathological changes in aged animals

Neuronal changes associated with ageing have been widely reported in the brains of old animals, although seldom with the consistency and magnitude of pathology that is suggested from the human neuropathological literature.

Gross atrophy of the aged brain in non-human mammals comparable to that generally seen in Alzheimer's disease (see above) is seldom reported, although a small (10%) reduction in whole brain volume in rats of approximately 30 months of age has been reported by Sabel & Stein (1981), and Johnson & Erner (1972) have estimated that total neuronal populations in cell suspensions of mouse brain decline by over 50% from youth to 29 months of age. Diamond *et al.* (1975) have made detailed morphometric measurements of the depth of neocortex in rats of different ages which suggested a significant decline in all areas from a peak at one month of age to approximately two years. However, this decline all took place within the first three months of life, with no further changes observed between three months and two years. By contrast, within the hippocampus of rats there was actually a progressive increase in the thickness of allocortical laminae over the two-year span, and a similar progressive increase was also seen in the width and depth of the whole diencephalon (Diamond *et al.*, 1975), although the opposite has been reported in the CA1 region of the hippocampus in aged primates (Brizzee *et al.*, 1980). In the large cross-species series studied by Dayan (1971), no gross morophological atrophy was seen in any species, and the only consistent loss of neurones was in the pyramidal cell layer of the cerebellum. Some additional cell loss in the neocortex was only seen in a few of the animals of some species.

Equally inconsistent results have been obtained in quantitative studies of the loss of neurones in particular areas of the brain or in discrete nuclei. Thus, for example, there is a considerable (13–18%) loss of cerebellar Purkinje cells in the rat and an increase in the number of Purkinje cells manifesting pathological features with age (Inukai,

1928). Similarly, there have been reports of a decline in neuronal cell density in both the neocortex (Kullenbeck, 1944; Mufson & Stein, 1980) and in subcortical sites, including the hypothalamus, septum, amygdala, and substantia nigra (Sabel & Stein, 1981). However, other studies have found no change in the density of neurons in a variety of cortical and subcortical sites (Brizzee *et al.*, 1968; Diamond *et al.*, 1977; Peng & Lee, 1979; Goldman & Coleman, 1981; Curcio & Coleman, 1982).

At the cellular level a variety of changes have been reported in neuronal morphology. Inukai (1928) first described three classes of pathological abnormality which develop with age in the Purkinje cells of the rat cerebellum: (i) 'pycnotic' changes in which the cells become shrunken and densely staining with methylene blue, until they finally disintegrate; (ii) accumulation of a 'mesh-like' appearance within the cytoplasm, reminiscent of neurofibrillary changes; and (iii) increased occurrence of abnormally shaped or multiple nuclei within the cells. Subsequent studies have had more refined techniques available to consider such changes with much greater precision. Thus, for example, cortical (Feldman, 1976) and olfactory (Hinds & McNelly, 1977) neurons have also been seen to undergo perikaryal atrophy, although the latter authors found the highest proportion of olfactory neurons with double nuclei to occur at around 3 months of age, and to decline subsequently. In addition to the somatic changes, substantial rearrangement or retraction has been observed in the dendritic domain of cortical and hippocampal neurons and in the organization of axo-dendritic contacts. For example, Vaughan (1976, 1977) has described at the ultrastructural level 'large whorled membranous bodies' forming in the dendrites of cortical pyramidal neurons, and a reduction in the density of the basal dendritic tree. In the cerebellum, Glick & Bondareff (1979) have reported an age-related decline in the density of synaptic contacts onto pyramidal neurons of aged rats.

An increase in neuroglia in the ageing brain has been widely assumed to be a corollary of neuronal loss, and complementary changes in glial cells have been reported. For example, increased numbers of microglia have been reported in the cerebral cortex (Vaughan & Peters, 1974) and hippocampus (Landfield *et al.*, 1981) of aged rats. Additionally, reactive astrocytes in the hippocampus and cere-bellum appear to show a progressive hypertrophy with age (Landfield *et al.*, 1978; Bjorklund *et al.*, 1985). However, such changes in all neuroglial populations are not always observed (Ling & Leblond, 1973; Vaughan & Peters, 1974; Bond-areff, 1977; Diamond *et al.*, 1977).

The most consistent age-related change in cere-bral pathology is lipofuscin accumulation: the pro-gressive appearance in many neurons of character-istic cytoplasmic lipid droplets. Lipofuscin is composed of a range of liposomal enzymes and other proteins, and ultrastructurally appears as dense granular structures within the neuronal cyto-plasm and in particular associated with the Golgi apparatus (Bondareff, 1977). The accumulation of lipofuscin granules with age is widespread. For example, Brizzee *et al.* (1974) found a progressive increase in the percentage of neurons containing lipofuscin granules in virtually all brain areas from very low levels in the youngest monkeys (three months) up to a mean of 47% of neurons in the oldest group (up to 20 years of age). Similar age related changes have been confirmed in many different species, and were relatively consistent across the 47 species of vertebrates compared by Dayan (1971). Although lipofuscin provides a clear marker of age-related changes in the ageing brain, its biochemical nature and development are still not well understood, although it is generally believed to be a residual product of some autopha-gic process.

By contrast with these age-related changes in animals which mimic many of the changes observed in normal human ageing, neuropatholo-gical features associated with Alzheimer's disease are either extremely rare or absent in aged animals. Thus Dayan (1971) did observe microvascular changes in the brains of old animals from several species, but neither neuritic plaques nor neurofi-brillary tangles were seen in any of his specimens. These two hallmark features have in fact been described in aged animals of several species. Neuritic plaques have been seen in the brains of aged dogs (Wisniewski *et al.*, 1970), rhesus monkeys (Kitt *et al.*, 1984, 1985; Struble *et al.*, 1985; Wisniewski *et al.*, 1973), squirrel monkeys (Walker *et al.*, 1987) and rats (Vaughan & Peters, 1981), that have not undergone any experimental infection or manipulation. The plaques in rhesus

monkeys have been associated with several differ-ent neurotransmitter systems, including ACh, NA, enkephalin, GABA and somatostatin (Kitt *et al.*, 1984, 1985; Struble *et al.*, 1984; Walker *et al.*, 1985, 1986). However, in all these studies the plaques differed from human senile plaques in not manifes-ting the presence of paired helical filaments (Wis-niewski *et al.*, 1973; Vaughan & Peters, 1981).In-tracellular tangles of paired helical neurofilaments have not been reported in untreated aged animals (but see Section 14.5 below).

14.3.2 Neurochemical changes in the aged brain

Unlike the neurofibrillary tangles characteristic of human ageing and dementia, neurochemical changes similar to those seen in human ageing have been widely observed in aged animals of all species investigated, and have been the subject of numer-ous reviews (e.g. Pradhan, 1980; McGeer, 1981; Algeri *et al.*, 1983). Biochemical studies of neuro-transmitter indices in the brains of aged animals have highlighted not only a decline in neurotrans-mitter levels associated with cell loss *per se*, but also a decline in synaptic function of viable neurons associated with changes in the efficiency of the presynaptic mechanisms of neurotransmitter synthesis, transport, storage, release and reuptake, and in the number or affinity of the postsynaptic receptors. Thus, many studies have investigated multiple markers of the functional integrity of a particular neurotransmitter system, rather than taking only a single measure of neurotransmitter concentration.

Of the classical transmitter systems, the most extensive studies have been in the monoaminergic systems. In the rodent brain, baseline levels of DA (Finch, 1973; Demarest *et al.*, 1980; Ponzio *et al.*, 1982) and DOPAC (Demarest *et al.*, 1980; Oster-burg *et al.*, 1981), tyrosine hydroxylase activity (Reis *et al.*, 1977), and DA uptake (Haycock *et al.*, 1977; Jonec & Finch, 1975) have all been reported to decrease, particularly in the neostriatum. Whereas such presynaptic changes are generally moderate (e.g. 15–30%), they can vary widely from 0 to 66% loss in different strains of rat (Algeri *et al.*, 1983; Gilad & Gilad, 1987). The number of D_1 and D_2 receptors also declines in the aged rat brain (Severson & Finch, 1980; Joyce *et al.*, 1986; Giorgi *et al.*, 1987). This is accompanied by reduced levels of DA-activated adenylate cyclase (Govoni *et al.*, 1977; Schmidt & Thornberry, 1978), a decline in glucose utilization (London *et al.*, 1981) and an impaired response to acute haloperidol treatment (Finch, 1982; Randall *et al.*, 1981; Trabucchi *et al.*, 1982), although the chronic development of D_2 receptor supersensitivity in response to nigrostriatal denervation was similar in mature and aged rats (Joseph *et al.*, 1981; Hirschhorn *et al.*, 1982).

Other studies have shown similar decreases of NA in the cortex and hypothalamus of aged animals (Westfall, 1987; Ponzio *et al.*, 1978; Estes & Simpkins, 1980), and parallel impairments in NA metabolism and catabolism (Finch, 1973), uptake and release (McIntosh & Westfall, 1987), and receptor binding (Greenberg & Weiss, 1978; Maggi *et al.*, 1979; Misra *et al.*, 1980; Leslie *et al.*, 1985), although the changes are often less dramatic than the decline in striatal DA markers.

By contrast, in the majority of studies, the concentration of 5-HT and the related markers tryptophan hydroxylase and 5-HIAA have been found to be much more stable with age (Finch, 1973; Pradhan, 1980; Ponzio *et al.*, 1982; Morgan, 1987). Although a few studies have reported small decreases in specific nuclei, such as the raphe, in aged animals (Pradhan, 1980), others have actually indicated age-related increases in 5-HT metabolism with age (Simpkins *et al.*, 1977). Similarly, 5-HT_1 receptor levels are also relatively stable with age, although 5-HT_2 receptors may show moderate declines (Finch & Morgan, 1984).

The majority of these studies have been conducted in aged rodents. However, a similar pattern of decline has recently been reported in aged primates. Thus, Goldman-Rakic & Brown (1981) conducted a systematic evaluation of monoamine content and synthesis in discrete regions of the rhesus monkey brain across four age bands. They found that the most substantial cortical decline was in DA content, particularly in the prefrontal cortex, whereas the depletion of NA was mild, and no significant depletions of 5-HT content were seen. Subcortically, significant DA depletions were seen in the caudate nucleus and hypothalamus, whereas DA levels in the brainstem actually increased. Catecholamine synthesis showed even greater ($> 50\%$) reductions in both cortical and subcortical regions of the aged monkey brain, whereas indoleamine synthesis was unaffected.

More recently, Lai *et al.* (1987) have reported parallel declines with ageing in the concentration of dopamine D_2 receptors in the caudate and putamen of rhesus monkeys.

Only in the last few years, with the recent interest in the 'cholinergic hypothesis', has similar attention been given to age-related changes in cholinergic systems in the brain. There is still scant information on ACh content itself (Pradhan, 1980), and greater attention has been given to levels of the synthesizing enzyme ChAT. While several studies have reported a significant decline in ChAT activity in the aged rodent cortex, striatum and other areas (McGeer *et al.*, 1971; Pradhan, 1980; Strong *et al.*, 1980), the decline in these cases is generally small (Bartus *et al.*, 1982) and many other studies find no significant decline with age (e.g. Ingram *et al.*, 1981; Reis *et al.*, 1977). However, there do appear to be substantial decreases in sodium-dependent high-affinity uptake mechanisms (Sherman *et al.*, 1981), ACh synthesis and release (Gibson *et al.*, 1981; Meyer *et al.*, 1984), and muscarinic receptor binding (James & Kanungo, 1976; Lippa *et al.*, 1980; Strong *et al.*, 1980).

In addition to the changes in the classical amine transmitters, more recent biochemical and immuno-histochemical studies have also identified age-related disturbances in GABA synthesis in the hippocampus (Ingram *et al.*, 1981; Lowy *et al.*, 1985) and in many peptidergic systems throughout the brain, including substance P, neurotensin, LHRH, somatostatin, met- and leu-enkephalin, ACTH, MSH, LPH, TRH and endorphin (reviewed by de Weid & van Ree, 1982; Banks & Kastin, 1986).

14.3.3 Behavioural changes in aged animals

All mnemonic and cognitive tests in animals involve observations of changes in behaviour in response to particular sets of training stimuli. It is therefore impossible to assess performance independently of the animals' sensory capacities, motivation and motor abilities, and changes in these factors need to be clearly appreciated and explicitly controlled for in tasks involving learning.

14.3.3.1 Motor deficits

Numerous studies have reported motor deficits in aged animals. At a simple level mice, rats and cats all show impairments in generalized locomotor activity, exploration and in the habituation of activity within and between test sessions (Smith & Dugal, 1965; Goodrick, 1965, 1975; Elias *et al.*, 1975; Brennan & Quartermain, 1980;Gage *et al.*, 1984*a, b*; Levine *et al.*, 1987). More precise assessment of the coordination and balance of young and old animals in these species indicate that the old groups show greater impairments in tests requiring coordinated control of motor behaviour (such as balance on a narrow beam) than in tests of simple reflexes (such as placing responses or negative geotaxis) (Wallace *et al.*, 1980*b*; Ingram *et al.*, 1981; Levine *et al.*, 1987).

A number of studies have attempted to associate motor impairments with changes in central motor systems including the basal ganglia. In particular, it has often been suggested that a decline in the dopaminergic innervation of the neostriatum may underlie some of the motor deficits. Many of the motor impairments seen in old rats are akin to those induced by central dopaminergic lesions in young rats (Marshall & Berrios, 1979; Gage *et al.*, 1983*b*). The time course of the stereotypic response to the dopaminergic stimulant amphetamine is markedly altered in aged rats (Hicks *et al.*, 1980), and old rats show a reduced sensitivity to dopaminergic drugs when tested for amphetamine-induced rotation (Joseph *et al.*, 1978), or haloperidol-induced catalepsy (Randall, 1980). Moreover, some age-related deficits in coordinated motor behaviour can be ameliorated by treatment with DA agonists (Marshall & Berrios, 1979) or by dopaminergic neural grafts into the neostriatum (Gage *et al.*, 1983*a*). Not all activational changes need be associated with the DA system, however, and other studies have implicated 5-HT systems in the hippocampus in the age-related changes in exploratory activity (Brennan *et al.*, 1981).

Zornetzer & Rogers (1983) raise the possibility that the focus on DA systems underlying age-related motor impairments may be misplaced. They emphasize that many of the tests of posture, balance and coordinated movement used to assess motor behaviour in aged animals also provide sensitive indices of cerebellar function, and they review a variety of anatomical and electrophysiological studies indicating a decline in the functional integrity of the cerebellar circuitry in senescence. In particular, they report impressive within-subject correlations between the impaired balance of old rats on a rotating rod and the decline in density of both Purkinje cell numbers and synaptic contacts in the molecular layer of the cerebellum. Their caution needs to be applied even more generally to an awareness of changes in the peripheral musculature of aged animals (Tuffery, 1971; Tauchi *et al.*, 1971).

14.3.3.2 Sensory deficits

When using training stimuli, such as visual or auditory discriminative stimuli, or footshock to maintain escape or avoidance responding, it is necessary to take account of possible changes over age in the animals' sensitivity. Whereas some studies have reported no differences between young and old rats in sensitivity to electric footshock (e.g. de Koning-Verest *et al.*, 1980; Lippa *et al.*, 1980), others have found marked differences. For example, Pare (1969) reported an increased shock detection threshold in old rats, although this covaried with body weight. Aged rats also show delayed reaction times to both footshock and loud auditory stimuli (Birren, 1955), and decreased startle responses to suprathreshold stimuli (Campbell *et al.*, 1980; de Koning-Verest *et al.*, 1980), although the relative contributions of motor and sensory factors to these changes are not clear. On the other hand, increased sensitivity to footshock (Gordon *et al.*, 1978) and a hotplate (Chan & Lai, 1982) have also been reported.

A decline in visual and auditory sensitivity with ageing is of course well recognized (e.g. Goetsch & Isaac, 1982), and is in large part attributable to a decline in the peripheral sense organs, receptors and afferent nerves. Cambell *et al.* (1980) used the fact that a weak but detectable stimulus preceding by 50–100 ms a more intense auditory startle stimulus attenuates the startle response, to determine detection thresholds in aged rats to stimuli of several modalities. They found that the burst of low intensity white noise that was necessary to inhibit auditory startle was relatively constant through the first two years of life, but then increased substantially during the succeeding months. Similarly, the rats' sensitivity to a 20-ms prepulse of light showed a similar decrease in the oldest age groups. The power of this approach to the determination of visual or auditory sensitivity is that both the auditory startle and the prestimulus inhibition effect are unconditioned, and the effect of the prepulse stimulus can be established with relative constancy across a range of startle stimuli and responses, independent of learning capacities.

14.3.3.3 Maze learning

The study of the role of age in the learning ability of rats has a venerable history extending back to the early days of experimental psychology (Liu, 1928). Most early studies employed maze learning tasks. For example, Stone (1929) reported that rats trained in a difficult sequential pattern maze reached a maximum learning rate at around 30–70 days of age. A high level of performance was maintained through two-thirds of the life span, but began to decline in early senescence around 20 months of age. In such studies, task difficulty was found to be an important variable (Elias & Elias, 1970; Kubanis & Zornetzer, 1981). Thus, Yerkes (1909) first reported that shock-motivated discrimination learning was deficient in aged mice only when the visual discrimination was difficult, whereas on a simple task the aged mice were if anything superior to young mice. Similar observations have been replicated by Goodrick (1972); not only was there a general decline with increasing age in rats' learning of a multiple choice maze, but additionally the age-related deficits became progressively greater as the number of choices increased from one to 14. By contrast, tasks involving learning of simple two-choice visual discriminations, whether based on shape, pattern or brightness, reveal few differences between young and old rats (Stone, 1929; Fields, 1953; Elias & Elias, 1976).

The last decade has seen a renewed interest in maze learning by old rats, not so much as a general tool for assessing discrimination or learning capacity, but with a particular focus on the spatial abilities of the animals. This focus arises from the recognition of the hippocampus and its central cholinergic innervation as a neural substrate for spatial mapping strategies used by rats and other mammals (O'Keefe & Nadel, 1978). Barnes (1979) first reported that aged rats were deficient in learning the location of a dark tunnel in order to escape from a brightly-lit exposed circular platform. This

deficit was significantly correlated with a decline in synaptic enhancement on granule cells of the dentate gyrus after high-frequency stimulation of the perforant path afferents to the hippocampal system. Gage et al., (1984b, c) confirmed similar spatial mapping deficits of aged animals in the Morris water maze, associated with a decline of metabolic activity in the septo-hippocampal pathway and frontal cortex. Moreover, the water maze impairment of old rats could be ameliorated following cholinergic replacement, by transplantation of embryonic septal cells to the hippocampus, strongly supporting the hypothesis that decline in the septo-hippocampal system underlay the functional deficits (Gage et al., 1984c; Gage & Bjorklund, 1986).

Rapp et al. (1987) have provided a detailed analysis of performance of old rats in variations of the Morris water maze task in an attempt to pinpoint the nature of the age-related impairment. They replicated the deficit observed in the original task, in which the rats must use distant spatial cues to infer the location of a submerged escape platform, but found no deficit when old animals were trained to locate a visible escape platform. Subsequently rats trained with the visible platform were switched to learning to locate a submerged platform in a novel location. The young animals alone showed an initial spatial bias to the original platform location, but they then learned to escape to the submerged platform more rapidly than the old rats. This pattern of results strongly suggested a specific spatial impairment in the group of old rats.

14.3.3.4 Passive avoidance memory

Maze learning studies illustrate general learning and cognitive impairments in old animals, whereas other studies have been interested in whether ageing is associated with explicit memory impairments (see Kubanis & Zornetzer, 1981, and Bartus et al., 1983, for reviews). Perhaps the most widely studied test of memory involves passive avoidance acquisition and retention by rodents. In passive avoidance training a rat or mouse is punished immediately it engages in some spontaneous behaviour. Typically, this involves administration of footshock following stepping off a raised platform onto a grid floor or entry from the brightly-lit into the dark compartment of a two-compartment test box ('step down' and 'step-through' passive avoidance, respectively). Then, at some later time, the animal is replaced in the same situation: memory for the previous aversive event is manifested by inhibition of the spontaneous behaviour – i.e. the animal stays where it is initially placed. An advantage of passive avoidance tests is that avoidance can generally be learned in a single trial, so that the time of acquisition can be determined precisely. Moreover, in contrast to many other maze learning tasks, passive avoidance tests require no long-term food-deprivation, which may be traumatic for aged rats.

Gold & McGaugh (1975; Gold *et al.*, 1981) first reported deficits in step-through passive avoidance in old rats. In comparison with young controls, one-year old rats showed no deficiencies when tested one day after training, but were deficient on one-, three- or six-week tests; two-year old rats showed deficits in retention within six hours of training. Subsequent studies from other laboratories have amply confirmed the basic deficit in passive avoidance by old rats (McNamara *et al.*, 1977; Lippa *et al.*, 1980; Dean *et al.*, 1981; Kubanis *et al.*, 1981; Zornetzer *et al.*, 1982).

Whereas it is clear that old animals manifest deficiencies in passive avoidance retention, it is more difficult to determine whether the deficit is attributable to an age-related decline in memory, as opposed to increasing difficulties in detecting the training stimuli or engaging in the response. While it is apparent that aged animals do have increasing sensory and motor impairments (see Sections 14.3.3.1 and 14.3.3.2), these do not provide a sufficient explanation for all of the deficits observed in passive avoidance tests. Thus, Bartus and colleagues showed that old rats that were deficient on 24-h passive avoidance retention did not differ in shock threshold sensitivity, in initial step-through latency or in retention when tested within one hour of training (Lippa *et al.*, 1980; Bartus *et al.*, 1983). However, their data did indicate a clear decline in performance at even the shortest retention times, which in all probability failed to reach significance only because of the small group sizes. Nevertheless, the more systematic studies of the time course of forgetting in old and young rats by Gold and colleagues (1981) indicate

a clear onset of deficits, with their training parameters, only at the longer retention intervals.

An alternative approach to assessing the rate of forgetting in long-term memory has been adopted by Thompson & Fitzsimons (1976). They trained rats to a common level of acquisition of a discrimination avoidance, and then retested animals eight days later. The number of trials to relearn the task to criterion increased monotonically with age. Here, the animals were at least matched for initial learning to a common criterion, but since the older animals were also slower to learn the initial brightness discrimination, their subsequent impairment might have been due to slower relearning rather than greater forgetting.

14.3.3.5 *Short and long-term memory*

As in the human memory literature (see Section 14.1.3), recent studies have attempted to disentangle more precisely the influence of ageing on discrete memory processes. The eight-arm radial maze was introduced by Olton & Samuelson (1976) as a convenient test of rats' working memory. The maze has a central platform and eight radiating arms, with a food well at the end of each arm. All wells are baited at the start of each trial and the rat is trained to enter all arms in order to collect the food. Thus, the animals require an efficient working memory of the arms already entered on the daily trial in order to achieve the optimal performance of visiting each arm only once. A decrease in accuracy is generally found on later choices in the trial, both as more information has to be recalled and as a progressively longer time passes since the initial choices. Within 5–10 days of training, normal young rats adopt the strategy, akin to optimal foraging, of visiting each arm just once to collect and consume the food there. This task has been found to be sensitive to hippocampal damage (Olton *et al.*, 1978).

Wallace *et al.* (1980*a*) trained food-deprived rats aged five, eight, 14 or 26 months on the eight-arm radial maze, and found that although all age groups showed substantial learning over 30 days of training, the older rats performed more poorly, in terms of the number of different arms chosen in the first eight trials, at all stages of training. However, when an enforced delay was inserted between the fourth and fifth choices, old and young rats showed a similar deterioration proportional to the length of the delay interval. This suggests that the deficit in aged rats was not attributable to a working memory deficit *per se*, but rather to a generalized impairment of sequential choice performance. Wallace *et al.* (1980*b*) tested and eliminated the possibilities that the performance deficit in the aged group was attributable to enhanced proactive interference from previous trials, or greater perseverative tendencies. They concluded that a decline in visual sensitivity to the spatial cues necessary for the guidance of task performance may have been the determining factor.

Using the same eight-arm radial maze task, Barnes *et al.* (1980) found a similar moderate deficit in the number of trials taken by 26-month-old rats to reach a criterion level of performance. These authors argued against the conclusion that the deficit of the old rats was due to a decline in visual acuity, since on the spatial task involving escape from an exposed circular platform, the old rats showed similar asymptotic performance even though different sets of external cues were used, which differed markedly in salience (Barnes, 1979; Barnes *et al.*, 1980). They conclude instead in favour of a spatial mapping impairment: 'that the deficit in the old animals is a failure to relate specific places to an integrated map of the environment' (Barnes *et al.* (1980), p. 37).

Nevertheless, in these studies the deficit of the old rats in the radial maze was relatively mild. In order to increase task difficulty, Ingram *et al.* (1981) increased the number of arms to 12, and found that in this harder version of the task the old group of rats was indeed substantially deficient. Moreover, the deficits in the individual old animals were correlated with the extent of decline in ChAT and GAD activity in the hippocampus, suggesting a neurochemical substrate for the age-related impairments. In another study, Bernstein *et al.* (1985) tested young and old mice of the C56BL/6J strain in the radial maze, and in this species found no deficits in the old mice. However, in this strain, there were also no age-related differences between young and old mice in central ChAT or GAD activity, although performance of the individual animals of all ages did correlate with changes in ChAT and GAD activity in particular neocortical areas.

In a further attempt to separate working memory from reference memory in aged rats, Lowy *et al.* (1985) modified a T-maze alternation task to involve a stable spatial discrimination in the stem of the maze, requiring long-term reference memory, prior to approach to the alternation choice point, a decision based on working memory. Both components of the task were impaired in the old rats, although only the working memory component correlated with any of the neurochemical indices, and in particular with hippocampal GAD activity.

The difficulty with all these tasks is that of separating the old animals' memory performance from their clear spatial deficits. In order to assess working memory directly, some task is required that will assess the decay of information from working memory, independently of the animals abilities to learn the general task demands. This problem has been addressed most directly in primates with variations on delayed response, matching and non-matching tasks, and only recently extended to other species.

14.3.3.6 Delayed response

Bartus *et al.* (1978, 1980) trained rhesus and cebus monkeys on an automated discrete-trial delayed response task. In this task, the monkey could see a flashing stimulus presented on one panel of a 3×3 array, but was prevented from responding immediately by an observation window. Then, after a variable delay (from 0–30 s), the window was opened and the animal was reinforced for pressing the previously illuminated panel. Aged monkeys (aged at least 18 years) did not differ from young controls when they were allowed to respond directly to the stimulus or when the zero-second delay was imposed, but showed progressively greater impairments at the longer delays. Bartus *et al.* (1978, 1980) argue that these data represent unequivocal evidence for a specific impairment in short-term memory in their aged monkeys, since any spatial, motivational, attentional, visual acuity or psychomotor deficits should also have been apparent in the zero-second delay condition.

A number of central systems may underlie these deficits. Bartus and colleagues (1982, 1983) have focussed on central cholinergic systems following the report of a beneficial effect of physostigmine on short-term memory in aged rhesus monkeys

(Bartus, 1979). Moreover, they found evidence that both the cholinesterase inhibitor physostigmine and the muscarinic agonist arecoline could improve the delayed performance of aged cebus monkeys, although the effective dose ranges were quite narrow and the optimal dose varied dramatically from monkey to monkey (Bartus *et al.*, 1980). Other cholinesterases and muscarinic agonists were also found to be effective, whereas no improvement was observed with dopaminergic, GABAergic or adrenergic agonists (Bartus *et al.*, 1983). By contrast, Arnsten & Goldman-Rakic (1985*a*) present convincing evidence that the alpha$_2$ adrenergic agonist clonidine could enhance performance of old rhesus monkeys on a related delayed response task, in each of the 13 rhesus monkeys they tested, and related this to NA function in the prefrontal cortex (Arnsten & Goldman-Rakic, 1987). As in monkeys with frontal lesions, the aged animals were also deficient in delayed alternation tests of spatial working memory in a delay-dependent manner, and the deficit was also ameliorated by clonidine (Arnsten & Goldman-Rakic, 1985*b*). Again, no beneficial effects of the DA agonists l-DOPA or apomorphine were observed.

By contrast with the studies in primates, there have been few studies of discrete-trial delayed response in aged rats. Wallace *et al.* (1980*b*) trained rats on the go/no go Konorski paradigm, in which two stimuli (a light or a tone) are separated by an interval. A response lever is presented with the second stimulus, and the rat is reinforced for responding if the two stimuli are the same (light-light or tone-tone) but not if they are different. On this task the old rats tended to perform more poorly at all delay intervals, including the zero-second delay, although the difference was not significant. However the task proved difficult for all rats, and even after as little as five seconds, young and old rats alike showed discrimination barely above chance levels. By contrast, Dunnett *et al.* (1988) trained six-, 15- and 24-month-old rats on either delayed response or delayed alternation in a two-lever discrete-trial task, and found that the oldest group was impaired on both versions. However, in this study the age-related deficits were delay dependent: the old rats performed close to 100% with short delays of zero to four seconds, but showed progressively greater impairments as

the delay intervals were lengthened to 24 s. This pattern of deficit in rats is similar to that observed by Bartus *et al.* (1978, 1980) in monkeys, and argues strongly for an explicit short-term working memory deficit. However, in contrast to the primate studies, Dunnett *et al.* (1988) failed to find any beneficial effects in the old rats of either physostigmine or arecoline over a wide range of doses. Other classes of pharmacological agents have not yet been tested on rodents in delayed-response tasks.

To summarize, it is clear that ageing in animals reproduces many of the neuropathological, neurochemical and behavioural changes observed in human ageing. Nevertheless, it remains the case that the association between the structural changes and their functional sequelae are largely unknown. Moreover, although the aged laboratory animal may provide a good model of natural ageing in humans, the specific pathology of Alzheimer's disease is only poorly approximated by the structural changes in the brains of aged animals; consequently, these animals provide a poor model for the specific conditions associated with dementia.

14.4 PHARMACOLOGICAL MODELS

An appealing strategy in developing animal models of dementia has been the use of pharmacological manipulations to induce symptoms similar to those observed in Alzheimer's disease and normal ageing. Investigators using this approach have typically attempted to induce learning and memory deficits in young subjects with drugs of known pharmacological activity. A main advantage of drug models is their facility of use. Restrictions on animal availability do not apply (e.g. as in the aged animal), and the time needed for animal preparation is much less than that required of other models (e.g. surgical procedures in lesion models). Additionally, each animal can generally act as its own control with appropriate use of a balanced schedule of drug administration. Thus, drug models have become a practical, inexpensive and widely used means of screening the relatively large number of animals that are often required for behavioural (including memory) assessment.

Another advantage of the drug model is its neurochemical specificity. Many pharmacological compounds are available which are specific to identified neurotransmitter systems and to specific receptor subtypes. The availability of drugs which could induce symptoms of dementia would help to uncover particular neurotransmitter systems involved in cognitive or memory dysfunction, and also point to possible avenues for treatment. The potentials of this approach are exemplified by the large literature attempting to relate the ACh transmitter system with deficits in memory performance and the age-related cognitive decline in Alzheimer's disease. As described earlier, it is well established that the classical neuropathological signs and the cognitive decline characteristic of Alzheimer's disease are correlated with degeneration of cholinergic inputs to the neocortex. With drugs models, it may then be possible to determine whether cholinergic dysfunction is directly related to memory deficits. In fact, there is considerable evidence suggesting that blockade of ACh muscarinic receptors with cholinergic antagonists leads to a disruption of memory function (Drachman & Leavitt, 1974; Bartus & Johnson, 1976; Spencer & Lal, 1983), and that this drug-induced memory deficit can be reversed with cholinergic agonists (Bartus *et al.*, 1978; see Salamone, this vol., Ch. 15). These data have been influential in stimulating the search for appropriate cholinergic agonists as treatments for dementia.

Although the pharmacological model has practical appeal for delineating neurotransmitter systems involved in memory dysfunction, can it approximate the neuropathological signs and memory impairments observed in Alzheimer's disease? In this section we examine critically the potential for drug models to serve as a basis for understanding and treating dementia, using the cholinergic antagonist scopolamine as a prototypic (and the most extensively studied) example. Scopolamine has been found to induce mnemonic impairments in a variety of species, including humans (see below). Three related issues need to be addressed: (i) Are the profiles of the amnesia induced by scopolamine in humans, non-human primates, and rodents similar to those observed in Alzheimer's disease? (ii) Does the mechanism of action of scopolamine mimic the known neuropathology of Alzheimer's disease? (iii) Does scopolamine present an adequate model for testing potential treatments (including cholinomimetics) for Alzheimer's disease?

14.4.1 Scopolamine effects in humans

There is substantial evidence that scopolamine has amnesic effects in humans. Memory deficits have been reported on a variety of tasks, including immediate and delayed free recall (Crow & Grove-White, 1973; Drachman & Leavitt, 1974; Ghonheim & Mewaldt, 1975; Petersen, 1977; Beatty *et al.*, 1986), serial learning (Sitaram *et al.*, 1978; Caine *et al.*, 1981), supraspan digit storage (Drachman & Leavitt, 1974; Drachman, 1977), the Brown-Petersen distractor task (Caine *et al.*, 1981; Beatty *et al.*, 1986), and dichotic listening (Drachman & Sahakian, 1980). However, the effects of scopolamine on some tasks have been controversial. For example, there have been both positive and negative reports that scopolamine may affect category naming (Drachman & Leavitt, 1974; Drachman, 1977; Caine *et al.*, 1981; Beatty *et al.*, 1986) and paired-associated learning (Ostfeld & Arguette, 1962; Crow & Grove-White, 1973; Beatty *et al.*, 1986). Possible explanations for these discrepancies could lie in either the dose of scopolamine or the route of drug administration used in the different studies.

There are also some cognitive tasks which appear to be resistant to the effects of scopolamine. For example, tasks similar to the standard digit-span test used in the WAIS, in which subjects are asked to repeat digit sequences of various lengths in their order of presentation, are unaffected by scopolamine administration (Ostfeld & Arguette, 1962; Drachman & Leavitt, 1974; Ghonheim & Mewaldt, 1975; Beatty *et al.*, 1986). Other behaviours that are not disrupted by scopolamine include the ability to produce semantic or rhyming word associations (Caine *et al.*, 1981), and delayed free recall when the to-be-remembered words are presented *before* the scopolamine injection (Ghonheim & Mewaldt, 1975; Petersen, 1977). Finally, Caine *et al.* (1981) found no drug effect on tasks of auditory attention and detection but did report a scopolamine-induced deficit on a task requiring recognition of the order of two tones.

Interpretations of the amnesic effects of scopolamine fall into three categories. Firstly, scopolamine might produce a general depression of the nervous system and thus may affect attentional processes as well as (or in contrast to) memory (Carlton, 1963, 1969; Warburton, 1972). Most investigators have discounted this possibility because tests of immediate memory (e.g., digit span) are spared (Drachman & Leavitt, 1974; Ghonheim & Mewaldt, 1975; Drachman, 1977). Moreover, no impairments have been found on test of auditory attention and detection (Caine *et al.*, 1981) or visual scanning (Crow & Grove-White, 1973). In contrast, Ostfeld & Arguette (1962) reported that scopolamine produced severe impairments on distraction (Stroop) and sustained attention (buzzer-press) tests, and they concluded that a main effect of scopolamine was on the maintenance of an attentive state. They suggest that immediate memory may be intact following scopolamine because the digit-span task does not require attention to be sustained for any length of time. A similar rationale might apply to the experiments by Warburton & Brown (1971; Brown & Warburton, 1971), in which scopolamine was found to influence the d-prime index of stimulus sensitivity in a signal detection task, in which subjects were required to sustain high levels of vigilance for accurate performance. Nevertheless, a degree of caution is needed in particular with the earlier Ostfeld & Arguette (1962) studies: they indicated that 30% of their subjects experienced hallucinations and 50% slept for brief periods, which might suggest very non-specific disruption of complex task performance. Although Ostfeld & Arguette (1962) employed doses of scopolamine that were in the same range as in other studies, the s.c. route of administration that they used might have produced unusual effects in drug metabolism.

A second possible mechanism that could be affected by scopolamine is the encoding and storage of new information. Drachman & Leavitt (1974) demonstrated that if the number of digits to be remembered is within the limits of short-term memory (i.e. 7 ± 2), scopolamine-treated subjects performed as well as saline-treated controls. Only if the number of digits exceeded the short-term memory span were severe deficits apparent. The lack of effect of scopolamine on the recency effect but the profound disruption of the primacy effect, in serial position curves of list recall, is consistent with this interpretation (Crow & Grove-White, 1973). Similarly, a severe disruption of performance on the Brown-Peterson distractor task by scopolamine also suggests a deficit in the encoding

of new information (Caine *et al.*, 1981; Beatty *et al.*, 1986).

The final, and perhaps most controversial mechanism, is the suggestion that scopolamine disrupts the retrieval of stored information (Drachman & Leavitt, 1974). This interpretation was based in part on the apparent deficit in scopolamine-treated subjects when asked to name members from a particular category. This task involves retrieval of information from memories stored *prior to* the administration of the drug (Caine *et al.*, 1981; Beatty *et al.*, 1986). However, the effects of scopolamine on retrieval have been questioned, on the basis of experiments in which the word lists to be remembered were given prior to scopolamine, and only recall was tested under the influence of the drug. Under such conditions, no impairments in memory were observed (Ghonheim & Mewaldt, 1975; Petersen, 1977). These data suggest that scopolamine does not primarily affect retrieval, and that a deficit in encoding and storing new information is at present the most parsimonious explanation for the amnesic effects of the drug.

14.4.2 Scopolamine-induced amnesia and Alzheimer's disease

So how similar are the behavioural deficits induced by scopolamine to those of Alzheimer's disease? A number of studies have made comparisons between the effects of the drug in young subjects with aged, but neurologically intact, subjects (see, e.g., Drachman & Leavitt, 1974) and reported remarkably similar profiles of impairment in the two groups. Thus, on the battery of tasks used by Drachman and colleagues, the only difference seen between scopolamine-treated young subjects and elderly subjects was that the elderly group was not impaired in the naming of objects in specified categories (Drachman, 1977; Drachman & Leavitt, 1974; Drachman & Sahakian, 1980). On all other tests the relative degree of impairment in the two groups was congruent. For example, both groups showed marked deficits on supraspan digit storage, free recall and performance IQ, only mild disruption of verbal IQ, and no impairment on digit span. Although these studies provide useful information on age-related deficits, as in the studies of ageing animals, the relevance to dementia, particularly of the more virulent early onset subclass, is not fully established. The mnemonic and other cognitive impairments in Alzheimer's disease are generally much greater in both range and extent than those reported in the neurologically intact elderly, and the issue of whether the difference is quantitative or qualitative is not as yet fully resolved (see below).

Experiments to evaluate both scopolamine-treated young subjects and Alzheimer's patients on the same battery of tests would provide the most direct comparison, and unfortunately only very few studies have attempted such a comparison with Alzheimer's or other dementia patients. In one recent study, Beatty *et al.* (1986) reported that the pattern of deficits observed in Alzheimer's disease might be quite different from those observed after scopolamine treatment. They found that when compared to age-matched controls demented patients showed deficits in immediate free recall, delayed recognition on the Brown-Peterson distractor task, verbal fluency, and symbol-digit associations. As discussed earlier, scopolamine treatment produced deficits only in free recall and on the Brown-Peterson task in this experiment. Moreover, there was some evidence that the patterns of errors were somewhat different between the two experimental groups. Thus, the Alzheimer's patients made a significant number of perseverative errors on the Brown-Peterson task, whereas the scopolamine-treated group did not.

These observations led Beatty *et al.* (1986) to conclude that scopolamine induces a pattern of memory deficits that are quite different from those seen in dementia. It is pertinent to note, however, that Beatty *et al.* (1986) did not directly compare the scopolamine-treated young subjects with the Alzheimer's patient group, but rather each had their respective young and elderly control groups. The control group for the Alzheimer's patients consisted of a group of neurologically intact subjects who were matched to the patients for both age and education. In particular, the scopolamine-treated young subjects showed a profile of deficits that was strikingly similar to the pattern of performance observed in the elderly patient controls, as reported previously by Drachman & Leavitt (1974). Thus, the drug administration regimes used in these studies produced behavioural effects that were more similar to the moderate decline associ-ated with normal ageing than to the greater impairment in Alzheimer's dementia. However, it is not clear whether the difference between the demented elderly and the drug-treated young subjects is indeed qualitative or simply a matter of degree. The demented patients were certainly more impaired quantitatively, both in showing more severe deficits on tasks where the drug-treated young subjects were deficient and in showing deficits on a wider range of tasks than were sensitive to the drug treatment. However, it is not clear that a similar pattern of deficits might not equally have been achieved with higher drug doses in the young subjects.

14.4.3 Scopolamine effects in non-human primates

Young monkeys injected with scopolamine have shown deficits on delayed colour matching tasks (Bohdanecky *et al.*, 1967; Glick & Jarvik, 1969), delayed spatial response tasks (Bartus & Johnson, 1976; Bartus, 1978), delayed non-matching to sample tasks (Aigner & Mishkin, 1986) and the retention and reversal of object discriminations (Ridley *et al.*, 1984). Bartus and colleagues have suggested that these amnesic effects are strikingly similar to those observed in the aged monkey (Bartus & Johnson, 1976; Bartus *et al.*, 1982; Dean & Bartus, 1985); thus, in parallel with the human drug studies, the scopolamine-treated young monkey may be regarded as a potential animal model for the study of age-related decline in cognitive performance, and perhaps also for the deficits associated with Alzheimer's dementia.

As in the human studies, a fundamental issue is whether the drug effects are primarily on mnemonic or other behavioural processes. It is clear that the drugs can affect non-mnemonic behaviours, which Jarvik *et al.* (1969) suggested could reflect perceptual disturbances. They reasoned that if scopolamine was *only* affecting memory processes, then its detrimental effects would be potentiated at longer retention intervals. Early studies that used a visual discrimination task suggested that processes other than memory were being affected by scopolamine (Bohdanecky *et al.*, 1967; Glick & Jarvik, 1969). In these studies, scopolamine-induced deficits on the memory task were apparent when retention was immediate (i.e., even in the zero-second delay condition) and the severity was not affected by increasing the retention interval. However, in a later study, Bartus & Johnson (1976) suggested that scopolamine also induced a deficit in memory independent of any disturbance in perceptual processes. They utilized a spatial memory task that had little discrimination component. On this task, monkeys did show a more severe memory impairment at longer delays, and this deficit was dose dependent. Moreover, Bartus (1978) found that this scopolamine-induced deficit could be reversed by the acetylcholinesterase inhibitor physostigmine, suggesting that the amnesic effect of scopolamine was due to blockade of the cholinergic system, and that it reflected a disturbance in memory specifically, rather than in arousal, attention or perception.

At least one study has addressed the issue of whether scopolamine affects memory processes in nonhuman primates at the level of encoding or retrieval of information. Ridley *et al.* (1984) administered scopolamine to marmosets prior to learning a new object discrimination, so that the drug was active during new learning. On the following day, the monkeys were tested for either retention or reversal of the discrimination. Under saline control treatment, the monkeys initially performed below chance on the reversal task (due to retention of the discrimination learned the previous day) whereas the scopolamine-treated animals performed at chance levels on the reversal task. From these observations the authors concluded that scopolamine disrupted the encoding of the newly learned information. In subsequent experiments, the drug was given just prior to the reversal task (i.e. 24 h after the new learning). In this case only mild impairments were observed. Moreover, if the marmosets were trained to a higher criterion of performance and then given scopolamine before the retention task, no impairment was seen. These experiments suggest that scopolamine has only a mild effect on retrieval of information from a discrimination task, and that its main effect, as in humans, may be on the encoding or storage of new information.

14.4.4 Scopolamine effects in rodents

The amnesic effect of scopolamine has been studied most extensively in rodents, and the topic has received several recent reviews (Spencer & Lal,

1983; Collerton, 1986; Flood & Cherkin, 1986). Historically, this area of research has been plagued by inconsistent findings, that may reflect differences in the species used, behavioural paradigms, drug dosage, or the timing of scopolamine injections. In fact, whereas most studies have reported impaired learning or memory, some investigators have actually reported faciliative effects (Suits & Isaacson, 1968; Flood & Cherkin, 1986). In cases where improvements have been found, motor influences of the drug may be a contributing factor (e.g. in two-way active avoidance tasks). Alternatively, Flood & Cherkin (1986) have suggested that improvements may be attributable to a presynaptic action of scopolamine at low doses enhancing ACh release via muscarinic autoreceptors (Hammer *et al.*, 1980; Marchi *et al.*, 1981).

Since the present interest in scopolamine is as a potential pharmacological model for dementia, we here focus on those behavioural paradigms which reveal scopolamine-induced deficits, and in particular on the studies that have varied the time of injection as an aid to identification of the underlying functional deficits. Scopolamine has been consistently found to impair performance in a wide range of tasks including classical conditioning (Moore *et al.*, 1976), spontaneous alternation (Douglas & Isaacson, 1966), passive avoidance (Meyers, 1965; Bammer, 1982), sensory discriminations (Milar *et al.*, 1978), spatial working memory (Eckerman *et al.*, 1980), and spatial delayed response tasks (Dunnett, 1985). (For comprehensive reviews, see Spencer & Lal, 1983, and Collerton, 1986). In contrast, reference memory tasks may be scopolamine resistant (Wirsching *et al.*, 1984). While an interpretation of deficits in terms of a disruption of memory processes, as first promoted by Deutsch (1971), has tended to predominate in the literature, many other hypotheses have been proposed to account for the apparent amnesic effects of the drug, including deficits in attention (Warburton, 1972; Cheal, 1981), stimulus control (Heise & Milar, 1984), internal inhibition (Carlton, 1969), and state dependent effects (Berger & Stein, 1969).

The most extensively-used paradigm for the effects of scopolamine on memory has been the simple passive avoidance task (see Section 14.3.3.4). Even with single-trial learning on this test, rats and mice typically show good retention when tested up to seven days later (Glick & Zimmerberg, 1972). Consequently, the effects of drugs on acquisition, consolidation and retrieval processes can be separated by administration of the drug prior to or following the single training trial, or prior to the retention test, respectively.

Anticholinergic drugs given prior to the initial training trial have consistently been shown to impair acquisition of one-trial passive avoidance tasks (Buresova *et al.*, 1964; Meyers, 1965; Glick & Zimmerberg, 1972; Glick *et al.*, 1973). By contrast, studies which have administered anticholinergic drugs immediately after the training trial (influencing consolidation of the memory trace) or prior to the test trial (influencing retrieval of the memory) have yielded variable results. For example, when scopolamine has been given during consolidation, some investigators have reported disruption in memory (Glick & Zimmerberg, 1972), while others have found no impairments (Buresova *et al.*, 1964). Likewise, scopolamine administration prior to the test trial has been reported to have either detrimental (Meyers, 1965) or neutral (Bohdanecky & Jarvik, 1967) effects on retrieval.

Whereas the passive avoidance task has the advantage of simplicity and temporal precision of the time at which the to-be-remembered information is acquired, it has other practical and interpretive disadvantages. A wide range of variability between animals generally requires large group sizes to achieve clear results. More critically, performance is subject to diverse sensory, activational, motivational and attentional factors that can confound the interpretation of specific memory capacity (see Section 14.3). For this reason a variety of other paradigms have been investigated that enable a greater degree of precision in resolving the underlying functional influence of the drug.

A number of studies have used delayed response tasks to suggest that scopolamine may disrupt functions other than short term memory (Warburton, 1972; Ksir, 1974; Heise *et al.*, 1975; Milar *et al.*, 1978; Dunnett, 1985). A memory interpretation was excluded since deficits are typically seen even at the shortest retention delays (i.e. one second or less). Rather, it was concluded that scopolamine may affect the initial discrimination or registration of the relevant information. It is nevertheless pos-

sible that these additional effects of the drug are masking an additional mnemonic deficit. Thus, Dunnett (1985) found that scopolamine affected response accuracy at the shortest delays but that the deficit was even greater at longer delays. This pattern of effects may suggest that scopolamine does produce short-term memory deficits, which are most apparent at the longer delays, in addition to non-mnemonic impairments that influence performance at all delay intervals.

The effects of scopolamine on encoding, storage and retrieval have recently received renewed interest, using variants of the radial arm maze (see Section 14.3.3.5). Eckerman *et al.* (1980) first showed that scopolamine administered prior to the start of the daily trial disrupted task performance, to a similar degree during the first four choices in each daily trial as during the second four. They concluded that scopolamine affected the encoding of information in working memory rather than memory storage because there was no acceleration of performance decrement over the increasing retention intervals. Moreover, Watts *et al.* (1981) showed that prior training on the task did not attenuate the working memory impairment induced by scopolamine, and so the results of Eckerman *et al.* (1980) cannot simply be attributed to the drug disrupting long-term reference memory processes associated with task acquisition.

In order to distinguish between storage and retrieval deficits, a recent variant of the radial maze task is to introduce an enforced four-to-five hour interval between the fourth and fifth choices (Godding *et al.*, 1982; Beatty & Bierly, 1986). Upon return to the maze, rats with extensive training will remember the initial four choices, and will only enter the four previously unvisited arms (Godding *et al.*, 1982). If drug injections are administered immediately after the fourth choice, at the beginning of the intra-trial interval, encoding and storage of the previous information is susceptible to disruption. Conversely, if the drug is administered just prior to choice five, information about the previous choices should already be stored and any drug effects would be attributable to an impairment in retrieval processes. Scopolamine injected immediately after the fourth choice had no effect on performance on choices five to eight (Beatty & Bierly, 1986; Godding *et al.*, 1982). However, when scopolamine was given just prior to choice five, performance was disrupted. There-fore, scopolamine may affect memory retrieval (in addition to the well established impairment in the encoding of new information), although percep-tual or attentional impairments provide equally plausible explanations of this effect.

A final issue that has been investigated in the radial maze, is the lack of effect of scopolamine on long-term, or reference, memory. This can be seen in a further variant of the radial maze task in which four arms are never baited. This information remains constant throughout training, and normal rats learn not to enter those arms. Wirsching *et al.* (1984) trained rats to an 87% correct criterion on this task before administering scopolamine. The drug induced the expected errors in working memory (i.e. re-entries into the arms that had contained food) but no deficit in reference memory (i.e. the rats continued to avoid those arms that never contained food). These data suggest that scopolamine may affect short-term working memory but not the memory for well-learned events stored in long-term memory.

In summary, several common threads emerge from a comparison of the effects of scopolamine in rats, monkeys and humans. First of all, scopola-mine appears to impair the encoding of new infor-mation. These effects are most apparent when the drug is given during the acquisition stage of learn-ing. A second area of agreement is the lack of impairment of retrieval of information in tasks that are well learned and, presumably, stored in long-term memory. By contrast there may be some effect of scopolamine on the retrieval of infor-mation stored in short-term memory. Finally, sco-polamine may also induce perceptual and atten-tional changes, which is an important consideration in interpreting the mnemonic effects of the drug.

One inconsistency between primate and rodent studies has been the discrepant results in delayed response tasks. Although in the monkey scopola-mine appears to have delay-dependent effects on performance, with either minimal or no disrup-tion at the shortest delays (e.g. Bartus & Johnson, 1976), a similar pattern has been hard to obtain in rodents (e.g., Heise, 1975; Dunnett, 1985). This discrepancy may reflect detailed differences in the task design, drug dose or the species used.

14.5 INDUCED CORTICAL NEUROPATHOLOGY

The classical neuropathological features of Alz-heimer's disease in particular, and to a lesser extent ageing in general, include a range of neuro-degenerative changes in the brain, such as the neuritic plaques, neurofibrillary tangles (NFT), granulovacuolar degeneration and Hirano bodies. Such neurodegenerative changes appear to be rare in aged non-human animals and one extensive series of observations on aged birds and mammals of a large number of different species has failed to find any evidence of either neuritic plaques or neurofibrillary tangle formation (Dayan, 1971). However, such degenerative changes have occa-sionally been identified, particularly in ageing pri-mates but also in rats and dogs (see Section 14.3.1). Price *et al.* (1987) have suggested that behavioural deficits might be systematically screened in colonies of aged monkeys, and then correlated with post mortem indices of brain neuropathology (see Section 14.4, above). However, this is not an efficient approach to pro-viding an animal model of Alzheimer's disease, not only because of the expense, but also because the occurrence of such neuropathology is relatively sparse and it will not be known which monkeys in a colony will develop specific neuropathological features.

As an alternative, a number of experimental treatments have been found that induce degener-ative changes in animals akin to particular features of the Alzheimer neuropathology, and which might provide a more effective approach to modelling specific aspects of the disease. Two such treatments will be considered here: NFTs induced by aluminium toxicity and neuritic plaques induced by infection with the scrapie virus. Both treatments induce an encephalopathy that can be fatal for the animals, indeed it is invariably so in the case of scrapie infection. Nevertheless, early in the course of the disease, or following sub-lethal doses of aluminium, these treatments provide interesting insights into the psychopathology of the human disease. We propose to describe these two models in some detail, because we suspect that they are likely to be initially less familiar to many psychopharmacologists than, for example, the pharmacological or NBM lesion models of dementia. Other cortical lesion models have recently been reviewed by Sarter (1987).

14.5.1 Aluminium toxicity

Aluminium is the third commonest element in the earth's crust, but is one of the few that appears not to be essential for the living organism. The neurotoxic effects of intracerebral administration of large doses of aluminium salts in rabbits was discovered serendipitously (Klatzo *et al.*, 1965; Terry & Pena, 1965), but the similarity of the induced pathology to Alzheimer's neurofibrillary degeneration was immediately apparent. The histological abnormalities were not seen in glial elements, myelin or blood vessels but were restricted to the neurons, and comprised focal clearings in neuronal cytoplasm and neurofibrillary tangles (Katzo *et al.*, 1965). These changes were apparent throughout the neuraxis, although they were less prominent in the neocortex and hippocampus than in subcortical, brainstem and spinal regions.

In Klatzo's experimental rabbits, aluminium did not appear to have any acute toxicity, and the animals recovered rapidly following the intracerebral injections. Neurofibrillary degeneration was generally apparent within three to four days of the aluminium injection. However the majority of animals developed unsteady gait and ataxia after one to two weeks, which was closely followed by the development of epileptic seizures and death within approximately four weeks. Thus, when employing intracerebral injection of aluminium into the adult animal, behavioural tests need to be conducted within a limited period of one to two weeks after surgical intervention.

Since intracerebral injection of aluminium can induce neurofibrillary degeneration similar to one feature of the Alzheimer neuropathology, is this treatment in animals associated with deficits in learning or memory performance that might be akin to human dementia? Crapper & Dalton (1973*a*) first investigated this question in cats. Short term memory was assessed by training the cats in a delayed response task. Aluminium salts were then injected into the hippocampus, and performance was tested on each subsequent day. After an initial asymptomatic period lasting several days, delayed response performance declined precipitously to change levels. Although

the cats did develop neurological motor impairments over one to three weeks, similar to the previous observation in rabbits (Latzo *et al.*, 1965), the delayed response deficits preceded the neurological signs in four of the six cats, and so cannot be considered as secondary to their dyskinesias. The same study also found that the aluminium treatments disrupted acquisition of a one-way active avoidance task, but had no influence whatsoever on performance of intracranial self-stimulation, on which stable high levels of responding were maintained even into advanced stages of the induced encephalopathy. Crapper & Dalton (1973*b*) went on to provide a more detailed histological analysis of the material from the cats, and found not only that the onset of neurofibrillary degeneration preceded the appearance of the behavioural deficits, but also that the avoidance learning deficit in each cat correlated highly with the extent of neurofibrillary degeneration in the hippocampus, entorhinal cortex and neocortex.

Subsequent studies have reproduced similar effects in a number of species. Thus, for example, Petit *et al.* (1980) infused aluminium tartrate into the lateral ventricles of rabbits and found substantial deficits after ten days in the initial learning, and three days later in the retention, of a step-down active avoidance task, which they attributed to a dying back of cortical dendrites as a result of aluminium-induced neurofibrillary tangles (NFT) reducing axoplasmic transport of essential cellular constituents. Similarly, Pendlebury *et al.* (1987) conditioned eye blink to a tone stimulus in rabbits either prior to or following i.c.v. injection of aluminium chloride. The aluminium treatment induced significant deficits in both acquisition and retention/performance of the conditioned response, which correlated with the extent of cortical neuropathology. Moreover, the unconditioned reflex was unaffected by the treatment, ruling out a simple motor interpretation of the deficit.

In contrast, it has proved more difficult to induce similar deficits by aluminium treatment in rats. King *et al.* (1975) injected aluminium chloride into the ventral hippocampus of rats using similar procedures to those which did induce learning and memory impairments in cats in the same laboratory (Crapper & Dalton 1973*a*). Even when brain concentrations of aluminium were five to six times the concentration that was effective in cats, the aluminium treated rats showed no progressive encephalopathy, neurofibrillary degeneration or any impairments in acquiring an active avoidance response in a shuttle box. Although correlations between neurofibrillary degeneration and behavioural deficits cannot establish a causative relationship, the parallel failure of aluminium to induce neurofibrillary pathology or learning deficits in rats indicates that some feature of the pathological changes rather than high brain concentrations of aluminium *per se* are important for the induction of the functional deficits.

In two studies, aluminium has been administered chronically to rats orally. Bowdler *et al.* (1979) intubated rats with aluminium salts and found that high serum levels of aluminium did induce changes in visuomotor coordination and activity, but not in learning to avoid footshock in a shuttle box. More recently, Thorne *et al.* (1987) fed rats either no, low, medium or high aluminium diets. Although the four groups of rats did not differ significantly in passive avoidance retention latencies, retention latencies did correlate with post mortem assays of aluminium content in the hippocampus. The difference between the two analyses was due to highly variable brain levels of aluminium in each treatment group. A similar pattern of results – i.e. no differences between the treatment groups but a significant correlation with hippocampal concentrations of aluminium – was also seen in black-white discrimination-reversal learning in a two-choice box (A. Sahgal, personal communication). Although the correlations with brain aluminium content were significant for these two behavioural measures of memory, they were quite small ($r = 0.30$ and 0.43, respectively). Moreover, neither of these studies presented any histological data to assess degenerative changes in these animals. In view of the previous hypothesis that induction of neurofibrillary degeneration rather than brain concentration of aluminium *per se* is necessary for the induction of functional deficits (King *et al.*, 1975), the absence of these data make the correlations difficult to interpret.

To summarize the current status of this animal model, intracerebral administration of aluminium can induce neurofibrillary degeneration in the brains of some species of experimental animals which is associated with deficits on simple tests of the animals' learning and/or memory capacity.

Rats appear to be relatively resistant to this means of inducing neurofibrillary degeneration, which may have the advantage of providing one means of addressing the mechanism by which aluminium is pathogenic, but has the major disadvantage that it is in this species (apart from primates and pigeons) that adequate tests for memory capacity have been most refined; a more sophisticated analysis of the nature of aluminium-induced psychopathology is still lacking. Nevertheless, the experimental observations in species other than rodents have raised a number of key questions related to the specificity and usefulness of the model: Is the neurofibrillary degeneration induced by aluminium similar to that seen in demented humans? Is aluminium pathogenic in humans and might this provide clues to the aetiology of Alzheimer's disease?

The observations in experimental animals led a number of investigators to consider whether aluminium toxicity might provide a contributory cause in human Alzheimer's disease. Thus, Crapper *et al.* (1976) assayed aluminium content in seven normal brains and the brains of ten cases of diagnosed Alzheimer's disease, with confirmed neurofibrillary degeneration. Abnormally high concentrations of aluminium were found in some samples (mean 28%) from the Alzheimer brains, and heightened aluminium levels were seen in parallel with the regional distribution of NFT pathology. A similar association has been suggested in 'dialysis dementia'. Post-mortem brain concentrations of aluminium from patients on long-term dialysis for renal failure were increased seven-fold in patients dying without dementia but 15-fold in association with dementia (Arieff *et al.*, 1979). Indeed, even in a large group of elderly normal subjects, high serum levels of aluminium have been associated with impaired long-term visuo-motor coordination and poor long-term memory (Bowdler *et al.*, 1979).

In contrast with this apparent association between performance and aluminium in a variety of conditions associated with ageing, the central association in Alzheimer's disease (Crapper *et al.*, 1976) has been found to be much smaller than originally suggested (Trapp *et al.*, 1978), and proved hard to replicate when care is taken to balance subjects for age (McDermott *et al.*, 1979; Shore & Whatt, 1983). Nevertheless, technical developments have improved the precision of measurement. For example, Perl & Brody (1980) combined scanning electron microscopy with X-ray spectroscopy to measure aluminium (and other trace metal) content in individual neurons. The majority of neurons containing neurofibrillary tangles showed foci of aluminium in the nuclear region, in both demented and elderly control tissue (although the number of such neurons were of course much higher in the demented brains), whereas adjacent neurons from either source that did not contain neurofibrillary degeneration were virtually free of detectable aluminium. In parallel, Candy *et al.* (1986; Edwardson *et al.*, 1986) have recently found that a high concentration of aluminium accumulates in the form of aluminosilicate at the core of senile plaques in Alzheimer's disease. Thus, there appears to be clear evidence that increased brain concentration of aluminium is associated with neurofibrillary degeneration and dementia, although whether it has any causative role is not demonstrated. More critical for a causative role of aluminium in the aetiology of dementia is the number of distinct differences between aluminium-induced and Alzheimer's neuropathology. In the experimental animal studies (e.g., Wisniewski *et al.*, 1967, 1970), NFT accumulation following aluminium treatment was much more pronounced in subcortical and spinal cord structures, and was relatively sparse in neocortex and hippocampus, which is the reverse of the topographic distribution of NFTs in demented patients (Tomlinson *et al.*, 1970). Secondly, the structure of the NFTs differ in the two situations. Aluminium-induced tangles are comprised of short, straight, soluble 10-nm diameter filaments (Terry & Pena, 1965; Crapper & Dalton, 1973b; Selkoe *et al.*, 1979), and are immunoreactive to antibodies recognizing components of normal neurofilaments (Selkoe *et al.*, 1979; Munoz-Garcia *et al.*, 1986). By contrast, Alzheimer NFTs are comprised of paired helical filaments (Wisniewski *et al.*, 1976), which do not react to the neurofilament antibodies but were immunoreactive to an antiserum raised against purified microtubules (Grundke-Iqbal *et al.*, 1979; Munoz-Garcia *et al.*, 1986). Perhaps the clearest demonstration that the tangles of Alzheimer's disease involve a different process from those induced by aluminium is provided by the observations that human foetal cortical neurons maintained in tissue, culture develop straight filament tangles when aluminium salts are added to the culture medium, but develop paired helical filaments when the cells are exposed to tissue extracts derived from the brains of Alzheimer's patients (DeBoni & Crapper, 1978), although this report has not been replicated (A. Sahgal, personal communication).

We must conclude then, that the aluminium-induced tangle is not the same as that which develops in Alzheimer's disease, and that although it may be a contributory factor, aluminium is not the aetiological cause of Alzheimer's disease (Khachaturian, 1987). Nevertheless, intracerebral administration of aluminium to experimental animals can still provide a model of the effects of tangle formation in cerebral neurons, in which intracellular transport and function are consequently disrupted. The value of the model has until recently been limited, for purposes of behavioural analysis, by the fact that the large majority of affected animals develop severe neurological complications, and die from status epilepticus within a relatively short time period. However, Wisniewski *et al.* (1980) have found that if the intracerebral injections of aluminium are given to infant rabbits at day 15 of age, they all develop neurofibrillary neuropathology, but somewhat more than half survive for many months, many without any detectable neurological impairments. These surviving pups nevertheless show severe impairments in a maze learning task (Rabe *et al.*, 1982). Improvements in the adult model have involved intracerebral administration of aluminium to adult rabbits in the form of a metallic suspension (Bugiani & Ghetti, 1983) or as a slurry (Wisniewski *et al.*, 1982), which provides a much slower progression of the neurological disorder, more akin to the insidious onset of the disease in humans, permitting study of the progression of the disorder over longer time periods. However, in at least one of these cases, the treatment was associated with additional motor neuron loss and muscle atrophy (Bugani & Ghetti, 1983). These recent developments of chronic animal models open the way for more detailed behavioural analyses. It is clear that aluminium-induced neurofibrillary degeneration results in deficits on relatively simple tests of learning and/or memory performance, but the precise nature of these impairments remains to be resolved adequately.

14.5.2 Scrapie viral infections

Scrapie is an infectious neurological disease that occurs naturally in sheep and goats, and which in its advanced stages also produces problems of motor coordination characterized by unsteady gait and ataxia. The replicating agent is believed to be an unconventional virus, which does not produce any inflammatory reaction and has a long incubation period before leading to neurodegenerative changes and inevitable death (Gaidusek, 1977; Kimberlin, 1984). The infectious nature of scrapie has been demonstrated by transmission by innoculation of tissue from an infected animal into a recipient, and for convenience most experimental studies have been conducted in infected mice or hamsters.

Following infection, the virus appears to replicate in a number of sites in the body, but the only pathological changes are seen in the central nervous system. The characteristic neuropathological features of infection are vacuolar degeneration ('spongiform encephalopathy') of grey matter, and variable amounts of neuronal loss and reactive gliosis, in particular in the pyramidal cells of the hippocampus (Dickinson *et al.*, 1983; Kimberlin, 1984). These features of the neuropathology bear closest resemblance to several human slow viral diseases, such as Creutzfeld-Jakob disease and Kuru (Gaidusek, 1977). However, of particular interest for Alzheimer's research has been the observation that some strains of scrapie induce neuritic and amyloid plaque formation in the brains of infected mice (Bruce, 1973; Fraser 1975; Wisniewski *et al.*, 1975; Bruce, 1981, 1984), and a loss in cortical choline acetyltransferase activity (McDermott *et al.*, 1978). The relative degree of neuropathology appears to depend on a number of factors, including the scrapie strain, the strain of mouse (determined by the allelic forms of the socalled *Sinc* gene), and the route of innoculation (Bruce *et al.*, 1976). The incidence of plaque formation is associated with particular combinations of scrapie strain and mouse genotype that give rise to long incubation periods (Bruce, 1984). Whereas many clinical neuropathological studies have suggested a correlation between cholinergic degeneration and cortical plaque counts (see Section 14.1.2), the potential independence of these signs is suggested by the fact that in the scrapie model the greatest ChAT deficit is obtained in a strain that shows no plaques (McDermott *et al.*, 1978).

The structure of scrapie-induced plaques is remarkably similar to that seen in Alzheimer's disease in their content and arrangement of amyloid fibrils and degenerating neuronal processes, and in their basic staining characteristics (Bruce & Fraser, 1975; Wisniewski *et al.*, 1975; Bruce, 1984). Recent biochemical and molecular biological studies indicate that the amyloid precursor proteins in Alzheimer's disease and scrapie are different. Nevertheless, similarities in the characteristics of these proteins and the close morphological parallels between the plaques in the two conditions suggest similar processes of plaque formation (Masters & Beyreuther, 1987), although this conclusion remains controversial. Whichever conclusion is drawn, the scrapie-induced amyloidogenesis potentially provides a good animal model for studying the functional sequelae of plaque-associated degeneration in the brain.

There are only very few investigations of the functional effects of scrapie infection. Several early studies identified changes in spontaneous motor activity (Savage & Field, 1965; Suckling *et al.*, 1976; McFarland *et al.*, 1980) and feeding (Outram, 1972) in infected mice. In the first study of the effects on learning, McFarland & Hotchin (1980) found that the mice showed marked deficits in two-way active avoidance learning by 150 days after injection, although they did not indicate whether clinical signs were apparent at this stage of the scrapie infection. More recently, Hunter *et al.* (1986) took advantage of the relatively stable incubation period that exists for any particular combination of scrapie and mouse strains by training groups of mice infected with each of four strains of scrapie in passive avoidance just prior to the expected appearance of overt clinical signs. Proper controls were included to demonstrate that there were no significant differences in spontaneous motor activity, initial training latencies or shock thresholds. By contrast, clear reductions in passive avoidance retention latency were seen in mice infected with two of the four strains, but only when trained with relatively low shock levels (0.45 mA). In other groups of mice trained at a higher shock level (0.6 mA), no differences were seen between infected and control animals. However, in this study, retention trials were terminated after a relatively short 120-s cut-off latency, and in all conditions where the infection did not produce significant differences every mouse in both the infected and control groups scored the full 120-s. Thus, even if the infection had produced any differences they would be completely masked by 'ceiling' effects.

Although amyloid plaques have been described sporadically in the aged brains of a variety of species (Wisniewski & Terry, 1973; Vaughan & Peters, 1981; Struble *et al.*, 1985), scrapie remains the only convenient experimental model of neuritic plaque formation in Alzheimer's disease. It is therefore surprising that there have not been any further published studies of the behavioural effects of scrapie infection. Moreover, as indicated above, scrapie induces amyloid plaque formation only under conditions in which there is a relatively long incubation period. By contrast, the two learning studies just described both used strains having short incubation periods associated with few or no plaques. Although such shorter periods are more convenient experimentally, the key variable of interest, i.e. the association between amyloid plaque formation and learning or memory deficits akin to dementia, is missing. One recent study has found that in long-incubation models in which plaques are found, the plaques can develop as early as one sixth the way through the incubation period (within 100 days) and were the earliest signs of infection (Bruce, 1981). This observation suggests that the plaque formation is a primary element in the neuropathology and not simply a secondary event to other aspects of degeneration. It also makes the model practical for investigation of the functional effects of plaque formation in the absence of other pathological changes, although attention will be needed to the site of plaque formation in cortical and subcortical sites. Indeed, the murine scrapie model, in which the combination of mouse and scrapie strains permits the systematic variation of amyloid plaques, cholinergic deficiency and vacuolar degeneration, could provide a powerful means to analyse the contribution of each of these neuropathological features associated with Alzheimer's disease to particular aspects of learning, memory and cognitive performance. However, at least to our knowledge, this potential of the scrapie model remains almost totally neglected.

14.6 SUBCORTICAL (NUCLEUS BASALIS) LESIONS

Whereas the focus of cortical models of ageing and Alzheimer's disease has been based on classical descriptions of neuropathological changes in the ageing and demented brain, more recently a second major focus has been on identified neurotransmitter systems of the isodendritic core. Functional activity in cholinergic, noradrenergic and other monoamine systems all decline in the ageing human and animal brain (see Sections 14.1.2 and 14.3.2), and, of these, the cholinergic decline provides the clearest neurochemical marker of dementia of the Alzheimer type. Following the formal proposal of the 'cholinergic hypothesis of geriatric memory dysfunction' (Bartus *et al.*, 1982; Coyle *et al.*, 1983), there has been a corresponding resurgence of attempts to study the functional effects of experimental lesions in the basal forebrain nuclei from which the neo- and allo-cortical cholinergic projections arise, and in particular the nucleus basalis magnocellularis (NBM).

The potential advantage of subcortical lesion models is that they offer a basis for studying the functional role of identified neurotransmitter systems (particularly the ascending cholinergic system) in learning and memory processes, with the goal of specifying the neural substrates of the various functional components of dementia. Although the pharmacological probes, such as scopolamine, offer a greater degree of neurochemical specificity, the anatomical loci of pharmacological effects are relatively poorly specified and are likely to include subcortical structures (such as the neostriatum or tegmental nuclei) that are believed not to have any major involvement in the genesis of dementia. Additionally, because the destructive effects of lesions can be stable and long-lasting, such models may provide the basis for studying the long-term dynamic and cumulative effects of prospective treatment strategies.

From the outset, several disadvantages of subcortical lesion models need to be born in mind. Firstly, in practical terms, the stereotaxic procedures involved are time consuming, so that large group sizes are the exception rather than the rule. This can be particularly disadvantageous for experimental programmes aimed at screening ranges of doses or different compounds for potential therapeutic action.

Secondly, because the lesion effects are relatively enduring (at least in comparison to most drug treatments), it is difficult to conduct experimental designs in which each animal serves as its own control, or in which the timing of treatment is important to the interpretation. This point is made clear when compared with the information obtained from the pharmacological strategy involving administration of scopolamine at different stages of the encoding, storage and retrieval process (see Section 14.4.4). In contrast, subcortical lesions can only readily be administered before or after the training, and the effects of anaesthetic and postoperative recovery do not permit precise timing of the surgical intervention. However, against this disadvantage is set the ease with which prolonged training or testing in complex learning paradigms can be assessed following lesion, without the disadvantage of repeated administration involved in pharmacological paradigms.

Finally, the focus of subcortical lesion studies is to mimic elements of the degenerative changes associated with ageing, in order to determine the extent to which particular dementia-related deficits (such as memory impairment) might be attributable to decline in specific neural subsystems. A major issue to be addressed in the studies to be reviewed here is the extent to which the cell damage afflicted by the lesion accurately and specifically affects the targeted cell groups. A major confounding feature of the subcortical lesion literature has been the relatively uncritical acceptance that because the target of a lesion was a particular cell group, any functional impairments caused by the lesion are attributable to damage of those cells. For example, if a lesion is targeted on the NBM, can one assume that resulting deficits are due to cortical cholinergic deafferentation? It is necessary to be convinced not only that cholinergic cells of the NBM were effectively destroyed, but also that the functional consequences were not attributable to damage of other non-cholinergic neurons or fibre systems within the vicinity of the NBM, nor to remote non-specific effects of the lesion procedure or neurotoxin.

14.6.1 Electrolytic/radiofrequency lesions

Kelly & Moore (1978) first showed that large electrolytic lesions of the globus pallidus could produce 50% or greater reductions in ChAT activity in the neocortex, with no parallel changes in glutamatergic activity. Subsequent studies quickly ascertained that the critical site was the ventral pallidum/substantia innominata area containing the magnocellular basal forebrain neurons, and not the more dorsal globus pallidus *per se* (Johnston *et al.*, 1979; Wenk *et al.*, 1980). In parallel, Johnston *et al.* (1979) demonstrated comparable cortical depletions after kainic acid lesions of the same basal forebrain sites, indicating that the critical damage was of neurons in the vicinity of the lesion placement rather than of ascending fibre systems originating from more caudal cholinergic cell groups. Wenk *et al.* (1984) went on to map systematically the effects of alternative lesion placements using the excitotoxin ibotenic acid to identify optimal coordinates in the lateral and caudal segments of the basal forebrain cholinergic cell groups in order to achieve maximum cortical ChAT depletions (of the order of 54% loss). Although these initial studies employed only short survival times of approximately one week, Pepeu *et al.* (1985) have shown that substantial and significant cortical cholinergic depletions remain up to six months after electrolytic lesions of the NBM (i.e. ventral pallidum), although some limited recovery does take place.

The first attempts to study the functional consequences of lesioning the cholinergic cells of the NBM employed the classic electrolytic lesion technique. LoConte *et al.* (1982a) made unilateral electrolytic lesions in the cholinergic magnocellular forebrain neurons of the ventral pallidum and observed a disruption of learning in both active and passive avoidance tasks. Although active avoidance learning in a shuttle box was impaired, the animals showed normal escape latencies when footshock was applied, which was taken to support a cognitive as opposed to motoric impairment induced by NBM destruction. In particular, the passive avoidance deficit was apparent even on the shortest retention test 30 min after the training trial, suggesting that the impairment may have been on some aspect of task acquisition rather than retention. This conclusion was supported by the further observation that lesioned animals pre-trained on the active avoidance task prior to surgery were unimpaired in their retention and subsequent performance of the response. Uni-lateral lesions were used in these studies because animals with bilateral electrolytic lesions of the NBM showed severe weight loss and died within a week (LoConte *et al.*, 1982*a, b*). In spite of these profound neurological impairments, the lesions produced only a moderate (18–40%) decline in spontaneous ACh output from the ipsilateral fronto-parietal neocortex. Nevertheless cortical electroencephalographic activity was substantially reduced across all frequency ranges (LoConte *et al.*, 1982*b*). These authors have therefore con-cluded that destruction of the basal forebrain cho-linergic system disrupts the cortical activation necessary for 'selective awareness' as revealed in the avoidance learning deficits.

Subsequent studies have confirmed that electro-lytic or radiofrequency lesions of the NBM disrupt acquisition of both active and passive avoidance tasks (Arendash *et al.*, 1985; Miyamoto *et al.*, 1985; Araki *et al.*, 1986). However, Miyamoto *et al.* (1985) found that when animals were pretrained in the active avoidance or passive avoidance con-tingencies prior to the lesion surgery, the lesioned animals were impaired on subsequent retention tests. This is at variance with the previous report of LoConte *et al.* (1982*a*), but may be related to the greater success of the later studies in keeping the animals alive after bilateral lesions, making it pos-sible to study animals with more profound impair-ments. The effective lesions in these studies appear to produce greater cholinergic depletions in poster-ior cortical sites than in more anterior areas (Pepeu *et al.*, 1985). Although the precise topography of cortical cholinergic projections is not clear, there appears to be a crude maintenance of anterior-posterior organization (Lehmann *et al.*, 1980; Wenk *et al.*, 1980; McKinney *et al.*, 1983). In line with this organization, and the optimal placements of ibotenate lesion for maximal cortical ChAT depletion (Wenk *et al.*, 1984), Miyamoto *et al.* (1985) found that lesions placed in posterior NBM produced greater passive avoidance deficits than lesions placed in more anterior basal forebrain sites. Subsequently, Araki *et al.* (1986) confirmed the effectiveness of posterior sites for inducing deficits in passive avoidance, and found that the

same principle applied to the active avoidance task also.

Several studies have gone beyond aversively conditioned (shock avoidance) tasks to consider the effects of NBM lesions in appetitive maze learning paradigms. Dubois *et al.* (1985) found that radiofrequency lesions of the NBM did not impair spatial discrimination learning in a T-maze, but did severely disrupt learning of the radial eight-arm maze task (see Section 14.3.4). Since both tasks involve learning of a spatial discrimina-tion, the difference could be related either to the greater task difficulty of the radial maze than the T-maze, or to the fact that the radial maze task involves working memory, whereas spatial dis-crimination tasks do not. Hepler *et al.* (1985*b*) used a modified T-maze to distinguish between these alternatives. In their task, the start arm of the maze was divided with one route permanently blocked. At the end of the open route, the rat confronted a second choice point between two side arms. On the first of each pair of trials, one side arm of the maze was blocked so that the rat was forced to enter the other arm. On the second trial of each pair, the rat had a free choice between side arms, but only the arm not visited on the first trial contained food. The rats therefore had to learn a stable spatial discrimination (reference memory) in the start arm, prior to reaching the choice point between side arms. On the second trial of each pair, the rat then had to employ its working memory to select the side arm not visited on the forced choice trial. Following radiofrequency lesions of the NBM, the rats had no difficulty whatsoever in making the first spatial discrimination (based on reference memory), but they were severely impaired in their ability to make the second spatial discrimination (based on working memory). Interestingly, radio-frequency lesions of the anterior cholinergic cells of the medial septal area produced an even greater impairment in the working memory component of this task, in parallel with the clearcut working memory deficits that have been frequently reported following lesions within hippocampal circuits (Olton *et al.*, 1978; Olton, 1983).

Thus, it is clear that electrolytic or radiofre-quency lesions of the NBM do disrupt animals' performance in learning and memory tasks. However, the major unresolved issue with such lesions is the extent to which the effects are truly attributable to destruction of the magnocellular cholinergic neurons of the basal forebrain. All such lesions certainly cause large holes in an area of the brain through which many regulatory pathways ascend from the brainstem, and involve the ventral pallidal/substantia innominata zone through which passes the outflow from the ventral striatal complex. Thus, Pepeu *et al.* (1985) note that in parallel with a 39% ChAT loss in the parietal cortex, the lesion also decreased 5-HT levels, although NA levels are unaffected. Of more concern, the posterior NBM lesions that produced the greatest disruption of active and passive avoidance learning were associated with substantial reductions of hippocampal NA, DA and 5-HT, as well as striatal DA loss (Araki *et al.*, 1986).

In addition although learning and memory tasks have been the primary focus of these studies, it is clear from the reports that the lesions induce more extensive neurological and motor impairments. Although LoConte *et al.* (1982*b*) found no changes in escape latencies, and therefore concluded that active avoidance deficits were cognitive rather than motor in nature, they reported substantial increases in exploratory activity in a Y maze. Similarly Dubois *et al.* (1985) found marked increases in nocturnal locomotor activity, abnormal exploratory behaviour in a hole box, and a profound disturbance of spontaneous hoarding behaviour. In addition, bilateral lesions induce profound impairments in eating and drinking, so that the animals require maintenance by intragastric intubation for several days or even weeks postoperatively if they are to be kept alive (Dubois *et al.*, 1985; Hepler *et al.*, 1985*b*). There must therefore be some concern about whether motivational impairments or the general debility of the animals may influence performance on some tasks, especially when they are tested after a relatively short postoperative period.

14.6.2 Specific cholinotoxic lesions

Clearly, what is required is a cholinergic neurotoxin that would enable selective destruction of cholinergic neurons analogous to the possibilities provided by selective monoaminergic neurotoxins such as 6-hydroxydopamine or 5,6- and 5,7-dihydroxytryptamine (Baumgarten *et al.*, 1978; Jonsson, 1980; Schultz, 1982). A number of

attempts have been made to derive such a toxin. Choline, the precuror in the synthesis of acetylcholine, is actively taken up into cholinergic terminals by a specific sodium-dependent high-affinity uptake system (Freeman & Jenden, 1976; Jope, 1979). Fisher & Hanin (1980) therefore proposed that either neurotoxic analogues of choline or an irreversible inhibitor of the high-affinity uptake system may provide a selective and specific means of lesioning cholinergic neurons.

14.6.2.1 Hemicholinium-3

One such inhibitor of the high affinity uptake system is hemicholinium-3 (HC3) (Birks & MacIntosh, 1961). As HC3 does not cross the blood–brain barrier, it requires intracerebral or intraventricular administration. Caulfield *et al.* (1981) reported that i.c.v. HC3 impaired passive avoidance learning in mice, and Ridley *et al.* (1987) have shown that similar administration in marmosets abolishes spatial discrimination learning. However, Fisher & Hanin (1986) suggest that this approach does not provide an effective lesion model, not only because the influence is not localized, but more particularly because it is reversible. In particular, Ridley *et al.* (1987) found no effect of HC3 two days after administration.

More recently, Hurlburt *et al.* (1987) have employed chronic infusion of HC3 delivered bilaterally into the NBM over 14 days by a subcutaneous Alzet minipump. This treatment resulted in a substantial cholinergic cell loss from the basal forebrain, lasting at least three months after surgery. Moreover, the HC3-infused rats showed substantial impairments in acquisition of a serial T-maze task. The histological report from this significant study demonstrated cholinergic cell loss from the basal forebrain and degenerating terminals in silver staining of the neocortex; unfortunately, however, little information was provided on the selectivity of the lesion and sparing of non-cholinergic neurons in the vicinity of the infusion. The possibility of additional non-specific damage is suggested by the hyperactivity, enhanced exploration and long-term body weight decline that was incidentally reported as occurring in many animals. Thus, although chronic HC3 administration may provide a promising technique for cholinergic depletion, there are at present too few studies to assess adequately its potential as a technique for selective cholinergic cell loss to model age-related impairments.

14.6.2.2 Ethylcholine mustard aziridinium ion

Much greater attention has been given to a second prospective cholinotoxin, the ethylcholine mustard aziridinium ion, commonly known as AF64A (Fisher & Hanin, 1986). Mantione *et al.* (1981) demonstrated loss of functional high-affinity choline uptake in the cortex and hippocampus of mice three days after i.c.v. injection of AF64A. Further studies indicated that this was associated with a decline of ChAT activity and steady state concentrations of acetylcholine in the neocortex, hippocampus and striatum (Fisher *et al.*, 1982), although only the hippocampal cholinergic loss persisted beyond 21 days after injection. By contrast to the cholinergic loss, hippocampal concentrations of 5-HT or NA did not change (see also Hortnagl *et al.*, 1987). This suggested that AF64A might provide a means of inducing specific and persistent central cholinergic hypofunction (Fisher & Hanin, 1986).

In a systematic series of studies, Walsh, Hanin and colleagues have reported both cognitive and non-cognitive deficits following i.c.v. injection of AF64A (Walsh *et al.*, 1984; Walsh & Hanin, 1986; Chrobak *et al.*, 1987). Thus, Walsh *et al.* (1984) demonstrated moderate deficits in passive avoidance tested 35 days after injection, and more substantial deficits in learning an eight-arm radial maze task tested 60–90 days after lesion. In the subsequent study, AF64A treated rats were selectively impaired in the working memory component of a paired trial alternation task in a T-maze, whereas the reference memory component for the simple task demands of running the maze stem for food was unaffected (Chrobak *et al.*, 1987). These learning deficits following i.c.v. AF64A have been replicated by other groups (Jarrard *et al.*, 1984; Pope *et al.*, 1985). Effects of central cholinergic hypofunction associated with i.c.v. AF64A in these studies were also observed in unconditioned tests. Thus, the AF64A treated rats were generally more active in tests of locomotor activity, and more reactive to aversive stimulation in the hot-plate test (Walsh *et al.*, 1984).

In these studies, i.c.v. administration of AF64A was reported to produce relatively selective depletions of ACh in the hippocampus (down 44–62% and neocortex (down 63%), with little influence on striatal concentrations of ACh, and producing no significant depletions of NA, 5-HT or DA in the cortex, striatum or hippocampus. When assessed biochemically, therefore, AF64A injections would appear to provide relatively selective cholinergic depletions. However, two causes for concern have been raised.

Firstly, as in pharmacological studies, i.c.v. administration provides little topographic specificity about which particular cholinergic cell groups are implicated in particular components of the animals' deficits (Fisher & Hanin, 1986). Although there are several studies reporting on the functional consequences of intrahippocampal (Bailey *et al.*, 1986; Blaker & Goodwin, 1987) or striatal (Sandberg *et al.*, 1984a; Stwertka & Olson, 1986) injection of AF64A, there are not to our knowledge any reports yet of the behavioural consequences following intracerebral injections in either the NBM cholinergic cell body region or into cortical terminal areas.

More critically, a number of authors have suggested substantial non-specific toxic effects of AF64A. McGurk *et al.* (1987) emphasize the general reliance on biochemical indices of selectivity by the proponents of AF64A, in the absence of even routine histopathological analysis. Thus, following intracerebral injection of AF64A into the caudate nucleus, Sandberg *et al.* (1984b) reported selective effects on striatal cholinergic neurons, whereas McGurk *et al.* (1987) could find no dose of AF64A that would reduce striatal ChAT activity without, at the same time, producing considerable non-specific tissue destruction. A similar profile of non-specific lesions and tissue necrosis was seen in the striatum by Stwertka & Olson (1986), and was associated with additional behavioural pathology (spontaneous and drug-induced circling), suggesting non-cholinergic damage by AF64A to dopaminergic terminals or to striatal output neurons.

Fewer attempts have been made to achieve selective cholinergic lesions of the NBM, but similar disagreement is found. Thus, whereas Kozlowski & Abrogast (1986) could identify no non-specific basal forebrain damage with doses that were effective in destroying AChE-positive NBM neurons and depleting cortical ChAT activity (albeit by only 10–15%), both Asante *et al.* (1983) and McGurk *et al.* (1987) reported that all effective doses of AF64A caused large areas of non-specific necrosis around the injection site.

These and other studies (e.g. Levy *et al.*, 1984; Villani *et al.*, 1986) indicate that considerable controversy still surrounds the utility of AF64A as a selective cholinergic toxin for intraparenchymal injection. Indeed, even after i.c.v. administration of AF64A, Jarrard *et al.* (1984) have observed substantial necrosis of the fimbria-fornix, similar to that seen with electrolytic lesions, that could fully account for the hippocampal cholinergic depletions observed in these animals. Consequently, in spite of the initial promise, AF64A does not appear to provide a readily available means of making selective lesions in basal forebrain cholinergic neurons. One reason for the discrepancies between studies might be that the selectivity of AF64A action is dependent on its fresh preparation and use within hours of synthesis. However, if this turns out to be the critical factor for its selective action, the use of AF64A lies outside the scope of all but a few research laboratories with access to specialized neurochemical facilities.

14.6.3 Excitotoxic lesions

The majority of studies of the effects of NBM lesions have pursued an intermediate strategy between the total non-specificity of electrolytic or radiofrequency lesions and the controversial cholinotoxins. The 'excitotoxic' class of amino acids appear to exert their neurotoxicity via action at the glutamate receptor, and these substances have the distinctive property of selectively destroying neuronal cell soma in the vicinity of the injection whilst sparing fibres of passage and afferent terminals (Kizer *et al.*, 1978; Watkins & Evans, 1981; Coyle, 1982).

14.6.3.1 Kainic acid

The first major toxin of this class to be widely utilized was kainic acid (McGeer *et al.*, 1978). As first noted by Johnston *et al.* (1979) injection of kainic acid into the ventral pallidum produced extensive loss of acetylcholinesterase (AChE)-positive cells in the NBM and of AChE-positive fibres in the neocortex. More detailed neurochemical measurements revealed significant 45–50% declines in ChAT activity, ACh concentration and synaptosomal choline uptake in the cortex, but no parallel changes in a variety of biochemical markers for glutamate, GABA, catecholamines, serotonin or histamine activity. Several behavioural analyses of rats with bilateral kainate lesions of the NBM went on to demonstrate significant impairments in passive avoidance (Friedman *et al.*, 1983; Lerer *et al.*, 1985), spatial reversal and alternation but not basic spatial discrimination learning (Lerer *et al.*, 1985; Beninger *et al.*, 1986a), and the working memory component of a radial maze task (Beninger *et al.*, 1986b). Kainate lesioned animals manifested learning deficits akin to those seen after scopolamine treatment (Beninger *et al.*, 1986a, b), and manifested an enhanced locomotor response to peripheral treatment with nicotine (Ksir & Benson, 1983), possibly suggesting an increased binding or affinity of cortical nicotinic receptors following deafferentation.

However, the major disadvantage of kainic acid as an excitotoxin is that it has frequently been found to produce extensive remote damage, perhaps via its epileptogenic side effects (Nadler *et al.*, 1978; Ben-Ari *et al.*, 1979, 1980). Thus, although kainate has not generally been reported to produce non-specific necrosis and holes in the NBM, widespread neuronal cell loss has generally been seen including not only adjacent nuclei such as the lateral hypothalamus, reticular nucleus of the thalamus and amygdala, but also more remote sites such as the piriform cortex and hippocampus (Dunnett *et al.*, 1987). Associated with the widespread damage, animals treated with kainic acid show pronounced postoperative aphagia, especially when lesioned bilaterally, and many die in spite of tube feeding (Lerer *et al.*, 1985).

14.6.3.2 Ibotenic acid

Consequently, more recent interest has focussed on other excitotoxins that act at different glutamate receptors (McLennan, 1983; Watkins, 1981), including ibotenic acid, n-methyl-d-aspartic acid, quinolinic acid and quisqualic acid, which appear to provide more focal neurotoxic effects than kainate (Dunnett *et al.*, 1987). Of these, ibotenic acid has been by far the most widely used.

Along with the other lesions already considered, ibotenate lesions of the NBM have been found to produce a substantial disruption of both passive and active avoidance learning (Flicker *et al.*, 1983; Fine *et al.*, 1985*a*; Hepler *et al.*, 1985*b*; Dunnet *et al.*, 1985, 1987; Everitt *et al.*, 1987; Miyamoto *et al.*, 1987). This does not imply that all learning is impaired, since taste aversion (Everitt *et al.*, 1987), spatial position habits (Murray & Fibiger, 1986), and stable routes through complex mazes (Knowlton *et al.*, 1985) are unimpaired. A number of studies have therefore focussed on attempts to unravel the precise nature of the learning deficit.

The first distinction to have received attention is the working/reference memory distinction. The performance of tasks dependent on working memory is disrupted by ibotenic acid lesions of the NBM, as assessed in serial spatial alternation in a T-maze (Murray & Fibiger, 1986), paired trial alternation in a T-maze (Salamone *et al.*, 1984; Hepler *et al.*, 1985*a*) and the standard eight-arm radial maze working memory task (Dubois *et al.*, 1985; Hepler *et al.*, 1985*b*; Miyamoto *et al.*, 1987). A study by Bartus *et al.* (1985) suggests that both the spatial and the memory elements might be critical in these tasks, since rats with similar lesions were not impaired in learning a radial maze with explicit intramaze cues unless delay intervals were interspersed.

The basic radial maze task has been manipulated in several ways to tease out its relative components. Thus, in a study by Kesner *et al.* (1987*a*), rats were first trained in the basic eight-arm maze task, and were then constrained to visit the arms in a set sequence. At the end of each trial, they were then given choices between pairs of arms and rewarded for selecting the one most recently visited. Whereas the rats with NBM lesions were relatively efficient at learning to make temporal-order decisions about recently visited arms, discrimination of the relative recency of earlier arms was severely impaired and did not deviate from chance (see also Kesner *et al.* (1986)). Murray & Fibiger (1985) employed a subset of never-baited arms to distinguish the working and reference memory components (see Olton, 1983, and Section 14.3.4). They found that NBM lesions selectively disrupted the reference memory component leaving the working memory component intact. This dissociation has been confirmed by Kesner *et*

al. (1987*b*). By contrast, Knowlton *et al.* (1985) came to apparently the opposite conclusion: disruption of the working but not the reference memory component. However, this latter study differed from the former two reports in that the basal forebrain lesions were far more extensive, with the ibotenate injections being directed to medial septum and diagonal band as well as NBM. Thus, Knowlton's observations could be attributable to the more anterior damage, since working memory performance in the radial maze is particularly sensitive to damage anywhere in the hippocampal system, including the medial septal nuclei (Olton, 1983; Olton *et al.*, 1978; Mitchell *et al.*, 1982).

If the NBM cortical system is critical for adequate performance on spatial reference memory tasks, then Morris' (1981) water maze task should provide another test sensitive to NBM damage. Several reports have confirmed disruption by ibotenic acid lesions of the NBM, in the acquisition, postoperative retention, and the precision of spatial navigation performance in this task (Dunnett *et al.*, 1985, 1987; Whishaw *et al.*, 1985; Miyamoto *et al.*, 1987). By contrast, Hagan *et al.* (1988) failed to replicate this result, in spite of achieving respectable 34–40% depletions in cortical ChAT levels, and suggested that the differences were attributable to the varying extent of non-specific damage produced by the ibotenic acid injections in the different studies. This is supported by the latter of the two studies by Dunnet *et al.* (1987), in which the degree of impairment in water maze acquisition produced by different toxins was more closely related to the extent of non-specific subcortical cell loss induced by each toxin, than to the extent of cortical ChAT depletion.

In addition to the variety of maze learning procedures, several studies have more recently employed operant tasks to analyse specific aspects of the lesioned animals' response selection. Dunnett (1985) and Etherington *et al.* (1987) employed operant versions of delayed matching to sample tasks, employing spatial to-be-remembered stimuli. In both studies, ibotenic acid lesions of the NBM disrupted performance of this short-term (working) memory task. However, matching performance was equally disrupted at the shortest delay intervals, suggesting a generalized impairment in the animals' ability to employ the conditional discrimination rule. A similar conclusion was reached by Everitt *et al.* (1987) who directly demonstrated that NBM lesions impaired both learning and post-operative retention of a visual conditional discrimination task that required no short-term memory mediation. Thus, ibotenic acid lesions of the NBM do not reproduce the delay-dependent pattern of impairment on short-term memory tasks that is seen in Alzheimer patients (Morris *et al.*, 1987; Sahakian *et al.*, 1988).

In a further study of temporal conditioning, Meck *et al.* (1987) employed an elegant paradigm for analysis of temporal discrimination in rats. They trained rats on a fixed interval operant schedule, in which the reinforcement was omitted on periodic trials. Well-trained rats increase their rate of responding up to the time when reinforcement would be delivered. On trials in which reinforcement is omitted the rate of responding peaks at the expected reinforcement delivery time and thereafter declines. This 'peak-interval' procedure can therefore be used to identify the animals' subjective expectancy of reinforcement. Meck *et al.* found that whereas septal or hippocampal lesions produced shortening of the rats' expected memory of reinforcement, NBM or frontal cortex lesions lengthened it. Thus, lesions in both components of the basal forebrain cholinergic system impaired rats' timing capacities, but in opposite directions.

Thus, numerous studies have, by now, demonstrated deficits induced by ibotenic acid lesions of the NBM in a variety of learning and memory tasks that range from the simplest passive avoidance tests of long-term retention to relatively sophisticated analyses of conditional spatial and temporal discrimination. It is appropriate, as in previous sections, to consider the degree of cholinergic specificity achieved by ibotenic acid lesions of the basal forebrain before focussing on the nature of the underlying impairments manifested by animals with such lesions and implications for our understanding of central cholinergic function.

In contrast to previous assumptions, Coffey *et al.* (1988) have recently suggested that ibotenic acid injected into the medial septal area or lateral geniculate nucleus of the thalamus can induce a substantial disruption of *en passage* axons. It is not known whether similar axon damage might also occur in the NBM; similar axonal damage was not seen at other injection sites, and the NBM itself was not considered in this study. Support for the axon-sparing properties of ibotenate lesions in the NBM come from comparisons of cholinergic and non-cholinergic markers in the neocortex, as mentioned above for kainate. In particular, Dubois *et al.* (1985) found no change in frontal cortical concentrations of NA or DA following ibotenic acid injection into the NBM, whereas a radiofrequency lesion of the same site depleted cortical NA by 56% and DA by 92%. In the same groups of animals, the greater loss of cortical ChAT was produced by the ibotenic acid (32%) than the radiofrequency (22%) lesions.

These biochemical observations are comparable with those in studies employing ibotenic acid lesions alone. Thus, the various studies described above have reported 26–49% decline in cortical ChAT activity following NBM lesions, 29–53% declines in cortical AChE activity, and changes of the same order in other presynaptic cholinergic markers in the cortex, such as synaptosomal high affinity choline uptake (Altman *et al.*, 1985; Bartus *et al.*, 1985; Watson *et al.*, 1985; El-Defrawy *et al.*, 1986; Miyamoto *et al.*, 1987; Mufson *et al.*, 1987). Although Wenk & Olton (1984) suggested that 39% depletions of cortical ChAT would recover to normal levels within 12 weeks after NBM lesions, all other studies that have considered the time course find significant deficits lasting well beyond this period (Bartus *et al.*, 1985, 1986; Fine *et al.*, 1985*b*, 1987; El-Defrawy *et al.*, 1986; Everitt *et al.*, 1987; Mufson *et al.*, 1987; Berman *et al.*, 1988). Postsynaptic binding studies have generally indicated no changes in binding affinity of muscarinic agonists and antagonists (Altman *et al.*, 1985; Bartus *et al.*, 1985; Watson *et al.*, 1985), although Watson *et al.* (1985) suggested that the number of muscarinic binding sites might be reduced by up to 25%. Conversely, non-cholinergic neurotransmitter systems that are either afferent or intrinsic to the neocortex do appear to be relatively spared, as has been shown for markers of various afferent catecholamine systems (Fine *et al.*, 1987; Miyamoto *et al.*, 1987), and for intrinsic amino acids (Murray & Fibiger, 1985) and neuropeptides (Fine *et al.*, 1987; Mufson *et al.*, 1987).

Whereas the axon sparing properties appear to be relatively robust, considerable caution is still required about the nature of the cells damaged at the locus of excitotoxin injection, and the extent of spread beyond the borders of the basal forebrain (Dunnett *et al.*, 1987; Everitt *et al.*, 1987; Abrogast & Kozlowski, 1988). Thus, for example, although the neurotransmitters of the majority of ventral pallidal and substantia innominata cells are at present unknown, subpopulations are believed to utilize substance P and met-enkephalin (Haber & Nauta, 1983). Biochemical changes in these transmitters have not been compared with loss of cholinergic markers from the basal forebrain following NBM lesions, but might be expected to be substantial. Where the postoperative health of the animals is reported, a general observation is that ibotenic acid lesions of the NBM cause severe regulatory deficits so that the animals require intubation for one to two weeks in order to be kept alive (Whishaw *et al.*, 1985; Miyamoto *et al.*, 1987). However, Dunnett *et al.* (1987) found that bilateral injections of quisqualic acid, that caused equivalent cortical ChAT depletions, produced only mild neurological and regulatory deficits, in comparison with ibotenic acid. This suggests not only that perhaps quisqualic acid should be the toxin of choice in such studies, but also that many of the deficits induced by ibotenic acid are not attributable to cholinergic cell loss. If this applies to the regulatory deficits, might it not also apply to the learning and memory impairments that are the primary focus of interest? Thus, for example, Everitt *et al.* (1987) employed a contingency analysis in an attempt to separate out the effects of ibotenic acid lesions in the NBM, and concluded that the extent of individual animals' impairments in learning the conditional discrimination task was more closely related to the extent of generalized cell damage in the dorsal and ventral pallidum than to the loss of magnocellular cholinergic neurons *per se*.

14.6.3.3 Quisqualic acid

As yet there are few studies of whether quisqualic acid lesions induce deficits on cognitive tests comparable with the wealth of studies using ibotenic acid, but the available data warrant caution. Thus, Etherington *et al.* (1987) found that ibotenic acid lesions of the NBM, associated with 17% cortical ChAT loss, produced substantial operant delayed matching to sample impairments, suggesting a

cognitive impairment. However, quisqualate lesions of the same site, associated with a three times higher depletion of cortical ChAT activity, produced no deficit whatsoever on the same task. Similarly, Etherington (1988) found that quisqualic acid lesions of the NBM did not impair paired-trial delayed alternation in the T-maze, in spite of cortical ChAT depletions that were greater than in any of the studies that saw disruption on a similar task with ibotenic acid (Salamone *et al.*, 1984; Hepler *et al.*, 1985a). Finally, in a study involving direct comparison between ibotenic acid, n-methyl-d-aspartic acid and quisqualic acid, we have found that deficits in the Morris water maze spatial navigation task are much less marked in animals with quisqualate than ibotenate lesions (Dunnett *et al.*, 1987). Thus, in most of these tests, it is unlikely that the animals' cognitive deficits result from the cholinergic cell loss that is observed, but rather from damage to some other population of basal forebrain cells that are sensitive to ibotenic acid and relatively resistant to quisqualic acid. In fact, in spite of its lack of behavioural specificity, the only test that has been found to be as sensitive to quisqualic acid as to ibotenic acid lesions of the NBM (with comparable cortical ChAT depletion) has been step-through passive avoidance retention (Dunnett *et al.*, 1987).

Nevertheless, certain cognitive (but not specifically mnemonic) deficits are seen following quisqualate lesions of the NBM. In a series of as yet unpublished studies, we have found that rats with quisqualate lesions of the NBM are unimpaired in postoperative retention of either delayed matching or non-matching to position in operant chambers, in contrast to the disruptive effect of ibotenic acid lesions on performance of the same task (Dunnett, 1985). In particular, analysis of performance of the lesioned rats at different short-term retention delay intervals reveals a normal forgetting function that is indistinguishable from that of control rats. However, the rats do have a quite specific deficit in learning the non-matching discrimination rule. This is true both in task acquisition, and in reversal tests. Conversely, Etherington (1988) has found that quisqualate lesions of the NBM did disrupt acquisition of paired trial matching but not paired trial alternation in the T-maze. The key distinction between operant chambers and maze tasks is that rats have a spontaneous tendency to alternate their responses in a spatial maze (Dember & Fowler, 1958), akin to inherited strategies for optimal foraging, whereas they tend spontaneously to repeat reinforced responses in an operant situation (Evenden & Robbins, 1984). These results therefore indicate that quisqualate lesions of the NBM disrupt not memory, but rather some particular aspect of discrimination learning, as suggested by Everitt *et al.* (1987). The relevant dimension is related to the animals' ability to change spontaneous or previously learned response strategies when reinforcement contingencies are modified. This is compatible with Kesner's notion of disruption of tasks dependent on 'effortful processing' (Di Mattia & Kesner, 1984), although that concept is difficult to specify with precision or independently of reference to whether the lesioned animals reveal deficits.

14.6.4 Nucleus basalis – cortical interactions

None of the techniques for making NBM lesions is sufficiently specific to interpret the precise functions of cortical cholinergic projections unambiguously. Nevertheless, two specific issues arise from these studies which can be directly addressed: (a) the topographic loss of cortical cholinergic innervation does not suggest involvement of this system in a specific, selective aspect of cognitive function; and (b) comparison of neuropathological consequences of cortical and NBM lesions suggests that the cortical degeneration may be the primary aetiological event in the development of Alzheimer's disease.

14.6.4.1 Topography of cortical cholinergic depletion

The first issue arises from consideration of the topographical distribution of cortical deafferentation induced by NBM lesions in rats. Both the biochemical depletion of cholinergic markers in the neocortex and the histochemical loss of ChAT and AChE staining, in the studies of the functional effects of NBM lesions described above, are generally restricted to the dorsolateral fronto-parietal cortex throughout its rostro-caudal extent. Medial and ventral cortical areas, including medial and orbital prefrontal, cingulate, temporal, entorhinal and piriform areas are spared in virtually all

reports of NBM lesions. This pattern of denervation has several implications.

Firstly, with loss of subcortical regulation of the neocortex over such an extensive area, we should not expect to find impairment in some specific cortical function, such as reference memory. Rather, diverse cortical areas with widely divergent functions lose subcortical cholinergic regulation. Secondly, the cortical areas that are most affected are the sensory and motor zones which involve the majority of the dorsolateral surface in rats (Hall & Lindholm, 1974; Donoghue & Wise, 1982). Therefore the motor and sensorimotor impairments induced by NBM lesions (Dunnett *et al.*, 1985; Whishaw *et al.*, 1985) may be direct consequences of the cortical deafferentation, rather than inconvenient non-specific effects. Thirdly, cognitive effects of the lesions (such as the matching/non-matching discrimination deficits), if due to cortical deafferentation, need to be considered in term of functions of the association areas that are deafferented, viz. the posterior parietal and superior temporal cortices, the functions of which remain obscure in the rat. Finally, attempting to model memory deficits associated with cortical dysfunction by NBM lesions in rats is probably misguided because the cortical areas most likely to be responsible (prefrontal and temporal association areas) are left relatively intact by the lesions.

These various implications have as an underlying theme the perspective that loss of a diffuse cortical cholinergic input is *a priori* more likely to be associated with disruption of generalized cortical regulation, manifested in a wide diversity of functional measures, than with disturbance of some particular cognitive function. In line with this perspective, Richardson & DeLong (1988) have recently emphasized a potential role for cortical cholinergic neurotransmission in the maintenance of cortical arousal, based on a review of electrophysiological and electroencephalographic observations. Although this perspective is not yet well specified, the approach is likely to prove more fruitful in the long run.

14.6.4.2 Primary and secondary neuropathological changes

Alzheimer's disease has been seen to involve neuropathological changes in both cortical and subcortical sites, and the causative relationship between these two levels remains controversial. Some versions of the cholinergic hypothesis, in which the functional changes observed in dementia are attributed to degeneration in subcortical cholinergic systems, imply that the classical cortical indices of Alzheimer's disease (senile plaques and neurofibrillary tangles) are also a postsynaptic degeneration that is secondary to loss of cholinergic afferents to the neocortex and hippocampus. The variety of types of correlative evidence that has been adduced in favour of the cholinergic hypothesis has already been considered extensively in the previous sections.

Conversely, other authors have argued that the sequence of causation progresses in the opposite direction. For example, Pearson *et al.* (1985) considered the distribution of neuritic plaques and neurofibrillary tangles in the cortex of Alzheimer patients. The occurrence of these neuropathological markers was most marked in layers III and V of the regions of association cortex that are most heavily interconnected with hippocampal and olfactory areas of the brain. Rogers & Morrison (1985) reported a virtually identical laminar distribution of senile plaques in Alzheimer's disease, and emphasized that the neurons in the most affected cortical laminae are dominated by their role in cortico-cortical associative relationships, and do not correspond to any known subcortical innervation pattern. They therefore concluded that the cholinergic deterioration plays no causative role in plaque aetiology, which rather derives from pathological events initiated in the neocortex. Likewise, Pearson *et al.* (1985; Pearson & Powell, 1987) conclude that the distribution of neuropathological changes most likely reflects a progressive spread of the disease, not via anterograde subcortical projections onto the neocortex, but by anterograde and retrograde degeneration via cortico-cortical connections originating in olfactory areas of the brain.

A similar regional distribution is seen in the brains of cases of Down's syndrome, all of whom develop the classical cortical neuropathology associated with Alzheimer's disease in middle age (Mann *et al.*, 1985b Coyle *et al.*, 1986). Coyle *et al.* (1986) note that in Down's syndrome, as in Alzheimer's disease, degeneration of cholinergic markers is seen in the cortex and NBM. Since the

cortical cholinergic markers are reduced to a much greater extent than cell loss and atrophy of NBM cell bodies, this evidence favours the view that the cholinergic degeneration progresses as a retrograde consequence of the cortical pathology.

To what extent have the experimental studies in animals contributed to the study of these issues? From one perspective, the neural degeneration in cortical and subcortical sites in some circumstances progress independently. Thus, in the various murine models involving infection with the scrapie virus, McDermott *et al.* (1978) found no close correspondence between the mouse/scrapie strain combinations that induced plaque development and those that induced cortical ChAT loss. Instead, each appeared independently and secondary to the experimental spongiform encephalopathy. However, in this model it is clear that the molecular nature of the scrapie-induced plaques are not the same as the senile plaques involved in human dementia (see Section 14.5.2).

Support for the primacy of the subcortical cholinergic systems in the development of cortical neuropathology has been derived from the association of AChE staining with cortical plaques and the observation that plaque-associated neurofibrils may be ChAT positive not only in humans but also in the case of the much rarer occurrence of plaques in aged monkeys (Kitt *et al.*, 1984; Struble *et al.*, 1982). However, several other transmitter-associated markers have also been associated with degenerative neurofibrils around plaques (Struble *et al.*, 1984; Kitt *et al.*, 1985; Morrison *et al.*, 1985; Walker *et al.*, 1985). Moreover, the direction of causation is ambiguous; it is equally as plausible that plaques in the cortex induce fibrillary degeneration of afferent terminals around the amyloid core (Mann, 1985), as that degenerating terminals underlie plaque formation.

Arendash *et al.* (1987) have recently reported that if NBM lesions are inflicted on rats at young maturity, neurofibrillary tangles can develop in the cortex as the animals age. Since no similar pathology has ever been reported in natural ageing in rats, this suggested that extensive subcortical cholinergic degeneration may be necessary to trigger the development of neocortical plaques during ageing, and suggests the primacy of the subcortical degeneration in the development of cortical pathology. However, concentrations of both somatostatin and neuropeptide Y increased dramatically in the cortex as a long-term consequence of the NBM lesions in the old rats, in contrast to the consistent decrease in these markers in Alzheimer's disease.

Given the remarkable nature of the observations of Arendash *et al.* (1987), caution needs to be applied until a replication is forthcoming. By contrast, there are now numerous studies indicating retrograde degeneration or atrophy of subcortical cholinergic cells in response to lesions of cortical (Pearson *et al.*, 1983*a*; Sofroniew *et al.*, 1983; Stephens *et al.*, 1986) or hippocampal (Powell, 1954; Kromer *et al.*, 1981; Gage *et al.*, 1986; Armstrong *et al.*, 1987) targets. The majority of these studies have involved gross lesion techniques that cause primary damage to the cholinergic axons and terminals as well as to the postsynaptic cortical targets. This lesion strategy provides no problem when the focus of the studies is the consideration of treatments, such as gangliosides or nerve growth factor (NGF), that might inhibit the retrograde degeneration associated with axotomy. For example, the development of retrograde degenerative changes in the cholinergic cells of the septum and nucleus basalis in response to axotomy or terminal lesion can be inhibited by postoperative administration of the ganglioside GM_1 (Cuello *et al.*, 1986; Gradowska *et al.*, 1986; Sofroniew *et al.*, 1986*b*; Stephens *et al.*, 1987) or of nerve growth factor (Hefti *et al.*, 1984; Haroutunian *et al.*, 1986; Hefti, 1986; Kromer, 1987). However, if the lesion procedure causes direct damage to the cholinergic terminals, it cannot be concluded with confidence that the retrograde degeneration in the septum or NBM is truly secondary to postsynaptic pathology in the target area.

To resolve this problem, Sofroniew & Pearson (1985) have applied the excitotoxins kainic acid and n-methyl-d-aspartic acid to the neocortical surface in order to lesion intrinsic cortical neurons selectively while leaving intact the glial cells, microvascular supply and afferent fibres and terminals in the denervated cortex. In this situation also they observed retrograde atrophy of the NBM cholinergic neurons. Finally, to confirm that these neurotoxins did not cause some subtle and undetected damage to the cholinergic terminals, Sofroniew *et al.* (1986*a*) transplanted embryonic cortex to the denervated cortex and found not only that the cholinergic terminals sprouted to reinner-

vate the grafts, but also that the provision of the alternative target prevented the retrograde degeneration of the cell soma in the NBM. Similar results have been observed in the septo-hippocampal circuitry, where cholinergic grafts can protect septal cholinergic cells from retrograde atrophy (Gage *et al.*, 1986).

These experimental studies appear to demonstrate that a variety of forms of cortical pathology can lead to secondary retrograde degeneration in the subcortical cholinergic afferents, and suggest that a similar sequence of events may underlie the degenerative changes in Alzheimer's disease, in particular when compared with the parallel development of changes in Down's syndrome (Mann, see above). Thus, subcortical cholinergic neurons may be dependent upon the availability of the target-derived growth factor for their survival. In particular, nerve growth factor has been proposed as the critical factor for cholinergic cells (Hefti & Will, 1987). It undergoes retrograde transport by cholinergic neurons from the cortex to the NBM (Seiler & Schwab, 1984) and from the hippocampus to the septum (Schwab *et al.*, 1979), and it promotes the survival of basal forebrain cholinergic neurons following axotomy, as described above. These observations have led several authors to propose that loss of such target-derived trophic factors following cortical denervation can provide a mechanism for the observed retrograde changes in subcortical cholinergic neurons in Alzheimer's disease as well as in the experimental animal models (Appel, 1981; Hefti, 1983).

It remains unclear whether a similar sequence of events is associated with natural ageing. In normal ageing it seems more plausible that neuropathological and neurochemical changes can develop in diverse cortical and subcortical sites independently, although secondary trans-synaptic changes will also become apparent as degeneration progresses. The heterogeneity of the age-related changes, and their capacity to occur independently (Gate *et al.*, 1984*b*), caution against a unitary interpretation of ageing. Nevertheless, Fischer *et al.* (1987) have recently observed that intracerebral injection of nerve growth factor in aged rats can inhibit both the progressive age-related decline in septal cholinergic cell atrophy and the development of maze learning deficits in the aged animals.

14.7 CONCLUSIONS

We have reviewed four main classes of animal models that have been developed for the study of the causes and progress of cognitive and memory changes associated with ageing and dementia.

Studies of ageing in rodents and primates (Section 14.3) constitute what Bartus *et al.* (1983) designate a 'Class A' model. The major advantage of such models is their face validity, and aged animals are seen to manifest similar neuropathological, neurochemical and behavioural changes to those associated with normal ageing in humans. The major disadvantages of ageing *per se* as an animal model for the human neuropathology are firstly that both neuropathological and behavioural changes are multidimensional so that causative relationships are extremely difficult to establish, and secondly that the specific cortical neuropathological changes associated with dementia in humans are only rarely and sporadically detectable in animals.

The other three classes of models that we have discussed all fall within the category of 'Class B' models (Bartus *et al.*, 1983), in which particular neuropathological features are artificially induced in young animals. Studies of the effects of pharmacological treatments in young animals, in particular using anticholinergic drugs (Section 14.4), reproduce many behavioural impairments associated with ageing. This strategy is easy and efficient to conduct, and has been widely used to screen novel drugs for cholinergic action. Nevertheless, attempts to characterize the behavioural disruption induced by scopolamine have not yet clearly resolved the functional role of central cholinergic systems in cognition. A major disadvantage of this strategy has been the lack of anatomical specificity provided by peripheral drug administration, where the compound is likely to influence multiple central systems.

Studies of the effects of encephalopathy induced by aluminium and by the scrapie virus (Section 14.5) have the major advantage of modelling specific neuropathological features associated with Alzheimer's disease, namely neurofibrillary tangles and neuritic plaques, respectively. Nevertheless the molecular structure of the abnormal neurofibrillary proteins in these two models is not the same as those observed in the human disease. Consequently, although these two models might provide information on the functional sequelae of particular types of cortical neuropathology, they are unlikely to provide specific insight into the aetiology of human dementia. Additionally, the behavioural effects of aluminium-induced and viral encephalopathies have still received only minimal attention.

The effects of subcortical lesions of the basal forebrain cholinergic systems (Section 14.6) have received detailed experimental attention in recent years. The major experimental difficulty with this approach has been the specificity with which selective lesions of the cholinergic neurons of the NBM can be achieved. Moreover, the weight of evidence suggests that subcortical degeneration in Alzheimer's disease is secondary to neuropathological changes in the neocortex, rather than vice versa.

Notwithstanding these reservations about the primacy of the subcortical cell loss in the developing functional impairments associated with ageing and dementia, the focus on NBM-cortical cholinergic systems that has been stimulated by the cholinergic hypothesis has generated a wealth of experimental information on the functional dependence of the neocortex on subcortical regulation. It is nevertheless the case that the precise nature of that regulation remains obscure.

In conclusion, none of the available animal models meets all the conditions necessary for a fully adequate model of human dementia. In combination, however, they have generated over the last decade substantial advances in our insight into the ways in which neurodegenerative changes contribute to the functional decline associated with normal and pathological ageing.

We thank Drs Arjun Sahgal, Moira Bruce, and Rachel Etherington for their expert comments and advice on parts of the manuscript. Our own studies on the effects of ageing and NBM lesions in rodents have been supported by the Mental Health Foundation and the Medical Research Council.

REFERENCES

Abrogast, R.E. & Kozlowski, M.R. (1988). Quantitative morphometric analysis of the neurotoxic effects of the excitotoxin, ibotenic acid, on the basal forebrain. *Neurotoxicology* **9**, 39–46.

Adolfsson, R., Gottfries, C.G., Roos, B.E. & Winblad, B. (1979). Changes in brain catecholamines in patients with dementia of the Alzheimer type. *British Journal of Psychiatry* **135**, 216–23.

Aigner, T.G. & Mishkin, M. (1986). The effects of physostigmine and scopolamine on recognition memory in monkeys. *Behavioral and Neural Biology* **45**, 81–7.

Alexander, D.A. (1973). Some tests of intelligence and learning for elderly psychiatric patients: a validation study. *British Journal of Psychiatry* **12**, 188–93.

Algeri, S., Calderini, G., Toffano, G. & Ponzio, F. (1983). Neurotransmitter alterations in aging rats. In D. Samuel, S. Algeri, S. Gershon, V.E. Grimm & G. Toffano (eds.), *Aging of the Brain*, pp. 227–43. Raven Press, New York.

Altman, H.J., Crosland, R.D., Jenden, D.J. & Berman, R.F. (1985). Further characterizations of the nature of the behavioral and neurochemical effects of lesions to the nucleus basalis of Meynert in the rat. *Neurobiology of Aging* **6**, 125–30.

Anderton, B.H., Breinburg, D., Downes, M.J., Green, P.J., Tomlinson, B.E., Ulrich, J., Wood, J.N. & Kahn, J. (1982). Monoclonal antibodies show that neurofibrillary tangles and neurofilaments share antigenic determinants. *Nature* **298**, 84–6.

Appel, S.H. (1981). A unifying hypothesis for the cause of amyotrophic lateral sclerosis, parkinsonism, and Alzheimer's disease. *Annals of Neurology* **10**, 499–503.

Arai, H., Kobayashi, K., Ichimiya, Y., Kosaka, K. & Iizuka, A. (1984). A preliminary study of free amino acids in the post mortem temporal cortex from Alzheimer-type dementia patients. *Neurobiology of Aging* **5**, 319–21.

Araki, H., Uchiyama, Y., Kawashima, K. & Aihara, H. (1986). Impairment of memory and changes in neurotransmitters induced by basal forebrain lesions in rats. *Japanese Journal of Pharmacology* **41**, 497–504.

Arendash, G.W., Strong, P.N. & Mouton, P.R. (1985). Intracerebral transplantation of cholinergic neurons in a new animal model for Alzheimer's disease. In: J.T. Hutton & A.D. Kenny (eds.), *Senile Dementia of the Alzheimer Type*, pp. 351–76. A.R. Liss, New York.

Arendash, G.W., Millard, W.J., Dunn, A.J. & Meyer, E.M. (1987). Long-term neuropathological and neurochemical effects of nucleus basalis lesions in the rat. *Science* **238**, 952–6.

Arendt, T., Bigl, V., Tennstedt, A. & Arendt, A. (1985). Neuronal loss in different parts of the nucleus basalis is related to neuritic plaque formation in cortical target areas in Alzheimer's disease. *Neuroscience* **14**, 1–14.

Arieff, A.L., Cooper, J.D., Armstrong, D. & Lazorowitz, V.C. (1979). Dementia, renal failure, and brain aluminium. *Annals of Internal Medicine* **90**, 741–7.

Armstrong, D.M., Terry, R.D., DeTeresa, R.M., Bruce, G., Hersh, L.B. & Gage, F.H. (1987). Response of septal cholinergic neurons to axotomy. *Journal of Comparative Neurology* **264**, 421–36.

Arnsten, A.F.T. & Goldman-Rakic, P.S. (1985*a*). Catecholamines and cognitive decline in aged nonhuman primates. *Annals of the New York Academy of Science* **444**, 218–34.

Arnsten, A.F.T. & Goldman-Rakic, P.S. (1985*b*). Alpha₂-adrenergic mechanisms in prefrontal cortex associated with cognitive decline in aged nonhuman primates. *Science* **230**, 1273–6.

Arnsten, A.F.T. & Goldman-Rakic, P.S. (1987). Noradrenergic mechanisms in age-related cognitive decline. In W.J. Wurtman, S.H. Corkin & J.H. Growden (eds.), *Alzheimer's Disease: Advances in Basic Research and Therapies*, pp. 275–82. Centre for Brain Sciences, Cambridge, MA.

Asante, J.W., Cross, A.J., Deakin, J.F.W., Johnson, J.A. & Slater, H.R. (1983). Evaluation of ethylcholine mustard aziridinium ion (ECMA) as a specific neurotoxin of brain cholinergic neurones. *British Journal of Pharmacology* **80**, 573P.

Baddeley, A. (1976). *The Psychology of Memory*. Harper & Row, New York.

Baddeley, A., Logie, R., Bressi, S., Della Sala, S. & Spinnler, H. (1986). Dementia and working memory. *Quarterly Journal of Experimental Psychology* **38A**, 603–18.

Bailey, E.L., Overstreet, D.H. & Crocker, A.D. (1986). Effects of intrahippocampal injections of the cholinergic neurotoxin AF64A on open-field activity and avoidance learning in the rat. *Behavioral and Neural Biology* **45**, 263–74.

Bammer, G. (1982). Pharmacological investigations of neurotransmitter involvement in passive avoidance responding: a review and some new results. *Neuroscience and Biobehavioral Reviews* **6**, 246–96.

Banks, W.A. & Kastin, A.J. (1986). Aging, peptides and the blood-brain barrier: implications and speculations. In T. Crook, R. Bartus, S. Ferris & S. Gershon (eds.), *Treatment Development Strategies for Alzheimer's Disease*, pp. 245–65. Mark Powley, Madison, Conn.

Barnes, C.A. (1979). Memory deficits associated with senescence: a neurophysiological and behavioral study in the rat. *Journal of Comparative and Physiological Psychology* **93**, 74–104.

Barnes, C.A., Nadel, L. & Honig, W.K. (1980). Spatial memory deficit in senescent rats. *Canadian Journal of Psychology* **34**, 29–39.

Bartus, R.T. (1978). Evidence for a direct cholinergic involvement in the scopolamine-induced amnesia in monkeys: effects of concurrent administration of physostigmine and methylphenidate with scopolamine. *Pharmacology, Biochemistry and Behavior* **9**, 833–6.

Bartus, R.T. (1979). Physostigmine and recent memory: effects in young and aged nonhuman primates. *Science* **206**, 1087–9.

Bartus, R.T. & Johnson, H.R. (1976). Short-term memory in the rhesus monkey: disruption from the anti-cholinergic scopolamine. *Pharmacology, Biochemistry and Behavior* **5**, 39–46.

Bartus, R.T., Fleming, D. & Johnson, H.R. (1978). Aging in the rhesus monkey: debilitating effects on short-term memory. *Journal of Gerontology* **33**, 858–71.

Bartus, R.T., Dean, R.L. & Beer, B. (1980). Memory deficits in aged cebus monkeys and facilitation with central cholinomimetics. *Neurobiology of Aging* **1**, 145–52.

Bartus, R.T., Dean, R.L., Beer, B. & Lippa, A.S. (1982). The cholinergic hypothesis of geriatric memory dysfunction. *Science* **217**, 408–17.

Bartus, R.T., Flicker, C. & Dean, R.L. (1983). Logical principles for the development of animal models of age-related memory impairments. In T. Crook, S. Ferris & R. Bartus (eds.), *Assessment in Geriatric Psychiatry*, pp. 263–99. Mark Powley, New Canaan, Conn.

Bartus, R.T., Flicker, C. Dean, R.L., Pontecorvo, M., Figueirdo, J.C. & Fisher, S.K. (1985). Selective memory loss following nucleus basalis lesions: long term behavioral recovery despite persistent cholinergic deficiencies. *Pharmacology, Biochemistry and Behavior* **23**, 125–35.

Bartus, R.T., Pontecorvo, M., Flicker, R.L., Dean, R.L. & Figueirdo, C. (1986). Behavioral recovery following bilateral lesions of the nucleus basalis does not occur spontaneously. *Pharmacology, Biochemistry and Behavior* **24**, 1287–92.

Baumgarten, H.G., Klemm, H.P., Lachenmeyer, L., Bjorklund, A., Lovenberg, W. & Schlossberger, H.G. (1978). Mode and mechanism of action of neurotoxic indoleamines: a review and a progress report. *Annals of the New York Academy of Science* **305**, 3–25.

Bayles, K.A. (1982). Language function and senile dementia. *Brain and Language* **16**, 265–80.

Beal, M.F., Benoit, R., Mazurek, M.F., Bird, E.D. & Martin, J.B. (1986). Somatostatin-28_{1-12}-like immunoreactivity is reduced in Alzheimer's disease cerebral cortex. *Brain Research* **368**, 380–3.

Beatty, W.W. & Bierley, R.A. (1986). Scopolamine impairs encoding and retrieval of spatial working memory in rats. *Physiological Psychology* **14**, 82–6.

Beatty, W.W., Butters, N. & Janowsky, D.S. (1986). Patterns of memory failure after scopolamine treatment: implications for cholinergic hypotheses of dementia. *Behavioral and Neural Biology* **45**, 196–211.

Beller, S.A., Overall, J.E. & Swann, A.C. (1985). Efficacy of oral physostigmine in primary degenerative dementia. A double-blind study of response to different dose levels. *Psychopharmacology* **87**, 147–51.

Ben-Ari Y., Tremblay, E., Ottersen, O.P. & Naquet, R. (1979). Evidence suggesting secondary epileptogenic lesions after kainic acid: pretreatment with diazepam reduces distant but not local brain damage. *Brain Research* **165**, 362–5.

Ben-Ari, Y., Tremblay, E., Ottersen, O.P. & Meldrum, B.S. (1980). The role of epileptic activity in hippocampal and 'remote' cerebral lesions induced by kainic acid. *Brain Research* **191**, 79–97.

Beninger, R.J., Jhamandas, K., Boegman, R.J. & El-Defrawy, S. (1986*a*). Effects of scopolamine and unilateral lesions of the basal forebrain on T-maze spatial discrimination and alternation in rats. *Pharmacology, Biochemistry and Behavior* **24**, 1353–60.

Beninger, R.J., Wirsching, B.A., Jhamandas, K., Boegman, R.J. & El-Defrawy, S. (1986*b*). Effects of altered cholinergic function on working and reference memory in the rat. *Canadian Journal of Physiology and Pharmacology* **64**, 376–82.

Benson, D.F., Kuhl, D.E., Hawkins, R.A., Phelps, M.E., Cummings, J.L. & Tsai, S.Y. (1983). The fluorodeoxyglucose ^{18}F scan in Alzheimer's disease and multi-infarct dementia. *Archives of Neurology* **40**, 711–14.

Benton, J.S., Bowen, D.M., Allen, S.J., Haan, E.A., Davison, A.N., Neary, D., Murphy, R.P. & Snowdon, J.S. (1982). Alzheimer's disease as a disorder of the isodendritic core. *Lancet* **i**, 456.

Berger, B.D. & Stein, L. (1969). An analysis of the learning deficits produced by scopolamine. *Psychopharmacology* **14**, 271–83.

Berman, R.F., Crosland, R.D., Jenden, D.J. & Altman, H.J. (1988). Persisting behavioral and neurochemical deficits in rats following lesions of the basal forebrain. *Pharmacology, Biochemistry and Behavior* **29**, 581–6.

Bernstein, D., Olton, D.S., Ingram, D.K., Waller, S.B., Reynolds, M.A. & London, E.D. (1985). Radial maze performance in young and aged mice: neurochemical correlates. *Pharmacology, Biochemistry and Behavior* **22**, 301–7.

Birks, R.I. & MacIntosh, F.C. (1961). Acetylcholine metabolism of a sympathetic ganglion. *Canadian Journal of Biochemistry and Physiology* **39**, 787–827.

Birren, J.E. (1955). Age differences in startle reaction time of the rat to noise and electric shock. *Journal of Gerontology* **10**, 437–40.

Bjorklund, H., Eriksdotter-Nilsson, M., Dahl, D., Rose, G., Hoffer, B.J. & Olson, L. (1985). Image analysis of GFA-positive astrocytes from adolescence to senescence. *Experimental Brain Research* **58**, 163–70.

Blaker, W.D. & Goodwin, S.D. (1987). Biochemical and behavioral effects of intrahippocampal AF64A in rats. *Pharmacology, Biochemistry and Behavior* **28**, 157–63.

Blessed, G., Tomlinson, B.E. & Roth, M. (1968). The association between quantitative measures of dementia and of senile change in the cerebral grey matter of elderly subjects. *British Journal of Psychiatry* **114**, 797–811.

Bohdanecky, Z. & Jarvik, M.E. (1967). Impairment of one-trial passive avoidance learning in mice by scopolamine, scopolamine methylbromide, and physostigmine. *International Journal of Neuropharmacology* **6**, 217–22.

Bohdanecky, Z., Jarvik, M.E. & Carley, J.L. (1967). Differential impairment of delayed matching in monkeys by scopolamine and scopolamine methyl bromide. *Psychopharmacologia* **11**, 293–9.

Bondareff, W. (1977). The neural basis of aging. In J.E. Birren & K.W. Schaie (eds.). *Handbook of the Psychology of Aging*, pp. 157–76. van Nostrand Reinhold, New York.

Bondareff, W, Mountjoy, C.Q. & Roth, M. (1981). Selective loss of neurones of origin of adrenergic projection to cerebral cortex (nucleus locus coeruleus) in senile dementia. *Lancet* **i**, 783–4.

Bowdler, N.C., Beasley, D.S., Fritze, E.C., Goulette, A.M., Hatton, J.D., Hession, J., Ostman, D.L., Rugg, D.J. & Schmittdel, C.J. (1979). Behavioral effects of aluminum ingestion on animal and human subjects. *Pharmacology, Biochemistry and Behavior* **10**, 505–12.

Bowen, D.M., Smith, C.B., White, P. & Davison, A.N. (1976). Neurotransmitter-related enzymes and indices of hypoxia in senile dementia and other abiotrophies. *Brain* **99**, 459–96.

Bowen, D.M., Allen, S.J., Benton, J.S., Goodhart, M.J., Haan, E.A., Palmer, A.M., Sims, N.R., Smith, C.C.T., Spillane, J.A., Esiri, M.M., Neary, D., Snowdon, J.S., Wilcock, G.K. & Davison, A.N. (1983). Biochemical assessment of serotonergic and cholinergic dysfunction and cerebral atorphy in Alzheimer's disease. *Journal of Neurochemistry* **41**, 266–72.

Brennan, M.J. & Quartermain, D. (1980). Age-related differences in within session habituation: the effect of stimulus complexity. *Gerontology* **20**, 71.

Brennan, M.J., Dallob, A. & Friedman, E. (1981). Involvement of hippocampal serotonergic activity in age-related changes in exploratory behavior. *Neurobiology of Aging* **2**, 199–203.

Brizzee, K.R., Sherwood, N. & Timiras, P. (1968). A comparison of cell populations at various depth levels in cerebral cortex of young adult and aged Long Evans rats. *Journal of Gerontology* **23**, 289–97.

Brizzee, K.R., Ordy, J.M. & Kaack, B. (1974). Early appearance and regional differences in intraneuronal and extraneuronal lipofuscin accumulation with age in the brain of a nonhuman primate (*Macaca mullata*). *Journal of Gerontology* **29**, 366–81.

Brizzee, K.R., Ordy, J.M. & Bartus, R.T. (1980). Localization of cellular changes within multimodal sensory regions in aged monkey brain: possible implications for age-related cognitive loss. *Neurobiology of Aging* **1**, 45–52.

Brown, K. & Warburton, D.M. (1971). Attenuation of stimulus sensitivity by scopolamine. *Psychonomic Science* **22**, 297–8.

Bruce, M.E. (1981). Serial studies of the development of cerebral amyloidosis and vacuolar degeneration in murine scrapie. *Journal of Comparative Pathology* **91**, 589–97.

Bruce, M.E. (1984). Scrapie and Alzheimer's disease. *Psychological Medicine* **14**, 497–500.

Bruce, M.E. & Fraser, H. (1975). Amyloid plaques in the brains of mice infected with scrapie: morphological variation and staining properties. *Neuropathology and Applied Neurobiology* **7**, 289–98.

Bruce, M.E., Dickinson, A.G. & Fraser, H. (1976). Cerebral amyloidosis in scrapie in the mouse: effect of agent strain and mouse genotpye. *Neuropathology and Applied Neurobiology* **2**, 471–8.

Bugiani, O. & Ghetti, B. (1983). Effects of prolonged exposure to aluminum on the nervous system. In D. Samuel, S. Algeri, S. Gershon, V.E. Grimm & G. Toffano (eds.), *Aging of the Brain*, pp. 271–5. Raven Press, New York.

Buresova, O., Bures, J., Bohdanecky, Z. & Weiss, T. (1964). Effects of atropine on learning, extinction, retention and retrieval in rats. *Psychopharmacology* **5**, 255–63.

Caine, E.D., Weingartner, H., Ludlow, C.L., Cudahy, E.A. & Wehry, S. (1981). Qualitative analysis of scopolamine-induced amnesia. *Psychopharmacology* **74**, 74–80.

Campbell, B.A., Krauter, E.E. & Wallace, J.E. (1980). Animal models of aging: sensory-motor and cognitive function in the aged rat. In D. Stein (ed.), *The Psychobiology of Aging*, pp. 201–26. Elsevier North Holland, Amsterdam.

Candy, J.M., Perry, R.H., Perry, E.K., Irving, D., Blessed, G., Fairbairn, A. & Tomlinson, B.E. (1983). Pathological changes in the nucleus of Meynert in Alzheimer's and Parkinson's diseases. *Journal of Neurological Science* **54**, 277–89.

Candy, M.J., Klinowski, K., Perry, R.H., Perry, E.K., Fairbairn, A., Oakley, A.E., Carpenter, T.A., Atack, J.R., Blessed, G. & Edwardson, J.A. (1986). Aluminosilicates and senile plaque formation in Alzheimer's disease. *Lancet* i, 354–7.

Carlsson, A. & Winblad, B. (1976). Influence of age and time interval between death and autopsy on dopamine and 3-methoxytyramine levels in human basal ganglia. *Journal of Neural Transmission* **83**, 271–6.

Carlton, P.L. (1963). Cholinergic mechanisms in the control of behaviour by the brain. *Psychological Review* **70**, 19–39.

Carlton, P.L. (1969). Brain acetylcholine and inhibition. In J.T. Tapp (ed.), *Reinforcement and Behavior*, pp. 287–327. Academic Press, New York.

Caulfield, M.P., Fortune, D.H., Roberts, P.M. & Stubley, J.K. (1981). Intracerebroventricular hemicholinium-3 (HC-3) impairs learning of a passive avoidance task in mice. *British Journal of Pharmacology* **74**, 865P.

Chan, S.H.H. & Lai, Y.-Y. (1982). Effects of aging on pain responses and analgesic efficacy of morphine and clonidine in rats. *Experimental Neurology* **75**, 112–19.

Chan-Palay, V. (1987). Somatostatin immunoreactive neurons in the human hippocampus and cortex shown by immunogold/silver intensification on vibratome sections: coexistence with neuropeptide Y neurons, and effects in Alzheimer-type dementia. *Journal of Comparative Neurology* **260**, 201–23.

Chan-Palay, V., Lang, W., Haesler, U., Kohler, C. & Yasargil, G. (1986). Distribution of altered hippocampal neurons and axons immunoreactive with antisera against neuropeptide Y in Alzheimer's-type dementia. *Journal of Comparative Neurology* **248**, 376–94.

Cheal, M.L. (1981). Scopolamine disrupts maintenance of attention rather than memory processes. *Behavioral and Neural Biology* **33**, 163–87.

Christie, J.E., Shering, A., Ferguson, J. & Glen, A.I.M. (1981). Physostigmine and arecoline: effects of intravenous infusions in Alzheimer presenile dementia. *British Journal of Psychiatry* **138**, 46–50.

Chrobak, J.J., Hanin, I. & Walsh, T.J. (1987). AF64A (ethylcholine aziridinium ion), a cholinergic neurotoxin, selectively impairs working memory in a multiple component T-maze task. *Brain Research* **414**, 15–21.

Coffey, P.J., Perry, V.H., Allen, Y., Sinden, J. & Rawlins, J.N.P. (1988). Ibotenic acid induced demyelination in the central nervous system: a consequence of a local inflammatory response. *Neuroscience Letters* **84**, 178–84.

Collerton, D. (1986). Cholinergic function and intellectual decline in Alzheimer's disease. *Neuroscience* **19**, 1–28.

Corkin, S. (1982). Some relationships between global amnesia and the memory impairments of Alzheimer's disease. In S. Corkin, K.L. Davis, J.H. Growden, E. Usdin & R.J. Wurtman (eds.), *Alzheimer's disease: A Report of Progress in Research*, pp. 149–64. Raven Press, New York.

Corkin, S., Growden, J.H., Sullivan, E.V. & Shedlack, K. (1981). Lecithin and cognitive function in aging and dementia. In A.D. Kidman, J.M. Tomkins & R.A. Westerman (eds.), *New Approaches to Nerve and Muscle Disorders: Basic and Applied Contributions*, pp. 229–49. Elsevier/North Holland, Amsterdam.

Corkin, S., Davis, K.L., Growden, J.H., Usdin, E. & Wurtman, R.J. (eds.) (1982). *Alzheimer's disease: A Report of Progress in Research*, pp. 133–9. Raven Press, New York.

Corkin, S., Growden, J.H., Nissen, M.J., Huff, F.J., Freed, D.M. & Sagar, H.J. (1987). Recent advances in the neuropsychological study of Alzheimer's disease. In R.J. Wurtman, S. Corkin & J.H. Growden (eds.), *Alzheimer's Disease: Advances in Basic Research and Therapies*, Center for Brain Sciences and Metabolism Trust, Cambridge, Mass.

Corsellis, J.A.N. (1976). Ageing and the dementias. In W. Blackwood & J.A.N. Corsellis (eds.), *Greenfield's Neuropathology*, 3rd edn, pp. 796–848. Arnold, London.

Coyle, J.T. (1982). Excitatory amino acid neurotoxins. In L.L. Iversen, S.D. Iversen & S.H. Snyder (eds.), *Handbook of Psychopharmacology, Volume 15: New techniques in Psychopharmacology*, pp. 237–69. Plenum Press, New York.

Coyle, J.T., Price, D.L. & DeLong, M.R. (1983). Alzheimer's disease: a disorder of cortical cholinergic innervation. *Science* **219**, 1184–90.

Coyle, J.T., Oster-Granite, M.L. & Gearhart, J.D. (1986). The neurobiologic consequences of Down syndrome. *Brain Research Bulletin* **16**, 773–87.

Craik, F.I.M. (1984). Age differences in remembering. In L.R. Squire & N. Butters (eds.), *Neuropsychology of Memory*, pp. 3–12. Guilford Press, New York.

Crapper, D.R. & Dalton, A.J. (1973a). Alterations in short-term retention, conditioned avoidance response acquisition and motivation following aluminum induced neurofibrillary degeneration. *Physiology and Behavior* **10**, 925–33.

Crapper, D.R. & Dalton, A.J. (1973b). Aluminum induced neurofibrillary degeneration, brain electrical activity and alterations in acquisition and retention. *Physiology and Behavior* **10**, 935–45.

Crapper, D.R., Krishnan, S.S. & Quittkat, S. (1976). Aluminum, neurofibrillary degeneration and Alzheimer's disease. *Brain* **99**, 67–80.

Crapper-McLachlan, D.R. (1987). Aluminum and Alzheimer's disease. *Neurobiology of Aging* **7**, 525–32.

Cross, A.J., Crow, T.J., Perry, E.K., Perry, R.H., Blessed, G. & Tomlinson, B.E. (1981). Reduced dopamine-beta-hydroxylase activity in Alzheimer's disease. *British Medical Journal* **282**, 93–4.

Crow, T.J. & Grove-White, I.G. (1973). An analysis of the learning deficit following hyoscine administration to man. *British Journal of Pharmacology* **49**, 322–7.

Cuello, A.C., Stephens, P.H., Tagari, P.C., Sofroniew, M.V. & Pearson, R.C.A. (1986). Retrograde changes in the nucleus basalis of the rat, caused by cortical damage, are prevented by exogenous ganglioside GM$_1$. *Brain Research* **376**, 373–7.

Curcio, C.A. & Coleman, P.D. (1982). Stability of neuron number in cortical barrels of aging mice. *Journal of Comparative Neurology* **212**, 158–72.

Dahl, D., Selkoe, D.J., Pero, R.T. & Bignami, A. (1982). Immunostaining of neurofibrillary tangles in Alzheimer's senile dementia with a neurofilament antiserum. *Journal of Neuroscience* **2**, 113–19.

Daitz, H.M. & Powell, T.P.S. (1954). Studies on the connections of the fornix system. *Journal of Neurology Neurosurgery and Psychiatry* **17**, 75–82.

Damasio, H., Eslinger, P., Damasio, A.R., Rizzo, M., Huang, H.K. & Demeter, S. (1983). Quantitative computed tomographic analysis in the diagnosis of dementia. *Archives of Neurology* **40**, 715–19.

D'Amato, R.J., Zweig, R.M., Whitehouse, P.J., Wenk, G.L., Singer, H.S., Mayeux, R., Price, D.L. & Snyder, S.H. (1987). Aminergic systems in Alzheimer's disease and Parkinson's disease. *Annals of Neurology* **22**, 229–36.

Davies, P. & Maloney, A.J.F. (1976). Selective loss of central cholinergic neurons in Alzheimer's disease. *Lancet* **ii**, 1403.

Davies, P., Katzman, R. & Terry, R.D. (1980). Reduced somatostatin-like immunoreactivity in cerebral cortex from cases of Alzheimer's disease and Alzheimer senile dementia. *Nature* **288**, 279–80.

Davis, K.L., Mohs, R.C. & Tinklenberg, J.R. (1979). Enhancement of memory by physostigmine. *New England Journal of Medicine* **301**, 946.

Dawbarn, D., Rossor, M.N., Mountjoy, C.Q., Roth, M. & Emson, P.C. (1986). Decreased somatostatin immunoreactivity but not neuropeptide Y immunoreactivity in cerebral cortex in senile dementia of Alzheimer type. *Neuroscience Letters* **70**, 154–9.

Dayan, A.D. (1971). Comparative neuropathology of ageing. Studies on the brains of 47 species of vertebrates. *Brain* **94**, 31–42.

Dean, R.L., Goas, J.A., Regan, B., Scozzafava, J. & Bartus, R.T. (1981). Age-related differences in the lifespan of the C57BL/6J mouse. *Experimental Aging Research* **7**, 427–51.

Dean, R.L. & Bartus, R.T. (1985). Animal models of geriatric cognitive dysfunction: evidence for an important cholinergic involvement. In J. Traber & W.H. Gispen (eds.), *Senile Dementia of the Alzheimer Type*, pp. 269–82. Springer-Verlag, Heidelberg.

DeBoni, U. & Crapper, D.R. (1978). Paired helical filaments of the Alzheimer type in cultured neurons. *Nature* **271**, 566–8.

de Koning-Verest, I.F., Knook, D.L. & Wolthuis, O.L. (1980). Behavioral and biochemical correlates of aging in rats. In D. Stein (ed.), *The Psychobiology of Aging*, pp. 177–99. Elsevier North Holland, Amsterdam.

Delabar, J.-M., Goldgaber, D., Lamour, Y., Nicole, A., Huret, J.-L., de Grouchy, J., Brown, P., Gadjusek, D.C. & Sinet, P.-M. (1987). Beta amyloid gene duplication in Alzheimer's disease and karyotypically normal Down syndrome. *Science* **235**, 1390–2.

Demarest, K.T., Riegle, G.D. & Moore, K.E. (1980). Characteristics of dopaminergic neurons in the aged male rat. *Neuroendocrinology* **31**, 222–7.

Dember, W.W. & Fowler, H. (1958). Spontaneous alternation behavior. *Psychological Bulletin* **55**, 412–28.

DeSouza, E.B., Whitehouse, P.J., Kuhar, M.J., Price, D.L. & Vale, W.V. (1986). Reciprocal changes in corticotrophin releasing factor (CRF)-like immunoreactivity and CRF receptors in cerebral cortex of Alzheimer's disease. *Nature* **319**, 593–5.

Deutsch, J.A. (1971). The cholinergic synapse and the site of memory. *Science* **174**, 788–94.

de Wied, D. & van Ree, J.M. (1982). Neuropeptides, mental performance and aging. *Life Sciences* **31**, 709–19.

Diamond, M.C., Johnson, R.E. & Ingham, C.A. (1975). Morphological changes in the young, adult and aging rat cerebral cortex, hippocampus and diencephalon. *Behavioral Biology* **14**, 163–74.

Diamond, M.C., Johnson, R.E. & Gold, M.W. (1977). Changes in neuron number and size and glia number in the young, adult and aging rat medial occipital cortex. *Behavioral Biology* **20**, 409–18.

Dickinson, A.G., Bruce, M.E. & Scott, J.R. (1983). The relevance of scrapie as an experimental model for Alzheimer's disease. *Banbury Report 15: Biological Aspects of Alzheimer's Disease*, pp. 387–98. Cold Spring Harbor.

DiMattia, B.V. & Kesner, R.P. (1984). Serial position curves in rats: automatic versus effortful information processing. *Journal of Experimental Psychology: Animal Behavior Processes* **10**, 557–63.

Donoghue, J.P. & Wise, S.P. (1982). The motor cortex of the rat: cytoarchitecture and microstimulation mapping. *Journal of Comparative Neurology* **212**, 76–88.

Douglas, R.J. & Isaacson, R.L. (1966). Spontaneous alternation and scopolamine. *Psychonomic Science* **4**, 283–4.

Drachman, D.A. (1977). Memory and cognitive function in man. Does the cholinergic system have a specific role? *Neurology* **27**, 783–90.

Drachman, D.A. & Leavitt, J. (1974). Human memory and the cholinergic system. *Archives of Neurology* **30**, 113–21.

Drachman, D.A. & Sahakian, B.J. (1980). Memory, aging and pharmacosystems. In D. Stein (ed.), *The Psychobiology of Aging: Problems and Perspectives*, pp. 348–68. Elsevier/North Holland, Amsterdam.

Dubois, B., Mayo, W., Agid, Y., LeMoal, M. & Simon, H. (1985). Profound disturbances of spontaneous and learned behaviors following lesions of the nucleus basalis magnocellularis in the rat. *Brain Research* **338**, 249–58.

Dunnett, S.B. (1985). Comparative effects of cholinergic drugs and lesions of nucleus basalis or fimbria-fornix on delayed matching in rats. *Psychopharmacology* **87**, 357–63.

Dunnett, S.B., Toniolo, G., Fine, A., Ryan, C.N., Bjorklund, A. & Iversen, S.D. (1985). Transplanatation of embryonic ventral forebrain neurons to the neocortex of rats with lesions of nucleus basalis magnocellularis – II. Sensorimotor and learning impairments. *Neuroscience* **16**, 787–97.

Dunnett, S.B., Whishaw, I.Q., Jones, G.H. & Bunch, S.T. (1987). Behavioural, biochemical and histochemical effects of different neurotoxic amino acids injected into nucleus basalis magnocellularis of rats. *Neuroscience* **20**, 653–69.

Dunnett, S.B., Evenden, J.L. & Iversen, S.D. (1988). Delay-dependent short-term memory deficits in aged rats. *Psychopharmacology* **96**, 174–80.

Eckerman, D.A., Gordon, W.A., Edwards, J.D., MacPhail, R.C. & Gage, M.I. (1980). Effects of scopolamine, pentobarbital, and amphetamine on radial arm maze performance in the rat. *Pharmacology, Biochemistry and Behavior* **12**, 595–602.

Edwardson, J.A., Klinowski, J., Oakley, A.E., Perry, R.H. & Candy, J.M. (1986). Aluminosilicates and the ageing brain: implications for the pathogenesis of Alzheimer's disease. In *Ciba Foundation Symposium Volume 121: Silicon Biochemistry*, pp. 160–79. Wiley, Chichester.

El-Defrawy, S.R., Boegman, R.J., Jhamandas, K., Beninger, R.J. & Shipton, L. (1986). Lack of recovery of cortical cholinergic function following quinolinic or ibotenic acid injections into the nucleus basalis magnocellularis in rats. *Experimental Neurology* **91**, 628–33.

Elias, P.K. & Elias, M.F. (1976). Effects of age on learning ability: contributions from the animal literature. *Experimental Aging Research* **2**, 165–86.

Elias, P.K., Elias, M.F. & Eleftheriou, B.E. (1975). Emotionality, exploratory behavior, and locomotion in aging inbred strains of mice. *Gerontology* **21**, 46–55.

Estes, K.S. & Simpkins, J.W. (1980). Age-related alterations in catecholamine concentrations in discrete preoptic area and hypothalamic regions in the male rat. *Brain Research* **194**, 556–60.

Etherington, R. (1988). Behavioural effects of cortical cholinergic deafferentation. Unpublished Ph.D. thesis, University of Cambridge.

Etherington, R., Mittleman, G. & Robbins, T.W. (1987). Comparative effects of nucleus basalis and fimbria-fornix lesions on delayed matching and alternation tests of memory. *Neuroscience Research Communications* **1**, 135–43.

Etienne, P., Robitaille, Y., Wood, P., Gauthier, S., Nair, N.P.V. & Quirion, R. (1986). Nucleus basalis neuronal loss, neuritic plaques and choline acetyltransferase activity in advanced Alzheimer's disease. *Neuroscience* **19**, 1279–91.

Evenden, J.L. & Robbins, T.W. (1984). Win-stay behaviour in the rat. *Quarterly Journal of Experimental Psychology* **36B**, 1–26.

Everitt, B.J., Robbins, T.W., Evenden, J.L., Marston, H.M., Jones, G.J. & Sirkia, T.E. (1987). The effects of excitotoxic lesions of the substantia innominata, ventral and dorsal globus pallidus on the acquisition and retention of a conditional visual discrimination: implications for cholinergic hypotheses of learning and memory. *Neuroscience* **22**, 441–69.

Feldman, M.L. (1976). Aging changes in the morphology of cortical dendrites. In R.D. Terry & S. Gershon (eds.), *Neurobiology of Aging*, pp. 221–7. Raven Press, New York.

Fields, P.E. (1953). The age factor in multiple-discrimination learning by white rats. *Journal of Comparative and Physiological Psychology* **46**, 387–9.

Finch, C.B. (1973). Catecholamine metabolism in the brains of ageing male mice. *Brain Research* **52**, 261–76.

Finch, C.B. (1982). Rodent models for aging processes in the human brain. In Corkin, S., Davis, K.L., Growden, J.H., Usdin, E. & Wurtman, R.J. (eds.), *Alzheimer's disease: A Report of Progress in Research*, pp. 249–56. Raven Press, New York.

Finch, C.B. & Morgan, D.G. (1984). Serotonin-2 (S-2) binding sites decrease in old mice, but are down-regulated to the same extent by sub-chronic amitriptyline at all ages. *Society for Neuroscience Abstracts* **10**, 16.

Fine, A., Dunnett, S.B., Bjorklund, A. & Iversen, S.D.

(1985*a*). Cholinergic ventral forebrain grafts into the neocortex improve passive avoidance memory in a rat model of Alzheimer's disease. *Proceedings of the National Academy of Sciences of the USA* **82**, 5227–30.

Fine, A., Dunnett, S.B., Bjorklund, A., Clarke, D.J. & Iversen, S.D. (1985*b*). Transplantation of embryonic ventral forebrain neurons to the neocortex of rats with lesions of nucleus basalis magnocellularis – I. Anatomical observations. *Neuroscience* **16**, 769–86.

Fine, A., Pittaway, K., de Quidt, M., Czudek, C. & Reynolds, G.P. (1987). Maintenance of cortical somatostatin and monoamine levels in the rat does not require intact cholinergic innervation. *Brain Research* **406**, 326–9.

Fischer, W., Wictorin, K., Bjorklund, A., Williams, L.R., Varon, S. & Gage, F.H. (1987). Amelioration of cholinergic neuron atrophy and spatial memory impairment in aged rats by nerve growth factor. *Nature* **329**, 65–8.

Fisher, A. & Hanin, I. (1980). Choline analogs as potential tools in developing selective animal models of central cholinergic hypofunction. *Life Sciences* **27**, 1615–34.

Fisher, A. & Hanin, I. (1986). Potential animal models for senile dementia of Alzheimer's type, with emphasis on AF64A-induced toxicity. *Annual Review of Pharmacology and Toxicology* **26**, 161–81.

Fisher, A., Mantione, C.R., Abraham, D.J. & Hanin, I. (1982). Long-term central cholinergic hypofunction induced in mice by ethylcholine aziridinium ion (AF64A) *in vivo. Journal of Pharmacology and Experimental Therapeuts* **222**, 140–5.

Flicker, C., Dean, R.L., Watkins, D.L., Fisher, S.K. & Bartus, R.T. (1983). Behavioral and neurochemical effects following neurotoxic lesions of a major cholinergic input to the cerebral cortex in the rat. *Pharmacology, Biochemistry and Behavior* **18**, 973–81.

Flood, J.F. & Cherkin, A. (1986). Scopolamine effects on memory retention in mice: a model for dementia? *Behavioral and Neural Biology* **45**, 169–84.

Fraser, H. & Bruce, M.E. (1973). Argyrophilic plaques in mice innoculated with scrapie from particular sources. *Lancet* **i**, 617.

Freeman, J.J. & Jenden, D.J. (1976). Source of choline for acetylcholine synthesis in brain. *Life Sciences* **19**, 949–62.

Friedman, E., Lerer, B. & Kuster, J. (1983). Loss of cholinergic neurons in the rat neocortex produces deficits in passive avoidance learning. *Pharmacology, Biochemistry and Behavior* **19**, 309–12.

Gage, F.H. & Bjorklund, A. (1986). Cholinergic grafts into the hippocampal formation improve spatial learning and memory in aged rats by an atropine sensitive mechanism. *Journal of Neuroscience* **6**, 2837–47.

Gage, F.H., Bjorklund, A., Stenevi, U. & Dunnett, S.B. (1983*a*). Intracerebral grafting in the aging brain. In W.H. Gispen & J. Traber (eds.), *Aging of the Brain*, pp. 125–37. Elsevier, Amsterdam.

Gage, F.H. Dunnett, S.B., Stenevi, U. & Bjorklund, A. (1983*b*). Aged rats: recovery of motor impairments by intrastriatal nigral grafts. *Science* **221**, 966–9.

Gage, F.H., Bjorklund, A., Stenevi, U., Dunnett, S.B. & Kelly, P.A.T. (1984*a*). Intrahippocampal septal grafts ameliorate learning impairments in aged rats. *Science* **225**, 533–6.

Gage, F.H., Dunnett, S.B. & Bjorklund, A. (1984*b*). Spatial learning and motor deficits in aged rats. *Neurobiology of Aging* **5**, 43–8.

Gage, F.H., Kelly, P.A.T. & Bjorklund, A. (1984*c*). Regional changes in brain glucose metabolism reflect cognitive impairments in aged rats. *Journal of Neuroscience* **4**, 2845–66.

Gage, F.H., Wictorin, K., Fischer, W., Williams, L.R., Varon, S. & Bjorklund, A. (1986). Retrograde cell changes in medial septum and diagonal band following fimbria-fornix transection: quantitative temporal analysis. *Neuroscience* **19**, 241–55.

Gaidusek, D.C. (1977). Unconventional viruses and the origin and disappearance of kuru. *Science* **197**, 943–60.

Ghonheim, M.M. & Mewaldt, S.P. (1975). Effects of diazepam and scopolamine on storage, retrieval and organisational processes in memory. *Psychopharmacology* **44**, 257–62.

Gibson, G.E., Peterson, C. & Jenden, D.J. (1981). Brain acetylcholine synthesis declines with senescence. *Science* **213**, 674–6.

Gibson, P.H. & Tomlinson, A.N. (1977). Numbers of Hirano bodies in hippocampus of normal and demented people with Alzheimer's disease. *Journal of Neurological Science* **33**, 199–206.

Gilad, G.M. & Gilad, V.H. (1987). Age-related reductions in brain cholinergic and dopaminergic indices in two rat strains differing in longevity. *Brain Research* **408**, 247–50.

Giorgi, O., Calderini, G., Toffano, G. & Biggio, G. (1987). D-1 dopamine receptors labelled with ^{3}H-SCH 23390 decrease in the striatum of aged rats. *Neurobiology of Aging* **8**, 51–4.

Glick, R. & Bondareff, W. (1979). Loss of synapses in the cerebellar cortex of the senescent rat. *Journal of Gerontology* **34**, 818–22.

Glick, S.D. & Jarvik, M.E. (1969). Amphetamine, scopolamine, and chlorpromazine interactions on delayed matching performance in monkeys. *Psychopharmacologia* **16**, 147–55.

Glick, S.D. & Zimmerberg, B. (1972). Amnesic effects of scopolamine. *Behavioral Biology* **7**, 245–54.

Glick, S.D., Mittag, T.W. & Green, J.P. (1973). Central cholinergic correlates of impaired learning. *Neuropharmacology* **12**, 291–6.

Godding, P.R., Rush, J.R. & Beatty, W.W. (1982). Scopolamine does not disrupt spatial working memory in rats. *Pharmacology, Biochemistry and Behavior* **16**, 919–23.

Goetsch, V.L. & Isaac, W. (1982). Age and visual sensitivity in the rat. *Physiological Psychology* **10**, 199–201.

Gold, P.E. & McGaugh, J.L. (1975). Changes in learning and memory during aging. In J.M. Ordy & K.R. Brizzee (eds.), *Neurobiology of Aging*, pp. 145–58. Plenum Press, New York.

Gold, P.E., McGaugh, J.L., Hankins, L.L., Rose, R.P. & Vasquez, B.J. (1981). Age dependent changes in retention in rats. *Experimental Aging Research* **8**, 53–8.

Goldman, G. & Coleman, P.D. (1981). Neuron numbers in locus coeruleus do not change with age in Fisher 344 rat. *Neurobiology of Aging* **2**, 33–6.

Goldman-Rakic, P.S. & Brown, R.M. (1981). Regional changes of monoamines in cerebral cortex and subcortical structures of aging rhesus monkeys. *Neuroscience* **7**, 177–87.

Goodrick, C.L. (1965). Social interactions and exploration of young, mature and senescent male rats. *Journal of Gerontology* **20**, 215–18.

Goodrick, C.L. (1972). Learning by mature-young and aged Wistar albino rats as a function of test complexity. *Journal of Gerontology* **27**, 353–7.

Goodrick, C.L. (1975). Behavioral rigidity as a mechanism for facilitation of problem solving for aged rats. *Journal of Gerontology* **30**, 181–4.

Gordon, W.C., Scobie, S.R. & Frankl, S.E. (1978). Age-related differences in electric shock detection and escape thresholds in Sprague-Dawley albino rats. *Experimental Aging Research* **4**, 23–35.

Govoni, S., Loddo, P., Spano, P.F. & Trabucchi, M. (1977). Dopamine receptor sensitivity in brain and retina of rats during aging. *Brain Research* **138**, 565–70.

Gradkowska, M., Skup, M., Kiedrowski, L., Calzolari, S. & Oderfeld-Nowak, B. (1986). The effect of GM$_1$ ganglioside on cholinergic and serotonergic systems in the rat hippocampus following partial denervation is dependent on the degree of fiber degeneration. *Brain Research* **375**, 417–22.

Greenberg, L.H. & Weiss, B. (1978). Beta-adrenergic receptors in aged rat brain: reduced number and capacity of pineal gland to develop sensitivity. *Science* **201**, 61–3.

Grundke-Iqbal, I., Johnson, A.B., Wisniewski, H.M., Terry, R.D. & Iqbal, K. (1979). Evidence that Alzheimer neurofibrillary tangles originate from neurotubules. *Lancet* **i**, 578–80.

Gurland, B.J. & Birkett, D.P. (1983). The senile and pre-senile dementias. In M.H. Lader (ed.), *Handbook of Psychiatry. 2. Mental Disorders and Somatic Illness*, pp. 128–46. Cambridge University Press, Cambridge.

Haber, S.N. & Nauta, W.J.H. (1983). Ramifications of the globus pallidus in the rat as indicated by patterns of immunohistochemistry. *Neuroscience* **9**, 245–60.

Hagan, J.J., Salamone, J.D., Simpson, J., Iversen, S.D. & Morris, R.G.M. (1988). Place navigation in rats is impaired by lesions of medial septum and diagonal band but not nucleus basalis magnocellularis. *Behavioural Brain Research* **27**, 9–20.

Hall, R.D. & Lindholm, E.P. (1974). Organization of motor and somatosensory neocortex in the albino rat. *Brain Research* **66**, 23–38.

Hammer, R., Berrie, C.P., Birdsall, N.J.M., Burgen, A.S.V. & Hulme, E.C. (1980). Pirenzepine distinguishes between subclasses of muscarinic receptors. *Nature* **283**, 90–2.

Hardy, J., Cowburn, R., Barton, A., Reynolds, G., Dodd, P., Wester, P., O'Carroll, A.-M., Lofdahl, E. & Winblad, B. (1987a). A disorder of cortical GABAergic innervation in Alzheimer's disease. *Neuroscience Letters* **73**, 192–6.

Hardy, J., Cowburn, R., Barton, A., Reynolds, G., Lofdahl, E., O'Carroll, A.M., Wester, P. & Winblad, B. (1987b). Region specific loss of glutamate innervation in Alzheimer's disease. *Neuroscience Letters* **73**, 77–80.

Haroutunian, V., Kanof, P.D. & Davis, K.L. (1986). Partial reversal of lesion-induced deficits in cortical cholinergic markers by nerve growth factor. *Brain Research* **386**, 397–9.

Haycock, J.W., White, W.F., McGaugh, J.L. & Cotman, C.W. (1977). Enhanced stimulus-secretion coupling from brains of aged mice. *Experimental Neurology* **57**, 873–82.

Hefti, F. (1983). Alzheimer's disease caused by a lack of nerve growth factor? *Annals of Neurology* **14**, 109–10.

Hefti, F. (1986). Nerve growth factor (NGF) promotes survival of septal cholinergic neurons after fimbrial transactions. *Journal of Neuroscience* **6**, 2155–62.

Hefti, F. & Will, B. (1987). Nerve growth factor is a neurotrophic factor for forebrain cholinergic neurons: implications for Alzheimer's disease. In R.J. Wurtman, S.R. Corkin & J.H. Growden (eds.), *Alzheimer's Disease: Advances in Basic Research and Therapies*, pp. 265–74. Centre for Brain Sciences and Metabolism, Cambridge, Mass.

Hefti, F., Dravid, A. & Hartikka, J. (1984). Chronic intraventricular injections of nerve growth factor elevate hippocampal choline acetyltransferase in adult rats with partial septo-hippocampal lesions. *Brain Research* **293**, 305–11.

Heise, G.A. (1975). Discrete trial analysis of drug action. *Federation Proceedings* **34**, 1898–903.

Heise, G.A. & Milar, K.S. (1984). Drugs and stimulus control. In L.L. Iversen, S.D. Iversen & S.H. Snyder (eds.), *Handbook of Psychopharmacology, Volume 18: Drugs, Neurotransmitters, and Behavior*, pp. 129–90. Plenum Press, New York.

Heise, G.A., Hrabrich, B., Lilie, N.L. & Martin, R.A. (1975). Scopolamine effects on delayed spatial alternation in the rat. *Pharmacology, Biochemistry and Behavior* **3**, 993–1002.

Heise, G.A., O'Connor, R. & Martin, R.A. (1976). Effects of scopolamine on variable intertrial interval spatial alternation and memory in the rat. *Psychopharmacologia* **18**, 38–49.

Henderson, A.S. & Huppert, F.A. (1984). The problem of mild dementia. *Psychological Medicine* **14**, 5–11.

Henke, H. & Lang, W. (1983). Cholinergic enzymes in neocortex, hippocampus and basal forebrain of non-neurological and senile dementia of Alzheimer-type brains. *Brain Research* **267**, 281–91.

Hepler, D.J., Olton, D.S., Wenk, G.L. & Coyle, J.T. (1985*a*). Lesions in nucleus basalis magnocellularis and medial septal area of rats produce qualitatively similar memory impairments. *Journal of Neuroscience* **5**, 866–73.

Hepler, D.J., Wenk, G.L., Cribbs, B.L., Olton, D.S. & Coyle, J.T. (1985*b*) Memory impairments following basal forebrain lesions. *Brain Research* **346**, 8–14.

Hicks, P., Strong, R., Schoolar, J.C. & Samorajski, T. (1980). Aging alters amphetamine-induced stereotyped gnawing and neostriatal elimination of amphetamine in mice. *Life Sciences* **27**, 715–22.

Hinds, J.W. & McNelly, N.A. (1977). Aging of the rat olfactory bulb: growth and atrophy of constituent layers and changes in size and number of mitral cells. *Journal of Comparative Neurology* **171**, 345–68.

Hirschhorn, I.D., Makman, M.H. & Sharpless, N.S. (1982). Dopamine receptor sensitivity following nigrostriatal lesion in the aged rat. *Brain Research* **234**, 357–68.

Hortnagl, H., Potter, P.E. & Hanin, I. (1987). Effect of cholinergic deficit induced by ethylcholine aziridinium (AF64A) on noradrenergic and dopaminergic parameters in rat brain. *Brain Research* **421**, 75–84.

Hubbard, B.M. & Anderson, J.M. (1981). A quantitative study of cerebral atrophy in old age and senile dementia. *Journal of Neurological Science* **50**, 135–45.

Hunter, A.J., Caulfield, M.P. & Kimberlin, R.H. (1986). Learning ability of mice infected with different strains of scrapie. *Physiology and Behavior* **36**, 1089–92.

Hurlburt, B.J., Lubar, J.F., Switzer, R., Dougherty, J. & Eisenstadt, M.L. (1987). Basal forebrain infusion of HC-3 in rats: maze learning deficits and neuropathology. *Physiology and Behavior* **39**, 381–93.

Hyman, B.T., Van Hoesen, G.W., Damasio, A.R. & Barnes, C.L. (1984). Alzheimer's disease: cell specific pathology isolates the hippocampal formation. *Science* **225**, 1168–70.

Ihara, Y., Abraham, C. & Selkoe, D.J. (1983). Antibodies to paired helical filaments in Alzheimer's disease do not recognize normal brain proteins. *Nature* **304**, 727–30.

Ingram, D.K., London, E.D., Reynolds, M.A., Waller, S.B. & Goodrick, C.L. (1981). Differential effects of age on motor performance in two mouse strains. *Neurobiology of Aging* **2**, 221–7.

Inukai, T. (1928). On the loss of Purkinje cells with advancing age from the cerebellar cortex of the albino rat. *Journal of Comparative Neurology* **34**, 1–31.

Isaacs, A.D. (1983). The senium. In M.H. Lader (ed.), *Handbook of Psychiatry, Volume 2: Mental Disorders and Somatic Illness*, pp. 88–97. Cambridge University Press, Cambridge.

Jacobs, R.W. & Butcher, L.L. (1986). Pathology of the basal forebrain in Alzheimer's disease and other dementias. In A.B. Scheibel & A.F. Weschler (eds.), *The Biological Substrates of Alzheimer's Disease*, pp. 87–100. Academic Press, New York.

James, T.C. & Kanungo, M.S. (1976). Alterations in atropine sites of the brain of rats as a function of age. *Biochemistry and Biophysics Research Communications* **72**, 170–5.

Jarrard, L.E., Kant, G.J., Meyerhoff, J.L. & Levy, A. (1984). Behavioral and neurochemical effects of intraventricular AF64A administration in rats. *Pharmacology, Biochemistry and Behavior* **21**, 273–80.

Jarvik, M.E., Goldfarb, T.L. & Carley, J.L. (1969). Influence of interference on delayed matching in monkeys. *Journal of Experimental Psychology* **81**, 1–6.

Johnson, H.A. & Erner, S. (1972). Neuron survival in the aging mouse brain. *Journal of Gerontology* **7**, 111–17.

Johnston, M.V., McKinney, M. & Coyle, J.T. (1979). Evidence for a cholinergic projection to neocortex from neurons in the basal forebrain. *Proceedings of the National Academy of Sciences of the USA* **76**, 5392–6.

Jonec, V. & Finch, C.B. (1975). Ageing and dopamine uptake by subcellular fractions of the C57BL/6J male mouse brain. *Brain Research* **91**, 197–215.

Jonsson, G. (1980). Chemical neurotoxins as denervation tools in neurobiology. *Annual Review of Neuroscience* **3**, 169–87.

Jope, R. (1979). High affinity choline transport and acetyl CoA production in brain and their roles in the regulation of acetylcholine synthesis. *Brain Research Reviews* **1**, 313–44.

Joseph, J.A., Berger, R.E., Engel, B.T. & Roth, G.S. (1978). Age-related changes in the nigrostriatum: a behavioral and biochemical analysis. *Journal of Gerontology* **33**, 643–9.

Joseph, J.A., Filburn, C.R. & Roth, G.S. (1981). Development of dopamine receptor denervation supersensitivity in the neostriatum of the senescent rat. *Life Sciences* **29**, 575–84.

Joyce, J.N., Loeschen, S.K., Sapp, D.W. & Marshall, J.F. (1986). Age-related regional loss of caudate-putamen dopamine receptors revealed by quantitative autoradiography. *Brain Research* **378**, 158–63.

Kang, J., Lemaire, H.-G., Unterbeck, A., Salbaum, J.M., Masters, C.L., Grzeschik, K.-H., Multhaup, G., Beyreuther, K. & Muller-Hill, B. (1987). The precursor of Alzheimer's disease amyloid A4 protein resembles a cell-surface receptor. *Nature* **325**, 733–6.

Kay, D.W.K., Bergmann, K., Foster, E.M., McKechnie, A.A. & Roth, M. (1972). Mental illness and hospital usage in the elderly: a random sample followed up. *Comparative Psychiatry* **11**, 26–35.

Kelly, P.H. & Moore, R.Y. (1978). Decrease of neocortical choline acetyltransferase after lesion of the globus pallidus in the rat. *Experimental Neurology* **61**, 479–84.

Kesner, R.P., Crutcher, K.A. & Measom, M.O. (1986). Medial septal and nucleus basalis magnocellularis lesions produce order memory deficits in rats which mimic symptomatology of Alzheimer's disease. *Neurobiology of Aging* **7**, 287–95.

Kesner, R.P., Adelstein, T. & Crutcher, K.A. (1987a). Rats with nucleus basalis magnocellularis lesions mimic mnemonic symptomatology observed in patients with dementia of the Alzheimer's type. *Behavioral Neuroscience* **101**, 451–6.

Kesner, R.P., DiMattia, B.V. & Crutcher, K.A. (1987b). Evidence for neocortical involvement in reference memory. *Behavioral and Neural Biology* **47**, 40–53.

Khachaturian, Z.S. (1987). Aluminum toxicity among other views on the etiology of Alzheimer's disease. *Neurobiology of Aging* **7**, 537–9.

Kidd, M. (1963). Paired helical filaments in electron microscopy of Alzheimer's disease. *Nature* **197**, 192–3.

Kimberlin, R.H. (1984). Scrapie: the disease and the infectious agent. *Trends in Neuroscience* **7**, 312–16.

King, G.A., De Boni, U. & Crapper, D.R. (1975). Effect of aluminum upon conditioned avoidance response acquisition in the absence of neurofibrillary degeneration. *Pharmacology, Biochemistry and Behavior* **3**, 1003–9.

Kitt, C.A., Price, D.L., Struble, R.G., Cork, L.C., Wainer, B.H., Becher, M.W. & Mobley, W.C. (1984). Evidence for cholinergic neurites in senile plaques. *Science* **226**, 1443–5.

Kitt, C.A., Struble, R.G., Cork, L.C., Mobley, W.C., Walker, L.C., John, T.H. & Price, D.L. (1985). Catecholaminergic neurites in senile plaques in prefrontal cortex of aged nonhuman primates. *Neuroscience* **16**, 691–9.

Kizer, J.S., Nemeroff, C.B. & Youngblood, W.W. (1978). Neurotoxic amino acids and structurally related analogs. *Pharmacological Review* **29**, 301–18.

Klatzo, I., Wisniewski, H. & Streicher, E. (1965). Experimental production of neurofibrillary degeneration. *Journal of Neuropathology and Experimental Neurology* **24**, 187–99.

Knowlton, B.J., Wenk, G.L., Olton, D.S. & Coyle, J.T. (1985). Basal forebrain lesions produce a dissociation of trial dependent and trial-independent memory performance. *Brain Research* **345**, 315–21.

Kopelman, M.D. (1985). Rates of forgetting in Alzheimer-type dementia and Korsakoff's syndrome. *Neuropsychologia* **23**, 623–38.

Kopelman, M.D. (1986). The cholinergic neurotransmitter system in human memory and dementia: a review. *Quarterly Journal of Experimental Psychology* **38A**, 535–73.

Kozlowski, M.R. & Abrogast, R.E. (1986). Specific toxic effects of ethylcholine nitrogen mustard on cholinergic neurons of the nucleus basalis of Meynert. *Brain Research* **372**, 45–54.

Kromer, L.F. (1987). Nerve growth factor treatment after brain injury prevents neuronal death. *Science* **235**, 214–16.

Kromer, L.F., Bjorklund, A. & Stenevi, U. (1981). Innervation of embryonic hippocampal implants by regenerating axons of cholinergic septal neurons in the adult rat. *Brain Research* **210**, 153–71.

Ksir, C. (1974). Scopolamine effects on two-trial delayed-response performance in the rat. *Psychopharmacology* **34**, 127–34.

Ksir, C. & Benson, D.M. (1983). Enhanced behavioral response to nicotine in an animal model of Alzheimer's disease. *Psychopharmacology* **81**, 272–3.

Kubanis, P. & Zornetzer, S.F. (1981). Age-related behavioral and neurobiological changes: a review with an emphasis on memory. *Behavioral and Neural Biology* **31**, 115–72.

Kubanis, P., Gobbel, G. & Zornetzer, S.F. (1981).

Age-related memory deficits in Swiss mice. *Behavioral and Neural Biology* **32**, 241–7.

Kullenbeck, H. (1944). Senile changes in the brain of Wistar Institute rats. *Anatomical Record* **88**, 441.

Lai, H., Bowden, D.M. & Horida, A. (1987). Age-related decreases in dopamine receptors in the caudate nucleus and putamen of the rhesus monkey (*Macaca mulatta*). *Neurobiology of Aging* **8**, 45–9.

Landfield, P.W. (1983). Mechanisms of altered neural function during aging. In W.H. Gispen & J. Traber (eds.), *Aging of the Brain*, pp. 51–71. Elsevier, Amsterdam.

Landfield, P.W., Waymire, J.C. & Lynch, G. (1978). Hippocampal aging and adrenocorticoids: quantitative correlations. *Science* **202**, 1098–101.

Landfield, P.W., Braun, L.D., Pitler, T.A., Lindsey, J.D. & Lynch, G. (1981). Hippocampal aging in rats: a morphometric study of multiple variables in semithin sections. *Neurobiology of Aging* **2**, 265–75.

Lehman, J., Nagy, J.I., Atmadja, S. & Fibiger, H.C. (1980). The nucleus basalis magnocellularis: the origin of a cholinergic projection to the neocortex of the rat. *Neuroscience* **5**, 1161–74.

Lerer, B., Warner, J., Friedman, E., Vincent, G. & Gamzu, E. (1985). Cortical cholinergic impairment and behavioral deficits produced by kainic acid lesions of rat magnocellular basal forebrain. *Behavioral Neuroscience* **99**, 661–77.

Leslie, F.M., Loughlin, S.E., Sternberg, D.B., McGaugh, J.L., Young, L.E. & Zornetzer, S.F. (1985). Noradrenergic changes and memory loss in aged mice. *Brain Research* **359**, 292–9.

Levine, M.S., Lloyd, R.L., Fisher, R.S., Hull, C.D. & Buchwald, N.A. (1987). Sensory, motor and cognitive alterations in aged cats. *Neurobiology of Aging* **8**, 253–63.

Levy, A., Kant, G.J., Meyerhoff, J.L. & Jarrard, L.E. (1984). Non-cholinergic neurotoxic effects of AF64A in the substantia nigra. *Brain Research* **305**, 169–72.

Ling, E.-A & Leblond, C.P. (1973). Investigation of glial cells in semithin sections. II. Variation with age in the numbers of the various glial cell types in rat cortex and corpus callosum. *Journal of Comparative Neurology* **149**, 73–81.

Lippa, A.S., Pelham, R.W., Beer, B., Critchett, D.J., Dean, R.L. & Bartus, R.T, (1980). Brain cholinergic dysfunction and memory in aged rats. *Neurobiology of Aging* **1**, 13–19.

Liu, S.Y. (1928). The relation of age to the learning ability of the white rat. *Journal of Comparative Psychology* **8**, 75–85.

LoConte, G., Bartolini, L., Casamenti, F., Marconcini-Pepeu, J. & Pepeu, G. (1982*a*). Lesions of cholinergic forebrain nuclei: changes in

avoidance behavior and scopolamine actions. *Pharmacology, Biochemistry and Behavior* **17**, 933–7.

LoConte, G., Casamenti, F., Milaneschi, E. & Pepeu, G. (1982*b*). Effects of magnocellular forebrain nuclei lesions on acetylcholine output from the cerebral cortex, electrocorticogram and behaviour. *Archives of Italian Biology* **120**, 176–87.

London, E.D., Nespor, S.M., Ohata, M. & Rapoport, S.I. (1981). Local cerebral glucose utilization during development and aging of the Fischer 344 rat. *Journal of Neurochemistry* **37**, 217–21.

Lowy, A.M., Ingram, D.K., Olton, D.S., Waller, S.B., Reynolds, M.A. & London, E.D. (1985). Discrimination learning requiring different memory components in rats: age and neurochemical comparisons. *Behavioral Neuroscience* **99**, 638–51.

Maggi, A., Schmidt, M.J., Ghetti, B. & Enna, S.J. (1979). Effect of aging on neurotransmitter receptor binding in rat and human brain. *Life Sciences* **24**, 367–74.

Mann, D.M.A. (1985). the neuropathology of Alzheimer's disease: a review with pathogenetic, aetiological and therapeutic considerations. *Mechanisms of Ageing and Development* **31**, 213–55.

Mann, D.M.A., Yates, P.O. & Marcyniuk, B. (1985*a*). Correlation between senile plaque and neurofibrillary tangle counts in cerebral cortex and neuronal counts in cortex and subcortical structures in Alzheimer's disease. *Neuroscience Letters* **56**, 51–5.

Mann, D.M.A., Yates, P.O. & Marcyniuk, B. (1986*b*). Some morphometric observations on the cerebral cortex and hippocampus in presenile Alzheimer's disease, senile dementia of Alzheimer type and Down's syndrome in middle age. *Journal of Neurological Science* **69**, 139–59.

Mantione, C.R., Fisher, A. & Hanin, I. (1981). The AF64A-treated mouse: possible model for central cholinergic hypofunction. *Science* **213**, 579–80.

Marchi, M., Paudice, P. & Raiter, M. (1981). Autoregulation of acetylcholine release in isolated hippocampal nerve endings. *European Journal of Pharmacology* **73**, 75–9.

Marycyniuk, B., Mann, D.M.A. & Yates, P.O. (1986). The topography of cell loss from locus coeruleus in Alzheimer's disease. *Journal of Neurological Science* **76**, 335–45.

Marshall, J.F. & Berrios, N. (1979). Movement disorders of aged rats: reversal by dopamine receptor stimulation. *Science* **206**, 477–9.

Masters, C.L. & Beyreuther, K. (1987). Neuronal origin of cerebral amyloidogenic proteins: their role in Alzheimer's disease and unconventional virus diseases of the nervous system. In *Ciba*

Foundation Symposium Volume 126: Selective Neuronal Death, pp. 49–64. Wiley, Chichester.

McDermott, J.R., Fraser, H. & Dickinson, A.G. (1978). Reduced cholineacetyltransferase activity in scrapie mouse brain. *Lancet* **ii**, 318–19.

McDermott, J.R., Smith, I., Iqbal, K. & Wisniewski, H.M. (1979). Brain aluminum in aging and Alzheimer's disease. *Neurology* **29**, 809–14.

McFarland, D.J. & Hotchin, J. (1980). Early behavioral abnormalities in mice due to scrapie virus encephalopathy. *Biological Psychiatry* **15**, 37–44.

McFarland, D.J., Baker, F.D. & Hotchin, J. (1980). Host and viral genetic determinants of the behavioral effects of scrapie encephalopathy. *Physiology and Behavior* **24**, 911–14.

McGeer, E.G. (1981). Neurotransmitter systems in aging and senile dementia. *Progress in Neurobiology* **8**, 111–19.

McGeer, E.G., Fibiger, H.C., McGeer, P.L. & Wickson, V. (1971). Aging and brain enzymes. *Experimental Gerontology* **6**, 391–6.

McGeer, E.G., Olney, J.W. & McGeer, P.L. (eds.) (1978). *Kainic Acid as a Tool in Neurobiology*. Raven Press, New York.

McGurk, S.R., Hartgraves, S.L., Kelly, P.H., Gordon, M.N. & Butcher, L.L. (1987). Is ethylcholine mustard aziridinium ion a specific cholinergic neurotoxin? *Neuroscience* **22**, 215–24.

McIntosh, H.H. & Westfall, T.C. (1987). Influence of aging on catecholamine levels, accumulation, and release in F-344 rats. *Neurobiology of Aging* **8**, 233–9.

McKinney, M., Coyle, J.T. & Hedreen, J.C. (1983). Topographic analysis of the innervation of the rat neocortex and hippocampus by the basal forebrain cholinergic system. *Journal of Comparative Neurology* **217**, 103–21.

McLennan, H. (1983). Receptors for the excitatory amino acids in the mammalian central nervous system. *Progress in Neurobiology* **20**, 251–71.

McMenemy, W.H. (1963). Aging and the dementias. In W. Blackwood, W.H. McMenemy, A. Meyer, R.M. Norman & D.S. Russell (eds.), *Greenfield's Neuropathology*, 2nd edn, Chapter 9. Arnold, London.

McNamara, M.C., Benignus, V.A., Benignus, G. & Miller, A.T. (1977). Active and passive avoidance in rats as a function of age. *Experimental Aging Research* **3**, 3–16.

Meck, W.H., Church, R.M., Wenk, G.L. & Olton, D.S. (1987). Nucleus basalis magnocellularis and medial septal area lesions differentially impair temporal memory. *Journal of Neuroscience* **7**, 3505–11.

Meyer, E.M., St. Onge, E. & Crews, F.T. (1984). Effects of aging on rat cortical presynaptic cholinergic processes. *Neurobiology of Aging* **5**, 315–17.

Meyers, B. (1965). Some effects of scopolamine on a passive avoidance response in rats. *Psychopharmacology* **8**, 111–19.

Milar, K.S., Halgren, C.R. & Heise, G.A. (1978). A reappraisal of scopolamine effects on inhibition. *Pharmacology, Biochemistry and Behavior* **9**, 307–13.

Miller, A.K.H., Alston, R.L. & Corsellis, J.A.N. (1980). Variation with age in the volumes of grey and white matter in the cerebral hemispheres of man: measurements with an image analyser. *Neuropathology and Applied Neurobiology* **6**, 119–32.

Misra, C.H., Shelat, H.S. & Smith, R.C. (1980). Effect of age on adrenergic and dopaminergic receptor binding in rat brain. *Life Sciences* **27**, 521–6.

Mitchell, S.J., Rawlins, J.N.P., Steward, O. & Olton, D.S. (1982). Medial septal area lesions disrupt theta rhythm and cholinergic staining in medial entorhinal cortex and produce impaired radial arm maze behavior in rats. *Journal of Neuroscience* **2**, 292–302.

Miyamoto, M., Shintani, M., Nagaoka, A. & Nagawa, Y. (1985). Lesioning of the rat basal forebrain leads to memory impairments in passive and active avoidance tasks. *Brain Research* **328**, 97–104.

Miyamoto, M., Kato, J., Narumi, S. & Nagawa, Y. (1987). Characteristics of memory impairment following lesioning of the basal forebrain and medial septal nuclei in rats. *Brain Research* **419**, 19–31.

Moore, J.W., Goodall, N.A. & Solomon, P.R. (1976). Central cholinergic blockade by scopolamine and habituation, classical conditioning, and latent inhibition of the rabbit's nictitating membrane response. *Physiological Psychology* **4**, 395–9.

Morgan, D.G. (1987). The dopamine and serotonin systems during aging in human and rodent brain. A brief review. *Progress in Neuro-Psychopharmacology and Biological Psychiatry* **11**, 153–7.

Morris, R.G. (1984). Dementia and the functioning of the articulatory loop system. *Cognitive Neuropsychology* **1**, 143–57.

Morris, R.G. & Kopelman, M.D. (1986). The memory deficits in Alzheimer-type dementia: a review. *Quarterly Journal of Experimental Psychology* **38A**, 575–602.

Morris, R.G., Evenden, J.L., Sahakain, B.J. & Robbins, T.W. (1987). Computer-aided assessment of dementia: comparative studies of neuropsychological deficits in Alzheimer-type dementia and Parkinson's disease. In S.M. Stahl, S.D. Iversen & E.C. Goodman (eds.), *Cognitive*

Neurochemistry, pp. 21–36. Oxford University Press, Oxford.

Morris, R.G.M. (1981). Spatial localisation does not require the presence of local cues. *Learning and Motivation* **12**, 239–49.

Morrison, J.H., Rogers, J., Scherr, S., Benoit, R. & Bloom, F.E. (1985). Neuritic plaques in Alzheimer's patients contain somatostatin immunoreactivity. *Nature* **314**, 90–2.

Moscovitch, M. (1982). A neuropsychological approach to perception and memory in normal and pathological aging. In F.I.M. Craik & S. Trehub (eds.), *Aging and Cognitive processes*, pp. 55–78. Plenum, New York.

Mountjoy, C.Q., Roth, M., Evans, N.J.R. & Evans, H.M. (1983). Cortical neuronal counts in normal elderly controls and demented patients. *Neurobiology of Aging* **4**, 1–11.

Mountjoy, C.Q., Rossor, M.N., Iversen, L.L. & Roth, M. (1984). Correlation of cortical cholinergic and GABA deficits with quantitative neuropathological findings in senile dementia. *Brain* **107**, 507–18.

Mufson, E.J. & Stein, D.G. (1980). Behavioral and morphological aspects of aging: an analysis of rat frontal cortex. In D.G. Stein (ed.), *The Psychobiology of Aging: Problems and Perspectives*, pp. 99–125. Elsevier/North Holland, New York.

Mufson, E.J., Kehr, A.D., Wainer, B.H. & Mesulam, M.-M. (1987). Cortical effects of neurotoxic damage to the nucleus basalis in rats: persistent loss of extrinsic cholinergic input and lack of transsynaptic effect upon the number of somatostatin-containing, cholinesterase-positive, and cholinergic cortical neurons. *Brain Research* **417**, 385–8.

Mulder, D.W. (1975). Organic brain syndromes associated with diseases of unknown cause. In A.M. Freedman, H.I. Kaplan & B.J. Sadock (eds.), *Comprehensive Testbook of Psychiatry*, 2nd edn, pp. 1086–93. Williams & Wilkins, Baltimore.

Munoz-Garcia, D., Pendlebury, W.W., Kessler, J.B. & Perl, D.P. (1986). An immunocytochemical comparison of cytoskeletal proteins in aluminum-induced and Alzheimer-type neurofibrillary tangles. *Acta Neuropathologica* **70**, 243–8.

Muramoto, O., Sugishita, M. & Ando, K. (1984). Cholinergic system and constructional apraxia: a further study of physostigmine in Alzheimer's disease. *Journal of Neurology, Neurosurgery and Psychiatry* **47**, 485–91.

Murray, C.L. & Fibiger, H.C. (1985). Learning and memory deficits after lesions of the nucleus basalis magnocellularis: reversal by physostigmine. *Neuroscience* **14**, 1025–32.

Murray, C.L. & Fibiger, H.C. (1986). Pilocarpine and physostigmine attenuate spatial memory

impairments produced by lesions of the nucleus basalis magnocellularis. *Behavioral Neuroscience* **100**, 23–32.

Nadler, J.V., Perry, B.W. & Cotman, C.W. (1978). Preferential vulnerability of hippocampus to intraventricular kainic acid. In E.G. McGeer, J.W. Olney & P.L. McGeer (eds.), *Kainic Acid as a Tool in Neurobiology*, pp. 219–38. Raven Press, New York.

Nagai, R., McGeer, P.L., Peng, J.H., McGeer, E.G. & Dolman, C.E. (1983). Choline acetyltransferase immunohistochemistry in brains of Alzheimer's disease patients and controls. *Neuroscience Letters* **36**, 195–9.

Norberg, A. & Winblad, B. (1981). Cholinergic receptors in human hippocampus – regional distribution and variance with age. *Life Sciences* **29**, 1937–44.

O'Keefe, J. & Nadel, L. (1978). *The Hippocampus as a Cognitive Map*, Oxford University Press, London.

Olton, D.S. (1983). Memory functions and the hippocampus. In W. Seifert (ed.), *Neurobiology of the Hippocampus*, pp. 335–73. Academic Press, New York.

Olton, D.S. & Samuelson, R.J. (1976). Remembrance of places passed: spatial memory in rats. *Journal of Experimental Psychology: Animal Behavior Processes* **2**, 97–116.

Olton, D.S., Walker, J.A. & Gage, F.H. (1978). Hippocampal connections and spatial discrimination. *Brain Research* **139**, 295–308.

Osterburg, H.H., Donahue, H.G., Severson, J.A. & Finch, C.E. (1981). Catecholamine levels and turnover during aging in brain regions of male C57BL/6J mice. *Brain Research* **224**, 337–52.

Ostfeld, A.M. & Arguette, A. (1962). Central nervous system effects of hyoscine in man. *Journal of Pharmacology and Experimental Therapeutics* **137**, 133–9.

Outram, G.W. (1972). Changes in drinking and feeding habits of mice with experimental scrapie. *Journal of Comparative Pathology* **82**, 415–27.

Palacios, J.M. (1982). Autoradiographic localization of muscarinic cholinergic receptors in the hippocampus of patients with senile dementia. *Brain Research* **243**, 173–5.

Palmer, A.M., Wilcock, G.K., Esiri, M.M., Francis, P.T. & Bowen, D.M. (1987). Monoaminergic innervation of the frontal and temporal lobes in Alzheimer's disease. *Brain Research* **401**, 231–8.

Pare, W.P. (1969). Age, sex and strain differences in the aversive threshold to grid shock in the rat. *Journal of Comparative and Physiological Psychology* **69**, 214–18.

Pearson, R.C.A. & Powell, T.P.S. (1987). Anterograde vs retrograde degeneration of the nucleus basalis medialis in Alzheimer's disease. In R.J. Wurtman,

S. Corkin & J.H. Growden (eds.), *Alzheimer's Disease: Advances in Basic Research and Therapies*, pp. 123–33. Center for Brain Sciences and Metabolism Trust, Cambridge. Mass.

Pearson, R.C.A., Gatter, K.C. & Powell, T.P.A. (1983*a*). Retrograde cell degeneration in the basal nucleus in monkey and man. *Brain Research* **261**, 321–6.

Pearson, R.C.A., Sofroniew, M.V., Cuello, A.C., Powell, T.P.S., Eckenstein, F., Esiri, M.M. & Wilcock, G.K. (1983*b*). Persistence of cholinergic neurons in the basal nucleus in a brain with senile dementia of the Alzheimer's type demonstrated with immunohistochemical staining for choline acetyltransferase. *Brain Research* **289**, 375–9.

Pearson, R.C.A., Esiri, M.M., Hiorns, R.W., Wilcock, G.K. & Powell, T.P.S. (1985). Anatomical correlates of the distribution of the pathological changes in neocortex in Alzheimer's disease. *Proceedings of the National Academy of Sciences of the USA* **82**, 4531–4.

Pendlebury, W.W., Beal, M.F., Kowall, N.W. & Solomon, P.R. (1987). Results of immunocytochemical, neurochemical, and behavioral studies in aluminum-induced neurofilamentous degeneration. In R.J. Wurtman, S.R. Corkin & J.H. Growden (eds.), *Alzheimer's Disease: Advances in Basic Research and Therapies*, pp. 529–33. Center for Brain Sciences and Metabolism, Cambridge, Mass.

Peng, M.T. & Lee, L.R. (1979). Regional differences of neuron loss of rat brain in old age. *Gerontology* **25**, 205–11.

Pepeu, G., Casamenti, F., Bracco, L., Ladinsky, H. & Consolo, S. (1985). Lesions of nucleus basalis in the rat: functional changes. In J. Traber & W.H. Gispen (eds.), *Senile Dementia of the Alzheimer Type*, pp. 305–15. Springer-Verlag, Berlin.

Perl, D.P. & Brody, A.R. (1980). Alzheimer's disease: X-ray spectrometric evidence of aluminum accumulation in neurofibrillary tangle-bearing neurons. *Science* **208**, 297–9.

Perret, E. & Birri, R. (1982). Aging, performance decrements, and differential cerebral involvement. In S. Corkin, K.L. Davis, J.H. Growden, E. Usdin & R.J. Wurtman (eds.), *Alzheimer's disease: A Report of Progress in Research*, pp. 133–9, Raven, New York.

Perry, E.K., Perry, R.H., Blessed, G. & Tomlinson, B.E. (1977). Necropsy evidence of central cholinergic deficits in senile dementia. *Lancet* **i**, 189.

Perry, E.K., Tomlinson, B.E., Blessed, G., Bergmann, K., Gibson, P.H. & Perry, R.H. (1978). Correlation of cholinergic abnormalities with senile plaques and mental test scores in senile dementia. *British Medical Journal* **ii**, 1457–9.

Perry, E.K., Cockray, G.J., Dimaline, R., Perry, R.H., Blessed, G. & Tomlinson, B.E. (1981). Neuropeptides in Alzheimer's disease, depression and schizophrenia: a post mortem analysis of vasoactive intestinal polypeptide and cholecystokinin in cerebral cortex. *Journal of Neurological Science* **51**, 465–72.

Petersen, R.C. (1977). Scopolamine induces learning failures in man. *Psychopharmacology* **52**, 283–9.

Petit, T.L., Biederman, G.B. & McMullen, P.A. (1980). Neurofibrillary degeneration, dendritic dying back, and learning-memory deficits after aluminum administration: implications for brain aging. *Experimental Neurology* **67**, 152–62.

Ponzio, F., Brunello, N. & Algeri, S. (1978). Catecholamine synthesis in brain of ageing rats. *Journal of Neurochemistry* **30**, 1617–20.

Ponzio, F., Calderini, G., Lomuscio, G., Vantani, G. & Toffano, G. (1982). Changes in monoamines and their metabolite levels in brain regions of aged rats. *Neurobiology of Aging* **3**, 23–9.

Pope, C.N., Englert, L.F. & Ho, B.T. (1985). Passive avoidance deficits in mice following ethylcholine aziridinium chloride treatment. *Pharmacology, Biochemistry and Behavior* **22**, 297–9.

Pradhan, S.N. (1980). Central neurotransmitters and aging. *Life Sciences* **26**, 1643–56.

Price, D.L., Cork, L.C., Struble, R.G., Kitt, C.A., Walker, L.C., Powers, R.E., Whitehouse, P.J. & Griffin, J.W. (1987). Dysfunction and death of neurons in human degenerative neurological diseases and in animal models. In *Ciba Foundation Symposium Volume 126: Selective Neuronal Death*, pp. 30–43. Wiley, Chichester.

Rabe, A., Lee, M.H., Shek, J. & Wisniewski, H.M. (1982). Learning deficit in immature rabbits with aluminum-induced neurofibrillary changes. *Experimental Neurology* **76**, 441–6.

Randall, P.K. (1980). Functional aging of the nigro-striatal system. *Peptides* **1**, Suppl. 1, 177–84.

Randall, P.K., Severson, J.A. & Finch, C.E. (1981). Aging and the regulation of striatal dopaminergic mechanisms in mice. *Journal of Pharmacology and Experimental Therapeuts* **219**, 675–700.

Rapp, P.R., Rosenberg, R.A. & Gallagher, M. (1987). An evaluation of spatial information processing in aged rats. *Behavioral Neuroscience* **101**, 3–12.

Reinikainen, K.J., Rieffinen, P.J., Halonen, T. & Laakso, M. (1987). Decreased muscarinic receptor binding in cerebral cortex and hippocampus in Alzheimer's disease. *Life Sciences* **41**, 453–61.

Reis, D.J., Ross, R.A. & Joh, T.H. (1977). Changes in the activity and amounts of enzymes synthesizing catecholamines and acetylcholine in brain, adrenal medulla and sympathetic ganglia of aged rats and mice. *Brain Research* **136**, 465–74.

Reisine, T.D., Yamamura, H.I., Bird, E.D., Spokes, E.

& Enna, S.J. (1978). Pre- and post-synaptic neurochemical alterations in Alzheimer's disease. *Brain Research* **159**, 477–81.

Richardson, R.T. & DeLong, M.R. (1988). A reappraisal of the functions of the nucleus basalis of Meynert. *Trends in Neuroscience* **11**, 264–7.

Ridley, R.M., Bowes, P.M., Baker, H.F. & Crow, T.J. (1984). An involvement of acetylcholine in object discrimination learning and memory in the marmoset. *Neuropsychologia* **22**, 253–63.

Ridley, R.M., Baker, H.F. & Drewett, B. (1987). Effects of arecoline and pilocarpine on learning ability in marmosets pretreated with hemicholinium-3. *Psychopharmacology* **91**, 512–14.

Rinne, J.O., Laakso, K., Lonnberg, P., Molsa, P., Paljarvi, L., Rinne, J.K., Sako, E. & Rinne, U.K. (1977). Brain muscarinic receptors in senile dementia. *Brain Research* **336**, 19–25.

Rochford, G. (1971). A study of naming errors in dysphasic and in demented patients. *Neuropsychologia* **9**, 437–43.

Rogers, J. & Morrison, J.H. (1985). Quantitative morphology and regional and laminar distributions of senile plaques in Alzheimer's disease. *Journal of Neuroscience* **5**, 2801–8.

Rossor, M.N. (1982). Neurotransmitters and CNS disease: dementia. *Lancet* **ii**, 1200–4.

Rossor, M.N. & Iversen, L.L. (1986). Non-cholinergic neurotransmitter abnormalities in Alzheimer's disease. *British Medical Bulletin* **42**, 70–4.

Rossor, M.N., Iversen, L.L., Johnson, A.J., Mountjoy, C.Q. & Roth, M. (1981). Cholinergic deficit in frontal cerebral cortex in Alzheimer's disease is age dependent. *Lancet* **ii**, 1422.

Rossor, M.N., Emson, P.C., Mountjoy, C.Q., Roth, M. & Iversen, L.L. (1982a). Reduced amounts of immunoreactive somatostatin in temporal cortex in senile dementia of Alzheimer type. *Neuroscience Letters* **20**, 373–7.

Rossor, M.N., Svendsen, C., Hunt, S.P., Mountjoy, C.Q., Roth, M. & Iversen, L.L. (1982b). The substantia innominata in Alzheimer's disease: an histochemical and biochemical study of cholinergic marker enzymes. *Neuroscience Letters* **28**, 217–22.

Rosser, M.N., Iversen, L.L., Reynolds, G.P., Mountjoy, C.Q. & Roth, M. (1984). Neurochemical deficit in early and late onset types of Alzheimer's disease is age dependent. *British Medical Journal* **288**, 361–4.

Roth, M. (1986). The association of clinical and neurobiological findings and its bearing on the classification and aetiology of Alzheimer's disease. *British Medical Bulletin* **42**, 42–50.

Roth, M., Tomlinson, B.E. & Blessed, G. (1967). The relationship between qualitative measures of dementia and of degenerative changes in the cerebral grey matter of elderly subjects. *Journal of the Royal Society of Medicine* **60**, 254–8.

Royal College of Physicians (1981). Organic mental impairment in the elderly: implications for research, education and the provision of services. *Journal of the Royal College of Physicians of London* **15**, 141–67.

Sabel, B.A. & Stein, D.G. (1981). Extensive loss of subcortical neurons in the aging brain. *Experimental Neurology* **73**, 507–16.

Sahakian, B.J., Joyce, E. & Lishman, W.A. (1987). Cholinergic effects on constructional abilities and on mnemonic processes. *Psychological Medicine* **17**, 329–33.

Sahakian, B.J., Morris, R.G., Evenden, J.L., Heald, A., Levy, R., Philpot, M. & Robbins, T.W. (1988). A comparative study of visuospatial memory and learning in Alzheimer-type dementia and Parkinson's disease. *Brain* **111**, 695–718.

Salamone, J.D., Beart, P.M., Alpert, J.E. & Iversen, S.D. (1984). Impairment in T-maze reinforced alternation performance following nucleus basalis magnocellularis lesions in rats. *Behavioural Brain Research* **13**, 63–70.

Sandberg, K., Sandberg, P.R. & Coyle, J.T. (1984a). Effects of intrastriatal injections of the cholinergic neurotoxin AF64A on spontaneous nocturnal locomotor behavior in the rat. *Brain Research* **299**, 339–43.

Sandberg, K., Hanin, I., Fisher, A. & Coyle, J.T. (1984b). Selective cholinergic neurotoxin: AF64A's effects in rat striatum. *Brain Research* **293**, 49–55.

Saper, C.B., German, D.C. & White, C.L. (1985). Neuronal pathology in the nucleus basilis and associated cell groups in senile dementia of the Alzheimer's type. *Neurology* **35**, 1089–95.

Sarter, M. (1987). Animal models of brain ageing and dementia. *Comparative Gerontology* **1**, 4–15.

Savage, R.D. & Field, E.J. (1965). Brain damage and emotional behaviour: the effects of scrapie on the emotional responses of mice. *Animal Behavior* **13**, 443–6.

Schlotterer, G., Moscovitch, M. & Crapper-McLachlan, D. (1983). Visual processing deficits as assessed by spatial frequency contrast sensitivity and backward masking in normal ageing and Alzheimer's disease. *Brain* **107**, 309–23.

Schmidt, M.J. & Thornberry, J.F. (1978). Cyclic AMP and cyclic GMP accumulation in vitro in brain regions of young, old, and aged rats. *Brain Research* **139**, 169–77.

Schultz, W. (1982). Depletion of dopamine in the striatum as an experimental model of parkinsonism: direct effects and adaptive mechanisms. *Progress in Neurobiology* **18**, 121–66.

Schwab, M.E., Otten, U., Agid, Y. & Thoenen, H. (1979). Nerve growth factor (NGF) in the rat

CNS: absence of specific retrograde axonal transport and tyrosine hydroxylase induction in locus coeruleus and substantia nigra. *Brain Research* **168**, 473–83.

Seiler, M. & Schwab, M.E. (1984). Specific retrograde transport of nerve growth factor (NGF) from neocortex to nucleus basalis in the rat. *Brain Research* **300**, 473–83.

Selkoe, D.J., Liem, R.K.H., Yen, S.-H. & Shelanski, M.L. (1979). Biochemical and immunological characterization of neurofilaments in experimental neurofibrillary degeneration induced by aluminum. *Brain Research* **163**, 235–52.

Selzer, B. & Sherwin, I. (1983). A comparison of clinical features in early and late onset primary degenerative dementia. *Archives of Neurology* **40**, 143–6.

Semple, S.A., Smith, C.M. & Swash, M. (1982). The Alzheimer disease syndrome. In S. Corkin, K.L. Davis, J.H. Growden, E. Usdin & R.J. Wurtman (eds.), *Alzheimer's disease: A Report of Progress in Research*, pp. 93–107. Raven, New York.

Severson, J.A. & Finch, C.E. (1980). Reduced dopaminergic binding during aging in the rodent striatum. *Brain Research* **192**, 147–62.

Sherman, K.A., Kuster, J.E., Dean, R.L., Bartus, R.T. & Friedman, E. (1981). Presynaptic cholinergic mechanisms in brain of aged rats with memory impairments. *Neurobiology of Aging* **2**, 99–104.

Shore, D. & Wyatt, R.J. (1983). Aluminum and Alzheimer's disease. *Journal of Nervous and Mental Disease* **171**, 553–8.

Simpkins, J.W., Mueller, C.P., Huang, H.H. & Meites, J. (1977). Evidence for depressed catecholamine and enhanced serotonin metabolism in aging male rats: possible relation to gonadotropin secretion. *Endocrinology* **100**, 1672–8.

Sitaram, N., Weingartner, H. & Gillin, J.C. (1978). Human serial learning: enhancement with arecoline and choline and impairment with scopolamine. *Science* **201**, 274–6.

Sjogren, J. (1952). Clinical aspects of morbus Alzheimer and morbus Pick. *Acta Psychiatrica Neurologica Scandinavica, Supplementum* **82**, 69–115.

Smith, L.C. & Dugal, L.P. (1965). Age and spontaneous running activity of male mice. *Canadian Journal of Physiology and Pharmacology* **43**, 852–6.

Sofroniew, M.V. & Pearson, R.C.A. (1985). Degeneration of cholinergic neurons in the basal nucleus following kainic acid or N-methyl-D-aspartic acid application to the cerebral cortex in the rat. *Brain Research* **339**, 186–90.

Sofroniew, M.V., Pearson, R.C.A., Eckenstein, F., Cuello, A.C. & Powell, T.P.S. (1983). Retrograde

changes in cholinergic neurons in the basal nucleus of the forebrain of the rat following cortical damage. *Brain Research* **289**, 370–4.

Sofroniew, M.V., Isacson, O. & Bjorklund, A. (1986a). Cortical grafts prevent atrophy of cholinergic basal nucleus neurons induced by excitotoxic cortical damage. *Brain Research* **378**, 409–15.

Sofroniew, M.V., Pearson, R.C.A., Cuello, A.C., Tagari, P.C. & Stephens, P.H. (1986b). Parenterally administered GM$_1$ ganglioside prevents retrograde degeneration of cholinergic cells of the rat basal forebrain. *Brain Research* **398**, 393–6.

Spencer, D.G. & Lal, H. (1983). Effects of anticholinergic drugs on learning and memory. *Drug Development Research* **3**, 489–502.

Stephens, P.H., Cuello, A.C., Sofroniew, M.V., Pearson, R.C.A. & Tagari, P.C. (1986). Effects of unilateral decortication on choline acetyltransferase activity in the nucleus basalis and other areas of the rat brain. *Journal of Neurochemistry* **45**, 1021–6.

Stephens, P.H., Tagari, P.C., Garofalo, L., Maysinger, D., Piotte, M. & Cuello, A.C. (1987). Neural plasticity of basal forebrain cholinergic neurons: effects of gangliosides. *Neuroscience Letters* **80**, 80–4.

Stone, C.P. (1929). The age factor in animal learning: II. Rats on a multiple light discrimination box and a difficult maze. *Genetic Psychology Monogographs* **6**, 125–201.

Strong, R., Hicks, P., Hsu, L., Bartus, R.T. & Enna, S.J. (1980). Age-related alterations in the rodent brain cholinergic system and behavior. *Neurobiology of Aging* **1**, 59–63.

Struble, R.G., Cork, L.C., Whitehouse, P.J. & Price, D.L. (1982). Cholinergic innervation of senile plaques. *Science* **216**, 413–15.

Struble, R.G., Kitt, C.A., Walker, L.C., Cork, L.C. & Price, D.L. (1984). Somatostatinergic neurites in senile plaques of aged non-human primates. *Brain Research* **324**, 394–6.

Struble, R.G., Price, D.L. Jr., Cork, L.C. & Price, D.L. (1985). Senile plaques in cortex of aged normal monkeys. *Brain Research* **361**, 261–75.

Stwertka, S.A. & Olson, G.L. (1986). Neuropathology and amphetamine-induced turning resulting from AF64A injections into the striatum of the rat. *Life Sciences* **38**, 1105–10.

Suckling, A.J., Bateman, S., Waldron, C.B., Webb, H.E. & Kimberlin, R.H. (1976). Motor activity changes in scrapie-affected mice. *British Journal of Experimental Pathology* **57**, 742–6.

Suits, E. & Isaacson, R.L. (1968). The effects of scopolamine hydrobromide on one-way and two-way avoidance learning in rats. *International Journal of Neuropharmacology* **7**, 441–6.

Tauchi, H., Yoshioka, T. & Kobayashi, H. (1971). Age change of skeletal muscles of rats. *Gerontology* **17**, 219–27.

Terry, R.D. & Davies, P. (1983). Some morphologic and biochemical aspects of Alzheimer's disease. In D. Samuel, S. Algeri, S. Gershon, V.E. Grimm & G. Toffano (eds.), *Aging of the Brain (Aging Volume 22)*, pp. 47–59. Raven Press, New York.

Terry, R.D. & Katzman, R. (1983). Senile dementia of the Alzheimer type. *Annals of Neurology* **14**, 497–506.

Terry, R.D. & Pena, C. (1965). Experimental production of neurofibrillary degeneration. II. Electron microscopy, phosphate histochemistry and electron probe analysis. *Journal of Neuropathology and Experimental Neurology* **24**, 200–10.

Terry, R.D., Peck, A., Deteresa, T., Schechter, R. & Haroupian, D.S. (1981). Some morphometric aspects of the brain in senile dementia of the Alzheimer type. *Annals of Neurology* **10**, 184–92.

Thompson, C.I. & Fitzsimons, T.R. (1976). Age differences in aversively motivated visual discrimination learning and retention in male Sprague-Dawley rats. *Journal of Gerontology* **31**, 47–52.

Thorne, B.M., Donohoe, T., Lin, K.-N., Lyon, S., Medeiros, D.M. & Weaver, M.L. (1986). Aluminum ingestion and behavior in the Long-Evans rat. *Physiology and Behavior* **36**, 63–7.

Tomlinson, B.E. & Kitchener, D. (1972). Granulovacuolar degeneration of hippocampal pyramidal cells. *Journal of Pathology* **106**, 165–85.

Tomlinson, B.E., Blessed, G. & Roth, M. (1968). Observations on the brains of non-demented old people. *Journal of Neurological Science* **7**, 331–56.

Tomlinson, B.E., Blessed, G. & Roth, M. (1970). Observations on the brains of demented old people. *Journal of Neurological Science* **11**, 205–42.

Tomlinson, B.E., Irving, D. & Blessed, G. (1981). Cell loss in the locus coeruleus in senile dementia of Alzheimer type. *Journal of Neurological Science* **49**, 419–28.

Tomonaga, M. (1974). Ultrastructure of Hirano bodies. *Acta Neuropathologica* **30**, 365–6.

Trabucchi, M., Spano, P.F., Govoni, S., Riccardi, F. & Bosio, A. (1982). Dopaminergic function during aging in rat brain. In E. Giacobini, G. Filogamo, G. Giacobini & A. Vernadakis (eds.), *The Aging Brain: Cellular and Molecular Mechanisms of Aging in the Nervous System (Aging Volume 20)*, pp. 195–201. Raven Press, New York.

Trapp, G.A., Miner, G.D., Zimmerman, R.L., Mastri, A.R. & Heston, L.L. (1978). Aluminum levels in brain in Alzheimer's disease. *Biological Psychiatry* **13**, 709–18.

Tuffery, A.R. (1971). Growth and degeneration of motor end-plates in normal cat hind limb muscle. *Journal of Anatomy* **110**, 221–47.

Tulving, E. (1972). Episodic and semantic memory. In E. Tulving & W.D. Donaldson (eds.), *Organization of Memory*, pp. 381–403. Academic Press, New York.

Vaughan, D.W. (1976). Membranous bodies in the cerebral cortex of aging rats: an electron microscope study. *Journal of Neuropathology and Experimental Neurology* **35**, 152–66.

Vaughan, D.W. (1977). Age related deterioration of pyramidal cell basal dendrites in rat auditory cortex. *Journal of Comparative Neurology* **171**, 501–16.

Vaughan, D.W. & Peters, A. (1974). Neuroglial cells in the cerebral cortex of rats from young adulthood to old age: an electron microscope study. *Journal of Neurocytology* **3**, 405–29.

Vaughan, D.W. & Peters, A. (1981). The structure of neuritic plaques in the cerebral cortex of aged rats. *Journal of Neuropathology and Experimental Neurology* **40**, 472–87.

Villani, L., Contestabile, A., Migani, P., Poli, A. & Fonnum, F. (1986). Ultrastructural and neurochemical effects of the presumed cholinergic toxin AF64A in the rat interpeduncular nucleus. *Brain Research* **379**, 223–31.

Walker, L.C., Kitt, C.A., Struble, R.G., Schmechel, D.E., Oertel, W.H., Cork, L.C. & Price, D.L. (1985). Glutamic acid decarboxylase-like immunoreactive neurites in senile plaques. *Neuroscience Letters* **59**, 165–9.

Walker, L.C., Kitt, C.A., Struble, R.G., Cork, L.C. & Price, D.L. (1986). Heterogeneity of neurites in senile plaques of aged rhesus monkeys. *Society for Neuroscience Abstracts* **12**, 272.

Walker, L.C., Kitt, C.A., Schwam, E., Buckwald, B., Garcia, F., Sepinwall, J. & Price, D.L. (1987). Senile plaques in aged squirrel monkeys. *Neurobiology of Aging* **8**, 291–6.

Wallace, J.E., Krauter, E.E. & Campbell, B.A. (1980*a*). Animal models of declining memory in the aged: short-term and spatial memory of the aged rat. *Journal of Gerontology* **35**, 355–63.

Wallace, J.E., Krauter, E.E. & Campbell, B.A. (1980*b*). Motor and reflexive behavior in the aging rat. *Journal of Gerontology* **35**, 364–70.

Walsh, T.I. & Hanin, I. (1986). A review of the behavioral effects of AF64A, a cholinergic neurotoxin. In A. Fisher, I. Hanin & C. Lachman (eds.), *Alzheimer's and Parkinson's Diseases*, pp. 461–7. Plenum Press, New York.

Walsh, T.J., Tilson, H.A., DeHaven, D.L., Mailman, R.B., Fisher, A. & Hanin, I. (1984). AF64A, a cholinergic neurotoxin, selectively depletes acetylcholine in hippocampus and cortex, and produces long-term passive avoidance and radial arm maze deficits in the rat. *Brain Research* **321**, 91–102.

Warburton, D.M. (1972). The cholinergic control of internal inhibition. In R.A. Boakes & M.S. Halliday (eds.), *Inhibition and Learning*, pp. 431–60. Academic Press, New York.

Warburton, D.M. & Brown, K. (1971). Scopolamine-induced attenuation of stimulus sensitivity. *Nature* **230**, 126–7.

Watkins, J.C. (1981). Pharmacology of excitatory amino acid receptors. In P.J. Roberts, J. Storm-Mathiesen & G.A.R. Johnston (eds.), *Glutamate: Transmitter in the Central Nervous System*, pp. 1–24. Wiley, London.

Watkins, J.C. & Evans, R.H. (1981). Excitatory amino acid transmitters. *Annual Review of Pharmacology and Toxicology* **21**, 165–204.

Watson, M., Vickroy, T.W., Fibiger, H.C., Roeske, W.B. & Yamamura, H.I. (1985). Effects of ibotenate-induced lesions of the nucleus basalis magnocellularis upon selective cholinergic biochemical markers in the rat anterior cerebral cortex. *Brain Research* **346**, 387–91.

Watts, J., Stevens, R. & Robinson, C. (1981). Effects of scopolamine on radial maze performance in rats. *Physiology and Behavior* **26**, 845–51.

Weingartner, H., Kaye, W., Smallberg, S.A., Ebert, M.H., Gillin, J.C. & Sitaram, N. (1981). Memory failures in progressive idiopathic dementia. *Journal of Abnormal Psychology* **90**, 187–96.

Wenk, G.L. & Olton, D.S. (1984). Recovery of neocortical choline acetyltransferase activity following ibotenic acid injection in the nucleus basalis of Meynert in rats. *Brain Research* **293**, 184–6.

Wenk, H., Bigl, V. & Meyer, U. (1980). Cholinergic projections from magnocellular nuclei of the basal forebrain to cortical areas in rats. *Brain Research Reviews* **2**, 295–316.

Wenk, G.L., Cribbs, B. & McCall, L. (1984). Nucleus basalis magnocellularis: optimal coordinates for selective reduction of choline acetyltransferase in frontal neocortex by ibotenic acid injections. *Experimental Brain Research* **56**, 335–40.

Whishaw, I.Q., O'Connor, W.T. & Dunnett, S.B. (1985). Disruption of central cholinergic systems in the rat by basal forebrain lesions or atropine: effects on feeding, sensorimotor behaviour, locomotor activity and spatial navigation. *Behavioural Brain Research* **17**, 103–15.

Whitehouse, P.J., Price, D.L., Clark, A.W., Coyle, J.T. & DeLong, M.R. (1981). Alzheimer's disease: evidence for selective loss of cholinergic neurons in the nucleus basalis. *Annals of Neurology* **10**, 122–6.

Whitehouse, P.J., Price, D.L., Struble, R.G., Clark, A.W., Coyle, J.T. & DeLong, M.R. (1982). Alzheimer's disease and senile dementia: loss of neurons in the basal forebrain. *Science* **215**, 1237–9.

Wilcock, G.K. & Esiri, M.M. (1982). Plaques, tangles and dementia: a quantitative study. *Journal of Neurological Science* **56**, 343–56.

Wilcock, G.K., Esiri, M.M., Bowen, D.M. & Smith, C.C.T. (1982). Alzheimer's disease: correlation of cortical choline acetyltransferase activity with the severity of dementia and histological abnormalities. *Journal of Neurological Science* **57**, 407–17.

Wilson, R.S., Kaszniak, A.W. & Fox, J.H. (1981). Remote memory in senile dementia. *Cortex* **17**, 41–8.

Wirsching, B.A., Beninger, R.J., Jhamandas, D., Boegman, R.J. & El-Defrawy, S.R. (1984). Differential effects of scopolamine on working and reference memory of rats in the radial maze. *Pharmacology, Biochemistry and Behavior* **20**, 659–62.

Wischik, C.M. & Crowther, R.A. (1986). Subunit structure of the Alzheimer tangle. *British Medical Bulletin* **42**, 51–6.

Wisniewski, H.M. & Terry, R.D. (1973). Reexamination of the pathogenesis of the senile plaque. In H.M. Zimmerman (ed.), *Progress in Neuropathology, Volume II*, pp. 1–26. Grune & Stratton, New York.

Wisniewski, H., Narkiewicz, O. & Wisniewski, K. (1967). Topography and dynamics of neurofibrillary degeneration in aluminum encephalopathy. *Acta Neuropathologica* **9**, 127–33.

Wisniewski, H.M., Terry, R.D. & Hirano, A. (1970). Neurofibrillary pathology. *Journal of Neuropathology and Experimental Neurology* **29**, 163–76.

Wisniewski, H.M., Ghetti, B. & Terry, R.D. (1973). Neuritic (senile) plaques and filamentous changes in aged rhesus monkeys. *Journal of Neuropathology and Experimental Neurology* **32**, 566–84.

Wisniewski, H.M., Bruce, M.E. & Fraser, H. (1975). Infectious etiology of neuritic (senile) plaques in mice. *Science* **190**, 1108–10.

Wisniewski, H., Narang, H.K. & Terry, P.D. (1976). Neurofibrillary tangles of paired helical filaments.

Wisniewski, H.M.. Sturman, J.A. & Shek, J.W. (1980). Aluminium chloride induced neurofibrillary changes in the developing rabbit: a chronic animal model. *Annals of Neurology* **8**, 479–90.

Wisniewski, H.M., Sturman, J.A. & Shek, J.W. (1982). Chronic model of neurofibrillary changes induced in mature rabbits by metallic aluminum. *Neurobiology of Aging* **3**, 11–22.

Yerkes, R.M. (1909). Modifiability of behavior in its relation to the age and sex of the dancing mouse. *Journal of Comparative and Neurological Psychology* **19**, 237–71.

Zornetzer, S.F. & Rogers, J. (1983). Animal models for assessment of geriatric mnemonic and motor deficits. In T. Crook, S. Ferris & R. Bartus (eds.), *Assessment in Geriatric Psychopharmacology*, pp. 301–22. Mark Powley, New Canaan, Conn.

Zornetzer, S.F., Thompson, R. & Rogers, J. (1982). Rapid forgetting in aged rats. *Behavioral and Neural Biology* **36**, 49–60.

Strategies for drug development in the treatment of dementia

JOHN D. SALAMONE

Dementing disorders such as Alzheimer's disease and senile dementia of the Alzheimer's type (SDAT) offer a major challenge to drug development researchers. Advances in medical science and improvements in nutrition and sanitation have greatly extended the average lifetime in most of the developed world. However, medical science does not yet offer protection against the threat to the quality of that extended life that is offered by the possibility of a crippling cognitive dysfunction. Within the last few years the medical and scientific communities have responded to this challenge. There has been a rapid increase in the amount of research into the anatomical, neurochemical and molecular basis of dementia. Most of the major pharmaceutical firms have initiated projects in dementia-related areas. Thus far, no drug has been tested that offers clinically significant relief of dementia symptoms. Nevertheless, enough progress has been made that one can begin to outline strategies for the development of such a drug, an exercise which will provide the focus of the present chapter.

15.1 DRUG DEVELOPMENT FOR DEMENTIA

In order to understand the process of drug development for dementia, it is important to consider the three following research areas: the clinical pathology of dementing disorders, the experimental psychobiology and psychopharmacology of learning and memory, and the clinical pharmacology of dementia. The physiological and behavioral pathologies present in the demented patient provide a starting point for the development of animal models of the disorder. The growing literature on the psychopharmacology of animal learning and memory also offers some clues as to drug categories with potential therapeutic utility. Both of these areas have been reviewed in the previous chapter (Dunnett & Barth, Chapter 14). However, it should be recognized that the clinical pharmacology of dementia is the area that will invariably yield the most important leads for drug development. A researcher could use impeccable logic and artful scientific ingenuity to construct an animal model of dementia for drug testing. Yet the model must be validated ultimately on the empirical criteria of acceptance or rejection of drugs in accordance with the clinical literature.

Dependence upon the clinical pharmacology for providing leads and validating models is a part of what makes drug development in the dementia area so difficult. Despite the large number of drugs that have been tested, there are no good reference compounds in dementia research comparable to diazepam for anxiety or the neuroleptics of schizophrenia (see Table 15.1 for a cursory overview of the clinical pharmacology of dementia). There has been some modest improvement in symptoms of dementia with administration of anticholinesterases in controlled laboratory conditions (see review by Becker & Giacobini, 1988). However, it is not certain whether this evidence is strong enough to warrant using these compounds as standards for the validation of models for the development of novel drugs. The lack of good reference compounds from the clinical literature means that novel drugs must be 'fed forward' through animal

Table 15.1. *Overview of drug effects in treatment of AD/SDAT*

Strategy	Drug	Reference	Effect
Acetylcholine	arecoline	Christie *et al.* (1981)	+
	choline	Thal *et al.* (1981)	−
	choline + THA	Summers *et al.* (1986)	+
	nicotine	Newhouse *et al.* (1988)	+
	physostigmine	Beller *et al.* (1985)	+
	physostigmine	Davis *et al.* (1983)	+
	physostigmine	Mohs & Davis (1982)	+
	physostigmine	Mohs *et al.* (1985)	−
	physostigmine	Thal & Fuld (1983)	+
	RS86	Wettstein & Speigel (1984)	−
Serotonin	L-tryptophan	Smith *et al.* (1984)	−
	zimelidine	Cutler *et al.* (1985*b*)	−
Catecholamines	Ritalin	Ban (1978)	
Somatostatin	L-363,586	Cutler *et al.* (1985*a*)	−
Vasopressin	DGAVP	Peabody *et al.* (1985)	−
Broad spectrum	minaprine	Passeri *et al.* (1985)	+
Nootropic	piracetam	Reisberg *et al.* (1982)	−
	piracetam + physostigmine	Serby *et al.* (1983)	+
	pramiracetam	Branconnier *et al.* (1983)	−

−, little or no effect on learning/memory performance
+, mild improvement of learning/memory performance

studies years prior to clinical testing and eventual feedback evaluation of the model that accepted those drugs. This delicate balance between the dependence upon the clinical literature and the lack of clinically useful drugs constitutes the major obstacle to drug development.

Another problem that must be discussed is the heterogeneous nature of dementia. There are several conditions that could lead to cognitive dysfunctions in middle aged or elderly patients, including multiple cerebral infarcts, drug or solvent toxicity, and a variety of degenerative diseases (Terry & Katzman, 1983). The most common forms of dementia are AD and SDAT, which share certain neuropathological features such as neurofibrillary tangles and neuritic plaques (Rothschild & Kasansin, 1936). There are also some differences between AD and SDAT, with the younger, classic Alzheimer's patients showing more severe and widespread neurochemical deficits (Rossor *et al.*,

1984). The major focus of this paper will be on describing the development of drugs for the treatment of Alzheimer's disease or SDAT, and for simplicity the two disorders will usually be considered together. However, disorders such as Alzheimer's disease, SDAT, or infarct dementia will be discussed individually when it is relevant to the particular point being raised.

In order to illustrate some of the possible strategies for drug development, I have constructed a hypothetical system that incorporates a variety of testing procedures. This system (Fig. 15.1) should not be considered as a specific proposal for a drug development program, but rather as a general outline that includes a number of possible approaches. There are two proposed stages of drug testing: neuropharmacological assays and animal modelling. Novel compounds and reference drugs can enter the system at either of these stages, although the system allows primarily for a linear

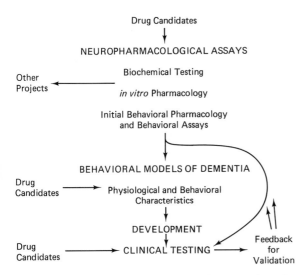

Drug Candidates

NEUROPHARMACOLOGICAL ASSAYS

Other Projects

Biochemical Testing

in vitro Pharmacology

Initial Behavioral Pharmacology and Behavioral Assays

BEHAVIORAL MODELS OF DEMENTIA

Drug Candidates

Physiological and Behavioral Characteristics

DEVELOPMENT

Drug Candidates

CLINICAL TESTING

Feedback for Validation

Fig. 15.1. Hypothetical system for development of novel drugs for dementia. The present system is designed to incorporate features of a neuropharmacologically-based project and a disorder-based project.

step-by-step passage through from biochemical screens to complex behavioral models. The present system was also constructed to illustrate a possible solution to the question of whether it is better to design a neuropharmacologically-based program (e.g. cholinergic) or a disorder-based program (e.g. dementia). Because there are advantages and disadvantages to both types of projects, the system outlined in Fig. 15.1 was designed to incorporate elements of both.

15.2 NEUROPHARMACOLOGICAL ASSAYS

Neuropharmacological testing would involve assessment of compounds based upon activity in biochemical or pharmacological procedures that are indicative of actions on particular neurotransmitter systems. The purpose of this stage is to generate compounds with a desired neuropharmacological profile (e.g. cholinergic agonist, serotonergic uptake blocker). The particular target profile is selected on the basis of several lines of evidence, including the neuroanatomical and neurochemical pathologies of the disorder, and psychopharmacological work with animals and humans. A series of biochemical and pharma-

cological assays indicative of activity in the desired system can be used to screen drug candidates.

There are several advantages to planning a drug development project with a neuropharmacological emphasis. Such an approach defines a biological activity that is relatively easy to determine. The development of 'memory-enhancing' drugs without reference to any specific pharmacological activity is a rather diffuse goal. However, the goal is much more well-defined and easily managed if one identifies a particular desired neurochemical activity. Such a tangible goal facilitates the organization of a drug synthesis program and the investigation of structure–activity relations. A drug synthesis program that is designed with reference to a neuropharmacological target has other, hidden advantages. The novel compounds that are generated for a dementia project will be well characterized in terms of their pharmacological profile, and thus may be useful for other projects as well. For example, a novel cholinergic agonist that has a long duration of action and central penetrability may turn out to be useless as a treatment for dementia. However, this same drug might prove to be an excellent non-narcotic analgesic.

A neuropharmacologically-based project is not without its disadvantages. Focussing on a specific pharmacological profile narrows the approach to treating the disorder. Without adequate clinical pharmacology to validate the chosen profile, a drug development program would be pouring resources into an area that may prove to be fruitless. It is also conceivable that functions as complex as learning, memory, attention, language, and praxis cannot be restored by a single drug with a narrow spectrum of biological activity. Despite these problems, it is sensible to include aspects of the neuropharmacological approach in any dementia project. The advantages of this approach are very powerful, and many of the disadvantages can be rectified by making changes in other parts of the project. In addition, there is a historical tradition in drug development research that supports a neuropharmacological orientation. To some extent, drug companies tend to pursue neuropharmacologically-defined projects because such projects have been successful in the past (e.g. dopamine antagonists for schizophrenia, dopamine agonists for Parkinson's disease, beta-blockers for high blood pressure).

15.2.1 Selection of desired pharmacological profile

Having chosen to pursue a strategy that is organized along neuropharmacological lines, the next issue that must be addressed is the determination of the target neurochemical profile. Typically, several lines of evidence are used to formulate a drug treatment strategy. In the case of Alzheimer's disease and SDAT, the researcher must rely heavily on information about the neurochemical pathologies, and the psychopharmacology of learning and memory (see Dunnett & Barth, Chapter 14 of this volume).

Considerable evidence indicates that there is a severe loss of cholinergic input to the neocortex and hippocampus in a variety of dementias including Alzheimer's disease, SDAT, and dementia in some Parkinson's patients (Bowen *et al.*, 1976; Davies & Maloney, 1976; Perry *et al.*, 1978). Animal studies have shown that anticholinergic drugs interfere with the performance of tasks involving learning, memory, or sensory discrimination processes in rats (Pazzagli & Pepeu, 1964; Ksir, 1974; Heise *et al.*, 1976; Milar, 1981; Spencer & Lal, 1983; Dunnett, 1985; Hagan *et al.*, 1986; Salamone *et al.*, 1987), monkeys (Hearst, 1964; Evans, 1975; Bartus & Johnson, 1976; Penetar & McDonough, 1983), and humans (Drachman & Leavitt, 1974; Drachman & Sahakian, 1980; Dunne & Hartley, 1985). Deficits in learning or memory tasks can also be produced by lesions that disrupt cholinergic projections to neocortex or hippocampus (Gaffan, 1974; Flicker *et al.*, 1983; Rawlins & Olton, 1983; Dunnett, 1985; Hepler *et al.*, 1985; Lerer *et al.*, 1985; Ridley *et al.*, 1985; Wishaw *et al.*, 1985; Salamone, 1986; Salamone *et al.*, 1984, 1987; Everitt *et al.*, 1987). Under some conditions, administration of cholinergic agonists or anticholinesterases has improved learning or discrimination performance (Bartus, 1979; Flood *et al.*, 1985; Haroutunian *et al.*, 1985). This evidence has led some researchers to suggest that cholinomimetic drugs may be useful as treatments for dementia (Bartus *et al.*, 1982; Coyle *et al.*, 1983; Smith & Swash, 1978). In some studies, cholinergic agonists or anticholinesterases have been reported to improve cognitive performance in demented patients (Christie *et al.*, 1981; Beller *et al.*, 1985; see also Table 15.1).

Despite the evidence cited above, it is premature at this time to emphasize acetylcholine (ACh) to the exclusion of other neurotransmitters. Neuropathological evidence clearly demonstrates that the degeneration observed in Alzheimer's or SDAT patients affects a number of brain systems, including noradrenaline (NA) and serotonin (5-HT) inputs to the cortex (Bondareff *et al.*, 1981; Arai *et al.*, 1984; Rossor *et al.*, 1984). Animal studies indicate that lesions of the ascending dorsal NA system can cause subtle deficits in complex learning or sensory discrimination tasks (Carli *et al.*, 1983; Everitt *et al.*, 1983; Tsaltas *et al.*, 1983). Electrophysiological data indicate that this NA system is involved in responses to arousing or stressful stimuli (Abercrombie & Jacobs, 1987). Therefore, it is possible that pharmacological modulation of NA activity may be useful as a treatment for some of the symptoms of dementia. The 5-HT uptake blockers zimeldine and alaproclate have been tested as possible treatments for dementia, but yielded little improvement in cognitive functions (Bergman *et al.*, 1983; Cutler *et al.*, 1985*a*).

There have been several investigations of neuropeptide involvement in the pathology of Alzheimer's disease and SDAT. Vasopressin, cholecystokinin and vasoactive intestinal peptide immunoreactivities were shown to be normal in Alzheimer's disease and SDAT patients (Rossor *et al.*, 1980, 1981; Perry *et al.*, 1981), but somatostatin immunoreactivity was reduced in neocortical tissue or CSF of patients with Alzheimer's disease, SDAT, or the dementia of Parkinson's disease (Davies *et al.*, 1980; Davies & Terry, 1981; Serby *et al.*, 1984). Beal *et al.* (1985) reported a decrease in numbers of somatostatin receptors in Alzheimer's disease and SDAT, and Armstrong *et al.* (1985) observed that 20–50% of the neuritic plaques of Alzheimers disease and SDAT patients contained somatostatin-positive profiles. The presence of somatostatin deficits in demented patients led to a clinical study using the somatostatin analogue L-363,586, but the drug had no clinically significant effect (Cutler *et al.*, 1985*b*).

In addition to the neurotransmitter deficits outlined above, it should be emphasized that Alzheimer's disease and SDAT are characterized by loss of intrinsic cortical neurons (Mountjoy *et al.*, 1983) and that hippocampal output as well as input is deteriorating (Hyman *et al.*, 1984). This loss of cortical circuitry poses difficult problems for the neuropharmacological approach to dementia drug therapy. Replacement of lost afferent neurotransmitter activity could be of little benefit if the target cells for that neurotransmitter are lost, or if the intracortical connections involved in cognitive functions are themselves degenerating. The problems posed by the loss of cortical neurons, and the consideration of neuromodulatory and information-carrying functions of neurochemicals depleted in Alzheimer's disease and SDAT will be discussed again later in this chapter.

Along with considering which neurotransmitter system to use as a focus for drug development, there is also the necessity of determining the best approach for pharmacological modulation of that system. One possible strategy is to use direct receptor agonists to serve as a replacement therapy for neurotransmitters lost from degenerating neurons. An advantage of this approach is that agonists can continue to exert effects upon receptors even if the presynaptic terminals undergo a severe or total degeneration as the disease progresses. The use of agonists with a selective profile would have an additional potential advantage of providing a differential stimulation of receptor populations. There have been recent advances in receptor selectivity in cholinergic (Hammer & Giachetti, 1984), noradrenergic (Aghajanian & Rogowski, 1983; Vizi, 1986), and serotonergic (Conn & Sanders-Bush, 1987) systems. Of course, the use of selective agonists as treatments is based on the assumption that the desired effect of enhanced cognitive function will be related to the stimulation of one subpopulation of receptors, whilst undesirable side effects would result from stimulation of a different subpopulation. So far, there is little direct evidence to support this position.

One disadvantage of direct receptor agonists is that they provide a tonic stimulation of the receptors rather than a phasic modulation of activity that corresponds to the physiological signal. If the neurotransmitter system being replaced is one that operates by finely tuned spatial and temporal modulation to convey specific information, then it would be impossible to replace or augment this information by administration of a direct agonist. For example, one could not restore organized muscular activity to a denervated muscle by systemic administration of a nicotonic cholinergic

agonist. Drugs that enhance the activity of functioning terminals by stimulating neurotransmitter synthesis (e.e., precursors), enhancing release, inhibiting breakdown, or providing a positive allosteric modulation of the receptors could serve to enhance existing levels of neurotransmitter activity, thereby enhancing the signal that is reduced by the neural degeneration. Of course, such an approach would be fruitless in cases where the degeneration is so severe that there is little or no neurotransmitter release left to enhance. The relative merits of direct vs. indirect stimulation of neurotransmission as a basis for therapy is an empirical question that should be addressed by basic scientists and clinical pharmacologists. However, it is a question that may have a very complex answer, that may vary from one transmitter system to another, and perhaps even from one patient to another.

15.2.2 Neuropharmacological testing

The types of tests to be used for assessing drug candidates depend upon the desired neurochemical profile. Biochemical assays for receptor binding, enzyme activity, or activation of second messengers can be used to offer relatively rapid assessment of the activity or selectivity of novel compounds. *In vitro* pharmacological preparations such as organ baths, electrophysiology, or release studies yield information regarding agonist/antagonist activity, efficacy, potency, and functional selectivity of drug candidates. In both cases the data provide important feedback to the synthetic chemists, enabling them to modify their synthesis program in accordance with the stated goals of the project. The methods described above reflect the familiar characteristics of the modern neuropharmacological approach to drug design.

The psychopharmacologist also has an important contribution to make at this stage of drug development. Behavioral actions of drugs that are indicative of activity in a particular transmitter system can be used as behavioral assays that complement and extend the biochemical and *in vitro* pharmacological assays. Observation of perioral movements (Rupniak *et al.*, 1985; Salamone *et al.*, 1986), analgesia (Dayton & Garrett, 1973), or parasympathetic activation (Salamone *et al.*, 1986) serve as *in vivo* indices of cholinergic activity.

Measurements of the head twitch or paw treading responses of 5-HT activity (Tricklebank *et al.*, 1985; Handley & Singh, 1986) offer an additional test of this type. The use of such tests is analogous to screening for neuroleptic activity by observing catalepsy or suppression of various motor activities (Janssen *et al.*, 1965).

In addition to providing behavioral signs of a particular neuropharmacological action, simple behavioral assays can yield information about the characteristics of drug action *in vivo* that could not be obtained from receptor binding assays or organ preparations. Because dementia is a central nervous system disorder, the ability of a drug to penetrate the blood–brain barrier is an absolutely essential feature of a drug candidate. A measure of central penetrability can be obtained by comparing the potency of a given compound for inducing central as opposed to peripheral behavioral effects. Other desirable characteristics of a drug treatment for dementia, such as long duration of action and oral activity, can also be assessed with simple behavioral assays. In this way, a profile of the *in vivo* pharmacology of a novel compound can be developed from tests that are simpler and more rapid than most tests of learning or memory.

In spite of the obvious utility of behavioral assays, such tests should be used cautiously. The drug actions being measured are not directly related to the primary therapeutic effect, but instead to undesirable side effects of the drugs. In the absence of clinical data clearly indicating that a certain neuropharmacological profile is required, the rejection of drugs based on poor performance in tests of movement or salivation could be a serious error. If the therapeutic effects of drugs for dementia turn out to be related to the activation of a particular receptor class that is different from the class responsible for the behavior in the simple behavioral assay, then limited use should be made of that behavioral assay.

15.2.3 Flexibility with the neuropharmacological approach

Thus far, the neuropharmacological approach to dementia drug therapy has been described in terms of a relatively narrow focus. A succession of questions was asked that emphasized a specific transmitter system, or even a specific subset of that system, for potential drug development. Although this focussed orientation has certain advantages, it is not the only approach that should be considered. In many demented patients, particularly the younger, more severely affected ones, there is a widespread deterioration of several transmitter systems (Rossor *et al.*, 1984). Also, it is probably unlikely that functions as complex as memory or attention are localized to a particular transmitter system. The implication of both the widespread pathology of dementia and the distribution of memory/cognitive functions to several different systems is that a drug with a broad spectrum of activity, or a drug combination, may prove to be more useful than one with a limited range.

Minaprine is a drug that was suggested as a treatment for dementia because of its action on a number of neurotransmitter systems (Passeri *et al.*, 1985). In this clinical study, minaprine was shown to have limited effect on patients with Alzheimer's disease and SDAT, but was somewhat better for treating infarct dementia patients (Passeri *et al.*, 1985). Adenosine receptor antagonists have been shown to stimulate release of several neurotransmitters (Daly, 1982). Drugs such as 3,4-diaminopyridine, which stimulate neurotransmitter release by promoting calcium uptake into nerve terminals, have also been offered as potential therapies for dementia (Peterson & Gibson, 1983).

If no single compound can be found that provides the desired combination of activities, perhaps a drug 'cocktail' could be devised. For example, stimulants such as methylphenidate can enhance vigilance in normal subjects under controlled laboratory conditions, but methylphenidate alone did not lead to a clinically significant improvement of symptoms in Alzheimer's disease or in SDAT patients (Ban, 1978). The anticholinesterase physostigmine has usually yielded positive effects that in Alzheimer's disease and SDAT patients are small, and have a narrow, inverted-U shaped dose-response curve (e.g. Mohs & Davis, 1982). It is possible that combining the two drugs would lead to a greater effect than either one alone, or at least that the therapeutic window of physostigmine could be improved. Using a drug combination might also allow clinicians to administer each at a dose lower than that required when given alone, thereby lowering the risk of side effects.

'Nootropic' or 'cerebroactive' compounds

represent a category of drugs that have been suggested for use in demented patients, but are outside the scope of the typical neuropharmacological project. This category includes drugs such as piracetam, aniracetam, flunarizine, oxiracetam, and the ergot derivatives found in Hydergine. Nootropic drugs are already widely prescribed to elderly patients in some countries, with Italy being the largest user among western countries (Spagnoli & Tognoni, 1983). The nootropics do not form a distinct pharmacological category, although there is some suggestion that many nootropics bear a structural resemblance to the folded conformation of GABA that is involved in some transport mechanisms (Nicolaus, 1982). The link connecting the various types of nootropics is that they are all said to enhance oxygen or glucose metabolism in the brain, and they demonstrate a facilitation of 'integrative activity' in the central nervous system (Nicolaus, 1982).

A thorough review of the nootropic literature would be difficult in a few paragraphs of this chapter. Perhaps the best way to summarize this area is to state that there is still some controversy concerning both the efficacy in animal tests and the therapeutic utility of these drugs (Spagnoli & Tognoni, 1983; Ennaceur & Delacour, 1987). In tests with rodents, the most consistent positive effects are usually observed if the nootropic drug is being used to reverse the effects of noxious stimuli such as hypoxia, electroconvulsive shock, or protein synthesis inhibitors (Gamzu & Perrone, 1981; Cumin et al., 1982). Treatment of demented patients with nootropic drugs alone has generally not led to any clinically significant improvement (Reisberg et al., 1981, 1982). There was some suggestion from the animal and human literature that giving a combined treatment with a nootropic and the acetylcholine precursor choline led to better positive effects than treatment with either drug alone (Bartus et al., 1982; Reisberg et al., 1982). However, Ennaceur & Delacour (1987) failed to observe a positive effect on delayed alternation or avoidance in rats with combined piracetam and choline treatment, and in fact found that the combination actually lowered performance. It remains possible that further drug development in this area will lead to compounds with greater efficacy. In addition, it is conceivable that nootropics are not suitable for treatment of Alzheimer's disease and SDAT patients, but that patients with other types of dementia would benefit.

In a dementia project that is largely oriented around a single pharmacological focus, the testing of combinations of drugs with wide activity spectra, or nootropics gives the project some breadth that is probably necessary. The neuropharmacological testing program can be expanded to include assays for several systems. However, it may be that a dementia project would not want to divert any effort away from the synthesis of drugs with the primary target profile. In this instance drugs developed from other sources that are outside the target profile could enter directly into assessment in animal models (see below).

15.2.4 Use of compounds accepted by neuropharmacological analysis

Typically, a drug development project would specify what neuropharmacological criteria would qualify a compound for acceptance or rejection, the criteria having been developed by some process similar to that described in the preceding sections. Regardless of the particular profile that is desired, the project is then faced with the question of what to do with compounds that meet with acceptance. According to the diagram in Fig. 15.1, the next logical step is to assess these compounds for effects in animal models of dementia. However, it should also be noted that the system described in Fig. 15.1 is not inflexible, and allows for the possibility of bypassing rigorous testing in animal models in favor of moving the drug candidate rapidly on to further development. Although this hurried passage from assays to toxicity testing and human volunteers is not be recommended for all disorders, it may well be advisable for dementia (see also Altman et al., Chapter 16 of this volume). Considering that the validation of animal models rests on some correspondence with the clinical literature, and that there are no clinically acceptable drugs for dementia, it can be argued that a novel drug with a promising profile should be rapidly passed on for clinical testing. This strategy would enable the empirical testing of various hypothesized treatments in human patients

Perhaps it is extreme to advocate either an expeditious passage of a novel drug to human subjects, or a laborious step-by-step process of extensive animal testing. It may be wise to use a compromise approach, with the best of an early series of drugs passing more quickly to possible human evaluation, but with work continuing in parallel on further detailed assessment in animal models, and development of back-up compounds. The procedure used to evaluate drug performance in human patients will not be discussed in this chapter (but see Altman *et al.*, in Chapter 16). The testing of drug candidates in animal models is the focus of the next section.

15.3 BEHAVIORAL MODELS

Animal models are used in drug development programs so that compounds that have passed through neuropharmacological testing, or have been generated from other sources, can be evaluated under conditions that are thought to bear some relation to the disorder. As stated at the outset of this chapter, the information that is most relevant to drug development research is whether or not the models predict performance in the clinical setting accurately. The present paucity of clinically active drugs means that the development of animal models of dementia is at a formative and challenging stage.

It is true that testing compounds in animal models that have not been properly validated may be a waste of time, and at worst may reject potentially useful compounds on a dubious basis. Nevertheless, there are compelling reasons why it is advisable to screen compounds using animal models. Testing novel drugs in humans is a very expensive and time consuming process. It is not feasible to develop simultaneously a large number of novel compounds and test them all in human subjects. For this reason, it is necessary to limit the field of possible candidates to as small a number as possible. Animal models provide a rational, though not necessarily empirically validated, way of filtering out compounds.

The preceding chapter (Dunnett & Barth, Chapter 14) provides a detailed discussion of various methods that offer potential animal models of dementia. The present discussion is meant to provide a brief overview of this topic with an emphasis on the way in which such tests can be integrated into a system of screens and models explicitly for the purpose of drug development. In most cases an animal model of dementia will itself consist of two models. One component would embody some of the physiological pathologies of the human disorder, such as lesion-induced neurotransmitter deficits, cortical damage, hypoxia, or aged animals. The other would include animal tests related to the behavioral pathologies of dementia, which essentially would be animal models of human functions such as learning, memory, or attention. In combination, the two models would yield a situation in which some physiological pathology produces a deficit in learning or memory in animals that is reversible with pharmacological treatment.

15.3.1 Criteria for evaluating animal models of dementia

Any animal model that is being used for drug development must be evaluated in terms of a number of criteria. One feature that drug researchers must deal with is the logistical characteristics of the particular model. The labor intensiveness of the behavioral procedures is a very important consideration, which may drive researchers to use automated tests that allow for large-scale screening with relatively few workers. The time it takes to obtain a useful characterization of each compound is another aspect of the logistics of animal modelling. It would be difficult for chemists to execute a competitive drug design program if it took several years to evaluate each drug. Rapid behavioral procedures related to learning or memory, such as passive avoidance, offer some advantages in terms of testing time.

Another set of criteria to be examined here concerns the validity of the model. According to Willner (1984; and Chapter 1 of this volume), animal models, like psychometric measures, can be evaluated in terms of predictive validity, face validity, and construct validity. Predictive validity refers to the extent to which the performance of the test predicts performance in the condition being modelled. Face validity is defined as the extent to which the model resembles the condition being modelled. The construct validity of a model reflects the soundness of the theoretical rationale upon which that model is based.

As outlined in the previous section of this chapter, animal models of dementia comprise both physiological models and behavioral models. Both sets of models can be analyzed individually in terms of the validation criteria listed above (see Dunnett & Barth, this vol., Ch. 14, for more detailed discussion). For example, various methods of achieving a cholinergic deficiency can be evaluated for their face validity compared to the pathology of Alzheimer's disease and SDAT. Scopolamine has been suggested as an animal model of the cholinergic deficiency in Alzheimer's disease and SDAT (Bartus, 1978; Spencer & Lal, 1983). Although this treatment does provide a cholinergic dysfunction that can lead to deficits on tests of memory, it has a relatively low face validity, due to its postsynaptic action. Hemicholinium is a pharmacological model that fares somewhat better in terms of face validity, because this drug depletes presynaptic ACh by inhibiting choline uptake (Caulfield *et al.*, 1981; Ridley *et al.*, 1987). Neurotoxic lesions of the nucleus basalis magnocellularis (see review by Salamone, 1986) reduce the cholinergic innervation of the neocortex, but the face validity of this preparation is somewhat marred by the spread of damage to areas other than the basal nucleus, and by the lack of hippocampal deficit. Fimbria–fornix lesions (Gaffan, 1974; Dunnett, 1985) transect the cholinergic projection to hippocampus, but also damage systems not impaired in Alzheimer's disease, and fail to affect the neocortical projection. Basal forebrain lesions that produce damage to cortical and hippocampal cholinergic projections (Wenk *et al.*, 1984), or injection of the cholinergic toxin AF64A (Walsh *et al.*, 1984), may produce a cholinergic depletion that more closely mimics that seen in Alzheimer's disease and SDAT. Yet as described above, the pathology of dementia can include other neurotransmitter deficits and loss of cortical neurons. Most cholinergic depletion models manifest only moderate face validity. Studies that combine lesions to a number of systems, or use aged rodents or primates that show some of the pathological features of Alzheimer's disease and SDAT, hold more promise as models with relatively high face validity (Struble *et al.*, 1985; Walker *et al.*, 1987).

Construct validity is most applicable to the evaluation of the behavioral models used. More precisely, it is important to recognize that the behavioral tests should measure something directly related to the constructs of 'learning', 'memory' or 'attention', as opposed to non-specific performance variables. Studies that investigate short-term memory by looking at time-related decreases in performance are examples of establishing construct validity in behavioral tests (Bartus & Johnson, 1976; Bartus, 1979; Penetar & McDonough, 1983; Dunnett, 1985; Salamone *et al.*, 1987). Attempts to describe behavioral deficits in terms of effects on 'working' or 'reference' memory also represent instances of understanding behavioral measures in relation to mnemonic constructs (Knowlton *et al.*, 1985; Murray & Fibiger, 1985).

However, considering that models of dementia are being discussed in the context of the development of drug treatments, it is of the utmost importance to validate the drug effects upon the composite physiological and behavioral model. Described in this manner, the condition being modelled is not the disorder itself, but rather the therapeutic effect of the drug on the disorder. Throughout this chapter it has been emphasized that the most important characteristic of an animal model that is used for drug development is the ability to reject or accept drugs in relation to their clinical efficacy. In fact, this represents the predictive validity of a model when the condition being modelled is the therapeutic effect of the drug.

The relative importance of predictive validity is best illustrated by considering that a model with low face validity and low construct validity could still be useful for developing novel drugs if it had a high predictive validity. The high predictive validity gives the researcher information about the drugs that pass through that model. More specifically, if a drug shows an action in a model that predicts therapeutic activity, this justifies the use of resources for further development. However, low predictive validity would limit the utility of a model for drug development, even if the model resembled the disorder and was based on a very sound theoretical rationale. It would be difficult to support a substantial commitment of resources to developing a drug on the basis of having passed through a model with low predictive validity, because there is little empirical basis for assuming that such a drug will work better than a drug that failed in the model.

In the specific case of dementia research, this discussion of the relative merits of predictive validity seems moot, because most of the results obtained from clinical pharmacology studies with a variety of drugs have been negative (Thal *et al.*, 1981; Dowson, 1982; Greenwald & Davis, 1983; Wettstein & Speigel, 1984; Cutler *et al.*, 1985*a*, *b*; Peabody *et al.*, 1985). The positive effects of physostigmine were small and variable from subject to subject (Mohs & Davis, 1982; Davis *et al.*, 1983; Thal & Fuld, 1983; but see Beller *et al.*, 1985, for more consistent results). This pattern does not easily provide a large enough range to generate correlations between clinical effects and the actions of drugs in animal models. In terms of validation criteria, the present state of dementia research is one in which scientists must rely on the less useful criteria of face validity and construct validity, guided only by minimal predictive information.

15.3.2 Evaluation of drug actions in animal models of dementia

Despite the problems stated above, scientists have begun to investigate the effects of drugs on animal models of dementia. This area of research is rapidly growing, and thus any chapter that attempts to detail recent developments is doomed to obsolescence before it is even published. For this reason, analysis will be restricted to a few studies that will serve as illustrations of how such research can be performed, and how the results can be evaluated in terms of a drug development project. The work to be examined includes studies on the effects of cholinomimetic drugs in reversing behavioral deficits produced by scopolamine, hemicholinium, nucleus basalis lesions, and in the aged primate.

As reviewed in previous sections of this chapter, scopolamine has been shown repeatedly to produce impairments in the performance of a variety of tasks involving learning, memory or sensory discrimination. In spite of the large volume of this work, relatively few studies have demonstrated the reversibility of these impairments with administration of cholinomimetic drugs. Glick & Zimmerberg (1972) showed that physostigmine administered to mice after the first trial in a passive avoidance test partially reversed the impairment produced by scopolamine that was administered before that trial. The impairment of visual frequency discrimination produced by scopolamine

(Salamone *et al.*, 1987) can be reversed by physostigmine (Salamone, in preparation).

Bartus (1979) compared the effects of physostigmine and the stimulant drug methylphenidate for reversal of the scopolamine-induced impairment of delayed response performance in rhesus monkeys. Using a careful dose-titration procedure, a mild impairment was produced by moderate doses of scopolamine. Doses of physostigmine that were just below those necessary for producing general behavioral deficits were capable of partially reversing the scopolamine deficit, but the particular dose of physostigmine varied across animals. Doses of methylphenidate that did not produce any effects when given alone potentiated the scopolamine deficit. The results of the Bartus (1979) study are particularly interesting when compared to the human clinical literature. The moderate and variable effects of physostigmine resemble the effects of this drug on Alzheimer's disease and SDAT patients (e.g. Davis *et al.*, 1983). The lack of positive results with methylphenidate also corresponds with the results of studies with human demented patients (Ban, 1978).

Hemicholinium is an inhibitor of choline uptake that reduces ACh synthesis, and has been shown to cause impairments in the performance of tests of learning or discrimination in rodents and primates (Caulfield *et al.*, 1981; Ridley *et al.*, 1987). Ridley *et al.* (1987) injected hemicholinium into the cerebral ventricles of monkeys and observed a deficit in the performance of a serial position reversal discrimination procedure. This deficit was reversed by the direct muscarinic agonists pilocarpine or arecoline. These results are very important, considering that arecoline has been shown to cause some modest improvement in the symptoms of human Alzheimer's disease and SDAT subjects (Christie *et al.*, 1981). Also, research using ACh depletion can begin to address the question of whether postsynaptic agonists can reverse the deleterious effects of presynaptic depletions.

Behavioral deficits produced by neurotoxic lesions of the nucleus basalis have been used to model the effects of cholinomimetic drugs on Alzheimer's disease and SDAT patients. Haroutunian *et al.* (1985) reversed the effects of nucleus basalis lesions in rats on a passive avoidance task by administering physostigmine after the acquisition trial. Murray & Fibiger (1985) observed that physostigmine improved performance of basalis-lesioned rats on a radial arm maze task. In a subsequent study, these researchers also reported that physostigmine and the direct muscarinic agonist pilocarpine enhanced the impaired performance of lesioned rats on a T-maze alternation task (Murray & Fibiger, 1986). Contrary to the results described above, Aigner *et al.* (1987) found that physostigmine did not improve performance of basalis-lesioned monkeys on a non-match-to-sample task. However, these subjects did not show a deficit produced by the lesion, so the lack of drug effect is very difficult to interpret. Ridley *et al.* (1986) did obtain a deficit in visual discrimination performance in marmosets after lesions of the nucleus basalis, and were able partially to reverse this deficit with administration of arecoline.

Bartus and his colleagues have conducted a series of studies examining the effects of cholino-mimetic drugs on aged primates (Bartus, 1979; Bartus *et al.*, 1980). Deficits on a delayed response task that were greater with longer delays were demonstrated in aged rhesus (Bartus, 1979) and cebus (Bartus *et al.*, 1980) monkeys. In either case, administration of physostigmine led to a small improvement in performance, with the optimal dose varying from subject to subject. Arecoline produced a more stable increase in performance than physostigmine in aged cebus monkeys (Bartus *et al.*, 1980). Injection of the ACh precursor choline chloride failed to improve test accuracy (Bartus *et al.*, 1980).

Several important features of the studies reviewed above should be emphasized. Physostigmine is capable of improving learning or memory performance in a variety of animal models of Alzheimer's disease and SDAT. The pattern of results emerging from the primate models is that the drug effects showed a modest correspondence with the clinical literature, because physostigmine yielded positive effects, but methylphenidate and choline did not. In addition, the therapeutic actions of physostigmine in the animal models had some face validity relative to the therapeutic effects shown in human clinical studies. The positive effects of physostigmine in the animal models tended to be small, with an inverted-U-shaped dose–response curve, and an optimal dose that varied across subjects. As stated above, these features are all present in human studies.

The type of data generated by physostigmine in animal models is both encouraging and disheartening. There is at least a tentative indication that the animal models being used presently can provide a basis for drug assessment. However, the results of drug studies in animal models also tend to confirm the observation that despite the mild positive effects one can obtain under ideal laboratory conditions, no drug has yet been found that is clinically useful. The variability of the dose–response curves across subjects and the toxicity and decrements in performance observed at high doses all pose difficulties for consistent clinical usage. Some of the positive results obtained with arecoline suggest that this drug should receive some additional scrutiny from clinicians.

The elusive nature of the actions of drugs in animal studies also poses some problems for drug development researchers. In evaluating the ability of some novel drug to reverse deficits in animal models of dementia, it is sometimes difficult to demonstrate convincingly that a positive result has been obtained. One often observes a pattern of results in which every animal shows some positive effect at some dose, but there is no statistically significant effect over all subjects at any one dose. If the researcher is interested in extracting some meaning from these data, a few options are available. After obtaining a dose–response curve for each animal, the researcher can attempt to replicate the effect at the optimal dose. In addition, each individual subject can be evaluated separately for statistical significance if there is a large body of data on each subject under the impaired control condition. Finally, one can use analysis of variance with orthogonal trend analysis to detect a significant quadratic trend that reflects the inverted-U-shaped dose–response curve. This procedure can be a powerful way of detecting drug effects of the type observed with physostigmine (unpublished observations).

It is also important to consider which control procedures are the most appropriate for evaluating positive actions of a drug. In clinical studies, it is desirable to show that a given drug has a significant positive effect relative to the disordered condition. Therefore, studies with animal models must use the impaired condition (e.g. scopolamine alone, or basalis lesions plus saline) as the control. It is inappropriate merely to demonstrate that treatment of a lesioned group renders that group no longer significantly different from normal control subjects. For example, Murray & Fibiger (1986) argued that physostigmine and pilocarpine produced positive effects, based partly upon the observation that basalis–lesioned animals treated with these drugs were no longer significantly different from saline-treated non-lesioned subjects. Such a procedure is dubious for several reasons. In cases when the lesion effect is small, the drug could reduce the significance of the effect merely by increasing the variability of the performance, or improving only a few subjects. Also, close examination of Figure 8 of Murray & Fibiger (1986) indicates that, although pilocarpine may have caused a slight improvement in the performance of lesioned subjects, saline may have caused a slight decrement in the performance of control subjects. Moreover, the assessment of a drug action by testing for a lack of significance between treated disordered subjects and normal subjects makes use of the classic statistical error of 'proving' the null hypothesis.

15.3.3 The use of animal models of dementia in a drug development project

Now that the results of some studies testing drugs in animal models of dementia have been discussed, it is useful to consider how animal models can be incorporated into a project for the development of drugs for dementia. Each of the models has some advantages and some limitations. Testing a drug for its ability to reverse a scopolamine-induced deficit in passive avoidance in rats is a logistically favorable procedure that is suited to a cholinergic project, but one that has low face validity, dubious construct validity, and uncertain predictive validity. The use of aged monkeys on complex tests that resemble human non-verbal performance tasks is a procedure with relatively high validity, but it is also rather expensive and time consuming. At present, there is not a single animal model of dementia that combines all the necessary features. For this reason, it is advisable to design a testing program that incorporates a number of model procedures.

An example of how a group of model procedures could be used is shown in Fig. 15.2. This figure contains two levels of behavioral models, with each

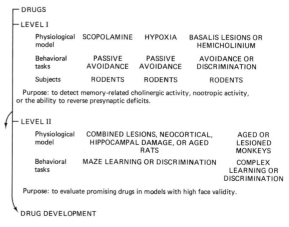

Fig. 15.2. Hypothetical organization of a system of behavioral models of dementia to be used for drug development.

level serving different functions. This first level would include tests designed to detect a number of possible activities that are related to hypothesized treatments. For example, compounds could be rapidly assessed for their ability to reverse scopolamine- or hypoxia-induced deficits in simple behavioral tests such as passive avoidance in rats or mice. There should also be tests of compounds in models that would focus on the ability of drugs to reverse deficits produced by presynaptic deficiency models such as hemicholinium or nucleus basalis lesions. At this first level, no single tests would be used as a screen to filter out drugs, but rather they would form a 'safety net' for the detection of one or more target activities. Some of these target activities should be directly related to the neuropharmacological strategy of the particular project. However, it is not necessary that all tests should be related to one particular neurotransmitter. Inclusion of models such as hypoxia would enable the researchers to assess compounds that are outside the target neurochemical profile, and would broaden the characterization of all the compounds tested.

Based upon the overall pattern of effects on these tests, drugs could be sent on for testing in the second level of behavioral models. This involves assessment of compounds in complex lesion models in rodents, or in aged or lesioned primates. At this level, the models would be expected to have a relatively high standard of face validity. In addition, the rather high logistical constraints of the complex models could be tolerated if used in the way described here. These models would not be expected to assess every compound leaving the chemists bench, but only those compounds with the more promising profiles, or perhaps only the better compounds from each group of analogues being tested. After assessment in the complex models, compounds with favorable profiles of activity can then be passed on for further development.

A drug development program is under some of the same constraints as a foraging animal. A rigorous cost/benefit analysis must guide each decision. The program must be willing to expend energy to obtain results, and to consider a variety of alternatives, just like an animal foraging for food. However, neither an animal in the wild nor a company in the market place can afford to spread itself too thinly by searching everywhere for its goal and wasting resources. The hypothetical testing program described in Figs. 15.1 and 15.2 is meant to provide some balanced consideration of all these factors. The efficiency of the neuropharmacological approach must be balanced against its lack of breadth. The logistical characteristics of complex testing procedures must be balanced against their validity. The ultimate organization of a drug development project must be influenced by a recognition of the limitations and constraints of the project, as well as the goals.

15.4 FUTURE RESEARCH

Considering that there are presently no clinically useful drugs for the treatment of dementia, the most fruitful period of dementia research is still yet to come. It is only in the last decade that there has been a dramatic explosion of research in this area, and drug researchers have only had the last few years to begin to formulate strategies for the treatment of dementia based on detailed scientific knowledge of the disorder. Within the next few years, there will probably be dozens of new drugs in laboratories around the world that will pass from drawing boards, computer screens and research laboratories to clinical assessment in demented patients. The clinical testing of novel compounds will enable the validation of the very hypotheses and model systems that first led to the development of those compounds.

Within the next few years, there is some need for closer cooperation between basic researchers and clinicians. The development of better tools for the early detection of dementia might be important for drug development, because it is possible that only those patients identified relatively early will be responsive to drug treatment. It is possible that the identification of clinical subtypes of Alzheimer's disease or SDAT will emerge, and that only certain subtypes will be helped by a particular drug. There is also some need for convergence between the methodologies used in experimental and clinical studies. Most animal testing programs compare many different drugs on a small range of tests. However, the clinical studies conducted thus far have not provided comparisons between a large number of drugs on a standard protocol. For example, the most useful way to indicate the relative merits of anticholinesterases compared to direct agonists is to assess a number of these compounds in a standard clinical battery in the same laboratory. For their part, animal researchers can focus more on tests that are functionally similar to human procedures. In fact, the use of computerized tests of nonverbal cognitive skills could enable the same tests to be performed on monkeys and humans.

Future research on the development of drugs for the treatment of dementia is liable to grow in directions that are quite different from those described here. In particular, it is likely that drug development programs will move away from the neurotransmitter-related projects described above. Basic research on the cellular and molecular basis of the dementias will probably shift the focus of attention to the factors leading to the onset of cellular degeneration. With the inception of such an approach, drug candidates will move from being neurotransmitter agonists to being stable analogues of neurotrophic factors. Several studies have already indicated that the peptide, nerve growth factor, can have growth-promoting effects in damaged or developing cholinergic systems (Hefti *et al.*, 1984; Mobley *et al.*, 1985; Will & Hefti, 1985). Ganglioside GM1 enhanced choline acetyltransferase activity in rats with electrolytic lesions of the nucleus basalis (Cassamenti *et al.*, 1985). In the future, cell culture studies on factors that promote cell survival or neurotransmitter synthesis may replace tests such as receptor binding assays as initial screens for potential drug candidates.

Fetal cell transplantation may also be considered as a future treatment for dementia, even as it is now being tested in Parkinsonian patients. Transplantation of embryonic ventral forebrain neurons was able to reverse deficits in the performance of memory tasks in rats following lesions of the fimbria–fornix (Dunnett *et al.*, 1982, 1988) or nucleus basalis (Dunnett *et al.*, 1985). These transplants could be used to replace some of the afferent cortical innervation that is lost in demented patients. It should be considered that transplants could be used ultimately to provide cells that synthesize neurotrophic factors, thereby helping to delay the course of the neuronal degeneration.

Considerable resources are currently being committed to the medical, academic and industrial research effort to find a treatment for the dementias. Although it is hoped that there will soon be a useful treatment, it is not certain that this will be the case. One must ask what will happen if, five or ten years into the future, there is still no major breakthrough in treatment methods. It should be recognized that dementia is a major health problem, that will only get worse because of our aging populations. Rather than considering dementia as a unitary disease that has a potential to be managed with a single treatment, it is much more useful to think of it as being a long-term research problem that will be approached in a variety of ways for many years to come. The relative amount of financial and logistical resources devoted to finding treatments for dementia ultimately must reflect the possible benefit such treatments will have for human society.

REFERENCES

Abercrombie, E.D. & Jacobs, B.L. (1987). Single unit response of noradrenergic neurons in the locus coeruleus of freely moving cats. 1. Acutely presented stressful and nonstressful stimuli. *Journal of Neuroscience* **7**, 2837–43.

Aghajanian, G.K. & Rogowski, M.A. (1983). The physiological role of alpha adrenoreceptors in the CNS: new concepts from single-cell studies. *Trends in Pharmacological Science* **1**, 315–17.

Aigner, T.G., Mitchell, S.J., Aggleton, J.P., Delong, M.R., Struble, R.G., Price, D.L., Wenk, G.L. & Mishkin, M. (1987). Effects of scopolamine and physostigmine on recognition memory in monkeys

with ibotenic acid lesions of the nucleus basalis of Maynert. *Psychopharmacology* **92**, 292–300.

Arai, H., Kosaka, K. & Iizuka, R. (1984). Changes of biogenic amines and their metabolites in postmortem brains from patients with Alzheimer-type dementia. *Journal of Neurochemistry* **43**, 388–93.

Armstrong, D.M., LeRoy, S., Shields, D. & Terry, R.D. (1985). Somatostatin-like immunoreactivity within neuritic plaques. *Brain Research* **338**, 71–9.

Ban, T.A. (1978). Vasodilators, stimulants, and anabolic agents in the treatment of geropsychiatric patients. In Lipton, M.A., DiMascio, A. & Killam, K.F. (eds.), *Psychopharmacology: A Generation of Progress*, pp. 1525–33. Raven Press: New York.

Bartus, R.T. (1978). Evidence for a direct cholinergic involvement in the scopolamine-induced amnesia in monkeys: effects of concurrent administration of physostigmine and methylphenidate with scopolamine. *Pharmacology, Biochemistry and Behavior* **9**, 833–6.

Bartus, R.T. (1979). Physostigmine and recent memory: effects in young and aged nonhuman primates. *Science* **206**, 1087–9.

Bartus, R.T. & Johnson, H.R. (1976). Short-term memory in the rhesus monkey: disruption from the anticholinergic scopolamine. *Pharmacology, Biochemistry and Behavior* **5**, 39–46.

Bartus, R.T., Dean, R.L. & Beer, B. (1980). Memory deficits in aged cebus monkeys and facilitation with central cholinomimetics. *Neurobiology of Aging* **1**, 145–52.

Bartus, R.T., Dean, R.L., Beer, B. & Lippa, A.S. (1982). The cholinergic hypothesis of geriatric memory dysfunction. *Science* **217**, 408–17.

Beal, M.F., Mazurek, M.F., Tran, V.T., Chattha, G., Bird, E.D. & Martin, J.B. (1985). Reduced numbers of somatostatin receptors in the cerebral cortex in Alzheimer's disease. *Science* **229**, 289–91.

Becker, R.E. & Giacobini, E. (1988). Mechanisms of cholinesterase inhibition in senile dementia of the Alzheimer's type: clinical, pharmacological and therapeutic aspects. *Drug Development Research* **12**, 163–95.

Beller, S.A., Overall, J.E. & Swann, A.C. (1985). Efficacy of oral physostigmine in primary degenerative dementia. *Psychopharmacology* **87**, 147–51.

Bergman, I., Brane, G., Gottfries, C.G., Jostell, K.G., Karlsson, I. & Svennerholm, L. (1983). Alaproclate: a pharmacokinetic and biochemical study in patients with dementia of Alzheimer type. *Psychopharmacology* **80**, 279–83.

Bondareff, W., Mountjoy, C.Q. & Roth, M. (1981). Selective loss of neurons of origin of adrenergic projection to cerebral cortex (locus coeruleus) in senile dementia. *Lancet* **1**, 783–4.

Bowen, D.M., Smith, C.B., White, P. & Davison, A.N. (1976). Neurotransmitter-related enzymes and indices of hypoxia in senile dementias and other abiotrophies. *Brain* **99**, 459–96.

Branconnier, R.J., Cole, J.O., Dessaub, E.C., Spera, K.F., Ghazvinian, S. & DeVitt, D. (1983). The therapeutic efficacy of pramiracetam in Alzheimer's disease: Preliminary observations. *Psychopharmacology Bulletin* 19, 726–30.

Carli, M., Robbins, T.W., Evenden, J.L. & Everitt, B.J. (1983). Effects of lesions to ascending noradrenergic neurones on performance of a 5-choice serial reaction task in rats: implications for theories of dorsal bundle function based on selective attention and arousal. *Behavioral Brain Research* **9**, 361–80.

Cassamenti, F., Bracco, L., Bartolini, L. & Pepeu, G. (1985). Effects of ganglioside treatment in rats with a lesion of the cholinergic forebrain nuclei. *Brain Research* **338**, 45–52.

Caulfield, M.P., Fortune, D.H., Roberts, P.M. & Stubley, J.K. (1981). Intracerebroventricular hemicholinium-3 (HC-3) impairs learning of a passive avoidance task in mice. *British Journal of Psychiatry* **74**, 865P.

Christie, J.F., Shering, A., Ferguson, J. & Glen, A.I.M. (1981). Physostigmine and arecoline: effects of intravenous infusions in Alzheimer presenile dementia. *British Journal of Psychiatry* **138**, 46–50.

Conn, P.J. & Sanders-Bush, E. (1987). Central serotonin receptors: effector systems, physiological roles and regulation. *Psychopharmacology* **92**, 267–77.

Coyle, J.T., Price, D.L. & Delong, M.R. (1983). Alzheimer's disease: a disorder of cortical cholinergic innervation. *Science* **219**, 1184–90.

Cumin, R., Bandle, E.F., Gamzu, E. & Haefely, W.E. (1982). Effects of the novel compound aniracetam (Ro 13-5957) upon impaired learning and memory in rodents. *Psychopharmacology* **78**, 104–11.

Cutler, N.R., Haxby, J.V., Narang, P.K., May, C. & Burg, C. (1985*a*). Evaluation of an analogue of somatostatin (L363,586) in Alzheimer's disease. *New England Journal of Medicine* **312**, 725.

Cutler, N.R., Haxby, J., Kay, A.D., Narang, P.K., Lesko, L.J., Costa, J.L., Ninos, M., Linoila, M., Potter, W.Z., Renfrew, J.W. & Moore, A.M. (1985*b*). Evaluation of Zimelidine in Alzheimer's disease: cognitive and biochemical measures. *Archives of Neurology* **42**, 744–8.

Daly, J.W. (1982). Adenosine receptors: targets for future drugs. *Journal of Medicinal Chemistry* **25**, 197.

Davies, P. & Maloney, A. (1976). Selective loss of central cholinergic neurons in Alzheimer's disease. *Lancet* **2**, 1403.

Davies, R. & Terry, R.D. (1981). Cortical somatostatin-like immunoreactivity in cases of Alzheimer's disease and senile dementia of the Alzheimer's type. *Neurobiology of Aging* **2**, 9–14.

Davies, P., Katzman, R. & Terry, R.D. (1980). Reduced somatostatin-like immunoreactivity in cerebral cortex from cases of Alzheimer disease and Alzheimer senile dementia. *Nature* **288**, 279–80.

Davis, K.L., Mohs, R.C., Rosen, W.G., Greenwald, B.S., Levy, M.I. & Horvath, T.B. (1983). Memory enhancement with oral physostigmine in Alzheimer's disease. *New England Journal of Medicine* **308**, 721.

Dayton, H.E. & Garrett, R.L. (1973). Production of analgesia by cholinergic drugs. *Proceedings of the Society for Experimental Biology and Medicine* **142**, 1011–13.

Dowson, J.H. (1982). Pharmacological treatment of chronic cognitive deficit: A review. *Comprehensive Psychiatry* **23**, 85–95.

Drachman, D.A. & Leavitt, J.L. (1974). Human memory and the cholinergic system: relationship to aging? *Archives of Neurology* **30**, 113–21.

Drachman, D.A. & Sahakian, B.J. (1980). Memory, aging, and pharmacosystems. In Stein, D. (ed.) *The Psychobiology of Aging: Problems and Perspectives*, pp. 347–68. Elsevier North Holland: Amsterdam.

Dunne, M.P. & Hartley, L.R. (1985). The effects of scopolamine upon verbal memory: evidence for an attentional hypothesis. *Acta Psychologica* **38**, 205–17.

Dunnett, S.B. (1985). Comparative effects of cholinergic drugs and lesions of nucleus basalis or fimbria–fornix on delayed matching in rats. *Psychopharmacology* **87**, 357–63.

Dunnett, S.B., Low, W.C., Iversen, S.D., Stenevi, U. & Bjorklund, A. (1982). Septal transplants restore maze learning in rats with fornix fimbria lesions. *Brain Research* **251**, 335–48.

Dunnett, S.B., Toniolo, G., Fine, A., Ryan, C.N., Bjorklund, A. & Iversen, S.D. (1985). Transplantation of embryonic ventral forebrain neurons to the neocortex of rats with lesions of nucleus basalis magnocellularis-II. Sensorimotor and learning impairments. *Neuroscience* **16**, 787–97.

Ennaceur, A. & Delacour, J. (1987). Effect of combined or separate administration of piracetam and choline on learning and memory in the rat. *Psychopharmacology* **92**, 58–67.

Evans, H. (1975). Scopolamine effects on visual discrimination: modifications related to stimulus control. *Journal of Pharmacology and Experimental Therapeutics* **195**, 105–13.

Everitt, B.J., Robbins, T.W., Gaskin, M. & Fray, P.J. (1983). The effects of lesions to ascending noradrenergic neurons on discrimination learning and performance in the rat. *Neuroscience* **10**, 397–410.

Everitt, B.J., Robbins, T.W., Evenden, J.L., Marston, H.M., Jones, G.H. & Sirka, T.E. (1987). The effects of excitotoxic lesions of the substantia innominata, ventral and dorsal globus pallidus on the acquisition and retention of a conditional visual discrimination: implications for cholinergic hypotheses of learning and memory. *Neuroscience* **22**, 441–70.

Flicker, C., Dean, R.L., Watkins, D.L., Fisher, S.K. & Bartus, R.T. (1983). Behavioral and neurochemical effects following neurotoxic lesions of a major cholinergic input to the cerebral cortex in the rat. *Pharmacology, Biochemistry and Behavior* **18**, 973–81.

Flood, J.F., Smith, G.E. & Cherkin, A. (1985). Memory enhancement: supra-additive effects of subcutaneous cholinergic drug combinations in mice. *Psychopharmacology* **86**, 61–7.

Gaffan, D. (1974). Recognition impaired and association intact in the memory of monkeys after transection of the fornix. *Journal of Comparative and Physiological Psychology* **29**, 577–88.

Gamzu, E. & Perrone, L. (1981). Pharmacological protection against hypoxic and electro brain shock disruption of avoidance retrieval in mice. *Federation Proceedings* **7**, 525.

Glick, S.D. & Zimmerberg, B. (1972). Amnesic effects of scopolamine. *Behavioral Biology* **7**, 245–54.

Greenwald, B.S. & Davis, K.L. (1983). Experimental pharmacology of Alzheimer's disease. In Mayeux, R. & Rosen, W.G. (eds.), *The Dementias*, pp. 87–101. Raven Press: New York.

Hagan, J.J., Tweedie, F. & Morris, R.G.M. (1986). Lack of task specificity and absence of post-training effects of atropine on learning. *Behavioral Neuroscience* **100**, 483–93.

Hammer, R. & Giachetti, A. (1984). Selective muscarinic receptor antagonists. *Trends in Pharmacological Sciences* **2**, 18–20.

Handley, S.L. & Singh, L. (1986). Neurotransmitters and shaking behaviour – more than a 'gut bath' for the brain? *Trends in Pharmacological Sciences* **8**, 324–8.

Haroutunian, V., Barnes, E. & Davis, K.L. (1985). Cholinergic modulation of memory in rats. *Psychopharmacology* **87**, 266–71.

Hearst, E. (1964). Drug effects on stimulus generalization gradients in the monkey. *Psychopharmacology* **6**, 57–70.

Hefti, F., Dravid, A. & Hartikka, J. (1984). Chronic intraventricular injections of nerve growth factor elevate hippocampal choline acetyltransferase activity in adult rats with partial septo-hippocampal lesions. *Brain Research* **293**, 305–11.

Heise, G.A., Conner, R. & Martin, R.A. (1976). Effects of scopolamine on variable intertrial interval spatial alternation and memory in the rat. *Psychopharmacology* **49**, 131–7.

Hepler, D.J., Olton, D.S., Wenk, G.L. & Coyle, J.T. (1985). Lesions of the nucleus basalis magnocellularis and medial septal area of rats produce qualitatively similar memory impairments. *Journal of Neuroscience* **5**, 866–73.

Hyman, B.T., Van Hoesen, G.W., Damasio, A.R. & Barnes, C.L. (1984). Alzheimer's disease: cell-specific pathology isolates the hippocampal formation. *Science* **225**, 1168–70.

Janssen, P.A.J., Niemegeers, C.J.E. & Schellekens, K.H.L. (1965). Is it possible to predict the clinical effects of neuroleptic drugs (major tranquilizers) from animal data? *Arzneimittel-Forschung* **15**, 104–17.

Knowlton, B.J., Wenk, G.L., Olton, D.S. & Coyle, J.T. (1985). Basal forebrain lesions produce a dissociation of trial-dependent and trial-independent memory performance. *Brain Research* **345**, 315–21.

Ksir, C.J. (1974). Scopolamine effects on two-trial delayed-response performance in the rat. *Psychopharmacology* **34**, 127–34.

Lerer, B., Warner, J., Friedman, E., Vincent, G. & Gamzu, E. (1985). Cortical cholinergic impairment and behavioral deficits produced by kainic acid lesions of the rat magnocellular basal forebrain. *Behavioral Neurosciences* **99**, 661–77.

Millar, K.S. (1981). Cholinergic drug effects on visual discriminations: a signal detection analysis. *Psychopharmacology* **74**, 383–8.

Mobley, W.C., Rutkowski, J.L., Tennekoon, G.I., Buchanan, K. & Johnston, M.V. (1985). Choline acetyltransferase activity in striatum of neonatal rats increased by nerve growth factor. *Science* **229**, 284–7.

Mohs, R.C. & Davis, K.L. (1982). A signal detectability analysis of the effect of physostigmine on memory in patients with Alzheimer's disease. *Neurobiology of Aging* **3**, 105–10.

Mohs, R.C., Davis, B.M., Johns, C.A., Mathe, A.A., Greenwald, B.S., Horvath, T.B. & Davis, K.L. (1985). Oral physostigmine treatment of patients with Alzheimer's disease. *American Journal of Psychiatry* **136**, 1275–7.

Mountjoy, C.Q., Roth, M., Evans, N.J.R. & Evans, H.M. (1983). Cortical neuronal counts in normal elderly controls and demented patients. *Neurobiology of Aging* **4**, 1–11.

Murray, C.L. & Fibiger, H.C. (1985). Learning and memory deficits after lesions of the nucleus basalis magnocellularis: reversal by physostigmine. *Neuroscience* **14**, 1025–32.

Murray, C.L. & Fibiger, H.C. (1986). Pilocarpine and

physostigmine attenuate spatial memory impairments produced by lesions of the nucleus basalis magnocellularis. *Behavioral Neuroscience* **100**, 23–32.

Newhouse, P.A., Sunderland, T., Tariot, P.N., Blumhardt, C.L., Weingartner, H., Mellow, A. & Murphy, D.L. (1988). Intravenous nicotine in Alzheimer's disease: a pilot study. *Psychopharmacology* **95**, 171–5.

Nicolaus, B.J.R. (1982). Chemistry and pharmacology of nootropics. *Drug Development Research* **2**, 462–74.

Passeri, M., Cunciotta, D., De Mello, M., Storchi, G., Roncucci, R. & Biziere, K. (1985). Minaprine for senile dementia. *Lancet* **1**, 824.

Pazzagli, A. & Pepeu, G. (1964). Amnesic properties of scopolamine and brain acetylcholine in the rat. *International Journal of Neuropharmacology* **4**, 291–9.

Peabody, C.A., Thiemann, S., Pigache, R., Miller, T.P., Berger, P.A., Yesavage, J. & Tinklenberg, J.R. (1985). Desglycinamide-9-arginine-8-vasopressin (DGAVP Organon 5667) in patients with dementia. *Neurobiology of Aging* **6**, 95–100.

Penetar, D.M. & McDonough, J.H. (1983). Effects of cholinergic drugs on delayed match-to-sample performance of Rhesus monkeys. *Pharmacology, Biochemistry and Behavior* **19**, 963–7.

Perry, E.K., Tomlinson, B.E., Blessed, G., Bergmann, K., Gibson, P.H. & Perry, R.H. (1978). Correlation of cholinergic abnormalities with senile plaque and mental test scores in senile dementia. *British Medical Journal* **2**, 1457–9.

Perry, R.H., Dockray, G.H., Dimaline, R., Perry, E.K., Blessed, G. & Tomlinson, B.F. (1981). Neuropeptides in Alzheimer's disease, depression and schizophrenia: a post mortem analysis of vasoactive intestinal peptide and cholecystokinin in cerebral cortex. *Journal of the Neurological Sciences* **51**, 465–72.

Peterson, C. & Gibson, G.E. (1983). Amelioration of age-related neurochemical and behavioral deficits by 3,4-diaminopyridine. *Neurobiology of Aging* **4**, 25–30.

Rawlins, J.N.P. & Olton, D.S. (1982). The septo-hippocampal system and cognitive mapping. *Behavioral Brain Research* **5**, 331–58.

Reisberg, B., Ferris, S.H. & Gershon, S. (1981). An overview of pharmacologic treatment of cognitive decline in the aged. *American Journal of Psychiatry* **138**, 593–600.

Reisberg, B., Ferris, S.H., Schneck, M.K., Corwin, J., Mir, P., Friedman, E., Sherman, K.A., McCarthy, M. & Bartus, R.T. (1982). Piracetam in the treatment of cognitive impairment in the elderly. *Drug Development Research* **2**, 275–480.

Ridley, R.M., Baker, H.F., Drewett, B. & Johnson,

J.A. (1985). Effects of ibotenic acid lesions of the basal forebrain on serial reversal learning in marmosets. *Psychopharmacology* **86**, 438–43.

Ridley, R.M., Murray, T.K., Johnson, J.A. & Baker, H.F. (1986). Learning impairment following lesion of the basal nucleus of Meynert in the marmoset. *Brain Research* **376**, 108–16.

Ridley, R.M., Baker, H.F. & Drewett, P. (1987). Effects of arecoline and pilocarpine on learning ability in marmosets pretreated with hemicholinium-3. *Psychopharmacology* **91**, 512–14.

Rossor, M.N., Fahrenkrug, J., Emson, P., Mountjoy, C., Iversen, L.L. & Roth, M. (1980). Reduced cortical choline acetyltransferase activity in senile dementia of Alzheimer's type is not accompanied by changes in vasoactive intestinal peptide. *Brain Research* **201**, 249–53.

Rossor, M.N., Rehfeld, J.F., Emson, P.C., Mountjoy, C.Q., Roth, M. & Iversen, L.L. (1981). Normal cortical concentration of cholecystokinin with reduced choline acetyltransferase activity in senile dementia of Alzheimer's type. *Life Science* **29**, 405–10.

Rossor, M.N., Iversen, L.L., Reynolds, G.P., Mountjoy, C.Q. & Roth, M. (1984). Neurochemical characteristics of early and late Alzheimer's disease. *British Medical Journal* **288**, 961–4.

Rothschild, D. & Kasansin, J. (1936). Clinicopathologic study of Alzheimer's disease: relationship to senile condition. *Archives of Neurology and Psychiatry* **36**, 293–321.

Rupniak, N.M.J., Jenner, P. & Marsden, C.D. (1985). Pharmacological characterization of spontaneous or drug-associated purposeless chewing movements in rats. *Psychopharmacology* **85**, 71–9.

Salamone, J.D. (1986). Behavioural functions of nucleus basalis magnocellularis and its relationship to dementia. *Trends in Neuroscience* **9**, 256–8.

Salamone, J.D., Beart, P.M., Alpert, J.E. & Iversen, S.D. (1984). Impairment in T-maze reinforced alternation performance following nucleus basalis magnocellularis lesions in rats. *Behavioral Brain Research* **13**, 63–70.

Salamone, J.D., Lalies, M.D., Channell, S.L. & Iversen, S.D. (1986). Behavioural and pharmacological characterization of the mouth movements induced by muscarinic agonists in the rat. *Psychopharmacology* **88**, 467–71.

Salamone, J.D., Channell, S.L., Welner, S.A., Gill, R., Robbins, T.W. & Iversen, S.D. (1987). Nucleus basalis lesions and anticholinergic drugs impair spatial memory and visual discrimination in the rat. In *Cellular and Molecular Basis of Cholinergic Function*, ed. M.J. Dowdall & J.N. Hawthorn, pp. 835–40. Chichester, England: Ellis Horwood.

Serby, M., Corwin, J., Rotrosen, J., Ferris, S.,

Reisberg, B., Freidman, E., Sherman, K., Bartus, R. & Jorden, B. (1983). Lecithin and piracetam in Alzheimer's disease. *Psychopharmacology Bulletin* **19**, 126–9.

Serby, M., Richardson, S.B., Twente, S., Siekierski, J., Corwin, J. & Rotrosen, J. (1984). CSF somatostatin in Alzheimer's disease. *Neurobiology of Aging* **5**, 187–9.

Smith, C.M. & Swash, M. (1978). Possible biochemical basis of memory disorder in Alzheimer's disease. *Annals of Neurology* **3**, 471–3.

Smith, D.F., Stomgren, E., Pertersen, H.N., Williams, D.G. & Sheldon, W. (1984). Lack of effect of tryptophan treatment in demented geronto-psychiatric patients. *Acta Psychiatrica Scandinavica* **70**, 470–7.

Spagnoli, A. & Tognoni, G. (1983). 'Cerebroactive' drugs: clinical pharmacology and therapeutic role in cerebrovascular disorders. *Drugs* **26**, 44–69.

Spencer, D.G. & Lal, H. (1983). Effects of anticholinergic drugs on learning and memory. *Drug Development Research* **3**, 489–502.

Struble, R.G., Price, D.L. Jr., Cork, L.C. & Price, D.L. (1985). Senile plaques in aged normal monkeys. *Brain Research* **361**, 267–75.

Summers, W.K., Majovski, L.V., Marsh, G.M., Tachiki, K. & Kling, V. (1986). Oral tetrahydroaminoacridine in long-term treatment of senile dementia, Alzheimer's type. *New England Journal of Medicine* **315**, 1241–5.

Terry, R.D. & Katzman, R. (1983). Senile dementia of the Alzheimer's type. *Annals of Neurology* **14**, 497–506.

Thal, L.J. & Fuld, P.A. (1983). Memory enhancement with oral physostigmine in Alzheimer's disease. *New England Journal of Medicine* **308**, 720.

Thal, L.J., Rosen, W., Sharpless, N.S. & Crystal, H. (1981). Choline chloride fails to improve cognition in Alzheimer's disease. *Neurobiology of Aging* **2**, 205–8.

Tricklebank, M.D., Forler, C., Middlemiss, D.N. & Fozard, J. (1985). Subtypes of the 5-hydroxy-tryptamine receptor mediating the behavioral responses to 5-methoxy-N,N-dimethyltryptamine. *European Journal of Pharmacology* **117**, 15–24.

Tsaltas, E., Preston, G.C. & Gray, J.A. (1983). Effects of dorsal bundle lesions on serial and trace conditioning. *Behavioral Brain Research* **10**, 361–74.

Vizi, E.S. (1986). Compounds acting on alpha 1- and alpha 2- adrenoreceptors: agonists and antagonists. *Medicinal Research Reviews* **6**, 431–49.

Walker, L.C., Kitt, C.A., Schwam, E., Buckwald, B., Garcia, F., Sepinwall, J. & Price, D.L. (1987). Senile plaques in aged squirrel monkeys. *Neurobiology of Aging* **8**, 291–6.

Walsh, T.J., Tilson, H.A., DeHaven, D.L., Mailman, R.B., Fisher, A. & Hanin, I. (1984). AF64A, a cholinergic neurotoxin, selectively depletes acetylcholine in hippocampus and cortex, and produces long-term passive avoidance and radial-maze deficits in the rat. *Brain Research* **321**, 91–102.

Welner, S.A., Dunnett, S.B., Salamone, J.D., Maclean, B. & Iversen, S.D. (1988). Transplantation of embryonic ventral forebrain grafts to the neocortex of rats with bilateral lesions of nucleus basalis magnocellularis ameliorates a lesion-induced deficit in spatial memory. *Brain Research* **463**, 192–7.

Wenk, G.L., Cribbs, B. & McCall, L. (1984). Nucleus basalis magnocellularis: optimal coordinates for selective reduction of choline acetyltransferase in frontal neocortex by ibotenic acid injections. *Experimental Brain Research* **56**, 335–40.

Wettstein, A. & Speigel, R. (1984). Clinical trials with the cholinergic drug RS 86 in Alzheimer's disease (AD) and senile dementia of the Alzheimer's type (SDAT). *Psychopharmacology* **84**, 572–3.

Whishaw, I.Q., O'Connor, W.T. & Dunnett, S.B. (1985). Disruption of central cholinergic systems in the rat by basal forebrain lesions or atropine: effects on feeding, sensorimotor behaviour, locomotor activity, and spatial navigation. *Behavioral Brain Research* **17**, 103–15.

Will, B. & Hefti, F. (1985). Behavioral and neurochemical effects of chronic intraventricular injections of nerve growth factor in adult rats with fimbria lesions. *Behavioral Brain Research* **17**, 17–24.

Willner, P. (1984). The validity of animal models of depression. *Psychopharmacology* **83**, 1–16.

Dementia: the role of behavioral models

HARVEY J. ALTMAN, SAMUEL GERSHON and HOWARD J. NORMILE

Recent advances in the fields of medicine and health care delivery have significantly improved life expectancy throughout the world. Particularly within many of the industrialized nations, this has also resulted in a significant shift in the demographics of the overall population. Now more than ever, society must seriously come to grips with the needs of the elderly. This is proving, however, to be a more formidable task than was originally thought. In fact, it would appear that the magnitude of the social and health-related needs of this population are becoming more, rather than less, acute with the passage of time.

16.1 THE IMPACT OF DEMENTIA ON HEALTH CARE-RELATED COSTS

The elderly have been found to present with a unique profile of medically-related needs and problems. One of these, dementia, accounts for more admissions and more hospital in-patient days than any other psychiatric illness of the elderly (Cummings & Benson, 1983; Khachaturian, 1985). It is estimated that each person with a dementing illness will require an average of seven years of care, either at home or in a residential care facility. In the United States, care provided at home is estimated to cost about $12,000 annually, for a total of $84,000 per person. This is a conservative figure, however, as many persons suffering from dementia spend their last few years in a nursing home at an average cost of $22,000 per year, and some spend from ten to 15 years in a nursing home, for a total cost of $220,000 to $330,000 (Michigan Department of Public Health,

1987). In 1986 alone, over $35 billion was spent in the U.S. on the care of the demented with another $39 billion for what has been referred to as 'indirect costs' (National Institute on Aging, 1986). The human cost is even less easily estimated because it is extremely difficult to put a dollar figure on the impact such care has on the caregiver or the patient's family.

16.2 DEMENTIA AS A CLINICAL SYNDROME

According to the DSM-III-R (American Psychiatric Association, 1987) the essential feature of dementia is an impairment in short- and long-term memory, associated with impairments in any one of the following: abstract thinking, judgement, personality or other disturbances of higher cortical function. In addition, the disturbance must be severe enough to interfere significantly with work or usual social activities or relationships.

Not all types/forms of dementia are, however, the same and a number of approaches have been offered as a means of differentiating them. In general, this has been achieved either by (1) focusing on the nature of the essential neuroanatomic structures thought to underlie a particular pattern of neurophysiological impairments, or by (2) focusing on the course of the neurodegenerative deterioration without specifically regarding the nature of the neuroanatomic structures actually involved. For example, when differentiation is based on an analysis of the major neuroanatomical structures thought to underlie dementia, two basic patterns of neuropsychological impairments can be identified – those involving primarily the cerebral cortex (e.g., Alzheimer's disease and Pick's disease) and those involving primarily subcortical structures such as the basal ganglia, thalamus and brainstem. Specific examples of the latter include: (1) the extrapyramidal syndromes (e.g., Parkinson's disease, Huntington's chorea, progressive supranuclear palsy, Wilson's disease, spinocerebellar degeneration and idiopathic basal ganglia calcification); (2) hydrocephalus; and (3) toxic or metabolic encephalopathies (e.g., systemic illness, endocrinopathies, deficiency states, drug intoxications, heavy metal exposure, industrial dementias and pseudodementias). A third or 'mixed' category, is frequently included to account for those dementia-producing states which appear to involve both cortical and subcortical structures (e.g., multi-infarct dementia, the infectious dementias, and a variety of miscellaneous dementias including post-traumatic, postanoxic and neoplastic) (Cummings & Benson, 1983).

This approach to the differential analysis of dementia-producing states is attractive because it also attempts to contrast the nature of the neuropsychological profiles idiosyncratic of people suffering from cortical vs. subcortical dementias. Specifically, people suffering from cortical dementias generally present with language and speech problems (aphasia with naming difficulties, impaired comprehension, and agnosia) on top of disturbances in memory and cognition. Visuospatial and constructional disturbances are also frequently encountered, as are disturbances in affect. Personality, on the other hand, is generally well preserved, as are most aspects of motor behavior such as posture, tone, movements and gait (at least until the later phases of the disease). The subcortical dementias differ in that language is preserved while motor activity is often markedly altered (posture – stooped/extended; tone – usually increased; movements – abnormal/tremor, chorea, asterixis; gait – abnormal). The memory and cognitive impairments of these two classes of dementias also appear to differ, although the differences in some cases may have more to do with interpretative biases than anything else. In spite of this potential shortcoming, an appreciation of such differences could be useful, diagnostically, to provide valuable insight into the locus and extent of the underlying pathology (for a more detailed discussion of this last point see Weingartner,1984; Wolkowitz *et al.*, 1985*a, b*; Berg, 1985). For example, some have suggested that the essential feature of the cortical dementias is an inability to encode information effectively (i.e., new learning). This encoding deficit is thought to be directly related to an impairment of episodic memory, the extent of which is directly related to an impairment of semantic or knowledge memory (Weingartner, 1984; see also Dunnett & Barth, this vol., Ch. 14). Others would disagree and suggest that the deficit should more appropriately be regarded as either a processing deficit at the acquisition phase (Davis & Mumford, 1984) or a retrieval deficit (Morris *et al.*, 1983). Evidence in support of an encoding deficit is derived from the observation that patients suffering from cortical dementias, such as Alzheimer's disease, generally have difficulty in finding words, naming, appreciating the meaning of ongoing events, and logical conceptual thinking (de Ajuriaguerra & Tissot, 1975). Support for the retrieval hypothesis is derived from data which indicate that demented subjects can sometimes be shown to perform more poorly than age-matched controls in tasks requiring forced recognition, but seem unimpaired when the tasks are shifted to cued recall (Morris *et al.*, 1983). Support for a processing deficit at the acquisition phase is also derived from comparisons of the performance of demented vs. non-demented age-matched controls in tasks involving cued vs. forced recognition. Instead, the differential improvement seen in demented subjects following cued recall is interpreted to be due to a stimulation of a memory trace that is already present in the brain, but inaccessible under normal conditions: the cue provides information as to the location of the trace and the memory can then be retrieved (Davis & Mumford, 1984).

It has been suggested, on the other hand, that the memory impairments associated with the subcortical dementias are predominantly retrieval deficits (Cummings & Benson, 1983). This hypothesis is again based on the observation that while people suffering from subcortical dementias have difficulty spontaneously retrieving old information, they are aided by clues and structure and can often be shown to have learned much of what they originally were unable to recall without prompting.

Some of the hypotheses generated to explain the

basis of specific failures in mnemonic processing should be viewed with considerable caution. For example, it is nearly impossible to know with any degree of certainty whether a subsequent memory failure is due to improper encoding (addressing), consolidation, storage or retrieval. It is often presumed that if the existence of the trace can be demonstrated, then the deficit is probably due to a failure in the retrieval process. However, again, it is still unclear whether the retrieval deficit is due to a failure of the retrieval mechanism itself, in initial encoding or in the mechanims responsible for laying down or delineating the path that information will eventually take when it is necessary to retrieve that information at a later date. Similar problems are encountered when attempting to describe the algorithm or strategy being used to solve various types of learning and memory problems (e.g., procedural vs. declarative, episodic vs. semantic, allocentric vs. egocentric, spatial vs. nonspatial). Again, the different taxonomic classifications offer important descriptors of how information is being viewed and processed by the brain. However, all of the terms fall short of completely describing or predicting how the information will eventually be handled by the organism. Unfortunately, an in depth analysis of this subject goes beyond the scope of this chapter. However, the reader is encouraged to consult a number of well written reviews that have been published on this subject (Weingartner, 1984; Squire & Zola-Morgan, 1985). The other dimension which may distinguish between cortical and subcortical dementias is cognition, which may best be described as dilapidated in people suffering from the subcortical dementias (McHugh & Folstein, 1979). That is, these individuals may accurately perform individual steps of a complex problem but fail to synthesize the elements or utilize strategies (mnemonics) to solve the problem (Cummings & Benson, 1983).

As indicated earlier, one can also attempt to differentiate the various dementia-producing states by focusing on the apparent 'course' of the specific neurodegenerative deterioration rather than on the specific structures affected or the differences in their respective neuropsychological profiles. For example, differentiation of dementia-producing states can be made according to whether the pathology underlying the dementia is progressive, as in

Alzheimer's disease, fixed, as in hypoxic-ischemic encephalopathy due to cardiac arrest, or due to a reversible cause, examples of which include intoxication (e.g., opiate analgesics, adrenocortical steroids), infection (e.g., leptomeningitis, encephalitis, syphillis, AIDS), metabolic disorders (e.g., tumorous diseases of the thyroid, parathyroid, adrenals and pituitary, diabetes), nutritional deficits (e.g., thiamine or folate deficiency), vascular disease (e.g., hypertension, arteriosclerosis), space-occupying lesions (e.g., chronic subdural hematoma, benign or malignant tumors of the brain) and depression (National Institutes of Health Consensus Development Conference Statement, 1987).

Clearly, improvements in diagnostic accuracy will facilitate the determination of appropriate treatment strategies as well as aid in refining the differentiation between the various types of dementing illnesses. However, improvements in diagnostic criteria only address a single aspect of the problem. Advances in our understanding of the etiology, pathophysiology and treatment of each form or subtype of dementing illness is still seriously needed. Much of this can and will be resolved at the clinical level. However, a significant proportion can not. It is here that basic research in animals may prove particularly relevant.

16.3 ANIMAL MODELS OF DEMENTING ILLNESS

The basic tenant underlying the use of animals for biomedical research has been that the structure and function of basic CNS systems in animals parallel, at least to some degree, those in man and should, therefore, be able to provide valuable insight into how such systems operate in man. Indeed, there are numerous examples of successful applications of this rationale in the literature. Within the fields of psychiatry and neurology, often-cited examples include the impact animal research has had on the resolution and treatment of psychosis (Haber *et al.*, 1977; Bunney, 1977; Kornetsky & Markowitz, 1978; Losonczy *et al.*, 1987) and Parkinson's disease (Langston, 1985; Chiueh *et al.*, 1985; Wagner *et al.*, 1986).

Research into the etiology, pathophysiology and potential treatment of dementia, and Alzheimer's disease in particular, have also relied heavily on data derived from research conducted at the animal level.

Unfortunately, research in animals has had only limited impact on our understanding of the etiology, pathophysiology and treatment of dementia. There may be a number of reasons why research in animals has failed to provide this insight including: (1) an apparent lack of acceptable models; (2) a fundamental lack of understanding of the neurobiological basis of learning and memory; (3) a lack of information regarding the importance of certain types of morphological or biochemical changes that appear to occur in association with certain types of dementing illnesses; and, possibly most important of all, (4) the various paradigms used to assess learning and memory performance, or the basic premise underlying the generalizability of basic CNS functioning at the animal level to that in humans, may not be entirely applicable. The only way animal research will be able to contribute more effectively to an overall understanding of the etiology, pathophysiology and treatment of Alzheimer's disease is by re-evaluating the merits of each of the above points and modifying future research strategies appropriately.

16.3.1 Lack of an acceptable model

Most of the work on models of dementia have focused on Alzheimer's disease. A number of models have been proposed involving experimentally-induced reductions in forebrain cholinergic markers and memory. A variety of procedures have been used, the most notable being lesions of the cholinergic basal forebrain (nucleus basalis magnocellularis (NBM), ventral and horizontal nucleus of the diagonal band of Broca and medial septal nucleus) (Johnston *et al.*, 1979; Altman *et al.*, 1985*a*; Knowlton *et al.*, 1985; Watson *et al.*, 1985; Kesner *et al.*, 1986; Bartus *et al.*, 1986; Berman *et al.*, 1988) and administration of AF64A into various brain regions (Fisher & Hanin, 1980; Mantione *et al.*, 1981; Fisher *et al.*, 1982; Chrobak *et al.*, 1987). Other models include the induction of neurofibrillary tangles following aluminum toxicity (Simpson *et al.*, 1985; Ghetti *et al.*, 1985; Crapper-McLachlan, 1986), and the production of abnormal amyloid plaque formation following Scrapie virus infestation (Prusiner, 1982; Wisniewski *et al.*, 1985;

Bassant *et al.*, 1987; McKinley & Prusiner, 1987).

Considerable debate centers on whether or not such experimental preparations as summarized above should appropriately be regarded as 'models' of Alzheimer's disease (see Dunnett and Barth, this vol., Ch. 14). However, by definition, an animal model does not have to mimic the clinical manifestation(s) of a particular behavioral syndrome in order to be applicable: it simply has to provide a basis on which to make reliable generalizations about that syndrome and/or the amelioration of its behavioral symptomatology (see, for example, Salamone, this vol., Ch. 15). Each of the models appears to comply, at least in principle, with the definition of what an appropriate model should be. For example, lesions of the cholinergic basal forebrain reliably reduce cholinergic functioning within the brain and impair learning and memory (Altman *et al.*, 1985*a*; Knowlton *et al.*, 1985; Watson *et al.*, 1985; Kesner *et al.*, 1986; Bartus *et al.*, 1986; Berman *et al.*, 1988). Aluminum administration does produce an accumulation of neurofibrillary tangle formations which, in at least one report, also resulted in impairments in learning and memory (Crapper & Dalton, 1973). Finally, infestation of the Scrapie virus does result in a significant increase in the accumulation of amyloid (Prusiner, 1982; Bassant *et al.*, 1987; Merz *et al.*, 1987; McKinley & Prusiner, 1987). On the other hand, each of the models listed above also has its shortcomings. For example, reductions in cholinergic functioning following lesions of the basal forebrain cholinergic system, administration of AF64A, or any other experimental procedure selective to the cholinergic system, fails to mimic adequately the global nature of the neuropathology associated with Alzheimer's disease. The neurofibrillary tangles produced following aluminum administration are not morphologically identical to those seen in Alzheimer's brains (Crapper-McLachlan, 1986). In addition, there is considerable controversy regarding the significance of elevations in the levels of aluminum or the presence of aluminum in tangle-bearing neurons to the etiology and pathogenesis of Alzheimer's disease (Crapper-McLachlan, 1986; Wisniewski *et al.*, 1986). Finally, viral-like proteins have, as yet, not been identified within Alzheimer's brains and the transmittability of Alzheimer's disease via a viral agent is still highly suspect, even in light of the recent report of neurodegenerative changes in the brains of rats injected with white blood cells from Alzheimer's patients (Manuelidis *et al.*, 1988).

In spite of these shortcomings, however, each of the models can still provide valuable insight into the possible etiology, pathophysiology and/or treatment of the behavioral symptomatology associated not only with Alzheimer's disease, but with dementia in general. The problem, if one really exists, centers on the apparent overemphasis initially placed on the ability of the models to account *completely* for all of the neuropathological or behavioral manifestations of the disease. Reports on the effects of scopolamine- and NBM lesion-induced effects on learning and memory serve as particularly good examples (see Dunnett & Barth, this vol., Ch. 14). Briefly, it has been argued that neither manipulation completely mimics either the pathology or the cognitive impairments associated with the disease. However, these experimental manipulations have provided essential evidence in support of an involvement of the cholinergic nervous system in learning and memory, and underscored the relevance of the cholinergic deterioration in Alzheimer's brains to the cognitive deficits of the disease. It is for these reasons that cholinergic models of Alzheimer's disease are still important and highly relevant. Specifically, they provide: (1) the basis for continued study into the neuroanatomical basis and specificity of the role of the cholinergic nervous system in learning and memory; and (2) a means with which to assess the efficacy of specific pharmacological manipulations to block or partially attenuate the behavioral deficits. Clearly, it would be most helpful to have an experimental preparation which more closely mimicked the sequelae of the disease. However, in its absence, we should not discount too hastily the value of the models presently available.

16.3.2 Lack of a clear understanding of the basic processes underlying learning and memory

Another major obstacle confronting the development of appropriate treatment strategies for the dementias centers around our present level of understanding regarding the neurobiological basis of learning and memory. It is becoming increasingly apparent that dementing illnesses such as Alzheimer's disease affect a number of neurotransmitter and neuropeptidergic systems, many of which have also been implicated as playing an important role in the processes underlying learning and memory. However, there is still a great deal of debate regarding the precise relationship(s) between the functioning of each of these neurotransmitter or neuropeptidergic systems and how information is processed by the brain. This last point is illustrated by the results in our laboratory on the effects of manipulations of serotonin (5-HT) on learning and memory in rodents. For example, pre-test or immediate post-training administration of a variety of 5-HT receptor antagonists facilitates retention of a previously learned one-trial inhibitory avoidance response in mice. On the other hand, administration of these same compounds prior to training appears to interfere with subsequent retention in this task (Altman & Normile, 1986, 1987). Interpretation of these results is complicated by the observations that the performance of rats trained in the Stone 14-Unit T-maze, a complex spatial discrimination task, is significantly enhanced following selective cytotoxic lesions of central presynaptic 5-HT nerve terminals (Altman *et al.*, 1985*b*, 1989) while long-term (i.e., chronic) administration of the 5-HT$_2$ receptor antagonist, ketanserin, impairs the performance of rats trained in this task (Normile *et al.*, 1988). Furthermore, while cytotoxic lesions of central presynaptic terminals enhance learning in rats trained in the Stone 14-Unit T-maze, the performance of similarly treated rats trained in an eight-arm radial-arm maze remains unaffected (Altman *et al.*, 1989). Similar behavioral inconsistencies following manipulations of cholinergic (Deutsch & Rogers, 1979), noradrenergic (Altman & Quartermain, 1983; Haycock *et al.*, 1977), dopaminergic (Altman & Quartermain, 1983; Haycock *et al.*, 1977), and a variety of peptidergic systems (van Ree *et al.*, 1978; Arnsten & Segal, 1979; Koob & Bloom, 1983; Sahgal, 1984; Altman *et al.*, 1987*a*) have frequently been reported.

These apparent differences in behavioral results have been attributed to many factors, not the least of which are differences in (1) drug specificity, (2) the nature of the task being used to assess drug or lesion effects, (3) the route of administration, and (4) the types of animals being used. There is also the intriguing possibility that some of these differences in behavioral response are actually due to inherent differences in the involvement of a specific modulatory system or area of the brain in learning and memory. Unfortunately, owing to the complexity of the systems involved and the global nature of the strategies often employed (e.g., whole animal), identification of these effects may be difficult if not impossible.

16.3.3 Clinical significance of changes in certain aspects of brain functioning in demented subjects

An additional factor which tends to complicate the development of an effective treatment strategy for dementia is lack of basic information regarding the relevance of a number of changes observed within the brain to the behavioral symptomatology associated with specific subtypes of dementing illnesses. For example, with respect to Alzheimer's disease, one of the most reliable, yet perplexing, observations is the consistent reduction in the neuropeptide somatostatin (Soininen *et al.*, 1984; Francis *et al.*, 1984; Cutler *et al.*, 1985; Gomez *et al.*, 1986; Beal *et al.*, 1986; Reinikainen *et al.*, 1987; Hoyer, 1987). Assuming that every structure and chemical within the brain has a functional correlate, does the reduction in somatostatin contribute to the cognitive impairment of Alzheimer's disease, or does it contribute to one or more of the other myriad changes in behavior or homeostasis that characterize this disease? This kind of question could be addressed at the animal level. For example, what effect(s) would alterations in the levels or turnover of somatostatin have on learning and memory in animals? Would stimulation of somatostatin neurotransmission attenuate naturally-occurring or experimentally-induced memory impairments? The answers to these and other questions would certainly assist in our understanding of how changes in this peptidergic system affect the processing of information by the brain. It should be noted, however, that stimulation of the somatostatin receptors with the somatostatin analogue L-363,586 has been tried clinically and not found to improve memory performance (Cutler *et al.*, 1985).

Serotonin is another example. For years, involvement of the serotonergic nervous system in the pathophysiology of Alzheimer's disease was not seriously regarded. Recently, however, significant reductions in the levels of 5-HT and its metabolite, 5-HIAA, as well as reductions in the density of 5-HT receptors in various regions of the Alzheimer's brain, have been described (Bowen *et al.*, 1979, 1983; Cross *et al.*, 1983, 1984, 1986; Palmer *et al.*, 1984; Arai *et al.*, 1984). In addition, significant reductions in the number of serotonergic perikarya and a significant increase in the incidence of neurofibrillary tangles within the raphe nuclei have been observed (Ishii, 1966; Bowen *et al.*, 1979, 1983; Mann & Yates, 1983; Cross *et al.*, 1984, 1986; Yamamoto & Hirano, 1985). Again, keeping in mind that the serotonergic nervous system clearly plays a role in learning and memory (Ögren, 1982; Normile & Altman, 1987*b*), are certain aspects of either the memory deficits or other behavioral deficits associated with Alzheimer's disease the result of alterations in the functional integrity of the serotonergic nervous system? Are the changes in this system idiosyncratic to the pathology associated with Alzheimer's disease, or are they, as is suspected to be the case for the adrenergic nervous system (Mann, 1983), general features of dementing illness as a whole? Again, answers to these types of questions would bring us closer to an understanding of the relevance of specific changes in neuronal functioning to the behavioral symptomatology associated with dementia, and provide the basis for the development of effective treatment strategies.

16.3.4 The basic premise

There may be yet another reason why research at the animal level has failed to provide clinicians with effective tools for treating the cognitive impairments associated with dementing illness, and it is a rather sobering one. The original premise upon which generalizations between animals and people are based may be faulty. That is, it may not be applicable in every sense to extrapolate results at the animal level to the clinical population. With respect to learning and memory, in general, the apparent lack of preclinical predictability of the clinical response is quite clear and extremely perplexing. Numerous examples are available in which pharmacological manipulations that reliably improve learning and memory in animals fail to affect such processes similarly in the clinic. The results derived from manipulations of the opiate system on learning and memory in animals and man serve poignantly to underscore this point. Naloxone and naltrexone, both opiate receptor antagonists, have repeatedly been shown to facilitate memory and/or attenuate experimentally induced amnesias in rats (Carrasio *et al.*, 1982; Gallagher,1982; Liang *et al.*, 1983). On the other hand, only one laboratory has successfully demonstrated such an enhancement of performance clinically (Reisberg *et al.*, 1983*a, b*); all subsequent attempts failed to replicate these initial observations (Blass *et al.*, 1983; Nasrallah *et al.*, 1985; Pomara *et al.*, 1985; Tariot *et al.*, 1985; Steigler *et al.*, 1985; Hyman *et al.*, 1985). Surprisingly, the apparent failure of naloxone and naltrexone to stimulate memory in people has not discouraged researchers at the animal level from continuing to explore the complex role the opiate system appears to play in the processes underlying learning and memory. Similar, though certainly not as dramatic, examples of the failure of basic psychopharmacological research to predict clinical efficacy can be derived from a comparison of the effects that vasopressin, adrenocorticotropic hormone, and their respective analogues, have had on the processes underlying learning and memory in animals and people (for extensive reviews on this subject see Wolkowitz *et al.*, 1985*a, b*; Altman *et al.*, 1987*a*; Hoch, 1987; Normile *et al.*, 1989).

In light of the magnitude, consistency and generality of many of the behavioral effects seen at the animal level, one must consider the possibility that either the original premise regarding the generalizability between animals and people may not be universally applicable or the types of animal learning and memory tasks are inappropriate. However, recent progress in modeling the biological basis of the amnestic syndrome using delayed matching- or non-matching-to-sample tasks in primates (Squire & Zola-Morgan, 1985), and in modeling list/order – primacy/recency components of the memory impairments of Alzheimer's patients through systematic behavioral analysis of the effects of selective lesions of the cholinergic basal forebrain and medial septal nucleus in rats (DiMattia & Kesner, 1984, 1989; Kesner, 1985, 1988; Kesner *et al.*, 1986) would appear to suggest that the problem may lie more with the choice of appropriate behavioral tasks than with the general ability to extrapolate data

derived at the animal level to the clinical setting. For example, in a series of studies, Kesner and colleagues have demonstrated that in a complex list learning task, rats with small (asymmetrical or unilateral) medial septal lesions, but not NBM or parietal cortical lesions, exhibit intact memory for the last item, but impaired performance on all other items of the list (Kesner *et al.*, 1984; DiMattia & Kesner, 1989), similar to that seen in human subjects with mild senile dementia of the Alzheimer's type (Kesner *et al.*, 1987). In contrast, moderately impaired Alzheimer's patients had difficulty in remembering any item within the list (Kesner *et al.*, 1987), which was very reminiscent of the observed performance of rats with large medial septal lesions, medium-sized dorsal hippocampal or parietal cortical lesions, but not NBM lesions (DiMattia & Kesner, 1984, 1989; Kesner *et al.*, 1984). Thus, cholinergic cells originating within the medial septum and projecting to the hippocampus could mediate item memory in both animals and people. On the other hand, in an order memory task, mildly impaired Alzheimer's patients exhibit intact first-item memory, but impaired memory for all other items (Kesner *et al.*, 1987), similar to rats with small NBM lesions (DiMattia & Kesner, 1984; Kesner *et al.*, 1984) or humans with parietal lesions (Saffran & Marin, 1975; Kesner *et al.*, 1986), but not rats with small medial septal or dorsal hippocampal lesions (DiMattia & Kesner, 1989; Kesner *et al.*, 1986). In contrast, moderately demented Alzheimer's subjects were impaired in remembering any of choice orders (Kesner *et al.*, 1987), similar to rats with large NBM or medial septal lesions (Kesner *et al.*, 1986). These data would appear to suggest that cholinergic cells within the NBM, which project to dorsolateral frontal and parietal cortices, may play an important role in mediating order memory in both animals and people.

16.4 FUTURE DIRECTIONS

In light of the above, therefore, what are our best courses of action? Clearly animal research has to continue. It still provides a valuable mechanism with which to screen potentially efficacious pharmacological strategies. Particularly promising examples include: (1) the development of longer-acting (e.g., tetrahydroaminoacridine-THA) (Bartus *et al.*, 1983; Summers *et al.*, 1986) or controlled-release formulations of cholinesterase inhibitors (Thal *et al.*, 1989) and new direct-acting cholinergic agonists such as AF102B (Fisher *et al.*, 1987*a*), RS-86 (Wettstein & Spiegel, 1984; Hollander *et al.*, 1987) and BM-5 (Nordstrom *et al.*, 1983); (2) the use of nerve growth factor (NGF) either to stimulate choline acetyltransferase activity (Hefti *et al.*, 1985; Mobley *et al.*, 1985) or to attenuate the effects of cholinergic neurodegeneration or destruction (Will & Hefti, 1985; Haroutunian *et al.*, 1986; Williams *et al.*, 1986; Fisher *et al.*, 1987*b*); (3) phosphatidylserine to retard membrane degeneration (Bruni & Toffano, 1982; Calderini *et al.*, 1983; Drago *et al.*, 1983; Delwaide *et al.*, 1986); (4) infusion pumps to deliver potentially efficacious pharmacological agents directly into the brain, thus bypassing peripheral drug effects and the blood–brain barrier (Harbough, 1987); and (5) the use of neuronal transplants or grafts to resupply degenerated areas of the brain (Gage *et al.*, 1984; Björklund, 1985; Gash, 1987).

More effort should, however, be directed toward resolving the theoretical impasse concerning the lack of preclinical to clinical predictability. Far too many clinical trials are still resulting in little to no clinically significant improvements in memory and/or cognition. The resolution to this impasse may, however, prove a formidable task in light of our present understanding of the neurobiological basis of learning and memory. However, its resolution lies at the heart of the usefulness of animal research to provide us with additional insight as to how to deal with or ameliorate the learning and memory impairments associated with dementing illness.

Animal research should also direct more attention to long-term drug effects. The learning and memory deficits associated with many of the dementing illnesses are permanent and, in many cases, progressive. Acute treatment strategies may fall short of actually addressing the needs of this population: a long-term drug program almost certainly will be necessary. However, little is known about the effects of long-term drug administration on the processes underlying learning and memory. This is due, in part, to a continued need to understand further the effects of long-term drug administration on the internal micropharmacology of the neuron and on the functional status of the

events revolving around interneuronal trans-synaptic communication. At the behavioral level, for example, it is possible that drugs which were ineffective or which even impaired performance at the acute level will prove effective when administered chronically. Illustrative of this line of reasoning are the results of Loullis *et al.* (1983), who administered physostigmine or scopolamine chronically to rats for two weeks and tested passive avoidance training and retention testing following a one-week washout period. In this study physostigmine impaired, while scopolamine enhanced, the subsequent retention of the previously learned avoidance response. These results stand in marked contrast to the behavioral effects produced by these drugs following acute administration and underscore the need for additional psychopharmacological studies of the chronic effects of various drug manipulations on learning and memory.

More attention should also be paid to (1) possible interactions between drugs that affect the same neurotransmitter or neuropeptide system, and (2) combinations of drugs which affect different systems of the brain. For example, recent experiments in mice show an improvement in retention test performance with cholinergic drug combinations (e.g., arecoline, oxotremorine, deanol, edrophonium and tacrine) at significantly lower doses than those needed to facilitate memory individually (Flood *et al.*, 1983, 1985). Furthermore, it has recently been reported that simultaneous administration of alaproclate and oxotremorine enhanced memory at significantly lower doses than were needed when either of the drugs was administered alone (Altman *et al.*, 1984, 1987*b*). Finally, interactions between adrenergic and serotonergic systems in the mediation of memory have also been reported (Normile & Altman, 1987*a*). These studies are important because they may lead to the development of multiple drug strategies in the treatment of dementing illness, and may also lead to the development of drugs or drug combinations which enhance memory at significantly lower doses, thus reducing the likelihood of producing adverse side effects.

Continued effort should also be directed towards identifying and validating acceptable animal models of the neuropathological and behavioral symptomatology of dementia. As indicated earlier in this and preceding chapters, a number of potentially useful animal models have already been offered.

At the clinical level more attention should be focused on the *exact* replications of experimental designs. Typically, clinical studies rarely replicate the designs of previous studies (for a detailed review of this subject see Brinkman & Gershon, 1983). Generally, either the same drug is compared against entirely different behavioral measures, or similar behavioral measures are compared across different types of drugs. This has, in the past, contributed to significant confusion regarding either the relative efficacy of particular drugs or classes of drugs, or the behavioral effects they have produced. In addition, care should be exercised in interpreting the results of those clinical studies which attempt to assess drug effects against a neuropsychological battery of tasks. Frequently, one study reports significant effects on performance in certain aspects of the battery while another study reports significant effects of the same or similar treatment on performance in entirely different aspects of the battery. Again, replicability is needed, which should limit future confusion in interpreting the results of clinical trials of drugs.

Finally, it is not always necessary for clinical drug trials to be based on the experimental outcomes of similar studies in animals. This point may be particularly relevant in light of the lack of preclinical predictability of clinical response addressed earlier. That is, in certain cases it may be more efficacious to go directly to the clinical population to assess the behavioral effects of a specific pharmacological agent, rather than to amass a clear rationale for its use at the animal level. This may actually expedite the assessment of the clinical efficacy of certain types of drugs.

In conclusion, animal studies provide an essential mechanism for elucidating the neuropathological and/or behavioral bases of neuropsychiatric disease as well as contribute to the development of effective treatment strategies. However, refinements in the theoretical constructs which form the basis of many of the experimental designs may be necessary. Clearly, the 'classical approach' to drug development has not been particularly rewarding with respect to the treatment of dementia. New approaches to the problem may need to be considered, the most important of which may be a reassessment of the ability of present behavioral paradigms to predict clinical response. Acute drug trials are important and clearly provide a rapid method of economically screening a wide range of potentially efficacious compounds. However, attention should also be directed toward an analysis of the chronic effects of these drugs. The same applies to the development and application of animal models. Care should be exercised not to overvalue or overemphasize the generalizability of such models. However, reliable model systems on which to base future experimental designs or assess pharmacological outcomes are essential to continued progress in this field. Consideration should also be given to the development of drug 'cocktails' or pharmacological agents possessing multiple pharmacological effects. The neuronal processes underlying learning and memory are extremely complex, involving a number of neuronal systems. Pharmacological agents or cocktails which can alter the functioning of more than one neuronal system may prove more efficacious than pharmacological manipulations which selectively alter the functioning of only one neuronal system. Finally, refinements in the design of future clinical studies may need to be considered with particular attention paid to precise replication of experimental designs.

REFERENCES

Altman, H.J. & Quartermain, D. (1983). Facilitation of memory retrieval by centrally administered catecholamine stimulating agents. *Behavioral Brain Research* **7**, 51–63.

Altman, H.J. & Normile, H.J. (1986). Enhancement of the memory of a previously learned aversive habit following pre-test administration of a variety of serotonergic antagonists in mice. *Psychopharmacology* **90**, 24–7.

Altman, H.J. & Normile, H.J. (1987). Different temporal effects of serotonergic antagonists on passive avoidance retention. *Pharmacology, Biochemistry and Behavior* **28**, 353–99.

Altman, H.J., Nordy, D.A. & Ögren, S.-O. (1984). Role of serotonin in memory: Facilitation by alaproclate and zimeldine. *Psychopharmacology* **84**, 496–502.

Altman, H.J., Crosland, R.C., Jenden, D.J. & Berman, F.R. (1985a). Further characterization of the nature of the behavioral and neurochemical effects of lesions to the nucleus basalis of Meynert in the rat. *Neurobiology of Aging* **6**, 125–30.

Altman, H.J., Normile, H.J. & Ögren, S.-O. (1985*b*). Facilitation of discrimination learning in the rat following cytotoxic lesions of the serotonergic nervous system. *Society for Neuroscience Abstracts* **11**, 874.

Altman, H.J., Normile, H.J. & Gershon, S. (1987*a*). Non-cholinergic pharmacology in human cognitive disorders. In *Cognitive Neurochemistry: Assessment Techniques and Research Strategies for Understanding the Pharmacological Basis of Human Neuropsychology*, eds. S. Iversen, L. Iversen & S.M. Stahl, pp. 346–71. Oxford: Oxford University Press.

Altman, H.J., Stone, W.S. & Ögren, S.-O. (1987*b*). Evidence for a possible functional interaction between serotonergic and cholinergic mechanisms in memory retrieval. *Behavioral and Neural Biology* **48**, 49–62.

Altman, H.J., Ögren, S.-O., Berman, R.F. & Normile, H.J. (1989). The effects of p-chloroamphetamine, a depletor of brain serotonin, on the performance of rats in two types of positively reinforced complex spatial discrimination tasks. *Behavioral and Neural Biology* **52**, 131–44.

American Psychiatric Association (1987). *Diagnostic and Statistical Manual of Mental Disorders*, 3rd edn – revised, pp. 103–7. Washington, D.C.: American Psychiatric Association.

Arai, H., Kosaka, K. & Iizuka, R. (1984). Changes of biogenic amines and their metabolites in postmortem brains from patients with Alzheimer-type dementia. *Journal of Neurochemistry* **43**, 388–93.

Arnsten, A. & Segal, D. (1979). Naloxone alters locomotion and interaction with environmental stimuli. *Life Sciences* **25**, 1035–42.

Bartus, R.T., Dean, R.L. & Beer, B. (1983). An evaluation of drugs for improving memory in aged monkeys: Implications for clinical trials in humans. *Psychopharmacological Bulletin* **19**, 168–84.

Bartus, R.T., Pontecorvo, M.J., Flicker, C., Dean, R.S.L. & Figieurido, J.C. (1986). Behavioral recovery following bilateral lesions of the nucleus basalis does not occur spontaneously. *Pharmacology, Biochemistry and Behavior* **24**, 1287–92.

Bassant, M.H., Court, L. & Cathala, F. (1987). Impairment of the cortical and thalamic electrical activity in scrapie-infected rats. *Electroencephalography and Clinical Neurophysiology* **66**, 307–16.

Beal, M.F., Mazurek, M.F., Svendsen, C.N., Bird, E.D. & Martin, J.B. (1986). Widespread reduction of somatostatin-like immunoreactivity in the cerebral cortex in Alzheimer's disease. *Annals of Neurology* **20**, 489–95.

Berg, L. (1985). Does Alzheimer's disease represent an exaggeration of normal aging? *Archives of Neurology* **42**, 737–9.

Berman, R.F., Crosland, R.D., Jenden, D.J. & Altman, H.J. (1988). Persisting behavioral and neurochemical deficits in rats following lesions of the basal forebrain. *Pharmacology, Biochemistry and Behavior* **29**, 581–6.

Björklund, A. (1985). Functional reactivation of the deafferinated hippocampus by embryonic septal grafts as assessed by measurements of local glucose utilization. *Experimental Brain Research* **58**, 570–9.

Blass, J.P., Drachman, D., Katzman, R. & Spar, J.E. (1983). Letter to the editor. *New England Journal of Medicine* **211**, 556.

Bowen, D.M., White, P., Spillane, J.A., Goodhardt, M.J., Curzon, G., Iwanoff, P., Meier-Ruge, W. & Davison, A.N. (1979). Accelerated ageing or selective neuronal loss as an important cause of dementia? *Lancet* **i**, 11–14.

Bowen, D.M., Allen, S.J., Benton, J.S., Goodhardt, M.J., Haan, E.A., Palmer, A.M., Sims, N.R., Smith, C.C.T., Spillane, J.A., Esira, G.K., Neary, D., Snowdon, J.S., Wilcock, G.K. & Davison, A.N. (1983). Biochemical assessment of serotonergic and cholinergic dysfunction and cerebral atrophy in Alzheimer's disease. *Journal of Neurochemistry* **41**, 266–72.

Brinkman, S.D. & Gershon, S. (1983). Measurement of cholinergic drug effects on memory in Alzheimer's disease. *Neurobiology of Aging* **3**, 139–45.

Bruni, A. & Toffano, G. (1982). Lysophosphatidylserine, a short-lived intermediate with plasma membrane regulatory properties. *Pharmacological Research Communication* **14**, 469–84.

Bunney, B.S. (1977). Central dopaminergic systems: Two *in vivo* electrophysiological models for predicting therapeutic efficacy and neurological side effects of putative antipsychotic drugs. In *Animal Models of Psychiatry and Neurology*, eds. Hanin, I. & Usdin, E., pp. 91–104. Oxford: Pergamon Press.

Calderini, G., Bonnetti, A.C., Battistella, A., Cress, F.T. & Toffano, G. (1983). Biochemical changes of rat brain membranes with aging. *Neurochemistry Research* **8**, 483–91.

Carrasio, M.A., Dias, R.D. & Izequierdo, I. (1982). Naloxone reverses retrograde amnesia induced by electroconvulsive shock. *Behavioral and Neural Biology* **34**, 352–7.

Chiueh, C.C., Burns, R.S., Markey, S.P., Jacobowitz, D.M. & Kopin, I.J. (1985). Primate model of Parkinsonism: Selective lesion of nigrostriatal neurons by 1-methyl-4-phenyl-1,2,3,6-tetrahydropyridine produces an extrapyramidal syndrome in Rhesus monkeys. *Life Sciences* **36**, 213–18.

Chrobak, J.J., Hanin, I. & Walsh, T.J. (1987). AF64A (ethylcholine aziridinium ion), a cholinergic neurotoxin, selectively impairs working memory in a multiple component T-maze task. *Brain Research* **414**, 15–21.

Crapper-McLachlan, D.R. (1986). Aluminum and Alzheimer's disease. *Neurobiology of Aging* **7**, 525–32.

Crapper, D.R. & Dalton, A.J. (1973). Alterations and short-term retention, conditioned avoidance response acquisition and motivation following aluminum induced neurofibrillary degeneration. *Physiology and Behavior* **10**, 925–33.

Cross, A.J., Crow, T.J., Johnson, J.A., Joseph, M.H., Perry, E.K., Perry, R.H., Blessed, G. & Tomlinson, B. (1983). Monoamine metabolism in senile dementia of the Alzheimer type. *Journal of Neurobiological Sciences* **60**, 383–92.

Cross, A.J., Crow, T.J., Johnson, J.A., Perry, E.K., Perry, R.H., Blessed, G. & Tomlinson, B.E. (1984). Studies on neurotransmitter receptor systems in cortex and hippocampus in senile dementia of the Alzheimer-type. *Journal of Neurobiological Sciences* **64**, 109–17.

Cross, A.J., Crow, T.J., Ferrier, I.N. & Johnson, J.A. (1986). The selectivity of the reduction of serotonin S-2 receptors in Alzheimer-type dementia. *Neurology of Aging* **7**, 3–7.

Cummings, J.L. & Benson, D.F. (1983). *Dementia: A Clinical Approach*. Boston: Butterworths.

Cutler, N.R., Haxby, J.V., Narang, P.K., May, C. & Burg, C. (1985). Evaluation of an analogue of somatostatin (L368,586) in Alzheimer's disease. *New England Journal of Medicine* **312**, 725.

Davis, P.E. & Mumford, S.J. (1984). Cued recall and the nature of the memory disorder in dementia. *British Journal of Psychiatry* **144**, 383–6.

de Ajuriaguerra, J. & Tissot, R. (1975). Some aspects of language in various forms of senile dementia (comparisons with language in childhood). In *Foundation of Language Development*, eds. E.H. Lenneberg & E. Lenneberg, pp. 137–51. New York: Academic Press.

Delwaide, P.J., Gyselynck-Mambourg, A.M., Hurlet, A. & Ylieff, M. (1986). Double-blind randomized controlled study of phosphatidylserine in senile demented patients. *Acta Neurologica Scandinavica* **73**, 136–40.

Deutsch, J.A. & Rogers, J.B. (1979). Cholinergic excitability and memory: Animal studies and their clinical implications. In *Brain Acetylcholine and Neuropsychiatric Disease*, eds. K.L. Davis & P.A. Berger, pp. 175–204. New York: Plenum Press.

DiMattia, B.V. & Kesner, R.P. (1984). Serial position curves in rats: Automatic vs effortful information processing. *Journal of Experimental Psychology* **10**, 557–63.

DiMattia, B.V. & Kesner, R.P. (1988). Spatial cognitive maps: Differential role of parietal cortex and hippocampal formation. *Behavioral Neuroscience* **102**, 471–80.

Drago, F., Toffano, G., Catalano Rossi Danielli, L., Continella, G. & Scapagnini, U. (1983). Phosphatidylserine facilitates learning and memory processes in aged rats. In *Aging of the Brain*, eds. D. Samuel, S. Algeri, S. Gershon, V.E. Grimm & G. Toffano, pp. 309–16. New York: Raven Press.

Fisher, A. & Hanin, I. (1980). Choline analogs as potential tools in developing selective animal models of central cholinergic hypofunction. *Life Sciences* **27**, 1615–34.

Fisher, A., Mantione, C.R., Abraham, D.J. & Hanin, I. (1982). Long term central cholinergic hypofunction induced in mice by ethycholine aziridinium ion (AF64A) *in vivo*. *Journal of Pharmacological & Experimental Therapeutics* **222**, 140–5.

Fisher, A., Heldman, R., Brandeis, I., Karton, Z., Pittel, S., Dacher, A., Levy, A. & Mizobe, E. (1987*a*). AF1092B: A novel putative M1 agonist reverses AF64A-induced cognitive impairments in rats. *Society for Neuroscience Abstracts* **13**, 675.

Fisher, A., Wictorin, K., Björklund, A., Williams, L.R., Varon, S. & Gage, F.H. (1987*b*). Amelioration of cholinergic neuron atrophy and spatial memory impairments in aged rats by nerve growth factor. *Nature* **329**, 65–8.

Flood, J.F., Smith, G.E. & Cherkin, A. (1983). Memory retention: Potentiation of cholinergic drug combinations in mice. *Neurobiology of Aging* **4**, 37–43.

Flood, J.F., Smith, G.E. & Cherkin, A. (1985). Memory enhancement: Supra-additive effect of subcutaneous cholinergic drug combinations in mice. *Psychopharmacology* **86**, 61–7.

Francis, P.T., Bowen, D.M., Neary, D., Palo, J., Wikstrom, J. & Olney, J. (1984). Somatostatin-like immunoreactivity in lumbar cerebrospinal fluid from neurohistologically examined demented patients. *Neurobiology of Aging* **5**, 183–6.

Gage, F.H., Björklund, A., Stenevi, U., Dunnett, S.B. & Kelly, P.A.T. (1984). Intrahippocampal septal grafts ameliorate learning impairments in aged rats. *Science* **225**, 533–6.

Gallagher, M. (1982). Naloxone enhancement of memory processes: Effects of other opiate antagonists. *Behavioral and Neural Biology* **35**, 375–82.

Gash, D.M. (1987). Neural implants: A strategy for the treatment of Alzheimer's disease. In *Alzheimer's Disease and Dementia: Problems, Prospects and*

Perspectives, ed. H.J. Altman, pp. 165–70. New York: Plenum Publishing Corp.

Ghetti, B., Musicco, M., Norton, J. & Bugiani, O. (1985). Nerve cell loss in the progressive encephalopathy induced by aluminum powder. A morphologic and semiquantitative study of the purkinje cells. *Neuropathology and Applied Neurobiology* **11**, 31–53.

Gomez, S., Puymirat, P., Valade, P., Davous, P., Rondot, P. & Cohen, P. (1986). Patients with Alzheimer's disease show an increased content of 15K dalton somatostatin precursor and a lowered level of tetradecapeptide in their cerebrospinal fluid. *Life Sciences* **39**, 623–7.

Haber, S., Barchas, P.R. & Barchas, J.D. (1977). Effects of amphetamine on social behaviors of Rhesus Macaques: An animal model of paranoia. In *Animal Models of Psychiatry and Neurology*, eds. Hanin, I. & Usdin, E., pp. 107–14. Oxford: Pergamon Press.

Harbough, R.E. (1987). Intracranial cholinergic drug infusion in patients with Alzheimer's disease. In *Alzheimer's Disease and Dementia: Problems, Prospects and Perspectives*, ed. H.J. Altman, pp. 157–64. New York: Plenum Publishing Corp.

Haroutunian, V., Kanof, P.D. & Davis, K.L. (1986). Partial reversal of lesion-induced deficits in cortical cholinergic markers by nerve growth factor. *Brain Research* **386**, 397–9.

Haycock, J.W., van Buskirk, R.B., Ryan, J.R. & McGaugh, J.L. (1977). Enhancement of retention with centrally administered catecholamines. *Experimental Neurology* **54**, 199–208.

Hefti, F., Hartikka, J., Gnahn, H., Heumann, R. & Schwab, M. (1985). Nerve growth factor increases choline acetyltransferase but not survival of fiber outgrowth of cultured fetal septal cholinergic neurons. *Neuroscience* **14**, 55–68.

Hoch, F.J. (1987). Drug influences on learning and memory in aged animals and humans. *Neuropsychobiology* **17**, 145–60.

Hollander, E., Davidson, M., Mohs, R.C., Horvath, T.B., Davis, B.M., Zemishlany, Z. & Davis, K.L. (1987). RS86 in the treatment of Alzheimer's disease: Cognitive and biological effects. *Biological Psychiatry* **22**, 1067–78.

Hoyer, S. (1987). Somatostatin and Alzheimer's disease. *Journal of Neurology* **234**, 266–7.

Hyman, B.T., Eslinger, P. & Damasia, A. (1985). Effect of naltrexone on senile dementia of the Alzheimer type. *Journal of Neurology, Neurosurgery and Psychiatry* **48**, 1169–71.

Ishii, I. (1966). Distribution of Alzheimer's neurofibrillary changes in the brain stem and hypothalamus of senile dementia. *Archives of Neuropathology* **107**, 385–99.

Johnston, M.V., McKinney, M. & Coyle, J.T. (1979).

Evidence for a cholinergic projection to neocortex from neurons in basal forebrain. *Proceedings of the National Academy of Sciences* **76**, 5392–6.

Kesner, R.P. (1985). Correspondence between humans and animals in coding of temporal attributes: Role of hippocampus and prefrontal cortex. In *Memory Dysfunctions: An Integration of Animal and Human Research from Preclinical and Clinical Perspectives*, eds. D.S. Olton, E. Gamzu & S. Corkin, pp. 122–36. New York: New York Academy of Sciences.

Kesner, R.P. (1988). Reevaluation of the contributions of the basal forebrain cholinergic system to memory. *Neurobiology of Aging* **9**, 609–16.

Kesner, R.P., Meason, M.O., Forsman, S.L. & Holbrook, T.H. (1984). Serial position curves in rats: Order memory for episodic spatial events. *Animal Learning and Behavior* **12**, 378–82.

Kesner, R.P., Crutcher, K.A. & Meason, M.O. (1986). Medial septal and nucleus basalis magnocellularis lesions produce order memory deficits in rats which mimic symptoms of Alzheimer's disease. *Neurobiology of Aging* **7**, 287–95.

Kesner, R.P., Adelstein, T. & Crutcher, K.A. (1987). Rats with nucleus basalis magnocellularis lesions mimic mnemonic symptomatology observed in patients with dementia of the Alzheimer's type. *Behavioral Neuroscience* **101**, 451–6.

Khachaturian, Z. (1985). Progress of research on Alzheimer's disease. In *Senile Dementia of the Alzheimer Type*, ed. J.T. Hutton & A.D. Kenny, pp. 379–84. New York: Alan R. Liss.

Knowlton, B.J., Wenk, G.L., Olton, D.S. & Coyle, J.T. (1985). Basal forebrain lesions produce a dissociation of trial-dependent and trial-independent memory performance. *Brain Research* **345**, 315–21.

Koob, G.F. & Bloom, F.E. (1983). Behavioral effects of opioid peptides. *British Medical Bulletin* **39**, 89–94.

Kornetsky, C. & Markowitz, R. (1978). Animal models of schizophrenia. In *Psychopharmacology: A Generation of Progress*, eds. M.A. Lipton, A. DiMascio & K.F. Killam, pp. 583–93. New York: Raven Press.

Langston, J.M. (1985). MPTP neurotoxicity: An overview and characterization of phases of toxicity. *Life Sciences* **36**, 201–6.

Liang, K.C., Messing, R.B. & McGaugh, J.L. (1983). Naloxone attenuates amnesia caused by amygdaloid stimulation: The involvement of a central opioid system. *British Medical Journal* **1**, 550.

Losonczy, M.F., Davidson, M. & Davis, K.L. (1987). The dopamine hypothesis of schizophrenia. In *Psychopharmacology: The Third Generation of Progress*, ed. Meltzer, H.Y., pp. 715–26. New York: Raven Press.

Loullis, C.C., Dean, L., Lippa, A.S., Meyerson, L.R., Beer, B. & Bartus, R.T. (1983). Chronic administration of cholinergic agents: Effects on behavior and calmodulin. *Pharmacology, Biochemistry and Behavior* **18**, 601–4.

Mann, D.M.A. (1983). The locus coeruleus and its possible role in aging and degenerative disease of the human central nervous system. *Mechanisms of Aging and Development* **23**, 73–94.

Mann, D.M.A. & Yates, P.O. (1983). Serotonin nerve cells in Alzheimer's disease. *Journal of Neurology, Neurosurgery and Psychiatry* **108**, 97–113.

Mantione, C.R., Fisher, A. & Hanin, I. (1981). AF64A-treated mouse: Possible model for central cholinergic hypofunction. *Science* **213**, 579–80.

Manuelidis, E.E., de Figueirdo, J.M., Kim, J.H., Fritch, W.W. & Manuelidis, L. (1988). Transmission studies from blood of Alzheimer's disease patients and healthy relatives. *Proceedings of the National Academy of Sciences (U.S.A.)* **13**, 4898–901.

McHugh, P.R. & Folstein, M.F. (1979). Psychopathology of dementia: Implications for neuropathology. In *Congenital and Acquired Cognitive Disorders*, ed. Katzman, R., pp. 17–30. New York: Raven Press.

McKinley, M.P. & Prusiner, S.B. (1987). Scrapie prions, amyloid placques, and a possible link with Alzheimer's disease. In *Alzheimer's Disease and Dementia: Problems, Prospects and Perspectives*, ed. H.J. Altman, pp. 75–86. New York: Plenum Publishing Corp.

Michigan Department of Public Health (1987). *Alzheimer's Disease and Related Conditions: Reducing Uncertainty*, vol. **1**, p. 1.

Mobley, W.C., Rutkowski, J.L., Tennekoon, G.I., Buchanan, K. & Johnston, M.V. (1985). Choline acetyltransferase activity in striatum of neonatal rats increased by nerve growth factor. *Science* **229**, 284–6.

Morris, R., Wheatley, J. & Britton, P. (1983). Retrieval from long-term memory in senile dementia: Cued recall revisited. *British Journal of Clinical Psychology* **22**, 141–2.

Nasrallah, H.A., Varney, N., Coffman, J.A., Bayliss, J. & Chapman, S. (1985). Effects of naloxone on cognitive deficits following electroconvulsive therapy. *Psychopharmacology Bulletin* **21**, 89–90.

National Institute on Aging (1986). *Progress Report on Alzheimer's Disease: Vol. III.* Washington, D.C.: National Institute on Aging.

National Institutes of Health Consensus Development Conference Statement, vol. **6**, no. 11. (1987). *Differential Diagnosis of Dementing Illness*, Washington, D.C.: National Institutes of Health.

Nordstrom, O., Alberts, P., Westlind, A., Unden, A. & Bartfai, T. (1983). Presynaptic agonist at muscarinic cholinergic receptors. *Molecular Pharmacology* **24**, 1–5.

Normile, H.J. & Altman, H.J. (1987*a*). Evidence for a possible interaction between noradrenergic and serotonergic neurotransmission in the retrieval of a previously learned aversive habit in mice. *Psychopharmacology* **92**, 388–92.

Normile, H.J. & Altman, H.J. (1987*b*). Serotonin, Alzheimer's disease and learning/memory in animals. In *Alzheimer's Disease and Dementia: Problems, Prospects and Perspectives*, ed. H.J. Altman, pp. 141–56. New York: Plenum Publishing Corp.

Normile, H.J., Galloway, M.P. & Altman, H.J. (1988). Effects of chronic serotonin receptor antagonist administration on acquisition of a complex spatial discrimination task in young and old rats. *Society for Neuroscience Abstracts* **14**, 725.

Normile, H.J., Altman, H.J. & Gershon, S. (1989). Issues regarding possible therapies using cognitive enhancers. In *Attention Deficit Disorder and Hyperkinetic Syndrome*, eds. T. Sagvolden & T. Archer, pp. 235–53. Oslo: L. Erlbaum Associates.

Ögren, S.-O. (1982). Central serotonin neurons and learning in the rat. In *Biology of Serotonergic Transmission*, ed. N.N. Osborne, pp. 317–34. Chichester: John Wiley & Sons.

Palmer, A.M., Sims, N.S., Bowen, D.M., Neary, D., Palo, J., Wikstrom, J. & Davison, A.N. (1984). Monoamine metabolite concentrations in lumbar cerebrospinal fluid of patients with histology verified Alzheimer's dementia. *Journal of Neurology, Neurosurgery and Psychiatry* **47**, 481–4.

Pomara, N., Roberts, R., Rhiew, H.B., Stanley, M. & Gershon, S. (1985). Multiple, single-dose naltrexone administration fails to affect overall cognitive functioning and plasma cortisol in individuals with probable Alzheimer's disease. *Neurobiology of Aging* **6**, 233–6.

Prusiner, S.B. (1982). Novel proteinaceous infectious particles cause scrapie. *Science* **216**, 136–8.

Reinikainen, K.J., Riekkinen, P.J., Jolkkonen, J., Kosma, V.-M. & Soininen, H. (1987). Decreased somatostatin-like immunoreactivity in cerebral cortex and cerebrospinal fluid in Alzheimer's disease. *Brain Research* **402**, 103–8.

Reisberg, B., Ferris, S.H., Anand, R., Mir, P., DeLeon, M.J. & Roberts, E. (1983*a*). Naloxone effects on primary degenerative dementia (PDD). *Psychopharmacology Bulletin* **19**, 45–7.

Reisberg, B., Ferris, S.H., Anand, R., Mir, P., DeLeon, M.J. & Roberts, E. (1983*b*). Effects of naloxone in senile dementia: A double-blind trial. *New England Journal of Medicine* **308**, 721–2.

Saffran, E.M. & Marin, O.S.M. (1975). Immediate memory for word lists and sentences in a patient with deficient auditory short-term memory. *Brain Language* **2**, 420–33.

Sahgal, A. (1984). A critique of the vasopressin-memory hypothesis. *Psychopharmacology* **83**, 215–28.

Simpson, J., Yates, C.M., Whyler, D.K., Wilson, H., Dewar, A.J. & Gordon, A. (1985). Biochemical studies on rabbits with aluminum induced neurofilament accumulations. *Neurochemical Research* **10**, 229–38.

Soininen, H.S., Jolkkonen, J.T., Reinikainen, K.J., Halonen, T.O. & Riekkinen, P.J. (1984). Reduced cholinesterase activity and somatostatin-like immunoreactivity in the cerebrospinal fluid of patients with dementia of the Alzheimer type. *Journal of the Neurological Sciences* **63**, 167–72.

Squire, L.R. & Zola-Morgan, S. (1985). The neuropsychology of memory: New links between humans and experimental animals. In *Memory Dysfunctions: An Integration of Animal and Human Research from Preclinical and Clinical Perspectives*, eds. D.S. Olton, E. Gamzu & S. Corkin, pp. 137–49. New York: New York Academy of Sciences.

Steigler, W.A., Mendelson, M., Jenkins, T., Smith, M. & Gay, T. (1985). Effects of naloxone in senile dementia. *Journal of the American Geriatrics Society* **33**, 155.

Summers, W.K., Majovski, L.V., Marsh, G.M., Tachiki, K. & Kling, V. (1986). Oral tetrahydroaminoacridine in long-term treatment of senile dementia, Alzheimer's type. *New England Journal of Medicine* **315**, 1241–5.

Tariot, P.N., Sunderland, T., Weingartner, H., Murphy, D.L., Cohen, M.R. & Cohen, R.M. (1985). Low- and high-dose naloxone in dementia of the Alzheimer's type. *Psychopharmacology Bulletin* **21**, 680–2.

Thal, L.J., Lasker, B., Sharpless, N.S., Bobotas, G., Schor, J.M. & Nigalye, A. (1989). Plasma physostigmine concentrations after controlled-release oral physostigmine. *Archives of Neurology* **46**(1), 13.

van Ree, J.M., Bohus, B., Versteeg, D.H.G. & De Wied, D. (1978). Neurohypophyseal principles and memory processes. *Biochemical Pharmacology* **27**, 1793–800.

Wagner, G.C., Jarvis, M.F. & Rubin, J.G. (1986). L-DOPA reverses the effects of MPTP toxicity. *Psychopharmacology* **88**, 401–2.

Watson, M., Vickroy, T.W., Fibiger, H.C., Roeske, W.R. & Yamamura, H.I. (1985). Effects of bilateral ibotenate-induced lesions of the nucleus basalis magnocellularis upon selective cholinergic biochemical markers in the rat anterior cerebral cortex. *Brain Research* **346**, 387–91.

Weingartner, H. (1984). Psychobiological determinants of memory failure. In *Neuropsychology of Memory*, eds. L.R. Squire & N. Butters, pp. 203–12. New York: Guilford Press.

Wettstein, A. & Spiegel, R. (1984). Clinical trials with the cholinergic drug RS86 in Alzheimer's disease (AD) and senile dementia of the Alzheimer's type (SDAT). *Psychopharmacology* **84**, 572–3.

Will, B. & Hefti, F. (1985). Behavioral and neurochemical effects of chronic intraventricular injections of nerve growth factor in adult rats with fimbria lesions. *Behavioral and Neural Biology* **17**, 17–24.

Williams, L.R., Varon, S., Peterson, G.M., Wictorin, K., Fischer, W., Björklund, A. & Gage, F.H. (1986). Continuous infusion of nerve growth factor prevents basal forebrain neuronal death after fimbria-fornix transection. *Proceedings of the National Academy of Sciences (U.S.A.)* **83**, 9231–5.

Wisniewski, H. M., Bruce, M. E. & Fraser, H. (1975). Infectious etiology of neuritic (senile) plaques in mice. *Science* **190**, 1108–10.

Wisniewski, H.M., Moretz, R.C. & Iqbal, K. (1986). No evidence for aluminum in etiology and pathogenesis of Alzheimer's disease. *Neurobiology of Aging* **7**, 532–5.

Wolkowitz, O.M., Tinklenberg, J.R. & Weingartner, H. (1985a). A psychopharmacological perspective of cognitive functions. I. Theoretical overview and methodological considerations. *Neuropsychobiology* **14**, 88–96.

Wolkowitz, O.M., Tinklenberg, J.R. & Weingartner, H. (1985b). A psychopharmacological perspective of cognitive functions. II. Specific pharmacologic agents. *Neuropsychobiology* **14**, 133–56.

Yamamoto, T. & Hirano, A. (1985). Nucleus raphe dorsalis in Alzheimer's disease: Neurofibrillary tangles and loss of large neurones. *Annals of Neurology* **17**, 573–7.

PART VII

Models of drug abuse and drug dependence

17

Animal models of drug abuse and dependence

ANDREW J. GOUDIE

17.1 THE CONCEPTS OF DEPENDENCE AND ABUSE

'There do not seem to be good grounds for treating the many types and forms of drug dependence as a unitary disorder. No absolute criteria can be specified and distinctions between use and abuse depend on the drug, the subject, the context in which they interact and the characteristics of the observer' (Kumar & Stolerman, 1977).

The widely used terms drug abuse and drug dependence have no universally accepted definitions. Furthermore, the literature on drug abuse and dependence contains references to many so-called explanatory constructs (such as craving, psychological dependence, compulsive behaviour, euphoria, drug-induced 'highs', and so on) which are often not defined clearly and which cannot be measured reliably. It is not clear whether abuse and dependence are statistical deviations from the norm, or whether they are genuine disorders with specific underlying etiologies. Attempts to clarify concepts in this area on the basis of theoretical discussions are not always profitable since they can lead into esoteric pastures in which one begins to ask whether it is possible to be 'dependent' on gambling, on ingestion of excessive amounts of food, on one's parents etc. (Kumar & Stolerman, 1977). Similarly, it is possible to ask whether animals could ever be said to 'abuse' drugs. Nevertheless, it is obvious that attempts to establish the validity of animal models of drug abuse and dependence will be fraught with difficulty unless some attempt is made to clarify and define the basic concepts. Thus this chapter begins with a general discussion of the nature of drug abuse and dependence.

A fundamental operational distinction can be drawn between the physical dependence potential of a drug and its abuse liability (Brady & Lukas, 1984; Brady & Fischman, 1985), the concept of abuse liability being related operationally to the ability of a drug to support self-administration (regardless of whether or not this is associated with dependence). This distinction between abuse and dependence is based largely on evidence indicating that physical dependence, which occurs largely as a *result* of prolonged drug usage, is not a necessary or even a sufficient condition for a drug to be repeatedly self-administered in a potentially mal-adaptive way (see Schuster & Johanson, 1981; Falk *et al.*, 1983; and Young & Herling, 1986, for reviews). Instead, it is suggested that experimental procedures can be developed to assess drug abuse liability, and that the actions of drugs in such procedures do not necessarily correlate with their ability to induce physical dependence (see also Sanger, Chapter 18 of this volume). For example, cocaine has a high abuse liability (Aigner & Balster, 1978; Griffiths *et al.*, 1978), but it does not induce a profound degree of physical dependence (Schuster & Johanson, 1981).

It is clear then that factors other than the ability to induce dependence are involved in excessive drug taking, although prior drug experience can under some circumstances increase drug intake (Falk *et al.*, 1983). Indeed, chronic drug intake is undoubtedly affected by both the abuse liability and dependence potential of the specific drug in question, although the relative roles of these two processes differ as a function of the duration of drug intake and as a function of the drugs studied. The situation is complicated further in cases of real-life drug taking in that polydrug abuse is common. There is even evidence that abuse of *one* drug type may be characterised by different syndromes of dependence in different groups of individuals, as suggested for the benzodiazepines (Cappell *et al.*, 1987). Thus, animal models of drug dependence and abuse can probably never do more than attempt to simulate the action of underlying behavioural and psychological *processes* in drug dependence. It is unrealistic to expect more from such models (Schuster, 1986; Johanson *et al.*, 1987).

17.2 CRITERIA FOR THE EVALUATION OF ANIMAL MODELS OF DRUG ABUSE/DEPENDENCE

The general criteria for evaluating animal models have already been outlined by Willner in the opening chapter of this book, (see also Willner, 1986). These criteria are briefly described here with specific reference to models of drug abuse and dependence. We then review a number of more common 'models' of drug abuse and dependence and, for each model, consider the extent to which it meets the various criteria.

In this chapter the term 'model' is used in a very general way – thus animal 'models' are taken to include all procedures that may provide insight(s) into basic behavioural, pharmacological and physiological processes involved in drug abuse/dependence. As noted by Willner (this volume, Chapter 1), models may also be used as bioassays and as screening tests. In the context of drug abuse research, the self-administration model (see Section 17.3.1) is often used to analyse the neurochemical systems involved in drug reinforcement (e.g. Wise, 1982) and as a screening test for potential drugs of abuse (e.g. Spyraki & Fibiger, 1981). However, the procedure is also used as a simulation to analyse basic processes involved in drug abuse (Johanson *et al.*, 1987). Similarly, the drug discrimination procedure (see Section 17.3.4) is used both as a bioassay to examine drug actions on receptors (Appel & Cunningham, 1986) and as a screening test for novel drugs with potential therapeutic applications (Lal & Emmett-Oglesby,

1983; Lal & Yaden, 1985). However, the drug discrimination procedure is also often conceptualised as an animal simulation of human 'subjective' responses to drugs. Such 'subjective' responses have often been thought to be closely linked to drug abuse potential and the drug discrimination procedure therefore represents a potential tool for analysing this aspect of drug abuse.

As noted above, it is unrealistic to expect that any model will cover all the complex interacting processes involved in the clinical syndrome(s) of drug abuse and dependence. Thus no model of drug abuse/dependence will fulfil all the validation criteria that may be applied to it. Nevertheless, it is legitimate to expect models to meet at least some of these criteria, as discussed below.

The notion of predictive validity relates initially to the question of whether specific models discriminate accurately between drugs that are abused by humans and those that are not. Subsequently, given the wide range of drugs that are abused by humans, it would be of considerable value for a model to allow ranking of drugs in terms of abuse potential and then to compare relative abuse potential with relative therapeutic efficacy so as to select out those drugs that have low abuse potential and high therapeutic efficacy in the same dose range (Griffiths *et al.*, 1979). However, if the constructs of dependence and abuse are themselves poorly defined, it may be difficult to determine whether or not false positives and negatives represent genuine failures of predictive validity. Furthermore, it will inevitably be difficult to *measure* the abuse/dependence potential of drugs in humans in order to determine whether rankings of abuse/dependence potential in animal models correlate with clinical rankings. It should therefore be clear that rigorous assessment of the predictive validity of animal models in this area will not be easy to achieve.

As far as face validity is concerned, McKinney & Bunney (1969) suggested that animal models of behavioural disorders should, ideally, resemble clinical syndrome(s) in terms of etiology, biochemistry, symptomatology and treatment. However, as noted by Willner (this volume, Chapter 1), this requirement for face validity assessment is probably too stringent, since little is known about the etiology and biochemical basis of drug abuse/dependence in humans. Although behavioural research in animals has provided insights into possible therapies for the treatment of dependence (Schuster, 1986), it is clear that, as in other areas in which attempts have been made to develop animal models of behavioural disorders, assessment of the face validity of animal models of abuse/dependence can only proceed in terms of similarities in symptomatology.

The notion of construct validation presupposes, by definition, that model(s) studied have sound theoretical bases (Willner, 1986). This is rarely the case for models of drug abuse/dependence. For example, it has been suggested (Stewart *et al.*, 1984) that chronic drug intake is rarely motivated by the fear of withdrawal, but rather results from the actions of drugs as positive reinforcers (incentives). The controversy as to whether drug taking is maintained by positive or negative reinforcement is clearly a fundamental one from the point of view of scientific theory. In many human situations, and in many animal models, these two processes may interact as determinants of chronic drug self-administration. Thus it is difficult to know whether one should be attempting to validate a model in terms of the constructs of abuse potential, dependence potential or both. The suggestion that positive reinforcement is of greater importance, in contrast to the more widely accepted notion that it is the alleviation of withdrawal which maintains persistent drug taking behaviour has been characterised by some clinicians as 'Theoretically interesting . . . but distant from the usual clinical situation' (O'Brien *et al.*, 1986, p. 331). Nevertheless, the fact that such a controversy exists at all, demonstrates the shaky theoretical grounds on which animal models of drug abuse/dependence are all based and highlights the considerable difficulties inherent in the construct validation of such models. Indeed, it is possible that the very existence of such a basic theoretical controversy may force clinicians and animal researchers alike to examine theoretical issues in carefully controlled empirical studies. The fact that difficulties are encountered at present in the construct validation of animal models of dependence may therefore prove ultimately to have a positive aspect.

17.3 MODELS OF ABUSE LIABILITY

17.3.1 Self-administration procedures

Following basic principles of operant conditioning (Skinner, 1953), if drug infusions contingent on a behaviour (such as lever pressing) lead to increases in the relevant behaviour, then drugs can be categorised as positively reinforcing stimuli (Schuster & Johanson, 1981; Woods, 1983). Early studies (Weeks, 1962; Deneau *et al.*, 1969) showed that a wide range of drugs are self-administered by animals intravenously or intragastrically. Thus drugs can be viewed as positive reinforcers and studied in the same way as other reinforcing stimuli (Kelleher & Goldberg, 1976). This conceptualisation of drug abuse defines the problem as a behavioural one (Johanson, 1978) and it allows analysis of the processes involved in terms of the operant conditioning 'technology' evolved over the years by behavioural scientists to study the actions of conventional reinforcing stimuli such as food and water (Morse, 1976). It is important to note that reinforcing stimuli are defined operationally in terms of their actions on behaviour, not in terms of their hedonic properties. Therefore, the finding that a drug is self-administered does not necessarily imply that the drug was self-administered for its euphoriant or other 'subjective' actions, although abused drugs often are euphoriants in humans and it has frequently been suggested that there is an important correlation between the 'subjective' effects of drugs and their tendency to be self-administered (Jasinski *et al.*, 1984; Johanson *et al.*, 1987).

The literature on the self-administration paradigm is now very extensive and an attempt to provide a comprehensive review of this area is beyond the scope of this chapter (see Brady & Lukas, 1984; Young & Herling, 1986, for reviews). Instead, we consider below some basic findings from animal self-administration studies which have important implications for conceptualisations of abuse. Studies conducted in recent years that have adopted operant procedures for analysing human drug self-administration (see Henningfield *et al.*, 1986, for a review) are not considered here. (Although such studies *are* clearly valuable in validating the self-administration paradigm as a model of drug abuse liability.)

17.3.1.1 The importance of schedules of reinforcement in drug self-administration

A major finding from animal self-administration studies is that the reinforcing properties of drugs are dependent to a considerable extent on the way their presentation is scheduled in relation to behaviour (Falk *et al.*, 1983; Young & Herling, 1986), just as is observed with more conventional reinforcing stimuli such as food and water (see Spealman & Goldberg, 1978, for review). Historical, experiential and environmental factors are known to be important determinants of the extent to which a particular drug will affect behaviour (Barrett, 1977; Barrett & Witkin, 1986). Such factors can also influence the extent to which a particular drug maintains self-administration in a particular experiment. Drug stimuli are not unique in this respect, perhaps the prototypical example of such an effect being the finding that, under the 'correct' experimental conditions, animals will self-administer electric shocks that under different conditions they work to avoid (see Morse & Kelleher, 1977, for review). There is now reliable evidence that such apparently paradoxical actions occur with drug stimuli also (Goldberg & Spealman, 1982; Corrigall *et al.*, 1986; Carr & White, 1986).

This point will be illustrated by reference to a study with cocaine (Spealman, 1979, 1985). Monkeys were initially trained to respond on one lever to self-administer cocaine under a Variable Interval (VI) 3-min schedule of reinforcement: the first response after a random period of time (on average 3 min) resulted in the presentation of an i.v. infusion of cocaine. Once animals had learned to self-administer cocaine reliably, the procedure was modified so that responses on a second lever resulted in the termination for one minute (under a concurrent Fixed Interval (FI) 3-min schedule), of the schedule of cocaine availability on the first lever. The development of responding for one monkey under this concurrent schedule is shown in Fig. 17.1.

After sufficient training (*c.* 10 sessions), the two schedules maintained distinctive rates and patterns of responding, these being the types typically associated with each of the schedules (Ferster & Skinner, 1957). When infusions of cocaine (on the VI lever) were suspended, responding declined on both schedules, indicating that the cocaine infu-

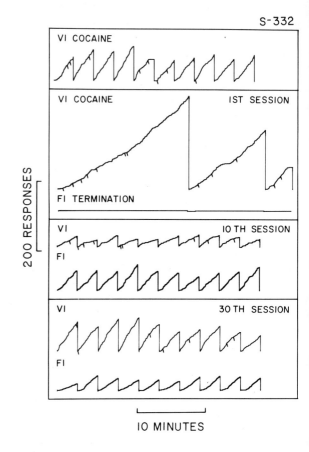

Fig. 17.1. Development of responding for one monkey (S-332) under a concurrent Variable Interval schedule of intravenous cocaine infusion and a Fixed Interval schedule of termination of access to cocaine. The horizontal axis shows time and the vertical axis shows cumulative responses. Diagonal marks in the upper record of each panel show infusions of cocaine. The first panel shows responding maintained by Variable Interval cocaine infusions. Subsequent panels show the 1st, 10th and 30th sessions after the introduction of the concurrent Fixed Interval schedule of cocaine termination. In the first panel the recorder reset every 3 min. In the subsequent panels both recorders reset at the start of the 1-min time-out period, during which the pens did not operate. Note the progressive development of characteristic scalloped pattern of Fixed Interval responding. Reproduced from Spealman R.D. In: Behavioral Pharmacology: The Current Status, Seiden L.S. & Balster, R.L. (eds.), pp. 23–8, A.R. Liss Inc. (1985), with permission of author and publisher.

sions maintained responding on both levers. Thus, responding was simultaneously maintained by cocaine infusions and by the opportunity to terminate the schedule of cocaine self-administration. These data indicate that the reinforcing actions of drugs are not simply a consequence of their pharmacological properties. Such apparently counterintuitive dual stimulus functions of drugs and other stimuli caution against the widely accepted belief that drugs are self-administered largely for their euphoriant or positive hedonic actions. If cocaine were self-administered simply because of its well known euphoriant actions, it is difficult to understand why animals should work to turn off access to a schedule of cocaine self-administration.

A number of other studies show that various drugs have different types of reinforcing actions dependent on the precise behavioural context in which they are studied. (See White *et al.*, 1977, Carr & White, 1986, and Goldberg & Spealman, 1982, for similar studies with opiates, stimulants and nicotine.) There is also evidence that the effects of morphine on brain neurochemistry are greater when the drug is self-administered than when the same drug doses are given passively (Smith *et al.*, 1980), indicating that the actions of drugs cannot readily be dissociated from the context in which they are self-administered.

A further example of this principle comes from studies with so-called second-order schedules of reinforcement. A second-order schedule is defined as one in which one schedule is a subunit of a more complex schedule, so that presentation of a primary reinforcer (such as a drug or food) occurs only after the completion of a number of consecutive schedule components. At the end of each single component the experimental subject is presented with a brief stimulus (the conditioned or secondary reinforcer) which has been paired repeatedly in the past with the primary reinforcer. Figure 17.2 shows cumulative records from a monkey responding under a second-order schedule in which completion of each 5-min FI component resulted in the presentation of a 2-s flash of a coloured light (indicated by a diagonal deflection on the cumulative record). The completion of each tenth FI unit, (i.e. after *c.* 50 min), resulted in the presentation of the light and also an i.v. infusion of cocaine (indicated by a downward deflection of the

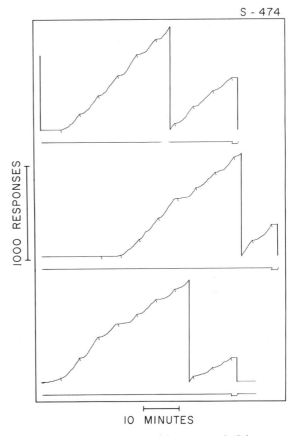

S - 474

1000 RESPONSES

10 MINUTES

*Fig. 17.2. Representative cumulative record of the performance from one monkey (S-474) in a second-order schedule in which every tenth completion (Fixed Ratio 10) of a five-minute Fixed Interval schedule produced an intravenous infusion of 100 μg/kg of cocaine. The Fixed Interval schedule resulted in the presentation of a yellow light for two seconds, as indicated on the cumulative records by short diagonal strokes. Note how the light presentations resulted in characteristic scalloped patterns of responding during each Fixed Interval component. Infusions of cocaine are shown for each record by downward deflections of the second (lower) pen recorder for a period of 100 s. The complete experimental session consisted of three sequences (from top to bottom) each of which terminated in a cocaine infusion. The recording pen reset to the base of the record after 1100 responses or at the end of each sequence. Reproduced from Goldberg S.R. et al., Federation Proceedings **34**, 1771–16 (1975), with permission of author and publisher.*

lower pen). Each cocaine injection was accompanied by a 100-s time-out period and each session terminated when animals had received three infusions of cocaine.

The second-order schedule clearly supported an extensive sequence of responding with relatively few infusions of cocaine, and a considerable number of responses (*c.* 1500) were emitted before the first presentation of cocaine. Furthermore, the light stimulus paired with cocaine infusion supported 'scalloped' patterns of responding typical of FI schedules of reinforcement (Fig. 17.2). An extensive body of experimental research has been conducted with second-order schedules of self-administration. Brief, drug-paired stimuli have been shown to modify the acquisition, maintenance and extinction of self-administration, (see Goldberg & Gardner, 1981, for review).

Despite the fact that the actions of drugs in supporting self-administration are attributable, at least in part, to factors other than molecular structures, it is nevertheless essential in assessing the predictive validity of the self-administration model to consider the range of drugs which do and do not maintain self-administration. Many different types of drugs maintain self-administration (see Young & Herling, 1986; and Sanger, this vol., Ch. 18 for review). It has frequently been suggested that drugs which animals self-administer are those which are prone to abuse by humans (Falk *et al.*, 1983), although there are exceptions to this rule. Animals do not self-administer hallucinogens and cannabinoids, but they do self-administer nomifensine and local anaesthetics which are not generally considered drugs of abuse in humans. Whether or not these are examples of 'false positives' and 'misses' in screening for abuse liability is difficult to decide since it is clear that whether or not a drug acts as a reinforcer depends on factors other than the drug studied. It may simply be the case that the experiments so far conducted in animals have been the 'wrong' ones, in terms of the procedures used (Woods, 1983). In contrast, Johanson *et al.* (1987) have suggested that drugs with the highest abuse liabilities are those that are self-administered under the widest range of conditions (see Sanger, this vol., Ch. 18 for further discussion of this issue).

17.3.1.2 Assessing the efficacy of drugs as reinforcers

Although the reinforcing actions of drugs can vary as a function of the procedures used, it is nevertheless reasonable to ask whether the self-administration paradigm allows predictions to be made as to the relative efficacy of drugs as reinforcers within specific procedures. As discussed above (Section 17.2), this issue is important in determining the predictive validity of any model.

The progressive-ratio procedure has perhaps been used most widely to assess the efficacy of reinforcing drugs. In this procedure, animals are trained to self-administer a drug until stable responding is obtained. Subsequently, the ratio of responses required for reinforcement is progressively increased. Animals reduce their response rate considerably as the ratio requirement increases, and a 'breaking point' is defined arbitrarily as the ratio at which response rate falls below some minimal criterion. Progressive-ratio studies with food as a reinforcer established that breaking points increase in a systematic manner as a function of the period of food deprivation or the concentration of a liquid reinforcer (Hodos, 1961; Hodos & Kalman, 1963). Similarly, breaking points increase as a function of the intensity of reinforcing electrical brain stimulation (Hodos, 1965). Thus it is assumed that the breaking point provides a measure of the 'strength' of the reinforcer in question.

Figure 17.3 shows an example of a progressive-ratio study conducted with a series of phenylethylamine-derived drugs (Griffiths *et al.*, 1978). In this study, five baboons were trained to respond for infusions of cocaine on a Fixed Ratio (FR) 160 schedule of reinforcement; each reward was followed by a three-hour time-out period when responses had no consequences. Thus animals could self-administer eight infusions each day. Subsequently, a number of drugs were substituted for cocaine and the ratio requirement was systematically increased over days (from FR 160 to FR 6000), until animals reached a breaking point at which they self-administered at most one infusion a day. The patterns of breaking points indicate that for all individual animals cocaine maintained the highest breaking points, followed progressively by diethylpropion, chlorphenter-

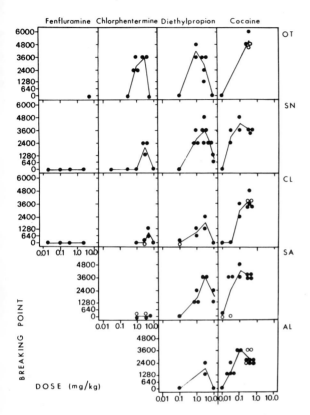

Fenfluramine Chlorphentermine Diethylpropion   Cocaine

Fig. 17.3. Breaking points determined for various drugs (fenfluramine, chlorphentermine, diethylpropion and cocaine) in five different baboons (OT, SN etc.). Each point represents a single breaking point observation. Lines represent the means of the breaking points determined at different doses of each drug. Filled circles refer to the first breaking point obtained for a drug dose, open circles refer to the breaking point obtained during the second exposure to a specific dose. Reproduced from Griffiths R.R. et al., Psychopharmacology 55, 5–13 (1978), with permission of author and publisher.

mine, and finally, fenfluramine which was not self-administered. Using similar procedures, Risner & Silcox (1981) demonstrated that, in dogs, cocaine maintained higher breaking points than amphetamine, whilst fenfluramine was again not self-administered. Indeed, there is reliable evidence from progressive-ratio studies that cocaine supports higher breaking points than a range of other drugs, including methylphenidate, secobarbital and nicotine, as well as all of the various drugs considered above (see Young & Herling, 1986, for review).

An alternative method of comparing the reinforcing actions of drugs involves the discrete-trial choice procedure (Johanson & Schuster, 1975, 1977; Johanson, 1978), which assesses the preference for one drug infusion over another when animals are required to make one of two mutually exclusive responses. Generally, these studies show that with cocaine higher doses are preferred over lower doses (at least over certain dose ranges), presumably because they are more 'efficacious' reinforcers (Griffiths *et al.*, 1979). Johanson & Schuster (1977) reported that in a choice procedure, cocaine was preferred over diethylpropion over a wide range of pairs of doses, suggesting that cocaine is a more efficacious reinforcer than diethylpropion, in agreement with the results from progressive-ratio studies (see Fig. 17.3). Choice procedures are of interest as measures of the reinforcing actions of drugs as they have often been used in human studies of drug reinforcement. For example, Griffiths *et al.* (1986) demonstrated, in a research ward study, that heavy coffee drinkers preferred caffeinated coffee over decaffeinated coffee when they had previously received recent extensive exposure to caffeine but not when they had been caffeine-free for some time. Thus the reinforcing action of caffeine depended upon whether or not subjects were caffeine tolerant/dependent. Such studies demonstrate the close links between animal and human studies of drugs as reinforcers and the way in which procedures used with animals can be adopted to investigate the reinforcing actions of drugs in humans.

It is clear that self-administration procedures can be used to assess comparative reinforcing efficacy of drugs. Studies with phenylethylamines (Griffiths *et al.*, 1979) indicate that drugs which are efficacious reinforcers (e.g. cocaine) are frequently prone to abuse, while drugs that do not act as effective reinforcers, are less prone to abuse (e.g. chlorphentermine) or are virtually completely abuse-free (e.g. fenfluramine). Whether or not such data provide strong evidence for a high decree of predictive validity of the self-administration procedure is difficult to assess without reliable data on the relative abuse potential of drugs in humans. However, enough evidence is available to support the proposition that drugs which show up as 'strong' reinforcers in animal studies are often prone to widespread abuse in humans.

17.3.1.3 Factors maintaining drug self-administration

It has been established unequivocally that drugs can act as reinforcers in animals that are not physically dependent on any drug (Pickens & Thompson, 1968; Deneau *et al.*, 1969; Johanson *et al.*, 1976; see also Sanger, this vol., Ch. 18). Nevertheless, although drugs can act as reinforcers in the absence of prior drug experience, such experience can facilitate their reinforcing actions (see, e.g., Winger & Woods, 1973). Thus there is undoubtedly a link between drug experience/dependence and subsequent drug self-administration (Yanagita, 1980), although it is difficult to dissociate the relative roles paid in maintenance of drug self-administration by positive reinforcing and negative (alleviation of withdrawal) reinforcing actions of drugs.

Common conceptualisations of drug abuse/dependence often suggest that drug self-administration is motivated predominantly by alleviation of a withdrawal syndrome that is associated with the development of physical dependence, despite the fact that it is clear that drugs can be repeatedly self-administered by animals in the absence of any overt signs of withdrawal. However, it is also clear that drug deprivation in dependent individuals can enhance drug intake (Cappell *et al.*, 1987), and that drug experience which is presumed to lead to tolerance/dependence can potentiate the reinforcing actions of drugs (see, e.g., Griffiths *et al.*, 1986). Thus there is controversy over the role of physical dependence in the maintenance of drug self-administration. One radical approach to this question is that proposed by Stewart *et al.* (1984), who argued that opiate and stimulant drugs act primarily as incentives (i.e. as positive reinforcers), rather than as negative reinforcing stimuli which alleviate the presumed aversive properties of withdrawal. A major line of evidence in support of this thesis is the finding that in animals which have been trained to self-administer a drug and then extinguished (so that operant responding declines substantially), non-contingent experimenter-administered injections can act as 'priming' stimuli and reinstate behaviour.

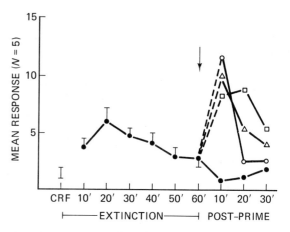

Fig. 17.4. *Mean number of responses (with standard errors) for five rats during 10-min periods. During the first period shown in the Figure (CRF = continuous reinforcement) infusions of cocaine (1.0 mg/kg) were available for each response. Subsequently, responding to cocaine was extinguished over a 60-min period – each response on the lever produced only an audible click. Note how during extinction, response rates first increased above those seen after cocaine infusion then decreased. At the time indicated at the arrow a non-contingent priming dose of cocaine was infused. The data for various priming doses of cocaine are shown separately – 0.0 mg/kg (filled circles), 0.5 mg/kg (open circles), 1.0 mg/kg (triangles) and 2.0 mg/kg (squares). Note how priming infusions of cocaine reinstated responding in a dose-related manner. Reproduced from de Wit H. & Stewart J., Psychopharmacology 75, 134–43 (1981), with permission of author and publisher.*

An example of such a 'priming' effect (de Wit & Stewart, 1981) is shown in Fig. 17.4. Rats were trained to self-administer cocaine (1.0 mg/kg per infusion). Once stable patterns of self-administration had developed, the effect of extinction was studied by disconnecting the infusion pump after a period of self-administration. When extinction was initiated responding increased initially, then it declined over the rest of the extinction period. After 60 min of extinction, the rats were given non-contingent 'priming' doses of cocaine. 'Priming' injections, reinstated responding in a dose-related fashion, despite the fact that responses did not lead to cocaine infusions. These data suggest that the presence of an abused drug in the body leads to drug-related behaviour and thus appear to show that 'relapse' is caused by the incentive properties of the drug rather than by the alleviation of withdrawal. 'Priming' effects such as those described above have been described with other CNS stimulants and with opiates (Stewart, 1984). There is evidence that 'priming' effects can be seen not only between drugs with similar pharmacological actions, but also, to some extent, between different classes of drugs (see Stewart *et al.*, 1984, for review). It is suggested that the incentive properties of drugs can be conditioned to environmental stimuli and that conditioned incentive stimuli 'prime' the drug-free individual to take drugs when detoxified (Stewart *et al.*, 1984).

Clearly, this conceptualisation of relapse, which emphasises the positive reinforcing actions of drugs, stands in stark contrast to the idea that relapse is due to withdrawal and to conditioned withdrawal (see Section 17.4.2 below). Stewart *et al.* (1984) suggest that animals and humans who take drugs learn to work for and seek out environments and objects associated with drugs and that such stimuli are potent determinants of the maintenance and reinstatement of drug-taking behaviour. Indeed, de Wit & Stewart (1981) demonstrated that stimuli (tones) associated repeatedly with cocaine infusions can themselves serve as conditioned 'priming' stimuli. It is not possible to discuss here in detail the question of whether relapse is mediated by withdrawal and conditioned withdrawal as opposed to incentive and conditioned incentive actions of drugs (see Hartnoll, this volume, Chapter 19, for further discussion of this question). However, studies such as those described above are important, in that they may have profound implications for the concepts of drug abuse and dependence. Obviously, such provocative studies could only have been conducted with the self-administration model. They therefore show the considerable value of simulations for analysing the basic processes involved in persistent drug intake.

17.3.1.4 *Validity of the self-administration model*

The findings described above show that the self-administration procedure has considerable face validity as an animal model of drug abuse, at least in terms of symptomatology, since it demonstrates in animals one of the defining characteristics of drug abuse – persistent self-administration of

drugs (Falk *et al.*, 1983). The model even gives insights into some of the processes potentially involved in the generation of persistence, as shown by the second-order schedule studies described. Assessment of the predictive validity of the model is difficult to achieve, since there is ample evidence that factors other than the drug itself act to maintain self-administration. Thus the finding that animals do not self-administer most hallucinogens may simply reflect the fact that the 'correct' experiments have not been done with these drugs. Despite this caveat (and with some exceptions – see Sanger, this vol., Ch. 18), there does seem to be good general agreement between the drugs that humans abuse and those that are self-administered by animals. Furthermore, the self-administration model allows drugs to be ranked in terms of their efficacy as reinforcers, and as a first approximation, those drugs which are 'strong' reinforcers in animals are common drugs of abuse in humans. It is difficult to assess the construct validity of the self-administration model because of the tenuous nature of the constructs of abuse and dependence. However, some of the findings from self-administration studies potentially have profound implications for the construct of dependence (e.g. with reference to the importance of withdrawal in dependence). The self-administration paradigm is clearly of considerable value for studying fundamental processes in drug abuse, although the conclusions that can be drawn from research in this are necessarily limited in the extent to which they relate to general conceptualisations of dependence and abuse.

17.3.2 Schedule-induced drug intake

It is often suggested that a defining feature of drug-abusing individuals is that they expend extensive effort and long periods of time engaged in behaviours associated with drug-taking (obtaining resources to buy drugs, obtaining drugs, etc. – see, e.g., Falk *et al.*, 1983; Sanger, 1986). It is believed that such behaviours are indulged in to the exclusion of more socially acceptable behaviours and are often detrimental to the health, finances and social relationships of the individuals concerned. Thus drug abuse has been characterised as 'compulsive' and 'excessive' (Falk *et al.*, 1983). It follows that animal models which seek to model aspects of the behaviour of abusing individuals should attempt to explain the apparently 'excessive' nature of their drug-related behaviour (Falk *et al.*, 1983).

As noted above, the ways in which drug stimuli are scheduled in relation to behaviour are criticial determinants of the amount of behaviour expended in obtaining a drug reward. Experimental studies have also shown, however, that in addition to supporting operants that lead to presentations of primary rewards, under some circumstances intermittent schedules of reinforcement also support substantial amounts of other behaviours that are not directly rewarded. These behaviours are typically referred to as schedule-induced, interim (as opposed to terminal), or adjunctive (Staddon, 1977). Such schedule-induced behaviours have received considerable attention from behavioural scientists interested in drug abuse because the behaviours observed appear to be 'compulsive' and 'excessive' and, at face value, similar to behaviours thought to occur in drug-abusing individuals.

The presumed 'excessive' nature of schedule-induced behaviour can best be illustrated by consideration of the prototypical example of such behaviour – schedule-induced polydipsia (S.I.P.). In the initial, chance discovery of S.I.P., Falk (1961) trained rats to lever press for small amounts of food reward delivered intermittently. Immediately after the presentation of each food pellet they went to an adjacent water spout and ingested a small amount of water, prior to the resumption of lever-pressing. Although the animals were not deprived of water, this 'post-pellet' drinking resulted in consumption of *c.* 90 ml of water in a three-hour session, the typical daily intake being less than a third of this. Thus, exposure to an intermittent schedule of food delivery resulted in ingestion, in a short period of time, of a substantial amount of water for which there was no physiological need.

The basic features of S.I.P. are now well established, as the phenomenon has been studied extensively (see Staddon, 1977, and Wallace & Singer, 1976, for reviews). They were described by Sanger (1986) as follows:

(i) S.I.P. is, almost by definition, 'excessive', as described above.

(ii) Drinking typically occurs immediately after delivery of food.

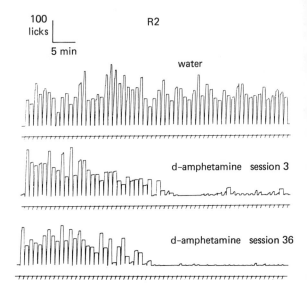

Fig. 17.5. *Drinking records from one rat in three separate 60-min sessions. Food pellets were delivered on a Fixed Time 60-s schedule as indicated by a downward deflection in the lower trace for each session. The upper pen shows the cumulative number of licks at a bottle containing water (top) or a solution of d-amphetamine in water (0.01 mg/ml). The upper pen reset at the delivery of each food pellet. During water ingestion reliable schedule-induced post-pellet polydipsia was recorded. During ingestion of amphetamine solutions reliable post-pellet drinking was observed initially but degenerated during the session. Reproduced from Sanger, D.J.,* Psychopharmacology **54**, *273–6 (1977), with permission of author and publisher.*

(iii) The amount of fluid consumed depends upon the specific schedule of food delivery utilised. For example, Falk (1966) examined the effects of FI schedules varying in duration from 2 to 300 s; the amount of drinking induced by these schedules of food reward increased as the interval increased from 2 to 180 s but then decreased in magnitude.

(iv) S.I.P. takes some sessions to reach asymptotic values.

(v) Polydipsia can also be induced with Fixed Time schedules in which food is presented intermittently without the subject having to make any operant response.

(vi) The more food-deprived subjects are, the greater the degree of polydipsia.

The upper panel of Fig. 17.5 (from Sanger, 1977) shows some of these features of S.I.P. in rats.

S.I.P. is indeed a striking phenomenon and it is not surprising that it has been suggested that 'excessive' behaviour thought to be characteristic of human drug abusers is related to 'excessive' behaviour seen in S.I.P. For example, Falk *et al.* (1983) suggested that

Both nature and society seldom provide an even flow of the commodities important to survival . . . Life in many ways is a set of complex intermittent schedules. The adjunctive behavior induced by such schedules would depend upon what behavior alternatives were available to the individual (Falk *et al.*, 1983, p. 72.)

It is therefore possible that, under the correct 'inducing' conditions, the intermittent flow of essential rewards could induce 'excessive' persistent drug-taking behaviour, if alternative sources of reinforcement are not available.

A number of authors (Gilbert, 1978; Falk, 1981; Falk *et al.*, 1983; Sanger, 1986) have noted that if this conceptualisation of drug taking behaviour is to be viable it is necessary to show:

(i)   That schedule-induction procedures can cause ingestion of drug solutions, rather than simply water, in laboratory animals.

(ii) That there is a class of behaviours which can be schedule-induced. In other words, it is necessary to show that the 'excessive' nature of S.I.P. is not simply a consequence of some complex, unique relationship between the motivational systems regulating food and water intake.

(iii) That schedule-induced behaviours occur in humans and that such behaviours are induced by a number of relatively non-specific stimuli.

We consider below the extent to which these three criteria have been met. It is now well established that the basic S.I.P. procedure can induce animals to self-administer large quantities of various drugs including ethanol, benzodiazepines, opiates, phencyclidine, barbiturates and nicotine (see Gilbert, 1978, for review). S.I.P. paradigms can also be used to induce animals to self-administer enough drug for physical dependence to develop. For example, Falk & Tang (1987) showed that it is possible to induce physical dependence on

the short-acting benzodiazepine midazolam with a S.I.P. procedure involving only brief (three hours daily) sessions of intermittent food presentation. The lower sections of Fig. 17.5 actually show schedule-induced self-administration of an amphetamine solution of sufficient concentration for the rat to ingest an oral dose of about 0.7 mg/kg, which would have had significant behavioural effects. Indeed, the schedule-induced ingestion of amphetamine declined during each session, probably as a consequence of a direct pharmacological action (drug-induced adipsia) of the ingested amphetamine (Sanger, 1977).

It is much less clear whether schedule-induced polydipsia is a unique example of a specific kind of 'excessive' behaviour, or whether it represents simply one example of a class of adjunctive behaviours. It is known that if reinforcers are administered to laboratory animals on intermittent schedules of reinforcement, it is possible to observe various behaviours between reinforcers which appear superficially to be schedule-induced. Amongst such behaviours are wheel-running, pica, eating, aggression, chewing, biting and gnawing, grooming and locomotion (see Roper, 1981, and Sanger, 1986, for reviews). The critical issue is to what extent such behaviours are truly schedule-induced. In order to show that a specific behaviour is scheduled-induced it is necessary to show that large amounts of the behaviour occur between reinforcers and, more critically, that the behaviour is dependent on the presence of the intermittent schedule. It is therefore necessary to compare amounts of purported schedule-induced behaviours with appropriate baseline conditions. Two baseline conditions proposed by Roper (1981) are (i) A massed-reinforcer condition in which the same total amount of reinforcer is freely available in a session of the same duration: and (ii) An extinction condition in which no reinforcer is available during the whole session. Only if the behaviour occurs at a higher rate under an intermittent schedule of reward than under these baseline conditions does Roper (1981) consider it to be genuinely schedule-induced. Many studies of purported schedule-induced behaviours fail to include comparisons with any baseline conditions (Roper, 1981). Roper (1981) and Roper & Crossland (1982) suggested that when many studies which claim to report schedule-induced behaviours other than drinking are studied in detail, only wood-chewing (in rats) and aggression (in pigeons) are truly schedule-induced. Other behaviours, such as wheel running, do occur at high rates between reinforcers, but the amount of behaviour recorded is not greater than that seen under the appropriate baseline control conditions. Even those behaviours that do meet the criteria for schedule-induction may differ in important ways from S.I.P. For example, schedule-induced chewing in rats, although apparently a good example of a schedule-induced behaviour, differs in that whilst most S.I.P. immediately follows consumption of the food pellet, most schedule-induced chewing is seen in the middle of the inter-reinforcer interval (Roper & Crossland, 1982). Schedule-induced chewing is also slow to develop and shows considerable variability over days and within sessions, in contrast to typical findings in S.I.P. studies (see Fig. 17.5, for example). Such data caution against the idea that there is a class of behaviours which are properly considered schedule-induced and which all have a common causal origin.

Since it has been difficult to prove that a number of different types of schedule-induced behaviours occur in animals, it is not surprising that the question as to whether or not schedule-induced behaviour occurs in humans remains a subject of controversy. Falk *et al.* (1983) have suggested that in humans environmental contingencies induce 'Excessive drug-taking, violence, exercise or even scientific and literary endeavour'. Similarly, it has been suggested that overeating, nail-biting and stereotyped movement patterns in humans may be schedule-induced. However, the evidence in support of such suggestions is at best tentative (see Sanger, 1986, for review). As noted by Sanger (1986), this may be in part because the reinforcement schedules controlling human behaviours are difficult to define and because of difficulties involved in conducting stringent experimental studies in humans. However, it remains true that a minority of the empirical studies that have been conducted in this area to date have used the necessary baseline control conditions. There must therefore be some doubt as to whether or not schedule-induced behaviours occur in humans.

Thus of the three criteria outlined above for the assessment of schedule-induced behaviours as a potential model of human drug taking, it *is* unequi-

vocally established that S.I.P. procedures can induce animals to self-administer large amounts of drug solutions which can lead to physical dependence. However, as noted by Sanger (1986): 'Simply to demonstrate that animals consume a drug solution under conditions where almost any fluid is consumed does not provide an adequate model of human drug taking'. As far as the two other criteria are concerned, there is doubt about the existence of a class of schedule-induced behaviours. If such a class of behaviours exists, its defining characteristics remain unclear. Finally, there is, as yet, little robust evidence to suggest that schedule-induced behaviours occur reliably in humans. All these considerations render somewhat tenuous the idea that the phenomenon of S.I.P. provides a sound foundation for an animal model of drug abuse.

However, despite these caveats, some of the most interesting research in this area in recent years has been concerned with another schedule-induced phenomenon – schedule-induced drug self-injection (Si.Si.). In the Si.Si. procedure animals are typically maintained at a reduced body weight and given food on a Fixed Time (F.T.) schedule for a short period each day, during which lever presses result in i.v. drug infusions. Thus the procedure is directly analogous to the S.I.P. procedure, except that instead of being given access to water, animals lever press to obtain drug infusions. Under these conditions animals self-administer significant amounts of a number of drugs, including nicotine, heroin, alcohol, and methadone (see Jefferys *et al.*, 1979; Singer *et al.*, 1982; and Sanger, 1986, for reviews). A critical finding of this work has been that the amount of drug self-administered is usually greater in animals exposed to both an F.T. schedule of food reward *and* a regime of food deprivation than in animals exposed to either food deprivation alone or simply to the drug self-administration schedule when not food deprived (e.g. Lang *et al.*, 1977). Jefferys *et al.* (1979) proposed that the Si.Si. paradigm provided a new way of conceptualising drug dependence since this paradigm allowed the study of *interactions* between pharmacological and environmental factors in drug abuse. They also proposed that the Si.Si. procedure allows experimenters to avoid a number of methodological problems associated with oral drug ingestion procedures, such as those arising from aversive tastes of drug solutions, and from the delay inherent in the slowness of drug absorption from the gut.

However, as noted by Sanger (1986), these factors have not limited the use of the S.I.P. paradigm as a method for inducing ingestion of large quantities of a number of drugs by the oral route. Whether or not the Si.Si. procedure really is an improvement on the S.I.P. procedure as a model of drug abuse is open to question at present. Furthermore, Sanger (1986) has suggested that Si.Si. behaviour may differ in critical ways from schedule-induced drinking. For example, in contrast to S.I.P., rates of responding do not always increase progressively over Si.Si. sessions. In contrast to S.I.P., Si.Si. does not therefore seem to be learned. Secondly, rates of responding are often such that self-administration does not occur reliably after every food pellet and thus does not appear to share the 'excessive/compulsive' feature of S.I.P. A further puzzling aspect of Si.Si. studies is the finding that there is no clear pharmacological pattern to Si.Si. studies (Sanger, 1986). Although for some drugs, such as nicotine and methadone, self-administration is greatest when animals are exposed to both an F.T. schedule and restricted food access, for other drugs, such as amphetamine, the F.T. schedule retards self-administration (Takahashi *et al.*, 1978) – the greatest amount of self-administration is seen in animals that are simply food deprived. In contrast, for cocaine, the presence of an intermittent schedule neither facilitates nor retards self-administration (Singer *et al.*, 1982). It is therefore possible to obtain different effects of procedural manipulations on self-administration of drugs such as amphetamine and cocaine which have many common pharmacological properties. In summary, while the Si.Si. procedure is an interesting development in self-administration studies, the relevance of this model to human drug abuse is unclear. This is perhaps not surprising given our poor understanding of the S.I.P. procedure itself.

In terms of the criteria used in this chapter to assess the validity of models of drug abuse, it is clear that the S.I.P. model has low predictive validity since it seems possible to induce animals to ingest almost any drug solution. The predictive validity of the Si.Si. model is less well established, since it has not been studied as intensively. With

regard to face validity, the S.I.P. model relies heavily on the idea that it models the putative 'excessive/compulsive' nature of drug abuse. However, it is legitimate to ask to what extent drug abuse in humans does in fact involve 'excessive/compulsive' behaviours? The description of drug-related behaviours in such terms involves a degree of subjectivity and is rarely supported by quantitative evidence. While it may be true that abusers of illegal agents such as opiates spend substantial periods of time engaged in drug-related behaviours, it is possible that abusers of *legal* drugs of abuse, such as the benzodiazepines, do not spend substantial amounts of time in drug procurement and ingestion. The face validity of the S.I.P. model of drug abuse may therefore be more apparent than real. In terms of construct validity, it should be apparent, from the discussion above, that the relevance of schedule-induced behaviours to drug abuse is not at all clear (Sanger, 1986). Claims that the Si.Si. procedure leads to a fundamental reconceptualisation of drug abuse and dependence (see, e.g., Jefferys *et al.*, 1979) seem somewhat premature, particularly when there are puzzling pharmacological patterns in the data generated in this procedure. In summary, both the S.I.P. and the Si.Si. models of drug abuse seem at best models that should not be accepted uncritically.

17.3.3 The conditioned place preference/aversion paradigm

The conditioned place preference/aversion (CPP/CPA) paradigm is now widely used to assess rewarding and aversive actions of drugs. The basic procedure consists of two (or more) different environments typically differentiated in terms of a number of stimulus modalities. One environment is consistently paired with drug treatment while the other is paired with saline treatment. Following training, animals are tested, typically when drug free, for their preference or aversion for the environment paired with the drug. There are numerous variations of the basic procedure (see Swerdlow *et al.*, 1988, and Spyraki, 1988, and Carr *et al.*, 1989 for reviews). All are based on the premise that animals learn to approach stimuli paired with rewards and to avoid stimuli paired with aversive agents. All the basic procedures essentially test the ability of drugs to enhance/diminish the incentive value of stimuli associated with drug treatment. These procedures are therefore probably related to operant second-order schedule studies which have shown that exteroceptive stimuli can exert powerful control over drug self-administration (see Section 17.3.1.1, above).

Since the notion of reward is central to many conceptualisations of drug abuse, it has been suggested that it is of considerable scientific value to have an alternative to the self-administration procedure to assess rewarding actions of drugs (Spyraki, 1988). Thus, implicit in the use of the CPP/CPA procedure is the assumption that it measures the same 'rewarding' property of a drug as the self-administration procedure. One would therefore expect that drugs which are self-administered should induce CPP and that the neural systems mediating CPP are the same as those that mediate rewarding actions of drugs in the self-administration paradigm. There is some evidence in support of both these claims, but neither is uncontroversial.

Consider first the question of which drugs reliably induce CPP. These include many agents that are self-administered by animals (and some that are abused by humans), including amphetamine, beta-phenylethylamine, cocaine, apomorphine, methylphenidate, nicotine, nomifensine, morphine, heroin, fentanyl, some enkephalins and endorphins, diazepam and other benzodiazepines (see review by Swerdlow et al., 1988). For some of the agents cited above contradictory data are found in the literature. For example, there is some controversy as to whether nicotine will support CPP (Clarke & Fibiger, 1987), although it should be noted that this drug supports self-administration only under restricted conditions (see Sanger, this vol., Ch. 18). Amongst agents that induce CPA are lithium, naloxone and naltrexone, kappa opiate agonists and pictrotoxin (Swerdlow et al., 1988). At face value, drugs that support CPP appear similar to those that support self-administration, while agents that induce CPA fail to support self-administration. Such data clearly support the idea that the CPP procedure detects drugs that are prone to self-administration. Indeed, Spyraki (1988) has suggested that the ability of the CPP paradigm to detect rewarding actions of diazepam (Spyraki et al., 1985; File,

1986) is indicative of the sensitivity of the procedure, because benzodiazepines are difficult to detect as drugs of abuse with the self-administration procedure. Furthermore, it is possible to demonstrate CPP induced by non-pharmacological rewards such as food (Spyraki et al., 1982a) and sucrose (White & Carr, 1985). Overall, there appear to be few 'false positives' and 'false negatives' in CPP/CPA studies (Swerdlow et al., 1988). Nevertheless, failures of predictive validity do occur. For example, studies with ethanol have not shown consistently that this drug induces CPP (Swerdlow et al., 1988). Indeed, Stewart & Grupp (1986) trained animals to self-administer an ethanol solution in one environment and water in another, but found that the animals developed an aversion for the ethanol-paired environment, despite the fact that they drank more ethanol than water. Ethanol, presumably, was rewarding, yet it resulted in paradoxical avoidance of the drug-paired environment. The causes of such surprising findings remain to be determined.

As far as the neurochemical systems involved in CPP are concerned, there is quite good evidence that dopaminergic (DA) systems, particularly the mesolimbic DA system, are involved in CPP induced by a number of drugs. Thus it is known that amphetamine-, heroin- (and food-) induced CPPs are blocked by haloperidol or by 6-OHDA lesions in the nucleus accumbens (Spyraki et al., 1982b, 1983). Significantly, Carr & White (1986) reported that amphetamine injected into the nucleus accumbens induced a CPP, but that no effect was seen in the nigrostriatal or other brain DA systems. Such findings directly implicate the mesolimbic DA system in CPP conditioning. Anomalous data have, however, been reported with cocaine: the CPP induced by i.p. cocaine injections is not blocked by haloperidol or by lesions of brain DA systems (Morency & Beninger, 1986). However, Spyraki et al. (1987) and Morency & Beninger (1986) have established that when cocaine is given i.v. or directly into the brain (i.c.v.) then the observed CPP *is* blocked by DA antagonists, suggesting that brain DA systems are involved in cocaine-induced CPP, but that peripherally injected cocaine can induce CPP by a DA-independent route. Morency and Beninger (1986) suggest that the local anaesthetic action of peripherally administered cocaine may cause CPP by alleviating the pain resulting from i.p. injections. Whatever the explanation, it appears that cocaine-induced CPP does involve a DA system, but that this effect can be masked by other actions of cocaine.

Other evidence suggests that CPP induced by opiates is stereospecific, can be blocked by naloxone and naltrexone, and is probably mediated by actions at mu rather than kappa receptors (Mucha & Herz, 1986). Endogenous opiate systems are therefore probably involved in CPP induced by opioids. Spyraki (1988) has argued that the mesolimbic DA system also plays a critical role in opiate-induced CPP, and is also involved in CPP induced by psychostimulants and by benzodiazepines. However, this conclusion is not universally accepted. Swerdlow *et al.* (1988) note that in some studies the CPP induced by the stimulants methylphenidate and nomifensine was not blocked by the DA antagonist haloperidol (Martin-Iverson *et al.*, 1985; Mithani *et al.*, 1986). Similarly, the finding that, haloperidol blocks heroin-induced CPP (Spyraki *et al.*, 1983), but has failed to modify heroin self-administration in some studies (e.g., Ettenberg *et al.*, 1982) suggests that differences may exist between neurochemical systems mediating CPP and self-administration. A review of this and related findings led Swerdlow *et al.* (1988) to the conclusion that:

The role of brain dopamine in the neural substrates of psychostimulant and opiate reinforcement as assessed by place conditioning varies greatly from the role of brain dopamine in drug reinforcement as predicted by other behavioural measures, particularly self-administration.

Thus it is not at present accepted unequivocally that the systems mediating self-administration are the same as those mediating CPP. Some of the controversy in this area may result from the nature of the CPP paradigm itself. Although originally introduced as a putatively simpler assay for reward than the self-administration procedure, it has become clear that the simplicity of the former is deceptive and that the results of CPP studies are dependent on factors other than the rewarding actions of drugs. This conclusion was highlighted initially by Swerdlow & Koob (1984) who showed that physical restraint, after injection of amphetamine *and* of saline, blocked amphetamine-

induced CPP. Restraint did *not* block conditioned stimulation in the amphetamine-paired environment. Thus it did not interfere with learning; it did, however, block spatial conditioning. Swerdlow & Koob (1984) suggested that, in unrestrained animals, a stimulant drug increases exploration of the drug-paired environment and so reduces environmental novelty relative to the vehicle-paired environment. When offered a choice, animals chose the drug-paired environment simply because it was less novel. This account of CPP raises the critical point that 'conditioned preferences' may be determined, at least to some extent, by drug-induced behaviours, rather than by rewarding actions of drugs. This analysis predicts that stimulants should induce CPP and sedatives should induce CPA. Many drugs which induce CPP (amphetamine, cocaine, opiates) are indeed stimulants at the relevant doses, while many agents which induce CPA are sedatives at the relevant doses (e.g. vasopressin, naloxone, naltrexone, haloperidol, pentobarbital – Swerdlow *et al.*, 1988). For many drugs locomotor stimulation follows an inverted U-shaped dose/response curve, falling off at high doses. Swerdlow *et al.* (1988) note that a similar inverted U-shaped dose–response curve is seen with respect to the ability of these drugs to induce CPP. Such data suggest that a close relationship exists between the actions of drugs as stimulants/sedatives and their ability to induce CPP/CPA.

In apparent conflict with this analysis are data obtained with benzodiazepines (Spyraki *et al.*, 1985; File, 1986) which show that sedatives induce CPP. However, as noted by Spyraki (1988), few studies have been conducted in which activity has been assessed *during* conditioning, which typically takes place in a novel environment in which 'sedatives' might enhance exploration. Thus it would be valuable in the future if CPP/CPA studies included assessments of drug effects on locomotor activity during conditioning trials.

The role of novelty in spatial preference studies has been emphasised by recent studies by Carr *et al.* (1988) and Vezina & Stewart (1987). Carr *et al.* (1988) demonstrated that, in a place preference apparatus, the amount of time spent in each environment depended in a complex way on the relative novelty of the two environments. Rats avoided a completely novel environment but showed preference for a relatively novel environment. Furthermore, if animals were repeatedly restrained in one environment but free to move about in the other, in a subsequent choice test they preferred the environment in which they had been restrained. On the basis of these data, Carr *et al.* (1989) have argued that the effect of restraint in the CPP procedure is to maintain the stimulus novelty of restraint-associated environments (see also Vezina & Stewart, 1987). In support of this analysis, they replicated Swerdlow & Koob's (1984) study, showing that restraint (in both environments) blocked amphetamine-induced CPP. However, they also showed that this effect was abolished if, before conditioning trials, animals were habituated to both environments. In contrast to the findings of Swerdlow & Koob (1984), restrained animals were able to learn an amphetamine-induced CPP provided they had been habituated to both environments prior to conditioning. These data support the conclusion that it is possible to induce CPPs which are not confounded by drug actions on locomotor activity. Nevertheless, the data of Carr *et al.* (1988) do demonstrate that relative novelty is a critical variable in CPP/CPA studies. In the majority of CPP studies, drug effects on reward and on activity have not been dissociated. Thus these measures of 'reward' may have been confounded to an unknown extent by drug actions on exploratory behaviour.

Indeed, as Carr *et al.* (1989) have suggested, data showing the importance of novelty in CPP studies emphasise the role of a further potential confounding factor in CPP studies – that of state dependency. Since most drugs induce state-dependent learning, and since, in CPP studies, animals are typically tested in the choice test when drug free, the drug-paired environment may be preferred in the choice test because it is effectively more novel, since this is the first time the drug-paired environment has been encountered in a drug-free state. Mucha & Iversen (1984) reported that morphine-induced CPP observed when animals were tested drug-free was of similar magnitude to that seen in animals tested under morphine, suggesting that, for this study at least, state-dependent learning was not a major determinant of choice behaviour (see also Spyraki *et al.*, 1985). However it remains possible that some CPP studies are confounded, to

an unknown extent, by state-dependent learning effects.

A further potential confounding factor in CPP studies is that in spatial conditioning studies subjects not only have to experience drug-induced affective responses; they also have to associate such changes with the relevant environment and then to recall this association. Thus spatial preference studies may also be confounded by actions of drugs on learning and recall (White & Carr, 1985), and an agent with potent amnestic actions might not be detected with CPP procedures. This could be of considerable clinical significance. For example, benzodiazepines differ in the extent to which they support spatial conditioning (File, 1986). It is thus possible that the CPP procedure could be used to detect benzodiazepines with low abuse potential. However, benzodiazepines have potent amnestic properties which could confound their actions in the CPP procedure and thus negate the value of the procedure as a measure of reward.

To what extent does the CPP procedure meet the various criteria by which an animal model of drug abuse should be assessed (see Section 17.2)? The model has a reasonable degree of predictive validity, although ethanol may be a "false negative". The CPP procedure has not as yet been used to compare drugs in terms of their relative abuse potential; it therefore is less useful in this respect than the self-administration model. Indeed, it is often difficult in CPP studies to generate typical graded dose–effect curves with any one drug (Carr *et al.*, 1989), so it is difficult to compare the relative efficacy even of different doses of the same drug. As far as face validity is concerned, the procedure makes no attempt to model the symptomatology of drug abuse. However, people may well seek out drug-associated stimuli and environments for their conditioned incentive effects (see Secton 17.4.2, below). Thus the model may possess face validity to an extent that is at present unknown. The CPP procedure gains most of its validation from the fact that it is a measure of drug-induced reward. As this concept is central to most conceptualisations of drug abuse, the CPP model clearly has some degree of construct validity. However, CPP procedures are not 'pure' measures of reward. Although they do provide an alternative to the self-administration procedure, it is apparent that the results of any one study may be confounded, to some extent, by drug actions on levels of locomotor activity, on memory, and as stimuli which may induce state-dependent learning.

17.3.4 Drug discrimination paradigms as models of 'subjective' drug effects

In recent years the drug discrimination (DD) bio-assay has become increasingly popular for investigating the actions of psychoactive drugs (Colpaert & Slangen, 1982, for reviews). In part, this is because it has often been suggested that the procedure allows experimenters to investigate animal models of human 'subjective' responses to drugs, which are often believed to be major determinants of abuse liability (Jasinski *et al.*, 1984). The DD procedure is therefore often considered to provide data on drug abuse potential (Falk *et al.*, 1983; Brady & Lukas, 1984; Colpaert, 1986). Although there are many variants of the procedure (Jarbe, 1987), the DD paradigm typically involves training animals to make an operant response to obtain a reward when treated with a specific drug (the so-called 'training drug'), but to make an alternative response when treated with a placebo inject-ion. After drug treatments, placebo-appropriate responses have no consequences; after placebo treatments, drug-appropriate responses have no consequences. To obtain reward, the subject must discriminate its internal state and select the appropriate response. Numerous studies have shown that many animal species can discriminate reliably between a wide range of drugs (Stolerman & Shine, 1985). Typically, once trained to discriminate a specific drug, animals are tested with this 'training' drug and with a variety of other drugs. They show dose-related selection of the drug-appropriate response option when tested with decreasing doses of the 'training' drug and they also show dose-related drug-appropriate responding (generalisation to the 'training' drug), when treated with drugs with similar profiles of action, but not when treated with drugs with different profiles of action. In the latter case, animals typically select the response option associated with placebo treatment.

The basic procedure can best be illustrated by consideration of a specific example of a DD study based upon drug effects on blood pressure. If normal rats are trained to discriminate the anti-hypertensive drug clonidine, they generalise to some antihypertensives but not to others (Bennett & Lal, 1982). However, Lal & Yaden (1985) reported that if genetically spontaneously hyper-tensive (SHR) rats were trained to discriminate clonidine, they generalised to a number of different antihypertensive agents with a variety of different mechanisms of action. In contrast, they selected the placebo response option after treatment with high doses of a number of different drugs. Doses of drugs which produced generalisation to clonidine in SHR rats correlated highly with doses of the same agents that reduced blood pressure. Col-lectively, these data suggest that, at least in SHR rats, the discriminative properties of clonidine are based on reductions in blood pressure. It is pre-sumed that in discriminating clonidine, animals learn to detect the 'subjective' effects of lowered blood pressure. Using similar procedures with other drugs, a number of experimenters have claimed to have developed DD assays based on hallucinogenic (Appel & Cunningham, 1986), anxiogenic (Emmett-Oglesby *et al.*, 1983), analge-sic (Weissman, 1976) and other 'subjective' actions of various drugs, many of which are mediated by actions on the central nervous system (Jarbe, 1987), and which may be related to drug abuse liability.

The mere presence of discriminative properties in animals cannot, however, be taken as evidence that any specific drug will be abused by humans since some drugs with discriminative properties in animals are not abused by humans (e.g. neuro-leptics, antihistamines; see Overton & Batta, 1977). A more sophisticated verison of the hypothesis that drug 'discriminability' may be related to abuse potential is based on the idea that only those drugs which have *potent* discriminative properties will be prone to abuse by humans. However, Overton & Batta (1977) reported a low ($+0.3$) correlation between the discriminability of drugs in rats (as measured by the number of sessions required to learn a DD task), and an index of drug abuse potential as assessed by the U.S. Drug Enforce-ment Agency Categorization (DEAC) of some 87 drugs. Although the correlation was statistically significant, it was nevertheless disappointingly low and suggested that 'discriminability' *per se* is not a good predictor of abuse potential. Of course, the validity of this analysis depends critically on the validity of the measures of 'discriminability' and abuse potential that were used. It is quite possible that abuse potential as assessed by DEAC may not be a good measure of drug abuse potential. More recent studies (reviewed in Jarbe, 1987) have also cast doubt on the very notion of 'discriminability'. For example, the speed at which a drug is discriminated in one procedure does not necessarily correlate with the speed at which the same drug is discriminated in a different procedure (Jarbe & Kroon-Jarbe, 1983). Validation of the assumed relationship between the abuse potential of drugs in humans and 'discriminability' of drugs in animals is therefore hampered by both the inadequacy of current methods for measuring the 'discriminability' of drugs in animals, and the questionable validity of current indices of abuse potential of drugs in humans.

However, despite these problems, it is widely believed that analysis of drug actions in the DD procedure will ultimately prove of value in assessing abuse potential. In support of this argument, Colpaert (1986) has argued that the clinical assessment of abuse potential in humans often relies on the assessment of similarities and differences in 'subjective' effects of drugs (e.g. Griffith *et al.*, 1975). In such procedures, (such as the Addiction Research Centre Inventory – Haertzen *et al.*, 1963), subjects are typically required to make behavioural (verbal) responses to indicate whether or not a drug has similar effects to reference compounds which are known to have abuse potential. At least at a superficial level, such procedures for assessing the 'subjective' effects of drugs in humans bear a formal similarity to those used to assess discriminative properties of drugs in animals. Both procedures presumably involve the attachment of specific behavioural responses to internal stimulus actions of drugs. For example, if a drug is perceived as amphetamine-like by humans with experience of amphetamine's effects, it is assumed that the drug will be prone to abuse of the amphetamine type. Similarly, if a drug generalises to amphetamine in the DD procedure, it is assumed that the drug may be prone to abuse of the amphetamine type (Porsolt *et al.*, 1982). The assumption is made that because of the formal similarity between the procedures they are both measuring the same drug action – that which leads to its 'subjective' effects.

Some attempts have been made to provide more formal validation of the idea that a close relationship exists between the actions of drugs as discriminative stimuli in animals and their 'subjective' actions in humans, by developing direct human analogues of animal DD procedures. Chait *et al.* (1984) successfully trained most of their human subjects to discriminate between amphetamine and placebo, by giving them one of two visually identical capsules on any one day and requiring them to contact the experimenters by telephone some hours later to report on its contents (subjects received a financial incentive for correct responses). The same group subsequently reported on the discriminative properties of various appetite-suppressant drugs in humans trained to discriminate amphetamine (Chait *et al.*, 1986*a*). When given phenmetrazine, subjects who did not know that they had received a novel drug reported that they had received amphetamine; in contrast, when subjects received fenfluramine, a non-stimulant appetite-suppressant, they made the placebo-appropriate response. Analyses of the 'subjective' actions of these agents indicated that amphetamine was consistently labelled as a stimulant, as was phenmetrazine. However, fenfluramine generally produced negligible effects on mood. These data show a close correlation between the effects of these drugs on mood and their actions as discriminative stimuli in humans. Furthermore, the profile of cross-drug generalisation seen in this study is identical to that which has usually been observed in a number of DD studies in various non-human species (see Schuster & Johanson, 1985, for review). Collectively, these data support the idea that DD procedures in animals measure drug effects related to their 'subjective' effects in humans.

However, in a further study, Chait *et al.* (1986*b*) reported that humans trained to discriminate amphetamine also generalised to the appetite-suppressants mazindol and phenylpropanolamine, basically in accordance with expectations from studies in animals that indicate that these agents show full or partial cross-generalisation (Schuster & Johanson, 1985). However, while phenylpropanolamine produced amphetamine-like effects on mood, mazindol did not, demonstrating a dissociation between the 'subjective' effects of drugs and their actions as discriminative stimuli. As noted by Chait *et al.* (1986*b*), it may simply be the case that procedures used to measure 'subjective' actions of drugs in humans are insensitive to similarities in their 'subjective' effects that really exist between the various agents, since it is difficult to measure mood reliably and accurately in humans. The dangers inherent in such arguments should, however, be obvious: they assume what they are supposed to prove. Similarly, in animal DD studies generalisation is usually, but not always, seen between hallucinogens, suggesting that the actions of these agents as discriminative stimuli may not depend on their 'subjective' effects. Colpaert (1986) has suggested that the 'subjective' effects of different hallucinogens may differ; thus the assumed link between 'subjective' actions of drugs and their discriminative properties may not be disproven by apparently contradictory data. The dangers inherent in such arguments are again clear: it becomes impossible ever to prove that discriminative properties of drugs do not covary with their 'subjective' effects; thus the basic premise is essentially irrefutable! Experimental attempts to prove the validity of this premise also seem beset by many potential methodological problems. For example, the pharmacological specificity of DD stimuli in animal studies is dose related. It is clearly a matter of considerable difficulty to select 'training' doses of drugs for use in human studies which will be functionally equivalent to 'training' doses typically used in animal studies. Hence any observed discrepancies between animal and human data could always be attributed plausibly to differences in the 'salience' of training doses.

Other data show that it is possible to dissociate discriminative properties of drugs from their tendency to maintain self-administration. For example, Woolverton & Balster (1982) demonstrated that while all local anaesthetics that were self-administered generalised to procaine, some local anaesthetics that generalised to procaine were not self-administered. Thus the link between discriminative properties and abuse potential is by no means clear. There are sufficient data which suggest that discriminative properties of drugs are not perfectly correlated with their 'subjective' properties, or with their tendency to be self-administered, for the DD procedure to be accepted only with caution at present as a model of abuse liability (see Griffiths *et al.*, 1985). Furthermore, as discussed above, the relationship between 'subject-

ive' properties of drugs and abuse potential is a complex one, as illustrated by studies which show that drugs can simultaneously have both aversive and positively reinforcing actions.

To summarise, the DD procedure has relatively low predictive validity as a model of abuse potential, since many drugs (some of which are not abused by humans) act as discriminative stimuli. Furthermore, the DD procedure does not allow ranking of drugs in terms of abuse potential. The model does have some predictive value in that it has a high degree of pharmacological specificity, and allows researchers to compare drugs with known drugs of abuse, and determine whether they have similar discriminative properties and similar mechanisms of action. However, there is, as yet, no unequivocal proof that simply possessing discriminative properties in common with abused drugs is necessarily indicative of abuse potential. As far as face validity is concerned, the model does not even attempt to reflect aspects of the symptomatology of drug abuse. The major criterion which the DD procedure attempts to fulfil as a model of abuse potential relates to construct validity. Specifically, the model suggests that analogues of humans' 'subjective' responses to drugs of abuse can be measured in animals and that these are important determinants of abuse potential. This hypothesis is not without its problems, as reviewed above. The possible link between 'subjective' effects in humans and self-administration in animals is also somewhat tenuous. It is clear that the validity of the DD procedure as an animal model of drug abuse is based largely on construct validation, and that such validation is controversial.

17.4 MODELS OF DEPENDENCE POTENTIAL

17.4.1 Behavioural tolerance

The development of tolerance is typically defined either as a progressive decrease in the effect of a constant drug dose with chronic treatment, or as the need to increase the dose during chronic treatment in order to obtain a specific effect of a drug (Corfield-Sumner & Stolerman, 1978; Bignami, 1979). Tolerance is probably best measured in terms of shifts in dose–effect curves resulting from chronic drug treatment; tolerance being associated with a shift to the right and/or a flattening of the dose–effect curve (Fernandes *et al.*, 1977, 1982; Schuster, 1978; Woods & Carney, 1978).

The rationale behind the categorisation of tolerance as a procedure for assessing drug dependence potential is based on the observation that acquisition of tolerance is often associated temporally with the development of withdrawal on discontinuation of drug treatment (Balster, 1985). Indeed, many different theoretical accounts of tolerance and dependence assume that these two processes are causally related (see, e.g., Solomon, 1977; Siegel & MacRae, 1984). This assumption appears reasonable since tolerance and dependence are both the product of chronic drug treatment, and since it is widely believed that tolerance and dependence both result from adaptive changes which minimise disruptive effects of drug-induced perturbations of normal physiological functioning (Balster, 1985). However, the evidence in support of the assumption that tolerance acquisition leads inevitably to the development of a withdrawal syndrome which results in increased drug intake is, remarkably, not as clear as one might expect (Cappell & Le Blanc, 1979; Alexander & Hadaway, 1982, but c.f. Cappell *et al.*, 1987). Furthermore, as discussed above, the conceptualisation of drug self-administration as motivated by the desire to alleviate withdrawal has been the subject of a robust theoretical critique emphasising the role of drugs as positive reinforcers (Stewart *et al.*, 1984).

There has also been increasing realisation in recent years that tolerance involves a range of different mechanisms of different levels of complexity, with different physiological, behavioural and cognitive substrates (see, e.g., Chen, 1968, 1979; Kalant *et al.*, 1971; Schuster, 1978; Demellweek & Goudie, 1982, 1983*a*, *b*; Siegel & MacRae, 1984; Vogel-Sprott *et al.*, 1984; Annear & Vogel-Sprott, 1985; Kalant, 1985; Goudie & Demellweek, 1986; Goudie & Griffiths, 1986). Tolerance is often categorised initially as being either dispositional or functional (Kalant *et al.*, 1971). Dispositional tolerance occurs as a consequence of alterations in drug absorption, distribution and metabolism, a protypical example being the increased metabolism of barbiturates which results from the induction of hepatic drug-metabolising enzymes. Functional tolerance is defined as tolerance arising from

changes in sensitivity of CNS structures. Thus functional tolerance can include changes at the level of receptors, in the synthesis and/or storage of neurotransmitters, in ion channels associated with neurotransmitter release, and so on. In contrast to dispositional and functional tolerance there is now good evidence that tolerance may also involve a number of different behavioural processes (Schuster *et al.*, 1966; Chen, 1979; Vogel-Sprott *et al.*, 1984; Goudie & Demellweek, 1986; Goudie & Griffiths, 1986). A detailed discussion of these different accounts of tolerance is beyond the scope of this chapter and more detailed reviews are available elsewhere (see, e.g., Goudie, 1988; Goudie & Emmett-Oglesby, 1989). Nevertheless, it is worth outlining some basic ideas behind these accounts of tolerance, since they are relevant to consideration of the fundamental issue of the relationship between tolerance and dependence.

One account of tolerance which emphasises the role of behavioural factors is based on the idea that organisms adjust their behaviour to cope with drug effects that interfere with efficient behavioural performance (Schuster *et al.*, 1966). There is good evidence from studies with a wide range of drugs, including stimulants (Demellweek & Goudie, 1983*a*, *b*; Wolgin *et al.*, 1987), sedatives (Branch, 1983), opiates (Sannerud & Young, 1986), and ethanol (Le Blanc *et al.*, 1973; Chen, 1979) that under some circumstances tolerance may involve the acquisition of learned strategies that minimise disruptive effects of drugs (see Wolgin, 1989, for review). There seems no reason why tolerance mediated by learned strategies should necessarily be associated with the development of a withdrawal syndrome. Therefore, the finding that tolerance develops to a drug in a specific assay does not necessarily tell us anything useful about the dependence potential of the drug.

A second account of tolerance which emphasises behavioural processes is the so-called compensatory classical conditioning hypothesis, as developed largely by Siegel (for reviews see Siegel & MacRae, 1984; Goudie & Demellweek, 1986; Siegel, 1989). According to this hypothesis, it is necessary to recognise that the circumstances under which drugs are administered usually constitute classical conditioning trials in which the effects of the drug act as an unconditioned stimulus (UCS) which triggers specific unconditioned responses (UCRs). Each drug administration is associated with so-called conditional stimuli (CSs), such as those associated with injection and/or oral administration of the drug and specific environmental cues, which reliably predict the presence of the drug in the body. Following repeated pairings of the drug UCS with the relevant CSs, a conditioned response (CR), develops in accord with the principles of Pavlovian conditioning. However, with drug UCSs it has been noted (Siegel & MacRae, 1984) that the CR is often in the direction opposite to the initial drug-induced UCR, in contrast to the more typical finding in Pavlovian conditioning studies that the CR resembles the UCR (see Eikelboom & Stewart, 1982; Siegel, 1988, for detailed discussions of this issue). Since the UCR to the drug is progressively offset by the gradually developing compensatory CR, with chronic treatment the net effect of the drug is progressively reduced and tolerance ensues. In contrast to the explanation of tolerance in terms of learned strategies outlined above, this account suggests that tolerance and dependence may be closely associated, since it is assumed that in the absence of the drug (i.e. in withdrawal) the presence of previously drug-predictive CSs will trigger the compensatory drug-opponent response which will effectively represent a withdrawal syndrome, the mirror-image of the drug's acute effects (Hinson & Siegel, 1980).

It is clear simply from consideration of these two 'behavioural' accounts of tolerance, that whether or not tolerance is associated with dependence is determined by the mechanism involved in the production of tolerance. If tolerance involves learned strategies, it will not be associated with dependence. In contrast, if tolerance is based on compensatory conditioning, it might be associated with dependence. As discussed in more detail elsewhere (Goudie, 1989), it is likely that in all behavioural procedures a number of mechanisms for producing tolerance will be involved and the relative importance of each mechanism may be critically determined by a number of experimental parameters. For example, there is now some evidence that tolerance mediated by compensatory conditioning may be of greatest significance, relative to other mechanisms, when relatively low doses of drugs are given infrequently (Baker & Tiffany, 1985; Goudie & Demellweek, 1986). This analysis

implies that demonstration that a drug produces tolerance in a specific behavioural or physiological assay tells us little of general significance about the tendency for that drug to induce dependence. Rather, it seems important to try to analyse tolerance studies in terms of the mechanisms involved (Goudie, 1989) and to determine the relative significance of these mechanisms under different experimental conditions.

What conclusions can we draw from the study of tolerance in terms of the validation criteria outlined for this chapter? As far as predictive validity is concerned, tolerance is a poor 'model' of dependence since almost all drugs, including drugs that are not prone to cause dependence, appear, under the right conditions, to result in the development of some form of tolerance. As far as face validity is concerned many drug-dependent individuals show tolerance to the drugs they are dependent upon, but so do many individuals who are not considered drug-dependent. Indeed, while tolerance may be a necessary condition for the development of dependence, it is not a defining characteristic of the dependent individual. As far as construct validity is concerned, the idea that tolerance can be a viable 'model' of dependence obviously depends on the underlying assumption that dependence arises as a result of the acquisition of tolerance and that dependence is a major determinant of frequent drug usage. While both these ideas are supported by some experimental evidence, it is also clear that tolerance is not necessarily associated with dependence, and that dependence is not necessarily the major determinant of frequent drug intake (see also Sanger, this vol., Ch. 18). Thus tolerance clearly has substantial shortcomings as a 'model' of dependence.

17.4.2 Conditioned withdrawal and conditioned compensatory responses

The idea that withdrawal could be conditioned to environmental stimuli was first put forward by Wikler (1948) as an explanation of relapse in detoxified addicts. Most of the research on this topic has been conducted with opiate addicts, but there is no reason why the basic principles cannot be extrapolated to other forms of dependence. Addicts who have been taking drugs for protracted periods of time will probably have experienced withdrawal repeatedly. If this withdrawal has frequently occurred in association with specific stimuli, addicts will have been classically conditioned over many trials so that when they are detoxified the stimuli will, by themselves, elicit conditioned withdrawal, which could cause a resumption of drug-taking. However, the fact that opiate addicts typically take drugs a number of times a day over a period of years, suggests that the opportunity also exists for conditioning of direct opiate actions, as opposed to the conditioning of withdrawal effects (O'Brien *et al.*, 1986). Thus it is to be expected that addicts will show both conditioned drug effects and conditioned withdrawal effects.

There is reliable evidence from animal studies that it is possible to condition withdrawal to environmental stimuli. The first study to demonstrate this with opiates appears to have been that of Irwin & Seevers (1956), who studied opiate-dependent monkeys that had been subjected to conditioning trials in which withdrawal was precipitated with an opiate antagonist. Some weeks after termination of opiate administration, injections of saline were capable of inducing indices of withdrawal. Subsequently, Wikler & Pescor (1967) demonstrated that the incidence of opiate withdrawal in rats was highest in environments in which withdrawal had been experienced previously. Conditioned withdrawal was observed months after cessation of opiate treatment and thus presumably occurred in the non-dependent animal. Subsequently, Goldberg, Schuster and colleagues investigated conditioned withdrawal phenomena systematically. In an initial study, Goldberg & Schuster (1967) reported that in morphine-dependent monkeys, the opiate antagonist nalorphine, which itself had no effects in non-dependent animals, produced suppression of operant behaviour and other signs of opiate withdrawal. A light/tone stimulus that was presented prior to nalorphine infusions initially had no effect on behaviour, but after repeated pairings it acquired the ability to suppress operant behaviour and to induce emesis and salivation when administered in association with infusions of saline. In a subsequent study, Goldberg & Schuster (1970) showed that it was possible to detect conditioned withdrawal some 120 days after cessation of chronic morphine treatment. Furthermore,

Goldberg *et al.* (1969) demonstrated, in animals trained to self-administer morphine, that after repeated pairings of a light with nalorphine, presentations of the light alone caused increases in morphine self-administration which were similar to those induced by nalorphine. Thus this series of studies established unequivocally that conditioned withdrawal can occur in monkeys, that such conditioned withdrawal can persist for months after the end of opiate treatment and that conditioned withdrawal can increase the rate of opiate self-administration.

A number of studies have shown that conditioned withdrawal can also develop in human (opiate) addicts. O'Brien *et al.* (1977) administered small doses of naloxone to methadone-maintained volunteers to induce a mild precipitated withdrawal syndrome. Injections of naloxone were paired with a tone/odour conditional stimulus. After repeated pairings, injections of saline in conjunction with the tone/odour stimulus resulted in a mild conditioned withdrawal syndrome, in the absence of naloxone injections.

In less-systematic studies, involving structured interviews with opiate addicts and postaddicts, O'Brien (1975) reported that, in a number of cases, it was possible to isolate specific situations which addicts considered capable of inducing illness, drug 'craving', or anxiety. Frequently, stimuli that elicited withdrawal-like effects involved environments associated with previous drug taking, seeing people use drugs, etc. Further evidence that has often been cited in support of the hypothesis that conditioned withdrawal may be an important determinant of relapse is the observation (Robins *et al.*, 1974; O'Brien *et al.*, 1980) that the relapse rate after detoxification of returning Vietnam veterans who were opiate addicts was substantially smaller (*c.* 18%) than that of typical addicts in the U.S.A. whose relapse rate has been estimated at around 80–90% (O'Brien *et al.*, 1986). There are, of course, many other possible explanations for these data.

All the evidence cited above suggests that conditioned withdrawal may be a determinant of relapse. However, the clinical relevance of the phenomenon is unclear. The conditioned withdrawal account of relapse is, of course, based on the premise that drug taking is determined predominantly by the desire to alleviate withdrawal.

As noted frequently above, this conceptualisation of drug abuse/dependence is not without its critics.

In a systematic test of the conditioned withdrawal account of relapse, McAuliffe (1982) interviewed 'street' opiate addicts and found that few reported conditioned withdrawal as a major determinant of repeated opiate use, citing instead euphoriant actions of opiates as the cause of continued drug use. Meyer & Mirin (1979) studied the behaviour of addicts in a research ward and reported that heroin self-administration was *not* associated with signs of conditioned withdrawal. In contrast, they suggested that heroin injections acted to increase craving for the drug when it was freely available. These data are obviously not readily reconciled with the conditioned withdrawal account of relapse. Clearly, the significance of this account of relapse is dependent upon the relative importance in long-term drug taking of positive and negative reinforcing actions of drugs. Even those who have investigated conditioned withdrawal in the animal laboratory or the clinic have argued that positive reinforcing and conditioned positive reinforcing actions of drugs may be important determinants of drug use (O'Brien *et al.*, 1986; Schuster, 1986). There is evidence, for example, that environmental stimuli repeatedly paired with opiate administration can, by conditioning, alleviate withdrawal from opioids (Lal *et al.*, 1976; Tye & Iversen, 1975). Likewise, it is clear from the conditioned place preference and second-order schedule self-administration studies described above that animals will readily seek out stimuli that have been repeatedly paired with drug administration. It is therefore reasonable to suggest that conditioned positive reinforcing actions of drugs might be just as, if not more, important, than conditioned withdrawal in determining drug use. Thus there is evidence (see O'Brien *et al.*, 1986, for review) that in human opiate addicts, injections of saline can have positive affective consequences which may be important in drug self-administration. Such conditioned positive effects of drugs may account for the behaviour known as 'needle freaking' in which addicts repeatedly self-administer injections which are not associated with any pharmacological consequences (Levine, 1974).

The conditioned compensatory response account of tolerance and withdrawal (see Section 17.4.1, above) clearly bears a close resemblance to Wikler's conditioned withdrawal account of relapse: it specifies in detail the mechanisms involved in the development of the conditioned responses thought to mediate relapse. As the conditioned compensatory response account of tolerance/dependence is based on the premise that drug taking is maintained by alleviation of withdrawal, all the criticisms of this position (see above) apply to this model. The conditioned compensatory response account of tolerance also suffers from the further problem that there is some controversy about the reliability of empirical demonstrations of conditioned compensatory responses (Goudie & Demellweek, 1986; see Siegel, 1989, for discussion). It is possible to develop models of tolerance based upon habituation processes, rather than on conditioned compensatory responses and such models can account for much of the data cited in support of the conditioned compensatory response account of tolerance (Baker & Tiffany, 1985). Clearly, further research is required before any definitive conclusions can be reached about the role of compensatory responses in relapse in dependent individuals.

How well do the notions of conditioned withdrawal and compensatory responses meet the criteria used throughout this chapter to evaluate models of abuse and dependence? As far as predictive validity is concerned, this is impossible to establish for conditioned withdrawal since most of the work in this area has involved opiates. The extent to which similar effects occur with other drugs is unclear. However, conditioned compensatory responses have been described with a range of drugs (see Goudie & Demellweek, 1986, for review). There therefore seems no reason to reject the idea that conditioned withdrawal and conditioned compensatory responses *might* be important determinants of relapse to opiates and other drugs. As far as face validity goes, it is possible to obtain evidence for conditioned withdrawal in human opiate addicts. Such symptomatology in addicts clearly increases the face validity of the conditioned withdrawal 'model' of dependence. As far as construct validity is concerned, the ideas of conditioned withdrawal and conditioned compensatory responses are basically explanatory constructs designed to account for an important facet of dependence – relapse. Thus studies on conditioned withdrawal actually modify the construct of dependence and cannot be validated by reference to it. Research on the conditioned withdrawal model is clearly important but at present the clinical significance of the phenomenon is unclear.

17.4.3 Dependence and behavioural dependence

Obviously, the most direct way of assessing dependence potential is to determine the effects of cessation of drug treatment. By definition, any change in behaviour or physiology associated with cessation of treatment is a withdrawal symptom. Such withdrawal symptoms are obviously not 'models', they are assays of dependence. There are numerous ways in which procedures for assessing dependence can vary. It is possible to assess the extent to which cessation of treatment leads to 'spontaneous' withdrawal; alternatively it is possible to make animals dependent on a standard drug (e.g. an opiate or a barbiturate) and then, using a substitution procedure, to determine the extent to which a test drug alleviates the effects of withdrawal from the standard compound. With the development of specific antagonists for opiates and benzodiazepines, it is also possible to assess 'precipitated' withdrawal induced by antagonists in animals treated chronically with the relevant drug. The study of 'precipitated' withdrawal has the substantial advantage that it is much less time consuming (see Sanger, this vol., Ch. 18, for further discussion).

In all dependence studies there are numerous ways in which chronic administration can be scheduled in terms of route, frequency and duration of dosing. There are also numerous behaviours and/or physiological indices, including responses of *in vitro* tissue preparations (Schulz & Herz, 1976), which may be used to detect withdrawal. A detailed discussion of the many methodological considerations involved in detecting dependence is beyond the scope of this chapter (see Katz & Valentino, 1986; Brady & Lukas, 1984). Nevertheless, it is relevant to discuss here studies of socalled 'behavioural dependence' (Schuster, 1968), since an evaluation of this concept reveals some of the difficulties encountered in defining and measuring dependence and withdrawal adequately.

Behavioural dependence studies are based on the belief that learned behaviours (typically using operant schedules) will be more sensitive in detecting withdrawal than gross behavioural or physiological indices. It is assumed that they detect withdrawal after treatment regimes that do not produce overt signs of withdrawal (Balster, 1985). There is evidence that withdrawal from ethanol (Ahlenius & Engel, 1974), tetrahydrocannabinol (Beardsley *et al.*, 1986), morphine (Ford & Balster, 1976; Steinfels & Young, 1981), cocaine (Woolverton & Kleven, 1988), and phencyclidine (Slifer *et al.*, 1984) all disrupt operant behaviour after treatment regimes which induce few gross overt signs of withdrawal. Thus a drug treatment regime may induce a behavioural withdrawal syndrome in the absence of more classical indices of withdrawal (Balster, 1985). However, as described above, abuse liability can be dissociated to some extent from dependence potential. It is, therefore, not clear exactly what we can usefully glean from subtle tests for withdrawal in terms of their ability to predict self-administration of a specific drug. While of considerable value in detecting subtle withdrawal effects, by definition, the clinical significance of studies of behavioural dependence is not at present clear.

The resolution of this question may depend upon the mechanism(s) by which behavioural dependence is/are produced. For example on termination of chronic treatment with an appetite suppressant drug, animals may show 'rebound' overeating in order to return to baseline body weight levels. By definition, such 'rebound' overeating resembles a withdrawal effect. However, since it probably arises from homeostatic adjustments to a period of reduced food intake, it is unlikely that such a withdrawal symptom will be associated with an increased tendency to self-administer the appetite suppressant. In many studies of behavioural dependence it is not at all clear what precise mechanism(s) cause disruption of operant behaviour. Some, or even all, forms of behavioural dependence may well be subtle manifestations of classical withdrawal syndromes (Balster, 1985). On the other hand, when subtle behaviours are used to monitor withdrawal, disruptions of behaviour may simply arise as a consequence of the state change that occurs when drug administration ends. Alternatively, withdrawal from many drugs is associated with apparent 'anxiety' in animals (Emmett-Oglesby *et al.*, 1983;

File *et al.*, 1987), such 'anxiety' could interfere with or punish ongoing operant behaviours. In effect, as with tolerance studies (see Section 17.4.1), unless something is known about the mechanisms underlying behavioural dependence, such studies do not allow assessment of the relevance of the observed behavioural dependence to drug self-administration. Mucha (1987) has recently raised a similar point with reference to classic somatic indices of opiate withdrawal in animals (weight loss, jumping, etc.), by showing that although opiate withdrawal induces a conditioned aversion to the place in which withdrawal occurs, individual somatic indices of withdrawal do not correlate consistently with the observed place aversion. Since it is presumed that the place aversion reflects an aversive property of opiate withdrawal which facilitates drug self-administration, it is clear that classical somatic indices of withdrawal may also be poor predictors of motivational properties of opiate withdrawal.

The predictive validity of behavioural dependence as a model of dependence potential is at present unknown. It would, for example, be of considerable interest to know if there are *any* drugs, particularly those that do not cause dependence in humans, that fail to induce behavioural dependence in animals. As far as face validity is concerned, behavioural dependence studies have high face validity, by definition. However, a puzzling feature of behavioural dependence studies is the finding that with repeated episodes of withdrawal, dependence diminishes progressively (Slifer *et al.*, 1984; Beardsley *et al.*, 1986). Thus 'tolerance' appears to develop to behavioural dependence. Whether or not such processes occur in dependent humans is a question of theoretical and clinical significance. At present, the absence of reports of such phenomena appears to limit the face validity of behavioural dependence as an animal model. As in tolerance studies, the construct validation of behavioural dependence is dependent largely on the assumption that drug dependence is critically involved in chronic drug abuse. As frequently noted above, this is not unquestioned. In summary, it is clear that further work is required before behavioural dependence phenomena can be considered to be validated models of dependence potential.

17.5 CONCLUSIONS

This concluding section attempts to highlight some theoretical issues which have arisen as a result of work on animal models of abuse and dependence.

One recurring theme throughout this chapter has been that animal models can only provide insights into some of the many processes involved in abuse and dependence. No single model adequately fulfils all the validation criteria described. In part, this is because the constructs of abuse and dependence are themselves not well defined and because reliable, widely-accepted measures of abuse/dependence potential of drugs in humans are not available. Indeed, research with animal models of dependence can result in significant reformulations of such constructs, as shown for example by emphasis on the idea that drugs may have significant abuse potential which does not derive simply from their ability to induce physical dependence. It is apparent that research has resulted in a significant reduction in emphasis on the idea that drug self-administration occurs only as a consequence of dependence. Although there is clearly not a consensus in the literature as to the relative significance of the positive rewarding actions of drugs and their negative reinforcing actions in alleviating withdrawal, this is not in itself a cause for dismay as it should cause basic researchers and clinicians to discard preconceived ideas and address this fundamental issue. Thus, in this area, the study of animal models may serve to influence the thinking of clinical researchers and therapists. Likewise, basic theoretical ideas derived from clinical research can be tested in animal models, as shown by research on the phenomenon of conditioned withdrawal. There should therefore be a two-way flow between research with animal models and related clinical research.

The fact that models of abuse and dependence do not meet all possible validation criteria does not in itself mean that these models are not useful for analysing basic processes in drug abuse/ dependence. For example, the conditioned place preference procedure in no way attempts to model any aspects of the behaviour of drug abusers. Nevertheless, it is clear that work in this area can provide potentially valuable insights into the process of reinforcement which is a central theoretical construct in drug abuse research.

Nevertheless, it must be acknowledged that it is not at all clear how the notion of reinforcement is related to other ideas that are considered fundamental in drug abuse research. The relationship between reinforcement, which is defined operationally in terms of response probability, and euphoria, which is often considered an important determinant of drug abuse, is not clear, as evidenced by studies which demonstrate that drugs can have both positive reinforcing and aversive actions at the same time (see Fig. 17.1). Although drugs such as cocaine undoubtedly have euphoriant actions in humans, it is clear that the actions of drugs on behaviour depend critically on the overall behavioural context in which they are self-administered. Such studies therefore illustrate the considerable power of models to force empirical investigation of behavioural processes which are thought to be well understood. Indeed, it is even possible that research of this type will force a reformulation of basic theoretical ideas such as that of reinforcement. Much of the research into animal models of drug abuse assumes implicitly that drug rewards are similar to other rewards, such as food and water, and that drug rewards work through the same neural and neurochemical systems. Indeed, there is good evidence from self-administration studies that drug rewards do maintain behaviour in much the same way as other stimuli. However, as recently reviewed by Liebman (1989), it is possible that, despite these similarities in effects on behaviour, rewarding stimuli may represent a heterogeneous set of events which do not act via a common neural system. In Liebman's terms, it remains to be determined whether or not reward will ultimately prove to be an efficient 'Natural fracture line' for the analysis of behaviour and of drug abuse. To date this has certainly been the case, as evidenced by much of the literature reviewed in this chapter. However, it remains possible that the notion of a global general reward system may ultimately prove illusory and that research will have to concentrate on more refined notions related more specifically to drug reward. It seems highly likely that, if theoretical developments such as these occur, they will be based in substantial part on the study of animal models of drugs as rewards, which may prove to have important influences on general neurobiological thinking.

I am grateful to my family for support during the writing of this chapter and to Professor M.W. Emmett-Oglesby for helpful, critical comments on it.

REFERENCES

Ahlenius, S. & Engel, J. (1974). Behavioural stimulation induced by ethanol withdrawal. *Pharmacology, Biochemistry and Behavior* **2**, 847–50.

Aigner, T.G. & Balster, R.L. (1978). Choice behavior in rhesus monkeys: Cocaine versus food. *Science* **201**, 534–5.

Alexander, B.K. & Hadaway, P.F. (1982). Opiate addiction: The case for an adaptive orientation. *Psychological Bulletin* **92**, 367–81.

Annear, W.C. & Vogel-Sprott, M. (1985). Mental rehearsal and classical conditioning contribute to ethanol tolerance in humans. *Psychopharmacology* **87**, 90–3.

Appel, J.B. & Cunningham, K.A. (1986). The use of drug discrimination procedures to characterize hallucinogenic drug actions. *Psychological Bulletin* **22**, 959–67.

Baker, T.B. & Tiffany, S.T. (1985). Morphine tolerance as habituation. *Psychological Review* **92**, 78–108.

Balster, R.L. (1985). Behavioral studies of tolerance and dependence. In *Behavioral Pharmacology: The Current Status*, ed. L.S. Seiden & R.L. Balster, pp. 403–18. New York: A.R. Liss Inc.

Barrett, J.E. (1977). Behavioral history as a determinant of the effects of d-amphetamine on punished behaviour. *Science* **198**, 67–9.

Barrett, J.E. & Witkin, J.M. (1986). The role of behavioral and pharmacological history in determining the effects of abused drugs. In *Behavioral Approaches to Drug Dependence*, ed. S.R. Goldberg & I.P. Stolerman, pp. 195–223. New York: Academic Press.

Beardsley, P.M., Balster, R.L. & Harris, L.S. (1986). Dependence on tetrahydrocannabinol in rhesus monkeys. *Journal of Pharmacology and Experimental Therapeutics* **239**, 311–19.

Bennett, D.A. & Lal, H. (1982). Discriminative stimuli produced by clonidine: An investigation of the possible relationship to adrenoceptor stimulation and hypotension. *Journal of Pharmacology and Experimental Therapeutics* **223**, 642–8.

Bignami, G. (1979). Methodological problems in the analysis of behavioural tolerance in toxicology. *Neurobehavioral Toxicology* **1**, Supplement 1, 179–86.

Brady, J.V. & Fischman, M.W. (1985). Assessment of drugs for dependence potential and abuse liability: An overview. In *Behavioral Pharmacology: The Current Status*, ed. L.S. Seiden & R.L. Balster, pp. 361–82. New York: A.R. Liss Inc.

Brady, J.V. & Lukas, S.E. (1984). Testing Drugs for Physical Dependence Potential and Abuse Liability. *National Institute on Drug Abuse Research Monograph Number 52*. Washington: U.S. Government Printing Office.

Branch, M.N. (1983). Behavioral tolerance to stimulating effects of pentobarbital: A within-subject determination. *Pharmacology, Biochemistry and Behavior* **18**, 25–30.

Cappell, H.D. & Le Blanc, A.E. (1979). Tolerance to, and physical dependence on, ethanol: Why do we study them? *Drug and Alcohol Dependence* **4**, 15–31.

Cappell, H., Busto, U., Kay, G., Naranjo, C.A., Sellers, E.M. & Sanchez-Craig, M. (1987). Drug deprivation and reinforcement by diazepam in a dependent population. *Psychopharmacology* **91**, 154–60.

Carr, G.D. & White, N.M. (1986). Anatomical dissociation of amphetamine's rewarding and aversive effects. *Psychopharmacology* **89**, 340–6.

Carr, G.D., Phillips, A.G. & Fibiger, H.C. (1988). Independence of amphetamine reward from locomotor stimulation demonstrated by conditioned place preference. *Psychopharmacology* **94**, 221–6.

Carr, G.D., Fibiger, H.C. & Phillips, A.G. (1989). Conditioned place preference as a measure of drug reward. In *The Neuropharmacological Basis of Reward*, ed. S.J. Cooper & J. Liebman, Monographs in Psychopharmacology, vol. 1, pp. 264–319. Oxford: Oxford University Press.

Chait, L.D., Uhlenhuth, E.H. & Johanson, C.E. (1984). An experimental paradigm for studying the discriminative stimulus properties of drugs in humans. *Psychopharmacology* **82**, 271–4.

Chait, L.D., Uhlenhuth, E.H. & Johanson, C.E. (1986a). The discriminative stimulus effects of d-amphetamine, phenmetrazine and fenfluramine in humans. *Psychopharmacology* **89**, 301–6.

Chait, L.D., Uhlenhuth, E.H. & Johanson, C.E. (1986b). The discriminative stimulus and subjective effects of phenylpropanolamine, mazindol and d-amphetamine in humans. *Pharmacology, Biochemistry and Behavior* **24**, 1665–72.

Chen, C.S. (1968). A study of the alcohol tolerance effect and an introduction of a new behavioural technique. *Psychopharmacologia* **12**, 433–40.

Chen, C.S. (1979). Acquisition of behavioral tolerance to ethanol as a function of reinforced practice in rats. *Psychopharmacologia* **63**, 285–8.

Clarke, P.B.S. & Fibiger, H.C. (1987). Apparent absence of nicotine-induced conditioned place preference in rats. *Psychopharmacology* **92**, 84–8.

Colpaert, F.C. (1986). Drug discrimination: Behavioral, pharmacological and molecular mechanisms of discriminative drug effects. In *Behavioral Analysis of Drug Dependence*, ed. S.R. Goldberg & I.P. Stolerman, pp. 161–93. New York: Academic Press.

Colpaert, F.C. & Slangen, J.L. (eds.) (1986). *Drug Discrimination: Applications in CNS Pharmacology*. Amsterdam: Elsevier Biomedical Press.

Corfield-Sumner, P.K. & Stolerman, I.P. (1978). Behavioural tolerance. In *Contemporary Research in Behavioural Pharmacology*, ed. D.E. Blackman & D.J. Sanger, pp. 391–448. New York: Plenum Press.

Corrigall, W.A., Linesman, M.A., D'Onofrio, R.M. & Lei, H. (1986). An analysis of the paradoxical effect of morphine on runway speed and food consumption. *Psychopharmacology* **89**, 327–33.

Demellweek, C. & Goudie, A.J. (1982). The role of reinforcement loss in the development of amphetamine anorectic tolerance. *I.R.C.S. Medical Science* **10**, 903–4.

Demellweek, C. & Goudie, A.J. (1983a). An analysis of behavioural mechanisms involved in amphetamine anorectic tolerance. *Psychopharmacology* **79**, 58–66.

Demellweek, C. & Goudie, A.J. (1983b). Behavioural tolerance to amphetamine and other psychostimulants. The case for considering behavioural mechanisms. *Psychopharmacology* **80**, 287–307.

Deneau, G., Yanagita, T. & Seevers, M.H. (1969). Self-administration of psychoactive substances by the monkey. *Psychopharmacologia* **16**, 30–48.

de Wit H. & Stewart, J. (1981). Reinstatement of cocaine-reinforced responding in the rat. *Psychopharmacology* **75**, 134–43.

Emmett-Oglesby, M.W., Spencer, D.G., Elmesallamy, F. & Lal, H. (1983). The pentylenetetrazol model of anxiety detects withdrawal from diazepam in rats. *Life Sciences* **33**, 161–8.

Ettenberg, A., Pettit, H.O., Bloom, F.E. & Koob, G.F. (1982). Heroin and cocaine intravenous self-administration in rats: Mediation by separate neural systems. *Psychopharmacology* **78**, 204–9.

Falk, J.L. (1961). Production of polydipsia in normal rats by an intermittent food schedule. *Science* **133**, 195–6.

Falk, J.L. (1966). Schedule-induced polydipsia as a function of fixed interval length. *Journal of the Experimental Analysis of Behavior* **10**, 199–206.

Falk, J.L. (1981). The place of adjunctive behavior in drug abuse research. In *Behavioral Pharmacology of Human Drug Dependence*, NIDA Research Monograph no. 37, ed. T. Thompson & C.E. Johanson. Washington, D.C.: U.S. Government Printing Office.

Falk, J.L. & Tang, M. (1987). Development of physical dependence on midazolam by oral self-administration. *Pharmacology, Biochemistry and Behavior* **26**, 797–800.

Falk, J.L., Dews, P.B. & Schuster, C.R. (1983). Commonalities in the environmental control of behavior. In *Commonalities in Substance Abuse and Habitual Behavior*, ed. P.K. Levison, D.R. Gerstein & D.R. Maloff, pp. 47–110. Lexington, U.S.A.: Lexington Books.

Fernandes, M., Kluwe, S. & Coper, H. (1977). The development of tolerance to morphine in the rat. *Psychopharmacology* **54**, 197–201.

Fernandes, M., Kluwe, S. & Coper, H. (1982). Development and loss of tolerance to morphine in the rat. *Psychopharmacology* **78**, 234–8.

Ferster, C.B. & Skinner, B.F. (1957). *Schedules of Reinforcement*. New York: Appleton-Century-Crofts.

File, S.E. (1986). Aversive and appetitive properties of anxiogenic and anxiolytic agents. *Behavioral Brain Research* **21**, 189–94.

File, S.E., Baldwin, H.A. & Aranko, K. (1987). Anxiogenic effects in benzodiazepine withdrawal are linked to the development of tolerance. *Brain Research Bulletin* **19**, 607–10.

Ford, R.D. & Balster, R.D. (1976). Schedule-controlled behavior in the morphine-dependent rat. *Pharmacology, Biochemistry and Behavior* **4**, 569–73.

Gilbert, R.M. (1978). Schedule-induced self-administration of drugs. In *Contemporary Research in Behavioral Pharmacology*, ed. D.J. Sanger & D.E. Blackman, pp. 289–323. New York: Plenum Press.

Goldberg, S.R. & Gardner, M.L. (1981). Second-order schedules: Extended sequences of behavior controlled by brief environmental stimuli associated with drugs self-administration. In *Behavioral Pharmacology of Human Drug Dependence*, ed. T. Thompson & C.E. Johanson, pp. 241–70, NIDA Research Monograph no 37. Washington, D.C.: U.S. Government Printing Office.

Goldberg, S.R. & Schuster, C.R. (1967). Conditioned suppression by a stimulus associated with nalorphine in morphine-dependent monkeys. *Journal of the Experimental Analysis of Behavior* **10**, 235–42.

Goldberg, S.R. & Schuster, C.R. (1970). Conditioned nalorphine-induced abstinence changes: Persistence in post morphine-dependent monkeys. *Journal of the Experimental Analysis of Behavior* **14**, 33–46.

Goldberg, S.R. & Spealman, R.D. (1982). Maintenance and suppression of behavior by intravenous nicotine injections in squirrel monkeys. *Federation Proceedings* **41**, 216–20.

Goldberg, S.R., Woods, J.H. & Schuster, C.R. (1969). Morphine: conditioned increase in self-administration in rhesus monkeys. *Science* **166**, 1306–7.

Goldberg, S.R., Kelleher, R.T. & Morse, W.H. (1975). Second-order schedules of drug injection. *Federation Proceedings* **34**, 1771–6.

Goudie, A.J. (1989). Behavioural procedures for assessing tolerance and sensitization. In *Neuromethods Series*, vol. 13, *Psychopharmacology*, ed. A.A. Boulton, G.B.Baker & A.J. Greenshaw, pp.565–621. Clifton, New Jersey: Humana Press Inc.

Goudie, A.J. & Demellweek, C. (1986). Conditioning factors in drug tolerance. In *Behavioral Basis of Drug Dependence*, ed. S.R. Goldberg & I.P. Stolerman, pp. 225–85. New York: Academic Press.

Goudie, A.J. & Emmett-Oglesby, M.W. (eds.) (1989). *Psychoactive Drugs: Tolerance and Sensitization*. Clifton, New Jersey: Humana Press Inc.

Goudie, A.J. & Griffiths, J.W. (1986). Behavioural factors in drug tolerance. *Trends in Pharmacological Sciences*, May Issue, pp. 192–6.

Griffith, J.D., Nutt, J.G. & Jasinski, D.R. (1975). A comparison of fenfluramine and amphetamine in man. *Clinical Pharmacology and Therapeutics* **18**, 563–70.

Griffiths, R.R., Brady, J.V. & Snell, J.D. (1978). Progressive-ratio performance maintained by drug infusions: Comparison of cocaine, diethylpropion, chlorphentermine and fenfluramine. *Psychopharmacology* **56**, 5–13.

Griffiths, R.R., Brady, J.V. & Bradford, J.V. (1979). Predicting the abuse liability of drugs with animal drug self-administration procedures: Psychomotor stimulants and hallucinogens. In *Advances in Behavioral Pharmacology*, vol. 2, ed. T. Thompson & P.B. Dews, pp. 163–208. New York: Academic Press.

Griffiths, R.R., Lamb, R.J., Ator, N., Roache, J.D. & Brady, J.V. (1985). Relative abuse potential of triazolam: Experimental assessment in animals and humans. *Neuroscience and Biobehavioral Reviews* **9**, 131–51.

Griffiths, R.R., Bigelow, G.E. & Liebson, I.A. (1986). Human coffee drinking: Reinforcing and physical dependence producing effects of caffeine. *Journal of Pharmacology and Experimental Therapeutics* **239**, 416–25.

Haertzen, C.A., Hill, H.E. & Belleville, R.E. (1963). Development of the Addiction Research Centre Inventory (ARCI): Selection of items that are sensitive to various drugs. *Psychopharmacologia* **4**, 155–66.

Henningfield, J.E., Lukas, S.E. & Bigelow, G. (1986). Human studies of drugs as reinforcers. In

Behavioral Analysis of Drug Dependence, ed. S.R. Goldberg & I.P. Stolerman, pp. 69–121. New York: Academic Press.

Hinson, R.E. & Siegel, S. (1980). The contribution of Pavlovian conditioning to ethanol tolerance and dependence. In *Alcohol Tolerance and Dependence*, ed. H. Rigter & J.C. Crabbe, pp. 181–99. Amsterdam: Elsevier.

Hodos, W. (1961). Progressive ratio as a measure of reward strength. *Science* **134**, 943–4.

Hodos, W. (1965). Motivational properties of long durations of rewarding brain stimulation. *Journal of Comparative and Physiological Psychology* **59**, 219–24.

Hodos, W. & Kalman, J. (1963). Effects of increment size and reinforcer volume on progressive-ratio performance. *Journal of the Experimental Analysis of Behavior* **6**, 387–92.

Irwin, S. & Seevers, M.H. (1956). Altered responses to drugs in the post-addict. (*Macaca mulatta*). *Journal of Pharmacology & Experimental Therapeutics* **116**, 31–2.

Jarbe, T.U.C. (1987). Drug discrimination learning. In *Experimental Psychopharmacology*, ed. A.J. Greenshaw & C.T. Dourish, pp. 433–81. New Jersey: Humana Press Inc.

Jarbe, T.U.C. & Kroon-Jarbe, E.R. (1983). Discriminability and procedure: Effects of cocaine and amphetamine. *Psychological Reports* **52**, 611–16.

Jasinski, D.R., Johnson, R.E. & Henningfield, J.E. (1984). Abuse lability assessment in human subjects. *Trends in Pharmacological Sciences* **5**, 196–200.

Jefferys, D., Oei, T.P.S. & Singer, G. (1979). A reconsideration of the concept of drug dependence. *Neuroscience & Biobehavioral Reviews* **3**, 149–53.

Johanson, C.E. (1978). Drugs as reinforcers. In *Contemporary Research in Behavioral Pharmacology*, ed. D.E. Blackman & D.J. Sanger, pp. 325–90. New York: Plenum Press.

Johanson, C.E. & Schuster, C.R. (1975). A choice procedure for drug reinforcers: cocaine and methylphenidate in the rhesus monkey. *Journal of Pharmacology and Experimental Therapeutics* **193**, 676–88.

Johanson, C.E. & Schuster, C.R. (1977). Procedures for the preclinical assessment of abuse potential of psychotropic drugs in animals. In: *Predicting Dependence Liability of Stimulant and Depressant Drugs*, ed. T. Thompson & K. Unna, pp. 203–30. Baltimore: University Park Press.

Johanson, C.E., Balster, R.L. & Bonese, K. (1976). Self-administration of psychomotor stimulant drugs: The effects of unlimited access. *Pharmacology, Biochemistry and Behavior* **4**, 45–51.

Johanson, C.E., Woolverton, W.L. & Schuster, C.R. (1987). Evaluating laboratory models of drug dependence. In *Psychopharmacology: The Third Generation of Progress*, ed. H.Y. Meltzer, pp. 1617–25. Raven Press: New York.

Kalant, H. (1985). Tolerance, learning and neurochemical adaptation. *Canadian Journal of Physiology and Pharmacology* **63**, 1485–94.

Kalant, H., Le Blanc, A.E. & Gibbins, R.J. (1971). Tolerance to, and dependence on, some nonopiate psychotropic drugs. *Pharmacological Reviews* **23**, 135–91.

Katz, J.L. & Valentino, R.J. (1986). Pharmacological and behavioral factors in opioiod dependence in animals. In *Behavioral Approaches to Drug Dependence*, ed. S.R. Goldberg & I.P. Stolerman, pp. 287–327. Orlando: Academic Press.

Kelleher, R.T. & Goldberg, S.R. (1976). Control of drug taking by schedules of reinforcement. *Pharmacological Reviews* **27**, 291–9.

Kumar, R. & Stolerman, I.P. (1977). Experimental and clinical aspects of drug dependence. In *Handbook of Psychopharmacology*, vol. 7, ed. S.D. Iversen, L.L. Iversen & S.H. Snyder, pp. 321–67. New York: Plenum Press.

Lader, M. & File, S.E. (1988). The biological basis of benzodiazepine dependence. *Psychological Medicine* **17**, in press.

Lal, H. & Emmett-Oglesby, M.W. (1983). Behavioral analogues of anxiety. *Neuropharmacology* **22**, 1423–41.

Lal, H. & Yaden, S. (1985). Discriminative stimuli produced by clonidine in spontaneously hypertensive rats: Generalization to antihypertensive drugs with different mechanisms of action. *Journal of Pharmacology and Experimental Therapeutics* **232**, 33–9.

Lal, H., Miksic, S. Drawbaugh, R., Numan, R. & Smith, N. (1976). Alleviation of narcotic withdrawal syndrome by conditional stimuli. *Pavlovian Journal of Biological Science* **11**, 215–62.

Lang, W.J., Latiff, A.A., McQueen, A. & Singer, G. (1977). Self-administration of nicotine with and without a food delivery schedule. *Pharmacology, Biochemistry and Behavior* **7**, 65–70.

Le Blanc, A.E., Gibbins, R.J. & Kalant, H. (1973). Behavioral augmentation of tolerance to ethanol in the rat. *Psychopharmacologia* **30**, 117–22.

Levine, D.G. (1974). Needle freaks: Compulsive self injections by drug users. *American Journal of Psychiatry* **131**, 297–300.

Liebman, J.M. (1989). Introduction. In *The Neuropharmacological Basis of Reward*, pp. 1–13 of *Monographs in Psychopharmacology*, ed. S.J. Cooper & J.M. Liebman. Oxford: Oxford University Press.

Martin-Iverson, M.T., Ortmann, R. & Fibiger, H.C.

(1985). Place preference conditioning with methylphenidate and nomifensine. *Brain Research* **332**, 59–67.

McAuliffe, W.E. (1982). A test of Wikler's theory of relapse: The frequency of relapse due to conditioned withdrawal sickness. *International Journal of the Addictions* **17**, 19–33.

McKinney, W.T. & Bunney, W.E. (1969). Animal models of depression: review of evidence and implications for research. *Archives of General Psychiatry* **21**, 240–8.

Meyer, R.E. & Mirin, S.M. (1979). *The Heroin Stimulus: Implications for a Theory of Addiction.* New York: Plenum Press.

Mithani, S., Martin-Iverson, M.T., Phillips, A.G. & Fibiger, H.C. (1986). The effects of haloperidol on amphetamine and methylphenidate induced conditioned place preference and locomotor activity. *Psychopharmacology* **90**, 247–52.

Morency, M.A. & Beninger, R.J. (1986). Dopaminergic substrates of cocaine-induced place conditioning. *Brain Research* **399**, 33–41.

Morse, W.H. & Kelleher, R.T. (1977). Determinants of reinforcement and punishment. In *Handbook of Operant Conditioning*, ed. W.K. Honig & J.E.R. Staddon, pp. 174–200. Englewood Cliffs, New Jersey: Prentice Hall.

Mucha, R.F. (1987). Is the motivational effect of opiate withdrawal reflected by common somatic indices of precipitated withdrawal? A place conditioning study in the rat. *Brain Research* **418**, 214–20.

Mucha, R.F. & Herz, A. (1986). Preference conditioning produced by opioid active and inactive isomers of levorphanol and morphine in rat. *Life Sciences* **38**, 241–9.

Mucha, R.F. & Iversen, S.D. (1984). Reinforcing properties of morphine and naloxone revealed by conditioned place preferences: A procedural analysis. *Psychopharmacology* **82**, 241–7.

O'Brien, C.P. (1975). Experimental analysis of conditioning factors in human narcotic addiction. *Pharmacological Reviews* **27**, 533–43.

O'Brien, C.P., Testa, T., O'Brien, T.J., Brady, J.P. & Wells, B. (1977). Conditioned narcotic withdrawal in humans. *Science* **195**, 1000–2.

O'Brien, C.P., Nance, E., Mintz, J., Meyers, A. & Ream, N. (1980). Follow up of Vietnam Veterans. Part 1: Relapse to drug use after Vietnam service. *Drug and Alcohol Dependence* **5**, 333–40.

O'Brien, C.P., Ehrman, R.N. & Ternes, J.W. (1986). Classical conditioning in human opioid dependence. In *Behavioral Approaches to Drug Dependence*, ed. S.R. Goldberg & I.P. Stolerman, pp. 329–56. New York: Academic Press.

Overton, D.A. & Batta, S.K. (1977). Relationship between abuse liability of drugs and their degree of discriminability in the rat. In *Predicting*

Dependence Liability of Stimulant and Depressant Drugs, ed. T. Thompson & K.R. Unna, pp. 125–36. Baltimore: University Park Press.

Pickens, R. & Thompson, T. (1968). Cocaine-reinforced behavior in rats: Effects of reinforcement magnitude and fixed ratio size. *Journal of Pharmacology and Experimental Therapeutics* **161**, 122–9.

Porsolt, R.D., Pawelec, C. & Jalfre, M. (1982). Use of the drug discrimination procedure to detect amphetamine-like effects of antidepressants. In *Drug Discrimination: Applications in CNS Pharmacology*, ed. F.C. Colpaert & J.L. Slangen, pp. 193–202. Amsterdam: Elsevier Biomedical Press.

Risner, M.E. & Silcox, D.L. (1981). Psychostimulant self-administration by beagle dogs in a progressive-ratio paradigm. *Psychopharmacology* **75**, 25–30.

Robins, L., Davis, D.H. & Goodwin, D.W. (1974). Drug use by U.S. army enlisted men in Vietnam: A follow-up on their return home. *American Journal of Epidemiology* **99**, 235–49.

Roper, T.J. (1981). What is meant by the term 'schedule-induced,' and how general is schedule induction? *Animal Learning and Behavior* **9**, 433–40.

Roper, T.J. & Crossland, G. (1982). Schedule-induced wood-chewing in rats and its dependence on body weight. *Animal Learning and Behavior* **10**, 65–71.

Sanger, D.J. (1977). d-Amphetamine and adjunctive drinking in rats. *Psychopharmacology* **54**, 273–6.

Sanger, D.J. (1986). Drug taking as adjunctive behavior. In *Behavioral Approaches to Drug Dependence*, ed. S.R. Goldberg & I.P. Stolerman, pp. 123–60. New York: Academic Press.

Sannerud, C.A. & Young, A. (1986). Modification of morphine tolerance by behavioral variables. *Journal of Pharmacology and Experimental Therapeutics* **237**, 75–81.

Schulz, R. & Herz, A. (1976). Aspects of opiate dependence in the myenteric plexus of the guinea pig. *Life Sciences* **19**, 1117–28.

Schuster, C.R. (1968). Variables affecting the self-administration of drugs by rhesus monkeys. In *Use of Non-human Primates in Drug Evaluation*, ed. H. Vagtborg, pp. 293–9. Austin, Texas: University of Texas Press.

Schuster, C.R. (1978). Theoretical basis of behavioral tolerance: Implications of the phenomenon for problems of drug abuse. In *Behavioral Tolerance: Research and Treatment Implications*, ed. N. Krasnegor, pp. 4–17. N.I.D.A. Research Monograph No. 18. Washington D.C.: U.S. Government Printing Office.

Schuster, C.R. (1986). Implications of laboratory research for the treatment of drug dependence. In

Behavioral Approaches to Drug Dependence, ed. S.R. Goldberg & I.P. Stolerman, pp. 357–86. New York: Academic Press.

Schuster, C.R. & Johanson, C.E. (1981). An analysis of drug-seeking behavior in animals. *Neuroscience & Biobehavioral Reviews* **5**, 315–23.

Schuster, C.R. & Johanson, C.E. (1985). Efficacy, dependence potential and neurotoxicity of anorectic drugs. In *Behavioral Pharmacology: The Current Status*, ed. L. Seiden & R.L. Balster, pp. 263–79. New York: A.R. Liss Inc.

Schuster, C.R., Dockens, W. & Woods, J.H. (1966). Behavioral variables affecting the development of amphetamine tolerance. *Psychopharmacologia* **9**, 170–82.

Siegel, S. (1989). Classical conditioning and habituation processes in drug tolerance and sensitization. In *Psychoactive Drugs: Tolerance and Sensitization*, ed. A.J. Goudie & M.W. Oglesby, Clifton, New Jersey: Humana Press Inc. (in press).

Siegel, S. & McRae, J. (1984). Environmental specificity of tolerance. *Trends in Neuroscience* **7**, 140–2.

Singer, G., Oei, T.P. & Wallace, M. (1982). Schedule-induced self injection of drugs. *Neuroscience & Biobehavioral Reviews* **6**, 77–83.

Skinner, B.F. (1953). *Science and Human Behaviour*. New York: Free Press.

Slifer, B.L., Balster, R.L. & Woolverton, W.L. (1984). Behavioral dependence produced by continuous phencyclidine infusion in rhesus monkeys. *Journal of Pharmacology & Experimental Therapeutics* **230**, 399–406.

Smith, J.E., Co, C., Freeman, M.E., Sands, M.P. & Lane, J.D. (1980). Neurotransmitter turnover in rat striatum is correlated with morphine self-administration. *Nature* **287**, 152–4.

Solomon, R.L. (1977). An opponent-process theory of acquired motivation: The affective dynamics of addiction. In *Psychopathology: Experimental Models*, ed. J.D. Maser & M.E.P. Seligman, pp. 66–73. San Francisco: Freeman.

Spealman, R.D. (1979). Behavior maintained by termination of a schedule of self-administered cocaine. *Science* **204**, 1231–3.

Spealman, R.D. (1985). Environmental factors controlling the control of behavior by drugs. In *Behavioral Pharmacology: The Current Status*, ed. L. Seiden & R.L. Balster, pp. 23–38. New York: A.R. Liss Inc.

Spealman, R.D. & Goldberg, S.R. (1978). Drug self-administration by laboratory animals: Control by schedules of reinforcement. *Annual Review of Pharmacology and Toxicology* **18**, 313–39.

Spryaki, C. (1988). Drug reward studied by the use of place conditioning in rats. In *The*

Psychopharmacology of Addiction, ed. M. Lader. New York: Academic Press (in press).

Spyraki, C. & Fibiger, H. (1981). intravenous self-administration of nomifensine in rats: Implications for abuse potential in humans. *Science* **212**, 1167–8.

Spyraki, C., Fibiger, H.C. & Phillips, A.G. (1982*a*). Attenuation by haloperidol of place preference conditioning using food reinforcement. *Psychopharmacology* **77**, 379–82.

Spyraki, C., Fibiger, H.C. & Phillips, A.G. (1982*b*). Dopaminergic substrates of amphetamine-induced place preference conditioning. *Brain Research* **253**, 185–93.

Spyraki, C., Fibiger, H.C. & Phillips, A.G. (1983). Attenuation of heroin reward by disruption of the mesolimbic dopamine system. *Psychopharmacology* **79**, 278–83.

Spyraki, C., Kazandjian, A. & Varonos, D. (1985). Diazepam-induced place preference conditioning: appetitive and antiaversive properties. *Psychopharmacology* **87**, 225–32.

Spyraki, C., Nomikos, G.G. & Varonos, D. (1987). Intravenous cocaine-induced place preference: attenuation by haloperidol. *Behavioral Brain Research* **26**, 57–62.

Staddon, J.E.R. (1977). Schedule-induced behavior. In *Handbook of Operant Behavior*, ed. W.L. Honig & J.E.R. Staddon, pp. 125–52. Englewood Cliffs, New Jersey: Prentice Hall.

Steinfels, G.F. & Young, G.A. (1981). Effects of narcotic abstinence on schedule-controlled behavior in dependent rats. *Pharmacology, Biochemistry and Behavior* **14**, 393–5.

Stewart, J. (1984). Reinstatement of heroin and cocaine self-administration in the rat by intracerebral application of morphine in the ventral tegmental area. *Pharmacology, Biochemistry and Behavior* **20**, 917–23.

Stewart, R.B. & Grupp, R.B. (1986). Conditioned place aversion mediated by orally self-administered ethanol in the rat. *Pharmacology, Biochemistry and Behavior* **24**, 1369–75.

Stewart, J., de Witt, H. & Eikelboom, R. (1984). Role of unconditioned and conditioned drug efects in the self-administration of opiates and stimulants. *Psychological Review* **91**, 251–68.

Swerdlow, N.R. & Koob, G.F. (1984). Restrained rats learn amphetamine-conditioned locomotion, but not place preference. *Psychopharmacology* **84**, 163–6.

Swerdlow, N.R., Gilbert, D.B. & Koob, G.F. (1988). Conditioned drug effects on spatial preference: Critical evaluation. In *Neuromethods: Psychopharmacology*, ed. A.A. Boulton, G.B. Baker & A.J. Greenshaw. New Jersey: Humana Press Inc. (in press).

Takahashi, R.N., Singer, G. & Oei, T.P.S. (1978). Schedule-induced self-inection of d-amphetamine by naive rats. *Pharmacology, Biochemistry and Behavior* 9, 857–62.

Tye, N. & Iversen, S.D. (1975). Some behavioural signs of morphine withdrawal blocked by conditional stimuli. *Nature* 255, 415–18.

Vezina, P. & Stewart, J. (1987). Morphine conditioned place preference and locomotion: The effect of confinement during training. *Psychopharmacology* 93, 257–60.

Vogel-Sprott, M., Rawana, E. & Webster, R. (1984). Mental rehearsal of a task under ethanol facilitates tolerance. *Pharmacology, Biochemistry and Behavior* 21, 329–31.

Wallace, M. & Singer, G. (1976). Schedule induced behavior: A review of its generality, determinants and pharmacological data. *Pharmacology, Biochemistry and Behavior* 5, 483–90.

Weeks, J.R. (1962). Experimental morphine addiction: Method for automatic intravenous injections in unrestrained rats. *Science* 138, 143–4.

Weissman, A. (1976). The discriminability of aspirin in arthritic and non-arthritic rats. *Pharmacology, Biochemistry and Behavior* 5, 583–6.

White, N.M. & Carr, G.D. (1985). The conditioned place preference is affected by two independent reinforcement processes. *Pharmacology, Biochemistry and Behavior* 23, 37–42.

White, N., Sklar, L. & Amit, Z. (1977). The reinforcing action of morphine and its paradoxical side effect. *Psychopharmacology* 52, 63–6.

Wikler, A. (1948). Recent progress in research on the neurophysiological basis of morphine addiction. *American Journal of Psychiatry* 105, 329–38.

Wikler, A. & Pescor, F.T. (1967). Classical conditioning of a morphine abstinence phenomenon, reinforcement of opioid-drinking behavior and 'relapse' in morphine-addicted rats. *Psychopharmacologia* 10, 255–84.

Willner, P. (1986). Validation criteria for animal models of human mental disorders: Learned helplessness as a paradigm case. *Progress in Neuro-Psychopharmacology & Biological Psychiatry* 10, 1–14.

Winger, G.D. & Woods, J.H. (1973). The reinforcing property of ethanol in the rhesus monkey: I. Initiation, maintenance and termination of intravenous ethanol-reinforced responding. *Annals of the New York Academy of Sciences* 215, 162–75.

Wise, R.A. (1982). Neuroleptics and operant behavior: The anhedonia hypothesis. *The Behavioral and Brain Sciences* 5, 39–87.

Wolgin, D.L. (1988). Instrumental conditioning factors in drug tolerance. In *Psychoactive Drugs: Tolerance and Sensitization*, ed. A.J. Goudie & M.W. Emmett-Oglesby, pp. 17–114. Clifton, New Jersey: Humana Press Inc.

Wolgin, D.L., Thompson, G.B. & Oslan, I.A. (1987). Tolerance to amphetamine: Contingent suppression of stereotypy mediates recovery of feeding. *Behavioral Neuroscience* 101, 264–71.

Woods, J.H. (1983). Some thoughts on the relations between animal and human drug-taking. *Progress in Neuro-Psychopharmacology & Biological Psychiatry* 7, 577–84.

Woods, J.H. & Carney, J. (1978). Narcotic tolerance and operant behaviour. In *Behavioral Tolerance: Research and Treatment Implications*, ed. N. Krasnegor, N.I.D.A. Research Monograph No. 18. Washington D.C.: U.S. Government Printing Office.

Woolverton, W.L. & Balster, R.L. (1982). Behavioral pharmacology of local anaesthetics: Reinforcing and discriminative stimulus effects. *Pharmacology, Biochemistry and Behavior* 16, 491–500.

Woolverton, W.L. & Kleven, M.S. (1988). Evidence for cocaine dependence in monkeys following a prolonged period of exposure. *Psychopharmacology* 94, 288–91.

Yanagita, T. (1980). Self-administration studies on psychological dependence. *Trends in Pharmacological Sciences*, February Issue, pp. 161–4.

Young, A. & Herling, S. (1986). Drugs as reinforcers: Studies in laboratory animals. In: *Behavioral Analysis of Drug Dependence*, ed. S.R. Goldberg & I.P. Stolerman, pp. 9–67. New York: Academic Press.

18

Screening for abuse and dependence liabilities

DAVID J. SANGER

18.1 INTRODUCTION

The non-medical use, misuse and abuse of psycho-tropic and other drugs is an all too prevalent aspect of modern life. Most of us need to be concerned with the problems presented by drug abuse, either as scientists, as drug users, as parents or simply as responsible citizens. However, the concern of scientists and other workers in the pharmaceutical industry is rather special. Like any industry, pharmaceutical companies seek to discover, develop and market successful products, in their case medicines. These new medicines should be effective in treating the disorders at which they are aimed, and hopefully more effective than available medicines, but should also, as far as possible, be devoid of side effects. As the medical, psychological and social problems produced by drug abuse can legitimately be considered as side effects when clinically useful drugs are abused, techniques are necessary which will effectively predict the abuse liability of new drugs and thus aid the development of novel drugs with low abuse potential.

Of course, drugs are abused in many ways and for many reasons, some having little relationship to the drugs' psychopharmacological effects. Furthermore, the definition of what is acceptable medical use and what constitutes harmless recreational use or unacceptable abuse may differ greatly in different societies, for cultural and political reasons. It is most unlikely, therefore, that any combination of preclinical and clinical tests will provide a perfect prediction of whether any particular drug will be abused. Nevertheless, as Goudie (Chapter 17, this volume) demonstrates clearly, there are now available a variety of laboratory procedures with which to investigate pharmacological and behavioural effects of drugs relevant to their abuse liability.

The purpose of the present chapter is to consider the extent to which results obtained in such tests correlate with actual patterns of abuse and thus provide data which can predict whether novel drugs are likely to be subject to abuse. Previous reviews of animal research have often pointed to what the reviewers have, justifiably, described as a close concordance between laboratory results and patterns of abuse (Kumar & Stolerman, 1977; Griffiths et al., 1980). However, there are also some anomalous findings which will be emphasised in the present review because of their importance in establishing the limits to the validity and reliability of preclinical tests as predictors of abuse liability.

18.2 EVALUATION OF ABUSE AND DEPENDENCE LIABILITIES – GENERAL CONSIDERATIONS

The decision to proceed with the development of any novel drug is always complex and difficult. No new drug will be a panacea and neither will it be completely devoid of undesirable effects. Therefore, a positive decision, with all its consequences for the investment of very costly resources, will be taken only if it is considered that the potential benefits, both medical and commercial, outweigh the potential hazards. It is the function of preclinical and clinical pharmacology to evaluate the potential medical benefits of a new drug, and the role of toxicology to investigate the potential risks.

Table 18.1. *General considerations for assessing the abuse and dependence liabilities of a novel compound*

(1) Chemical structure and structural relationship to known drugs of abuse
(2) Physical characteristics of the potential drug preparation
(3) Pharmacokinetic characteristics
(4) Proposed clinical use
(5) Withdrawal syndrome
(6) Self-administration and stimulus properties

Drug dependence and drug abuse are clearly undesirable aspects of any drug's pharmacological profile and thus might, strictly speaking, be considered to fall within the domain of toxicology, and, more specifically, that of the emerging discipline of behavioural toxicology. However, the assessment of abuse and dependence liability requires a considerable amount of specialised knowledge in psychopharmacology and is thus probably best dealt with by preclinical and clinical scientists trained in this subject.

Table 18.1 presents a list of considerations which will need to be taken into account, at different stages of a drug's development, in order to assess effectively the drug's potential for abuse and dependence. Some of these factors involving physical dependence, drug self-administration and stimulus properties will be dealt with in greater detail in later sections. It seems appropriate however to begin the present discussion with some general considerations.

It seems obvious that if a novel compound has a chemical structure closely related to known drugs of abuse, it may itself have potential for abuse. The assumption behind this statement, of course, is that chemical structure is an indication of pharmacological activity and indeed there are many examples of structure–activity relationships in pharmacology. However, there are also many instances where small alterations in chemical structure have given rise to significant and unpredictable changes in pharmacological and toxicological properties. Therefore, although it is known that certain groups of molecules, such as those with opiate-like or amphetamine-like structures, have

been associated with considerable abuse, chemical structure is only a very limited predictor of abuse.

The abuse liability of a potentially abusable drug can in some circumstances be minimised by the thoughtful design of the physical and pharmacokinetic characteristics of the final, clinically-used preparation. Thus, for example, the most serious problems of opioid drug abuse occur when drugs such as heroin are self-administered by the intravenous route. Drugs which have very low solubility in water are only taken in this way with great difficulty, however, so their abuse potential may be relatively low. Nevertheless, it would be dangerous to assume that such drugs have no abuse potential as there have been many examples of the ingenuity and sophistication of drug abusers in preparing drugs suitable for self-administration.

A particular example of how the abuse potential of drug preparations can be lowered is provided by attempts to combine opioid analgesics with the opioid antagonist naloxone. Preparations of opioids prepared for oral use as analgesics have often been injected by the intravenous route by drug abusers. To attempt to stop this practice several drug companies have carried out research with preparations of analgesics combined with naloxone. The ingenious idea behind such preparations relies on the fact that naloxone has low oral bioavailability due to first-pass metabolism. Therefore, a preparation of naloxone with an opioid agonist, taken orally, would give rise to the desirable pharmacological properties of the agonist, such as analgesia. However, if injected intravenously, naloxone should antagonise the pharmacological activity of the agonist including the euphoria and other subjective changes for which the drug was presumably being abused.

In order for such combinations to be effective considerable amounts of preclinical and clinical research are necessary. Legros and colleagues (1984) described a series of experiments in which the analgesic properties of combinations of naloxone with pentazocine were investigated in rats. They found that when given by intravenous or subcutaneous injection, a dose of naloxone one hundredth of the dose of pentazocine was necessary to block pentazocine's analgesic activity. However, when both drugs were given orally, a dose of naloxone one fifth that of the agonist was needed. These researchers concluded, therefore, that a combination of pentazocine with naloxone in a dose ratio of 100:1 might be clinically valuable.

Hoffmeister (1986) also carried out studies of combinations of naloxone with opioid agonists, in this case codeine, pentazocine, buprenorphine and tilidine, but using an intravenous self-administration procedure with monkeys. It had previously been shown that opioid antagonists have aversive properties in animals physically dependent on agonists and also in naive animals (Hoffmeister & Wuttke, 1973a; Downs & Woods, 1975). The purpose of Hoffmeister's experiments, therefore, was to investigate whether animals would respond to avoid infusions of combinations of the agonists with naloxone. He found that they would do so indicating that such a combination should be completely devoid of intravenous abuse potential. Of course, such findings need to be confirmed by careful clinical research (see Vanacker *et al.*, 1986; Preston *et al.*, 1988).

Another consideration concerns the proposed clinical use of the new drug. If a drug is aimed at an indication which is very common and will involve widespread self-medication, such as the treatment of cold symptoms or mild pain, it is clear that the product must be completely devoid of abuse potential. However, if the drug is to be used under much more limited or restricted conditions, such as in surgical anaesthesia, a certain level of abuse liability is clearly acceptable as the availability of the drug should never be such as to give rise to abuse. This is an important point as it illustrates the fact that abuse and dependence liabilities do not necessarily preclude the development of a novel drug. It may be considered that the therapeutic advantages of a new drug are great and that its clinical use can be controlled even though evidence suggests that it could give rise to abuse under certain circumstances. In such a case, it is likely that industrial decision-makers and clinicians would consider that the drug should be developed and marketed and this would be allowed by regulatory authorities. Of course, the conditions under which such a drug was manufactured, marketed, distributed and used would be subject to very strict controls.

Table 18.2. *Correlation between analgesic activity of opiates and their addiction liability in humans*

Drug	Rank order	
	Analgesia (tail flick)	Addiction liability
Ketobemidone	1	2
Methadone	2	1
Isomethadone	3	4
Morphine	4	3
Meperidine	5	5
D-propoxyphene	6	7
Codeine	7	6

Modified from Harris (1986).

18.3 PHARMACOLOGICAL PROFILES OF ABUSED DRUGS

Drugs which give rise to problems of abuse and dependence come from a variety of pharmacological classes. They include centrally-acting analgesics such as morphine and heroin, psychomotor stimulants such as cocaine and amphetamine, depressants and sedatives such as the barbiturates and ethanol, dissociative anaesthetics such as phencyclidine, and drugs which induce distortions of perception such as mescaline and LSD. Clearly the pharmacological profiles presented by these diverse substances in the animal laboratory vary widely. However, within each pharmacological category there are certain actions which may indicate that particular drugs have abuse liability.

It has been known for many years that while opioid drugs have great therapeutic value as analgesics they also give rise to physical dependence and are abused. Medicinal chemists have synthesised many thousands of opiate-related molecules, in attempts to produce drugs which were effective analgesics but which had little abuse or dependence liability. However, until recently, analgesic potency seemed to be closely associated with abuse liability. Table 18.2 is taken from a paper by Harris (1986) and shows older data illustrating the strong correlation between the abuse liability of different opioid drugs and their potency as analgesics in the animal laboratory. Although certain of these compounds, such as codeine, seem to have low abuse potential, such drugs also have low potency as analgesics. To a certain extent the position changed when it was discovered that some compounds, particularly nalorphine, which were opiate antagonists in the laboratory, had analgesic properties in the clinic (Lasagna & Beecher, 1954). This led to the discovery and development of several drugs, including pentazocine, butorphanol, propiram and buprenorphine, in which abuse liability has been greatly reduced though not completely eliminated.

The pharmacological activities of such compounds can now be understood in terms of partial agonism at a single opioid receptor or mixed agonist/antagonist properties at different receptor subtypes. Indeed, the discovery of opioid receptor subtypes (Martin et al., 1976) has again raised the possibility of a complete dissociation between analgesic properties and abuse liability. Although a full discussion of this point is beyond the scope of the present chapter, it has been shown that compounds such as ketazocine and U-50,488H, which are kappa receptor agonists, have analgesic properties without having psychopharmacological properties indicating abuse liability in animals and people (Kumor et al., 1986a). However, it is not yet clear whether such findings will lead to the development of better strong analgesics as it has been reported that kappa agonists also produce dysphoric psychomimetic subjective changes in people which might preclude their clinical use (Kumor et al., 1986b; Pfeiffer et al., 1986).

Analgesic activity in animals, therefore, only provides a strong indication of abuse liability within the limited pharmacological category of compounds acting at the mu opiate receptor (i.e. morphine-like opioids). Similarly, many other pharmacological activities are seen in some, but not all, drugs of abuse and, therefore, do not provide predictors of a compound's abuse potential. However, there is one behavioural activity which is produced by many abused drugs from several categories and this is the ability to increase levels of locomotor activity in rodents. Thus, increased locomotion in rats or mice has been reported with barbiturates and benzodiazepines (Marriott & Spencer, 1965), ethanol (Read et al., 1960), mu but not kappa opiate agonists (Iwamoto, 1981; Sanger, 1983), phencyclidine

(Chen et al., 1959), hallucinogens such as mescaline (Lush, 1974) and LSD (Kabes & Fink, 1972), and cannabis (Miller & Drew, 1974) as well as with psychomotor stimulants such as amphetamines and cocaine (Simon et al., 1972; Lyon & Robbins, 1975). Of course, drugs from some of these classes also reduce locomotion at higher doses, through sedation or muscle relaxation, and the mechanisms, both behavioural and neurochemical, through which locomotor activity can be increased by drugs are very varied. Thus barbiturates and benzodiazepines are believed to be more likely to increase exploratory activity in a novel environment because of a decrease in fear or anxiety associated with novelty (Marriott & Smith, 1972), whereas the opposite may be the case with amphetamine (e.g. Kumar, 1969). No single test procedure, therefore, is likely to pick out all potential drugs of abuse without picking out drugs which would not be subject to abuse. Certainly, it could not be suggested that locomotor stimulant activity is directly related to abuse liability particularly as there are some drugs which can produce very clear amphetamine-like increases in locomotion in rodents but which have not been associated with abuse; two examples are nomifensine (Braestrup & Scheel-Kruger, 1976) and bupropion (Cooper et al., 1980). Nevertheless, if a drug does increase locomotion in the animal laboratory, such a compound is usually treated with some caution.

18.4 PHYSICAL DEPENDENCE

As noted in the preceding chapter, a fundamental distinction is usually drawn between psychological (or psychic) and physical (or physiological) dependence although this dichotomy is not without its problems (Brady & Fischman, 1985). Psychological dependence is usually defined in terms of drug craving or repeated drug self-administration and thus is clearly associated with what is normally referred to as drug abuse. Physical dependence is generally defined as the state of an organism when a drug-withdrawal syndrome occurs. This is a very precise, operational definition which thus makes physical dependence relatively easy to investigate in the laboratory (at least in theory). However, as Katz & Valentino (1986) have pointed out, it can lead to circular arguments. Thus if physical dependence is defined

Table 18.3. *Reasons why physical dependence and abuse are not necessarily associated*

(1) Avoidance of a withdrawal syndrome cannot explain the initiation of drug taking
(2) Some abused drugs do not produce physical dependence
(3) Some drugs give rise to a withdrawal syndrome but are not abused
(4) Some abusers of dependence-producing drugs are not physically dependent

as the occurrence of withdrawal signs and symptoms it is not logical to explain withdrawal as being due to physical dependence.

The usual stereotype of a drug abuser involves both compulsive drug taking and physical dependence and it is often assumed that drug taking behaviour is maintained by the motivation to escape or avoid the withdrawal syndrome. Research both with animals (e.g. Goldberg *et al.*, 1971) and humans (e.g. Cappell *et al.*, 1987) has shown that drug taking can be maintained by such conditions. However, it is now clear that physical dependence is far from being a cause or even a necessary correlate of drug abuse (see also Goudie, this vol., Ch. 17).

Table 18.3 presents several arguments for the maintenance of a clear distinction between psychological and physical dependence. It is clear that although avoidance of an aversive withdrawal syndrome may contribute towards the maintenance of drug taking in individuals who have been using the drug long enough for a state of dependence to have developed, it cannot be involved in the initiation of drug self-administration. Furthermore, a number of drugs which are widely self-administered and cause problems of abuse, such as cocaine, do not generally give rise to a withdrawal syndrome. It has also been established that many users of heroin are not, in fact, physically dependent on the drug because of their patterns of use which may involve low doses and/or infrequent self-administration (Harding & Zinberg, 1983). Finally, as pointed out by Haefely (1986), many drugs produce marked changes in biological mechanisms which might be expected to give rise to processes of adaptation. Withdrawal of such a drug would then be expected to unmask a syndrome produced by rebound

mechanisms. Such phenomena have been observed with many drugs which are not subject to abuse. For example, clonidine, a widely used cardiovascular drug, can give rise to a withdrawal syndrome consisting of both cardiovascular and behavioural symptoms (Jarrott & Lewis, 1987). Although clonidine does have behavioural activity, principally sedation, and has been used successfully in the treatment of withdrawal symptoms in opiate abusers (Gold *et al.*, 1980), it is not, itself, abused.

It is not meant to imply by these considerations that physical dependence is a phenomenon with which the industrial, preclinical pharmacologist need not be concerned. Clearly, the development of severe withdrawal syndromes, as occurs with opioids, barbiturates and ethanol, can represent a significant side effect which may preclude the development of a novel drug. Therefore, laboratory studies are necessary in order to assess the physical dependence potential of novel compounds.

The development of physical dependence can be, and often is, investigated in rodents (Cheney & Goldstein, 1971; Essig, 1966) but the most detailed information on physical dependence with opioid-like and barbiturate-like drugs has been gathered in non-human primates. In several laboratories, colonies of physically-dependent monkeys are maintained and withdrawal syndromes have been studied in great detail (Seevers, 1936; Katz & Valentino, 1986). In order to assess whether a new compound is likely to produce physical dependence of either the opioid or barbiturate types the first procedure which is often carried out involves substituting the new compound for the maintenance drug in dependent animals. This procedure is usually referred to as the single dose suppression test as its purpose is to investigate whether a novel compound will suppress the withdrawal signs seen after chronic administration of another agent. Thus, if a compound suppresses withdrawal signs in morphine-dependent monkeys it may indicate that this compound might have opioid-like properties itself, including the production of physical dependence (Woolverton & Schuster, 1983a).

Although the single dose suppression test is rapid and can give an early indication of a compound's potential to produce physical dependence, it is not definitive. Certain compounds may suppress withdrawal signs without themselves producing physical dependence. Clonidine, for example, suppresses some of the symptoms produced by withdrawal from opioids but is not a drug which, in clinical use, produces problems of dependence. Furthermore one of the major signs of withdrawal from barbiturates or alcohol is the production of convulsions which, it might be expected, would be antagonised by many anticonvulsant drugs regardless of whether they give rise to dependence.

The only definitive test for physical dependence, therefore, involves the repeated administration of a compound and its withdrawal under conditions in which animals can be closely observed for behavioural and physiological signs. In a typical experimental procedure (cf. Yanagita, 1973) monkeys are given twice-daily administrations of the test compound for several weeks. The initial dose is chosen as one which gives rise to marked behavioural effects, such as sedation, and if tolerance occurs to this effect the dose is systematically increased throughout the study. With opioid-like and barbiturate-like drugs this leads to the administration of very high doses. The repeated drug administration is finally stopped and the animals are observed frequently over the succeeding days for signs of withdrawal. Typically, rating scales are used to assess the severity of withdrawal which will, of course, vary depending upon the dose and number of drug administrations given (e.g. Blasig *et al.*, 1973). In a variation of this technique, with dependence-producing drugs which act through specific receptor mechanisms, antagonists can be used to precipitate a withdrawal syndrome in dependent animals. Thus naloxone has been frequently used in this way with opioid-dependent animals (Saelens *et al.*, 1971; Wei, 1973) and Ro 15-1788 has been shown to precipitate withdrawal signs in animals receiving repeated administration of benzodiazepines (Cumin *et al.*, 1982; Lukas & Griffith, 1982). It should be noted, however, that signs of precipitated withdrawal are not always identical to those of spontaneous withdrawal (Lukas & Griffiths, 1982; Katz & Valentino, 1986).

Using such procedures, it is possible to establish whether a novel compound will give rise to physical dependence and, if so, whether the withdrawal syndrome is similar to that of a known dependence-producing drug such as an opioid or a barbiturate. What is more difficult to assess,

however, is the relevance of such dependence to the proposed clinical use of the new drug. As noted above, animal experiments often use frequent administration of high doses in attempts to study dependence. If such doses, or the frequency with which they are administered, far surpass what would be considered clinically relevant it may be difficult to extrapolate the results of the laboratory studies to the clinic. In such cases, further animal research, using more relevant conditions, such as lower doses, would clearly be indicated.

An illustration of this point is provided by the present concern over physical dependence produced by benzodiazepines (Lader, 1984). It is clear that high doses of benzodiazepines, taken for long periods give rise to a withdrawal syndrome in humans and in laboratory animals (Hollister *et al.*, 1961; Ryan & Boisse, 1983). What has been considerably more controversial, however, has been the extent to which these phenomena are relevant to the appropriate clinical use of these drugs (Woods *et al.*, 1987). It now seems clear that some patients may experience difficulties in trying to stop benzodiazepine medication under normal clinical conditions (Owen & Tyrer, 1983) and animal studies have also shown that some physical dependence can develop after relatively short-term drug administration (Lukas & Griffiths, 1984). Thus laboratory studies have been of value in defining the precise conditions under which dependence on benzodiazepines can develop and in investigating the mechanisms involved (e.g. Little *et al.*, 1987).

A further example of how care is necessary in interpreting the results of animal studies is provided by research with opioid agonist/antagonists such as buprenorphine. Buprenorphine has both agonist and antagonist effects in the laboratory (Cowan *et al.*, 1977) and, as would be expected, can either suppress or precipitate withdrawal signs in morphine-dependent monkeys (Lewis *et al.*, 1983). However, when animals were given repeated injections of buprenorphine and the drug subsequently withdrawn, few withdrawal signs were observed (Cowan, 1974; Dum *et al.*, 1981). Nevertheless, Dum and colleagues (1981) found that in rats which had been treated chronically with buprenorphine and then switched to morphine, naloxone did precipitate a withdrawal syndrome. The authors claimed that this finding indicated that buprenorphine, like other opioids, did give rise to physical dependence. The explanation for these results probably lies in the finding that buprenorphine dissociates very slowly from its receptor and thus no withdrawal signs are observed after spontaneous withdrawal. In any event, such results, although of theoretical interest, indicate that physical dependence produced by buprenorphine can be demonstrated only under conditions which have little relevance to the drug's clinical use. Indeed, clinical testing of buprenorphine did not indicate a withdrawal syndrome (Jasinski *et al.*, 1978) and buprenorphine has been shown to have considerable value in the treatment of heroin abusers (Jasinski *et al.*, 1978; Mello & Mendelson, 1980). Nevertheless, there are reports that buprenorphine is sometimes abused (see Hartnoll, Chapter 19, this volume).

18.5 SELF ADMINISTRATION

One of the defining features of human drug abuse is the repeated self-administration of a drug. Many studies therefore have investigated the possibiity that animals can be induced to self-administer drugs repeatedly under conditions where they appear to be taking the drug voluntarily. Although techniques have been developed in which animals choose to consume drug solutions orally (e.g. Stolerman & Kumar, 1970; Sanger, 1986) the most successful techniques involve the surgical placement of intravenous catheters through which animals are able to self-administer a drug solution by making an arbitrary response such as pressing a lever (Weeks, 1962; Thompson & Schuster, 1964). Intragastric self-administration has also been studied although it seems to give rise to less reliable results (Woolverton & Schuster, 1983*b*).

The technology for studying intravenous self-administration in animals is now well developed. Many drugs have been tested under standard conditions and the determinants of this behaviour have been reviewed on a number of occasions (Schuster & Thompson, 1969; Johanson, 1978; Spealman & Goldberg, 1978; Young & Herling, 1986). Therefore, the ability to maintain intravenous self-administration in animals is probably the most important information in evaluating the extent to which a new drug is liable to abuse.

Two basic techniques are usually employed in evaluating a novel compound. These are, self-injection under conditions of unlimited access (e.g. Johanson *et al.*, 1976; Balster & Woolverton, 1982), and substitution of a novel compound for a compound already being self-administered under conditions of limited access (Griffiths *et al.*, 1981; Johanson, 1987). Experiments using unlimited access simply connect the response lever and observe the animals to see whether they will initiate or maintain self-administration. Continuous reinforcement schedules in which every response produces a drug infusion are frequently used although the schedule may be changed if self-administration occurs. Other manipulations which may be carried out involve changing the drug concentration or administering repeated, non-contingent drug infusions to investigate whether this will facilitate the acquisition of drug self-administration or suppress ongoing self-administration. Substitution procedures involve the establishment of stable patterns of self-administration with a standard drug, usually during daily sessions of several hours duration, and then the substitution of the novel compounds. Standard drugs frequently used in such experiments include cocaine, codeine and pentobarbital. After substitution animals are observed to determine whether self-administration continues at similar or different levels or whether the behaviour is suppressed or gradually extinguishes. It is also usual in such studies to substitute the drug vehicle for the standard drug from time to time.

Both unlimited access and substitution procedures have advantages and disadvantages. Unlimited access procedures, for example, can reveal differences in patterns of self-administration between different drugs and also can indicate whether repeated self-administration can have toxic consequences. With some drugs, cocaine, for example, rhesus monkeys have been reported to self-administer lethal doses (Johanson *et al.*, 1976). However, it is difficult with such procedures to provide precise evaluations of the relative levels of self administration maintained by different drugs. Substitution procedures, on the other hand, provide exactly such information. The rate of self-administration maintained by different doses of a novel compound can be compared with response rates maintained by standard drugs in the same animals. A potential disadvantage of this method, however, is that the extent to which some drugs are self-administered may depend upon the reference drug used to establish responding (Bergman & Johanson, 1985). However, in general, studies of both unlimited access and limited access substitution procedures produce similar results (Balster & Woolverton, 1982).

When a novel compound is studied in a self-administration experiment the investigators will normally seek to answer two experimental questions. Firstly, the results should indicate whether the compound supports self-administration under standard conditions. If it does, the second question will concern the rates and patterns of self-administration of the new drug in comparison with those of known drugs of abuse. On the basis of these experimental results it should be possible to assess the extent to which the new drug might be subject to abuse.

In order to address effectively this issue, consideration has to be given to the reliability and validity of self-administration procedures (see also Goudie, this vol., Ch. 17). A number of reviews have considered this matter (e.g. Schuster & Thompson, 1969; Griffiths *et al.*, 1980) and have pointed out that most of the drugs abused in humans are self-administered in animals whereas a number of psychoactive drugs which are not subject to abuse also fail to maintain self-administration in animals. Thus, in addition to its apparent face and construct validity, intravenous self-administration also seems to have predictive validity.

However, as the predictive value of any laboratory test is of major importance in drug screening, it is worth considering this matter in greater depth. Table 18.4 presents a simplified summary of the results of many studies of drug self-administration in animals, mainly non-human primates. Drugs have been placed in three categories according to whether they have been shown consistently to maintain self-administration, are active in certain limited conditions, or are not self-administered. It is clear, that, overall, there does seem to be a correlation between self-administration in animals and abuse. Thus the drugs which have been shown most unequivocally to maintain self-administration, opioids (Griffiths & Balster, 1979), psychomotor stimulants (Griffiths *et al.*, 1979), barbiturates (Ator & Griffiths, 1987) and phencyclidine

Table 18.4. *Intravenous drug self-administration in animals*

(1) *Drugs which unequivocally support self-administration*
    Opioids: morphine, heroin, codeine
    Stimulants: cocaine, amphetamine
    Barbiturates: pentobarbital, secobarbital
    Dissociative anaesthetics: phencyclidine,
      ketamine
    Ethanol

(2) *Drugs which support self-administration under certain limited conditions*
    Nicotine
    Caffeine
    Benzodiazepines: diazepam, triazolam

(3) *Drugs which do not support self-administration*
    Antidepressants
    Neuroleptics
    Cannabis
    Hallucinogens: LSD, mescaline
    Some anorectics: fenfluramine,
      phenylpropanolamine
    Some anxiolytics: buspirone

(Marquis & Moreton, 1987) are precisely those drugs which have been most widely used in a non-clinical context and whose abuse has led to social and medical problems. However, other drugs which have had wide recreational use and may, under some circumstances be considered drugs of abuse, including cannabis and LSD, are not self-administered by animals.

The evaluation of drugs which maintain varied and equivocal levels of self-administration in animals also present difficulties. Thus, nicotine and caffeine are two of the most frequently used drugs in humans, although they are not usually considered drugs of abuse. Nicotine will maintain self-administration in animals but under conditions which are much more restricted than those which apply for more powerful stimulants such as cocaine and amphetamine (Henningfield & Goldberg, 1983). Similar results have been reported with caffeine (Griffiths *et al.*, 1979; Hoffmeister & Wuttke, 1973*b*). Self-administration of benzodiazepines has also been reported although results vary both within and between laboratories (Ator & Griffiths, 1987). Although benzodiazepines are very widely prescribed they seem rarely to be used for recreational purposes and are not usually considered to be drugs of abuse although there are some reports of intravenous abuse (e.g. Stark *et al.*, 1987). Clincal studies have also indicated that, under conditions where subjects choose to take amphetamine they do not show preferences for diazepam over placebo (Johanson & Uhlenhuth, 1982).

Many findings emphasise the validity of the self-administration procedure. However, there are also a small number of more anomalous results. It has already been pointed out that widely used and abused hallucinogenic drugs are not self-administered by experimental animals. There are also several drugs which are self-administered in the laboratory but are widely used or have been widely tested in humans and appear to have low abuse potential. Included in this group are the anorectic agent mazindol and the antidepressants nomifensine and bupropion. These drugs have pharmacological profiles similar to amphetamine-like psychomotor stimulants (including increased locomotor activity), and are also self-administered (Wilson & Schuster, 1976; Johanson, 1986; Spyraki & Fibiger, 1981). However, in humans they appear neither to produce amphetamine-like subjective effects, nor to be abused (Hamilton *et al.*, 1983; Miller & Griffith, 1983). There are also some findings with opioids which seem difficult to explain. For example, the kappa agonists ketocyclazocine and ethylketocyclazine, which have dysphoric effects in humans, are not self-administered by monkeys but are active in similar tests with rats (Woods *et al.*, 1982; Collins *et al.*, 1984).

Findings such as these, may, of course, have quite straightforward explanations including, perhaps, differences in metabolism between humans and lower animals. However, for the moment such results remain anomalous and indicate that some drugs self-administered by laboratory animals may be without abuse liability, and vice versa (this point is also discussed by Goudie, Chapter 17 of this volume).

18.6 DISCRIMINATIVE STIMULUS PROPERTIES

One of the most frequently employed methods in behavioural pharmacology in recent years has involved the establishment of drugs as discriminative stimuli (Winter, 1978; Jarbe, 1987). There are a variety of reasons for the use of such methods including their success in drug classification (Barry, 1974) and in analysing biochemical mechanisms (e.g. Cunningham *et al.*, 1987). Another factor, however, is the belief that, as drugs are functioning as internal stimuli, a parallel can be drawn between drug-induced discriminative stimulus control in animals and subjective drug effects in humans. As some researchers (e.g. Jasinski *et al.*, 1984) believe that subjective effects, and particularly drug-induced euphoria, are a primary determinant of abuse liability, it is possible that the stimulus properties of drugs seen in laboratory animals provide an indication of subjective effects in humans and thus of abuse liability (Schuster *et al.*, 1981).

This argument has probably been developed most systematically with opioids (Colpaert, 1986). There are many studies in rats, pigeons, squirrel monkeys and rhesus monkeys showing that opioids can be placed in several categories according to their discriminative stimulus properties. One of these categories includes morphine, heroin and other drugs which produce similar subjective effects in humans and are subject to abuse. A different discriminative stimulus is usually produced by drugs such as ketocyclazocine which also give rise to quite different subjective effects in people (Holtzman, 1983). This classification also corresponds closely to the sub-classification of opioid receptors although some differences between results obtained with different species have been reported (Herling & Woods, 1981; Holtzman, 1985).

With opioids therefore, drug discrimination may provide an indication as to whether drugs will produce morphine-like subjective effects in people and be subject to abuse. Extensive research has also been carried out to investigate the discriminative stimulus properties of psychomotor stimulants (Young & Glennon, 1986) and of sedatives and anxiolytics (Schuster *et al.*, 1981; Sanger, 1988) although there has been less discussion of the relevance of these studies to the evaluation of abuse liability. However, there are a number of results which indicate that the correlations between stimulus properties and abuse liabilities in these drugs classes is not strong.

Many experiments have studied the stimulus properties of barbiturates and benzodiazepines (Colpaert *et al.*, 1976; Sanger & Zivkovic, 1987). With a few notable exceptions (Ator & Griffiths, 1983, 1987; Spealman, 1985), the results of these studies have shown marked similarities between the stimulus properties of these two groups of drugs so that animals trained to discriminate a benzodiazepine from placebo show drug-appropriate responding after administration of a barbiturate, and vice-versa. However, it is not at all clear that these two groups of drugs have similar subjective effects or abuse liabilities in humans. With more recently developed anxiolytics and sedatives the situation is also quite complex. Thus, even among substances which displace benzodiazepines from their binding sites there are compounds which show discriminative stimulus properties similar to those of benzodiazepines (e.g. zopiclone, CL 218,872: Sanger *et al.*, 1985; Ator & Griffiths, 1986) whereas with some other drugs such as zolpidem there is only a partial overlap of stimulus properties (Depoortere *et al.*, 1986; Sanger & Zivkovic, 1986, 1987). The novel anxiolytic, busirone, whose mechanims of action seems to have nothing in common with those of barbiturates or benzodiazepines, also has discriminative stimulus properties distinctively different from those of other anxiolytics (Hendry *et al.*, 1983; Cunningham *et al.*, 1987).

With psychomotor stimulants the relationship between stimulus properties and abuse liability also seems unclear. A number of studies have shown that abused drugs such as cocaine and amphetamine substitute for each other (Colpaert *et al.*, 1978; Jarbe, 1981) although there may be subtle differences between the discriminative stimuli produced by these drugs (D'Mello & Stolerman, 1977). However, there are also a number of demonstrations that the amphetamine cue generalises partially or completely to a variety of other drugs, some without apparent abuse liability. Thus Fig. 18.1 shows that rats trained to discriminate d-amphetamine from saline showed dose-related responding on the lever associated with amphetamine, after administration of nomifensine and bupropion, confirming the results of previous studies (Jones *et al.*, 1980; Porsolt *et al.*, 1982). As noted earlier, neither of these compounds seems to exert amphetamine-like effects in humans. The

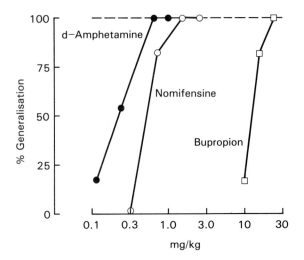

Fig. 18.1. Results of generalisation tests in rats trained to discriminate d-amphetamine (0.5 mg/kg) from saline in a standard, two-lever, FR 10 drug discrimination procedure (Sanger, unpublished). After increasing doses of the training drug and of the antidepressants, nomifensine and bupropion, rats showed dose-related responding on the amphetamine-associated lever. The measure of generalisation is the percentage of animals choosing to respond on the amphetamine lever. The amphetamine-like stimulus properties of nomifensine and bupropion correspond with their pharmacological profiles in rodents but apparently not with their effects in humans.

anorectics mazindol and phenylpropanolamine also produce amphetamine-like stimulus properties (Huang & Ho, 1974; Holloway *et al.*, 1985; Woolverton *et al.*, 1986; Evans & Johanson, 1987) although phenylpropanolamine is not self-administered by rhesus monkeys or baboons (Woolverton *et al.*, 1986; Lamb *et al.*, 1987). In relation to these drugs the series of studies reported by Chait and his colleagues (Chait *et al.*, 1984, 1985, 1986, 1987) are of considerable theoretical importance. These researchers trained human subjects to discriminate amphetamine from placebo and found, in confirmation of the results of animal experiments, that the stimulus produced by amphetamine generalised to mazindol and phenylpropanolamine. However, these drugs did not produce subjective effects identical to those of amphetamine and are not known to be abused to any significant degree. Such results therefore indicate that even in human subjects stimulus properties, subjective effects and abuse liabiity to not necessarily co-vary, at least with this drug class (for further discussion of these results see Goudie, this vol., Ch. 17, and Schuster & Johanson, 1985). Thus, although the stimulus properties of different drugs in laboratory animals may provide important information about the overall pharmacological profile of new compounds they do not seem to have a very high degree of validity for the prediction of abuse potential.

18.7 PHARMACOLOGICAL TREATMENTS FOR DRUG ABUSE AND DEPENDENCE

Few would deny that drug abuse and dependence represent a sociomedical problem which is immensely costly in both social and economic terms. Scientific and medical research, therefore, needs to deal not only with the development of novel drugs without abuse and dependence liabilities but also should search for suitable treatments for these problems.

One approach to the treatment of drug abuse and physical dependence involves a search for drugs which will have useful therapeutic activity. At first it might seem paradoxical that drugs should be used to treat a problem brought on by other drugs. Indeed, Naranjo (1985) suggests that there has been a certain amount of prejudice against this approach which has led to a lack of research both within and outside the pharmaceutical industry. However, there is no logical reason why drugs cannot be used effectively to deal with drug abuse and dependence and, in fact, pharmacological treatments are already widely used although not always to great effect. Jaffe (1987) notes, for example, that over a hundred different drugs or drug combinations have been described as being useful in treating the alcohol withdrawal syndrome.

Table 18.5 presents several ways in which drugs can be used to deal with problems of abuse and dependence with some examples of current practice. A new drug can be subsituted for a drug on which a patient is physically dependent with the purpose usually, of allowing a more manageable detoxification. Methadone maintenance, for example, has become a standard method of treatment of heroin addicts (Dole & Nyswander,

Table 18.5. *Ways in which drug therapy can be used in treating drug dependence*

(1) *Substitution of a pharmacologically similar but medically more acceptable preparation for maintenance and detoxification*
    e.g. Methadone for heroin addicts, nicotine chewing gum for smokers.

(2) *Alleviation of withdrawal symptoms*
    e.g. Benzodiazepines for alcohol withdrawal, clonidine for heroin withdrawal.

(3) *Blockade of rewarding properties of abused drugs*
    Abused drugs become aversive, or there is a specific pharmacological antagonism of their actions, or there may be a general blockade of pharmacological reinforcement. e.g. disulfiram for alcoholics, naltrexone for opiate abusers.

1965). Another current example is provided by the use of chewing gum or other products containing nicotine as substitutes for cigarettes as aids in reducing cigarette smoking (Russell *et al.*, 1980). The use of such preparations is based on the belief that many smokers are physically dependent on nicotine and are unable to give up smoking because of withdrawal symptoms. Substitution of nicotine-containing chewing gum should thus block the withdrawal signs and would allow the dependent subject to give up smoking and then, presumably, to become gradually detoxified by reducing chewing gum consumption over a period of time.

Another approach to the pharmacological treatment of drug abuse involves drugs which block the rewarding action of abused drugs and thus produce a decrease in drug self-administration. At least in theory, this can be done in several ways. A preparation can be used to make the abused drug aversive, as with disulfiram in alcoholics. Alternatively, a drug treatment can be used which blocks all the pharmacological effects of the abused drug, such as naltrexone in heroin abusers. Finally, it might be possible to find drugs which specifically block the rewarding effects of abused drugs. The first two of these approaches have been used with some success. They do suffer from one particular difficulty, however, which is that compliance tends to be poor. Thus, when a drug abuser discovers that a preferred drug of abuse no longer has any effect this can lead either to the extinction of the drug-taking behaviour or, alternatively, to the treatment drug no longer being taken (Jaffe, 1987).

In order to evaluate the potential of drugs as treatments for drug abuse or physical dependence it seems logical to make use of the same laboratory tests as are used in evaluating abuse and dependence potential. Thus, for example, physical dependence can be established in animals with an opiate and the extent to which withdrawal signs are alleviated by an experimental drug investigated. As might be expected in this case, there seems to be a good concordance between laboratory and clinical studies as with clonidine which suppresses some opiate withdrawal signs in both animals (Katz, 1986) and humans (Gold *et al.*, 1978). Similarly, in order to establish whether it is possible to affect drug taking with a pharmacological treatment, the self-administration of a drug of abuse needs to be established in the laboratory and the effects of potential treatments investigated. A number of such studies have been carried out and this work has led to some suggestions for treatments. However, there are many associated difficulties which have not, as yet, led to any breakthroughs in the treatment of drug abusers.

As an example of this type of research we can consider a recent series of studies concerned with the possibility that drugs which inhibit the reuptake of serotonin can be useful treatments of alcohol abuse (Amit *et al.*, 1984). Many methods have been used to attempt to induce alcohol drinking in laboratory animals and the extent to which these methods provide models of alcohol abuse has been hotly debated (e.g. Lester & Freed, 1973; Mello, 1973, 1985). It is possible, however, to induce animals to consume more of an alcohol solution than of another fluid, usually water. In one such study Rockman *et al.* (1979) found that the antidepressant drug zimeldine reduced alcohol intake, but not total fluid intake, of rats which had both water and an alcohol solution available and were drinking more alcohol than water under control conditions. Further experiments showed that only antidepressant drugs with the ability to inhibit serotonin reuptake had this effect on alcohol consumption. It was therefore proposed that serotonin might play an important role in the reinforcing properties of alcohol and that such drugs might be clinically useful in treating alcohol abuse (Amit *et al.*, 1984). On the basis of this laboratory research clinical studies were carried out and did indeed provide evidence that zimeldine can reduce alcohol intake in humans (Naranjo *et al.*, 1984, 1985; Amit *et al.*, 1985).

Zimeldine, however, was withdrawn from the market and the clinical trials in alcohol abusers discontinued, although it is likely that other drugs with a similar mechanism of action are currently being evaluated for this clinical use. In addition, there now appears to be some doubt as to the specificity of zimeldine's effect on alcohol intake in rats (Gill & Amit, 1987). It remains to be seen, therefore, whether this area of research will lead to an effective treatment for alcohol abuse and thus provide a clear example of a behavioural model successfully predicting clinical application.

18.8 CONCLUSION

Laboratory screening tests and animal models can never predict drug action in humans with complete accuracy. This is particularly true in psychopharmacology. In psychotropic drug research there are many examples of failures to predict therapeutic utility accurately and of faulty assessment of behavioural side effects. It can reasonably be argued that the assessment of dependence and abuse liabilities is one of the areas of behavioural pharmacology which has made most progress in recent years, as tests are now available which, in general, are both reliable and valid. However, as the present discussion and the preceding chapter (Goudie, this vol., Ch. 17) have indicated, there are also a number of apparently anomolous findings. Although techniques such as self-administration, precipitated withdrawal and drug discrimination can predict abuse and dependence liabilities with a level of accuracy impressive in behavioural science these methods are far from perfect. Therefore, the industrial pharmacologist concerned with psychotropic drug development needs to be cautious. Certain findings, such as high levels of intravenous self-administration in monkeys or intense withdrawal symptoms after relatively low doses, would certainly preclude the further development of most types of drugs, but things are rarely this simple. In order to decide whether a new drug is worthy of development all the factors described in the present chapter must be taken into account along with the potential therapeutic value and overall toxicological profile of the compound. All therapeutic drug use involves a degree of clinical and social risk. It is therefore the task of industrial and clinical scientists, together with representatives of government and society, to evaluate the acceptability of the abuse and dependence liabilities of any novel drug in relation to the advance in therapy which it represents.

REFERENCES

Amit, Z., Sutherland, A., Gill, K. & Ogren, S.O. (1984). Zimeldine: a review of its effect on ethanol consumption. *Neuroscience and Biobehavioral Reviews* **8**, 35–54.

Amit, Z., Brown, Z., Sutherland, A., Rockman, G., Gill, K. & Selvagii, N. (1985). Reduction in alcohol intake in humans as a function of treatment with zimeldine: implications for treatment. In *Research Advances in New Psychopharmacological Treatments for Alcoholism*, eds. C.A. Naranjo & E.M. Sellers, pp. 189–98, Amsterdam: Elsevier.

Ator, N.A. & Griffiths, R.R. (1983). Lorazepam and pentobarbital drug discrimination in baboons: cross-drug generalization and interaction with Ro 15-1788. *Journal of Pharmacology and Experimental Therapeutics* **226**, 770–82.

Ator, N.A. & Griffiths, R.R. (1986). Discriminative stimulus effects of atypical anxiolytics in baboons and rats. *Journal of Pharmacology and Experimental Therapeutics* **237**, 393–403.

Ator, N.A. & Griffiths, R.R. (1987). Self-administration of barbiturates and benzodiazepines: A review. *Pharmacology, Biochemistry and Behavior* **27**, 391–8.

Balster, R.L. & Woolverton, W.L. (1982). Unlimited access intravenous drug self-administration in rhesus monkeys. *Federation Proceedings* **41**, 211–15.

Barry, H. (1974). Classification of drugs according to their discriminable effects in rats. *Federation Proceedings* **33**, 1814–24.

Bergman, J. & Johanson, C.E. (1985). The reinforcing properties of diazepam under several conditions in the rhesus monkey. *Psychopharmacology* **86**, 108–13.

Blasig, J., Herz, A., Reinhold, K. & Zieglgansberger, S. (1973). Development of physical dependence on morphine in respect to time and dosage and quantification of the precipitated withdrawal syndrome in rats. *Psychopharmacology* **33**, 19–38.

Brady, J.V. & Fischman, M.W. (1985). Assessment of drugs for dependence potential and abuse liability: an overview. In *Behavioral Pharmacology: The Current Status*, eds. L.S. Seiden & R.L. Balster, pp. 361–82, New York: Alan R. Liss.

Braestrup, C. & Scheel-Kruger, J. (1976). Methylphenidate-like effects of the new antidepressant drug nomifensine (HOE 984). *European Journal of Pharmacology* **38**, 305–12.

Cappell, H., Busto, V., Kay, G., Naranjo, C.A., Sellers, E.M. & Sanchez-Craig, M. (1987). Drug deprivation and reinforcement by diazepam in a dependent population. *Psychopharmacology* **91**, 154–60.

Chait, L.D., Uhlenhuth, E.H. & Johanson, C.E. (1984). An experimental paradigm for studying the discriminative stimulus properties of drugs in humans. *Psychopharmacology* **82**, 272–4.

Chait, L.D., Uhlenhuth, E.H. & Johanson, C.E. (1985). The discriminative stimulus and subjective effects of d-amphetamine in humans. *Psychopharmacology* **86**, 307–12.

Chait, L.D., Uhlenhuth, E.H. & Johanson, C.E. (1986). Human drug discrimination: d-amphetamine and other anorectics. In *Problems of Drug Dependence*, 1985. National Institute of Drug Abuse Research Monograph 67, pp. 161–7, Washington, D.C.

Chait, L.D., Uhlenhuth, E.H. & Johanson, C.E. (1987). Reinforcing and subjective effects of anorectics in normal human volunteers. *Journal of Pharmacology and Experimental Therapeutics* **242**, 777–83.

Chen, G., Ensor, C.R., Russell, D. & Bohner, B. (1959). The pharmacology of 1-(1-phenylcyclo-hexyl) piperidine HCl. *Journal of Pharmacology and Experimental Therapeutics* **127**, 241–50.

Cheney, D.L. & Goldstein, A. (1971). Tolerance to opioid narcotics: time course and reversibility of physical dependence in mice. *Nature* **232**, 477–8.

Collins, R.J., Weeks, J.R., Cooper, M.M., Good, P.I. & Russell, R.R. (1984). Prediction of abuse liability of drugs using i.v. self-administration by rats. *Psychopharmacology* **82**, 6–13.

Colpaert, F.C. (1986). Drug discrimination: behavioral, pharmacological and molecular mechanisms of discriminative drug effects. In *Behavioral Analysis of Drug Dependence*, eds. S.R. Goldberg & I.P. Stolerman, pp. 161–93, London: Academic Press.

Colpaert, F.C., Desmedt, L.K.C. & Janssen, P.A.J. (1976). Discriminative stimulus properties of benzodiazepines, barbiturates and pharmacologically related drugs: relation to some intrinsic and anticonvulsant effects. *European Journal of Pharmacology* **37**, 113–23.

Colpaert, F.C., Niemegeers, C.J.E. & Janssen, P.A.J. (1978). Discriminative stimulus properties of cocaine and d-amphetamine, and antagonism by haloperidol: a comparative study. *Neuropharmacology* **17**, 937–42.

Cooper, B.R., Hester, T.J. & Maxwell, R.A. (1980). Behavioral and biochemical effects of the antidepressant bupropion (Wellbutrin): evidence for selective blockade of dopamine uptake in vivo. *Journal of Pharmacology and Experimental Therapeutics* **215**, 127–34.

Cowan, A. (1974). Evaluation in non-human primates: evaluation of the physical dependence capacities of oripavine-thebaine partial agonists in patas monkeys. In *Advances in Biochemical Psychopharmacology*, vol. 8, eds. M.C. Brande, L.S. Harris, E.L. May, J.P. Smith & J.E. Villarreal, pp. 427–8, New York: Raven Press.

Cowan, A., Doxey, J.C. & Harry, E.J.R. (1977). The animal pharmacology of buprenorphine, an oripavine analgesic agent. *British Journal of Pharmacology* **60**, 547–54.

Cumin, R., Bonetti, E.P., Scherschlicht, S. & Haefely, W.E. (1982). Use of the specific benzodiazepine antagonist, Ro 15-1788, in studies of physiological dependence on benzodiazepines. *Experientia* **38**, 833–4.

Cunningham, K.A., Callahan, P.M. & Appel, J.B. (1987). Discriminative stimulus properties of 8-hydroxy-2-(di-n-propylamino) tetralin (8-OHDPAT): implications for understanding the actions of novel anxiolytics. *European Journal of Pharmacology* **138**, 29–36.

Depoortere, H., Zivkovic, B., Lloyd, K.G., Sanger, D.J., Perrault, G., Langer, S.Z. & Bartholini, G. (1986). Zolpidem, a novel non-benzodiazepine hypnotic. I – Neuropharmacological and behavioral effects. *Journal of Pharmacology and Experimental Therapeutics* **237**, 649–58.

D'Mello, G.D. & Stolerman, I.P. (1977). Comparison of the discriminative stimulus properties of cocaine and amphetamine in rats. *British Journal of Pharmacology* **61**, 415–22.

Dole, V.P. & Nyswander, M.A. (1965). A medical treatment for diacetylmorphine (heroin) addiction. *Journal of the American Medical Association* **193**, 646–50.

Downs, D.A. & Woods, J.H. (1975). Fixed-ratio escape and avoidance-escape from naloxone in morphine-dependent monkeys: effects of naloxone dose and morphine pretreatment. *Journal of the Experimental Analysis of Behavior* **23**, 415–27.

Dum, J., Blasig, J. & Herz, A. (1981). Buprenorphine: demonstration of physical dependence liability. *European Journal of Pharmacology* **70**, 293–300.

Essig, C.F. (1966). Barbiturate withdrawal convulsions in white rats. *Neuropharmacology* **5**, 103–7.

Evans, S.M. & Johanson, C.E. (1987). Amphetamine-like effects of anorectics and related compounds in pigeons. *Journal of Pharmacology and Experimental Therapeutics* **241**, 817–25.

Gill, K. & Amit, Z. (1987). Effects of serotonin uptake blockade on food, water, and ethanol consumption in rats. *Alcoholism: Clinical and Experimental Research* **11**, 444–9.

Gold, M.S., Redmond, D.E. & Kleber, H.D. (1978). Clonidine blocks acute opiate withdrawal symptoms. *Lancet* **ii**, 599–602.

Gold, M.S., Pottash, A.L.C., Sweeney, D.R. & Kleber, H.D. (1980). Clonidine: a safe, effective, and rapid nonopiate treatment for opiate withdrawal. *Journal of the American Medical Association* **243**, 343–6.

Goldberg, S.R., Hoffmeister, F., Schlichting, U. & Wuttke, W. (1971). Aversive properties of nalorphine and naloxone in morphine-dependent rhesus monkeys. *Journal of Pharmacology and Experimental Therapeutics* **179**, 268–76.

Griffiths, R.R. & Balster, R.L. (1979). Opioids: Similarity between evaluations of subjective effects and animal self-administration results. *Clinical Pharmacology and Therapeutics* **25**, 611–17.

Griffiths, R.R., Brady, J.V. & Bradford, L.D. (1979). Predicting the abuse liability of drugs with animal drug self-administration procedures: psychomotor stimulants and hallucinogens. In *Advances in Behavioral Pharmacology*, vol. 2, eds. T. Thompson & P.B. Dews, pp. 163–208, New York: Academic Press.

Griffiths, R.R., Bigelow, G.E. & Henningfield, J.E. (1980). Similarities in animal and human drug-taking behavior. In *Advances in Substance Abuse*, vol. 1, ed. N.K. Mello, pp. 1–90, Greenwich, Conn.: JAI Press.

Griffiths, R.R., Lukas, S.E., Bradford, L.D., Brady, J.V. & Snell, J.D. (1981). Self-injection of barbiturates and benzodiazepines in baboons. *Psychopharmacology* **75**, 101–9.

Haefely, W. (1986). Biological basis of drug-induced tolerance, rebound and dependence. Contribution of recent research on benzodiazepines. *Pharmakopsychiatrie* **19**, 353–61.

Hamilton, M.J., Smith, P.R. & Peck, A.W. (1983). Effects of bupropion, nomifensine and dexamphetamine on performance, subjective feelings, autonomic variables and electroencephalogram in healthy volunteers. *British Journal of Clinical Pharmacology* **15**, 367–74.

Harding, W.M. & Zinberg, N.E. (1983). Occasional opiate use. In *Advances in Substance Abuse*, vol. 3, ed. N.K. Mello, pp. 27–61, Greenwich, Connecticut: JAI Press.

Harris, L.S. (1986). Nathan B. Eddy memorial award lecture. In *Problems of Drug Dependence, 1985*. National Institute on Drug Abuse Research Monograph 67, pp. 4–13, Washington, D.C.

Hendry, J.S., Balster, R.L. & Rosecrans, J.A. (1983). Discriminative stimulus properties of buspirone compared to central nervous system depressants in rats. *Pharmacology, Biochemistry and Behavior* **19**, 97–101.

Henningfield, J.E. & Goldberg, S.R. (1983). Nicotine as a reinforcer in human subjects and laboratory animals. *Pharmacology, Biochemistry and Behavior* **19**, 989–92.

Herling, S. & Woods, J.H. (1981). Discriminative stimulus effects of narcotics: evidence for multiple receptor-mediated actions. *Life Sciences* **28**, 1571–84.

Hoffmeister, F. (1986). Negative reinforcing properties of naloxone in the non-dependent rhesus monkey: influence on reinforcing properties of codeine, tilidine, buprenorphine, and pentazocine. *Psychopharmacology* **90**, 441–50.

Hoffmeister, F. & Wuttke, W. (1973a). Negative reinforcing properties of morphine-antagonists in naive rhesus monkeys. *Psychopharmacologia* **33**, 247–58.

Hoffmeister, F. & Wuttke, W. (1973b). Self-administration of acetylsalicylic acid and combinations with codeine and caffeine in rhesus monkeys. *Journal of Pharmacology and Experimental Therapeutics* **186**, 266–75.

Hollister, L.E., Motzenbecker, R.F. & Degan, R.O. (1961). Withdrawal reactions from chlordiazepoxide ('Librium'). *Psychopharmacologia* **2**, 63–8.

Holloway, F.A., Michaelis, R.C. & Huerta, P.L. (1985). Caffeine–phenylethylamine combinations mimic the amphetamine discriminative cue. *Life Sciences* **36**, 723–30.

Holtzman, S.G. (1983). Discriminative stimulus properties of opioid agonists and antagonists. In *Theory in Psychopharmacology*, vol. 2, ed. S. J. Cooper, pp. 1–45, London: Academic Press.

Holtzman, S.G. (1985). Discriminative stimulus properties of opioids that interact with mu, kappa and PCP/Sigma receptors. In *Behavioral Pharmacology: The Current Status*, eds. L.S. Seiden & R.L. Balster, pp. 131–47, New York: Allan R. Liss.

Huang, J.T. & Ho, B.T. (1974). Discriminative stimulus properties of d-amphetamine and related compounds in rats. *Journal of Pharmacology and Experimental Therapeutics* **2**, 669–73.

Iwamoto, E.T. (1981). Locomotor activity and antinociception after putative mu, kappa and sigma opioid receptor agonists in the rat: influence of dopaminergic agonists and antagonists. *Journal of Pharmacology and Experimental Therapeutics* **217**, 415–60.

Jaffe, J.H. (1987). Pharmacological agents in treatment of drug dependence. In *Psychopharmacology: The Third Generation of Progress*, ed. H.Y. Meltzer, pp. 1605–16, New York: Raven Press.

Jarbe, T.U.C. (1981). Cocaine cue in pigeons: time course studies and generalization to structurally related compounds (norcocaine, WIN 35,428 and 35,065-2) and (±)-amphetamine. *British Journal of Pharmacology* **73**, 843–52.

Jarbe, T.U.C. (1987). Drug discrimination learning. In *Experimental Psychopharmacology*, eds. A.J. Greenshaw & C.T. Dourish, pp. 433–79, Clifton, New Jersey: Humana Press.

Jarrott, B. & Lewis, S.J. (1987). Discontinuation syndrome in rats chronically treated with centrally acting α-adrenoceptor agonists. *Trends in Pharmacological Sciences* **8**, 244–6.

Jasinski, D.R., Pevnick, J.S. & Griffiths, J.D. (1978). Human pharmacology and abuse potential of the analgesic buprenorphine. *Archives of General Psychiatry* **35**, 501–16.

Jasinski, D.R., Johnson, R.E. & Henningfield, J.E. (1984). Abuse liability assessment in human subjects. *Trends in Pharmacological Sciences*, May, 196–200.

Johanson, C.E. (1978). Drugs as reinforcers. In *Contemporary Research in Behavioral Pharmacology*, eds. D.E. Blackman & D.J. Sanger, pp. 325–90, New York: Plenum Press.

Johanson, C.E. (1986). Stimulant depressant report. In *Problems of Drug Dependence*, 1985. National Institute of Drug Abuse Research Monograph 67, pp. 98–104, Washington, D.C.

Johanson, C.E. (1987). Benzodiazepine self-administration in rhesus monkeys: estrazolam, flurazepam and lorazepam. *Pharmacology, Biochemistry and Behavior* **26**, 521–6.

Johanson, C.E. & Uhlenhuth, E.H. (1982). Drug preferences in humans. *Federation Proceedings* **41**, 228, 233.

Johanson, C.E., Balster, R.L. & Bonese, K. (1976). Self-administration of psychomotor stimulant drugs: the effects of unlimited access. *Pharmacology, Biochemistry and Behavior* **4**, 45–51.

Jones, C.N., Howard, J.L. & McBennett, S.T. (1980). Stimulus properties of antidepressants in the rat. *Psychopharmacology* **67**, 111–18.

Kabes, J. & Fink, S. (1972). A new device for measuring spontaneous motor activity – Effects of lysergic acid diethylamide in rats. *Psychopharmacology* **23**, 75–85.

Katz, J.L. (1986). Effects of clonidine and morphine on opioid withdrawal in rhesus monkeys. *Psychopharmacology* **88**, 392–7.

Katz, J.L. & Valentino, R.J. (1986). Pharmacological and behavioral factors in opioid dependence in animals. In *Behavioral Analysis of Drug Dependence*, eds. S.R. Goldberg & I.P. Stolerman, pp. 287–327, London: Academic Press.

Kumar, R. (1969). Exploration and latent learning: differential effects of dexamphetamine on components of exploratory behaviour in rats. *Psychopharmacology* **16**, 54–72.

Kumar, R. & Stolerman, I.P. (1977). Experimental and clinical aspects of drug dependence. In *Handbook of Psychopharmacology*, vol. 7, eds. L.L. Iversen, S.D. Iversen & S.H. Snyder, pp. 321–67, New York: Plenum Press.

Kumor, K., Su, T.P., Vaupel, B., Haertzen, C., Johnson, R.E. & Goldberg, S. (1986*a*). Studies of kappa agonist. In *Problems of Drug Dependence*, 1985. National Institute of Drug Abuse Research Monograph 67, pp. 18–25, Washington, D.C.

Kumor, K.M., Haertzen, C.A., Johnson, R.E., Kocher, T. & Jasinski, D. (1986*b*). Human psychopharmacology of ketocyclazocine compared with cyclazocine, morphine and placebo. *Journal of Pharmacology and Experimental Therapeutics* **238**, 960–70.

Lader, M. (1984). Benzodiazepine dependence. *Progress in Neuropsychopharmacology and Biological Psychiatry* **89**, 85–95.

Lamb, R.J., Sannerud, C.A. & Griffiths, R.R. (1987). An examination of the intravenous self-administration of phenylpropranolamine using a cocaine substitution procedure in the baboon. *Pharmacology, Biochemistry and Behavior* **28**, 389–92.

Lasagna, L. & Beecher, H.K. (1954). The analgesic effectiveness of nalorphine and nalorphine-morphine combinations in man. *Journal of Pharmacology and Experimental Therapeutics* **112**, 356–63.

Legros, J., Khalili-Verasteh, H. & Margetts, G. (1984). Pharmacological study of pentazocine-naloxone combination: interest as a potentially non abusable oral form of pentazocine. *Archives Internationales de Pharmacodynamie* **271**, 11–21.

Lester, D. & Freed, E.X. (1973). Criteria for an animal model of alcoholism. *Pharmacology, Biochemistry and Behavior* **1**, 103–7.

Lewis, J.W., Rance, M.J. & Sanger, D.J. (1983). The pharmacology and abuse potential of buprenorphine: A new antagonist analgesic. In *Advances in Substance Abuse*, vol. 3, ed. N.K. Mello, pp. 103–54, Greenwich, Connecticut: JAI Press.

Little, H.J., Nutt, D.J. & Taylor, S.C. (1987). Kindling and withdrawal changes at the benzodiazepine receptor. *Journal of Psychopharmacology* **1**, 35–46.

Lukas, S.E. & Griffiths, R.R. (1982). Precipitated withdrawal by a benzodiazepine receptor antagonist (Ro 15-1788) after 7 days of diazepam. *Science* **217**, 1161–3.

Lukas, S.E. & Griffiths, R.R. (1984). Precipitated diazepam withdrawal in baboons: Effects of dose and duration of diazepam exposure. *European Journal of Pharmacology* **100**, 163–71.

Lush, I.E. (1974). A comparison of the effect of mescaline on activity and emotional defecation in seven strains of mice. *British Journal of Pharmacology* **55**, 133–9.

Lyon, M. & Robbins, T.W. (1975). The action of CNS stimulant drugs – a general theory concerning amphetamine effects. In *Current Developments in Psychopharmacology*, vol. 2, eds. W.B. Essman & L. Valzelli, pp. 80–163, New York: Wiley.

Marquis, K.L. & Moreton, J.E. (1987). Animal models of intravenous phencycinoid self-administration. *Pharmacology, Biochemistry and Behavior* **27**, 383–9.

Marriott, A.S. & Smith, E.F. (1972). An analysis of drug effects in mice exposed to a simple novel environment. *Psychopharmacology* **24**, 397–406.

Marriott, A.S. & Spencer, P.S.J. (1965). Effects of centrally acting drugs on exploratory behaviour in rats. *British Journal of Pharmacology* **25**, 432–41.

Martin, W.R., Eades, C.J., Thompson, J.A., Huppler, R.E. & Gilbert, P.E. (1976). The effects of morphine- and nalorphine-like drugs in the non-dependent and morphine-dependent chronic spinal dog. *Journal of Pharmacology and Experimental Therapeutics* **197**, 517–32.

Mello, N.K. (1973). A review of methods to induce alcohol addiction in animals. *Pharmacology, Biochemistry and Behavior* **1**, 89–101.

Mello, N.K. (1985). Animal models of alcoholism: contributions of behavioral pharmacology. In *Behavioral Pharmacology: The Current Status*, eds. L.W. Seiden & R.L. Balster, pp. 383–401, New York: Alan R. Liss.

Mello, N.K. & Mendelson, J.H. (1980). Buprenorphine suppresses heroin use by heroin addicts. *Science* **207**, 657–9.

Miller, L.L. & Drew, W.G. (1974). Cannabis: review of behavioral effects in animals. *Psychological Bulletin* **81**, 401–17.

Miller, L. & Griffith, J. (1983). A comparison of bupropion, dextroamphetamine, and placebo in mixed-substance abusers. *Psychopharmacology* **80**, 199–205.

Naranjo, C.A. (1985). Drug treatments for alcoholism: the need for innovation. In *Research Advances in New Psychopharmacological Treatments for Alcoholism*, eds. C.A. Naranjo & E.M. Sellers, pp. 1–9, Amsterdam: Elsevier.

Naranjo, C.A., Sellers, E.M., Roach, C.A., Woodley, D.V., Sanchez-Craig, M. & Sykora, K. (1984). Zimeldine-induced variations in ethanol intake in non-depressed heavy drinkers. *Clinical Pharmacology and Therapeutics* **35**, 374–81.

Naranjo, C.A., Sellers, E.M., Wu, P.H. & Lawrin, M.O. (1985). Moderation of ethanol drinking: role of enhanced serotoninergic neurotransmission. In *Research Advance in New Psychopharmacological Treatments for Alcoholism*, eds. C.A. Naranjo & E.M. Sellers, pp. 171–86, Amsterdam: Elsevier.

Owen, R.T. & Tyrer, P. (1983). Benzodiazepine dependence: A review of the evidence. *Drugs* **25**, 385–98.

Pfeiffer, A., Branti, V., Herz, A. & Emrich, H.M. (1986). Psychotomimesis mediated to K opiate receptors. *Science* **233**, 774–6.

Porsolt, R.D., Pawelec, C. & Jalfre, M. (1982). Use of the drug discrimination procedure to detect amphetamine-like effects of antidepressants. In *Drug Discrimination: Applications in CNS Pharmacology*, eds. F.C. Colpaert & J.L. Slangen, pp. 193–202, Amsterdam: Elsevier Biomedical Press.

Preston, K.L., Bigelow, G.E. & Liebson, I.A. (1988). Buprenophine and naloxone alone and in combination in opioid-dependent humans. *Psychopharmacology* **94**, 484–90.

Read, G.W., Cutting, W. & Furst, A. (1960). Comparison of excited phases after sedatives and tranquilizers. *Psychopharmacology* **1**, 346–50.

Rockman, G.E., Amit, Z., Carr, G., Brown, Z. & Ogren, S.O. (1979). Attenuation of ethanol intake by 5-hydroxytryptamine uptake blockade in laboratory rats. I. Involvement of brain 5-hydroxytryptamine in the mediation of the positive reinforcing properties of ethanol. *Archives Internationales de Pharmacodynamie* **241**, 245–59.

Ryan, G.P. & Boisse, N.R. (1983). Experimental induction of benzodiazepine tolerance and physical dependence. *Journal of Pharmacology and Experimental Therapeutics* **226**, 100–7.

Russell, M.A.H., Raw, M. & Jarvis, M.J. (1980). Clinical use of nicotine chewing gum. *British Medical Journal* **1**, 1599–622.

Saelens, J.K., Granat, F.R. & Sawyer, W.K. (1971). The mouse jumping test – a simple screening method to estimate the physical dependence capacity of analgesics. *Archives Internationales de Pharmacodynamie* **190**, 213–18.

Sanger, D.J. (1983). Opiates and ingestive behaviour. In *Theory in Psychopharmacology*, vol. 2, ed. S.J. Cooper, pp. 75–113, London: Academic Press.

Sanger, D.J. (1986). Drug taking as adjunctive behavior. In *Behavioral Analysis of Drug Dependence*, eds. S.R. Goldberg & I.P. Stolerman, pp. 123–60, London: Academic Press.

Sanger, D.J. (1988). Discriminative stimulus properties of anxiolytic and sedative drugs: pharmacological specificity. In *Transduction Mechanisms of Drug Stimuli*, eds. F.C. Colpaert & R.L. Balster, pp. 73–84, Berlin: Springer.

Sanger, D.J. & Zivkovic, B. (1986). The discriminative stimulus properties of zolpidem, a novel imidazopyridine hypnotic. *Psychopharmacology* **89**, 317–22.

Sanger, D.J. & Zivkovic, B. (1987). Discriminative stimulus properties of chlordiazepoxide and zolpidem. Agonist and antagonist effects of CGS 9896 and ZK 91296. *Neuropharmacology* **26**, 499–505.

Sanger, D.J., Joly, D. & Zivkovic, B. (1985). Behavioral effects of non-benzodiazepine anxiolytic drugs: a comparison of CGS 9896 and zopiclone with chlordiazepoxide. *Journal of Pharmacology and Experimental Therapeutics* **232**, 831–7.

Schuster, C.R. & Johanson, C.E. (1985). Efficacy, dependence potential and neurotoxicity of anorectic drugs. In *Behavioral Pharmacology: The Current Status*, eds. L.S. Seiden & R.L. Balster, pp. 263–79, New York: Alan R. Liss.

Schuster, C.R. & Thompson, T. (1969). Self-administration of and behavioral dependence on drugs. *Annual Review of Pharmacology and Toxicology* **9**, 483–502.

Schuster, C.R., Fischman, M.W. & Johanson, C.E. (1981). Internal stimulus control and subjective effects of drugs. In *Behavioral Pharmacology of Human Drug Dependence*, National Institute on Drug Abuse Research Monograph 67, pp. 116–29, Washington, D.C.

Seevers, M.H. (1936). Opiate addiction in the monkey. I. Methods of study. *Journal of Pharmacology and Experimental Therapeutics* **56**, 147–56.

Simon, P., Sultan, Z., Chermat, R. & Boissier, J.R. (1972). La cocaïne, une substance amphétaminique? Un problème de psychopharmacologie expérimentale. *Journal de Pharmacologie* **3**, 129–42.

Spealman, R.D. (1985). Discriminative stimulus effects of midazolam in squirrel monkeys: comparison with other drugs and antagonism by Ro 15-1788. *Journal of Pharmacology and Experimental Therapeutics* **235**, 456–62.

Spealman, R.D. & Goldberg, S.R. (1978). Drug self-administration by laboratory animals: control by schedules of reinforcement. *Annual Review of Pharmacology and Toxicology* **18**, 313–39.

Spyraki, C. & Fibiger, H.C. (1981). Intravenous self-administration of nomifensine in rats: implication for abuse potential in humans. *Science* **212**, 1167–8.

Stark, C., Sykes, R. & Mullin, P. (1987). Temazepam abuse. *Lancet*, 802–3.

Stolerman, I.P. & Kumar, R. (1970). Preferences for morphine in rats: validation of an experimental model of dependence. *Psychopharmacologia* **17**, 137–50.

Thompson, T. & Schuster, C.R. (1964). Morphine self-administration, food-reinforced and avoidance behaviors in rhesus monkeys. *Psychopharmacologia* **5**, 87–94.

Vanacker, B., Vandermeersch, E. & Tomassen, J. (1986). Comparison of intramuscular buprenorphine and a buprenorphine/naloxone combination in the treatment of post-operative pain. *Current Medical Research and Opinion* **10**, 139–44.

Weeks, J.R. (1962). Experimental morphine addiction: method for automatic intravenous injections in unrestrained rats. *Science* **138**, 143–4.

Wei, E. (1973). Assessment of precipitated abstinence in morphine-dependent rats. *Psychopharmacologia* **28**, 35–44.

Wilson, M.C. & Schuster, C.R. (1976). Mazindol self-administration in the rhesus monkey. *Pharmacology, Biochemistry and Behavior* **4**, 207–10.

Winter, J.C. (1978). Drug-induced stimulus control. In *Contemporary Research in Behavioral Pharmacology*, eds. D.E. Blackman & D.J. Sanger, pp. 209–37, New York: Plenum Press.

Woods, J.H., Young, A.M. & Herling, S. (1982). Classification of narcotics on the basis of their reinforcing, discriminative and antagonist effects in rhesus monkeys. *Federation Proceedings* **41**, 221–7.

Woods, J.H., Katz, J.L. & Winger, G. (1987). Abuse liabilities of benzodiazepines. *Pharmacological Reviews* **39**, 251–419.

Woolverton, W.L. & Schuster, C.R. (1983a). Behavioral and pharmacological aspects of opioid dependence: mixed agonist-antagonists. *Pharmacological Reviews* **35**, 33–52.

Woolverton, W.L. & Schuster, C.R. (1983b). Intragastric self-administration in rhesus monkeys under limited access conditions: methodological studies. *Journal of Pharmacological Methods* **10**, 93–106.

Woolverton, W.L., Johanson, C.E., de la Garza, R., Ellis, S., Seiden, L.S. & Schuster, C.R. (1986). Behavioral and neurochemical evaluation of phenylpropanolamine. *Journal of Pharmacology and Experimental Therapeutics* **237**, 926–30.

Yanagita, T. (1973). An experimental framework for evaluation of dependence liability of various types of drugs in monkeys. *Bulletin on Narcotics* **25**, 57.

Young, A.M. & Herling, S. (1986). Drugs as reinforcers: studies in laboratory animals. In *Behavioral Analysis of Drug Dependence*, eds. S.R. Goldberg & I.P. Stolerman, pp. 9–67, London: Academic Press.

Young, R. & Glennon, R.A. (1986). Discriminative stimulus properties of amphetamine and structurally related phenalkylamines. *Medicinal Research Reviews* **6**, 99–130.

19

The relevance of behavioural models of drug abuse and dependence liabilities to the understanding of drug misuse in humans

RICHARD HARTNOLL

Two key concepts used to organise much of the psychopharmacological material on psychoactive drug-taking are 'abuse liability' and 'dependence potential'. (See Goudie, Sanger, this volume, Chapters 17, 18). This chapter discusses the relevance of laboratory-based behavioural models in general, and of these two concepts in particular, to the understanding of drug use as it is found in human societies. The main argument put forward is that human drug taking involves complex patterns of behaviours that can only be understood in terms of interactions between many different levels of explanation, ranging from the biochemical to the historical, economic and political. Inevitably, the laboratory scientist is concerned with drug–organism interactions, and primarily drug–animal interactions, under very specific, controlled conditions, and at a limited number of levels of explanation. This focus is both necessary and useful for the tasks at hand, such as the elucidation of the chemical and neural mechanisms underlying the actions of different drugs, or the development and screening of new drugs. However, it will be argued that models generated through laboratory investigations are seriously limited in the extent to which they can be taken as broader explanations of the wide variety of phenomena that are observed in the development, maintenance and remission of human drug-taking. Similarly, the contributions of such models to the fields of social policy or clinical interventions are likely to be restricted to an adjunctive rather than a primary role.

19.1 THE SOCIAL CONTEXT OF DRUG ABUSE

Whilst some authors recognise that drugs are abused 'for many reasons, some having little relationship with the drug's psychopharmacological effects' (Sanger, this vol., Ch. 18) or that legal and cultural processes play a central role in determining what is drug abuse (or even, what is a drug) there is a tendency to acknowledge these confounding factors and then to proceed to ignore them. The underlying thrust of this chapter is to reverse this emphasis. It is suggested that drug abuse and dependence in humans are better understood from the starting perspective of human beings and of human societies, and that whilst animal models can make a contribution, they play only a secondary role in advancing that understanding. In short, such models overemphasise the drugs themselves and their effects on behaviour via neurochemical substrates, and they greatly understate the importance of factors that influence drug taking external to the drug–organism interaction, in particular, the ability of humans, both individually and collectively, to interpret or reinterpret almost any drug effect as desirable or undesirable, and the powerful influence of social context and social learning.

This is not to undervalue the understanding of drug actions. This is clearly a necessary link in the chain of explanation, but it is far from sufficient. Further, the process of trying to relate behavioural models to human behaviour is valuable in itself – it compels rigorous analysis of what is meant by 'abuse' and 'dependence', and of what would constitute an appropriate model. The value of models can be evaluated not only in terms of their validity but also in terms of the richness of the hypotheses that they generate and the clarity of thought that they demand.

19.1.1 Definitional issues

'Abuse liability' is defined operationally in terms of the degree to which a drug will maintain self-administration; 'dependence potential' in terms of the probability that repeated administration will lead to physical dependence, usually characterised as the development of tolerance and/or withdrawal symptoms on cessation of administration (see Goudie, this vol., Ch. 17).

In the human context, there is no comparable definition of 'drug abuse' that is either clearly specified or even vaguely agreed. Rather, the definition of 'abuse' depends on the culture, the point in history and the perspective of the definer. Thus from a legalistic point of view, all use of illicit drugs is abuse, whereas regular consumption of alcohol (which fulfills the criterion of maintaining self-administration) is not, except in Muslim countries where it is prohibited. Using Western legalistic criteria, the rank order of drugs in terms of the extent to which they are abused might be: cannabis, amphetamines, psychedelics, cocaine, heroin. A clinician, on the other hand, defines abuse in terms of the probability of harmful consequences rather than illegality, and might rank order drugs in terms of their frequency of abuse as tobacco, alcohol, tranquillisers, opiates and stimulants. In contrast, an epidemiologist describing the distribution of nonmedical psychoactive drug using behaviour in a population, regardless of legality or health consequences, might produce a list headed by caffeine (coffee, tea, soft drinks), alcohol, tobacco, cannabis, tranquillisers, and, some way behind, amphetamines, psychedelics, cocaine and opiates.

These rank orders are listed in Table 19.1, along with a modified version of Table 18.4 from Sanger (this vol., Ch. 18), based on the criterion of the extent to which drugs support self-administration in animals. These are not intended to be precise rankings, but it is clear that attempts to validate models of abuse liability based on self-administ-

ration paradigms against crude 'prevalence' rates in human societies encounter serious problems when deciding against which criteria to set the models. There are also serious questions concerning why certain drugs that do not easily support self-administration are nonetheless commonly used. These issues are taken up later in this chapter.

Similarly, the definition of drug dependence in humans is not clearly agreed. The history of attempts by the World Health Organisation to establish such definitions reveals a process of constantly shifting conceptualisation that seems to reflect changes in human and scientific values as much as it does the phenomena themselves (W.H.O., 1957; Eddy *et al.*, 1965; Young, 1971; Zinberg, 1984).

A further definitional issue is whether the drugs and the way they are administered in laboratory settings are comparable to the forms and routes of usage by people. Most studies have involved either oral, intraperitoneal or intravenous routes of administration. Whilst this may be appropriate for morphine, it is clearly inappropriate for cigarettes, for example, since it is very difficult to persuade animals to smoke, and quite difficult to persuade human smokers that other ways of taking nicotine are satisfactory. Similarly, the substances used in real life may not be comparable to the pure drugs administered in experimental situations. For example, the average purity of street heroin reported in some American cities has sometimes been as low as two to three per cent, yet continued self-administration and dependence are regularly observed, along with high levels of goal directed behaviour such as crime or extraordinarily persistent drug-seeking (Hanson *et al.*, 1985). What maintains behaviour in these cases? If it is assumed to be the drug, then it appears that the dose–effect relationships that are central to many self-administration models (Koob *et al.*, 1987) break down.

Finally, there are definitional problems in relation to which behavioural paradigms are most appropriate to human behaviour. This is illustrated, for example, by the debate over whether self-administration is maintained by positive reinforcement or by avoidance of withdrawal, and by uncertainty over whether various drug behaviours are best described in terms of operant or classical conditioning. There may be a much

Table 19.1. *'Abuse potential' of different drugs according to different criteria for 'abuse', listed in descending order*

Legal	Clinical	Epidemiological	Self-administration
Cannabis	Tobacco	Caffeine	Cocaine and amphetamines
Amphetamines	Alcohol	Alcohol	Opioids
Psychedelics	Tranquillisers and hypnotics	Tobacco	Barbiturates
Cocaine	Opioids	Cannabis	Dissociative anaesthetics
Heroin	Stimulants	Tranquillisers	Alcohol
		Amphetamines	Nicotine
		Psychedelics	Caffeine
		Opioids	Benzodiazepines
		Cocaine	Cannabis
		Psychedelics	

larger overlap, and thus blurring, between stimulus-led and response-led behaviours than is implied in traditional texts.

19.1.2 Levels of explanation

The concept of 'level of explanation' is both powerful and necessary in all areas of scientific endeavour; the study of drug use, abuse and dependence is no exception. The example of heroin usage in the United Kingdom will serve as an illustration.

From the latter 1970s, heroin use and dependence expanded in parts of the U.K., the most immediate explanation being a marked increase in the supply and availability of heroin. Among factors that contributed to this were international events such as the Iranian revolution, and later the expansion of heroin production in Pakistan/Afghanistan. For some years previously, Iran had experienced high levels of heroin (not opium) use and addiction, amongst younger urban males. After the Shah fell, many Iranians fled to London, and some brought substantial amounts of heroin with them. The availability of heroin increased noticeably, the price fell, and consumption increased (Lewis *et al.*, 1986; Hartnoll, 1987). Subsequently, when supply from Iran diminished, the illicit market had developed sufficiently for alternative sources to emerge, most notably the north-west frontier region of Pakistan and Afghanistan, where heroin production had increased for a variety of political and economic reasons, including the war. Thus a vital dimension of any understanding of drug abuse is the national and international political and economic context of production and supply (McCoy, 1972; Henman *et al.*, 1985; Whitaker, 1987; Hartnoll, 1989). In some developing countries, illegal drugs are a major source of much-needed hard currency. For peasants, the earnings from cannabis, opium or coca may far exceed those from other crops. Frequently, production and international traffic are closely related to civil war, ideological conflict and the arms trade. By definition, this level of analysis cannot be simulated in behavioural models. However, in the real world, these are powerful factors that help shape drug-using opportunities and behaviour.

Increased supply and availability were not, however, sufficient explanations of increased heroin use in Britain. There were also a number of domestic factors, at the level of social and subcultural change, that ensured a receptive demand (Hartnoll, 1987; Dorn & South, 1987; Pearson, 1987*a*). These included a trend towards the use of heroin by routes other than injection, especially sniffing. In particular, for people who were used to sniffing amphetamine sulphate or cocaine, heroin became just another powder. This shift in route of administration weakened the taboo against heroin, broadened the 'susceptible' population, and enabled people to distance themselves from the image of the 'junkie' and to use heroin in the (mistaken) belief that it was not possible to become dependent. This in turn faciliated an increased 'social acceptability' of heroin as a recreational drug amongst people who already had experienced a variety of other illegal drugs.

The shift in the image of heroin occurred at a time when many of the subcultural boundaries which had previously defined, and thus helped limit, illegal drug use, had broken down. In the late 1960s and early 1970s, cannabis, and subsequently LSD, were the drugs of choice of a range of people who identified, to some extent at least, with the value of the 'counter-culture'. In contrast, subcultural attitudes towards heroin, as expressed for example in the alternative press of the time, were highly negative (Hartnoll & Mitcheson, 1973). With the disintegration of those subcultural groupings and social identities, the 'protective' negative attitudes towards heroin were weakened. By the late 1970s heroin had become another substance in the cocktail cabinet of illegal drug users, rather than something in a different category. As part of these subcultural changes, some dealers who before had supplied only cannabis and perhaps LSD and cocaine, started selling a wider variety of drugs, including heroin. There were probably several reasons for this, including the economics of supplying different drugs vis-á-vis the risks involved, and the heavy criminal activity that had become part of the cannabis market by the late 1970s. As a result, heroin became available beyond the previously rather limited networks of heroin addicts. As the market expanded, in terms of supply, distribution and use, so it became more criminalised and institutionalised (Hartnoll, 1987). Patterns of heroin use, and the expansion of the illicit market, were also influenced by a shift in treatment policy over the 1970s, away from maintaining addicts and towards a more abstinence-oriented approach (Stimson, 1987). Thus people wishing to continue using heroin often turned to the illicit market.

Socio-economic conditions constitute a further dimension to the social level of analysis. Some, though not all, areas where heroin use has increased most markedly are poor urban communities where other social problems such as bad housing, unemployment and crime are also concentrated (Parker *et al.*, 1987; Wilkinson *et al.*, 1987). This is not the place to discuss the often over-simplified issue of drug use and unemployment (see Pearson, 1987*b*). However, it is worth noting Pearson's comment that the intermittent pattern of 'recreational' heroin use (e.g. weekend only use) noted by some observers (e.g., Zinberg,

1984) was not found amongst unemployed heroin users in some inner-city areas of Northern England, because, he suggests, 'every day is a weekend', i.e. the lack of a demarcation between work and leisure takes away an important social structure that enables individuals to regulate behaviours such as drug use. Others (for example Prebble & Casey, 1969; Hanson *et al.*, 1985) have suggested that, in North America at least, the lifestyle of inner city addicts should be seen as a 'job' that fulfills for them what work fulfills for the employed.

The brief moments of euphoria after each administration of a small amount of heroin constitute a small fraction of their daily lives. The rest of the time they are aggressively pursuing a career that is exacting, challenging, adventurous and rewarding. They are always on the move and must be alert, flexible and resourceful . . . The heroin user walks with a fast, purposeful stride, as if he is late for an important appointment – indeed he is. He is hustling, . . . trying to sell stolen goods, avoiding the police, looking for a dealer, . . . looking for someone who cheated him . . . He is, in short, taking care of business. For them, . . . the quest for heroin is the quest for a meaningful life, not an escape from life. And the meaning does not lie, primarily, in the effects of the drug on their minds and bodies; it lies in the gratification of accomplishing a series of challenging, exciting tasks, every day of the week. (Prebble & Casey, 1969, pp. 2–3.)

This search for meaning, Prebble & Casey then link to the social context of street users.

. . . If anyone can be called passive in the slums, it is not the heroin user, but the one who submits to and accepts these conditions. The career of the heroin user serves a dual purpose for the slum inhabitant; it enables him to escape, not from purposeful activity, but from the monotony of an existence severely limited by social constraints, and at the same time it provides a way for him to gain revenge on society for the injustices and deprivation he has experienced. (pp. 21–2.)

The levels of explanation exemplified so far are largely concerned with understanding global or population changes in drug taking and thus fall outside the immediate frame of reference of behavioural models. The remainder of this chapter focusses on perspectives and evidence regarding human drug taking at psychological levels of explanation, and on the relevance of animal models. However, it must be emphasised that psychological models, whether behavioural of psycho-pathological, do not exist unaffected by the levels of explanation discussed above. The availability of drugs, the cultural and subcultural attitudes to them, and the social circumstances of users and potential users, all overlap with and interact with the psychological determinants of drug use.

19.2 PSYCHOLOGICAL APPROACHES

What sorts of psychological approaches are appropriate to understanding drug abuse and dependence in humans? This begs the question of how the phenomena should be conceptualised. As emphasised earlier, this will involve a multidimensional approach.

A recent trend in psychological theorising on drug abuse and drug dependence has been to emphasise the parallels between drug taking and other forms of appetitive behaviour, and between drug dependence and other patterns of compulsive activity (Falk, 1983; Orford, 1985; Peele, 1985). Thus Orford (1985) examines compulsive eating and gambling, along with drug and alcohol use under the same rubric of 'excessive appetites', and concludes that all can be usefully approached in terms of a conflict between appetitive behaviour and restraint. The source of both the attractions of any given behaviour and its restraint lie not only in the activities or drugs themselves, but also in individual experiences and expectations and in socially defined norms and sanctions.

Similarly, Peele argues that addictive behaviour is not rigidly determined by the properties of drugs, nor even restricted to drugs, but rather that people become addicted to experiences. 'The addictive experience is the totality of effect produced by an involvement; it stems from pharmacological and physiological sources, but takes its ultimate form from cultural and individual constructions of experience.' (Peele, 1985, p. 98). In the more extreme cases, addiction may become a 'dysfunctional attachment to an experience that is acutely harmful to the person, but one that is an essential part of the person's ecology . . . (that they) cannot relinquish.' (Peele, 1985, p. 97). This state is seen as the result of a dynamic social learning process in which the experience is rewarding because it ameliorates urgently felt needs, despite long-term damage.

First and foremost, addictive experiences are potent modifiers of mood and sensation, in part because of their direct pharmacological action or physical impact, and in part because of their learned and symbolic significance . . . In the U.S., for example, both heroin and one-to-one love relationships have extreme cultural weight placed upon what are inherently affecting experiences. (Peele & Brodsky, 1975, quoted in Peele, p. 85).

This line of theorising should not be seen as an esoteric attempt to force theories of drug abuse and dependence upon a wide range of non-drug phenomena. Rather, it should be seen as the reverse – an attempt to conceptualise and define drug abuse and addictive behaviour as phenomena that are to be understood within the same framework of explanation as other human behaviour and sentiment, having major cultural, social and cognitive dimensions, as well as the strictly behavioural and pharmacological. Failure to do so may well lead to the assumption, implicit in some laboratory-derived models, that abuse and dependence are primarily properties of the drugs, when in fact these are characteristics of different people, who use drugs in different ways, for different purposes, with different expectations, and in different circumstances.

A broad and varied body of evidence indicates that not all, and often only a minority of individuals in a group or community use drugs, and even fewer continue to use them, even in situations of easy availability and high prevalence (e.g. Johnston *et al.*, 1988). However, despite a large literature, especially on personality differences, relatively few individual factors emerge that hold strongly across different samples and situations. (Orford, 1985). This is especially true of studies of the initiation into drug use, or of recreational/experimental use, where the main (though relatively weak) 'personality' predictors relate to broad tendencies such as 'social precocity', 'nonconformity', or 'rebelliousness' (Jessor & Jessor, 1977; Kandel, 1978; Oetting & Beauvais, 1988). In these studies of youthful drug taking the attitudes and behaviour of friends account for more of the variance of the outcome variable than do individual factors.

Studies of drug users who become more heavily involved in drug taking, and who are more likely to be seen in hospitals, clinics or prisons, suggest a stronger association between individual factors and dependent or problematic drug use. Many psychiatric papers report high proportions of addicts in various diagnostic categories (Hawks *et al.*, 1969; James, 1971; Cockett, 1971; Barnes & Noble, 1972; Becket, 1974). Others refer to much higher than expected levels of difficulties in childhood (truancy, learning difficulties, separation from parents, distant, non-communicating parents, alcoholism in the family) and of other markers of possible psycho-social disruption (suicide attempts, depression, social isolation or low self-esteem) in samples of treated or institutionalised addicts (see Fazey, 1976, and Plant, 1981, for reviews). However, it is hard to assess the role of individual differences in dependence for various methodological reasons.

Firstly, treated or institutionalised addicts cannot be assumed to be representative of all drug users who continue to take drugs regularly. There is evidence that regular heroin users who are not in treatment tend to manifest more adequate social and interpersonal functioning and lower levels of depression than their peers who do seek treatment (Hartnoll & Power, 1989). Thus, drug users in treatment are self-selected samples in which individuals with more severe personality difficulties and fewer coping mechanisms are very likely to be overrepresented. Studies based on such samples are liable to overstate the significance of individual characteristics in continued drug use, and may as much reflect factors involved in help-seeking as in continued drug use.

Secondly, almost all studies of the individual characteristics of regular drug users have been conducted retrospectively, often after many years of drug use. It is thus difficult to establish whether the observed characteristics are a result of excessive drug use and of lifestyles that are associated with illegality and social deviance, or whether they are enduring individual characteristics that predated drug use, or an uncertain mixture of the two. Ex-addicts who have ceased drug use for some years often show dramatic improvements in both social functioning and in emotional and personal functioning, suggesting that their 'addicted personality' was largely a function of their drug-using lifestyle, though these changes are also consistent with a maturational hypothesis (Winick, 1962). In other cases, it appears that relatively objective markers of pre-drug use problems, such as separation, divorced parents, difficulties at school or at home, do indeed distinguish people who experience problems with drug use from those who do not (see review by Plant, 1981).

Thirdly, even if there are pre-existing differences between 'problem drug users' and non-problem users, or between users and non-users, it is not always obvious that those differences explain drug use, or problematic use. Thus the personality or early-experience factors that may be predictive of problematic use are the same sorts of variables as have been considered predictors of other 'problems' such as delinquency, emotional disturbances and personality disorders (Bowlby, 1951; McCord & McCord, 1959; Glueck & Glueck, 1959; Robins *et al.*, 1971; Rutter & Madge, 1976). Individual characteristics identified as predictors of problem drug use or drug dependence may well be no more than clusters of factors that are associated with individuals who are, in a broad sense, more 'vulnerable', and who are more likely to experience difficulties in one or more areas of their lives, and who have fewer resources for coping with those difficulties when they do arise. Whether or not those difficulties involve drugs may have as much or more to do with availability, immediate circumstances and the wider social milieu as with the individual. Similarly, the personal attributes that to some extent distinguish users from non-users (e.g. non-conformity, etc.) may also be predictors of conventional success, for example in terms of 'leadership' or 'adventurousness'. Which outcome obtained may depend crucially on opportunities for realising personal attributes (Merton, 1957).

These methodological problems should not be taken as a reason for rejecting the notion that certain people are, by virtue of who they are and how they have grown up, more or less likely to use drugs and more or less likely to become dependent if they do use them. The psychiatric literature may have tended, through exclusion of social and other levels of explanation, to have overemphasised the role of personality and individual differences in explaining drug use. However the challenge remains for behavioural models to take account of individual differences and experiences in drug abuse and drug dependence.

It becomes clear from the account so far that a vital component of a suitable conceptualisation of human drug taking is at the level of social-psychological models. Drug taking is essentially a social activity in which both initiation and continued use are strongly influenced by peer group values and social networks. For example, Fraser & George (1988) have described how the closure of a public house that was the social hub of a network of heroin addicts in a Southern English town, and their failure to find an alternative, disrupted the social functioning of the group to such an extent that the great majority ceased heroin use within a year. Other studies (see de Alarcon, 1969; Rathod, 1972) have shown the central significance of friends in the decision to initiate drug use. Thus, models derived from studies of conformity and, above all, social learning theory are essential (Bandura, 1977).

A further major component of a satisfactory model, that has been the focus of increasing attention, is found in cognitive approaches (e.g. Wilson, 1987). These emphasise the importance of expectations of drug effects and drug withdrawal, and the need for people to make sense of, attribute meaning, and attach value to activities such as drug taking or to states such as intoxication or abstinence.

A forerunner of this perspective was Lindesmith (1947) who argued that the defining characteristic of addiction was not a physical dependence resulting from regular administration of a drug, but rather, the recognition by the user that withdrawal symptoms were linked to cessation of use. This implied that patients regularly given morphine in hospital could not be considered addicted unless they made this cognitive connection. A further implication, developed later, was that once this link was made, the person would start to anticipate withdrawal symptoms, which arouse anxiety and serve to intensify those symptoms (Wikler, 1973; Siegel, 1983; MacRae *et al.*, 1987). This is supported by evidence that hospital patients given regular morphine without knowing what it was report less severe symptoms on cessation than 'addicts', and that they interpret the symptoms as influenza-like discomfort rather than as withdrawals (Lindesmith, 1947; Zinberg, 1974). Conversely, studies of placebo effects also serve to emphasise the role of expectations in regard to withdrawal (or avoidance thereof) (Lasagna *et al.*, 1954; Vuchinich & Tucker, 1980).

Another example of the importance of expectations comes from the alcohol field. A prominent theory of excessive drinking, and one that has considerable face validity, is the tension-reduction hypothesis, i.e. drinking reduces anxiety and other symptoms of stress, thereby reinforcing the behaviour (Conger, 1956; see Cappell & Herman, 1972, and Hodgson *et al.*, 1979, for reviews). However, careful phenomenological studies of heavy drinkers suggest that increasing quantities of alcohol may actually increase tension and subjective feelings of dysphoria (Warren & Raynes, 1972). Further exploration of subjects' accounts indicated that they continued drinking in the belief that they would feel better and attributed the negative sensations to not having drunk enough (Mello & Mendelsohn, 1978; Hodgson *et al.*, 1979; Goldman *et al.*, 1988). Alternative hypotheses have been put forward to explain this and other phenomena observed in drug and alcohol use, which are discussed in detail by Orford (1985). Increasingly, however, it is becoming apparent that drug-taking behaviour involves a complex interplay between the drug, the person and his history and behaviours in which many cues, both external and internal, assume a significance that is not merely the result of conditioning but is also mediated by a person's belief structures and expectations of drug effects. Furthermore, these expectations are not constructed by individuals in isolation, but are formed under the influence of significant others and within the options offered by the prevailing wider culture.

A good example of the importance of significant others in shaping expectations of drug effects was provided by Becker (1953) who described how at a time of severe social disapproval of marijuana use, experienced users helped novice users (a) to overcome anxieties about using a drug that was popularly depicted as dangerous, (b) to learn how to use the drug, and, most importantly for present purposes, (c) to learn how to recognise certain effects of the drug and how to interpret those effects as pleasurable. It is not self-evident that marijuana, or other drugs, are inherently pleasurable. First-time users may report either little effect, or negative sensations such as dizziness, nausea, etc. Similarly, volunteers with no previous experience of heroin may report that they dislike the effects. (One do-it-yourself theory for the prevention of cigarette smoking is to give children a cigarette in the hope that they cough, splutter and feel sick enough never to try it again). Peele & Brodsky (1975) have described the ways in which wider beliefs about drug consumption may shape and give meaning to individuals' experiences of drug effects and thence to their behaviour. It is no accident that young males in cultures such as the north of England working class, where hard drinking is a symbol of masculinity, feel and behave differently when they drink from their counterparts in Mediterranean wine-producing countries, or from vicars at a sherry party.

Other psychological perspectives, which cannot be discussed in any detail but which can contribute valuable understanding of drug taking and dependence are adaptive models, developmental models, and approach/avoidance or conflict models.

Adaptive models examine drug use in terms of the different functions which this use may serve for different individuals, or for the same individual in different contexts (Alexander & Hadaway, 1982; Alexander, 1988). The tension-reduction hypothesis (Cappell & Herman, 1975) of excessive drinking is one example of an adaptive model. The career of the heroin user discussed by Prebble & Cassey (1969) offers another example. Orford (1985) has emphasised the diversity of functions that drug use may fulfil. Whilst it might be possible to set up experimental paradigms to model some of these functions, it is difficult to see how, for example, rebellion against society's values could be modelled in the laboratory.

Developmental approaches examine drug use in terms of the stages of people's lives at which it tends to occur (Winick, 1962; Alskne *et al.*, 1967). In adolescence, drug use, along with other behaviours, often constitutes part of a process of experimentation, of testing adult values, and of establishing an independent identity. For many, illegal drug use diminishes or ceases upon the assumption of adult reponsibilities (O'Donnell *et al.*, 1976; Johnston *et al.*, 1978). Even amongst those who become opiate dependent and whose drug use is prolonged into adulthood, there is a tendency for some, perhaps as many as half, to end it for a variety of reasons – sometimes related to treatment, but often not (Winick, 1962; Stimson & Oppenheimer, 1982; Waldorf, 1983; Biernacki,

1986). Similarly with alcohol, per capita consumption falls after the age of 30. These life-cycle changes are of course not uniformly observed and probably vary from culture to culture. However it is likely that they relate closely to self-conceptions that are considered appropriate at different stages of life. A sixty-year-old grandmother sniffing glue is logically possible but unlikely.

Finally, Orford (1985) has argued that drug use is best understood wihin the broad framework of a conflict between appetitive behaviour and various source of restraint.

19.3 ANIMAL MODELS

We turn now to the question of how some of the models of drug abuse and drug dependence discussed in the two accompanying chapters (Goudie, Sanger, this vol., Ch. 17, 18) contribute to the understanding of drug misuse outlined above.

19.3.1 Abuse potential and self-administration

It was noted earlier that there are important discrepancies between the extent to which drugs support self-administration and various other crude indicators of human drug misuse. Thus in this respect there is disagreement with the conclusions of Goudie and of Sanger (this vol., Ch. 17, 18). There are two questions here. First, why are some drugs (tobacco, cannabis, caffeine) which do not support self-administration, or other drugs (e.g. alcohol, benzodiazepines, psychedelics) that do not easily support it, nonetheless widely used? Second, why are other drugs that clearly maintain self-administration (especially opioids and certain stimulants) not more widely used?

In answer to the first question, it is possible that some drugs do not support self-administration because the appropriate experimental conditions have not (yet) been found. Such an argument, however, comes perilously close to a most unscientific position in appearing to imply that these drugs must support self-administration and that therefore any disconfirming evidence can be ignored. It is more satisfactory to draw on various approaches outlined in the previous section. Thus the work of Becker (1953) and others suggested that people learn from others and the world around them how to reinterpret as desirable behaviours that were initially experienced either as neutral or even aversive. To understand why individuals should do this, it is useful to turn to the social psychological perspectives, described earlier, that emphasise social conformity (e.g. adolescents starting to smoke, advertising, role modelling, etc.), social ritual (e.g. drinking morning coffee) and reinforcement of membership of an 'in group'; and the search for meaning (e.g. use of psychedelics to change consciousness). There are thus important patterns of drug use where the self-administration model has neither face, predictive nor construct validity. It is difficult to see how these social psychological constructs could be modelled in animals.

In answer to the second question, it can, of course, be argued that some of the differences in Table 19.1 are the result of major historical differences in the production, distribution and use of certain drugs. Thus alcohol, tobacco and caffeine have been legal, accepted and widely marketed in much of the world for centuries, whereas heroin and cocaine are esssentially drugs of the twentieth century and have been subject to strict controls and social disapproval. It is quite conceivable that, as predicted by self-administration models, if heroin and cocaine were widely and freely available, then their use would become more widespread. The role played by supply in increased heroin use in the U.K., described earlier, would be consistent with this. The recent 'scare' about rapidly escalating use of cocaine, especially 'crack' (a form of smokable freebase cocaine) in the United States following widespread increases in availability also gives superficial support to this position.

However, to take the example of cocaine, closer examination of the evidence suggests that the abuse potential of cocaine is neither as high nor as clearcut under 'normal' conditions. Surveys of cocaine use in general populations show that most people who use cocaine do so occasionally rather than repetitively or compulsively, even under circumstances of ready availability, as in many parts of the United States (see Clayton, 1985; O'Malley *et al.*, 1988). For example, to quote the conclusions of Alexander & Erickson (1989)

Most North Americans never use it and most of those who do try it only a few times. Of those who become regular users, most do not become addicted. Of those who become addicted, most return to moderate use or to abstinence without treatment. Of those whose addiction is serious enough to require treatment, most suffered from severe problems before they first used cocaine – they would have been susceptible to other addictions had cocaine not been available (p. 1).

Similarly, in parts of Peru, where cocaine is widely available and cheap (often as cocaine paste – a smokable freebase form of cocaine that constitutes an intermediate stage in the production of cocaine hydrochloride from coca leaves), it has been estimated that only one in ten of users progresses to regular, compulsive use, despite suggestions that cocaine is particularly reinforcing when smoked rather than sniffed (Jeri, 1987).

Studies that have focussed on cocaine users contacted outside institutional settings all indicate that the majority control their cocaine use, often consciously, and do not display compulsive patterns of use (Morningstar & Chitwood, 1983; Siegel, 1985; Erickson *et al.*, 1987; Cohen, 1989). The reasons most people who used cocaine did not become compulsive users included perceptions of risks, choice to spend money in other ways, alternative demands on time and priorities, and perception of cocaine as a leisure-time drug. It was further suggested that compulsive patterns of cocaine use are more likely to arise in more vulnerable people in more vulnerable situations (Alexander & Erickson, 1989).

What, then, is to be made of laboratory studies demonstrating that cocaine has a high abuse potential, and that under conditions of unlimited access, monkeys and rats may self-administer to the point of convulsions and in some cases death (Johanson, 1984; Johanson & Schuster, 1975).

There are two, related, points here. Firstly, the experimental conditions under which these results were obtained involved individual, isolated animals, restrained in small cages, implanted with a catheter through which the drug was administered, and given few possibilities of responding apart from pressing either a food lever or a cocaine lever. These conditions bear little relationship to any 'normal' environment. Rats and monkeys are gregarious, inquisitive animals. Unrestrained, in a larger, more stimulating environment, and with the company of other animals, it is likely that self-stimulation rates would fall. Studies with morphine suggest similar conclusions. Thus Alexander *et al.* (1981) report that 'the readiness to consume opiates displayed by caged animals does not hold for rats living in an environment that resembles the animal's natural setting, even after the rats have been habituated to drug use' (Peele, 1985, p. 84).

A further implication is that these studies were not only failing to model 'normal' animal behaviour, but that they were also seriously deficient as models of human behaviour, apart, perhaps, from that seen in the most extreme cases of social and sensory deprivation. Thus a human placed in solitary confinement with minimal human contact, little knowledge or hope of release, and faced only with the alternative of prison food, might well self-administer cocaine to the point of starvation or convulsions if given unlimited access. (A further important but often ignored factor is that cocaine is a potent appetite suppressant.) Other self-administration schedules (e.g., intermittent) do not show such dramatic results, though progressive ratio schedules may, in more modest fashion, offer useful indicators of the relative 'abuse potential' of different drugs within the same broad category (e.g. stimulants). However, as is noted below, even this is not without its problems.

Whilst cocaine has been used as an example, it is clear that similar considerations must also apply to other drugs. Thus, self-administration studies on caged, isolated animals have doubtful validity as models of how humans behave in varied, stimulating social and physical environments (Peele, 1985). Though the former Chief Inspector of the Drugs Branch at the Home Office, after visiting some of the Scottish housing estates where heroin prevalence was high, is on record as concluding 'God, if I lived here, I'd turn on to heroin too' (Spear, 1988).

19.3.2 Schedule-induced models

Goudie (this vol., Ch. 17) emphasises the importance of the particular parameters of reinforcement schedules in determining whether or not, and to what extent, drugs will support self-administration. Thus Spealman (1979) has reported that certain schedules of cocaine delivery can simultaneously maintain self-administration and maintain responses to terminate cocaine delivery. Woods *et al.* (1987) also note that 'the effects of a schedule of cocaine delivery upon patterns of self-administration may produce low rates of responding while only moments later produce extraordinarily high rates . . .' (p. 56). Similarly, Falk (1983) notes that

schedule-induced drug overindulgence remains strictly a function of current induction conditions. Even with a long history of schedule-induced drinking, termination of the schedule . . . produces an immediate fall in alcohol intake to a control level . . . Once again we have a picture of a reputedly enticing molecule failing to take over behaviour in spite of chronic binging. (p. 389).

This points to the importance of looking at the conditions under which schedules support or do not support behaviour. It also calls attention to consistent findings across many studies that intermittent reinforcement schedules and second-order schedules exert powerful control over behaviour. Thus a number of detailed studies of the microbehaviour of drug misusers, mainly heroin addicts, have been carried out by urban anthropologists (Prebble & Casey, 1969; Weppner, 1977; Hanson *et al.*, 1985). It might be useful to attempt to develop a behavioural analysis in terms of intermittent reinforcement schedules and higher-order stimuli.

An example drawn from a recent ethnographic study of street addicts in four U.S. cities (Hanson *et al.*, 1985) describes how the daily lives of young, male urban heroin users were structured around a routine which started with 'hitting the street' – meeting up with other users at known hang-out points – to hear the latest news on drug availability, police activity, etc., followed by several hours of 'hustling' to raise money in time to 'score' by lunchtime or early afternoon. After using and relaxing during the afternoon, the evening became a time for socialising. Although some retained a small amount of heroin to use upon awakening in the morning, most did not, and thus used only once per day. Yet despite the relatively short duration of action of heroin, few users reported serious withdrawal symptoms. This may well have reflected the low purity of the heroin. However, they persisted in daily self-administration and, during a relatively set period of the day, in highly motivated goal-directed behaviour. Thus, it apparently was not physical dependency that primarily accounted for continued heroin use, but rather, the reinforcement schedules involved, and the presence of second-order stimuli (other users, drug talk, scoring sites, etc.) At first look, place and time-ordered schedules also appear relevant.

The problem, however, is that real-life situations may be too complex and too varied for satisfactory description and analysis in terms of reinforcement schedules (cf. Falk, 1983; Peele, 1985). Thus it is likely that the conditions of reinforcement are determined in part by the physical characteristics of the environment, in part by external structures imposed by the social milieu, in part by the individual's various previous experiences, and in part by a more active process in which individuals create and interpret for themselves the significance of events in their lives.

The implications for future behavioural work are that it may be valuable to take detailed naturalistic descriptions of drug-users' behaviour as a basis for generating further hypotheses to guide research. This line of work may be potentially useful for clinicians in terms of bringing to bear a sharper tool for examining questions such as 'What particular aspects of a client's lifestyle help maintain drug use?' This in turn may point to more focussed strategies for helping people to identify situations, people or other stimuli (including their own ideation) that increase the likelihood of drug use, and to develop alternative responses. Some preliminary work in this direction is discussed by Saunders & Allsop (1987).

19.3.3 Drug discrimination models

The drug discrimination paradigm holds a certain attraction. If a drug that is chemically similar to amphetamines maintains the same dose-related pattern of discriminant responding, then there appears good reason to suspect that it has actions and effects in common with amphetamines and might thus be liable to similar patterns of misuse. It should be noted, however, that this is a screening procedure that classifies drugs in terms of their degree of similarity to 'marker' drugs. It is not, in itself, a model of behaviour, nor, as Goudie and Sanger (this vol., Ch. 17, 18) both indicate, do results using this method correlate highly with other measures of abuse potential.

The question of whether the drug discrimination paradigm models human 'subjective' responses to drugs is well discussed in the two accompanying chapters. Here it should be added that, as dis-

cussed earlier, the subjective responses of humans are strongly influenced by the culture and by the immediate social environment, as well as by their expectations. Thus for example, when the author was involved in extensive fieldwork amongst a variety of heavy drug users in London in the 1970s, there were clear examples of different subgroups of users, some based on friendship networks, others on geographical locations, where subjective perceptions and responses to the same drugs varied considerably between groups. In some groups, methylphenidate (Ritain) was more sought after, and more expensive, than dextroamphetamine (Dexedrine), or dipipanone (Diconal) more highly prized than heroin. Others, whose primary drug was amphetamine, neither used nor liked cocaine (despite one double-blind laboratory study suggesting the two drugs cannot be discriminated – Fischman *et al.*, 1976). Yet others found intravenous barbiturates a satisfactory substitute for heroin. These local variations in drug preference and patterns of use are common. Recently, there have been reports from cities such as Glasgow of heroin addicts switching to injecting Temazepam (Stark *et al.*, 1987; Sakol *et al.*, 1989). In other cities, such as Manchester (Strang, 1985), Dublin (O'Connor *et al.*, 1988), and Edinburgh (Robertson & Bucknall, 1986) the misuse of buprenorphine (Temgesic), which is usually considered to have a low abuse potential, has increased significantly amongst opiate users.

Thus whilst drug-discrimination models may hold promise in the laboratory for preliminary screening and classification purposes, caution is needed in making extrapolations to human drug users. Similarly, taste-discrimination studies would not necessarily have predicted the commercial success of certain international fast-food hamburger chains.

19.3.4 Dependence liability

As with the notion of 'abuse potential', so with 'dependence liability' it is essential to employ an approach that broadens beyond a narrow focus on drug and drug–organism interactions. Thus, cross-cultural studies of alcohol use emphasise the large variations that are observed between different social groups in terms of the pattern and level of consequences that arise (Bales, 1946; MacAndrew

& Edgerton, 1969). Similarly, the use of heroin, a drug which is unambiguously classified in laboratory studies as having high abuse-dependence liabilities, displays far greater variability in the real world. A clear example is provided by Robins' study of Vietnam veterans returning to the United States (Robins *et al.*, 1975). About half of the men used heroin at some time whilst in Vietnam, and about 20% became addicted. Once home, half of those who had been addicted used heroin, some regularly, in the United States (and 84% of these reported finding a supplier quickly). Despite this, only 14% of Vietnam-addicted veterans became readdicted.

Other studies of heroin users in the community rather than in treatment or institutional settings also point to the return to non-dependent heroin use by former addicts (Harding *et al.*, 1980; Waldorf, 1983), and to non-compulsive patterns of heroin use in a variety of settings. Thus Zinberg and his collegues describe how social sanctions and rituals play an important role in enabling some heroin users to maintain control over their heroin use by, for example, only using in certain circumstances (e.g. with friends or at weekends), setting limits on expenditure on drugs, never using alone or with addict-dealers, not injecting, and so on (Zinberg, 1984). They emphasise that although users learned these strategies gradually over the course of their drug-using careers, 'virtually all . . . required the assistance of other controlled users to construct appropriate rituals and social sanctions out of the folklore and practices of the diverse subcultures of drug takers.' (Zinberg *et al.*, 1977). This association with other controlled users 'redefines what is considered a highly deviant activity . . . as an acceptable social behaviour within the group,' but also 'provides the necessary reinforcement for avoiding compulsive use.'

Studies such as these, together with the evidence noted above (Section 19.3.3) concerning observations that 'addicts' may sometimes use drugs from different pharmacological categories interchangeably (e.g. benzodiazepines for heroin), or that drugs with supposedly low abuse potential (e.g. buprenorphine) may be widely misused in certain circumstances, raise questions about the validity and scientific value of 'dependence liability'. Recent clinical thinking is moving away from characterising dependence in terms of specific pharma-

cological categories, towards concepts such as a drug- or alcohol-dependence syndrome (Edwards, 1986; Babor *et al.*, 1987), in which the focus is on a cluster of behaviours manifested by the person (such as narrowing of drug-taking repertoire, increased salience of drug-seeking behaviour, rapid reinstatement of dependence after abstinence, as well as tolerance and use of drugs to avoid withdrawal symptoms). As with abuse potential, a useful development for behavioural scientists might be to look at how careful clinical descriptions of the behaviours of drug- or alcohol-centred patients might be modelled.

Nevertheless, models of behavioural tolerance and conditioned withdrawal raise interesting possibilities, although they should not be overstretched as explanations of dependence and relapse. For example the importance of non-pharmacological conditioning factors in the development of tolerance (Hinson & Siegel, 1982) leads to suggestions that the rate at which tolerance develops in clinical settings may be minimised through interspersing placebo sessions (partial reinforcement) between drug administrations, even if overall doses are the same (MacRae *et al.*, 1987). Siegel (1983) has suggested that drug-associated environmental cues may elicit conditioned responses (seen as drug-compensating conditional responses) which are interpreted as withdrawal symptoms, which in turn elicit craving and thence, perhaps, relapse in abstinent addicts. Although there are documented cases of this occurring, it should not be taken as a universal phenomenon. Another area where the concept of behavioural tolerance may have clinical implications concerns the likelihood that opiate overdoses will be fatal. Thus heroin-experienced rats are more likely to die from a high dose of heroin if it is administered in a novel environment where there are no cues that elicit compensatory conditioned responses in anticipation of the drug effect (Siegel *et al.*, 1979, 1982). There is also supportive evidence from humans (Siegel, 1984). However, it is important to emphasise that these phenomena are unlikely to be explicable in a purely classical conditioning paradigm. As Goudie & Griffiths (1986) and Goudie & Demellweek (1986) have emphasised, a variety of mechanisms are likely to be involved, and the interaction of these is in urgent need of empirical study. Similarly, reviews of the factors associated with relapse underline the importance of integrating other psychological perspectives, such as those used in the study of decision-making and coping repertoires, with the more narrowly-defined behavioural models discussed above (Marlatt & Gordon, 1985; Saunders & Allsop, 1987).

19.4 CONCLUSIONS

The broad conclusion drawn here is that behavioural models have a role to play when set within a broader framework of explanation that includes, above all, cultural, social and situational factors. Failure to do this can have serious consequences, not only in terms of supporting a distorted scientific understanding of drug taking and drug dependence, but also in terms of encouraging both clinicians and policy makers (to the extent that either are aware of the findings) to give too great an emphasis on the drugs and too little on the circumstances in which and the reasons for which drugs are used. A good example has been the use of the results from self-administration studies to conclude that 'cocaine is the most reinforcing and addictive drug known to man' implying that few can resist the urge to become instantly addicted and thus justifying sweeping political action directed at cocaine producers, and far-reaching police measures to suppress demand. The omission of the qualifier, that such results obtain only 'in certain, often extreme conditions of sensory, motor and social deprivation' makes a crucial difference to how the data ought to be interpreted.

Subject to this, behavioural models are valuable for a number of reasons. They are useful for the screening of new drugs, especially if several different paradigms are used, such as progressive ratio schedules and drug discrimination procedures, together with variations in the schedules employed. However, even here they are not infallible, as the examples of the misuse of temazepam and buprenorphine (Temgesic) show. Models are further valuable for increasing our understanding of the neural and chemical mechanisms involved in drug actions. This in turn is important both at a purely scientific level, and at the level of helping to develop drugs that may be useful adjuncts in the treatment of drug dependence. Examples include the use of the tricyclic antidepressant, despiramine, or experimental work with the dopamine antago-

nists amantadine and bromocryptine in the treatment of high dose, chronic cocaine use. (Kleber & Gawin, 1987; Kleber, 1988). However, the key word here is 'adjunct'. Experience from clinical practice underlines that the determinants of successful treatment are wide-ranging and variable, and include the nature of the relationship between treated and treater, the functions that drug use plays for individuals and their motives for change, the alternatives that are open to individuals, the type and extent of post-treatment support, and so on.

At a behavioural level, the models are useful for directing the attention of clinicians to the importance of lifestyle-related schedules and drug-associated cues in maintaining drug use or in prompting relapse. Further studies in this area would be very helpful. They may further provide an analytic tool for sharpening the focus of therapeutic interventions and relapse prevention. Finally, because the models covered in the last three chapters of this book are explicitly and in many cases rigorously related to a broader body of psychological theory, they fulfill a valuable function in forcing one to examine critically the different analytic frameworks that are used to 'explain' and subsequently to intervene in drug misuse and drug dependence.

There are probably many implications for future behavioural research. Three are selected here. Firstly, it would be useful to look at ways in which some of the social dimensions of drug taking might be modelled. Too many studies are based on isolated, caged animals who can make few if any alternative responses; there seems no reason in principle why some of the relevant social variables, for example company, or the degree of stimulation in the environment, might not be included in future studies. It would also be worth exploring how some of the cognitive aspects of drug use might be modelled, though this might prove more difficult. Secondly, there is clearly a need to examine more systematically the circumstances under which reinforcement schedules support drug-using behaviour to a greater or lesser extent, with a particular focus on the role of second-order schedules and drug-associated cues. Thirdly, laboratory experimenters could benefit from attention to careful studies of how different human drug users actually behave in different circumstances,

515

both in naturalistic settings and when they present for treatment, since eventually it is against this yardstick that behavioural models must be judged.

REFERENCES

Alexander, B.K. (1988). The disease and adaptive models of addiction: a framework evaluation. In: Peele, S. (ed.) *Visions of Addiction*, ed. S. Peele, pp. 45–66. Lexington Books: Lexington.

Alexander, B.K. & Erickson, P.G. (1989). Gaining perspective on cocaine: Addictive liability and effects of moderate use. *The Journal of the Addiction Research Foundation (Toronto)*, February 1st, 1989.

Alexander, B.K. & Hadaway, P.F. (1982). Opiate addiction: The case for an adaptive orientation. *Psychological Bulletin* **92**, 367–81.

Alexander, B.K., Beyersten, B.L., Hadaway, P.F. & Coambs, R.B. (1981). Effects of early and later colony housing on oral ingestion of morphine in rats. *Pharmacology, Biochemistry and Behavior* **15**, 571–6.

Alskne, M., Lieberman, L. & Brill, D. (1967). A conceptual model of the life cycle of addiction. *International Journal of the Addictions* **2**, 221–40.

Babor, T.F., Cooney, N.L. & Lauerman, R.J. (1987). The dependence syndrome concept as a psychological theory of relapse behaviour: an empirical evaluation of alcoholic and opiate addicts. *British Journal of Addiction* **82**, 393–405.

Bales, R.F. (1946). Cultural differences in rates of alcoholism. *Quarterly Journal of Studies on Alcohol* **6**, 480–99.

Bandura, A. (1977). *Social Learning Theory*. Prentice Hall: Englewood Cliffs, N.J.

Barnes, G.G. & Noble, P. (1972). Deprivation and drug addiction – a study of a vulnerable sub-group. *British Journal of Social Work* **2**, 299–311.

Becker, H. (1953). Becoming a marijuana user. *American Journal of Sociology* **59**, 235–42.

Beckett, H.D. (1974). Hypotheses concerning the etiology of heroin addiction. In: *Addiction*, ed. P.G. Bourne, pp. 37–54. Academic Press: London.

Biernacki, P. (1986). *Pathways from Heroin Addiction: Recovery Without Treatment*. Temple University Press: Philadelphia.

Bowlby, J. (1951). *Maternal Care and Mental Health*, World Health Organisation: Geneva.

Cappell, H. & Herman, C. (1972). Alcohol and tension reduction: A review. *Quarterly Journal of Studies on Alcohol* **33**, 33–64.

Clayton, R.R. (1985). Cocaine use in the United States: In a blizzard or just being snowed? In: *Cocaine Use in America: Epidemiologic and Clinical Perspectives*, ed. N.J. Kozel & E.H. Adams,

Research Monograph 61, pp. 8–34. National Institute on Drug Abuse: Rockville, MD.

Cockett, R. (1971). *Drug Abuse and Personality in Young Offenders*. Butterworth: London.

Cohen, P. (1989). *Cocaine Use in Amsterdam in Nondeviant Subcultures*. University of Amsterdam: Amsterdam.

Conger, J.J. (1956). Reinforcement theory and the dynamics of alcoholism. *Quarterly Journal of Studies on Alcoholism* **17**, 296–305.

de Alarcon, R. (1969). The spread of heroin abuse in a community. *United Nations Bulletin on Narcotics* **21**, 17–22.

Dorn, N. & South, N. (eds.) (1987). *A Land Fit for Heroin?* MacMillan Education: London.

Eddy, N.B., Halbach, H., Isbell, H. & Seevers, M.H. (1965). Drug dependence: Its significance and characteristics. *Bulletin of the World Health Organisation* **32**, 721–33.

Edwards, G. (1986). The alcohol dependence syndrome: a concept as a stimulus to enquiry. *British Journal of Addiction* **81**, 171–83.

Erickson, P., Adlaf, E., Murray, G. & Smart, G. (1987). *The Steel Drug: Cocaine in Perspective*. Lexington Books: Toronto.

Falk, J.L. (1983). Drug dependence: Myth or motive? *Pharmacology, Biochemistry and Behavior* **19**, 385–91.

Fazey, C. (1976). *The Aetiology of Non-medical Drug Use*. UNESCO: Paris.

Fischman, M.W., Schuster, C.R., Resnekov, L., Schick, J.F.E., Krasnegor, N.A., Fennell, W. & Freedman, D.X. (1976). Cardiovascular and subjective effects of intravenous cocaine administration in humans. *Archives of General Psychiatry* **33**, 983–9.

Fraser, A. & George, M. (1988). Changing trends in drug use: an initial follow-up of a local heroin using community. *British Journal of Addiction* **83**, 655–63.

Glueck, S. & Glueck, E.T. (1959). *Predicting Delinquency and Crime*. Harvard University Press: Harvard.

Goldman, M., Brown, S. & Christiansen, B. (1988). Expectancy theory – thinking about drinking. In *Psychological Theories of Drinking and Alcoholism*, ed. H.J. Blane & K.E. Leonard. Guilford Press: New York.

Goudie, A.J. & Demellweek, C. (1986). Conditioning factors in drug tolerance. In: *Behavioural Analysis of Drug Dependence*, ed. S.R. Goldberg & I.P. Stolerman, pp. 225–85. Academic Press: London.

Goudie, A.J. & Griffiths, J.W. (1986). Behavioural factors in drug tolerance. *Trends in Pharmacological Sciences*, May, 192–6.

Hanson, B., Beschner, G., Walters, J.M. & Bovelle, E.

(1985). *Life with Heroin*, Lexington Books: Lexington.

Harding, W.M., Zinberg, N.E., Stelmack, S.M. & Barry, M. (1980). Formerly-addicted-now-controlled opiate users. *International Journal of the Addictions* **15**, 47–60.

Hartnoll, R.L. (1987). Patterns of drug taking in Britain. In *Drug Misuse: A Reader*, ed. T. Heller, M. Gott & C. Jeffery, pp. 13–18. John Wiley in association with the Open University: London.

Hartnoll, R.L. (1989). The international context. In: *Drugs and British Society*, ed. S. MacGregor, pp. 36–51. Routledge: London.

Hartnoll, R.L. & Mitcheson, M.C. (1973). Attitudes of young people towards drug use. *United Nations Bulletin on Narcotics* **25**, 9–24.

Hartnoll, R.L. & Power, R. (1989). Why most of Britain's drug users are not looking for help. *Druglink* **4**, 8–9.

Hawks, D., Mitcheson, M., Ogborne, A. & Edwards, G. (1969). Abuse of methylamphetamine, *British Medical Journal* **2**, 715–21.

Henman, A., Lewis, R. & Malyon, T. (1985). *Big Deal: The Politics of the Illicit Drug Business*. Pluto Press: London.

Hinson, R.E. & Siegel, S. (1982). Nonpharmacological bases of drug tolerance and dependence, *Journal of Psychosomatic Research* **26**, 495–503.

Hodgson, R.H., Stockwell, T.R. & Rankin, H.J. (1979). Can alcohol reduce tension? *Behavior Research and Therapy* **17**, 459–66.

James, I.P. (1971). The changing pattern of narcotic addiction in Britain, 1959–1969. *International Journal of the Addictions* **6**, 119–34.

Jeri, F.R. (1987). The impact of cocaine production and abuse in developing countries. Paper presented at scientific meeting on cocaine, 14–16 January 1987 (Luxembourg, Commission of the European Communities, Directorate General for Employment, Social Affairs and Education Health and Safety Directorate).

Jessor, R. & Jessor, S. (1977). *Problem Behavior and Psycho-Social Development: A Longitudinal Study of Youth*. Academic Press: New York.

Johanson, C.E. (1984). Assessment of the dependence potential of cocaine in animals. In: *Cocaine: Pharmacology, Effects and Treatment of Abuse*, ed. J. Grabowski, Research Monograph 50, pp. 54–71. National Institute on Drug Abuse: Rockville, MD.

Johanson, C.E. & Schuster, C.R. (1975). A choice procedure for drug reinforcers: cocaine and methylphenidate in the rhesus monkey. *Journal of Pharmacology and Experimental Therapeutics* **193**, 676–88.

Johnston, L., O'Malley, P. & Eveland, L. (1978). Drugs and delinquency: A search for causal connections. In *Longitudinal Research on Drug Use*, ed. D.B.

Kandel, pp. 137–56. Hemisphere: Washington, D.C.

Kandel, D. (ed.) (1978). *Longitudinal Research on Drug Use: Empirical Findings and Methodological Issues*. Hemisphere: Washington D.C.

Kleber, H.D. (1988). Epidemic cocaine abuse: America's present, Britain's future? *British Journal of Addiction* **83**, 1359–71.

Kleber, H.D. & Gawin, F.H. (1987). Pharmacological treatments of cocaine abuse. In *Cocaine: A Clinician's Handbook*, ed. A.M. Washton & M.S. Gold, pp. 118–34. John Wiley & Sons: Chichester.

Koob, G.F., Vaccarino, F.J., Amalric, M. & Swerdlow, N.R. (1987). Neural substrates for cocaine and opiate reinforcement. In *Cocaine: Clinical and Biobehavioral Aspects*, ed. S. Fisher, A. Raskin & E.H. Uhlenhuth, pp. 80–108. Oxford University Press: Oxford.

Lasagna, L., Mosteller, F., von Felsinger, J.M. & Beecher, H.K. (1954). A study of the placebo response. *American Journal of Medicine* **16**, 770–9.

Lewis, R.J., Hartnoll, R.L., Bryer, S., Daviaud, E. & Mitchenson, M.C. (1985). Scoring smack: the illicit heroin market in London, 1980–83. *British Journal of Addiction* **80**, 281–90.

Lindesmith, A. (1947). *Opiate Addiction*. Principa: Bloomington.

Macandrew, C. & Edgerton, B. (1969). *Drunken Compartment: A Social Explanation*. Aldine: Chicago.

Macrae, J.R., Scoles, M.T. & Siegel, S. (1987). The contribution of Pavlovian conditioning to drug tolerance and dependence. *British Journal of Addiction* **82**, 371–80.

Marlatt, G.A. & Gordon, J. (1985). *Relapse Prevention*. Guilford Press: New York.

McCord, W. & McCord, J. (1959). *Origins of Crime: A New Evaluation of the Cambridge–Somerville Study*. Columbia University Press: New York.

McCoy, A.W. (1972). *The Politics of Heroin in South East Asia*. Harper Row: New York.

Mello, N.K. & Mendelson, J.H. (1978). Alcohol and human behaviour. In: *Handbook of Psychopharmacology*, vol. 12, ed. L.L. Iversen, S.D. Iversen & S.H. Snyder. Springer-Verlag: Berlin.

Merton, B.K. (1957). *Social Theory and Social Structure*. Collier MacMillan: Toronto.

Morningstar, P. & Chitwood, D. (1983). *The Patterns of Cocaine Use – An Interdisciplinary Study*. National Institute on Drug Abuse: Rockville MD.

O'Connor, J.J., Moloney, E., Travers, R. & Cambell, A. (1988). Buprenorphine abuse among opiate addicts. *British Journal of Addiction* **83**, 1085–7.

O'Donnell, J.A., Voss, H., Clayton, R., Slating & Room, R. (1976). *Young Men and Drugs: A*

Nationwide Survey Research Monograph 5. National Institute on Drug Abuse: Rockville, MD.

Oetting, E.R. & Beauvais, F. (1988). Common elements in youth drug abuse: peer clusters and other psychosocial factors. In: *Visions of Addiction*, ed. S. Peele, pp. 141–61. Lexington Books: Lexington.

O'Malley, P., Johnston, L. & Buchman, J.G. (1988). *Illicit Drug Use, Smoking and Drinking by America's High School Students, College Students and Young Adults, 1975–1987*. National Institute on Drug Abuse: Rockville, MD.

Orford, J. (1985). *Excessive Appetites: A Psychological View of Addictions*. John Wiley & Sons: Chichester.

Parker, H., Newcombe, R. & Bakx, K. (1987). The new heroin users: prevalence and characteristics in Wirral, Merseyside. *British Journal of Addiction* **82**, 147–58.

Pearson, G. (1987*a*). *The New Heroin Users*. Basil Blackwell: Oxford.

Pearson, G. (1987*b*). Social deprivation, unemployment and patterns of heroin use. In: *A Land Fit for Heroin?* ed. N. Dorn & N. South, pp. 62–94. MacMillan Education: London.

Peele, S. (1985). *The Meaning of Addiction*. Lexington Books: Lexington.

Peele, S. & Brodsky, A. (1975). *Love and Addiction*. Taplinger: New York.

Plant, M. (1981). What aetiologies: In: *Drug Problems in Britain: A Review of Ten Years*, ed. G. Edwards & C. Busch, pp. 246–80.

Academic Press: London.

Prebble, E. & Casey, J.H. (1969). taking care of business – The heroin user's life on the streets. *International Journal of the Addictions* **4**, 1–24.

Rathod, N.H. (1972). The use of heroin and methadone by injection in a new town. *British Journal of Addiction* **67**, 113–21.

Robertson, J.R. & Bucknall, A.B.V. (1986). Buprenorphine: dangerous drug or overlooked therapy. *British Medical Journal* **292**, 1465.

Robins, L.N., Murphy, C.E., Woodruff, R.A. & King, L.J. (1971). Adult psychiatric status of black schoolboys. *Archives of General Psychiatry* **24**, 338–45.

Robins, L.N., Helzer, J.E. & Davis, D.H. (1975). Narcotic Use in Southeast Asia and afterward. *Archives of General Psychiatry* **32**, 955–61.

Rutter, M. & Madge, N. (1976). *Cycles of Disadvantage*. Heinemann Educational Books: London.

Sakol, M.S., Stark, C.R. & Sykes, R. (1989). Buprenorphine and Temazepam abuse by drug takers in Glasgow – an increase. *British Journal of Addiction* **84**, 439–40.

Saunders, B. & Allsop, S. (1987). Relapse: a

psychological perspective. *British Journal of Addiction* **82**, 417–29.

Siegel, R.K. (1985). New patterns of cocaine use; changing doses and routes. In: Kozel, N. & Adams, E. (eds.) *Cocaine Use in America: Epidemiologic and Clinical Perspectives*. Research Monograph 61, ed. N. Kozel & E. Adams, pp. 204–20. National Institute on Drug Abuse: Rockville, MD.

Siegel, S. (1983). Classical conditioning, drug tolerance, and drug dependence. In: *Research Advances in Alcohol and Drug Problems*, vol. **7**, ed. Y. Israel, F.B. Glaser, H. Kalant, R.E. Popham, W. Schmidt & R.G. Smart, pp. 207–46. Plenum: New York.

Siegel, S. (1984). Pavlovian conditioning and heroin overdose: Reports by overdose victims. *Bulletin of the Psychonomic Society* **22**, 428–30.

Siegel, S., Hinson, R.E. & Krank, M.D. (1979). Modulation of tolerance to the lethal effect of morphine by extinction. *Behavioral and Neural Biology* **25**, 257–62.

Siegel, S., Hinson, R.E.,. Krank, M.D. & McCully, J. (1982). Heroin 'overdose' death; the contribution of drug-associated environmental cues. *Science* **216**, 436–7.

Spealman, R.D. (1979). Behaviour maintained by termination of a schedule of self-administered cocaine. *Science* **204**, 1231–3.

Spear, H.B. (1988). Conversation with H.B. Spear. *British Journal of Addiction* **83**, 473–82.

Stark, C.R., Sykes, R.L. & Mullin, P.T. (1987). Temazepam abuse. *Lancet* **2**, 802–3.

Stimson, G.V. (1987). British drug policies in the 1980s: a preliminary analysis and suggestions for research. *British Journal of Addiction* **82**, 477–89.

Stimson, G.V. & Oppenheimer, E. (1982). *Heroin Addiction: Treatment and Control in Britain*. Tavistock: London.

Strang, J. (1985). Abuse of buprenorphine. *Lancet* **2**, 725.

Vuchinich, R. & Tucker, J. (1980). A critique of cognitive labelling explanations of the emotional and behavioural effects of alcohol. *Addictive Behaviours* **5**, 179–88.

Waldorf, F. (1983). Natural recovery from opiate addiction: Some social-psychological processes of untreated recovery. *Journal of Drug Issues* **13**, 237–80.

Warren, G.H. & Raynes, A.E. (1972). Mood changes during three conditions of alcohol intake. *Quarterly Journal of Studies on Alcohol* **33**, 979–89.

Weppner, R.S. (ed.) (1977). *Street Ethnography*. Sage: London.

Whitaker, B. (1987). *The Global Connection: The Crisis of Drug Addiction*. Jonathan Cape: London.

W.H.O. (1957). World Health Organisation Expert Committee on Mental Health, 1957, *Addiction producing drugs:* 7th report of the W.H.O. Expert Committee, W.H.O. Technical Report Series 116. World Health Organisation: Geneva.

Wikler, A. (1973). Dynamics of drug dependence: implications of conditioning theory for research and treatment. *Archives of General Psychiatry* **28**, 611.

Wilkinson, J., Lawes, G., Unell, I., Bradbury, J. & Maclean, P. (1987). Problematic drug use and social deprivation. *Public Health* **101**, 165–8.

Wilson, G.T. (1987). Cognitive processes in addiction. *British Journal of Addiction* **82**, 343–53.

Winick, C. (1962). Maturing out of narcotic addiction. *United Nations Bulletin on Narcotics* **14**, 1–7.

Woods, J.H., Winger, G.D. & France, C.P. (1987). Reinforcing and discriminative stimulus effects of cocaine: Analysis of pharmacological mechanisms. In: *Cocaine: Clinical and Behavioural Aspects*, ed. S. Fisher, A. Raskin & E.H. Uhlenhuth, pp. 21–65. Oxford University Press: Oxford.

Young, J. (1971). *The Drugtakers: The Social Meaning of Drug Use*. MacGibbon & Kee: London.

Zinberg, N.E. (1974). The search for rational approaches to heroin use. In: *Addiction*, ed. P.G. Bourne. Academic Press: New York.

Zinberg, N.E. (1984). *Drug Set and Setting*. Yale University Press: New Haven.

Zinberg, N.E., Harding, W.M. & Winkeller, M. (1977). A study of social regulatory mechanisms in controlled illicit drug users. *Journal of Drug Issues* **7**, 117–33.

Index